Rec'd 8/30/80

$ 21. 50

McGraw-Hill Yearbook of Science and Technology 1979 REVIEW

1980 PREVIEW

1979

COMPREHENSIVE COVERAGE OF THE IMPORTANT EVENTS OF THE YEAR AS COMPILED BY THE STAFF OF THE McGRAW-HILL ENCYCLOPEDIA OF SCIENCE AND TECHNOLOGY

McGraw-Hill **Yearbook of**

Science and Technology

McGRAW-HILL BOOK COMPANY

NEW YORK   ST. LOUIS   SAN FRANCISCO
AUCKLAND   NEW DELHI
BOGOTA   PANAMA
HAMBURG   PARIS
JOHANNESBURG   SAO PAULO
LONDON   SINGAPORE
MADRID   SYDNEY
MEXICO   TOKYO
MONTREAL   TORONTO

On preceding pages:

Left. Part of diatom (magnified 3000 times).

Right. Closeup of a fecal pellet (magnified 4000 times) of a small zooplankton; the pellet was gathered during a sediment trap experiment at 5300 meters in the Sargasso Sea.

(Scanning electron micrographs from S. Honjo, The scanning electron microscope in marine science, Oceanus, 21(3):19–29, Summer 1978)

The Library of Congress cataloged the original printing of this title as follows:

McGraw-Hill yearbook of science and technology. 1962–
   New York, McGraw-Hill Book Co.

     v. illus. 26 cm.
    Vols. for 1962–   compiled by the staff of the McGraw-Hill encyclopedia of science and technology.

     1. Science–Yearbooks. 2. Technology–Yearbooks. I. McGraw-Hill encyclopedia of science and technology.
Q1.M13   505.8   62-12028
Library of Congress   (10)

Table of Contents

Editorial Advisory Board

Dr. Neil Bartlett
Professor of Chemistry
University of California, Berkeley

Dr. Richard H. Dalitz
Department of Theoretical Physics
University of Oxford

Dr. Freeman J. Dyson
The Institute for Advanced Study
Princeton, NJ

Dr. George R. Harrison
Dean Emeritus, School of Science
Massachusetts Institute of Technology

Dr. Leon Knopoff
Institute of Geophysics and Planetary Physics
University of California, Los Angeles

Dr. H. C. Longuet-Higgins
Royal Society Research Professor,
  Experimental Psychology
University of Sussex

Dr. Alfred E. Ringwood
Director, Research School of Earth Sciences
Australian National University

Dr. Arthur L. Schawlow
Professor of Physics
Stanford University

Dr. Koichi Shimoda
Department of Physics
University of Tokyo

Dr. A. E. Siegman
Director, Edward L. Ginzton Laboratory
Professor of Electrical Engineering
Stanford University

Prof. N. S. Sutherland
Director, Centre for Research on Perception
  and Cognition
University of Sussex

Dr. Hugo Theorell
The Nobel Institute
Stockholm

Lord Todd of Trumpington
University Chemical Laboratory
Cambridge University

Dr. George W. Wetherill
Director, Department of Terrestrial Magnetism
Carnegie Institution of Washington

Dr. E. O. Wilson
Professor of Zoology
Harvard University

Dr. Arnold M. Zwicky
Professor of Linguistics
Ohio State University

Editorial Staff

Sybil P. Parker, *Editor in Chief*

Jonathan Weil, *Staff editor*
Betty Richman, *Staff editor*

Joe Faulk, *Editing manager*
Olive H. Collen, *Editing supervisor*

Patricia W. Albers, *Senior editing assistant*
Judith Alberts, *Editing assistant*
Thomas Siracusa, *Editing assistant*
Sharon D. Balkcom, *Editing assistant*

Edward J. Fox, *Art director*
Richard A. Roth, *Art editor*
Ann D. Bonardi, *Art production supervisor*
Cynthia M. Kelly, *Art/traffic*

Consulting Editors

Contributors

A list of contributors, their affiliations, and the articles they wrote will be found on page 427.

Preface

The 1980 *McGraw-Hill Yearbook of Science and Technology*, continuing in the tradition of its 18 predecessors, presents the outstanding achievements of 1979 in science and technology. Thus it serves as an annual supplement to the *McGraw-Hill Encyclopedia of Science and Technology*, updating the basic information in the fourth edition (1977) of the Encyclopedia.

The Yearbook contains articles selected from hundreds of topics which appeared in the forefront of scientific progress during the past year. These articles report on the topics that were judged by the 69 consulting editors and the editorial staff as the most significant developments of 1979. Each article is written by one or more specialists who are actively pursuing research or are authorities on the subject being discussed.

The Yearbook is organized in three independent sections. The first section, a preview of 1980, includes seven feature articles with comprehensive, expanded coverage of topics that have broad interest and future significance. The second section, photographic highlights, presents photographs considered noteworthy because of the unusual nature of the subject or the advanced photographic technique. The third section comprises 140 alphabetically arranged articles on the year's research in many disciplines, including space technology, clinical pathology, building construction, nuclear power, invertebrate architecture, and word processing.

The *McGraw-Hill Yearbook of Science and Technology* provides the general public, students, teachers, librarians, and the scientific community with information needed to keep pace with scientific and technological progress throughout the world. For the past 19 years, the Yearbook has served this need through the ideas and efforts of the consulting editors and the contributions of eminent international specialists.

SYBIL P. PARKER
Editor in Chief

McGraw-Hill Yearbook of Science and Technology 1979 REVIEW

1980 PREVIEW

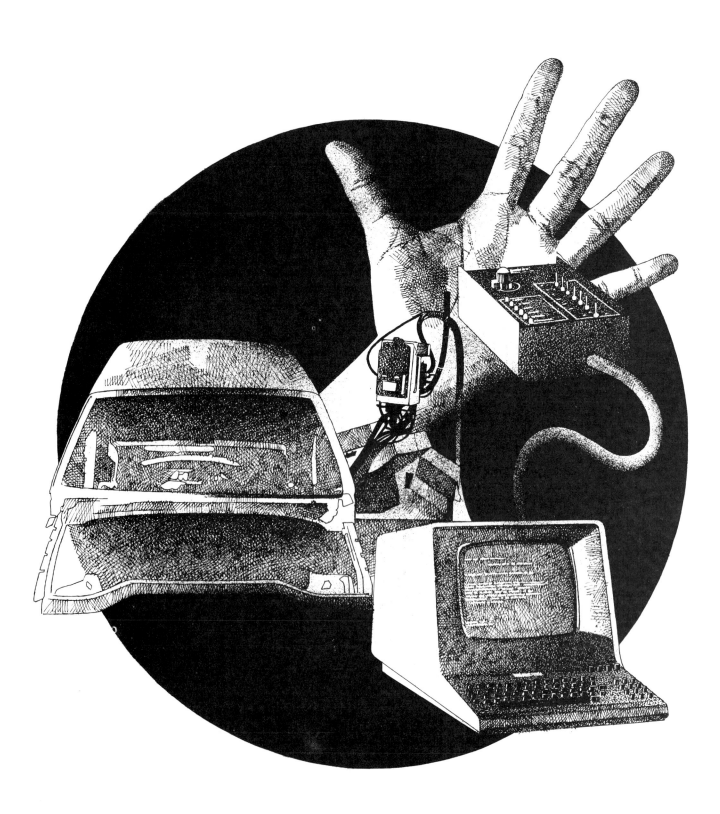

The industrial robot is a programmable mechanism designed to move and do work within a certain volume of space. The robot differs from a parts-transfer mechanism in that the action patterns can easily be changed by software changes in a controlling computer or sometimes by adjusting the mechanism. Because of this feature the industrial robot has been called off-the-shelf automation, and fills the gap between hard automation — that is, mechanisms designed for a specific job — and human labor.

The industrial robot was first developed in the United States in the late 1950s. After a slow beginning, its use in industry has accelerated rapidly, and robots of all configurations are presently in extensive use in most mechanized societies. The largest user is the automobile industry, where robots are performing a wide variety of tasks from spot-welding bodies to assembling instrument panels. Other users of robots span the manufacturing industry from woodworking (applying polish to fine furniture) to the aerospace industry.

Gordon I. Robertson received a doctorate in electrical engineering from London University in 1965. Until recently he was the research leader of the Advanced Automation Department at Western Electric Engineering Research Center, Princeton, NJ, where he worked on a variety of computer-controlled equipment.

CLASSIFICATION

Industrial robots may be divided into four classifications based on their capital cost and capabilities: pick-and-place, lightweight electric, heavy duty, and special-purpose. These are described below, and examples of their use given.

Pick-and-place robots. The pick-and-place robot is the simplest and cheapest of the four classifications. Bordering on a transfer mechanism, it is classified as an industrial robot because of its ability to be reconfigured for different tasks with a minimum of effort. A pick-and-place robot is characterized as one that moves between predetermined limits, which are set by hard stops or by microswitches, and is usually pneumatically powered, although larger models may be hydraulically powered. Because of its simple construction and low basic cost, this robot can be

Industrial Robots

obtained in various configurations, depending upon the nature of the task to be performed. Very simple models may have only one or two independent axes of motion, whereas more complicated models may have up to five or six axes, including rotatable and twisting wrists.

Figure 1 shows a basic pick-and-place robot working in conjunction with an X-Y table. Each independently movable axis or "degree of freedom" is powered by a small air cylinder. Because the limits of motion are determined by preset mechanical stops, there are only two stationary positions for each axis, namely, in or out, up or down. Tailoring the pick-and-place robot to its task involves adjusting the stops so that the end positions are appropriate, then programming the control system to move to these positions in the correct sequence.

To maintain the low-cost concept of the machine, the control system is simple. Its function is to energize valves which admit air or hydraulic fluid to the cylinders of the machine in the proper sequence. In addition, the control system has a limited capability of interacting with the outside world, such as recognizing the closing of microswitches or activating external relays. Any simple programmable sequencer is suited for this purpose; low-cost programmable controllers, drum switch-

es, or pneumatic logic systems have all been used with success.

The pick-and-place robot is used for application where low initial cost is an important consideration. Because of its simple construction and control system, its capabilities are limited, and it is usually used for parts transfer between two predetermined positions.

Although most robots in this classification are designed for lightweight part-transfer applications, the pick-and-place robot can be made much larger and more powerful to enable it to handle heavier loads, while still maintaining the inherent simplicity, low cost, and ease of maintenance of the pick-and-place concept. Figure 2 shows a hydraulically powered pick-and-place robot unloading a machine tool. This is a popular application of this type of robot, because the path to be followed is simple and without variations.

An additional advantage provided by hydraulic power is the ability of stopping at intermediate positions between the hard stops. This is accomplished by placing a microswitch at the desired position to control the hydraulic servo valve. Because hydraulic fluid is incompressible, the robot remains at the intermediate position. This feature, together with the long reach and 150-lb (68-kg) load-handling capacity, make the hydraulically

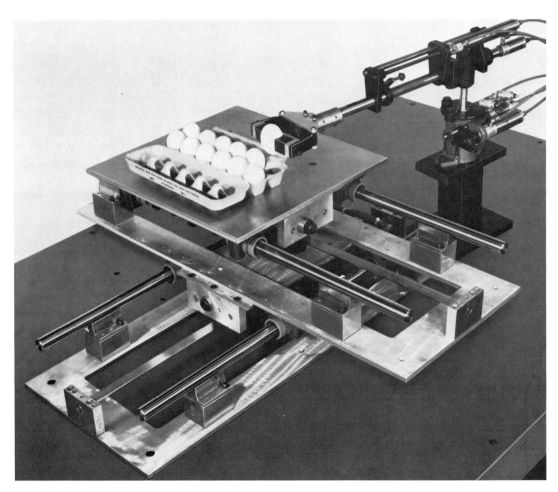

Fig. 1. Pick-and-place robot working in conjunction with an X-Y table. (Auto-Place, Inc.)

Fig. 2. PRAB hydraulic pick-and-place robot unloading machine flanges. (*PRAB Conveyors, Inc.*)

powered pick-and-place robot an attractive option for many jobs.

Lightweight electric robots. The lightweight electric robot resembles a small pick-and-place robot in its size and load-handling capacity. It is, however, a considerably more sophisticated machine, and is also much more expensive. In contrast to the pick-and-place robot, the lightweight electric machine can be driven to any position in its operating space.

The most popular configuration of this robot is shown in Fig. 3. The dimensions of the arm are usually chosen around the dimensions of the human arm. This is a convenient size since it permits many jobs which are traditionally done by human beings to be performed by the robot, and minimizes the space required to operate it. The robot shown in Fig. 3, for example, has dimensions between the major joints, shoulder to elbow and elbow to wrist, of 17 in. (43 cm). This gives it a reasonable working space of 30 in. (76 cm) radius from the base. The machine can also handle loads that a seated operator can conveniently work with—in the neighborhood of 5 lb (2.3 kg).

Theoretically, six independent axes of motion provide all the flexibility required. Three of these axes are needed to position the hand at the appropriate point in space, while the other three axes are used to orient it. However, in many cases only five independent axes are used, three to position the hand and two for orientation. Five axes slightly limit the ability to position an object, but still retain 90% of the capability of the arm while reducing the cost and complexity significantly.

Each joint, or axis, of the robot is controlled by a closed-loop servo system and is equipped with an electric motor, a suitable gear train, some sort of position feedback mechanism, and usually a feedback tachometer and electric brake. The motor, position feedback, and velocity feedback (tachom-

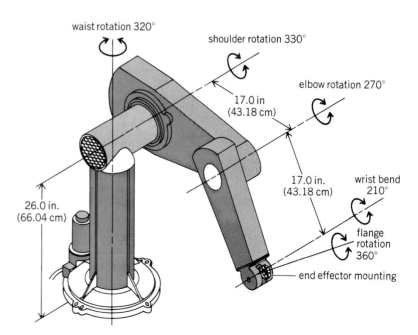

Fig. 3. Most popular configuration of a lightweight electric robot, showing the human-sized proportions. (*Unimation, Inc.*)

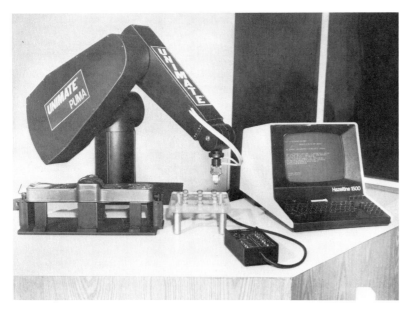

Fig. 4.   Unimate PUMA robot installing lights in an automobile panel. (Unimation, Inc.)

eter) form part of a servo loop which is driven by a control computer. The electric brake locks the joint while it is not in motion.

The robot is "taught" by moving it to the appropriate positions and recording those positions in the computer. The simplest way to do this is to release all the brakes and move it manually. The computer is then instructed to remember the coordinates of the joints for that position. A more sophisticated method is to use a teach box which, when plugged into the controlling computer, enables the operator to position the arm by pressing certain buttons. The teach box also incorporates the ability to inch the arm by small amounts, which aids in precise positioning. Having placed the arm in the desired position, the operator keys a switch which causes the computer to read and store the positions of all the joints. The robot is taught by repeatedly defining a series of stored positions. Upon playback, the computer will move the arm through the taught positions and thus mimic the

motion defined by the operator. Advanced methods of teaching computer-controlled robots involve the use of a "high-level language."

The lightweight electric robot costs 5–10 times as much as the basic pick-and-place robot. However, because of its ability to move objects to any place within its working space and the large memory capacity of its control computer, it is a much more versatile machine. Moreover, the computer can be programmed to handle complex decisions, and thus cause the robot to react to the external environment.

The principal market for the lightweight electric robot is in tasks such as assembly of small components typically performed by human operators. Figure 4 shows a lightweight electric robot set up to install lights in an automobile dash panel. Figure 5 is an artist's illustration of a series of lightweight electric robots at work on an assembly line. Each arm occupies approximately the same volume of space as a human operator would, and is designed so that if the robot fails it can be removed from the assembly line and its function performed by a human. Thus the entire assembly line is not dependent upon the robot.

Heavy-duty robots. Presently the heavy-duty robot is the workhorse of the industry. Although at least one model is driven electrically, most heavy-duty robots are hydraulically powered because of their high torque requirements. As the name implies, these robots are capable of handling loads in the 200–500-lb (90–230-kg) range, depending on the model. With a reach of approximately 6 ft (1.8 m), the heavy-duty robot considerably exceeds human capabilities.

Several different configurations of heavy-duty robots are available, as shown in Figs. 6, 7, and 8. The robot in Fig. 6 operates in cylindrical coordinates—the three primary axes to move the wrist are a rotation, a vertical motion, and a radial motion of the boom. The robot of Fig. 7, on the other hand, operates in spherical coordinates. The appropriate three axes are rotation and elevation of the turret and extension of the boom. The third robot, shown in Fig. 8, uses an articulated configuration. There are no significant differences in performance between the three configurations, except that the articulated machine has an advantage in terms of reaching awkward positions.

Hydraulically powered heavy-duty robots are equipped with a hydraulic pump mounted either in the base or at a remote location. Each joint of the robot contains a hydraulic motor and position and velocity feedback elements, as is the case with the lightweight electric robot. A brake, however, is not necessary because the joint can be locked by closing the hydraulic valve.

The robot is controlled by a built-in computer, and all models come equipped with a teach box or similar device for instructing the robot. By means of controls on the teach box, the arm is positioned to appropriate points, and by pressing a button these points are recorded. On playback, the arm moves smoothly through the recorded positions. Alternatively, some models are equipped with sophisticated controls that allow continuous motion for special tasks such as arc welding or work-

Fig. 5.   Assembly line of the future at General Motors. PUMA robots and humans work side by side to assemble components. (*Unimation, Inc.*)

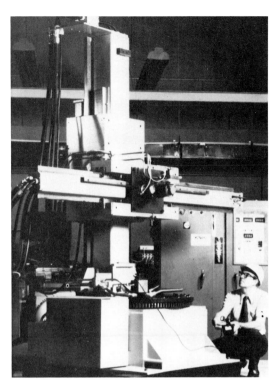

Fig. 6. The PRAB-AMF Versitran robot is an example of a cylindrical-coordinate heavy-duty machine. (*PRAB Conveyors, Inc.*)

ing on articles on a moving conveyor belt.

The heavy-duty robot is used for tasks that are hazardous, hot, strenuous, or otherwise difficult for humans to perform. Its usefulness is limited only by imagination; its versatility allows it to perform many different tasks, ranging from loading and unloading machine tools to TIG (tungsten – inert gas) welding of truck bodies. A major air-frame manufacturer is using an articulated heavy-duty robot to drill holes in aircraft wings. In this case, the robot replaces jobs previously done by three or four workers using large templates. A common use is spot-welding auto bodies, where the reprogrammability of the robot allows it to weld several different models in succession. Figure 9 shows the production line of a major car manufacturer with seven heavy-duty robots at work. This line is known as the "turkey farm" because of the appearance and action of the robots.

Special-purpose robots. One of the major advantages of an industrial robot is that it is a general-purpose mechanism which is tailored to a specific task by software changes. The concept of a special-purpose robot would appear to invalidate this advantage. There are, however, some tasks which are sufficiently different from those performed by other robots, and sufficiently broad in scope, that special-purpose robots have been designed and built to fill their requirements. Special-purpose robots presently on the market have been designed for spray painting and for welding.

Figure 10 shows a robot designed expressly for spray painting. It is hydraulically powered, but it differs from the heavy-duty hydraulic robot in its

lighter construction and, more importantly, in its control method. In contrast with the control system of other robots, which concentrate on accuracy of the end positions and are unconcerned with the path between those end positions, the control system for a spray-painting robot must accurately follow a predetermined path.

To program this machine, the operator, a skilled spray painter, holds a spray gun attached to the end of a special teaching arm (on some models the spray gun on the robot arm is used for teaching); he then sprays the article in the manner necessary to achieve a good coat of paint, and the computer continuously records the positions of all the robot joints. On playback, it reproduces the exact motions of the skilled painter and so duplicates the spray pattern.

Besides the obvious advantage of reducing labor costs, the spray-painting robot improves throughput and consistency and realizes paint savings of up to 50%. Because the robot repeats spray patterns with high consistency, problems such as heavy edges and sags are eliminated. In addition, it can function in a spray booth with a ventilation level unhealthy for people.

The spray-painting robot is not limited to applications of paint. Any material which is applied by spraying is suitable, and uses range from spraying car bodies to applying ablative coatings to the space shuttle.

A second type of special-purpose robot is designed for on-site arc welding, and is a light-weight portable unit which is attached to the job to be welded. Whereas the spray-painting robot is useful for repetitive jobs, the welding robot is designed expressly for single welding jobs in cramped or poorly ventilated environments which are difficult for humans.

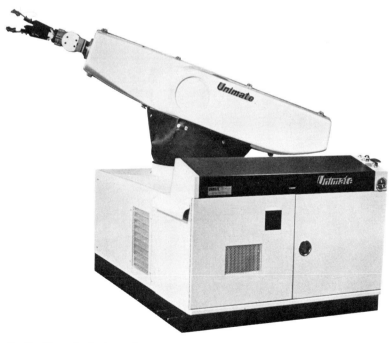

Fig. 7. The spherical-coordinate heavy-duty robot is typified by the Unimate Model 2000. (*Unimation, Inc.*)

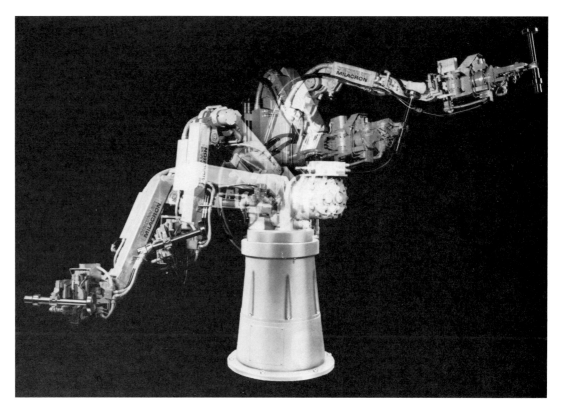

Fig. 8.   Cincinnati: Milacron T³ (The Tomorrow Tool) robot has articulated configuration. (*Cincinnati: Milacron, Inc.*)

Fig. 9.   Multi-robot automobile spot-welding line. (*Unimation, Inc.*)

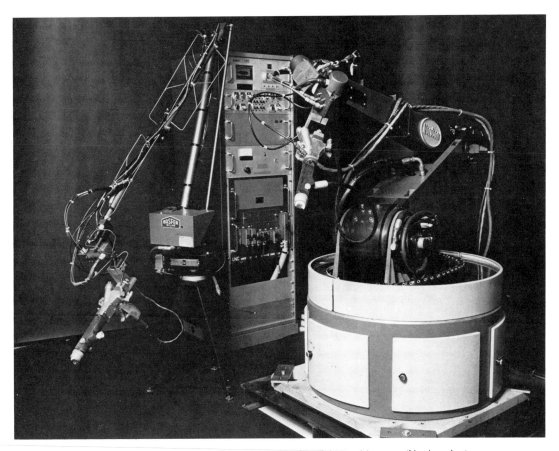

Fig. 10.   The Nordson spray-painting robot shown beside its lightweight teaching arm. *(Nordson, Inc.)*

Figure 11 shows the robot laying a weld on a ship's hull. The machine has five axes of motion, and is electrically powered. Like the spray-painting robot, its control system follows the path defined by the skilled operator; however, it has additional capabilites which allow the weld to be "weaved" and the welding wire to be automatically fed. Programming is accomplished by using a teaching wheel attachment. The operator attaches the wheel and then rolls it along the seam to be welded. All movements made while the record button is held down are memorized by the robot. The operator then leaves the area, and the robot lays down the weld. Transverse speed, wire speed rate, voltage, and weave conditions are programmed from the control panel.

ANATOMY OF A ROBOT

An industrial robot is an engineering system with many interacting elements. In order to understand the elements and to appreciate their interaction, the design criteria for a hypothetical machine will be reviewed.

Design criteria. The first step in designing a robot is to select the mission that it is to accomplish. From this, certain mechanical parameters are specified, namely, (1) the size and configuration of the robot, (2) the required positioning accuracy, (3) the load that is to be handled, and (4) the speed at which it is to be moved. For the purpose of discussion, let it be assumed that the robot is to have a reach of 5 ft (1.5 m) and have an articulated construction. The required accuracy is to be 0.050

in. (1.27 mm) at the hand, and the machine should be able to handle a 25-lb (11.3-kg) load and be fast enough to move between any two positions in 2 s or less. Figure 12 shows the initial design layout and key parameters.

After determining the configuration, the mass and moments of inertia of the links are calculated. To do so, the construction method of each link should be determined by considering the size and accuracy requirements. With a reach of 5 ft (1.5 m) and a total accuracy of 0.050 in. (1.27 mm), a rigid construction technique is necessary. On the other hand, a high speed is required, so that the construction cannot be too heavy. Assume that a cast aluminum construction is selected.

From the mass and moments of inertia of the links, it is possible to determine the motor torque required to move them at their rated speeds. The acceleration and maximum velocity of each joint is determined from point-to-point time of 2 s. The torque required to accelerate the joint is then found and added to the torque required to hold the link against gravity. From these numbers the type of motor for each joint is selected. In choosing the motor, attention must also be paid to the total mass of the motor, because three motors on the wrist and one on the elbow are carried on the arm and add to the weight. The motor must also be capable of sufficiently fine control so that the arm can be positioned with the necessary accuracy. Let it be assumed that direct-drive hydraulic motors are the optimum in this case.

Finally the feedback elements for each joint

Fig. 11. The Unimate Apprentice robot laying a weld on a ship's hull. (*Unimation, Inc.*)

must be chosen. The primary feedback determines the position of the joint; the secondary feedback (tachometer) gives information on the velocity of the joint, and is used for stabilizing the control loop. Three types of position feedback elements are available—precision potentiometers, resolvers, or optical encoders. Because of space and precision requirements and the fact that a digital output is an advantage, incremental optical encoders have been selected.

Velocity feedback is usually provided by a tachometer. However, tachometers, being small

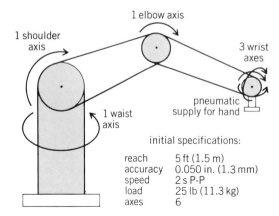

Fig. 12. Initial sketch and specifications for a hypothetical robot.

generators, must be driven from reasonably high-speed shafts. Because direct-drive hydraulic motors have been selected, a tachometer is not suitable. Velocity feedback will therefore be provided by digitally differentiating the position feedback signal.

The final element to be considered is the hand, or end effector. The hand is usually the only element of the robot that is changed to suit the task. A suitable mounting flange for a custom-made hand and a means of actuating it will therefore be provided. A two-state hand motion is all that is usually required, and so an air supply for a pneumatically operated hand will be available at the end of the wrist.

Interface and control computer. The sophistication of the control system of a robot greatly affects its versatility. With the increasing capabilities and the decreasing cost of microprocessors, systems that would have been prohibitively expensive 10 years ago are now possible.

The control system consists of two parts—the interface and the computer control. The function of the interface is to receive control commands from the computer and convert them into appropriate outputs to drive the motors. It is also responsible for closing the feedback servo loop to make sure that the joints follow the commanded positions.

The data from the computer to the interface are in the form of a stream of position instructions for each joint of the robot arm. Depending on the degree of sophistication of the control system, these position commands can represent either successive end positions for a particular motion or incremental points along a path to be followed.

If the design robot must be capable of complex motions, the second type of control system is selected, which enables path control to be maintained. Position data are provided by the computer, typically 60 times a second. Since there are six joints to be controlled, six position commands are transmitted in sequence every 1/60 s. In the interface, these position commands are separated (demultiplexed) and passed to the appropriate servo controllers.

Figure 13 is a block diagram of the interface. Note that the position feedback is differentiated and used to provide velocity damping. The output from the servo amplifier also has a null detector, which provides information to the computer as to when the final position has been reached.

The function of the control computer is to provide the stream of data defining the incremental position changes of all the joints in the robot. It must also establish, by checking the null detectors on the outputs of the servo amplifiers, that the robot arm has arrived at a particular position before initiating the next sequence of action. The computer also generates all appropriate timing pauses and provides necessary outputs for ancillary equipment, such as incrementing feeding mechanisms or signaling that a task is completed.

The easiest way for a computer to provide this data is to play back a prerecorded sequence of events. The spray-painting robot shown in Fig. 10 is controlled in just such a manner. This method does, however, unnecessarily restrict the utility of

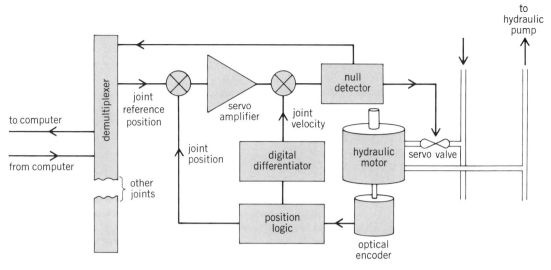

Fig. 13.   Interface diagram for one of the six joints of the hypothetical robot.

a general-purpose robot. The presence of the computer on a robot enables it to be a decision-making, or adaptively controlled, machine. Even simple adaptive control can greatly increase the versatility of the robot. For example, with appropriate input on its position, the robot can track and operate on a moving object such as a conveyor belt. To do this, it must first establish the required position of the hand at every instance and then calculate the appropriate angles for all the joints in real time.

A more sophisticated version of adaptive control enables the robot to generate paths that are not specified by the programmer. One example would be if the machine were equipped with vision (described below). In this case the path to be taken depends upon the input from the vision unit, and cannot be determined in advance. With an adaptive control program, the computer generates its own path and provides the appropriate stream of data to the interface.

FUTURE DEVELOPMENTS

Industrial robot technology is just beginning what will surely be a long and successful career. The necessary mechanical components have been available for a number of years, but it is the introduction of small, very capable computers that has spurred the development of the technology. The advent of the microprocessor and the continuing drop in price of electronics indicate a bright future for all aspects of automation technology.

High-level languages. Side by side with the development of hardware must come the development of suitable software. Since the major purpose of the industrial robot is to provide off-the-shelf automation that can be utilized by a typical engineer in a factory, some form of easy-to-use high-level language must be developed. Several research organizations are presently studying this concept, and there is at least one high-level language on the market. However, there are no universally accepted language and no agreed-upon constraints from which to construct such a language.

Two major schools of thought exist. In one, the language should be small and self-contained, since the robot should not need access to any other computer in order to be instructed or to execute the instructions. In the other school of thought, the computer should be integrated into a large, general-purpose, computer-aided manufacturing (CAM) system. With such integrated control, the robot would be capable of generating its own motions based solely upon the design of the object to be handled, with little or no input from the engineer. The second type of high-level language is considerable interest to the aerospace industry.

The table shows a section of a program written in the first type of high-level language. The arm is instructed to move by commands such as MOVE,

Fig. 14.   The remote center compliance (RCC) unit greatly simplifies tight-tolerance assembly tasks. Diameter is 2 in. plus (50 mm plus). (*Charles Stark Draper Laboratories, Inc.*)

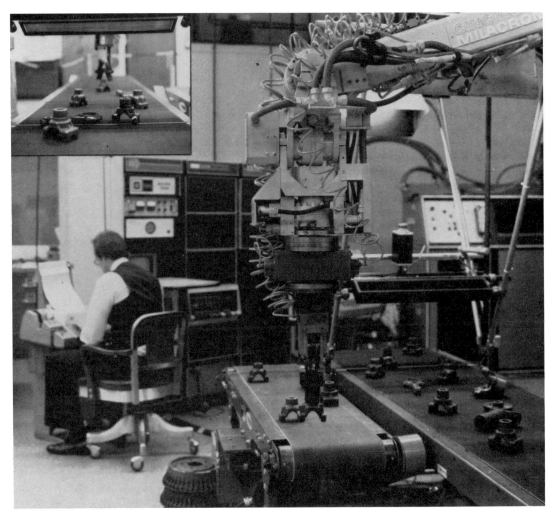

Fig. 15.   Using Consight, a robot acquires randomly placed items on a conveyor belt. (*General Motors, Inc.*)

DRAW, and GO TO. These commands cause the arm to move in different manners to the appropriate point. The exact coordinates of the points are not specified at the time of writing the program, but may either be defined later or even be made programmable, that is, the computer can calculate the coordinates of the point and insert them into the program.

More sophisticated programming techniques will ultimately enable robots to be programmed by voice in a conversational manner. This technique falls under the classification of artificial intelligence, and is being actively studied in several universities in the United States. Practical implementation is, however, several years into the future.

Sample instructions of a robot high-level language

SPEED 50	Run at 50% speed
COARSE	Set coarse tolerances
WAIT 51	Wait for switch 51
MOVE P3	Move to position P3
SET N = N + 1	Increment counter N
IF N = 4 GO TO 30	If counter = 4, branch to instruction 30

Assembly aids. A major area of interest for robot manufacturers is automatic assembly of mechanisms such as small electric motors, brake cylinders, and typewriter components. Although the level of sophistication and dexterity required is higher than is available from most present-day robots, several industry- and government-sponsored laboratories are developing techniques to assist the machine.

One fundamental aspect of assembly is the fitting of tight-tolerance objects, for instance, a bearing into its housing. A human fits a bearing by invoking delicate and complicated feedback sensation from the fingertips, which are entirely lacking in the robot. Moreover, the resolution of the robot arm is usually not capable of sufficiently fine adjustments to insert the bearing. Researchers at the Charles–Stark Draper Laboratory have studied this problem and produced an ingenious solution. A device known as a remote center compliance unit, shown in Fig. 14, is attached to the end of the arm, which by means of levers and springs has the effect of changing the apparent point at which force is applied to the bearing. In this manner, although the bearing is in fact being pushed into the housing, it is mechanically identical to a bearing which is being pulled into the housing from inside. Using this device, robots can insert bearings into housings with 0.0005 in. (12.7 μm) clearances in 0.5 s.

Vision. A limitation of present robots is the fact that they position their hand by deadreckoning. The object to be handled is expected to be in the appropriate position, which often requires elaborate feed mechanisms to dispense and properly place the object. Since parts feeders must be tailored individually for each part, this is not in keeping with the concept of off-the-shelf automation. To overcome this limitation, several research facilities are developing elaborate feedback systems, most of them based upon vision.

Compared with human vision, machine vision is rudimentary and poor in resolution. It is, however, capable of locating an object within a certain known area and of describing the identity, position, and orientation of that object. This information can be passed to the computer controlling the robot, which is then able to acquire the object.

There are several different approaches to machine vision. Figure 15 illustrates one approach, in which a bar of light projected across the conveyor belt illuminates a strip of the object as it passes through it. The outline of the object is matched against known templates, and the object identified by finding the closest match. A second approach is used in the demonstration shown in Fig. 16. Here a television camera (not shown) recognizes objects by establishing certain parameters, such as area and perimeter, associated with each object. In this demonstration the robot is selecting the letters that spell out its name and arranging them in the proper order.

CONCLUSIONS

The advent of low-cost microcomputers has turned the industrial robot from a mechanical nov-

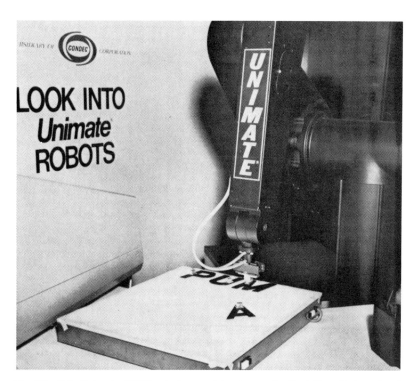

Fig. 16. This Unimate PUMA robot uses vision to acquire the letters that spell out its name, and arranges them in order. (*Unimation, Inc.*)

elty into a significant automation aid. The basic mechanism, with relatively simple control programs, has demonstrated its reliability and utility in various industries, and future developments in the fields of tactile and visual feedback systems will generate a considerable impact on the economics of automation.

[GORDON I. ROBERTSON]

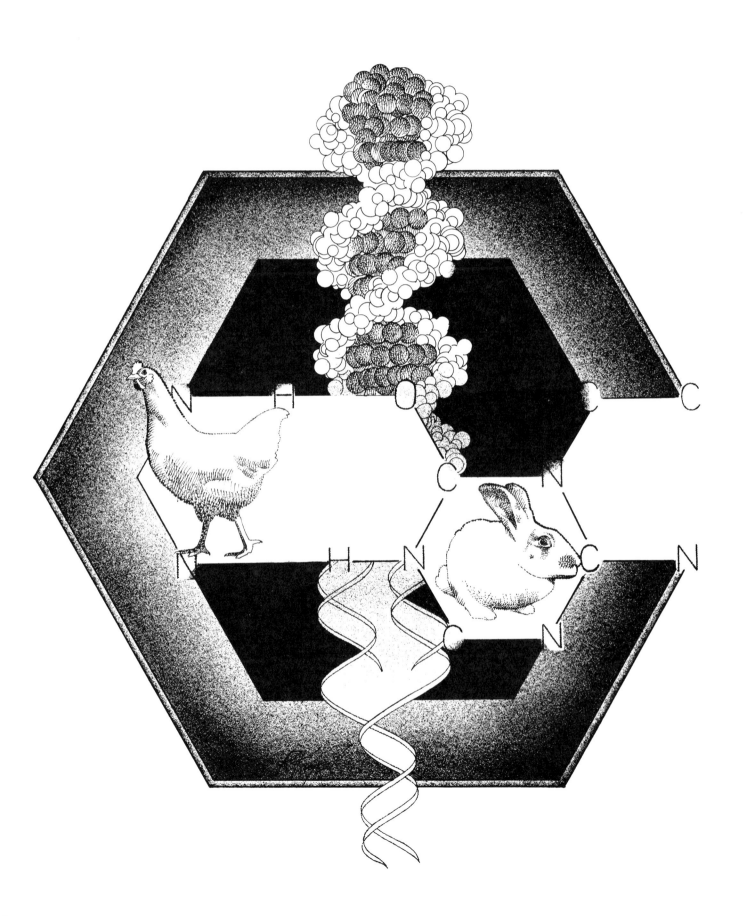

The genetic information which determines the phenotypic characteristics of an organism is stored in the macromolecule deoxyribonucleic acid (DNA). DNA is a double-stranded polymer of four different deoxyribonucleotides, which are distinguished from one another by their purine or pyrimidine base substituents, namely, the purines adenine (A) and guanine (G) and the pyrimidines thymine (T) and cytosine (C). The two strands of DNA are held together by pairing between the complementary bases A and T, and G and C. The information that specifies a given protein is encoded in the linear sequence of these four nucleotides in one of the DNA strands (called the sense strand). A single gene codes for one protein (or to be more specific, one polypeptide chain since some proteins consist of a complex of several different polypeptides). Through the genetic code, groups of three adjacent nucleotides—called codons—specify the amino acids which are to be joined in sequence to provide a given protein. The genetic code also specifies the signal (ATG, or rarely GTG) to commence protein synthesis and to terminate synthesis after the last amino acid of the protein has been specified (TAA, TAG, or TGA). The genetic code is identical in all organisms studied, from bacteria to humans.

The information stored in the DNA is not translated directly into protein. To produce the template for protein synthesis, the DNA is first transcribed into messenger ribonucleic acid (mRNA). The mRNA constitutes only a few percent of the total RNA mass in the cell. The major cellular RNAs are the structural RNAs of the ribosome, which is the enzyme complex responsible for protein synthesis. The ribosomes attach to the mRNA; they read the genetic information and polymerize the amino acids in the correct order according to the sequence of codons in the mRNA. A bacterial cell does not express all of its genes at one given time. Rather, it selectively

Richard A. Flavell is head of the laboratory of gene structure and expression at the National Institute for Medical Research, London. His work involves the study of the structure and expression of eukaryotic genes, in particular in relation to inherited disease of humans.

Eukaryotic Genes

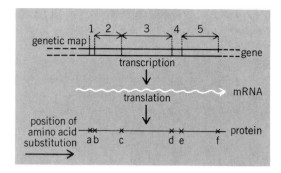

Fig. 1. The bacterial gene is colinear with its protein product. The position of mutations was deduced by measuring the distance between each respective mutation (1, 2, 3, and so on). The position of the amino acid substitutions which resulted from a given mutation was determined by protein sequencing.

transcribes only those genes that specify the proteins required by the cell at that moment. It represses the transcription of unnecessary genes, for example, those that code for proteins that are normally responsible for the synthesis of essential constituents which at that moment are provided by the environment. If the supply of the compound in the immediate environment is exhausted, the transcriptional repression of the gene in question stops and the mRNA is synthesized. In this way the protein composition of a bacterial cell is tailored to the needs of the organism at any moment.

The sequence of codons in the mRNA for a protein is colinear with the sequence of amino acid present in the protein. For a long time it was assumed that the nucleotide sequence of the gene is also colinear with the corresponding amino acids of its protein product. Direct proof of this was obtained for bacterial genes in the 1960s when the position of amino acid substitutions in the α-polypeptide chain of mutant forms of the enzyme tryptophan synthetase were proved to correspond with the positions of the respective mutations within the trpA gene, which codes for this protein (Fig. 1). Subsequently, the entire nucleotide sequence of a number of genes of bacterial viruses has been determined. Here the nucleotide sequence of a gene has been shown to be precisely colinear with the amino acid sequence of the protein product.

DIFFERENCES BETWEEN PROKARYOTES AND EUKARYOTES

In the last decade emphasis has shifted from the prokaryote to the eukaryote. A number of differences exists between prokaryotes and eukaryotes, which have influenced the way that progress has been made in the two systems. Of course, the main distinction between the cells of prokaryotes and eukaryotes is the possession by the eukaryote of a cell nucleus, containing the chromosomes and transcription machinery, together with several other cell organelles in the cytoplasm. Eukaryotic cells are in general much larger than those of prokaryotes and, of course, higher eukaryotes are multicellular organisms. In these creatures a division of labor takes place between the different organs of the body; these organs in turn consist of different cell types, each one expressing a characteristic complement of genes whose products enable the organ-specific functions to be performed.

The complexity of the developmental organiza-

tion of the higher eukaryote is reflected in the immense genome size characteristic of this group. Mammals contain a haploid genome (a single chromosomal DNA complement, present in a single gamete) of about 3×10^9 base pairs (bp), some thousand times more complex than a bacterial genome. Since a single gene is only about $1000-2000$ bp, it follows that a single gene constitutes only one part in a million of the total genome. In relatively "unspecialized" cells, a large number of different genes are being expressed, probably about 10,000 different genes being active at a given time. However, in certain tissues, a cell type is so specialized that only a few genes are expressed. As an example, consider the production of red blood cells. These cells have as their function the transport of oxygen in blood; the oxygen transporter is the protein hemoglobin, which is by far the major protein in red cells. The red cell can therefore be considered to be a bag full of hemoglobin where all luxuries—unnecessary for oxygen transport—have been disposed of. The red cell is derived from a stem cell, which generates the red cell precursors. These precursors undergo a limited number of cell divisions and, during this process, synthesize large amounts of globin mRNA. This mRNA, and its protein product hemoglobin, accumulates during the maturation of the red cell precursors; ultimately the mRNA concentration reaches the astonishing value of about 1% of the total cellular RNA in reticulocytes, the penultimate stage of red cell production. Finally, as the level of hemoglobin reaches its required value, the globin mRNA level drops to zero in the formation of erythrocytes from reticulocytes.

STUDY OF EUKARYOTIC GENES

Knowledge of the molecular biology of bacteria was a result of the powerful combination of the two disciplines of bacterial genetics and biochemistry.

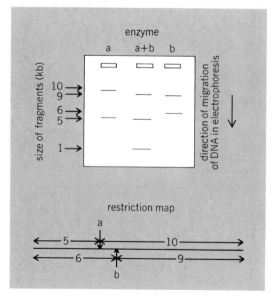

Fig. 2. Mapping of the cleavage sites for restriction enzymes on a DNA molecule. The DNA is digested with enzymes a and b; the size of the DNA fragments (kb= kilobase pairs) generated is determined by gel electrophoresis.

Unfortunately, the genetics of higher organisms is not at the advanced level that bacterial genetics had reached in the 1960s, mainly due to the long generation time of higher eukaryotes (at best days and at worst years) compared with bacteria (minutes or hours). Consequently, new biochemical approaches have been necessary for the elucidation of eukaryotic gene structure.

NEW DNA TECHNOLOGY

A plethora of new techniques have become available in the last decade; the most important of these are considered below, and then the discoveries which followed the use of these techniques.

DNA fragmentation with restriction enzymes. Bacterial restriction enzymes cut double-stranded DNA at specific sites. By now a large number of different restriction enzymes have been identified from many different types of bacteria. Different enzymes have different cleavage sites. For example, the enzyme EcoRI (isolated from *Escherichia coli*) cleaves double-stranded DNA at the sequence

$$\downarrow$$
GAATTC
CTTAAG
$$\uparrow$$

(the arrows indicate the point where cleavage occurs), whereas the enzyme BamHI (from *Bacillus amyloliquifaciens* HI) cleaves at

$$\downarrow$$
GGATCC
CCTAGG
$$\uparrow$$

In a DNA molecule which contains equal amounts of A, T, G, and C, such sequences will occur on average once every 4096 nucleotides: DNA can therefore be cut into relatively large pieces. The large number of different restriction enzymes makes it possible to dissect a given stretch of DNA into almost any desired segment. By using combinations of restriction enzymes, maps of the restriction enzyme cleavage sites can be constructed (Fig. 2). When sufficiently detailed, a restriction enzyme map can serve as a "fingerprint" to identify the region of DNA which the researcher is examining—a given DNA segment will show a highly characteristic pattern of restriction enzyme cleavage sites.

Recombinant DNA technology. In bacteria the major DNA component is the chromosomal DNA. In addition to this, however, small circular DNA molecules, called plasmids, which can replicate autonomously are found. In the last few years it has become possible to isolate these plasmid DNA molecules, to cut them open in laboratory cultures by using a restriction enzyme, and to insert a piece of foreign DNA into the plasmid by joining its ends with the ends of the plasmid DNA to reform a circle (Fig. 3). As long as the insertion of the foreign DNA into the plasmid does not inactivate essential plasmid genes, this recombinant DNA is still capable of replication in bacteria. The recombinant DNA is introduced into the bacterium under conditions whereby only one plasmid molecule is taken up; by using a selective system—such as a plasmid-coded resistance to an antibiotic—the plasmid-containing bacteria can be selected for and

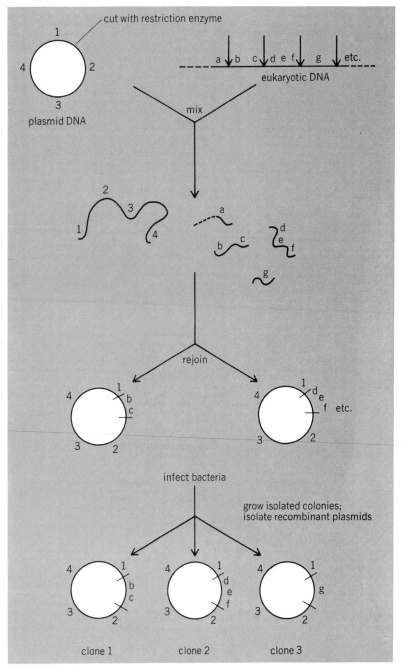

Fig. 3. Cloning of eukaryotic DNA by recombinant DNA techniques. A plasmid DNA molecule, capable of autonomous replication in *Escherichia coli*, and the eukaryotic DNA are digested with a restriction enzyme, the two samples mixed, and the ends rejoined by the enzyme polynucleotide ligase. This generates all possible recombinant molecules. Bacteria are then transformed with the DNA mixture under conditions where only one DNA molecule enters. The transformed bacteria can then be selected for by an antibiotic resistance marker present on the plasmid: nontransformed bacteria are killed by the antibiotic—the transformants are resistant; they grow up into individual colonies derived from one bacterium; DNA is then extracted from these bacteria and analyzed. In the figure three bacterial colonies (clones) are shown to contain different recombinant DNAs.

grown on agar as colonies derived from a single bacterium. Bulk growth of these bacteria can provide large amounts of the recombinant DNA, originally derived from a single DNA molecule.

A similar technology is available when the cloning vehicle is a bacteriophage (a bacterial virus) DNA instead of plasmid DNA. In this case, a nonessential region of the phage DNA is replaced by foreign DNA as described above for plasmids.

Molecular hybridization. As stated above, DNA is double-stranded; the complementary strands are held together by base pairing where A binds to T, and G to C. This base pairing can be broken, for example, by heating, at which the strands separate as two single-stranded DNA molecules (DNA denaturation). If the correct conditions are provided, the reaction can be reversed and a duplex DNA molecule is formed by the reassociation of the two single strands (Fig. 4). This reverse reaction works equally well with an RNA + DNA mixture, where a DNA-RNA hybrid forms. This technique has been of vital importance in molecular biology. Because the reassociation reaction is highly specific, it is possible to determine whether a DNA preparation contains a gene which specifies a given mRNA. This is usually done by fixing DNA to nitrocellulose filter paper (single-stranded, but not duplex, DNA sticks avidly to nitrocellulose) and then placing the filter in a solution containing the RNA (or a synthetic DNA copy of the RNA; see below) under reassociation conditions. If the mRNA binds to the filter-bound DNA to form a DNA-RNA hybrid, the gene of interest must be present.

Gel electrophoresis–filter hybridization. E. Southern at the University of Edinburgh devised an ingenious combination of restriction enzyme techniques and filter hybridization to detect the DNA fragments containing the genes of higher organisms. This principle of the method as adapted by A. J. Jeffreys and R. A. Flavell is shown in Fig. 5. where this method is applied to the detection of the rabbit β-globin gene. Rabbit DNA is cleaved with a restriction endonuclease; since mammalian DNA is so complex (3×10^9 bp), this generates about 10^6 different fragments, one or two of which

contain the rabbit β-globin gene. These fragments are separated on the basis of their size by agarose gel electrophoresis, then denatured in place in the gel, and transferred and fixed onto the nitrocellulose filter under conditions where the distribution of DNA on this filter perfectly mirrors that found in the gel. On hybridizing the filter with a radioactive DNA copy of the rabbit β-globin mRNA, this radioactive probe binds to the filter-bound DNA fragments which contain the rabbit β-globin gene. The radioactive β-globin gene hybrids can be detected after removal of the nonhybridized probe by exposure of the filter to x-ray film (the radiation causes blackening of the film, which indicates the original position of the β-globin gene fragments in the original gel).

Rapid DNA/RNA sequencing techniques. Until the mid-1970s, the sequencing of nucleic acids was the exclusive privilege of few selected enthusiasts who were willing to invest the considerable amount of effort required in exchange for a sequence of a few hundred nucleotides. More recently, however, rapid methods have been developed independently by F. Sanger and colleagues at Cambridge University and A. Maxam and W. Gilbert at Harvard University, which make possible the determination of a sequence thousands of nucleotides long in the space of months. Though these methods were initially available only for DNA, adaptations of the technology have also made rapid sequencing of RNA a reality.

EUKARYOTIC mRNAs

The fact that a single gene constitutes only one part in a million of the total DNA has, until recently, greatly impeded work on eukaryotic genes. However, because a suitable cell containing but a single copy of a given gene usually contains many more, indeed sometimes thousands, of copies of the transcripts of the gene (that is, mRNAs), work on eukaryotic mRNAs preceded work on the genes themselves. The structure of eukaryotic genes has therefore been approached via their mRNAs. To isolate a mRNA coding for a single protein, investigators have used specialized cell types, such as red blood cell precursors. In adult animals, hemoglobin is a tetrameric protein consisting of two molecules, each of two nonidentical subunits (α_2, β_2). These subunits are encoded by two mRNAs which are similar in size (about 600–700 nucleotides in total), and are in most senses typical eukaryotic mRNAs. That is, they are monocistronic (code only for a single polypeptide) and contain at the 3'-end of the mRNA a covalently attached poly(A) tail (a terminal sequence where all the nucleotides are A's) of about 50–150 nucleotides long. This poly(A) tail is not encoded in the DNA; instead, it is added to the 3'-end of the precursor RNA to globin mRNA by the terminal addition of adenosinemonophosphate (AMP) residues.

The presence of the poly(A) tail has greatly facilitated the isolation of eukaryotic mRNAs. This is achieved by binding the mRNA to a matrix-bound poly(dT). The poly(A) tail of the mRNA is complementary to the immobilized (dT) residues; the mRNA is, therefore, retained on the matrix by the formation of a poly(A)-poly(dT) hybrid. Since the other major cell RNAs do not have poly(A), they are not retained to any significant degree by the

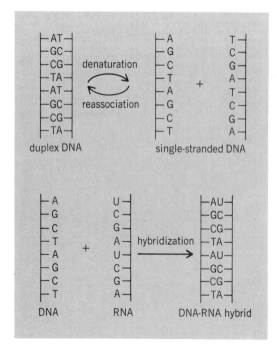

Fig. 4. Schematic representation of denaturation and renaturation of DNA and formation of a DNA-RNA hybrid.

matrix. To purify the globin mRNA free of any other nonglobin poly(A)-containing RNAs, a size selection is carried out whereby only RNAs of 600–700 nucleotides are collected. Thus, in a simple two-step procedure, essentially pure globin mRNA can be isolated.

An alternative way to obtain specific mRNAs is to make use of viral systems. Certain viruses make use of the cellular enzyme systems and divert the cell's energies to the production of viral products. For example, late in the infection cycle of human adenovirus, 50% of the RNA synthesized is of the order of a dozen mRNA species which code for the so-called late viral proteins (mainly structural proteins of the virus). Thus these specific adenovirus mRNAs are again present at high levels in an otherwise unspecialized cell and can therefore be isolated with little difficulty.

Molecular cloning of a synthetic globin gene. For certain problems, the presence of the inevitable minor contaminants in globin mRNA preparations is a difficulty. In addition, the development of the recent technology discussed above has shown that DNA is a much better experimental material than RNA. For these and other reasons, several laboratories set out to amplify DNA copies of globin mRNA as recombinant DNA. This is achieved by copying the globin mRNA into a single-stranded complementary DNA (cDNA), using an enzyme reverse transcriptase. The single-stranded cDNA is converted to a double-stranded DNA, after removing the globin mRNA, by a similar reaction. The resultant double-stranded cDNA is then inserted into a bacterial plasmid.

The availability of large amounts of cloned mRNA copies has made possible a highly detailed analysis of eukaryotic mRNAs. In the case first analyzed, this has made possible the complete sequence determination of the rabbit β-globin mRNA. This RNA contains, in fact, 56 nucleotides at the "leading" end which do not code for the β-globin protein; it is thought that this RNA sequence plays a role in the initiation of protein synthesis. Thereafter follow 438 nucleotides which code for the β-globin protein, finally terminated by 95 noncoding nucleotides, of as yet unknown function, and the terminal poly(A) sequence. Perhaps most important, however, is the fact that the technology developed for globin cDNA cloning is now applicable to practically every mRNA.

Elucidation of eukaryotic genes. Armed with powerful new techniques and considerable knowledge of the sequence organization of eukaryotic mRNAs, several groups have analyzed the structure of eukaryotic genes.

The first shock came from a series of studies on the sequence arrangement of the mRNAs of human adenovirus. Several independent groups at the Cold Spring Harbor Laboratory and S. Berget, C. Moore, and P. Sharp at the Massachusetts Institute of Technology showed that the late mRNAs of adenovirus 2 consist of a transcript of noncontiguous segments of the adenovirus genome; three common "leader" segments encoded in the left-hand end of the adenovirus DNA are covalently attached to a protein-coding mRNA "body," which is a transcript of a sequence far downstream to the leaders. This was most clearly demonstrated in experiments where the adenovirus mRNA-DNA

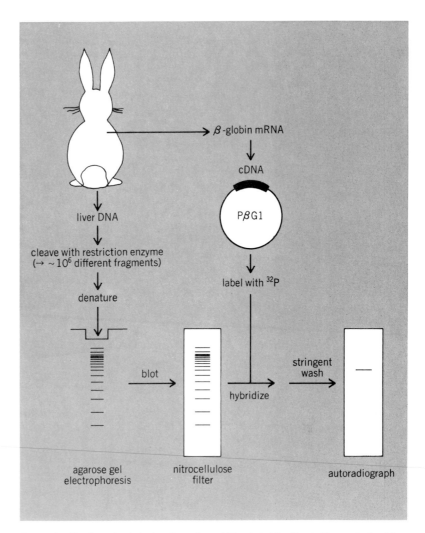

Fig. 5. Application of gel electrophoresis and filter hybridization to the analysis of the rabbit β-globin gene.

hybrid was examined in the electron microscope (Fig. 6). It can be seen that the mRNA forms a hybrid with four noncontiguous regions of the adenovirus DNA. The intervening DNA sequences not present in the mRNA therefore loop out in the DNA-RNA hybrid. A comparison of the position of the three leader and body sequences for several different late adenovirus mRNAs was even more startling: each different mRNA was a mosaic consisting of the same three leaders, coupled to a discrete body segment.

One more fact provided the key to a compelling model for adenovirus transcription. The late region of adenovirus has been shown to be transcribed as a giant pre-mRNA by J. Darnell and colleagues at Rockefeller University. The starting point of RNA synthesis that they determined for this RNA was the same as the site of the left-most leader sequence. This led to the proposal that transcription of the major late region of adenovirus occurs by synthesis of a single large RNA. A given mRNA is obtained by excision of the intervening sequences between each leader sequence, followed by a splicing reaction whereby the genetically noncontiguous (leader$_1$-leader$_2$-leader$_3$-body) mosaic RNA is created (see Fig. 7 for a schematic representation of this model applied to the globin gene system).

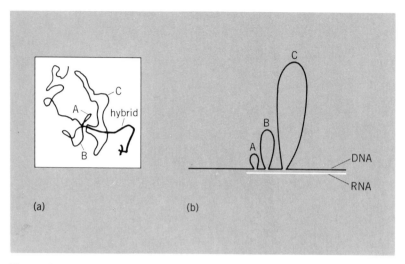

Fig. 6. Electron microscopic visualization of the DNA-RNA hybrid formed between adenovirus DNA and a spliced "late" mRNA. (a) Tracing of the original micrograph (from S. Berget et al., Spliced segments at the 5' terminus of Adenovirus 2 late mRNA, Proc. Nat. Acad. Sci. U.S., 74:3171–3175, 1977). (b) Schematic interpretation of the hybrid structure.

SPLIT GENES

In the summer of 1977, the structure of the rabbit and mouse β-globin genes was shown to have a similar type of arrangment to that described above (Fig. 7). The mouse β-globin gene structure was examined by electron microscopy of DNA-mRNA hybrids and some restriction enzyme mapping. The DNA-RNA hybrids showed that the mouse β-globin gene contained a 500–600-bp sequence, not present in the globin mRNA. The position of this interruption, however, was within the coding regions of the mouse β-globin gene, whereas the leader sequences described for the

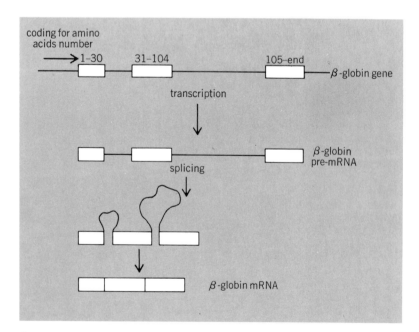

Fig. 7. Model for the expression of a split gene applied to the rabbit β-globin gene as example. The gene consists of three coding segments, separated by two intervening sequences. Transcription is thought to start at the beginning of the first coding segment. It is not yet clear where transcription terminates.

adenovirus mRNAs do not contain sequences that code for protein. In a parallel study, the same observation was made for the rabbit β-globin gene by restriction mapping on total genomic DNA as described above. Here, a restriction map was obtained which showed the rabbit β-globin mRNA to be encoded in at least two noncontiguous regions, separated by a 600-bp intervening sequence present somewhere within the region coding for amino acids 101–120 of the β-globin polypeptide chain. Similar discontinuities were found for the chicken ovalbumin gene, mouse immunoglobin genes, and genes in the SV40 and polyoma viruses.

The discontinuous structure of the rabbit β-globin gene was shown to be a constant feature in all tissues examined, including those actively making globin mRNA and germ-line cells destined to pass on the β-globin gene to progeny rabbits. It seems highly probable, therefore, that the discontinuous β-globin gene had to be transcribed as a larger pre-mRNA which should be processed in a cleavage-plus-splicing operation to remove the intervening sequences in the way described above for adenovirus. A β-globin pre-mRNA has been described by a number of workers, and electron microscopy of hybrids between the globin pre-mRNA and the globin gene has confirmed the presence of the intervening sequence transcripts in globin pre-mRNA. With these observations and those on adenovirus, a major puzzle in the mode of RNA synthesis in the cell nucleus has been solved. In the eukaryote, mRNAs are made as large precursor RNAs containing both the intervening sequences and the mRNA coding sequences; the processing of the precursors involves removal of the internal intervening sequences rather than extensive processing at the termini as was originally thought (Fig. 7).

In the time that has passed since these initial discoveries, the structures of a large number of eukaryotic genes have been elucidated in considerable detail. As an example, Fig. 7 shows the structure of the rabbit β-globin gene, which has been the subject of Jeffreys and Flavell's attention. Detailed DNA sequence information and restriction mapping have shown that the rabbit β-globin gene consists in fact of three mRNA coding segments, separated by two intervening sequences of 126 and 573 bp, respectively, which split the coding sequences at the positions corresponding to amino acids 30–31 and 104–105 of the 146-amino-acid-long β-globin polypeptide chain.

Compared with the β-globin gene, the structure of the chicken ovalbumin seems baroque. P. Chambon's group at Strasbourg and B. O'Malley's group at Houston have both unraveled the complex structure of this gene; spanning 8000 bp of chicken DNA, it consists of seven intervening sequences and eight mRNA coding regions! A large number of other genes have also been examined; although several eukaryotic genes are split, this is by no means always the case. The table shows that there are also several colinear genes in eukaryotes; too few data are available to determine which type of gene is in the majority.

Origin. An important point which has to be resolved is the origin of split genes. Two models can be envisaged (Fig. 8). One possibility is that genes arose as colinear coding sequences and that split genes arose later by the insertion of DNA sequences into the gene. Alternatively, split genes might be the primordial gene type; further evolution by recombination could reassort the gene segments to create novel combinations of protein-coding sequences, some of which could offer selective advantages to the organism. Colinear genes could evolve from split genes by a precise deletion of the intervening sequence or sequences.

At present there is little evidence to distinguish these two alternatives. It is clear that split genes are an extremely old phenomenon. It is believed that the genes coding for α- and β-globin are both derived from a common ancestral gene by a gene duplication event which must have occurred about 500,000,000 years ago. Both α- and β-globin genes contain two intervening sequences at identical homologous positions in the coding sequences. It must therefore be concluded that these two intervening sequences were also present in the same position in the progenitor globin gene. Though this is consistent with the idea that split genes evolved first, it is no proof. It will be of interest to see if the plant leg-hemoglobin which diverged from animal globins perhaps 1,000,000,000 years ago, also has the intervening sequences.

Also of interest is recent work in the two rat insulin genes. A. Efstratiadis, W. Gilbert, and colleagues have shown that while one of these genes has a single intervening sequence, the second gene contains an additional second intervening sequence. It is, of course, not clear from this observation whether the first gene has lost an intervening sequence or the second gene has gained one; examination of the structure of insulin genes in related species should answer this question, however. These data do at least suggest that precise excision or insertion of an intervening sequence can occur which at first sight supports the insertion sequence model—it is, however, also compatible with the second model (by precise excision of the second intervening sequence).

Significance. Why genes are split is an inevitable question related to the origin of split genes described above. If, for example, split genes are

A list of eukaryotic split and colinear genes

SPLIT GENES
 Drosophila 28S rRNA
 Adenovirus SV40 and polyoma
 RNA tumor viruses
 α-, β-, γ-, and δ-globins
 Ovalbumin, conalbumin, ovomucoid,
  and lysozyme (chicken)
 Immunoglobulins
 Silk fibroin
 Yeast and *Xenopus* tRNAs
 Yeast mitochondrial rRNA, and cytochrome *b*

COLINEAR GENES
 Histones (sea urchin, *Dictyostelium*)
 Yeast cytochrome *c*
 Dictyostelium actin
 Yeast mitochondrial ATPase subunit
 Yeast and *Xenopus* tRNAs
 Xenopus rDNA
 Xenopus and yeast 5S RNA

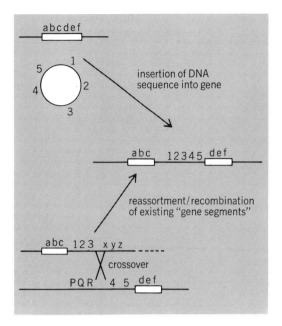

Fig. 8. Two models for the evolutionary origin of split genes.

the primordial gene type, their existence might be a consequence of the way life first evolved on Earth: the prolongation of the split-gene configuration until modern times would further suggest that during evolution this structure has provided a selective advantage to the eukaryote.

The evolutionary advantages of a split-gene configuration seem on paper to be significant. It can be envisaged that initially short coding regions could direct the synthesis of short polypeptides; these might associate in solution to generate a protein with a function more complete than that of its separate constituents. If these "gene" units could be assorted at the DNA level by the type of recombination events illustrated in Fig. 8, in a genetic background where a splicing enzyme is capable of removing the transcript of the intervening sequence, then a mRNA directing the synthesis of a single polypeptide containing both functional domains would be created. Further recombination provides the basis for continuous reshuffling of these gene units; yet if these recombinations occur within the intervening sequences, and if it is assumed that the sequence information required for splicing is common to most genes and is localized at the junction of the intervening and coding sequences, it follows that in most cases a functional mRNA would be produced which would encode a novel protein.

In fact, studies of protein structure have revealed that proteins are built up of functional domains. Considerable homology can be seen in the three-dimensional structure and protein sequence surrounding such functional structures as the active center of a dehydrogenase enzyme or regions of an antibody molecule, and this suggests that short homologous domains are evolutionarily related. Recently the structure of the immunoglobin genes has been worked out, notably by S. Tonegawa in Basel and P. Leder at the National Institutes of Health. Strikingly, the domain structure of

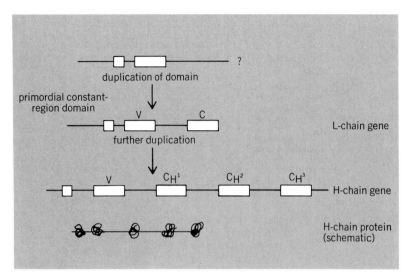

Fig. 9.   Model for the evolution of the antibody genes.

the immunoglobin light (L) and heavy (H) chains is precisely reflected in the gene structure. The active L-chain gene consists of three coding segments and two intervening sequences; the intervening sequences separate the constant and variable region domains and the variable domain from the prepeptide region (which is removed from the precursor immunoglobulin during its secretion by the lymphocyte), respectively. The H-chain gene is very similar except that now the number of constant region protein domains has increased from one to three; the gene structure reflects this by the presence of two extra coding regions and two intervening sequences of the same size as these of the L-chain.

A model for the evolution of immunoglobulin genes is shown in Fig. 9. Here it is assumed that a single domain-coding gene segment becomes duplicated to generate an L-chain-type gene and that later the constant-domain "gene" becomes duplicated further to give the H-chain gene. (If such a

system of domain-coding sequence duplication can occur, researchers might expect to find proteins which contain a known domain of one given protein fused to an unrelated series of domains to give a protein with a novel function, yet which retains part of the properties of the original protein. Such examples might, for example, be a segment of the α-subunit of the major histocompatibility antigen which contains a region homologous to an H-chain constant domain linked to other unrelated sequences; or the complement component C1q which contains a protein domain homologous to the repeating unit of collagen. Although these gene structures have not as yet been worked out, these examples — of which there must be more — suggest how new proteins can be created by the reassortment of available subgenic domain-coding structures.) Indeed, evidence for the existence of a single-domain gene is found in β_2-microglobulin — a polypeptide homologous to a constant region domain, which serves as the β subunit of the major histocompatibility antigen present on the surface of most cells.

Multiple use of genetic information. The above discussion emphasized the role that split genes might play in the evolution of the eukaryote. Are there other possible ways in which split genes might be advantageous to the organism? Three additional possibilities have been raised. The first is that split genes open the possibility to use the same piece of genetic information for several different gene products without duplicating the DNA sequence. An example of this is found in the papovaviruses SV40 and polyoma, where the region expressed early in infection codes for two proteins called T and t. These proteins share a common N-terminal amino acid sequence. The C-terminal regions differ, however, because the splicing reaction creates two different mRNAs by linking the N-terminal RNA segment to a different C-terminal segment in both cases (Fig. 10). Interestingly, the splicing reaction which generates T mRNA eliminates the codon which specifies the stop signal of t, and the result is a mRNA specifying a larger protein with presumably different properties. An example of the multiple use of noncoding leader sequences is found in the three late leader sequences of adenovirus discussed above. Multiple use of genetic information might be predominantly a viral phenomenon. While strong selection is exerted on the genetic information content of viruses (for example, the viral DNA must fit into a viral head), there is no indication that a similar selection occurs on the nuclear DNA content of higher organisms. In the latter case it might, therefore, be expected that gene (or subgene segment) duplication would fulfil the type of role discussed here and in the preceding section.

Gene expression and recombination. A plausible idea which remains to be discussed is that the fact that genes are split provides the cell with at least two levels at which to regulate gene expression: transcription and the processing of the pre-mRNA by splicing. For example, let it be imagined that the cell has 100 different transcriptional regulators and 100 different splicing systems. The number of different regulatory possibilities is therefore $100 \times 100 = 10^4$. This has obvious theoret-

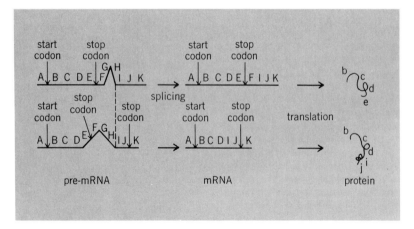

Fig. 10.   Multiple use of genetic information in papovaviruses. The "early" region of SV40 and polyoma viruses code for two major mRNAs (in polyoma there is actually a third, which is omitted here for simplicity). These are transcribed from the same region, and by different splicing pathways two different mRNAs are produced from a common precursor.

ical advantages over a system based exclusively on transcriptional regulation, where to achieve the same effect 10^4 transcriptional regulators would be required. Furthermore, it could be imagined that intervening sequences themselves could play an important regulatory role, either at the DNA level by serving as sites for DNA-binding proteins, or at the RNA level, where the (excised) intervening sequence transcript could serve a regulatory role. The weight of evidence (as yet admittedly not very much) does not favor this possibility. Such a regulatory role should require that selection at the level of the primary sequence of the intervening sequence should occur. Yet a comparison of the mouse and rabbit β-globin genes clearly showed that much greater differences exist between the primary sequence of the homologous intervening sequences of these genes than exist between the corresponding protein-coding sequences. Apparently, therefore, there has been less selection on these sequences.

The rapid evolutionary divergence of intervening sequences has suggested to some workers that split genes might have lower frequencies of recombination between homologous chromosomes at meiosis than would be expected for colinear genes; the reduced homology between the intervening sequences of duplicated genes could suppress the frequency of unequal crossing-over that occurs between the diverged duplicated genes, since it is believed that recombination is mediated primarily via base pairing of homologous segments of chromosomes and since it seems reasonable that a minimal length of homology will be required to start a successful crossing-over event.

FUTURE RESEARCH

This article shows that in the last few years much has been learned about the sequence organization of eukaryotic genes; this has in turn done much to influence ideas on the way humans have evolved. Essentially nothing has been learned, however, about how the expression of the bewildering number of genes is capable of building organisms of such exquisite complexity as a human. It is this type of question that originally promoted the swing from research on prokaryotes to eukaryotes; it is likely that the new technology described here will make possible answers to at least parts of this compound question in suitable experimental model systems by, say, 1985.

[RICHARD A. FLAVELL]

A plant, like any other living entity, may become diseased. A disease is epidemic when it intensifies and spreads over a large population of plants. Alternatively, it may be stated that an agent of disease, often called the pathogen, causes an epidemic in a plant population, commonly indicated as the host.

During an epidemic, a population of pathogens multiplies at the cost of a population of host plants; the plants are injured, their yield decreases, and considerable economic losses are suffered. Epidemiology is a quantitative science. In describing epidemics, therefore, numbers are important, including numbers of pathogenic units, numbers of plants affected, and numbers of kilograms of produce and dollars lost per hectare of crop.

An epidemic is a complex biological phenomenon. Growth and interaction of the pathogen and the host are affected by environmental factors (variables). The crop manager, when intervening by means such as chemical treatment, can become a third interactive component in an epidemic. Plant disease epidemics can be considered as dynamic systems that develop according to the conditions dictated by the environment (in the form of microclimate), by the characteristics of host and parasite (in the form of resistance and pathogenicity, respectively), and under human influence (in the form of plant nutrition and pesticide usage). Before the advent of computers, the complexity of and the dynamic interaction in epidemics rendered them difficult to analyze and comprehend. Currently, with the development of simulation as an investigative tool and increased access to computers, these same features, complexity and interaction, make plant disease epidemics suitable subjects for systems studies.

This article discusses the role of computer simulation in the study of plant disease epidemics. The rationale for applying simulation in epidemiology will be examined; so too will the encompassing framework of systems

P. S. Teng is assistant professor in epidemiology at the University of Minnesota. Born in Malaysia, he received his university education in New Zealand, and has spent time in both countries working on epidemics that cause significant crop loss.

J. C. Zadoks is a senior lecturer in plant pathology at the Agricultural University of Wageningen, the Netherlands. He has published numerous papers on the ecological aspects of plant diseases, and has served in many missions to developing countries.

Computer Simulation of Plant Disease Epidemics

analysis and computer modeling. Successful development of a simulation model is a notable achievement, but is not the end point in systems studies. The article will discuss how models can be used to further understanding of epidemics, to guide research efforts, and to "manage" disease. It will be necessary first to explain simulation and epidemiology separately, and then to show how they relate to each other.

SIMULATION AND EPIDEMIOLOGY

As an investigative technique, simulation is not hindered by the rigidity encountered in more established quantitative approaches for systems studies, such as linear optimization. Broadly speaking, simulation means replication of the essence of a system without actually attaining reality itself. A working definition of simulation is provided by Robert Shannon: "the process of designing a model of a real system and conducting experiments with this model for the purpose either of understanding the behavior of the system or of evaluating strategies for the operation of the system." The process of building a model is an indispensable part of simulation. A simulation model acts as a substitute of the real system, but has instead a computer core as its medium of operation. It is possible to simulate an epidemic without computers, but the computer has accelerated progress by eliminating the need for repetitive calculations.

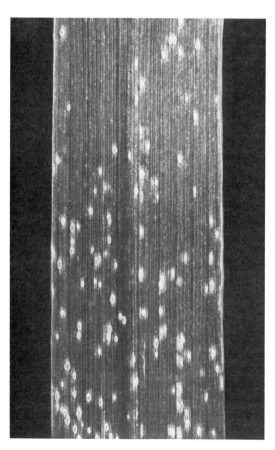

Fig. 1. Uredia (the small spots) of *Puccinia recondita* on a wheat leaf.

System. A system, in its simplest sense, is a limited section of reality, consisting of parts united by regular interaction to perform a specified function. In an epidemic, these parts are the pathogen and the host, which interact over time in the presence of environmental variables. In natural systems, it is common that the system under study is part of a hierarchy. Therefore, in practice, an upper limit is applied to define the extent of the system under investigation. What is inside the conceptual boundary is called the system proper, and what is outside is called the system environment. The system proper contains the main elements of interaction, the host and the pathogen. The system environment contains the variables which drive the system, for example, sunshine, wind, and rain. In some models, a crop manager's action is considered to be an interactive element of the system; in others, it is handled as a driving force.

Model. To study such a system, a model is constructed to represent the system in as many respects as feasible. A model is a set of mathematical equations that completely describes the system. Whereas the system description may be qualitative, the model is quantitative. Computer simulation models represent systems by means of symbols that are translated into electrical bits of activity. Computer simulation models can display many of the dynamic characteristics of biological systems, and the display may be in the form of tables and graphics projected on a television screen.

Systems approach. Underlying computer simulation is a scientific philosophy, popularly called the systems approach, which proposes a holistic view of systems. It states that systems will not be properly understood by studies limited to their components alone. This is because a change in one aspect of a system may well produce changes in other parts of the system so that the whole is more than the sum of its parts. The systems approach has been variously called systems analysis, systems simulation, and systems research. Regardless of its name, the approach remains a quantitative one, where some of the following elements are present: measurement, analysis, modeling, simulation, and system experimentation.

PLANT DISEASE EPIDEMIC AS A SYSTEM

As an example an economically important disease (whose pathogen has a relatively simple life cycle) will be considered: brown leaf rust of wheat caused by the fungus *Puccinia recondita* f. sp. *tritici*. The fungus mainly attacks the foliage and manifests itself on susceptible plants as small, rusty brown pustules called uredia (Fig 1). Although the fungus undergoes many morphological changes during the season, only the uredial stage is capable of causing significant damage to the wheat crop. During a single cropping season, many repetitive cycles of this uredial stage occur if weather conditions are favorable. Each of the infection cycles comprises many activities which are individually affected by different environmental factors.

Figure 2 shows what happens from the time a spore is formed until it becomes a fruiting unit, capable of producing more spores. Spores are produced in the pustules (labeled 1), each pustule being capable of producing up to 800 spores per day, at the peak of its life, and up to 10,000 in its lifetime. Pustules do not produce spores for an indefinite period; they die after passing through an infectious period. Spores ripen and become detached (labeled 2) from their stalks by wind or leaf movement. They become airborne, and after an interval are deposited. Spores deposited on the soil, other crops, or nonsusceptible wheat plants are lost. Only spores deposited on leaves of susceptible wheat can perpetuate the epidemic. Even so, spores on a suitable leaf surface may fail to encounter favorable conditions for germination, and especially those on the uppermost leaf layer may be killed by sunshine. For germination, rust spores require free moisture on the leaf surface, and this is commonly provided by dew droplets or light rain. Germinated spores put out germ tubes that penetrate the leaf (labeled 3). If the colonization is successful, the germ tube develops into fungal mycelium (labeled 4). The fungus grows within the leaf and builds up reserve material (labeled 5). During this period, the fungus is not visible to the observer, and epidemiologists have called this period of growth the latent period. Its duration depends on host, fungus, and environmental factors, of which ambient temperature is the most important. Not all latent infections will become pustules, since host plants have regulatory mechanisms that act against the invasion by the fungus. If the germ tube from the original spore is successful, after a latent period it becomes another uredium that erupts on the leaf surface and starts to produce further spores. And so the process repeats itself. Every infection cycle leads to a multiplication of the number of pustules by, say, tenfold. An epidemic of *P. recondita* in a wheat crop consists of a sequence of partly overlapping infection cycles; it is a polycyclic epidemic (Fig. 3).

In Fig. 3 the unit used for denoting the amount of disease is the percentage severity, which is the percentage of leaf area covered by disease. This is derived from considering the area of each uredium and its associated diseased area. The sum of rusted leaf area and area dead from disease gives the total diseased area. This, expressed as a percentage of the total leaf area, gives a measure called disease severity. The epidemic of brown leaf rust on wheat, monitored regularly over the lifetime of a crop, commonly shows an S shape when plotted as disease severity against time. For practical plant pathologists, this S-shaped curve is the abstract manifestation of an epidemic observed in a field of wheat. To the epidemiologist, this curve is the net sum of the interactions of host and pathogen and of all the complex processes whereby environment affects the growth of pathogen and host.

SIMULATION PROCESS

The construction of simulation models in epidemiology goes through a sequence of steps (Fig. 4).

Identification of problem and objectives. The objective for building a model has to be considered since it affects the structure of the model in terms of scope and detail. A clear objective enables a neat definition of the conceptual boundary and main components of the system and its environ-

ment. Here, it is worthwhile to digress a bit and elaborate on some terms used in simulation. The model is quantitative. At any point in time the state of the system is expressed in measurable units, the state variables of the system; examples of state variables are leaf area, number of latent infections, number of infectious pustules, number of spores produced, and number of germinated

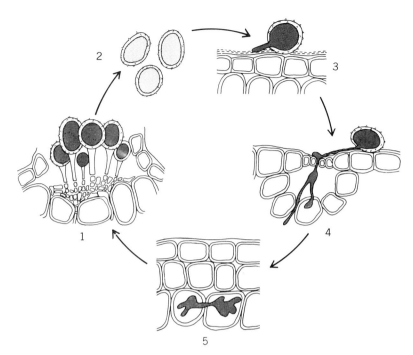

Fig. 2. Generalized infection cycle of the uredial stage of a leaf rust.

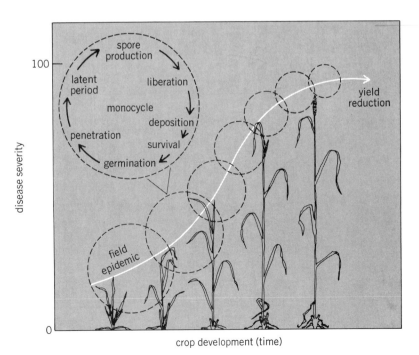

Fig. 3. Brown leaf rust epidemic on wheat and its components. Each of the rust generations, called a monocycle, is represented by a broken circle, and itself comprises numerous activities (left upper circle). Duration of a rust generation varies with prevailing environmental conditions, as shown by the different sizes of the circles.

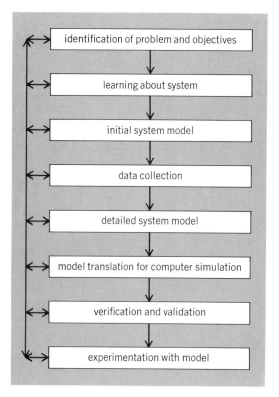

Fig. 4. Steps in the simulation process.

spores. The magnitude of the state variables changes over time according to the effect of driving variables that originate from the environment; examples of driving variables are temperature, dew, and wind. Variables change from one state to the next at rates determined by rate equations and are modulated by the driving variables. The particular rate of change of a state variable at any time is a function of driving variables, but also of state variables and of constants called parameters. In a particular combination of a virulent race of brown leaf rust and a susceptible variety of wheat, 100 germinated spores that become incipient colonies will result in approximately 20 pustules: the ratio of 0.20 can be used as a parameter. Parameters may, in their turn, become variables when this ratio changes with the age of the leaf.

Learning about the system. The next step in the simulation process involves learning about the host-pathogen system so that an initial model can be formulated. Epidemiologists endeavoring to build a simulation model have some prior knowledge of the system they are interested in. An initial system model is not necessarily a formalized model. The infection cycle of Fig. 2 is a good example of a pictorial system model. The learning phase enables the modeler to draw on experience and on published literature concerning the system. It also reveals what information is lacking. Thus this phase of system analysis is in itself a valuable exercise. It often leads to the disappointing conclusion that much of the knowledge about a particular disease is irrelevant for understanding the epidemics of that disease and for modeling them. Therefore the epidemiologist has to collect data about the effects of certain driving variables on some

state variables. For example, whereas there may be information on how temperature (driving variable) affects the number of germinated spores (state variable) after a period of moisture, there may be no information about the effects of light intensity, or of the conditions under which spores have been formed, on spore germination. Conducting experiments to provide data for simulation is time-consuming and resource-demanding. These are good reasons why, when studying a new disease, it will prove profitable to adopt a systems approach from the outset to make sure that experimentation relates to problem clarification.

Developing the system model. Development of a computer simulation model is an ongoing affair. The model progresses toward a better representation of the real system according to the incoming information. When the epidemiologist has collected enough information for the first detailed attempt at simulating the epidemic, the next step is to translate this information into a form suitable for the computer. Decisions have to be made on the computer language to be used, on the clockwork of the system, on how feedback mechanisms are to be built into the model, on methods of handling files of driving variables, and even on the question of whether or not the epidemiologist is competent to handle computer simulation. The general-purpose computer language FORTRAN has been widely used in simulating plant disease epidemics. Special simulation languages are available, such as CSMP (Continuous System Modeling Program) and GASP-IV. The availability of such special simulation languages means that plant pathologists wanting to simulate epidemics need not be experts in programming. It must be stressed that self-activity of the epidemiologist in simulation is important for a deeper understanding of the epidemic process.

A prelude to coding the model for computer simulation, and a procedure which many biologists find useful, is the drawing of a flow diagram showing how the various components of the system relate to each other. Such a flow diagram may take the form of block diagrams and relational diagrams or of a formalized flow chart. Some symbols have, through usage, become accepted as representations of system elements (Fig. 5).

Verification and validation. On completion of model translation, preliminary results at least can be expected from the simulation. Rarely will a model be successful at the first try; more likely, some errors have been made in coding the information for the computer. Model debugging, that is, getting rid of any "bugs" in the computer program, is a time-consuming affair but essential. The software in modern computer operating systems usually has facilities for identifying syntax or format "bugs" in a program, but the logical error is more difficult to find. An example of a logical error is disease severity assuming negative values or values exceeding 100%! The aspect of simulation concerned with ensuring that the model behaves in the manner intended by the epidemiologist might be called model verification. Common sense is the major guide in the verification phase. The process of checking the behavior of the simulation model against that of a field epidemic, given the same set of parameters and driving variables, is here called validation of the model. Criteria for establishing model validity are not well developed in epidemiology, but it is possible to collect field data on disease and to compare the simulated curve with the disease progress curve as observed in the field.

Experimentation. When at long last the epidemiologist has constructed a simulation model that satisfactorily depicts the pathosystem, the model may be used as an experimental tool. Here the computer becomes a powerful medium, since the epidemiologist can explore the effect of a change in some properties of the pathosystem on the development of the epidemic. This can be done by using many computer crop seasons. Replication can be introduced into simulation experiments by the use of random numbers and probability distribution, and results generated by computer simulation can be subjected to statistical analysis as if they were results from field experiments.

Although in Fig. 4 the design process of a simulation model was represented by discrete boxes, these steps do not need to follow one another strictly. The double-headed arrows indicate that one step can influence a previous one and that the process of developing a simulation model is in itself a dynamic process, with feedback and interaction.

SIMULATION OF BROWN LEAF RUST EPIDEMICS

The following example is given of a simulation of the pathosystem formed by wheat *(Triticum aestivum)* and brown leaf rust *(P. recondita)*. Such a simulation can be achieved at several levels of complexity, with increasing biological detail.

Growth of cereal rust epidemics approximately follows the logistic equation (1), where x is disease

$$\frac{dx}{dt} = r \cdot x \cdot (1-x), \ldots \qquad (1)$$

intensity or the fraction of host tissue that is infected (the total host material being assigned the value 1), t is time, and r is a proportionality factor called the apparent infection rate. In graphics the logistic equation is represented by a sigmoid (S-shaped) curve when time is plotted on the abscissa and disease on the ordinate (Fig. 3). The logistic equation has some defects, corrected by J. E. Vanderplank, considered by many the father of quantitative epidemiology, to become Eq. (2), where x_t

$$\frac{dx_t}{dt} = R_c \cdot (x_{t-p} - x_{t-i-p}) \cdot (1-x_t) \ldots \qquad (2)$$

is the proportion of diseased host material at time t, p is latent period (the interval between deposition of the spore on host material and eruption of the new pustule), i is the infectious period (the time length that a pustule is capable of spore production), and R_c is a proportionality factor called the corrected basic infection rate. Equations (1) and (2) are differential equations in which the time interval for solution is infinitesimally small. The curve described by a differential equation can be obtained by integrating the equation. While this is simple for Eq. (1), it has not yet been achieved for Eq. (2), in which more biological concepts are contained. In simulation, numerical methods are used for integration when an equation is too complex for

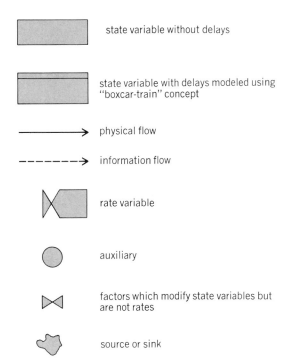

Fig. 5. Symbols used in flow diagrams.

state variable without delays

state variable with delays modeled using "boxcar-train" concept

physical flow

information flow

rate variable

auxiliary

factors which modify state variables but are not rates

source or sink

analytical solution. The discrepancy between numerical and analytical integration can be ignored here; the utility of numerical integration more than compensates for any loss in precision. Epidemiologists conducting computer simulations are seldom

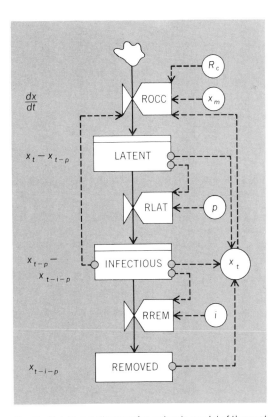

Fig. 6. Relational diagram for a simple model of the rust epidemic on wheat: model A.

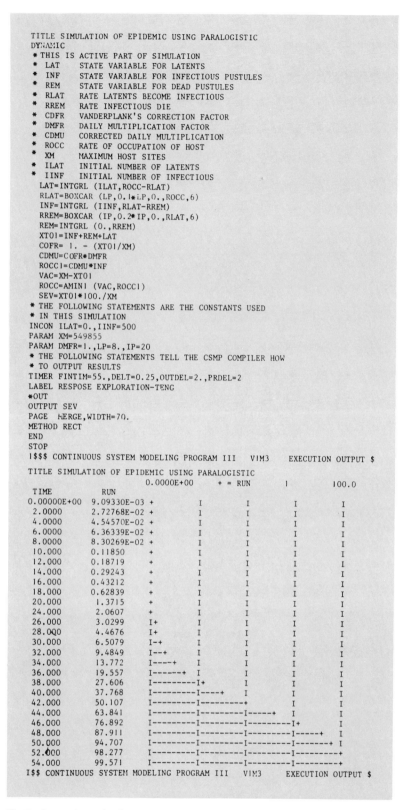

```
TITLE SIMULATION OF EPIDEMIC USING PARALOGISTIC
DYNAMIC
* THIS IS ACTIVE PART OF SIMULATION
*  LAT     STATE VARIABLE FOR LATENTS
*  INF     STATE VARIABLE FOR INFECTIOUS PUSTULES
*  REM     STATE VARIABLE FOR DEAD PUSTULES
*  RLAT    RATE LATENTS BECOME INFECTIOUS
*  RREM    RATE INFECTIOUS DIE
*  CDFR    VANDERPLANK'S CORRECTION FACTOR
*  DMFR    DAILY MULTIPLICATION FACTOR
*  CDMU    CORRECTED DAILY MULTIPLICATION
*  ROCC    RATE OF OCCUPATION OF HOST
*  XM      MAXIMUM HOST SITES
*  ILAT    INITIAL NUMBER OF LATENTS
*  IINF    INITIAL NUMBER OF INFECTIOUS
   LAT=INTGRL (ILAT,ROCC-RLAT)
   RLAT=BOXCAR (LP,0.1*LP,0.,ROCC,6)
   INF=INTGRL (IINF,RLAT-RREM)
   RREM=BOXCAR (IP,0.2*IP,0.,RLAT,6)
   REM=INTGRL (0.,RREM)
   XTO1=INF+REM+LAT
   COFR= 1. - (XTO1/XM)
   CDMU=COFR*DMFR
   ROCC1=CDMU*INF
   VAC=XM-XTO1
   ROCC=AMIN1 (VAC,ROCC1)
   SEV=XTO1*100./XM
* THE FOLLOWING STATEMENTS ARE THE CONSTANTS USED
* IN THIS SIMULATION
INCON ILAT=0.,IINF=500
PARAM XM=549855
PARAM DMFR=1.,LP=8.,IP=20
* THE FOLLOWING STATEMENTS TELL THE CSMP COMPILER HOW
* TO OUTPUT RESULTS
TIMER FINTIM=55.,DELT=0.25,OUTDEL=2.,PRDEL=2
LABEL RESPOSE EXPLORATION-TENG
*OUT
OUTPUT SEV
PAGE MERGE,WIDTH=70.
METHOD RECT
END
STOP
I$$$ CONTINUOUS SYSTEM MODELING PROGRAM III    VIM3    EXECUTION OUTPUT $
TITLE SIMULATION OF EPIDEMIC USING PARALOGISTIC
                            0.0000E+00     + = RUN    I        100.0
  TIME         RUN
0.00000E+00  9.09330E-03 +         I         I         I         I
  2.0000     2.72768E-02 +         I         I         I         I
  4.0000     4.54570E-02 +         I         I         I         I
  6.0000     6.36339E-02 +         I         I         I         I
  8.0000     8.30269E-02 +         I         I         I         I
 10.000      0.11850     +         I         I         I         I
 12.000      0.18719     +         I         I         I         I
 14.000      0.29243     +         I         I         I         I
 16.000      0.43212     +         I         I         I         I
 18.000      0.62839     +         I         I         I         I
 20.000      1.3715      +         I         I         I         I
 24.000      2.0607      +         I         I         I         I
 26.000      3.0299      I+        I         I         I         I
 28.000      4.4676      I+        I         I         I         I
 30.000      6.5079      I-+       I         I         I         I
 32.000      9.4849      I--+      I         I         I         I
 34.000     13.772       I-----+   I         I         I         I
 36.000     19.557       I-------+ I         I         I         I
 38.000     27.606       I---------I+        I         I         I
 40.000     37.768       I---------I-----+   I         I         I
 42.000     50.107       I---------I---------+         I         I
 44.000     63.841       I---------I---------I-----+   I         I
 46.000     76.892       I---------I---------I---------I+        I
 48.000     87.911       I---------I---------I---------I------+  I
 50.000     94.707       I---------I---------I---------I-------+ I
 52.000     98.277       I---------I---------I---------I---------I
 54.000     99.571       I---------I---------I---------I---------+
I$$ CONTINUOUS SYSTEM MODELING PROGRAM III    VIM3    EXECUTION OUTPUT $
```

Fig. 7. Computer code of model A in CSMP, and results of a simulation.

the addition of "newly diseased" to "already diseased" tissue.

Equation (2), represented by the relational diagram of Fig. 6, can be extended to Eq. (3), where x_m

$$\frac{dx_t}{dt} = R_c \cdot (x_{t-p} - x_{t-i-p}) \cdot \left(1 - \frac{x_t}{x_m}\right) \qquad (3)$$

is the maximum possible amount of disease. Disease intensity at any time (x_t) is the proportion of diseased host tissue.

Model A. For a simple model (Fig. 6) the total area of diseased host tissue can be considered as being the sum of the area covered by all rust pustules. Hence x_m and x_t can be expressed in terms of numbers of diseased sites, and disease intensity will be the fraction of diseased sites divided by the total number of sites. Examination of Eq. (3) shows that the rate of increase of disease (dx_t/dt) is proportional to the amount of infectious host tissue $(x_{t-p} - x_{t-i-p})$ and the fraction of uninfected host tissue $(1 - x_t/x_m)$. In the relational diagram of Fig. 6, diseased tissue is partitioned into three components: latent, infectious, and removed. The last has pustules which, having passed through an infectious period, are no longer capable of spore production, that is, of contributing to an increase in disease. The proportionality factor R_c is a multiplier to calculate the amount of disease that can result during one time interval from disease present at the beginning of that interval. If the solution interval is 1 day, it can be considered as a daily multiplication factor, or the number of daughter pustules per mother pustule per day.

Between the time that the fungus successfully penetrates a wheat leaf and the time it reemerges as a sporulating pustule, and between this time and when it stops producing spores, there are delays called latent period and infectious period. A way has to be found to introduce these delays in the computer code for simulation. Delays occur in many biological systems; for example, the larval instars of an insect life cycle can be considered a sequence of delays. The programming of delays is beyond the scope of this article.

The simple epidemic as defined in Fig. 6 can be simulated in CSMP using the program of Fig. 7, which also shows the resulting epidemic curve. The CSMP statements code the relationships shown in Fig. 6 for computer action. Statements labeled with an asterisk are nonexecutable statements which serve to make the program comprehensible to others. The first executable statement in the program is LAT = INTGRL (ILAT, ROCC — RLAT), which means that the value of the state variable LAT at any time is obtained as an integral of the initial number of latent infectious ILAT and the rate at which host tissue is infected (ROCC) minus the rate at which latent infections become infectious (RLAT). Parameters—that is, the properties of the system that are considered constant for this simulation—are coded by using statements that start with PARAM. The result of such a simulation is shown in the lower half of the figure; where the first column is the time in steps of four units and the second column is amount of disease in percentage severity. The S-shaped epidemic growth curve is obtained by interconnecting the plus signs in the figure.

interested in the mechanics of the integration procedure. It is sufficient to note that a choice of numerical integration methods is available. The integration procedure can be thought of as a cumulative procedure. The S-shaped curve results from

Model B. Until now, the amount of host tissue was kept constant during the simulation. Increased realism is obtained by expanding the model, using the variable host leaf area as measured in the field. In Fig. 8, host tissue is partitioned into two components: LIVE and DEAD. The amount of each of these at any time is governed by the rate of leaf growth (RLG) and the rate of leaf death (RLD). The rate at which live host tissue is attacked by disease (ROCC) is affected by the daily multiplication factor DMFR and a correction factor CF. The correction factor, derived by subtracting from LIVE the number of latent and infectious pustules, represents the amount of uninfected host tissue left and is equivalent to $(1 - x_t)$ in Eq. (1). The rate at which latent infections become infectious (RLAT) and the rate at which infectious pustules die (RREM) are influenced by the rate of leaf death. Rusts are obligate parasites, requiring living host tissue for their existence. Any pustule located on dead leaf tissue is dead too. The number of new infections per day varies according to the values of CF and DMFR. As disease increases, the value of CF decreases, so that it slows the rate of increase of the epidemic. The DMFR can be kept constant during a simulation, in which case it is treated as a parameter, or it can be made a variable. Model B can be used to experiment with a variable DMFR, which is a reflection of the rate of spore production.

Plant breeders and epidemiologists are aware that as resistance of wheat to brown rust increases, the number of spores produced per day declines, that is, the DMFR is reduced. Figure 9 shows that the bigger the DMFR, the larger the rate of increase of epidemics, as indicated by a shift of the curves to the vertical. This bears out what has been observed by plant pathologists in the field. The figure shows results of three constant values of DMFR: 2.0 (run 1), 3.0 (run 2), and 4.0 (run 3). Simulation time shown in the output is in steps of three units, up to 48.0 units. Disease severity is expressed as a percentage; for example, at 48.0 (last two rows in figure), severity values for DMFR = 2.0, 3.0, and 4,0 are 13.698, 48.682, and 83.366 respectively.

Model C. Figure 10 is a relational diagram of a detailed model of an epidemic of brown leaf rust. The figure shows the dynamic situation of a uredial monocycle, which, when integrated over time, becomes an epidemic. In a comparison to model B in Fig. 8, the most obvious difference is further partitioning of the daily multiplication factor DMFR into six state variables described by the processes from spore production to penetration. The rate of occupation is influenced by an additional auxiliary, the lesion size (LS). The connection between INFectious and SPORE PRODUCTION is a wavy line to indicate that the change in state is not a simple physical flow of matter but involves a change in dimensional property. The flow from SPORE PRODUCTION to PENETRATION is a flow of matter through a set of valves which reduces the flow in a stepwise manner.

When a wheat crop is at the flowering stage, it generally has four or five leaf layers, numbered from the top downward. Leaf 1, the flag leaf, is the youngest leaf, and leaves lower down progressively increase in age. The epidemic on the flag leaf layer will now be considered.

The daily multiplication factor (DMFR) is further divided into the epidemiological processes of spore production, liberation, deposition, survival, germination, and penetration (Figs. 2 and 3). Each of these component processes can be affected by the microclimate of the flag leaf, with physical factors acting as driving variables. The flag leaf area depends on the development of the crop and is expressed by two state variables, LIVE and DEAD. The number of sites available for infection by the pathogen is influenced by the size of a lesion (LS) determined from laboratory experiments. In Fig. 10 the rate at which living leaf tissue is occupied (ROCC) is now equivalent to the DMFR of the previous model. The ROCC cannot be a constant, because it depends on the outcomes of other processes, each of which is affected by different driving variables. Depending on the value of ROCC, a certain number of latent infections is established each day. These remain for an interval, the latent period, the length of which is in part determined by ambient temperature. Because biological entities exhibit inherent variability, not all germ tubes established on one day will simultaneously become erupted pustules when the latent period has expired. More likely, 80% of one lot of latent infections emerges at time t, another 10% at time $t + 1$, another 5% at time $t + 2$, and so on, that is, in a

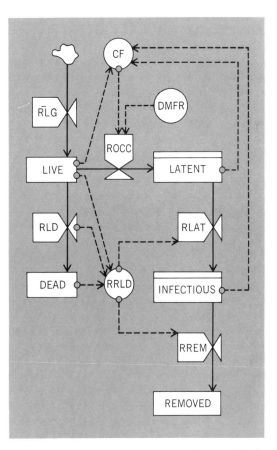

Fig. 8. Relational diagram of brown leaf rust on wheat: model B, with variable host tissue and constant daily multiplication factor.

dispersed fashion. This dispersed eruption pattern can be measured in laboratory experiments for a particular combination of rust race and wheat cultivar and can be accounted for in the simulation program.

The number of infectious pustules at any day is described by the state variable INF. The ability of a pustule to produce spores varies with the age of the pustule. Spore production peaks after a few days and then declines gradually during the infectious period. Apart from age, temperature affects spore production. Thus, on a certain day, it is necessary to know how many pustules there are in a certain age group, and how temperature affects their spore-producing capacity. Anyone familiar with statistical design will recognize this situation as a classical two-factor experimental design with interaction. To obtain data for a model, an experiment in a controlled environment would have to be conducted where diseased plants are kept under different temperatures and sampled for spores at frequent intervals.

Spores produced in any time interval are liberated, dispersed by wind, and then deposited. When the epidemics on several leaf layers are simulated simultaneously, each leaf layer may be regarded as a subsystem. The spores produced in one subsystem can enter another one to contribute to the epidemic there. Also, spores may leave the system forever, or they may enter the system from another crop. This part of the monocyclic epidemic, in which spore behavior depends entirely on physical factors, is one of the most difficult to simulate. Recently workers at the Rothamsted Experimental Station, in the United Kingdom, and the Agricultural University, in the Netherlands, have independently developed models for spore dispersal, based on physical laws and biological properties of crop and spore. It is now possible to simulate the spore dispersal pattern in a wheat crop with reasonable accuracy.

Spores that land on the leaf surface may or may not encounter favorable conditions for germination. The model must account for the fraction of spores that survives exposure to unfavorable conditions. Wheat rust spores require free moisture to germinate, but the rate of germination is governed by temperature. Germination, penetration, and colonization can be determined in the laboratory. Resulting parameters can be applied in the model. Natural conditions, however, are far more complex than those in controlled environments. Therefore a field check of parameters is advocated. In nature each of these parameters is a variable subject to effects of physical driving forces, leaf layer, leaf age, race characteristics of the rust, cultivar characteristics and nitrogen status of the host plant, and pesticide applications. This enumeration could easily lead to the conclusion that the position of the modeler-epidemiologist is hopeless. The contrary is true! Much work has to be done, but then the epidemiologist can be of great help to the modern plant breeder, who is interested in partial resistance. Partial resistance, just enough resistance for the needs of the crop manager but not more, is thought to be more stable than total resistance which, by genetic change in the rust population, can be suddenly and completely invalidated.

After the behavior of partial resistance (a promising future development) is studied in the laboratory, the field, and the model, the studies must also incorporate the effects of plant nutrition by fertilizer application, of water management, of fungicides, and so on. For every item, the whole modeling procedure must be repeated. This is, however, by no means a futuristic projection, since it is technically feasible to incorporate these effects into a single model.

SIMULATION AND DISEASE MANAGEMENT

For the plant pathologist, simulation is a means toward an end, and not an end in itself. The construction of a simulation model is an excellent way to improve one's understanding of an epidemic process. Eventually the model may become a tool for disease management or, better, for pathosystem management.

Long-term problems. Simulation models can be used for analyzing long-term effects of certain management policies, embracing many years. Effects of various degrees of partial resistance, multiline varieties, variety mixtures, fields of different varieties plotted in a mosaic pattern, sowing and harvest dates, split applications of nitrogen fertil-

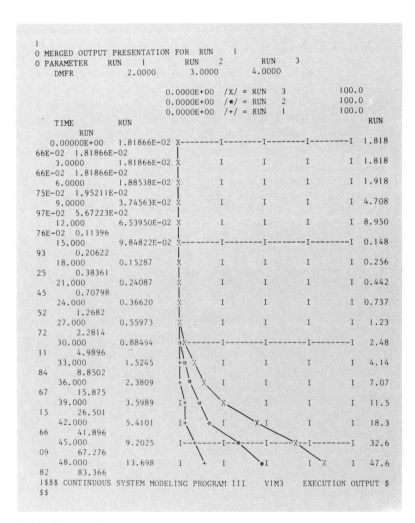

Fig. 9. Effect of different daily multiplication factors on epidemic development explored by using the wheat brown rust model B.

izer, and pesticide usage can be studied in an analytic and maybe even a predictive way. Very little has been done yet. Apparently science is waiting for a new breed of epidemiologists who can tackle large-scale and long-term problems without losing contact with reality.

Data bank. In the tactics of day-to-day pathosystem management, the simulation model is but one out of a set of tools. In the Netherlands, the project EPIPRE aims at supervised control of pests and diseases in wheat. One major tool is a data bank in which participating crop managers register their fields. Besides the core data, to be entered into the data bank once per year, regular updates on crop development, disease and pest observations, nitrogen application, and pesticide usage are registered.

Decision system. Another tool is a decision system based on various thresholds in disease severity—the damage, action, and warning thresholds—in which an economic criterion is incorporated. Care is taken of governmental regulations, such as deadlines for chemical treatments. The character of the pesticide, broad-spectrum or selective, is considered. A handy simplification is the grouping of variety-race combinations in a few classes: dangerous, intermediate, and not dangerous. Several times a week, the computer is told to update its information and to produce instructions for the participating crop managers. Following the decision system, the computer selects the appropriate recommendation from a set of preprogrammed advices. The recommendation is printed by the computer, and subsequently mailed to the crop manager.

Growth rates. Daily updates of weather data are obtained from the Royal Netherlands Meteorological Institute over a computer-to-computer telephone line. Weather data are used to determine the growth rates of epidemics in various danger classes for several diseases. Simulation models are needed only for general growth rates. When these general growth rates are known, the necessary projections into the future can be made for every registered field. Application of the simulation model to every separate field is far too expensive.

Value of the model. When the simulation model has satisfied the scientific curiosity of the researcher, it becomes an instrument, a tool. For pathosystem management, several other tools are

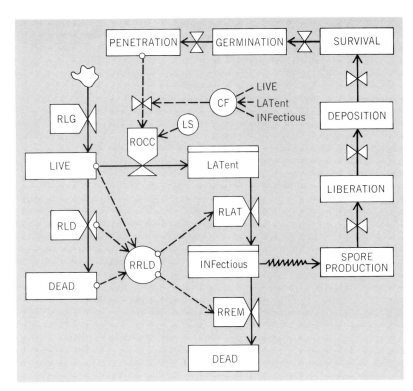

Fig. 10. Relational diagram of a detailed model describing leaf rust on wheat: model C.

needed. In fact, pathosystem management without simulation is perfectly feasible. Simulation not incorporated in disease management is a somewhat sterile exercise. If applied, simulation models are modest but useful arms in the tactical weaponry for plant disease management.

THE FUTURE

Plant disease epidemics continue to depress world food production, and the modern trend toward monocultures and intensive land use is exacerbating this outcome. The systems approach, and the multidisciplinary technology that it commands, augurs well for future control of disease by management. The same technology that helped put humans on the Moon has only just been harnessed to tackle an age-old problem, that of reducing crop loss and wastage.

[P. S. TENG; J. C. ZADOKS]

The ecological niche of an organism is the set of environmental conditions under which the particular functions of the organism could be expected to assure its survival. It comprises both the set of conditions where the organism lives (often termed the habitat of the organism) and the functional role of the organism in the ecosystem. Recent works in niche theory has enabled ecologists to develop predictions and actual applications. Historically, niche theory has arisen from the resolution of two different concepts of the ecological niche.

HISTORY OF THE NICHE CONCEPT

In 1917 Joseph Grinnell developed the concept of the ecological niche. Grinnell considered the type of vegetation (or habitat) in which the California thrasher lived, described the bird's food and feeding habits, noted the manner in which the coloration of the bird blended with its habitat, and discussed the bird's behavior. In his concluding remarks, Grinnell noted that both the behavior and the physical structure of the California thrasher determined the range over which the species was found, namely, the chaparral vegetation in California. Later, in 1928, Grinnell defined the niche as "the ultimate distributional unit within which each species is held by structural and instinctive limitations." In this definition Grinnell emphasized the manner in which an organism fitted into its environment and the factors that determined the geographical range of a species.

Also in 1928, Charles Elton redirected the definition of the ecological niche. Elton defined the niche of an animal as "its place in the biotic environment, *its relation to food and enemies* [italics in original text]." Elton further elaborated that, by studying the niches of animals, one might determine how different animal communities resembled each other. Thus the niches or roles of rabbits or hares in North America might be similar to the niches of the agouti and viscacha in South

Herman H. Shugart is a senior ecologist in the Environmental Sciences Division, Oak Ridge National Laboratory, and teaches in the botany department at the University of Tennessee. A graduate of the University of Arkansas and the University of Georgia, he is interested in ecological community dynamics and mechanistic modeling of complex ecological systems.

The Ecological Niche

America, the niches of animals like the hyrax, springbok, or mouse deer in Africa. This redefinition of the ecological niche differed from that of Grinnell in its emphasis on the food and function of an animal and its reduced emphasis on the manner in which a species' behavior or physical aspects might determine the species' distribution.

Also, the Italian mathematician Vito Volterra developed a mathematical model of species competing one with the other for food, and a Soviet scientist, G. F. Gause, went to work in the laboratory to obtain data to verify Volterra's theory. Gause's experimental results (Fig. 1) validated Volterra's theoretical predictions that if two species are similar in their food habits, one will eliminate the other. Exactly which species will survive depends on the way they interact and, in some cases, on the numbers of each of the species at the beginning of the experiment. This theoretical (from Volterra's mathematics) and experimental (from Gause's laboratory) result has come to be called the competitive exclusion principle. This principle is that no two species can occupy the same niche, or that complete competitors cannot coexist. Since Volterra's derivations referred to competing for food and because Gause referred to Elton's definition of the niche in publishing his work, the competitive exclusion principle was generally taken to apply to niches in the sense of Elton's definition. It is ironic in this regard that Grinnell in 1917 had stated, "It is, of course, axiomatic that no two species regularly established in a single fauna have precisely the same niche relationships."

By the 1940s, the term niche was a well-developed concept in ecology, and papers describing the niche relationships of different animals appeared in various scientific journals. The different interpretations of Grinnell and Elton caused some

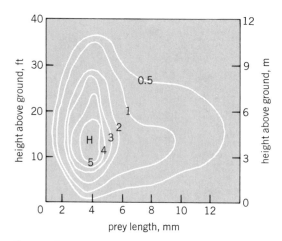

Fig. 2. Realized niche of the blue-gray gnatcatcher (*Poplioptera caerulea*). The contour lines map the feeding frequencies (percentage of the bird's total diet) on insects at different heights above the ground. *(From R. H. Whittaker, S. A. Levin, and R. B. Root, Niche habitat and ecotope, Amer. Natural., 107:321–338, 1973)*

confusion. With the experimental results from Gause and others (as well as mathematical results from Volterra and A. J. Lotka), the Eltonian definition with its emphasis on feeding habits and function began to prevail. By the 1950s, Grinnell's concept of the niche was thought of as a more detailed description of an animal's habitat—sometimes called the place niche.

The most recent addition to the development in the theory of the ecological niche has been the formulation of a mathematical or geometrical model of the niche by G. Evelyn Hutchinson, and this reformulation is the basis for the niche theory in ecology today. Hutchinson had published parts of his ideas on the niche in the 1940s, but most ecologists would cite a concluding paper given in 1957 at the Cold Springs Harbor Symposium on Biology. Hutchinson noted the difference between Grinnell's and Elton's definitions of the niche and included aspects of both concepts in his redefinition. In Hutchinson's definition, important features of the environment (called relevant environmental variables) were considered to be dimensions of an abstract space. The niche of an organism was the portion of this space with conditions under which the organism could be expected to live and reproduce. Since the abstract space formed by the relevant environmental variables could have any number of dimensions (one for each environmental variable), this definition is sometimes called the niche hyperspace concept. The total amount of niche space which was available in any particular ecosystem was termed the biotope.

NICHE HYPERSPACE CONCEPT

Hutchinson considered the set of all conditions under which an organism could potentially live to be the fundamental niche. Usually the set of conditions under which an organism actually lives is smaller, and is termed the realized niche. Figure 2 illustrates the part of the realized niche for a small bird, the blue-gray gnatcatcher. In this example,

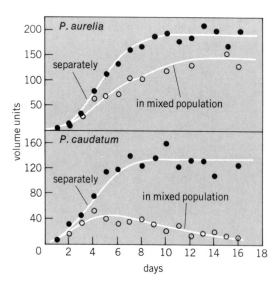

Fig. 1. Growth of two species of *Paramecium* cultivated separately and together. In mixed cultures, *P. aurelia* increases more slowly than in pure cultures. *Paramecium caudatum* is greatly reduced in its growth in the presence of *P. aurelia* and after a time might be expected to become extinct. *(From G. F. Gause, The Struggle for Existence, Williams and Wilkins, 1934)*

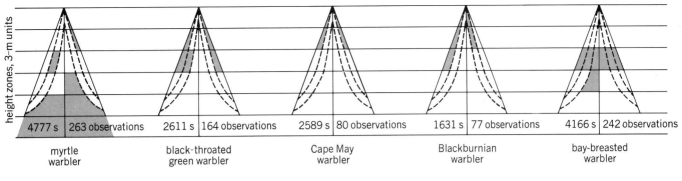

= concentrated foraging activities

Fig. 3. Parts of trees in a coniferous forest used by five species of warblers (*Dendroica*). The right side of each tree represents use based on the total number of birds observed; the left side represents use based on the total amount of time that birds were observed. (*Copyright 1958 by the Ecological Society of America; from E. R. Pianka, Evolutionary Ecology, Harper and Row, 1978*)

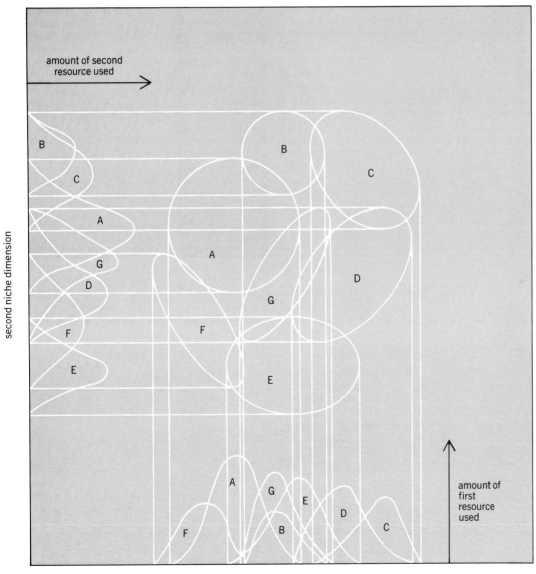

Fig. 4. Niche axes for seven hypothetical species. Although species A through F overlap considerably in their use of each dimension taken separately, there is very little overlap in each of the species when both niche axes are considered. (*From R. M. May, Theoretical Ecology, W. B. Saunders, 1976*)

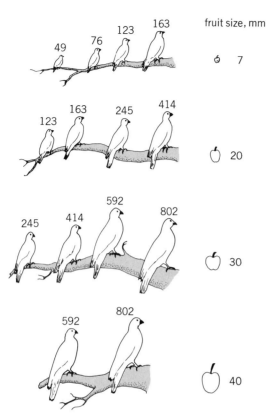

fruit size, mm

Fig. 5. Schematic representation of niche relations among eight species of fruit pigeons found in the New Guinea lowland rainforest. The weights of the pigeons in grams are written above each bird. Each fruit tree attracts up to four consecutive members of this size sequence of pigeons. Trees with large fruits attract increasingly larger pigeons. In any particular fruit tree the smaller pigeons feed on the smaller branches. *(From J. M. Dramond, Distribution ecology of New Guinea birds, Science, 179:767, 1973)*

the niche has two dimensions (height above the ground and the size of the insects that the animal catches), and it appears that the gnatcatcher spends a large percentage of its time feeding 10–15 ft (3–5 m) off the ground on insects 3–4 mm in length. Presumably some other bird (or other type of animal) that occurred in the same locations as the gnatcatcher but fed on a different size of insect or searched for insects at a different height in the trees and shrubs would not compete with the gnatcatcher. The advantage of the hyperspace niche definition is that one can equate competition (or the potential for competition) with the amount of overlap which occurs in the niche hyperspace.

Most studies, for example the classic study by R. H. MacArthur (Fig. 3), have found that similar species of animals in fact usually do not occupy the same parts of the niche hypervolume. But, because two organisms may overlap in any single dimension of the niche hyperspace (Fig. 4), scientists consider several geometrical aspects of the niche: niche dimension—the number of axes in the niche hyperspace; niche width—the size of the niche; and niche overlap—the amount of the niche hyperspace potentially shared by two or more species.

Niche dimension. Although the number of dimensions making up the niche hyperspace can be very large in theory, in practice most organisms can be separated one from another with relatively few dimensions (see table). Geometrically, once one has considered enough niche dimensions to separate the niches of two species in a niche hyperspace, the addition of other niche dimensions cannot cause the two niches to overlap. There are some general patterns of dimensionality evident in the niche studies in the table: (1) Habitats tend to be more important than foods. In turn, foods are usually more important than temporal separations. (2) Animals that feed upon other animals often are separated by being active at different times of the day. Thus temporal separations tend to be important for predators. (3) Terrestrial poikilotherms (commonly called coldblooded animals) portion food by being active at different times of the day. (4) Vertebrates tend to be separated less frequently by seasonal activity than do lower animals. (5) The food types dimensions tend to be less important for animals that feed on relatively small items.

Niche width. Niche width in terms of the hyperspace model amounts to a geometric concept much like the ecological constructs of generality and specialization. An organism with a broad niche (the niche is a relatively large volume of the niche hyperspace) would be able to survive under a wide array of conditions, and would be termed an ecological generalist. An organism with a narrow niche would be restricted in the set of conditions that allowed its survival, and would be considered a specialist. One area of study that has developed from niche theory has been the attempt to determine whether communities of organisms are dominated by specialists or generalists. Often these studies have provided some surprising results. For example, a widespread common animal may be a generalist that is adapted to live in a wide variety of conditions or it may be a specialist on some set of conditions that happen to occur very often. Organisms of the first sort are potential pest or weed species if introduced to a new area. Species of the latter sort could be extirpated over large areas by changes in the environment.

Niche overlap. One of the more appealing aspects of the niche hyperspace model as it was initially formulated by Hutchinson was the manner in which the overlaps of species niche volumes could be equated with competition or at least the potential for competition. Most studies of detailed ways of niche differentiation in plants and animals (see table) have found that there is very little overlap in the niche spaces in most animals, particularly when the relationships of these animals are studied in detail. This result has been variously interpreted. Some ecologists take the view that what is measured in most studies is the realized niche of species, and it is the overlaps of fundamental niches that relate to competition. Other ecologists claim that competition rapidly conditions populations to select their environments so that they do not overlap in the niche hyperspace in any way. The latter interpretation has been supported by experiments involving the removal of one of two similar species and noting that there is

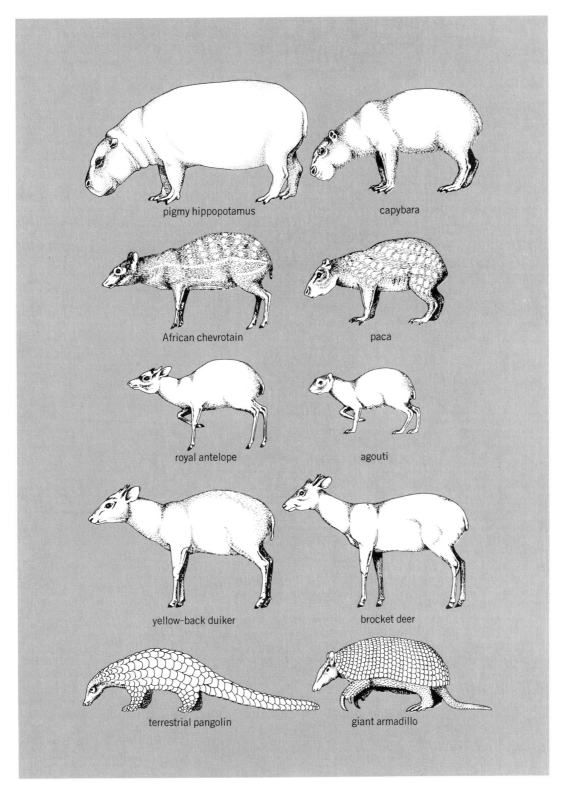

Fig. 6. Morphological convergence among African (left) and South American (right) rainforest mammals. Each pair of animals is drawn to the same scale. (*From B. J. Meggers, E. S. Ayensu, and W. D. Duckworth, Tropical Forest Ecosystems in Africa and South America: A Comparative Review, Smithsonian Institution Press, 1973*)

little change in the niche volume of the other. However, most classic interpretations of species distributions, as well as observations of new species invading islands, tend to be considered to favor the former interpretation. This is an area of research at present, and there seems to be a need for clever experiments with field populations as well as for some theoretical reformulation.

Ecological differences between similar species

	Consumers			Rank and description of resource dimensions*¶			Time	
Group and location	No. of species	No. of genera	Food	Macro-habitat†	Microhabitat‡	Food type§	Day	Year
SIMPLE ORGANISMS:								
Slime molds, forest, eastern North America	4	2	Bacteria	X		1-TS		
Paramecium, near Ann Arbor, MI	5	1	Organic minutia	1-Wat				1
Triclads, shallow littoral zone of lakes, Britain	4	3	Invertebrates		2 (X)-Depth	1-TH; X-S		
Nematodes, psammolittoral, Gulf of Mexico	46	?	Invertebrates, plants	1-Hor	1-Depth	3-F		
Rotifers, small lake, central Sweden	5	1	Flagellates		1-Depth	1[1]-S[1]		
Tubificid oligochaetes, Toronto Harbor	3	3	Bacteria	X		1-TS[12]		
Polychaetes, soft bottom, Beaufort, NC	5	3	Deposit feeder		1-Sediment type 2-Vertical zone			
Chaetognaths, Agulhas Current, Indian Ocean	18	4	Mostly copepods	2-Hor	2-Depth	1[1]-S		
MOLLUSKS:								
Gastropods, shallow water, Florida	8	6	Invertebrates			1-TS		2
Conus, Hawaii	25	1	Polychaetes, fish, gastropods	3-For	2-Substrate	1-TSH	4	
Conus, Pacific atolls	17	1	Mostly polychaetes		2-Substrate	1-TSH		
CRUSTACEA								
Crabs, intertidal bench, Tasmania	11	9	Algae, inverte-brates, detritus	1-Hor	1-As macrohabitat[10] 2-Cover 5-Vertical zone 1-Shell shape[s] 2-Shell weight[s]	3-TH		3[b]
Hermit crabs, intertidal, San Juan Islands, WA	3	1	Detritus		3-Bed and tidepool type[s,n]			
Diaptomus copepods, Clarke Lake, Ontario	3	1	Plant, animal particles		1-Depth	1[1]-S	X	1
Diaptomus copepods, Saskatchewan ponds	7	1	Plant, animal particles	1-Geo	4 (X)-Depth	2[1]-S		2
Amphipods, marine sand beaches, Georgia	5	5	Mostly detritus, algae, protozoa	1-Hor	2-Depth in sand	2-S[1] 5-TH		2[b,1]
Crustaceans, cave streams, West Virginia	4	3	Leaves, microorganisms		1[s]-Riffles or type pool	X		
INSECTS:								
Grasshoppers, prairie, northeastern Colorado	14	11	Grasses, forbs			1-TS; 1-TS		2
Melanoplus grasshoppers, grasslands,	3	1	Mostly grasses	2-Veg		1-TS		2
Termites, savanna-woodland, western Africa	5	1	Grasses	1[n]-Veg		4-TS; 4-S	3	2
Psocids, larch trees, Britain	9	5	Bark algae, fungi	4-Alt	2-Twig condition	1-THP		3
Butterflies, lowland rain forest, Costa Rica	12	7	Decaying fruit		X-Microclimate	X-B	1[n]	
Carabid beetles, fen, England	8	2	Mainly scavengers	1-Veg		A-TH; A-S		2[f]
Whirligig beetles, Michigan	3	1	Predators, scavengers	1-Lat	1-Lake size			3(X)[b]
Euglossa bees, Panama	19	1	Nectar	2-Veg	3[1]-Microclimate	1[1]-S	X	
Ants, Colorado	4	3	Animals	1-Veg	3-Type log or cover	2[1]-S		
Ants, Colorado	5	2	Seeds	2-Veg	3-Type log or cover	1[1]-S		
Megarhyssa wasps, beech-maple, Michigan	3	1	Parasitoids		1[1]-Depth of food[1] X-Leaf type[s]	X	2	X
Wasps, *Neodiprion*, Quebec	11	9	Parasitoids	4-Veg	1[1]-Depth of host	1-TL; 1[1]-S 4-TS		
OTHER ARTHROPODS:								
Millipedes, maple-oak forest, central Illinois	7	7	Leaf litter, decaying wood		1-Position in log, litter	X-F		
Mites, deciduous forest, central Maryland	7, 9	1	Invertebrates	1-Veg	1-Depth in soil	1[1]-S		4
Water mites, ponds, central New York	20	1	Parasites	3-Wat		1-TS; 2-P		4

*Footnotes appear on page 40.

Ecological differences between similar species (cont.)

Group and location	No. of species	No. of genera	Food	Macro-habitat[†]	Microhabitat[‡]	Food type[§]	Day	Year
FISH:								
Stream fish, dry season, Panama, moist tropics	12	12	Animals, plants	2-Str	2-Depth	1-TH	3	
River fish, River Endrick, Scotland	5	4	Arthropods, algae	1-Str		1-TSH		
Lake fish, eastern Ontario	17	15	Animals, plants	1-Hor[s]	5-Depth[5]	1-TH 5[s]-S[1,7]	3	3
Intertidal fish, Brittany	13	12	Invertebrates, algae	1-Hor	1-As macrohabitat	2-TH		
SALAMANDERS:								
Desmognathus, Appalachians	5	2	Arthropods	1-Aqu 2-Alt		2-S[1] A-TH		X
Triturus, ponds, England	3	1	Invertebrates	2?-Alt	2-Temperature	1-S, 4-TH		
FROGS:								
Tropical *Rana*, streamsides, rain forest, Borneo	3	1	Small animals	1?-Wat	1-Distance from stream	3-S 3-TH		X
Temperate *Rana*, northeastern North America	6	1	Small animals	2-Aqu 3-Lat		3[1]-S		1[b]
LIZARDS:								
Ameiva teids, Osa, Costa Rica	3	1	Arthropods, fruit	1-Veg		2-S[1]	2[1]	
Ctenotus skinks, desert, Australia	7	1	Arthropods		4-Plant cover[6,9]	3-S[3]; 1-TH	5[d]	2
Cnemidophorus whiptails, Trans-Pecos	5	1	Arthropods	1-Veg		2-TH		
Cnemidophorus, south and central New Mexico	4	1	Arthropods	1-Veg		2-TH; 2-S[1]	5[7]	4[b]
Anolis, Bimini	4	1	Arthropods	5-Veg	1-Structural habitat	2-S[3]; 2-TH	2	
Anolis, Jamaica	6	1	Arthropods, fruit	3-Veg	1-Structural habitat	1[1]-S	4	
Anolis, Puerto Rico	10	1	Arthropods, fruit	2-Veg	1-Structural habitat	3[3]-S	4?	
Phyllodactylus geckos, Sechura desert, Peru	4	1	Mostly arthropods	2-Alt 3-Soi	3-Plant species 5-Foraging substrate	1-TH 6-S	X	
OTHER REPTILES:								
Sternothaerus turtles, southeastern United States	4	1	Mollusks, arthropods	1-Wat		3 (X)-TH	1	
Garter snakes, Michigan	3	1	Animals	2-Veg		1-TH		3
BIRDS:								
Alcids, Olympic Peninsula	6	6	Fish, invertebrates	1-Dis	2-Feeding depth	2-TH		X[b]
Alcids, St. Lawrence Island, Alaska	3	2	Invertebrates	4-Dis	X-Feeding depth	1-S[2] 1-TH[5]	3	X[b]
Terns, Christmas Islands	5	4	Fish, invertebrates	2-Dis		1-S[1,2] A-TH	2	
Sandpipers, tundra, Alaska	4	1	Insects	1-Veg		1-TS 3-S		
Herons, Lake Alice, Fla.	4	4	Animals		2-Feeding place	1-THS;3-S		
Ducks, Medway Island, Britain	7	3	Plants, animals	3-Veg	2-Feeding method	1-THP		
Hummingbirds, Arima Valley, Trinidad	9	8	Nectar, insects	2-Veg	1-Kind of plant	2-S[1,2]		
Flycatchers, deciduous forest, southern West Virginia	5	5	Insects	1-Veg	4-Feeding height 3-Vegetation density	1[1]-S[1,2] A-TH		
Flycatchers, deciduous forest, eastern United States	5	4	Insects	1-Veg		2-S[1] X-TH		
Titmice, broadleaved woods, Britain	5	1	Insects, seeds	1-Veg	2-Structural habitat	2-S		
Vireos, New World	17	1	Insects	2-Alt 1-Geo	2-Foliage layer	4[1]-S		
Warblers, boreal forests, Vermont	5	1	Insects	X	1-Part of tree 2-Feeding style	X-TH	X	3
Icterids, channeled scabland, Washington	4	4	Insects (for nestlings)	1-Veg	1-As macrohabitat	2-TH		
Tanagers, Trinidad	10	5	Fruit, insects	2-Veg	1-Structural habitat	2-THS		
Honeycreepers, Trinidad	5	4	Fruits, insects	3-Veg	1-Structural habitat	2-THS		
Finches, southeastern United States	5	5	Seeds, insects	1-Veg	1-Height	3 (A)-S		
Finches, near Oxford, England	10	5	Seeds, insects, buds	3-Veg	4-Ground, air, or foliage	1-TSH[5,2] 1-S[2]		5

(continued)

Ecological differences between similar species (cont.)

	Consumers			Rank and description of resource dimensions*¶				
Group and location	No. of species	No. of genera	Food	Macro-habitat†	Microhabitat‡	Food type§	Time — Day	Time — Year
BIRDS (cont.):								
Geospiza finches, central Galápagos Islands	5	1	Seeds, fruit, buds, insects	2-Veg		1-S, H		
Camarhynchus finches, central Galápagos Islands	4	1	Insects	3-Veg	1-Structural habitat 1-Foraging method	3-S		
Grassland birds, ten sites, New World	2-4	2-4	Seeds, insects	2-Veg	3-Vertical	1¹-S	X	X
Foliage gleaners, oak woods, California	5	3	Insects		3-Substrate 4-Foraging layer	1-S 1-TH		
Upland birds, broadleaved woods, Britain	22	17?	Insects	2-Veg	1-Structural habitat A-holes^s	2-TH		A
MAMMALS:								
Pocket gophers, Colorado	4	3	Plants	1-Soi		X-FT		
Chipmunks, Sierra Nevada, CA	4	1	Seeds, fruits	1-Alt				
Rodents, deserts, North America	10	3	Seeds	1-Veg 3-Soi	1-Foliage height	X?-S	X	
Peromyseus, Ozarks of Missouri	3	1	Insects, seeds	1-Veg		X-THP		
Giant rats, western Malaysia	4	3	Plants, insects	1-Alt 3-Veg	2-Height in trees	3-THP		
Carnivores, Serengeti, Africa	7	5	Large animals	2-Veg		1-S^{1,22}	3	
Bats, central Iowa	8	6	Insects	2-Veg			1	
Bats, Central American lowlands	31	21	Animals, plants	2-Veg		1¹-S; 2-TH		

*Numbers in the columns denote importance rank of the decision concerning the ecological differences between similar species. The "rank" abbreviations are as follows: A, thought to be strictly an aspect of separation on another dimension; b, breeding time; f, feeding time; i, indicator used; n, thought not to be necessarily related to resource partitioning; x, known not to be an important dimension.

†Alt, altitude; Aqu, aquatic-terrestrial gradient; Dis, foraging distance from land; For, marine formation; Geo, geographic; Hor, aquatic horizontal zone; Lat, latitude; Soi, soil; Str, stream size or part of stream; Veg, vegetation type; Wat, size of type of water body.

‡s, used as shelter; structural habitat, food or perch substrate in vegetation.

§B, artificial baits; F, feeding type; H, hardness; S, size; T, taxonomic category [letters following T refer to species of food (S), higher taxonomic category (H), life stage of host (L), part of an individual prey (P)].

¶Indicators: 1, body size; 2, bill size; 3, head size; 4, ovipositor length; 5, bill shape; 6, body temperature; 7, mouth shape; 8, body form; 9, hindleg length; 10, hair quantity; 11, group size; 12, nutrient-utilization ability.

SOURCE: T. W. Schoener, Resource partitioning in ecological communities, *Science*, 185:27-39, 1974.

ECOLOGICAL COMMUNITIES

One of the prominent goals of ecologists in using niche theory is to develop a fundamental understanding of the manner in which plant and animal communities are constructed. The topics that generally arise in these considerations include the following.

Limiting similarity. The question arises as to how similar in their niches can two species be and still be able to survive in the presence of one another. Because the differences in animals in the same communities are often regular (Fig. 5), some scientists believe in a limit to the amount of niche overlap allowed in a given assemblage of species.

Maximum diversity. Since only a fixed number of niches of a given size could be packed into a given niche volume (biotope), one might expect that in any given ecosystem there should be some maximum number of species present and no more. Other, more elaborate, geometrical interpretations of the ways niches of species could be packed into the biotope of a given ecosystem have led to theoretical models for predicting total diversity in a given environment, as well as the abundances of these species.

Ecological convergence. Sometimes species on different continents, but in the same sort of ecosystems, will look almost identical, even though they are not at all related to one another, being in different families or orders (Fig. 6). This observation was one of the facts that led to Elton's definition of the niche as the role that different animals played in different communities. The existence of such convergence would indicate that certain configurations in niche space occur in a regular and perhaps predictable manner in ecosystems of a given sort.

APPLICATIONS OF NICHE THEORY

The potential of niche theory in solving practical ecological problems has begun to be realized over the past few years. The biological control of pest organisms has made considerable use of niche theoretical ideas in designing techniques for controlling pest animals without the dependence on chemical control agents. For example, in Australia scarab beetles have intentionally been introduced from Mexico, Africa, and Sri Lanka because the native fauna of marsupial dung could not utilize the niche provided by cattle dung. Accumulating cattle dung provides an egg-laying site for the Australian bush-fly (a pest insect). Such ecologi-

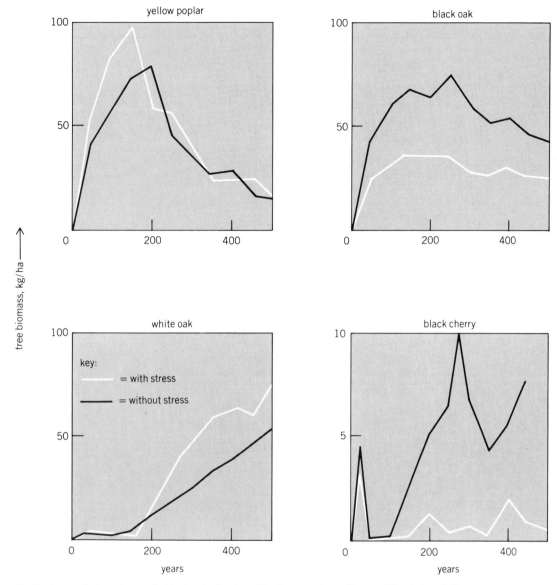

Fig. 7. Amount of wood expected in developing forest for four species, with and without air pollution.

cal engineering is an outgrowth of studies on the ecological niches of the animals involved.

Management for endangered species has lately taken the approach of developing an understanding of the requirements of the animal (its niche) and then using this information to determine what factors might be the species in danger of extinction. Often the important factors involve the habitat required by these species.

Finally, using niche theory, one can develop mathematical models and use them to predict the effects of changes in the environment on ecosystems. Figure 7 is an example of predictions from such a model. Such computer models based on niche attributes of species can help assess the effects of inadvertent changes caused by humans in the world environment.

FUTURE DEVELOPMENTS IN NICHE THEORY

Future work on niche theory could give a more basic understanding of competition and of factors controlling diversity of ecosystems. The plant ecologists have only recently begun to use niche theory, and one can expect an exciting decade as niche theory developed for animals is applied to plants. With the increasing pressure of human development on natural ecosystems, one should expect the ecological engineering of communities based on a well-developed theory of the ecological niche to become more common. As niche theory is applied, it will be tested. It is the success and failure of the present theory in real applications that will influence the direction of this area in the future.

[HERMAN H. SHUGART]

Only recently in human history did scientists learn of the presence of airborne microorganisms. For centuries since the time of Aristotle, it was widely believed that flies, mites, and molds were generated spontaneously in decaying animal and vegetable matter. In 1680 Anton van Leeuwenhoek's microscope at last rendered visible the world of minute organisms. But even the invention of the microscope did not end the controversy over spontaneous generation, and more time passed before airborne microbes, the infecting agents, were discovered.

Their presence was finally established by Louis Pasteur. In 1861 he designed a very simple air-sampling device consisting of a 0.5-cm-diameter tube, a guncotton plug, and a pump. He sampled the air in Paris, trapping airborne particles in the guncotton, which was then dissolved in an alcohol-ether mixture. The particles settled, the liquid was decanted, and the remaining deposit, when examined under a microscope, revealed bacteria, molds, and yeasts.

This discovery soon prompted others to study the atmosphere to find out what else was there, creating a flurry of activity in the late 1800s. Glass dishes full of nutrient media (petri dishes) and microscope slides were exposed in all sorts of locations — on mountaintops, in city streets, inside and outside buildings, and in sewers. Soon a picture of airborne microorganisms began to emerge. They appeared to be everywhere, and in greater concentrations in cities than in the country, in lowlands than on mountaintops, and during the day than at night.

When airplanes and balloons made it possible to explore higher in the atmosphere, interest in airborne microorganisms increased even more. Even Charles Lindbergh exposed hand-held petri dishes in flight. It was not until the mid-1930s, however, that F. C. Meier of the U.S. Department of Agriculture introduced the word aerobiology to describe research involving microbial life in the air.

Robert L. Edmonds left Australia in 1966 to attend the University of Washington (Ph.D., 1971). After several years at the University of Michigan, he returned to the University of Washington in 1973 as an assistant professor of forestry and became associate professor in 1979. He has contributed numerous papers on aerobiology, forest pathology, and soil microbiology to scientific publications.

Aerobiology

In 1942 the American Association for the Advancement of Science published the proceedings of a symposium on aerobiology, and the new discipline was launched. Thanks to the work of many researchers, but in particular, Philip H. Gregory of the Rothamsted Agricultural Experimental Station in England, interest in aerobiology has continued. In 1970, as part of the International Biological Program, aerobiology emerged as a viable area of multidisciplinary research.

Because aerobiology encompasses pathology, allergology, entomology, air-pollution effects, palynology, phytogeography, and meteorology, it is somewhat difficult to define. Simply, aerobiology is an interdisciplinary study of the biological components of the atmosphere, with emphasis on their release, transport, atmospheric interactions, and deposition, and on how they affect plant, animal, and human health. Aerobiology encompasses both indoor (intramural) and outdoor (extramural) environments. Although the array of microbes may not be so varied inside as outside, they can certainly

Types and sizes of airborne microbes*

Type	Diameter
Viruses	$0.015-0.45\,\mu\text{m}$
Bacteria	$0.3-10.0\,\mu\text{m}$
Algae	$0.5\,\mu\text{m}-1\,\text{cm}$
Fungus spores	$1.0-100\,\mu\text{m}$
Lichen fragments	$1.0\,\mu\text{m}-1\,\text{cm}$
Protozoa	$2.0\,\mu\text{m}-1\,\text{cm}$
Moss spores	$6.0-30.0\,\mu\text{m}$
Fern spores	$20.0-60.0\,\mu\text{m}$
Pollen	$10.0-100.0\,\mu\text{m}$
Plant fragments, minute seeds, insects, spiders, and so on	Greater than $100\,\mu\text{m}$

*From R. L. Edmonds, Aerobiology: Ecology of the atmosphere, *Biol. Dig.*, 4(5):11–24, 1978.

be found inside all buildings. Many microbes come from the outside through windows and doors, but some are generated inside.

MICROBES IN THE ATMOSPHERE

The types of airborne microbes and their sizes are shown in the table. The smallest microbes are the viruses, although single viruses are rarely found in the atmosphere. Mostly they occur on "rafts" of organic debris which protect them from drying and exposure to deadly ultraviolet radiation. Bacterial cells may also be carried in this way, or they may occur in clumps of cells, for most bacteria grow in colonies in close association with one another. Even fungus spores may occur in clumps or chains, although they commonly occur singly.

The shapes and sizes of airborne fungus spores vary greatly, as seen in Fig. 1. The reason for such variability is unknown. Some are shaped aerodynamically, and others are not. Many are heavily pigmented, whereas others are transparent. Pigmentation almost certainly helps protect spores against exposure to ultraviolet radiation. Spores of mosses and ferns, and pollen grains of higher plants, may also be airborne; some are illustrated in Fig. 1. Not all pollen grains, however, are dispersed by wind; many are dispersed by insects and birds.

As the diameter of organisms increases above 100μm, they begin to fall out very rapidly from the atmosphere. Yet many organisms of large size are dispersed in the atmosphere. Some have developed special features enabling them to be more buoyant than would be anticipated. For example, the first-instar larva of the gypsy moth, a tree defoliator, is about 2 mm long but has long hairs all over its body that provide increased flotation (Fig. 2). In addition, it may carry a length of silken thread, which further assists buoyancy. This airborne organism is, of course, visible with the naked eye. Generally, larger insects require powered flight, although some insects, such as aphids, use both passive and active dispersion.

IMPORTANCE OF AEROBIOLOGY

Why should there be concern about airborne microbes? It is known that certain human airborne diseases, including pneumonia, influenza, smallpox, tuberculosis, and numerous others caused by viruses, bacteria, and fungi, can be spread through

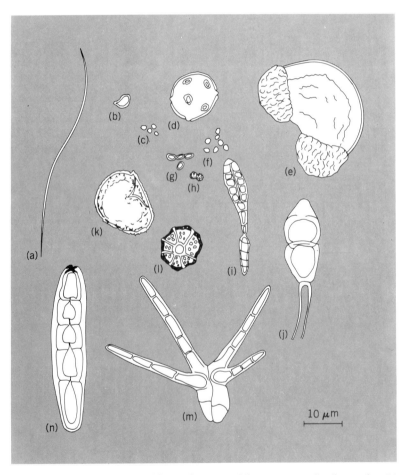

Fig. 1. Some representative airborne fungus and fern spores and pollen grains. (a) *Claviceps purpurea*, ascospore; (b) *Amanita rubescens*, basidiospore; (c) *Penicillium cyclopium*, conidia; (d) *Plantago* species, plantain pollen; (e) *Pinus sylvestris*, pine pollen; (f) *Aspergillus fumigatus*, conidia; (g) *Penicillium chrysogenum*, conidia; (h) *Aspergillus niger*, conidia; (i) *Alternaria* species, conidia; (j) *Puccinia graminis*, uredospore; (k) *Pteridium acquilinum*, fern spore; (l) *Epicoccum* species, conidium; (m) *Tetraploa arista*, conidium; (n) *Helminthosporium* species, conidium. (*From R. L. Edmonds, Aerobiology: Ecology of the atmosphere, Biol. Dig., 4(5):11–24, 1978*)

Fig. 2. The long hairs on this gypsy moth larva help it to float on the wind. (*Courtesy of M. L. McManus, U.S. Forest Service*)

the atmosphere. The influenza epidemic of 1918–1919 resulted in 550,000 deaths in the United States alone. There is also some evidence that the infectious agent of Legionnaires' disease may be airborne. Many people, as many as 30,-000,000 in the United States, suffer from hay fever or asthma. This type of allergy is commonly caused by airborne pollen and fungus spores. The chief culprit is the ragweed (*Ambrosia* species) pollen grain (Fig. 3). There are other interesting examples of hypersensitivity reactions caused by aeroallergens. One recent example is New Orleans epidemic asthma, which is an allergy to local airborne pollen and spores, and affects thousands of people. Other cases have been documented involving airborne spores associated with air-conditioning and humidifier systems, and airborne allergenic algae may also be produced from the surfaces of nutrient-enriched lakes.

New environmental problems have recently emerged, among them, the generation of microbial aerosols as a result of sewage sludge and wastewater applications on forest and crop land. It is not yet known what impact, if any, these applications will have on human health.

But what of animal and plant diseases? Although many animal diseases are spread by direct contact, as are human diseases, some animal diseases are airborne. The virus-caused foot-and-mouth disease is a classic example. And of course, there are many examples of airborne spread of virus diseases through insect vectors, such as equine encephalitis.

Many plant pathogens, many of which are fungi,

spread in the atmosphere via spores, some of which are capable of traveling hundreds of kilometers before finally infecting susceptible plants. Crop loss caused by these diseases costs many millions of dollars each year. The most notorious example of an airborne plant pathogen is the wheat stem rust fungus *(Puccinia graminis)*. Every year clouds of spores move from Mexico to Canada across the Great Plains after overwintering in warmer southern climates. This fungus can cause massive losses to wheat, and the only reason wheat can be grown successfully today is because disease-resistant strains have been developed. Corn blight (Fig. 4) and coffee rust are other examples of airborne diseases. Breeding for resistance, however, has a limit, for fungi readily produce mutant strains; thus it is constantly necessary to breed new disease-resistant plants.

Many insects have an airborne stage in their life cycle. The larvae of tree-defoliating insects, such as the gypsy moth, the Douglas fir tussock moth, and the spruce budworm, are airborne. The gypsy moth has defoliated millions of hectares of hardwood trees in the northeastern United States. In addition, many crop viruses are transmitted by aphids and leafhoppers.

Interest in airborne pollen is not restricted to aeroallergens. For plant geographers and palynologists, determining how far airborne pollen can travel is an aid in explaining past and present plant distribution patterns.

So far this discussion has focused on the atmosphere as a medium for passive transport. Airborne organisms, however, may interact with other components of the atmosphere in flight. It is now known that airborne organic particles and certain bacteria act as condensation nuclei for ice and rain. Microorganisms are also sensitive to pollu-

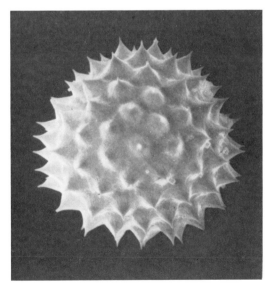

Fig. 3. Photomicrograph of a ragweed pollen grain, about 20 μm in diameter. The cause of ragweed allergy is not the spikes but, rather, a water-soluble extract. (*From R. L. Edmonds, Aerobiology: Ecology of the atmosphere, Biol. Dig., 4(5):11–24, 1978*)

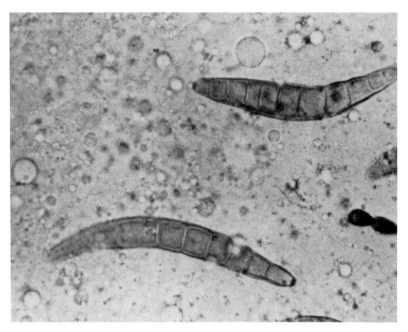

Fig. 4. Airborne spore of the southern corn leaf blight fungus. This fungus caused large losses to the corn crop in the United States in the early 1970s.

tant gases such as sulfur dioxide and ozone.

Finally, many nutrient elements and carbon are transported through the atmosphere. Thus, nutrient cycling, particularly on the global scale, can be considered to be in the realm of aerobiology.

HOW MICROORGANISMS ENTER THE ATMOSPHERE

Some microorganisms have developed many elaborate and sophisticated release mechanisms to ensure that propagules are injected into the atmo-

Fig. 5. Larvae of the gypsy moth hanging on silken threads from oak leaves. (*Courtesy of M. L. McManus, U.S. Forest Service*)

sphere. Other organisms, particularly the smaller ones such as viruses and bacteria, are liberated passively. Among passive mechanisms are the blowing or washing of microorganisms from plant material or soil surfaces. Indoor disturbance of dust releases many microorganisms. Recently, spraying operations and ventilation towers associated with sewage treatment have been found to generate large quantities of bacterial aerosol. Wave action also creates clouds of airborne bacteria.

Bacteria and viruses also may become airborne when humans cough and sneeze. Even ordinary talking liberates a considerable microbial aerosol. However, because aerosols generated in this way travel only a short distance before they are inhaled, it is sometimes difficult to determine whether diseases spread in this manner are airborne or contact diseases.

The fungi do not rely solely on passive release mechanisms, although common molds (or conidial fungi) do release their spores passively. The familiar green color of *Penicillium* mold on food is due to the spores which are produced on the surface of the fungal mycelium.

Mold spores are released either as a result of water splash (slime spores) or by wind action (dry spores). The stronger the air current, the greater the rate of blowoff for dry spores, with dry air being a more effective dispersal force than moist air.

Active discharge occurs in the Ascomycetes (truffle-type fungi) and Basidiomycetes (mushroom or puffball fungi). Ascomycetes produce ascospores which are shot for distances up to 50 cm from the fruiting bodies by the release of hydrostatic pressure. This is sufficient to project the spores into turbulent air, with the consequent possibilities of significant dispersal. Many ascospores are released with the onset of rain.

Basidospores are borne on the surface of mushroom gills or pores or in puffballs. They drop into turbulent air through the space between the gills, or are shot into the air in the case of puffballs. It has been estimated that 1.8×10^9 spores can be released from an 8-cm-diameter cap of the common field mushroom. A giant puffball may contain as many as 7×10^{12} spores. Spore discharge in mushrooms generally occurs under damp conditions, and most mushrooms are observed only in the cool, damp spring and fall months.

Not all fungus spores are released at the same time of day, however. Some are released in the early morning, and others in the midafternoon; others peak at night. Many colorless spores are released at night, since they are susceptible to damage by ultraviolet radiation. Many pigmented spores, particularly those of the molds, are released during daylight hours, since they are more protected against radiation.

Many spore-release mechanisms operating in the fungi are also present in moss and fern spores. Algae are generally released passively by wind blowing across land surfaces or by wave and wind action in oceans and lakes.

Many plants produce wind-dispersed pollen. Commonly, plants with wind-dispersed pollen are the initial colonizers on disturbed sites; among these plants are ragweed or long-lived plants such

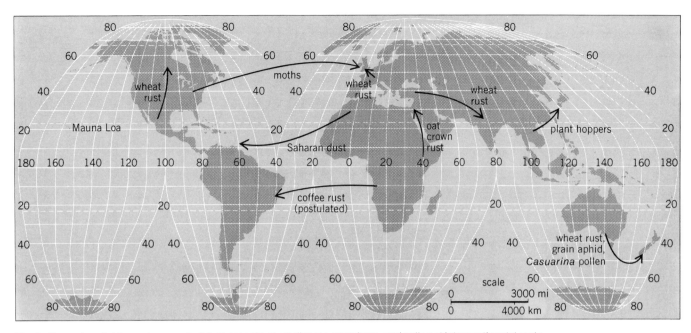

Fig. 6. Examples of airborne transport of dust, insects, plant disease organisms, and pollen at intercontinental scale.

as trees, particularly temperate conifers and hardwoods. Clouds of pollen are commonly produced in the spring months.

Although insects are mostly considered to be actively flying organisms, many smaller insects such as aphids, leafhoppers, and the larvae of flies, butterflies, and moths move randomly in the atmosphere with the air currents. Many of these insects have developed interesting mechanisms for projecting themselves into the airstream. For example, gypsy moth larvae, after hatching in May, hang from leaves on single silken threads (Fig. 5) which may be as long as 4 m. When a gust of wind breaks the thread, the larvae start their journey in the atmosphere. If conditions are not right, they do not fall.

DISPERSAL IN THE ATMOSPHERE

This topic can be considered in two parts: outdoor (extramural) environments and indoor (intramural) environments.

Outdoors. If one stands in one place for any length of time, it will be noted that the wind does not blow steadily from the same direction, even on windy days but, rather, fluctuates. Wind speed also changes unpredictably. This continuous fluctuation, known as turbulence, is the most important factor controlling the dispersion of microorganisms. Turbulent eddies are created by mechanical obstructions in the airflow, such as mountains and buildings, and by surface heating. Greatest dispersion occurs on sunny, windy days when turbulent eddies are large, whereas the smallest amount of dispersion occurs on clear, still nights or under inversion conditions.

The level of aerial transport of microorganisms can be microscale, mesoscale, or macroscale. Many organisms are dispersed only on the microscale (time scale of less than one hour and distances of only a few hundred meters). In fact,

fragile organisms can travel only short distances before they die.

Hardy spores such as those of the wheat rust and pollen grains may be transported at the mesoscale (days and a few hundred kilometers).

Macroscale transport is of sufficient scope to involve global circulation (Fig. 6). Intercontinental transport across oceans is common. Saharan dust has been sampled in Miami. It has also been postulated that coffee rust spores moved across the Atlantic from Africa to Brazil in the early 1970s. A counter suggestion is that although this is possible, it is unlikely, and the fungus really arrived by jet aircraft. Wheat rust spores also move at the macroscale.

Indoors. Although some microorganisms such as pollen can move from outdoor to indoor environments, the most important source of aerosols within intramural spaces is the occupants. The hospital is an indoor environment especially conducive to the spread of airborne pathogens. Hospitals contain a reservoir of individuals infected with pathogenic microorganisms which are potentially transmissible to other persons, including patients, hospital personnel, and visitors.

Even without artificial ventilation, the chimney effect in buildings can distribute aerosols throughout a large air volume in a short time, particularly in tall buildings. Ventilating systems are a mixed blessing. Such systems accelerate the distribution of particles, often in mystifying ways. For example, aerosols generated when teeth are drilled can be rapidly transported throughout an entire dental suite. On the other hand, mechanical ventilation systems have the capacity to reduce microbial contamination through filtration, and by correct filtration, a sterile environment can be created. However, microorganisms can also grow and reproduce in ventilating ducts, sometimes ahead of the filtration system.

Fig. 7. Gravity slide sampler. After a period of exposure, the microscope slide is examined under a microscope. (*Courtesy of W. R. Solomon, University of Michigan*)

SURVIVAL OF MICROBES IN THE ATMOSPHERE

Many physical and chemical properties of the atmosphere affect the survival of airborne microbes, particularly temperature, moisture, radiation, and gases.

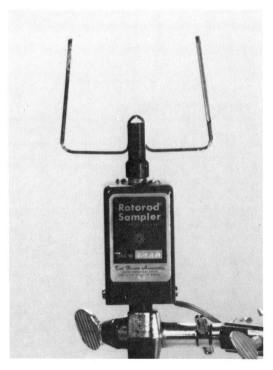

Fig. 8. Rotorod sampler. Depending on rod configuration and coating material, it can collect inert inorganic dust, viable bacteria, or pollen or spore material from the air. (*Ted Brown Associates, Los Altos Hills, CA*)

Atmospheric interactions. Survival decreases sharply below about −20°C and above 50°C. Bacteria tend to be more sensitive than fungi, and relative humidity significantly influences temperature effects.

Increased exposure to ultraviolet radiation also lowers the viability of airborne organisms. Dark pigments provide protection. Others use "rafts" of organic matter for protection.

Insects respond strongly to temperature, moisture, and pressure gradients. However, because they are active, they can restrict their activity to times which are most favorable for dispersal and survival.

Many gases affect the survival of airborne microbes. In general, exposure to pollutant gases such as nitrous oxide, sulfur dioxide, and ozone decreases microbial survival.

Although in most cases the atmosphere has the greatest influence on airborne microbes, sometimes this role is reversed. Many small organic particles and bacteria, in particular, *Pseudomonas syringae*, can act as ice condensation nuclei. They can thus have a large influence on precipitation processes and rainfall patterns. In fact, it has been postulated that the recent African drought resulted in part from overgrazing and a consequent reduction of soil organic matter, followed by a lack of small airborne organic particles and bacteria because of reduced decomposition.

Deposition of microorganisms. To have a major influence on animal, plant, and human systems, airborne microbes must be deposited on or in a substrate. The five common deposition mechanisms are: sedimentation, impaction or turbulent deposition, electrostatic or thermal deposition, precipitation scavenging, and respiratory deposition.

Sedimentation is an important mechanism of deposition in indoor environments. Outdoors, however, it is less important because of turbulence. Microorganisms moving in turbulent air are impacted on surfaces because of their inertia. Many plant foliage pathogens have large spores which are impacted efficiently.

Electrostatic or thermal deposition is not thought to be important. However, many microbes possess positive charges, and close to surfaces, this may be quite important in assisting deposition. An example of thermal deposition is the black particles that adhere to walls near heating vents.

Precipitation scavenging of aerosols is either by rainout or washout. Rainout occurs when extremely small particles (< 0.04 μm in diameter) are carried by molecular diffusion onto cloud droplets or ice crystals. Washout occurs when falling raindrops or snowflakes collide with and retain large particles. For particles greater than 20 μm in diameter, washout is very efficient. Below 20 μm, washout efficiency decreases rapidly, and it is negligible for particles less than 2 μm in diameter. Particles between 0.04 and 2 μm are not easily removed by the precipitation process.

The factors that influence deposition in animal and human respiratory systems are essentially those that govern impaction in any situation. Impaction in this case is in the respiratory system. This system can be likened to a tree of tubes which branches again and again. The alveoli (bubblelike structures at the ends of the branches) are analogous to leaves. Bronchioles are small branches that connect to one of two large branches (the bronchi) which in turn connect to the traches (the trunk). Inhaled air passes the nose and sinus labyrinth and the glottal opening before entering the trachea.

Each of the above structures acts as an impactor. The result is that, in ordinary breathing, few particles less than about 5 μm in diameter ever reach the glottal area; those that do seldom reach the bronchioles. Particles in the $1-3$-μm-diameter range are collected by the nasal sinuses and throat and are eventually swallowed.

AIR-SAMPLING TECHNIQUES

Air-sampling techniques, the first of which were devised over a century ago, are nearly as diverse as the disciplines included in the field of aerobiology. This diversity is largely due to the great variety of airborne organisms and the various reasons for sampling them. Air sampling may be qualitative, when an investigator wishes to determine what organisms are present; or it may be quantitative, when it is important to know concentrations. Some samplers are useful only for qualitative purposes.

The choice of a sampling method not only is a function of the sampling purpose but, more important, is dependent on the characteristics of the airborne particles, that is, their size, shape, surface structure, and density.

Types of samplers. Although air-sampling techniques are varied, there are only a few basic principles used in the design of samplers, for example, gravitational settling, impaction, suction, filtration, impingement into liquids, and electrostatic and thermal precipitation.

Gravity samplers. Exposure of a horizontal surface on which particles can settle is the simplest method of sampling. However, because of air turbulence, it is impossible to compute concentrations, and thus this method is qualitative. The most familiar sampler of this type is the Durham or gravity slide sampler (Fig. 7).

Impaction samplers. Airborne microbes possess inertia such that, when a particle approaches an obstacle, it will be impacted rather than flowing around it if the inertial force is great enough. Impaction efficiency is a function of wind speed, particle characteristics, and collector size. Acceptable efficiency can be obtained only for certain combinations of these variables. The efficiency of retention is also important, and a sampling surface must be coated with a good adhesive to ensure adequate retention.

Static impaction samplers such as glass cylinders have been used, but they are not recommended for field use under light or variable wind speeds. Rotating impaction samplers have proved to be more useful and reliable for obtaining quantitative samples of large spores and pollen. Collection efficiency is largely independent of wind speed. The rotorod (Fig. 8) is the best-known of this type; particles are collected on the surfaces of two upright metal arms mounted on a small motor which revolves at 2500 rpm. Rotoslide, rotobar, and rotodisk models with or without retracting arms, which

shield surfaces from wind impaction when they are not rotating, have also been developed.

Because of the high collection efficiency of these devices, they become overloaded if operated for prolonged periods, and thus are usually designed to operate sequentially or intermittently.

Suction samplers. Suction samplers draw air into an entrance, usually with a vacuum pump. Within the sampler, filtration, impaction, liquid impingement, and electrostatic and thermal precipitation can be used for collecting the material. Some samplers do not collect the material but measure concentration by optical methods.

There are certain problems in obtaining a representative air sample, particularly for large particles, and the number sampled may be different from that actually in the air. Isokinetic sampling is desirable, and is illustrated in Fig. 9a. Figure 9b illustrates a case when airspeed is less than sampling speed, and oversampling results. Undersampling occurs when airspeed exceeds sampling speed. Most suction samplers, however, are designed so that even an approximation to isokinetic sampling is impossible when the rapid fluctuations in wind speed and directions in the atmosphere are taken into account. Suction samplers are not normally suitable for sampling larger airborne particles. Entrance efficiencies are better for smaller particles, and are improved if the sampler is vaned into the wind.

An example of a suction sampler is the Hirst spore trap, designed for spores which are impacted on a microscope slide moving at 2 mm/hr. A later model, the Burkard 7-day trap (Fig. 10), is similar but collects spores on a circular drum.

Cascade impactors are suction samplers involving a number of orifices and impaction surfaces in series, each smaller than the preceding ones, so that large particles impact on the first stage, and smaller ones on succeeding stages. Many models are available. One of the most commonly used is the Andersen sampler (Fig. 11), in which particles are deposited on petri dishes of nutrient media for culture.

Filters. Filter samplers, particularly the high-volume samplers, which can sample up to 1415 liters/min, generally are used for sampling nonbiological air pollutants. Filter samplers with a smooth surface, such as molecular membrane types and

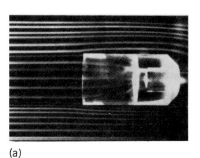 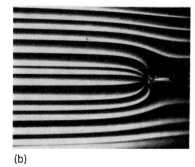

(a)   (b)

Fig. 9. Airflow in a suction sampler. (a) Isokinetic sampling; smoke moving in a laminar flow enters the sampler without disturbance. (b) Distortion of the laminar of the smoke; in this case, oversampling occurs. (From K. R. May, in P. H. Gregory and J. L. Monteith, eds., Airborne Microbes, Cambridge University Press, p. 80, 1967) some glass fiber filters, can be used for microscopic examination.

Liquid impingers. Liquid impingers are commonly used for fragile microorganisms including bacteria. Quantification is by plating and counting colonies. The samplers are not well adapted for low concentrations and are limited to use above 4.5°C.

Electrostatic, thermal, and optical samplers. Electrostatic and thermal precipitation and optical counters are generally not used in microbiological sampling.

Aircraft samplers. Pollen and spores have also been sampled from aircraft. Most of these samplers have nonisokinetic entrances, but some isokinetic samplers for use on light aircraft have been developed.

Insect samplers. Most of these devices can be classed as moving or rotating nets, tow nets, sticky barriers, and suction traps.

Choice of samplers. Particles <5 μm in diameter and not requiring culture are best sampled by suction samples. Anisokinetic conditions do not introduce serious errors for these small particles. Particles between 5 and 15 μm in diameter are not sampled very efficiently by either suction or impaction samplers. It is generally important to determine the collection efficiency for the particles being considered. Above a diameter of 15 μm, particles are best sampled by rotating impactors using continuous, intermittent, or sequential operations.

It is also important to consider the sampling location, sampling season, sampling period, various

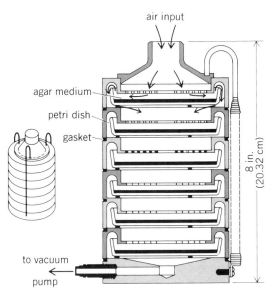

Fig. 11. Andersen sampler, a typical cascade impactor. *(From R. L. Edmonds, Aerobiology: The Ecological Systems Approach, Dowden, Hutchinson, and Ross, Stroudsburg, PA, 1979)*

stains for microscopic analysis, and culture techniques. Identification of airborne particles requires considerable training and experience, extensive consultation of the literature, reference collections, and the assistance of specialists.

AEROBIOLOGY IN THE FUTURE

Aerobiology is now an established interdisciplinary field of science. Airborne microorganisms affect the lives of humans in many ways, particularly because a large number of important diseases of animals, humans, and plants are spread in the air.

Many of these diseases are currently controlled by chemical means, or by selective breeding in the case of plants and animals. But in many cases a single method of control is not adequate. Microorganisms have a far greater capacity to mutate to more pathogenic forms able to resist chemicals and infect "resistant" individuals than plants and animals have to produce resistant strains. In addition, breeding has begun to reduce the supply of resistant genes.

The science of aerobiology can help in disease and insect control. If the time when spores or airborne diseases or insects alight on plant surfaces is known, chemical controls can be used most effectively.

An understanding of airflow in homes, places of work, and hospitals has made it possible to design systems which can prevent cross-infection. Identification of the source of the problem is also as important as the control.

Mathematical modeling and computer assistance are beginning to help in the area of disease prediction and the storage and integration of data. This whole area, although full of promise, needs much further development. Adequate monitoring systems for airborne diseases also need to be set up on a worldwide basis.

Fig. 10. Burkard seven-day spore trap, a typical suction sampler which is vaned into the wind. *(From R. L. Edmonds, Aerobiology: The Ecological Systems Approach, Dowden, Hutchinson, and Ross, Stroudsburg, PA, 1979)*

Although much of the initial work in aerobiology concentrated on diseases, other areas of interest have emerged, for example, the role of microorganisms and small organic particles in atmospheric precipitation processes and in nutrient cycling, particularly on a global basis. Although the time and space scales of palynological research are far larger than in most aerobiological problems, the explanation of plant distribution patterns can be aided by an understanding of modern pollen rain.

Finally, there remain some fascinating questions to be resolved: Do microorganisms reproduce in the atmosphere, and do they exist in outer space?

[ROBERT L. EDMONDS]

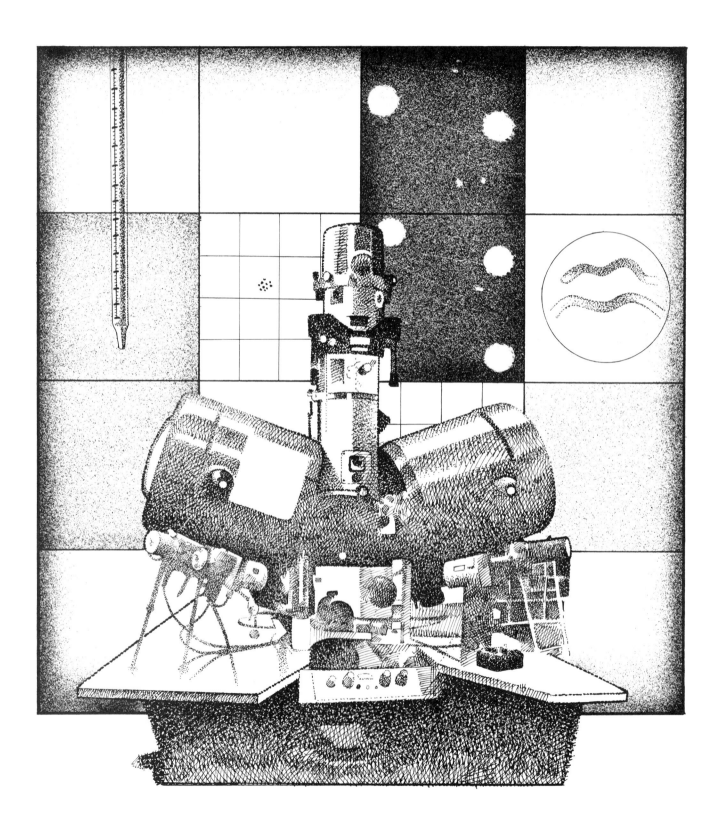

"Only within very narrow boundaries can man observe the phenomena which surround him; most of them escape his senses, and mere observation is not enough. To extend his knowledge, he has had to increase the power of his organs by means of special appliances . . . ." (From An Introduction to the Study of Experimental Medicine, 1865, by Claude Bernard, the French physiologist who was the first to discuss the importance of the principles of experimentation in medicine)

Richard J. Roman is a research fellow in the National Biotechnology Resource in Electron Probe Microanalysis, Department of Physiology, Harvard Medical School. He received his doctorate in pharmacology from the University of Tennessee in 1977.

Claude Lechene is director of the National Biotechnology Resource in Electron Probe Microanalysis and is a visiting professor of physiology at Harvard Medical School. Trained in mathematics and thermodynamics at the University of Paris, he received his M.D. degree there in 1965.

Electron Probe Micro-analysis in Health Research

The electron probe is increasingly filling an important void in the special appliances of physiologists. Whereas electron microscopy has allowed anatomists to describe the fine structure of cells in the nanometer range, and molecular biologists are studying the structure and function of genes, the elemental composition of the cells and the local distribution of chemical elements within the intracellular and extracellular compartments of tissues have not been determined. Part of the problem in defining cell composition has been the lack of methods for measuring the elemental content (10^{-15} to 10^{-18} mole) of minute samples of biologic fluids and the concentration of elements within and around cells whose volume is in the picoliter (10^{-12} liter) to attoliter (10^{-18} liter) range. Traditional microchemical methods for elemental analysis using flame photometry or atomic absorption spectrophotometry have minimal detectable amounts in the range of 10^{-9} to 10^{-12} mole. Further, these methods are destructive and require relatively large sample volumes, that is, milliliters to microliters. On the other hand, electron probe microanalysis can be used to measure the concentration of any element in the periodic table with atomic number of 5 (boron) or greater in a sample volume of an attoliter, provided the concentration of that element is greater than 10^{-4} M. Minimum detectable amounts as low as 10^{-20} g for iron have been reported. Theoretically, there is no limitation to increasing the sensitivity of the method. The advantages of electron probe microanalysis over all other techniques for measuring elemental composition of samples are, first, the required sample volumes are five orders of magnitude smaller than other methods, and second, the technique is nondestructive. Simultaneous or repeated measurements can be made under various assay conditions, and the concentration of every element in the periodic table above atomic number 5 can be determined in the same sample.

BASIC PRINCIPLES

Electron probe microanalysis is a spectrometric technique which utilizes the property of all chemical elements to emit x-rays of discrete characteristic wavelengths when excited by high-energy radiation. By identifying the wavelength of the x-ray photons emitted at any characteristic wavelength (or energy), one can qualitatively analyze the elemental content of a sample. By determining the intensity of radiation at any characteristic wavelength, one can quantitate the amount of that element contained within the excited sample volume.

X-ray production. In electron probe microanalysis the energy for the excitation of x-ray emission is provided by high-speed electrons bombarding a sample. There are two electron-beam—sample interactions which lead to the production of x-rays (Fig. 1). Occasionally, an incident-beam electron with sufficient energy may dislodge a K, L, or M inner-shell electron from an atom in the sample, and leave that atom in an excited or ionized state. The ionized atom returns toward its ground state when a higher-energy outer-shell electron fills the vacancy in the inner electron shell. This process is accompanied by the release of energy through the emission of an x-ray photon. Since electrons are in discrete orbitals at defined energy levels, the emitted x-ray photon has an energy equal to the difference between the energy level of the outer-shell electron which filled the electron orbital vacancy and its new energy level in the inner shell. The relationship between the wavelength and the energy of an x-ray photon is given by the formula $\lambda = hc/eE = 12.4/E$, where h is Planck's constant, c is the velocity of light, E is the energy of the x-ray given in kiloelectronvolts, e is the charge of an electron, and λ is the x-ray wavelength given in angstroms.

From this formula it can readily be seen that removal of an inner-shell electron by the incident-electron-beam excitation will lead to the emission of a photon whose characteristic energy or wavelength is limited to only a few possibilities, depending on the atomic structure of the element under study. Thus, under electron-beam excitation, each atom of a single element in a sample emits x-rays at a few well-defined wavelengths or produces radiation in discrete characteristic x-ray lines which can be separated and quantitated in an electron probe.

The other potential interaction of an incident-beam electron and the atoms of the sample leading to photon emission is scattering of incident-beam electrons by the nucleus (Fig. 1). An incident electron loses a variable amount of energy or is decelerated to various degrees by the positive electrical field of the nucleus. This loss of energy by beam electrons results in the production of x-ray photons with equal energy. Since the energy lost from any incident-beam electron can range continuously from zero up to the total energy of that electron, x-ray photons of many different energies or wavelengths are produced. This process produces a background x-ray spectrum called a continuum, white radiation, or bremsstrahlung. Unlike the production of characteristic radiation, which is specific to one element, the production of background or white radiation arises from the contribution of all atoms of all elements in a sample. A useful consequence of this phenomenon is that the intensity of white radiation emitted is proportional to the total mass of the sample excited by the electron beam. This signal can be measured and can provide an index of sample mass.

X-ray detection. In electron probe microanalysis the number of photons emitted by a sample at characteristic wavelengths can be determined by using either wavelength-dispersive or energy-dispersive spectrometry. In wavelength-dispersive spectrometry the emitted x-ray photons from the sample are collimated and directed onto a crystal. Only photons with wavelengths λ given by Bragg's law are diffracted. Bragg's law is expressed as $n\lambda = 2d \sin \Theta$, where n is an integer, d is the interplanar crystal spacing, and Θ is the incident angle at which the x-ray photon strikes the crystal. By adjusting the angle between the crystal and the sample, or by the use of different crystals with different interplanar spacing, x-ray photons characteristic for any element of interest can be selectively diffracted. Selectively diffracted photons are directed at the thin window of a gas-flow proportional counter for detection. The counter consists of a cylinder with a central wire along its axis through which an argon-methane gas mixture continually flows. The wire has an applied positive potential of approximately 1500 V relative to the cylinder. When an x-ray photon enters the proportional counter, it collides with and ionizes gas molecules. The negatively charged electrons released from the gas molecules are accelerated toward the positive potential of the central wire and gain sufficient energy to ionize other gas molecules. Electrons released from these secondary ionizations also move toward the central wire. In this way, one x-ray photon can produce a large number of electrons moving toward the central wire, and can generate an electric current impulse. The amplitude of this signal is proportional to the number of

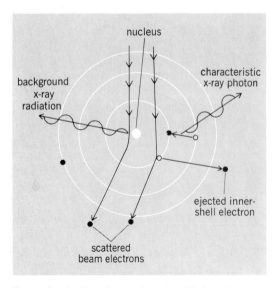

Fig. 1. Production of x-rays in a simplified model of the atom.

gas molecules ionized, which, in turn, is dependent on the energy of the original photon. The electrical signal from the proportional counter is passed through a preamplifier, an amplifier, and a pulse-height analyzer. The number of pulses of a given amplitude within a voltage window defined by the pulse-height analyzer is displayed on a digital counter as counts per second (cps). The cps are proportional to the amount of element of interest in the volume excited by the electron beam.

Detector. In energy-dispersive spectrometry, all x-rays of every wavelength emitted by the sample are directed on a solid-state detector. The detector consists of two metal electrodes with an applied potential, separated by a silicon crystal semiconductor doped with lithium. X-ray photons entering the semiconductor create electron-hole pairs by ionization of molecules within the detector. The number of pairs formed is proportional to the energy of the incident photon. Electron-hole pairs within the detector crystal are mobile and move in response to the applied electrical field and produce a current signal across the detector. More simply stated, x-ray photons create a current signal by increasing the conductivity of the detector. The signal is amplified greatly, transformed into a voltage pulse, and passed into a multichanneled analyzer in which the voltage pulses are separated according to their amplitude and stored in a computer memory. An energy spectrum, that is, the number of pulses at various voltages, can be displayed on a cathode-ray tube or fed into a computer. This voltage spectrum is directly related to the number of photons entering the detector at various energies. With the use of computer programs this complex signal can be unraveled, and amplitudes of various characteristic peaks determined. The major advantage of energy-dispersive spectrometry for analyzing x-ray emission from a sample is that the entire spectrum is displayed simultaneously, thus allowing instantaneous recognition of all the elements in the excited volume of sample. On the other hand, this technique has several important drawbacks. First, the amount of background signal is higher, and the peak-to-background ratio at characteristic wavelengths much lower with energy-dispersive spectrometry than with wavelength-dispersive crystal spectrometers. An additional disadvantage of energy-dispersive spectrometers is the relative lack of energy resolution. A crystal wavelength spectrometer has an energy resolution of 10 eV, whereas an energy-dispersive detector can at best resolve peaks separated by 150 eV. Practically, this means that although the characteristic x-ray photons of magnesium and sodium in a sample are unambiguously resolved by using wavelength-dispersive spectrometry, this may not be possible with energy-dispersive systems.

THE INSTRUMENT

In most electron probes the source of electrons for sample excitation is provided by a hot-filament electron gun (Fig. 2). The filament is a small-diameter wire biased positive with respect to a cathode shield (Wehelt cylinder) that has a central hole. Current is passed through the wire to raise the filament temperature. Electrons, emitted from the

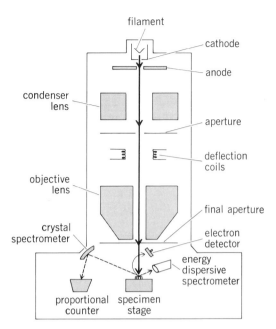

Fig. 2. Schematic representation of an electron probe microanalyzer.

filament, are focused as they pass through the central hole of the cathode while being accelerated to energies of 8–50 keV by high-voltage positive potential applied to the anode. The column of the electron probe is evacuated to at least 10^{-5} torr

Fig. 3. Cameca MS-46 electron probe microanalyzer with four wavelength-dispersive spectrometers. The instrument has a computer-controlled automated stage and spectrometer tuning.

$(1.3 \times 10^{-3}$ Pa) with a diffusion pump, since the electron beam would be absorbed or scattered by even a small amount of air. A condenser lens defocuses the electron beam and controls the amount of electrons passing through the first aperture. The diameter of the electron beam that strikes the sample is controlled by an objective electronic lens. The incident electron beam can be kept static or can scan the sample in a raster pattern if defection coils are added to the column. The specimen is positioned on a stage which moves in the x,y,z directions.

In the instrument shown in Fig. 3, the stage movements are automated and under computer control. The location of different areas of a sample, or different samples on the same sample holder, can be entered into computer memory, the coordinates recalled, and the stage movement controlled by computer, so that many samples can be sequentially brought under the electron beam and assayed repeatedly. The samples can be of any thickness—from bulk samples, opaque to electrons, to ultrathin $(0.2\text{-}\mu\text{m})$ samples through which incident electrons pass relatively unimpeded. Under electron-beam bombardment, the sample emits x-ray photons. The x-rays can be collimated, diffracted by a wavelength-dispersive crystal spectrometer, and counted or analyzed by using an energy-dispersive system with a solid-state detector. Typically, several spectrometers can be attached to an instrument. In the instrument depicted in Fig. 3 there are four wavelength-dispersive spectrometers; the newer instrument shown in Fig. 4 has three wavelength-dispersive spectrometers and one energy-dispersive spectrometer.

Several additional devices are usually incorporated in electron probes to allow correlation of sample morphology and elemental composition. A light microscope (500×) proves to be an invaluable help in selecting an area to be analyzed and in observing changes in the sample during analysis. Three types of electron signals are available from a sample excited by an electron beam. These signals can be detected and provide valuable information concerning the structure of the sample. Low-energy "secondary electrons," which emerge primarily from the specimen's surface, can be accelerated toward a grid with an applied positive potential and directed onto an electron detector. The electron detector can be a solid-state device, similar to that used in an energy-dispersive spectrometer, or a scintillator-photomultiplier detector. The voltage signal from the electron detector can be used to form an image indicative of surface variations and local density of the sample. Backscattered electrons are incident-beam electrons scattered by the specimen which emerge from the specimen surface with relatively high energy. The number of backscattered electrons varies with average density or atomic number of a region in a sample. Backscattered electrons can also be measured with an electron detector, and their distribution displayed to form an image of the sample. Finally, the specimen current, or the current absorbed by regions of the sample, can also be displayed on a cathode-ray tube to form an image. A different approach is used to correlate this image loosely with the average density within sample regions than is used with backscattered or secondary electrons. By detecting the available electron signals, the elec-

Fig. 4.   Cameca Camebax electron probe microanalyzer with transmission electron microscope attachment. This instrument has three wavelength-dispersive spectrometers and one energy-dispersive spectrometer.

tron probe, at worst, can produce scanning electron microscopy images of a sample and, at best, with ultrathin sections and additional attachments, can be used to form conventional transmission electron micrographs of the region of sample being analyzed.

Many instruments are available from a variety of manufacturers. Choice of an instrument should be made on the basis of superior analytical ability and the capability to handle a variety of sample types.

SAMPLE PREPARATION AND APPLICATIONS

Electron probe microanalysis is the method of choice whenever one needs to know the elemental composition of samples of ultrasmall volumes. Although electron probe microanalysis is widely used in metallurgy and material sciences, the application of the technique has been slow in the health sciences. Part of the problem concerning the slow incorporation of electron probe microanalysis in biologic studies has revolved around methods of sample preparation. In metallurgy, elements are fixed in position within a sample; in biology, the distribution of diffusible elements in a sample is of utmost importance. Sample preparation of biologic tissues for electron probe measurements must be geared to preserve the distribution of elements within the living sample. To date, electron probe microanalysis has been successfully applied to three classes of biologic samples: liquid droplets, isolated cells, and tissue. In the following sections, methods of sample preparation are briefly outlined, and some of the areas in which electron probe microanalysis has provided, or should provide in the future, unique information for health research are highlighted.

Liquid droplets. In liquid droplet microprobe analysis, fluid samples of nanoliter and picoliter size are collected from a living system by using glass micropipets. A few examples of this type of sample are tubular fluid collected from the lumen of individual mammalian nephrons, intracellular fluid from a large cell, fluid from a blastocyte grown in tissue culture, or endolymph and perilymph collected from the ear of an organism. After collection, the sample is prepared for electron probe analysis, as described below and schematically outlined in Fig. 5. The samples are ejected from the micropipets under oil onto siliconized glass concavity slides which are kept immersed in a bath of saline paraffin oil so that the samples do not evaporate. Aliquots of as little as 5 picoliters are taken up with a volumetric micropipet and deposited under oil onto the surface of a beryllium block. Hundreds of samples can be deposited on the same support. After both standard solutions of a known composition and samples of an unknown composition are in place on the support, the overlying layer of oil is washed off the surface of the block with xylene. The block is then placed in a bath of cold isopentane (−160°C) to freeze the liquid droplet samples. The support is then transferred to a freeze-dryer apparatus (−70°C; vacuum 10^{-5} torr or 1.3×10^{-3} Pa). Freeze-drying of the liquid droplets results in the formation of dried spots of very uniform, small (less than 1 μm) salt crystals on the surface of the beryllium support (Fig. 6). A liquid sample of 20 picoliters, salt concentration of

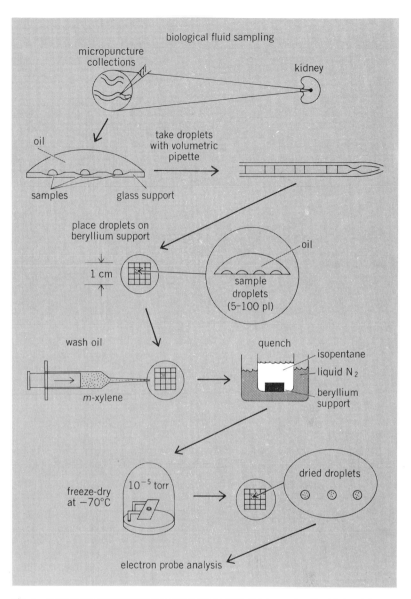

Fig. 5. Schematic representation of the procedure for preparing liquid droplet samples for electron probe microanalysis. (From J. V. Bonventre, K. Blouch, and C. Lechene, Biological Sample Preparation for Electron Probe Analysis: Liquid Droplets and Isolated cells, in M. A. Hayat, ed., X-ray Microscopy in Biology, University Park Press, in press)

20 mg/ml, typically gives rise to a dry spot of 100 μm diameter and less than 1 μm thick. The similar geometry of each dried sample tends to minimize possible difficulties in quantitative analysis of the sample by the electron probe due to nonuniform penetration of the electron beam or absorption of emergent x-rays by the sample itself.

In quantitative electron probe microanalysis of liquid droplets, each spot is excited by the electron beam of constant diameter and current density for a fixed time. X-ray intensities measured from unknown samples are recorded and compared to those of the standards. Routinely, an electron probe analyzes for seven elements of biologic interest (Na, K, Ca, Mg, Cl, P, S) in picoliter-droplet samples. Standard curves, relating x-ray counts versus elemental content for a series of picoliter-size aliquots of standard solutions, are

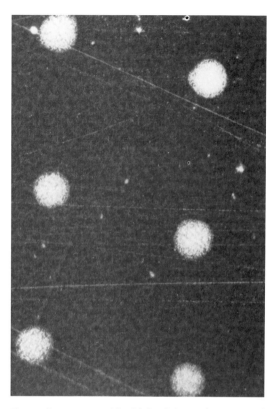

Fig. 6.   Appearance of liquid droplet samples prepared for electron probe microanalysis. Each spot of dried salt crystals is 40 μm in diameter and represents the salt content of 31.1 picoliters of fluid. (From C. Lechene, Electron Probe Microanalysis of Picoliter Liquid Samples, in T. Hall, P. Echlin, and R. Kaufman, eds., Microprobe Analysis as Applied to Cells and Tissues, Academic Press, 1974)

presented in Fig. 7. As can be seen, the curves are linear for all elements over the concentration range likely to be encountered in biologic samples. For all elements tested, the correlation coefficient of linear regression is essentially 1. Reproducibility of electron probe microanalysis measurements of liquid droplet samples of single or mixed salt solutions is excellent. Generally, the standard error of repeated measurements of x-ray intensity from single samples is less than 2% of the mean.

It should be noted that x-ray microanalysis as described gives the average amount of the element of interest within a sample without discrimination as to the form of that element. In biologic specimens, sulfur, not sulfate, and phosphorus, not phosphate, concentrations are measured. For example, the sulfur signal generated from a sample might be due to sulfur atoms incorporated in amino acids, organic acids, and proteins, as well as sulfur, in inorganic sulfate molecules. Additional analysis with available methods, such as ultramicroprecipitation of sulfur in the form of sulfate with barium chloride, is necessary to determine what fraction of sulfur in the original sample was in organic and inorganic forms.

The method of liquid droplet sample preparation outlined above must be modified if x-ray photons are to be analyzed by using energy-dispersive rather than wavelength-dispersive spectrometry, because the use of beryllium support would lead to a

prohibitively high background. To avoid this problem, techniques are available for preparing the liquid droplets on formvar or colloidion membranes. These membranes are extremely thin, and contribute little to the background signal.

Electron probe microanalysis of liquid droplet samples has increasingly been used in health research. Indeed, as more scientists become aware of this method, whereby they can analyze more elements in smaller volumes than ever before, application of the technique to new problems should continue to grow rapidly.

In renal, reproductive, exocrine, digestive, and auditory physiology and in biochemistry, electron probe microanalysis of very small liquid droplet samples has produced original information not obtainable by the use of any other microanalytical technique.

Isolated cells. In preparing isolated cells for analysis, the goals are: (1) to separate the cells from each other, (2) to maintain their ionic content without any leak or redistribution, and (3) to ensure that the cells are not coated with a layer of dried medium that would interfere with the analysis.

Blood cells, intestinal cells, bacteria, sperm, and large protozoans have all been analyzed after a cell suspension has been placed on a substrate and the cells have dried in air. Cells have also been analyzed after fixation with various media. The use of these methods may modify the distribution of elements between or within the extracellular fluid environment and the intracellular compartments of the cells, and thus interpretation of the results becomes difficult. To avoid ionic translocation, some researchers have frozen cell suspensions and freeze-dried the preparation. However, freeze-drying may also cause disruptions in the cells, for water vapor escapes from the cells under the influence of the vacuum.

One seemingly satisfactory method for studying isolated cells is to spray a cell suspension onto a support. This method has been used to study the composition of isolated human red blood cells. Cells were suspended in an isotonic acid sucrose solution and sprayed onto carbon disks and allowed to dry in air. The cells dried almost immediately after being deposited on the support, as only a small quantity of solution remained adherent to the cells after atomization. In addition, the adherent layer of medium which coated the cells contained no ions to interfere with the chemical elemental analysis of the cells. Sprayed red blood cells ready for electron probe analysis are illustrated in Fig. 8. It was shown that Na, K, Fe, and S signals were measured only when the electron beam was located over cells, and no signal could be detected in adjacent areas. This result indicates that the cells have completely retained their cellular content. Absolute quantification of the elemental concentrations within red blood cells can be obtained by using as standards cells which have been preloaded with solutions of known composition by the use of agents which reversibly modify red cell membrane permeability. This spray technique to prepare cells for electron probe microanalysis can be used to study innumerable problems in the health sciences. For example, using preloaded cells as standards, one can compare the chemical elemental composition of normal red blood cells with that of cells damaged by disease or drugs.

Tissue. The promise of electron probe microanalysis is to provide the physiologist with a tool to learn how ions are distributed between and within cells. This promise is being realized; routine and general methods for preparing tissues for electron probe that unequivocally maintain the ionic distribution of ions as in living tissue are being documented. It is immediately obvious that the usual histological and electron microscopic preparative methods that maintain the morphologic appearance of tissue (for example, fixation, staining, and embedding) are not applicable. Diffusible elements within the tissue are extracted when they are contacted by aqueous embedding media and translocated, as discussed previously, during ultramicrotomy and freeze-drying processes.

It should be noted, however, that in special applications of electron probe microanalysis of tissue, that is, when elements under study are bound or precipitated in tissue, or when the elemental localization within tissue is not critical, the sample may be prepared by using traditional electron microscopy techniques. These methods have been successfully applied to the study of the composition of insoluble pathologic inclusion in tissues. Another special application of electron probe microanalysis is in cytochemistry. Reaction products precipitated within tissues could be recognized, even if they were invisible, if they carried elemental tags which could be identified by the electron probe.

Electron probe microanalysis of diffusible elements in tissue samples kept frozen and hydrated is probably the most promising general method for tissue analysis. Electron microprobe studies of frozen hydrated tissues in cells of Malpighian tubules and salivary glands, in frog retina, and in frozen hydrated rat kidneys already indicate that concentrations of ions as in living tissues are maintained in this preparation and that electron probe analysis at a spatial resolution of 1 μm^3 or in volumes of 10^{-15} liter is possible. Two important applications of electron probe microanalysis to tissue samples should be mentioned. Studies have been done on epithelial transport in freeze-dried sections of frog skin and on the elemental basis for the mechanism of muscle contraction in ultrathin freeze-dried cryosections. Although the use of freeze-dried sections was adequate for electron probe studies of these highly structured, dense tissues, a general application of these preparative methods to all types of tissues is probably not feasible.

In one general system that is currently used to analyze elemental composition within frozen hydrated tissue, the tissue is first quench-frozen in solid nitrogen. The tissue is then placed in a liquid nitrogen bath (−196°C) for trimming and cutting. The sample is cut with a diamond-embedded, circular lapidary saw. The cutting of the tissue with this saw in liquid nitrogen produces a large, flat surface suitable for microprobe analysis. After cutting, the tissue is transferred, via an interlock chamber maintained at liquid nitrogen tempera-

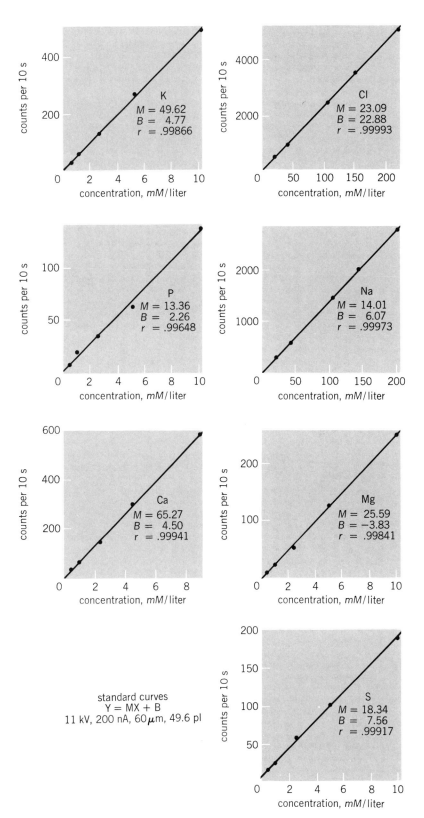

Fig. 7. Set of electron probe microanalysis standard curves relating K, Cl, P, Na, Ca, Mg, and S concentration versus x-ray counts per 10 s for a series of 49.6-picoliter aliquots of standard solutions and a blank. The line equating x-ray counts per 10 s versus elemental concentration is given by the formula $y = mx + b$, where m is the slope of the line and b is the y intercept; R = regression coefficient. The accelerating voltage was 11 keV, and the beam current was 200 nA. The beam diameter was 60 μm. (From J. V. Bonventre, K. Blouch, and C. Lechene, Biological Sample Preparation for Electron Probe Analysis: Liquid Droplets and Isolated Cells, in M. A. Hayat, ed., X-ray Microscopy in Biology, University Park Press, in press)

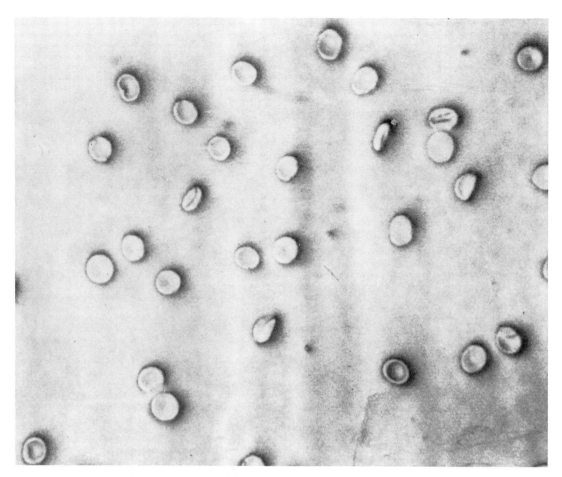

Fig. 8. Appearance of red blood cells prepared for electron probe microanalysis by spraying the cells onto a carbon support. *(From C. P. Lechene, C. Bronner, and R. J. Kirk, Electron probe microanalysis of chemical elemental content of single human red blood cells, J. Cell. Physiol., 90:120, 1977)*

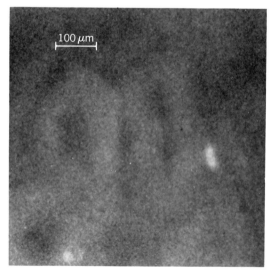

Fig. 9. Appearance of the surface of a frozen hydrated rat kidney prepared for electron probe microanalysis. The kidney was sectioned with a diamond lapidary saw in a liquid nitrogen bath. Note that the lumen and cell wall of the tubules and the peritubular space around the individual tubules are visible. Tubular diameter is approximately 25 μm. Magnification 160×.

ture, to the frozen stage of the Cameca Cambax microprobe (Fig. 4). In the probe column, the sample and tissue holder are kept cold (−190°C) by the circulation of liquid nitrogen through the stage of the instrument. With the light microscope or scanning electron microscope, the area to be analyzed is visualized. In bulk kidney samples, prepared as described, open tubular structures, glomeruli, and capillaries are recognizable (Fig. 9), and x-ray signals can be analyzed from various regions of the kidney. The problems of developing methods of accurate quantitation and determining the most effective electron microprobe conditions for analysis of frozen tissue remain to be completely solved, and studies concerning the stability of frozen hydrated samples under electron-beam excitation to be completed, before this type of analysis will be accepted as routine in health research.

SUMMARY

Electron probe microanalysis is a new and exciting spectrometric method for measuring the entire chemical element content of samples in volumes that are orders of magnitude smaller than can be used in other techniques. Electron probe microanalysis of millimolar concentration of elements in picoliter-volume liquid biologic samples is easy and routine. Electron probe microanalysis has already provided important information concerning the composition of various biologic fluids and cells which cannot be obtained by any other method. Analysis of the in-place concentrations of ions within frozen hydrated tissues is feasible and currently under active development in several laboratories. Electron probe microanalysis is already transforming the study of physiology, just as the electron microscope has done in anatomy. For the future, it is hoped that the electron probe will prove to be the "special appliance" of biologists that will lead them into an age in which cell function can be described in terms of a precise knowledge of cellular composition and ionic compartmentalization within and around cells.

[RICHARD J. ROMAN; CLAUDE LECHENE]

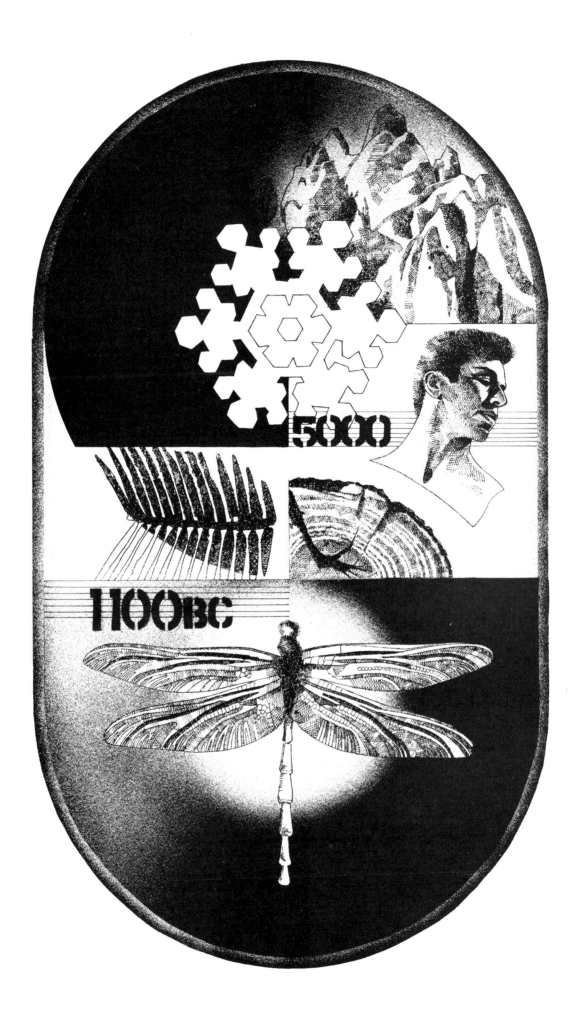

With increasing knowledge of the interacting physical and chemical mechanisms which affect the environment, there has emerged a strong desire to understand the factors which bring about climatic change. Through a better perception of these forces, scientists may be able to predict the impact of modulations in these controlling elements on global climate. Correct forcasting of future climatic variations with the adoption of appropriate procedures should make it possible to minimize the negative effects of such changes. The record of past climates has been examined with the expectation that these studies will provide information on the complex interactions between the internal and external factors which generate modern climate change. These studies should also give a better understanding of the processes by which these changes are amplified through responses in the Earth's atmosphere-ocean-ice system.

The science of paleoclimatology encompasses the various scientific disciplines concerned with Earth's past climates and associated climate phenomena. Paleoclimatology involves the examination of climates in the recent past as well as those which existed millions of years ago. Through use of the information acquired from such fields as glaciology, paleontology, paleogeography, isotopic geochemistry, and sedimentology, maps are produced depicting climate at specified time intervals during the geologic past. The compilation and evaluation of many maps, each detailing a specific time period, make it possible to follow the evolution of climate, to examine the variations, and to consider the factors responsible for such climate changes. Although much research has been devoted to mapping past climates within various time scales ranging from months to thousands of years, paleoclimatology is still in a relatively young stage of development.

Climatic studies of the last 100,000 years have produced estimates on a wide variety of

Joseph J. Morley is a research associate at Columbia University's Lamont-Doherty Geological Observatory and specializes in paleoclimatology and micropaleontology. Since receiving his Ph.D. in geology from Columbia, he has studied past climates as reflected in variations in the distribution pattern of microorganisms to acquire a better understanding of the processes that determine the present-day climate.

Paleo-
climatology

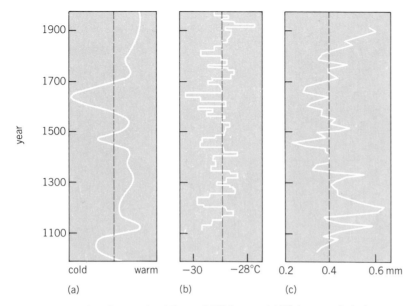

Fig. 1.   Three climatic records of the past 1000 years. (a) Winter severity index compiled from a 50-year moving average of Paris and London historical records. (b) Air temperatures over Greenland Ice Cap calculated from oxygen isotope measurements in an ice core. (c) Tree ring widths for 20-year mean tree growth (slow growth indicates low temperatures) of bristlecone pines from California.

variables, such as fluctuations in sea level, ice volume, air temperatures, precipitation, vegetation, and oceanic temperatures and salinities. In some instances, where the data base is adequate, initial analyses on the changes in these climatic

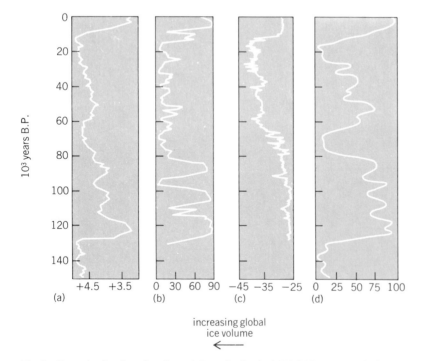

increasing global
ice volume
⟵

Fig. 2.   Records showing climatic variations for the last 150,000 years. (a) Global ice volume changes as reflected in variations in oxygen isotopes in benthic foraminifera in a deep-sea core from the eastern equatorial Pacific. Composition of $\delta^{18}O$ is expressed as per mil (°/oo) deviation from standard belemnite from Pee Dee Formation (PDB). (b) Percentages of arboreal pollen in core from Greece. (c) Air temperatures (°C) over Greenland Ice Cap calculated from oxygen isotope measurements in Camp Century ice core. (d) Percentages of polar assemblage in samples from a North Atlantic deep-sea core.

variables through time reveal what appear to be cycles in the climatic record. Possible internal and external forces responsible for these apparent periodic changes in climate are being investigated, and attempts are being made to determine the specific importance of each parameter relative to the total number of variables influencing climatic change.

DATING AND ANALYZING CLIMATIC RECORDS

In the effort to map and monitor past climates, accurate identification of the time period under study is crucial. This identification process can be accomplished in a relatively straightforward manner when examining the climatic record contained in annual tree rings, ice cores with their annual record of snow and ice accumulation, and varved sediments with their alternating layers designating annual or semiannual time intervals. The constant decay rates of radiocarbon, potassium/argon, and uranium/thorium have permitted the dating of climatic records which do not possess annual signatures. Additional specific time levels are provided by indirect methods such as first and last appearances of specific plants or animals, distinctive characteristic variations in the oxygen isotope record, and periodic magnetic reversals preserved in marine and nonmarine sequences. Since a variation in any one of these parameters occurs worldwide at essentially the same instant in geologic time, it provides the necessary control for compiling and examining climatic records.

Because of the numerous and complex processes which appear to influence climatic change locally, regionally, and globally, paleoclimatologists have depended heavily on advanced scientific equipment and techniques. The advent of the computer has greatly accelerated paleoclimatic research and made possible the use of complex statistical methods. One application of this advanced technology is in the generation of estimates of past climatic factors by relating abundance variations in marine fauna or flora to specific climatic variables such as ocean temperature and salinity. Computers are capable of absorbing huge data bases on these and other influential variables and generating complex environmental models for specific past time periods which can be compared with today's world. Satellites have provided a fast and accurate method for gathering detailed measurements of specific variables on a global scale. A recent study by G. J. Kukla and H. J. Kukla utilized this sophisticated instrumentation to acquire information on Northern Hemisphere interannual changes in area of snow and ice cover.

EXAMINATION OF PAST CLIMATES

Studies of past climates have concentrated on constructing time series (a series of statistical data collected at regular intervals of time) of varying lengths and sampling intervals from days and months through millions of years, depending on the available data. For example, instruments at various sites have recorded climatic variables such as temperature and precipitation for much of the last 100 years. These records are of such different lengths and composition, with each station not necessarily monitoring identical variables, that data comparisons are extremely difficult. Investi-

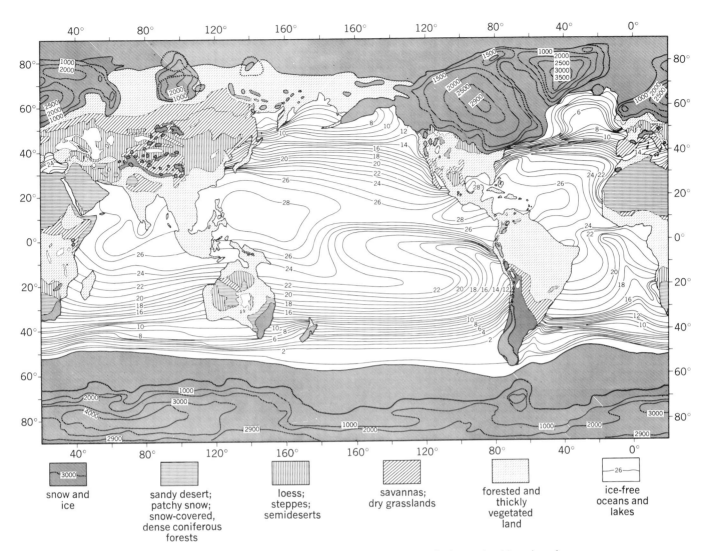

Fig. 3. Sea-surface temperatures, ice extent, ice elevation, and continental albedo and vegetation patterns for average August 18,000 years B.P. Contour intervals are 1°C for sea-surface temperatures and 500 m for elevation. Continental outlines reflect a sea level lowering of 85 m. (*From CLIMAP Project Members, The surface of the ice-age earth, Science, 191:1131–1137, 1976*)

gators have, however, calculated global mean values of some of these parameters. M. I. Budyko, for instance, in his efforts to study the effects of solar radiation variations on climate, has carefully sorted through the large quantity of instrumental data and constructed a reliable temperature index map for the Northern Hemisphere.

Last-1000-year record. In extending the climatic record back in time, H. H. Lamb has collected and cataloged historical narratives describing European weather (mostly from the regions of London and Paris) for major portions of the last 1000 years. Utilizing varied, generally inconsistent sources, Lamb devised a winter severity index which presents the data in a manner which can be easily compared with more standardized climatic records. Other scientists studying climate changes over this 1000 years have concentrated their efforts on examining the annual climatic variations in a particular region. V. C. LaMarche, Jr., correlated the width of individual tree growth rings to annual California temperatures. This tree-ring chronology resembles the historical data gathered by Lamb and others in that the resulting values of

various climatic variables can be identified with individual years. Another annual record of climate is contained in the Greenland Ice Cap, where seasonal air temperatures can be estimated from individual snow and ice layers. Measurements within each layer of the changes in the ratio of oxygen-18 to oxygen-16, a ratio which varies with changing temperatures, are used to derive the average yearly air temperatures over the ice cap.

A comparison and cross-checking of these three very different annual records (historical, tree-ring, ice core) show that all three contain a similar pattern, permitting the reconstruction of a reasonably consistent picture of Northern Hemisphere temperature variations for the last 800 years (Fig. 1). Although temperatures have fluctuated over the last several hundred years, periods dominated by colder or warmer than average temperatures are identifiable. During the Middle Ages (1100–1400), for example, temperatures averaged above normal. The Northern Hemisphere experienced a cold period from 1400 to approximately 1800. This interval is often referred to as the Little Ice Age. Abnormally low temperatures were accompanied by a major areal expansion of most glaciers, the southerly extension of pack ice in the North Atlantic, and a reduction in harvested grain and other crops throughout the British Isles and northern Europe. Since the mid to late 19th century, temperatures have risen, reaching a maximum in the 1940s. Most climatic records show a general cooling trend during the last 3 decades, with today's temperatures midway between the levels recorded in the 1940s and the cooler values of the 1880s.

Last-100,000-year record. Paleoclimatic studies involving time periods older than the last several hundred years have had to rely heavily on indirect evidence. Few historical sources exist beyond the year 1000 which describe climate-related events, and many of these reports are of questionable quality. As a result, other ways have been devised to acquire information concerning climate changes which occurred thousands to millions of years ago. Fluctuations in global ice volume during the last 1,000,000 years are recorded by changes in the oxygen-18 to oxygen-16 ratio in shells of microfossil foraminifera in deep-sea sediments. Through use of multivariate statistical techniques, abundance variations of foraminifera and other marine microfossils (coccoliths, diatoms, radiolaria) can provide quantitative estimates of past oceanographic parameters, such as sea-surface temperature and salinity. Since over two-thirds of the Earth's surface is covered by water,

these oceanographic values are extremely important in the development of a complete picture of past climate. Information on variations in sea level also provides data on climate changes, with relatively low sea levels occurring during cold (glacial) intervals, when large volumes of water are trapped in the polar ice caps, and relatively high sea stands coincident with warmer (interglacial) periods. Changes in the distribution of specific species of trees or plants can be detected by examining their abundance variations with respect to the total pollen population in samples from peat bog and lake borings. Changes in vegetation patterns can be related to environmental changes which most likely reflect terrestial climatic variations.

These indirect methods yield information on variations in ice volume, sea level, terrestrial vegetation patterns, sea-surface temperatures, and other oceanographic data which together provide a good composite of global climate changes through time. Although these indirect methods are of lower resolution and not as precise as instrumental data, they permit the extension of the climate record for thousands of years beyond that which can be mapped by using exclusively historical data.

Last glacial and interglacial. Figure 2 shows the climatic records developed through these diverse indirect methods, for a time span encompassing approximately the last 150,000 years. The oxygen-18 to oxygen-16 ($\delta^{18}O$) ratio was calculated from shells of a specific species of benthic (bottom-dwelling) foraminifera taken from a deep-sea sediment core (V19-29) in the eastern equatorial Pacific. Higher ratios of oxygen-18 to oxygen-16 indicate increased global ice volume. The generalized tree pollen curve was produced from studies on a long borehole from northeastern Macedonia (Greece). High percentages of pine and oak (arboreal) generally reflect warmer climates. The third data set shows the variations in air temperatures over the Greenland Ice Cap as recorded in an ice core from Camp Century. The final graph is a plot of the variations in the polar foraminiferal assemblage in a deep-sea core (K708-6) from the North Atlantic. These four climatic patterns display some major similarities even though they are from widely different geographical regions and were developed by using different methods. All four climatic records show a warm (interglacial) interval beginning about 10,000 years before present (B.P.) and extending through to the present. Another warm interval encompasses the period approximately 125,000 to 75,000 years B.P. In between these two interglacial periods is a long very cold (glacial) interval.

Detailed examinations of these and other climatic records covering the last 150,000 years reveal that the warmest temperatures within the present interglacial period (10,000 years B.P. to today) occurred during a 2000-year time span between 7000 and 5000 years B.P. Temperatures since this warm interval (sometimes referred to as the hypsithermal) have gradually declined, with three major cold excursions centered about 5300, 2800, and 350 years B.P. (Fig. 8, temperature trends for past 10,000 years).

Ice age Earth. Recently a major effort was exerted to produce a detailed climatic reconstruction

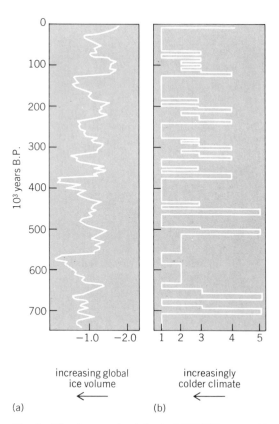

increasing global
ice volume
←

increasingly
colder climate
←

(a)   (b)

Fig. 4. Climate records of the last 700,000 years. (a) Global ice volume changes as reflected in variations in oxygen isotopes in foraminifera from a central equatorial Pacific deep-sea core. (b) Variations in soil types from loess (1) to savannas (5) as recorded at Brno, Czechoslovakia.

at a specific time during the last glacial period. The resulting map of global climate 18,000 years B.P. (Fig. 3) depicts sea-surface temperatures, ice extent, ice elevation, continental albedo, and vegetation. This data base was generated as part of an extensive research effort by many scientists in a project called CLIMAP (Climate: Long Range Investigation Mapping and Prediction).

Micropaleontologists derived estimates of sea-surface temperatures by relating faunal and floral variations to ocean temperatures. Approximately 250 deep-sea core samples of sediment deposited on the sea floor 18,000 years B.P. were analyzed in this phase of the reconstruction. The area covered by permanent ice sheets 18,000 years B.P. was determined from a widespread survey of existing literature, with ice elevations defined by a combination of ice margin limits and ice flow line characteristics. Radioisotope dating on submerged coastal terraces provided estimates on how much sea level had fallen to compensate for the large volume of water trapped in the form of ice in polar regions. Pollen studies from numerous land sites provided data on variations in vegetation patterns, and additional information for defining the limits of continental glaciers and ice sheets.

The full glacial world differed markedly from today's interglacial world, with huge ice sheets reaching thicknesses of up to 3 km covering a large portion of North America, the polar seas, and parts of northern Eurasia. In the Southern Hemisphere, the winter sea ice expanded to encompass a much larger area than today. This additional ice cover was accompanied by a lowering of sea level by a minimum of 85 m. Changes in vegetation included increased areal coverage of grasslands, steppes, and deserts, with a reduction in forested regions.

Sea-surface temperatures were an average 2.3°C cooler 18,000 years B.P., with increased upwelling in the eastern equatorial Atlantic and Pacific oceans and transport of cooler waters equatorward along the western coasts of Africa, Australia, and South America. The largest changes in sea-surface temperatures occurred in the subarctic and subantarctic regions, where temperatures declined 7 to 10°C from their present values. Temperatures over land masses averaged 6.5°C cooler than today.

Although these results show that major climate changes (glacial to interglacial) occur at approximately the same time in both hemispheres, detailed studies indicate that maximum glacial conditions occurred at different times in different areas. For example, the Labrador and Scandinavian ice sheets reached their maximum extent approximately 22,000 years B.P., whereas the Cordilleran Ice Sheet, centered in western North America, did not begin to retreat until 14,000 years B.P. Besides the differing times of maximum extent of major ice sheets, studies have also shown that the retreat (melting) of the glaciers was not constant, with various readvances of the ice margins occurring several times during deglaciation.

Modeling the last glacial maximum. Detailed reconstructions of past climates, such as that reproduced for 18,000 years B.P., provide basic and comprehensive data which are vital to understanding the specific processes that produce climate and cause its fluctuations and changes. The information in these climatic reconstructions serves as

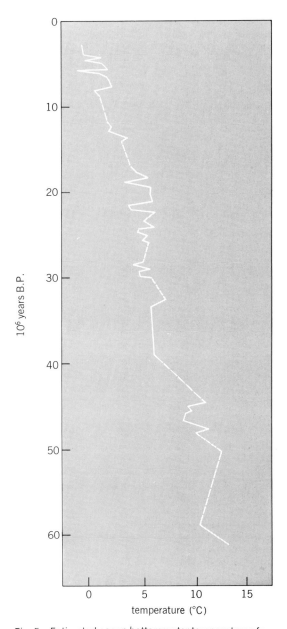

Fig. 5. Estimated ocean bottom water temperatures for the last 60,000,000 years based on oxygen isotope variations in benthic foraminifera from a site in the South Atlantic.

basic boundary conditions for the general circulation modelers. By insertion of data on extent and elevation of permanent ice, global patterns of sea-surface temperature, and continental geography, in addition to albedo measurements (amount of incoming solar radiation reflected by the Earth's surface), general circulation models can be developed which simulate numerically a three-dimensional atmosphere. Comparison of this past environment with that of today may assist in the identification of what particular physical processes are important in determining climate in addition to the sequence of events which create the observed climatic variations.

Last warm climate maximum. A majority of climatic records of the last 150,000 years indicate

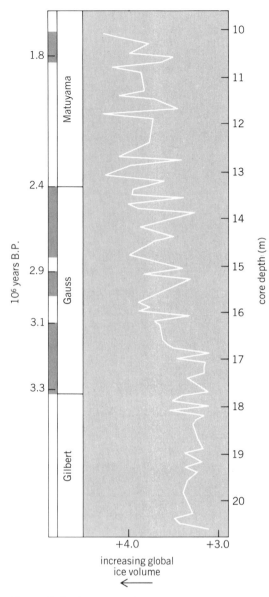

Fig. 6. Global ice volume changes as measured in oxygen isotope ($\delta^{18}O$) variations in a central equatorial Pacific deep-sea core showing development of Northern Hemisphere ice sheets approximately 3,200,000 years B.P. and initiation of large-scale interglacial-glacial climate oscillations about 2,500,000 years B.P.

mean glacial to mean interglacial can be as short as 2000 years, with a 5000- to 10,000-year span between full glacial and full interglacial.

Last-1,000,000-year record. Figure 4 shows two climatic records covering the last 700,000 years. The oxygen isotope ($\delta^{18}O$) curve in a deep-sea core (V28-239) from the central equatorial Pacific monitors variations in global ice volume, whereas the soil section from Brno, Czechoslovakia, records variations in vegetation patterns. Both curves show a climatic pattern alternating between full glacial (cold) and interglacial (warm), with approximately eight full glacial-interglacial climatic cycles during the last 700,000 years. Each of these cycles includes a major sequence of expansion and contraction of Northern Hemisphere ice sheets and warm through cold extremes in temperature. The sharp terminations marking the transition between these two climatic extremes (full glacial to full interglacial and full interglacial to full glacial) suggest that the climate changes during this transitional stage occurred very rapidly in geologic terms (within 5000 to 10,000 years).

Evolution of glacial-interglacial cycles. Global climate has not always oscillated between full glacial and full interglacial regimes. Information on climatic variations millions of years ago is not nearly as detailed or complete as that documented for the last several hundred thousand years. A combination of data gathered from studies of land sections and marine sediment sequences serves as the major source for describing the climatic pattern over this much longer time span. Evidence from this data base indicates that a substantially warmer climate persisted throughout much of the period from 100,000,000 to 65,000,000 years B.P. During this time, no ice caps were present in either hemisphere. The transition from this warm interval to the present regime of fluctuations between interglacial and glacial conditions occurred very gradually over many millions of years.

Figure 5 illustrates this slow cooling of the global climate as reflected in bottom water temperatures. These temperatures were derived from oxygen isotope analyses on benthic foraminifera from Deep Sea Drilling Site 357 in the South Atlantic Ocean. This general gradual cooling trend can best be explained through a combination of movement of the continents into near-polar positions (continental drift and sea-floor spreading) and development of new ocean circulation patterns, with the creation of intense circumpolar circulation in the Southern Hemisphere and the cessation of circumequatorial circulation.

Seasonal sea ice formation in the high latitudes of the Southern Hemisphere 39,000,000 years B.P. contributed to reducing bottom water temperatures by approximately 5°C. A permanent ice cap formed in Antarctica about 14,000,000 years B.P. In the Northern Hemisphere, mountain glaciers did not begin forming until approximately 10,000,000 years B.P. A rapid decline in temperatures around 5,000,000 years B.P. coincided with a large increase in ice volume and a drop in sea level, greatly restricting circulation between the Mediterranean Sea and the Atlantic Ocean. The final cooling event occurred with the development of the Northern Hemisphere ice sheets between

that the last glacial period lasted approximately 60,000 years. These records also show that this cold interval was preceded by a warmer (interglacial) period of approximately 50,000 years. Only during one short interval within this interglacial, however, did temperatures reach values that were as warm or warmer than today. This 10,000-year warm interval is sometimes referred to as the Eemian, with the cooler stages of the interglacial (115,000–75,000 years B.P.) designated early glacial or preglacial. Scientists have concluded, from detailed studies of the climate changes within the last interglacial (Eemian plus early glacial) and within the transitional stage between the last interglacial and glacial periods, that the time from

3,200,000 and 2,500,000 years B.P., producing the present glacial-interglacial climatic pattern. Figure 6 shows this final cooling phase as recorded in variations in oxygen isotope (δ^{18}O) composition of benthic foraminifera from central equatorial Pacific marine sediments in deep-sea core V28-179. This isotopic record indicates that glacial-interglacial fluctuations commenced about 3,200,000 years B.P. and reached maximum amplitudes, similar to those recorded in younger climatic records, 2,500,000 years B.P.

POSSIBLE CAUSES OF CLIMATE CHANGE

The reconstruction of past climates using all available information produces an overall view of climate change from which can be extracted some answers as to the possible factors which initiate or contribute to changes in climate. A partial list of probable factors affecting global climate is presented in Fig. 7, along with the time period over which variations in each factor would most likely cause climate changes. Variations in solar radiation, an external forcing function, may create the broad low-amplitude climate changes observed in the record. It has been postulated that major climatic cycles on the order of tens of millions of years may be produced by changes in the Earth's rate of rotation. The interactions of the magnetic fields of various celestial bodies may account for relatively long climatic variations on the order of tens of thousands of years.

The changing geometry of continental land masses (continental drift) and ocean basins (sea-floor spreading) with respect to the polar regions is another possible cause of climatic modifications. This particular factor was used in an earlier section to explain the gradual cooling of global climate over the last 60,000,000 years. It is generally accepted that the positioning of land masses with respect to ocean basins is a major factor in determining ocean circulation patterns. Such circulation patterns affect the properties of ocean bottom water, whose changes influence global climate. The amount and extent of continental ice sheets and sea ice also affect the composition of ocean bottom water. This brief description of what may create changes in ocean bottom water is but a single example of the complex interactions and responses of the hydrosphere (ocean) and cryosphere (ice) to variations in other factors (in this case, the changing geometry of land masses and ocean basins). These complex interactions in the air-water-ice system produce the final global climate product.

Analyses from tree-ring data indicate the presence of a drought cycle with a period of about 22 years. It has been suggested that cyclic variations in sunspot activity may be responsible for the periodic droughts.

Examination of longer climatic records, several hundred thousand years in length, has revealed cycles with periods of approximately 100,000, 41,000, and 23,000 years. These periods match those for variations in the Earth's orbital parameters of eccentricity, obliquity, and precession, indicating that these climatic cycles may have been caused by periodic variations in the geometry of the Earth's orbit.

Changes in vegetation patterns alter the amount of incoming solar radiation reflected by the Earth's surface (albedo). An increase in albedo should bring about a lowering of global temperatures. The influx into the Earth's atmosphere of large quantities of volcanic ash, smoke from extensive forest fires, or gases such as carbon dioxide from the burning of fossil fuels is also thought to produce changes in climate on a regional and possibly a global scale. An examination of the possible effects of all these factors which may modify climate indicates that a complex combination of mechanisms probably determines climate and any variations in it.

FUTURE CLIMATE

A review of the present and past climates should serve as a basis for predicting any possible future changes in climate. The climatic record for the past million years has been dominated by periods during which temperatures were colder than today. As Fig. 8 illustrates, temperatures have reached or exceeded today's warm values during less than 10% of the last 700,000 years. Today's relatively warm temperatures have persisted for the last 10,000 years. The last time temperatures were comparable to today's was during the 10,000-year interval commonly referred to as the Eemian. Since the present warm interval equals the length of the last warm period (Eemian), it has been predicted that the future climate will be colder and drier. It is difficult to determine the amount of time which will elapse during transition from the present interglacial to a colder period; however, previous studies indicate that this transition from interglacial to glacial could be quite rapid.

Even within today's relatively warm period, the Earth has experienced intervals characterized by colder temperatures. Since the 1880s global temperatures have risen, reaching a maximum in the 1940s, but have declined since then reaching a level today midway between the warm 1940s and the colder 1880s temperatures.

On a shorter time scale, it appears that climate on a year-to-year basis will be highly variable.

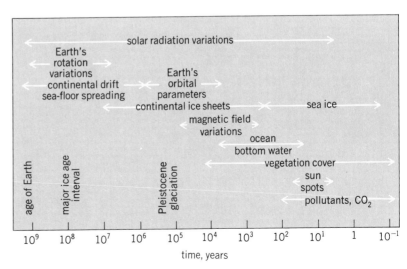

Fig. 7. Partial list of factors which may cause climate changes and the time scale during which variations in these factors would affect global climate.

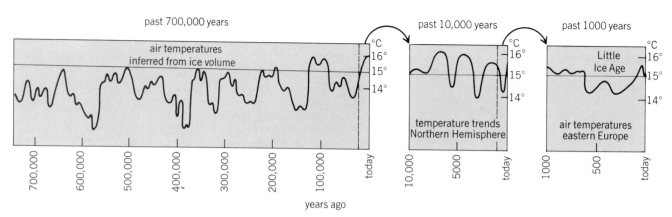

past 700,000 years

past 10,000 years

past 1000 years

years ago

Fig. 8. Chart showing the generalized temperature record for the past 700,000 years. (*Modified from S. W. Mat-* *thews, What's happening to our climate, Nat. Geogr., 150: 576–615, 1976*)

Evidence shows that the mild climate that in general existed during the early 1970s, with relatively cool summers and mild winters, is not typical. Future yearly climatic conditions should resemble those observed during the mid to late 1970s, with widely ranging temperatures and highly variable weather.

Even with the advances recently made in paleoclimatology which provide a better understanding of climate, forecasting the weather continues to be extremely difficult. Nevertheless, the need persists to predict climate for years, decades, even centuries into the future to take maximum advantage of its variations. In this prediction process, the specific importance of each factor which influences climate must be gaged, and the response of the Earth's atmosphere-ocean-ice system to changes in these parameters must be accurately measured.

[JOSEPH J. MORLEY]

Photographic Highlights

These photographs have been chosen for their scientific value and current relevance. Many result from advances in photographic and optical techniques as humans extend their sensory awareness with the aid of the machine, and others are records of important natural phenomena and recent scientific discoveries.

Jupiter as viewed by *Voyager 1* at a distance 17,500,000 mi (28,400,000 km) from the planet. Two of the satellites are visible: Io, seen against Jupiter's disk, and Europa, to the right of the planet. *(NASA)*

Venus clouds viewed from above the north pole. The collar cloud, seen in a roughly crescent-shaped pattern around the pole, is 1000 mi (1600 km) across at its widest part and rises 10 mi (16 km) above the main cloud deck. *(NASA)*

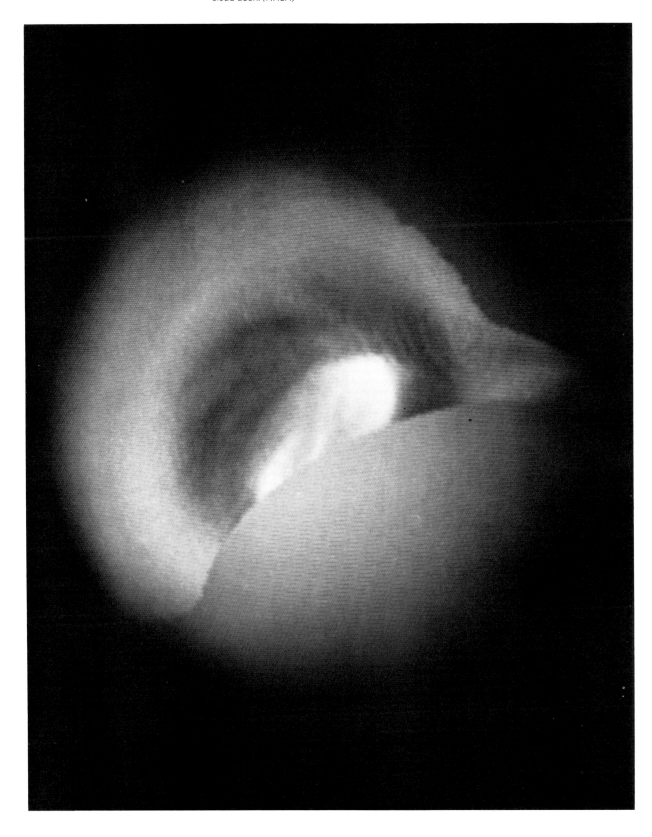

Four aspects of Venus as seen by the Pioneer orbiter. *(a)* Image taken on Dec. 30, 1978, from 26,600 mi (42,800 km) direcly over the equator and *(b)* image on Jan. 10, 1979, from 29,700 mi (47,800 km). *(c)* Three-quarter disk view taken on Jan. 14, 1979, from 40,000 mi (64,400 km) above the surface and 22° south of the equator. *(d)* Image taken on Jan. 18, 1979, from 40,000 mi (64,400 km) altitude. *(NASA)*

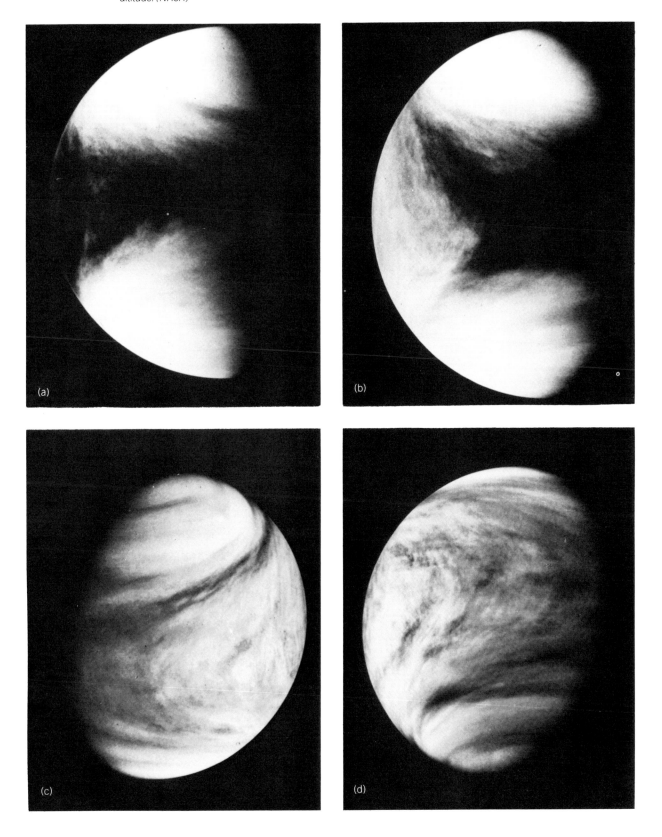

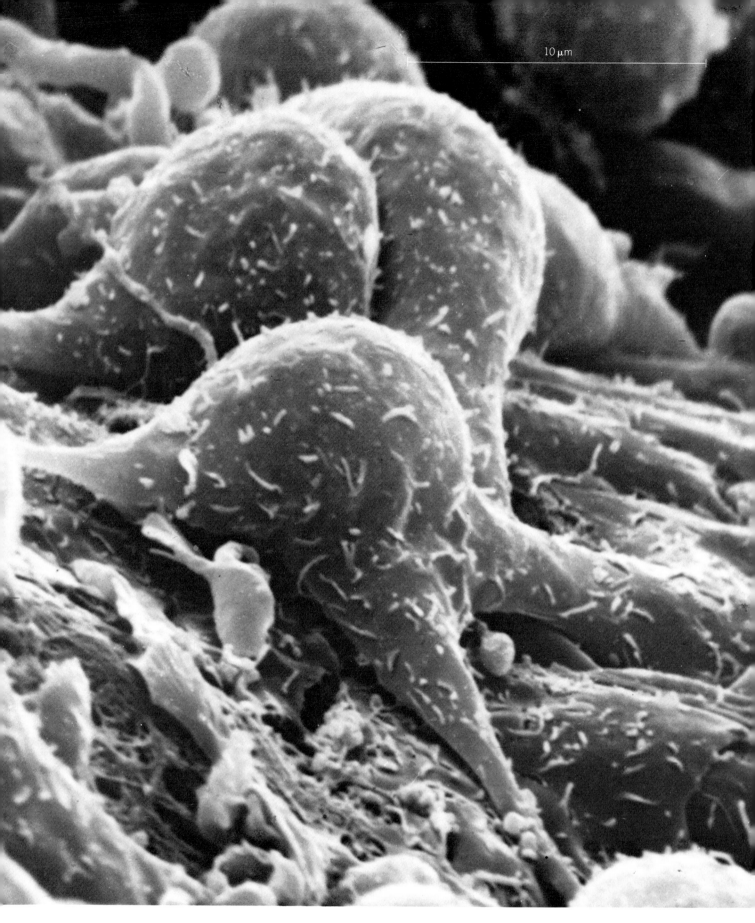
10 µm

Opposite page. Metastatic melanoma cells invading the wall of a blood vessel by attachment to the endothelial cells of the vein and insertion of their processes into fissures that form as endothelial cells retract. *(Courtesy of Garth L. Nicolson, University of California, Irvine)*

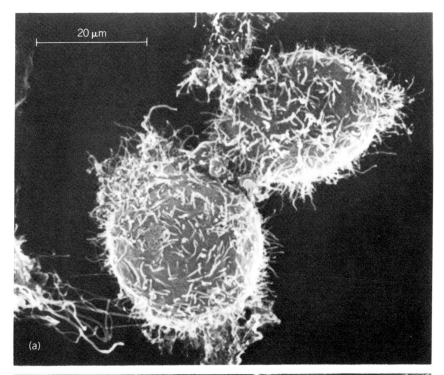

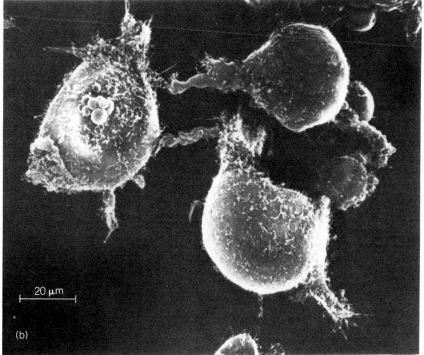

Scanning electron micrographs of fibroblastic cells with *(a)* high and *(b)* low agglutinability. *(From I. Pastan and M. Willingham, Cellular transformation and the morphologic phenotype of transformed cells, Nature, 274:645–649, Aug. 17, 1978)*

100 μm

A Primitive Crustacean, Cephalocarida

10 μm

(Above) Tip of the tentacles and (left) spines. *(Photographs by S. Honjo, back and front covers of Oceanus, vol. 21, no. 3, Summer 1978)*

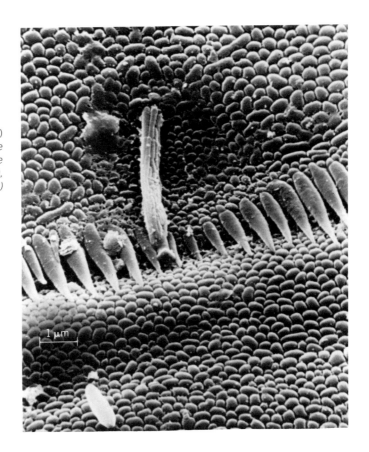

(Right) Surface skeleton and (below) part of the mouth. *(From S. Honjo, The scanning electron microscope in marine science, Oceanus, 21 (3):19–29, Summer 1978)*

1 μm

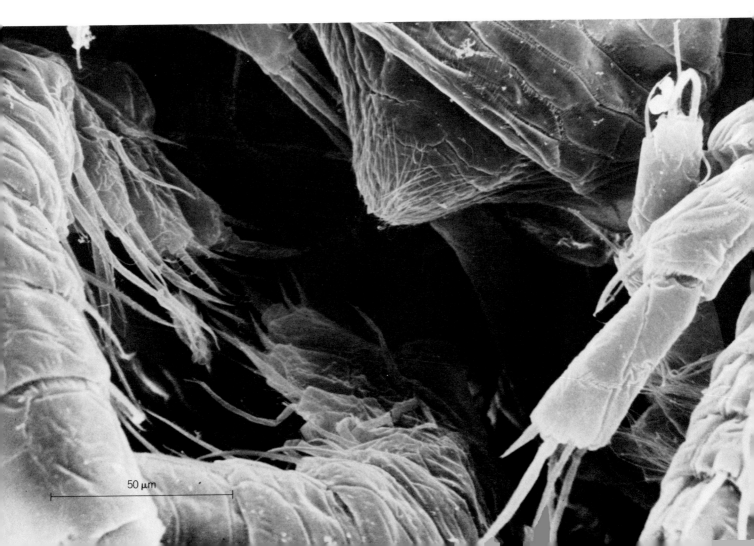

50 μm

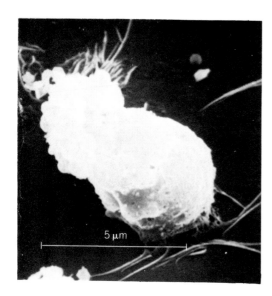

Scanning electron micrograph
of an S49 lymphoid cell
with a prominent mycoplasma cap.

S49 lymphoblastoid cells with
mycoplasma caps. *(a)* Thin
section of a lymphocyte-containing
cap associated with a uropod.
(b) Uropod showing the
membrane vesicles (arrows).

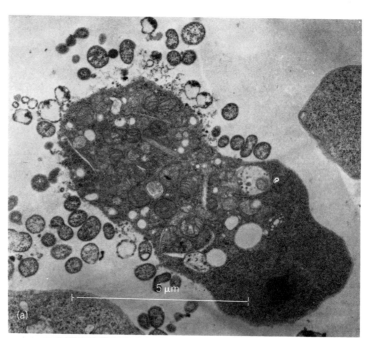

(a)

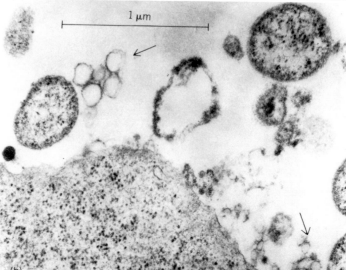

(b)

*From E. J. Stanbridge and
R. L. Weiss, Mycoplasma
capping on lymphocytes,
Nature, 276:583–587,
Dec. 7, 1978*

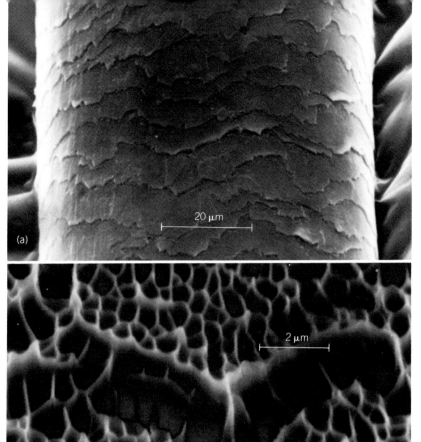

(a)

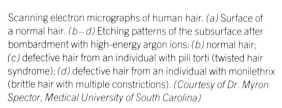
Scanning electron micrographs of human hair. *(a)* Surface of
a normal hair. *(b–d)* Etching patterns of the subsurface after
bombardment with high-energy argon ions: *(b)* normal hair;
(c) defective hair from an individual with pili torti (twisted hair
syndrome); *(d)* defective hair from an individual with monilethrix
(brittle hair with multiple constrictions). *(Courtesy of Dr. Myron
Spector, Medical University of South Carolina)*

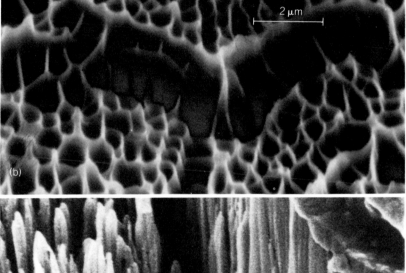

(b)

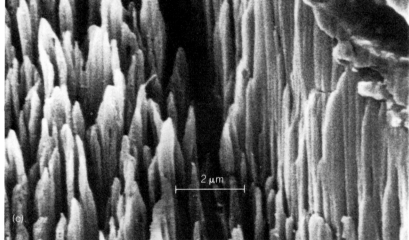

(c)

(d)

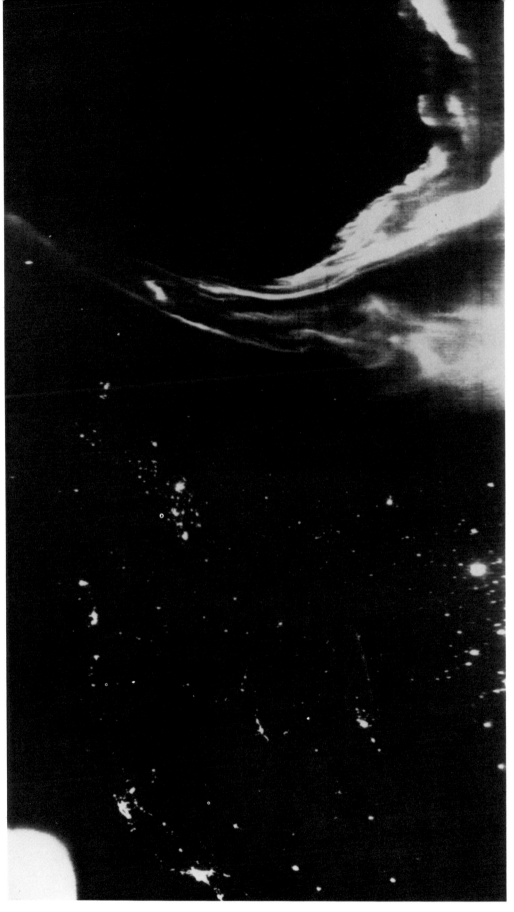

An auroral breakup over North America taken by satellite. The lights in the lower part are cities of the United States and Canada. Minneapolis can be seen near the right and San Francisco and Los Angeles in the lower left. *(From A. D. Johnstone, Pulsating aurora, Nature, 274:119–126, July 13, 1978)*

Silica fibers grown on a platinum wire as seen through a scanning electron
microscope. *(From Mater. Eng., 89 (2):72, February 1979)*

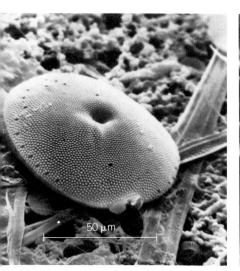

Dinoflagellate

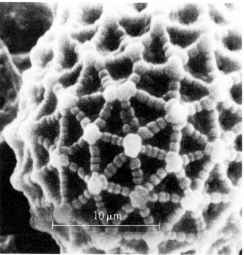

Unicellular plankton

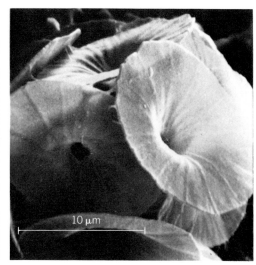

Trumpet-shaped coccolith

Suspended Particles in the Open Sea

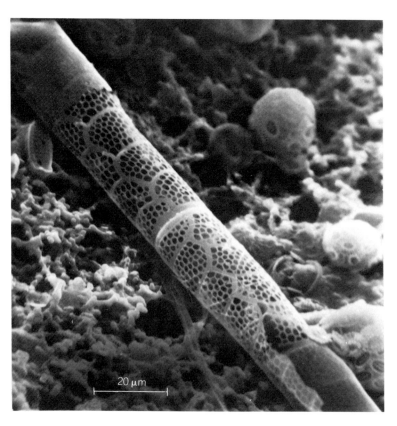

Diatom

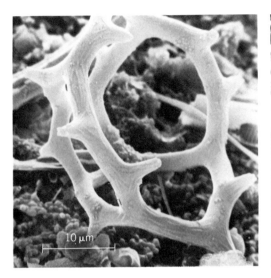

Silicaflagellate skeleton

Diatom

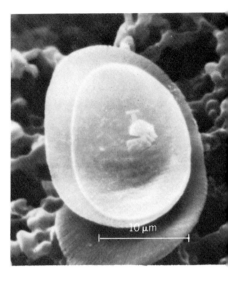

Spiral coccolith

Acantharia skeleton

Micrographs on both pages from S. Honjo, The scanning electron microscope in marine science, Oceanus, 21 (3):19–29, Summer 1978

Metallographic examination of iron produced by the Haya smelting process, an advanced African technology dating from 1500 to 2000 years ago. *(a)* Bloom from a 1976 smelt showing the growth interface between iron (white) and slag (gray). *(b)* Scanning electron micrograph of an intercrystalline fracture in a traditional bloom. *(c)* Photomicrograph of the same traditional bloom showing the long undistorted Neüman bands. *(d)* Etched micrograph of the 1976 bloom, showing the wide local range of carbon content. *(From P. Schmidt and D. H. Avery, Complex iron smelting and prehistoric culture in Tanzania, Science, 201:1085–1089, Sept. 22, 1978; copyright 1978 by the American Association for the Advancement of Science)*

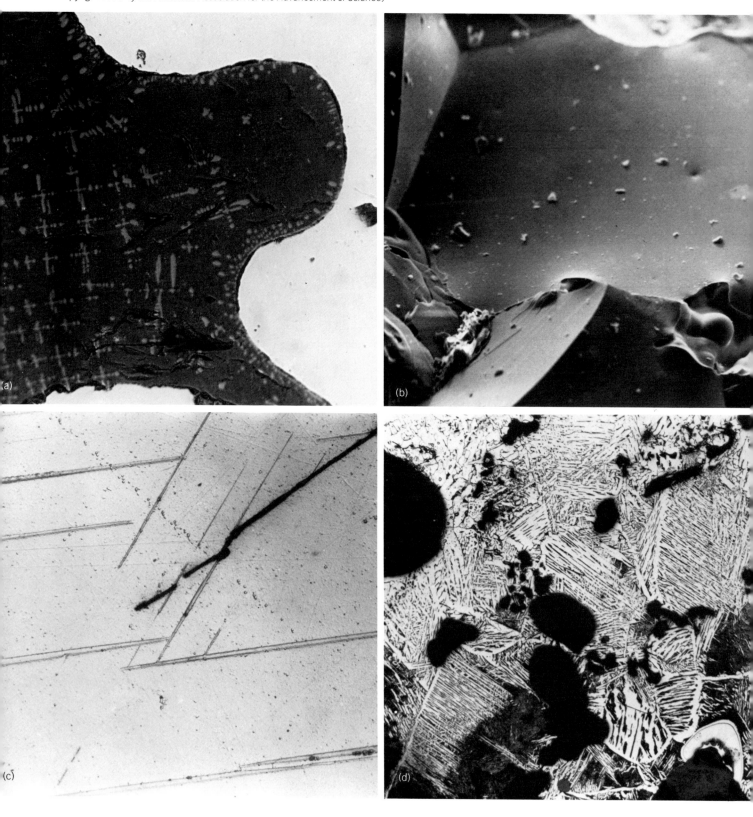

Color phases and wax secretion of the desert beetle *Cryptoglossa verrucosa.*
(a) Light blue (low humidity) and black (high humidity) color phases. *(b)* Wax-secreting tubercles of a black-phase beetle. *(c)* Spreading wax filaments from a single tubercle in response to low humidity. *(d)* Scraped cuticle surface showing the boundary layer created by the wax meshwork. *(e)* Individual wax filaments. *(From N. F. Hadley, Wax secretion and color phases of the desert tenebrionid beetle Cryptoglossa verrucosa (Le Conte), Science, 203:367–369, Jan. 26, 1979; copyright 1979 by the American Association for the Advancement of Science)*

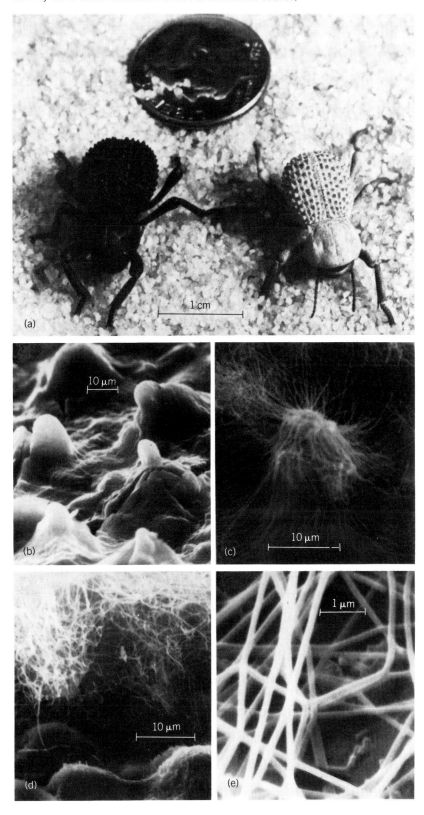

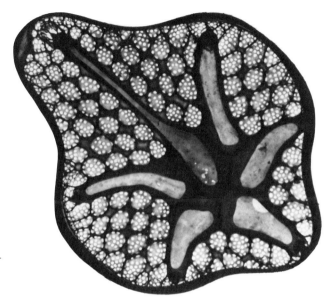

Shadowgraph of a diatom *(Thalsas-siosira)* valve, produced by the transmission electron microscope. *(From S. Honjo, The scanning electron microscope in marine science, Oceanus, 21 (3):19–29, Summer 1978)*

10 μm

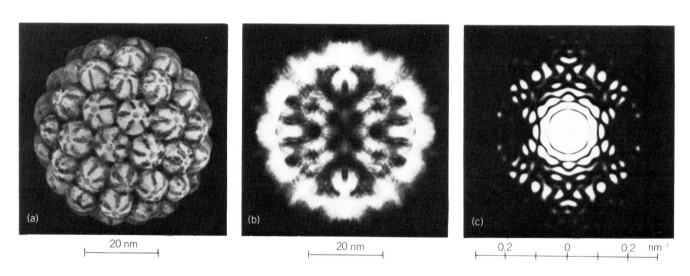

(a)   20 nm

(b)   20 nm

(c)   0.2   0   0.2   nm⁻¹

Crystal structures of a polyoma virion and capsid. *(a)* Model of the capsid viewed along a twofold axis. *(b)* Radiographic projection of the model. *(c)* Optical diffraction pattern of the projection. *(From W. Saunders and D. L. D. Caspar, Structural Biology Laboratory, Rosenstiel Basic Medical Sciences Research Center, Brandeis University)*

A-Z

Actinide elements

Uranium(IV) tetrakisborohydride, the first actinide borohydride to be prepared, was synthesized during World War II as part of the search for a volatile uranium compound to be used for the separation of ^{235}U from ^{238}U. Subsequently, $Th(BH_4)_4$ was prepared, and very recently, borohydrides were prepared for the other members of the first five elements of the actinide series: $Pa(BH_4)_4$, $Np(BH_4)_4$, and $Pu(BH_4)_4$. The heavier actinide borohydrides, $Np(BH_4)_4$ and $Pu(BH_4)_4$, resemble much more closely $Hf(BH_4)_4$ and $Zr(BH_4)_4$ in their physical and structural properties rather than their actinide neighbor in the periodic table, $U(BH_4)_4$. The tripositive actinide ions form tris-borohydride complexes, but these materials appear to be intractable, polymeric solids.

Preparation. All five actinide borohydrides, $An(BH_4)_4$, may be prepared by the solvent-free reaction of the anhydrous tetrafluoride with $Al(BH_4)_3$, as in the reaction below. The reaction

$$AnF_4 + 2Al(BH_4)_3 \rightarrow An(BH_4)_4 + 2AlF_2BH_4$$

was carried out at room temperature for the three lighter actinides, but for neptunium and plutonium borohydrides it was necessary to carry out the synthesis at 0°C. One of the important properties characteristic of the metal(IV) borohydrides is their volatility, which enables them to be purified from the reaction mixture by sublimation (or distillation). The vapor pressures of the complexes increase with higher atomic number of the actinide element.

Properties. Uranium borohydride is moderately volatile and forms dark green, lustrous crystals which in larger aggregates appear almost black. Thorium borohydride is a white crystalline solid which is much less volatile than the uranium compound. The protactinium compound is an orange solid with a volatility intermediate between the thorium and uranium borohydrides. These three complexes appear stable for long periods of time at room temperature. Neptunium borohydride is a dark green liquid which melts at 14.2°C. It must be kept in a greaseless storage tube at low temperature since it decomposes in the liquid phase fairly rapidly at 25°C. Plutonium borohydride is a bluish-black liquid with properties very similar to $Np(BH_4)_4$, but it decomposes more rapidly in the liquid phase. The uranium and thorium (and presumably protactinium) borohydrides react slowly with dry air at room temperature, whereas neptunium and plutonium borohydrides inflame violently when exposed to air. All react violently with water or water vapor. The physical properties of the actinide borohydrides are listed in the table, along with those of $Hf(BH_4)_4$ for comparison.

Structures. The actinide metal atom forms two types of bonds with the borohydride group: a double-hydrogen-bridge bond and a triple-hydrogen-bridge bond. The first type of bond results in the linking of two actinide atoms, whereas in the latter type of bond the fourth hydrogen atom forms a terminal bond with the boron atom. These bonding schemes are illustrated in Fig. 1.

The boron atoms in uranium borohydride are connected to the uranium atom by a combination of four double- and two triple-hydrogen-bridge bonds (Fig. 2). The double-hydrogen-bridge bonds link pairs of uranium atoms together to form a three-dimensional molecular network, while the

ACTINIDE ELEMENTS

(a)

(b)

Fig. 1. Types of hydrogen-bridge bonds in actinide borohydrides: (a) double-hydrogen-bridge bond; (b) triple-hydrogen-bridge bond with a terminal hydrogen atom.

Physical properties of the actinide borohydrides and hafnium borohydride*

Compound	Color	Stability at 20°C	Melting point, °C	Vapor pressure, mmHg/°C†	M⁴⁺ radius, nm	Solid	Gas-phase molecular symmetry	Density, g/cm³
$Th(BH_4)_4$	White	Stable	203, decomposes	0.05/130	9.72	Polymeric	T_d?	2.56
$Pa(BH_4)_4$	Reddish orange	Stable	Decomposes		9.35	Polymeric	T_d?	2.63
$U(BH_4)_4$	Dark green	Very slowly decomposes	Decomposes	0.3/34	9.18	Polymeric	T_d	2.71
$Np(BH_4)_4$	Bluish green	Decomposes	14	10/25	9.03	Monomeric	T_d	2.21
$Pu(BH_4)_4$	Bluish black	Decomposes	~14	~10/25	8.87	Monomeric	T_d	2.20
$Hf(BH_4)_4$	White	Very slowly decomposes	29	15/25	7.9	Monomeric	T_d	1.86

*From R. H. Banks et al., *J. Amer. Chem. Soc.*, 100:1957–1958, 1978.
†1 mmHg = 133.3 pascals.

triple-hydrogen-bridge bonds link the uranium atom to terminal boron atoms. This type of bonding results in a coordination of 14 hydrogen atoms about each uranium atom. Powder x-ray diffraction data show the Th and Pa borohydrides are isostructural with uranium borohydride. By contrast, $Hf(BH_4)_4$ is monomeric and has four triple-hydrogen-bridge bonds linking the boron atoms to the metal atom for a coordination of 12 hydrogen atoms about the hafnium atom. As can be seen from the table, the hafnium complex has a much higher vapor pressure at room temperature than the early actinide complexes. Since uranium borohydride is monomeric in the vapor phase, the increased volatility of $Hf(BH_4)_4$ is due to very weak intermolecular forces between neighboring complexes in the solid state as compared to the polymeric $U(BH_4)_4$. The greater volatility of neptunium and plutonium borohydrides as compared to $U(BH_4)_4$ suggested that these new compounds might also be monomeric in the solid state.

This has now been shown to be the case, as a single-crystal x-ray diffraction study of neptunium borohydride has been completed. The structure is shown in Fig. 3. The neptunium atom is connected to each of four boron atoms by a triple-hydrogen-bridge bond. This molecular structure is the same as in $Hf(BH_4)_4$, except that the packing of the complexes in the unit cell is in a tetragonal arrangement for $Np(BH_4)_4$ [and $Pu(BH_4)_4$] whereas it is in a cubic arrangement for $Hf(BH_4)_4$. The evidence to date suggests that for M(IV) ions with ionic radii less than that of the uranium(IV) ion, the metal borohydride complex will be monomeric in the solid state

Actinide borohydride complexes. In the original report of the preparation of uranium borohydride, an attempt was made to prepare this compound by reaction of UF_4 with $LiBH_4$ in the presence of diethyl ether. A green, ether-soluble product was formed, but it was not possible to obtain a solvent-free product. $Th(BH_4)_4$ has been synthesized by the reaction of $ThCl_4$ with $LiBH_4$ with diethyl ether as the solvent. Here, an etherate complex was formed as an intermediate, but the solvent could be completely removed by vacuum distillation. Treatment of $Th(BH_4)_4$ with $LiBH_4$ in diethyl ether resulted in the isolation of the salts $LiTh(BH_4)_5$ and $Li_2Th(BH_4)_6$. Etherates were again formed as intermediates, but the diethyl ether was easily removed.

The monomeric actinide borohydrides are extremely soluble in noncoordinating solvents such as *n*-pentane, while $U(BH_4)_4$ is slightly soluble and $Pa(BH_4)_4$ and $Th(BH_4)_4$ are insoluble. Again this solubility behavior appears related to the strength of the intermolecular forces between complexes in

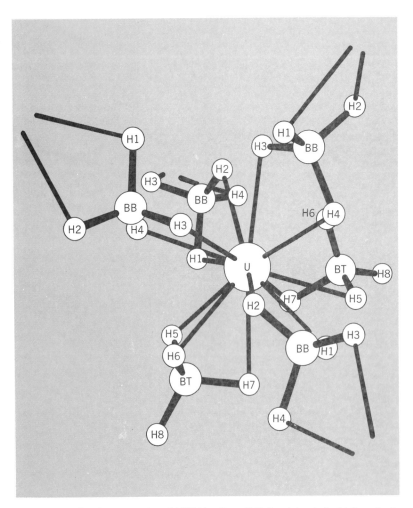

Fig. 2. Coordination geometry of $U(BH_4)_4$. *(From E. R. Berstein et al., 14-Coordinate uranium (IV): The structure of uranium borohydride by single crystal neutron diffraction, Inorg. Chem., 11:3009–3016, 1972; copyright 1972 by the American Chemical Society)*

the solid state, with the polymeric complexes being the least soluble. However, in coordinating solvents such as ethers, even the polymeric actinide borohydrides dissolve and appear to form complexes with the solvent. In the case of uranium borohydride, some of these green, volatile, etherate complexes have been isolated and structurally characterized.

Uranium borohydride etherates. The molecular structures of the $U(BH_4)_4 \cdot O(CH_3)_2$ and $U(BH_4)_4 \cdot O(C_2H_5)_2$ are related to that of uranium borohydride. In these etherate complexes there are infinite linear chains of alternating uranium and boron atoms (two per uranium atom) joined by double-hydrogen-bridge bonds. The remaining three borohydride groups are attached to the uranium atom by triple-hydrogen-bridge bonds. One ether molecule is associated to each uranium by the oxygen atom. The coordination about each uranium atom is again 14, as in $U(BH_4)_4$, that is, 1 oxygen atom and 13 hydrogen atoms. In the dimethyl ether adduct, successive ether molecules along the chain are trans, whereas in the diethyl ether adduct they are all cis. Uranium borohydride forms a complex with two molecules of tetrahydrofuran, $U(BH_4)_4 \cdot 2OC_4H_8$. This complex is monomeric, with a distorted octahedral arrangment about the uranium atom of four triple-hydrogen-bridge bonds and two tetrahydrofuran groups trans to each other coordinated through the oxygen atoms. Again as found in uranium borohydride, the coordination about the uranium atom is 14.

The only other characterized complex of uranium borohydride is the n-propyl ether complex, $U(BH_4)_4 \cdot O(C_3H_7)_2$. This compound was more volatile than the other etherate complexes and proved to be an unsymmetrical dimer. One uranium atom is bonded to two oxygen atoms from two n-propyl-ether groups trans to each other, and to four borohydride groups, one of which has a double-hydrogen-bridge bond to the other uranium atom. The other three borohydride groups form triple-hydrogen-bridge bonds to the first uranium atom. The second uranium atom has the double-hydrogen-bridge bond plus four triple-hydrogen-bridge bonds. The first uranium atom has the unexpected coordination of 13, while the second uranium atom has the usual coordination of 14.

Other actinide borohydride compounds. Some borohydride complexes of the type R_3MBH_4 or $R'M(BH_4)_2$, where M = Th or U and R = cyclopentadienyl or bis-trimethylsilylamido- and R' = cyclooctatetraenyl, have been characterized. In these cases the borohydride groups have all formed triple-hydrogen-bridge bonds to the metal atom.

A number of new borohydride complexes of the actinide elements have been synthesized and characterized in recent years. The high volatility of the $An(BH_4)_4$ complexes and their solubility in organic solvents make them useful as starting materials for the synthesis of organometallic compounds on a small scale. The reaction chemistry of these materials is now under active investigation.

For background information see ACTINIDE ELEMENTS in the McGraw-Hill Encyclopedia of Science and Technology. [NORMAN M. EDELSTEIN]

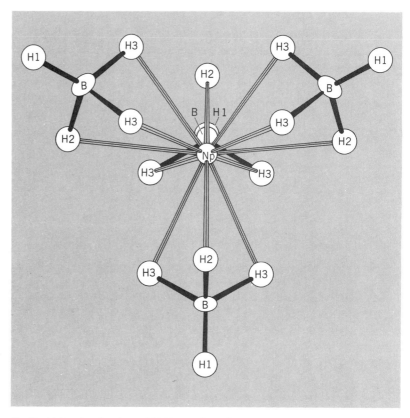

Fig. 3. Molecular structure of $Np(BH_4)_4$. *(From R. H. Banks et al., Volatility and molecular structure of neptunium (IV) borohydride, J. Amer. Chem. Soc., January 1980)*

Bibliography: R. H. Banks et al., *J. Amer. Chem. Soc.*, 100:1957–1958, 1978; T. J. Marks and J. R. Kolb, *Chem. Rev.*, 77:263–293, 1977; R. R. Rietz et al., *Inorgan. Chem.*, 17:653–663, 1978; H. I. Schlesinger and H. C. Brown, *J. Amer. Chem. Soc.*, 75:219–221, 1953.

Aflatoxin

Aflatoxins are a series of toxic secondary metabolites produced by the mold *Aspergillus flavus* and the related species *A. parasiticus*. This class of toxins was discovered in 1960 in England during an investigation of moldy peanut meal associated with a mycotoxicosis in turkeys. The toxicosis was subsequently named turkey X disease. When the same toxic principle was found in peanuts from Uganda contaminated with *A. flavus*, it was named aflatoxin after this parent organism.

There are 17 known members of this series; 11 are either chemically synthesized or biochemically altered derivatives of the 6 major aflatoxins. The six are designated aflatoxins B_1, B_2, G_1, G_2, M_1, and M_2 (see illustration), according to their ability to fluoresce blue (B series) or green (G series) on thin-layer chromatography plates or from their presence in mammalian milk after ingestion of B_1- and B_2-contaminated feed (M series). The six major aflatoxins derive their importance from their carcinogenic and hepatotoxic properties and their occurrence in natural products.

Case histories of diseases which resemble aflatoxicoses have been reported since 1935; for exam-

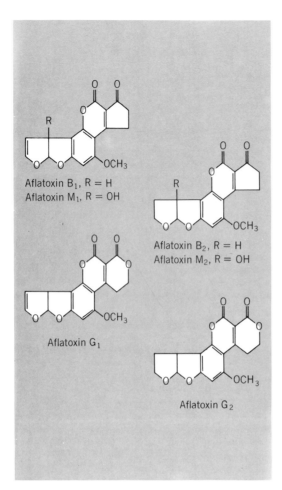

Structures of the major aflatoxins. (*From R. D. Stubble-field, in J. C. Touchstone and J. A. Sherma, eds., Densitometry in Thin Layer Chromatography: Practice and Applications, copyright 1979 by John Wiley & Sons; used with permission*)

ple, there have been reports of a high incidence of hepatomas in hatchery-raised trout in Italy, France, Japan, and the United States. In 1952 hogs fed moldy corn containing *A. flavus* and the fungus *Penicillium rubrum* developed liver diseases and hemorrhaging. A 1964 report described the isolation of aflatoxin from 40-year-old peanuts.

Biological activity. Aflatoxins are among the most carcinogenic and hepatotoxic natural products known, and they are highly toxic to most animals. Toxicity varies among different animals as well as among the six major aflatoxins. Day-old ducklings are very susceptible and are used extensively in biological confirmatory tests for aflatoxicoses. The LD_{50} (lethal dose for 50% of sample) values for aflatoxin B_1 in ducklings range from 12 to 18 μg/50 g administered orally. Aflatoxin B_1 is the most toxic, and G_2 least toxic, in the order: $B_1 \approx M_1 > G_1 > M_2 > B_2 > G_2$. Although nearly as toxic as B_1, aflatoxin M_1 is less carcinogenic.

The general toxicity of aflatoxins makes it imperative that they be absent from food designated for human consumption. Feeding test data from work with primates were extrapolated to humans to establish a reasonable dose response. It was hypothesized that a human regularly ingesting 1.7

mg/kg body weight of aflatoxin might develop serious liver damage in a few months, although a single dose at higher concentrations might be tolerated. However, other parameters such as species, sex, health, and age must be taken into account. Case histories involving Africans and people of southeastern Asia reveal that these areas have the highest incidence of aflatoxicoses. Body tissues of 22 out of 23 Thai children who died of encephalopathy and fatty degeneration of the viscera, a disease similar to Reye's syndrome, had high concentrations of aflatoxins in the liver, bile, and gastrointestinal contents. Cases elsewhere involve such symptoms as bile duct proliferation. When sublethal doses were present, there were manifestations of acute poisoning as well as chronic poisoning, and aflatoxin M_1 was found in the urine and feces. In recognition of this, the World Health Organization has recommended that protein supplement foods contain no more than 30 parts per billion (ppb) of aflatoxin.

Mode of action. The chief target organ of aflatoxins in animals is the liver. The toxins, especially B_1, caused marked alterations in nucleic acid and protein syntheses when administered in acute doses. The large amount of information on actinomycin D (a pigment produced by *Actinomyces* species) and the similarity in mode of action of aflatoxin B_1 and actinomycin D have been useful in elucidating the biochemistry of B_1. It is hypothesized that the initial action of B_1 is interaction with DNA transcription by direct binding, thereby preventing DNA from acting as a primer and causing an impairment of DNA and RNA synthesis through inhibition of the polymerase responsible for the respective syntheses. A more precise explanation of the mode of action of B_1 is not yet possible, because certain similar biochemical responses can be elicited with compounds having structural features unlike those of B_1. Thus, an exact portrayal of sequential biochemical events requires additional research.

Biosynthesis. Aflatoxins are biosynthesized by species of the genus *Aspergillus* after the logarithmic growth phase of the mold. The ring carbons are derived from acetate, with the aromatic ring methoxy carbon coming from methionine. When malonate was added to a synthetic aflatoxin production medium containing acetate, more malonate than acetate was incorporated in the toxin, suggesting that malonate may be incorporated without a preliminary decarboxylation to acetate. Zinc is essential for aflatoxin production as well as for the production of many other secondary metabolites. A deficiency of zinc in *A. parasiticus* NRRL 3240 increased the biosynthesis of lipid and decreased aflatoxin formation.

It has been suggested that there is a preliminary common pathway from which aflatoxin and sterigmatocystin are derived. The fact that sterigmatocystin is a toxic metabolite of *A. versicolor* suggests that it and *A. flavus* have similar biochemical synthetic pathways. An example of possible symbiosis is found when water is added to the unsaturated furanoid ring of B_1 by acid-producing molds, thereby forming aflatoxin B_{2a}.

Analysis and detection. Since aflatoxins fluoresce strongly under ultraviolet light (366 nm), early aflatoxin analyses were done by visual examination of fluorescing spots on thin-layer chromatography plates containing sample extracts and known standards. The fluorescence intensities of the individual aflatoxins are as follows: $B_2 > G_2 > B_1 > G_1$. Aflatoxin B_1 and M_1 fluoresce to nearly the same degree; M_2 is approximately 1.5 times stronger. Aflatoxin B_1 can be detected visually in amounts as low as 0.2–0.5 ng, and although the human eye can detect low-level differences in fluorescence, a fluorodensitometer is used because it is more rapid and reasonably precise and thus superior to any other known method. Currently, transmission densitometry is used for aflatoxin detection and measurement, although it does not provide conclusive evidence for positive identification.

Present status. Many common foods have been shown to contain mycotoxins, so foods are monitored routinely for aflatoxins. Manufacturers and government laboratories in the United States and many other countries can detect aflatoxin B_1 at levels of 10 μg/kg to assure consumer safety. Intensive research is in progress to develop better methods to detect, remove, or detoxify aflatoxin contaminants from infected foodstuffs that might enter the human food chain. Solvent extraction, drying, and physical removal of contaminants have been used, as has ammoniation treatment, to successfully lower the levels of aflatoxin in corn and cottonseed meal. The reactions involved in ammoniation are unknown.

Future research on aflatoxin will seek solutions to its mode of action, biosynthesis, methods of detoxification, and its effect on human beings. The aflatoxin B_1 molecule was synthesized in 1966.

[LOWELL L. WALLEN]

Bibliography: W. H. Butler, in I. F. H. Purchase (ed.), *Mycotoxins*, 1974; L. A. Goldblatt (ed.), *Aflatoxin*, 1974; J. V. Rodricks, C. W. Hesseltine, and M. A. Mehlman (eds.), *Mycotoxins in Human and Animal Health*, 1977; R. D. Stubblefield, in J. C. Touchstone and J. A. Sherma (eds.), *Densitometry in Thin Layer Chromatography: Practice and Applications*, 1979.

Agriculture, soil and crop practices in

A new approach to the interpretation of leaf or plant analyses has recently been published by E. R. Beaufils as part of a diagnosis and recommendation intergrated system (DRIS). This holistic system identifies the nutritional factors limiting crop production, thereby increasing the chances of obtaining higher crop yields resulting from improved fertilizer recommendations. This approach has the advantages that it takes nutritional balance into account, can be used over a range of tissue age irrespective of cultivar, and is universally applicable for a particular tissue of a given crop. Although the DRIS approach involves the study of all known yield factors, the discussion here will be confined to making a diagnosis utilizing a simple example of three nutrients (N, P, and K) in corn leaves.

Optimum N-P-K values. Indices which measure the deviation from optimum composition are used in determining the order of the nutrients in their limiting importance on yield. Their magnitude estimates the degree to which the crop requires a given nutrient. The optimum composition mentioned above is that of a population of high-yielding crops obtained from a survey of the particular industry over space and time. The indices are calculated from Eqs. (1)–(3), which simply give the relative

$$N \text{ index} = +\left[\frac{f(N/P) + f(N/K)}{2}\right] \quad (1)$$

$$P \text{ index} = -\left[\frac{f(N/P) + f(K/P)}{2}\right] \quad (2)$$

$$K \text{ index} = +\left[\frac{f(K/P) - f(N/K)}{2}\right] \quad (3)$$

where $f(N/P) = 68\left(\dfrac{N/P}{10.04} - 1\right)$ when N/P > 10.04

or $f(N/P) = 68\left(1 - \dfrac{10.04}{N/P}\right)$ when N/P < 10.04

$f(N/K) = 45\left(\dfrac{N/K}{1.49} - 1\right)$ when N/K > 1.49

or $f(N/K) = 45\left(1 - \dfrac{1.49}{N/K}\right)$ when N/K < 1.49

$f(K/P) = 41\left(\dfrac{K/P}{6.74} - 1\right)$ when K/P > 6.74

or $f(K/P) = 41\left(1 - \dfrac{6.74}{K/P}\right)$ when K/P < 6.74

N, P, and K = nutrient contents as a percentage of dry matter

deviations of the nutrients in a sample under diagnosis from the optimum values. Thus the sum of indices is always zero. The most negative index indicates the greatest requirement by the plant.

The use of these indices in making diagnoses, and their advantages over the critical value approach, will now be illustrated by using data from experiments. The data in Table 1 are for corn grown in experiments in the United States and South Africa in which yield responses to applied N, P, and K were obtained. The indices will be used to show that DRIS can diagnose correctly the nutrient limiting yield in the field.

Diagnosing requirements. A diagnosis is made by selecting a particular plot (as, for United States data, 0-0-0), diagnosing the most limiting nutrient by using the indices (in this case, P), and selecting a treatment in which P is added (0-20-0). The yield increase from 28 to 49% indicates that a correct diagnosis has been made, but now K has become the most required nutrient. The K requirement is satisfied in treatment 0-20-50, resulting in an increase in yield to 55%. This process is continued until the highest yield is reached, showing that the DRIS indices are able to make valid diagnoses. The same is true for the South African data. In the case of the United States data, the critical values at tasseling (N = 3.0%, P = 0.25%, and K = 1.9%) would be capable only of making valid diagnoses up to the 0-20-50 treatment. This illustrates that the DRIS approach can often diagnose requirements which are not picked up by the critical value approach. The data in Table 1 also illustrate the general applicability of the norms.

Time of sampling. To show that diagnosis can be validly made irrespective of plant age, data from a corn experiment sampled at different times after germination will be used (Table 2). Irrespec-

Table 1. Diagnosis of N-P-K requirements of corn using data from factorial experiments in which responses to all three nutrients were obtained (leaf samples taken at tasseling)

Treatment, kg/ha			Leaf composition, %			DRIS indices			Corn yield, % of maximum
N	P	K	N	P	K	N	P	K	
United States									
0	0	0	2.80	0.21	2.20	7	−22	15	28
0	20	0	3.20	0.28	1.00	31	13	−44	49
0	20	50	2.93	0.28	2.60	−6	−9	15	55
0	40	50	2.60	0.26	2.44	−9	−8	17	60
100	40	50	3.16	0.33	2.45	−5	−1	6	75
200	40	50	3.40	0.34	2.40	−1	−1	2	100
South Africa									
0	0	0	2.78	0.20	2.00	12	−23	11	24
0	20	0	3.02	0.32	0.75	36	40	−76	48
0	20	50	2.48	0.26	2.22	−9	−4	13	78
25	20	50	2.81	0.26	2.19	−1	−8	9	88
25	40	50	2.64	0.28	1.85	−3	2	1	92
50	40	50	2.85	0.30	1.95	−2	2	0	100

Table 2. Diagnosis of N-P-K requirements of corn at different times after germination

Days after germination	Leaf composition, %			DRIS indices		
	N	P	K	N	P	K
28	3.43	0.35	2.19	1	2	−3
42	3.27	0.36	2.03	−1	7	−6
56	3.06	0.39	1.91	−7	17	−10
77	2.81	0.38	1.43	−5	28	−23
84	2.79	0.40	1.30	−5	37	−32
91	2.65	0.45	1.44	−18	46	−28
98	2.34	0.40	1.14	−16	52	−36
105	2.30	0.37	0.92	−5	55	−50

tive of the time of sampling, the DRIS indices diagnose that K is more required than N which is more required than P (K>N>P). Using the critical value norms presented above, diagnosis would have been possible only at the 77-day stage, whereas with the DRIS approach diagnosis can be made at any stage, thus increasing flexibility of the DRIS approach in diagnostic work.

Type of cultivar. That the type of cultivar being used does not have a major effect on the DRIS diagnosis is illustrated in Table 3, where the same order of requirement (N>P>K) is obtained irrespective of cultivar. Furthermore, in the case of cultivars D, E, and F the critical value approach would not have been able to make a diagnosis. Although the above discussion has been limited to diagnosing the N, P, and K requirements of corn, the DRIS approach has been successfully applied to other crops such as rubber, soybeans, sugarcane, potatoes, wheat, and sorghum. In all cases,

Table 3. Effect of cultivar on diagnosis using the DRIS approach

Cultivar	Leaf composition, %			DRIS indices		
	N	P	K	N	P	K
A	2.87	0.30	2.47	−8	−3	11
B	2.92	0.31	2.58	−9	−3	12
C	2.85	0.33	2.56	−13	2	11
D	3.07	0.31	2.32	−3	−2	5
E	3.04	0.33	2.33	−6	2	4
F	3.15	0.35	2.52	−8	2	6

improved diagnostic precision has been obtained by using the DRIS approach. The results presented have demonstrated the flexibility of the approach as far as N, P, and K are concerned. Norms for other elements have been published for some crops, and hopefully these principles will be applied to other crops and nutrients.

For background information *see* AGRICULTURE, SOIL AND CROP PRACTICES IN in the McGraw-Hill Encyclopedia of Science and Technology.

[MALCOLM E. SUMNER]

Bibliography: E. R. Beaufils, *Fert. Soc. S. Afr. J.*, 1:1-30, 1971; E. R. Beaufils, *Univ. Natal Soil Sci. Bull.*, no. 1, 1973; M. E. Sumner, *Agron. J.*, 69: 226–230, 1977; M. E. Sumner and E. R. Beaufils, *Proc. S. Afr. Sugar Tech. Ass.*, 51:131–137, 1977.

Animal evolution

Taxonomic diversity refers to the number of kinds of organisms that occur in an ecosystem. During the past several decades, paleontologists have devoted considerable attention to documenting and explaining patterns of change in taxonomic diversity that have occurred over geologic time. Among the variety of explanations that have been proposed, equilibrium models of diversification (that is, of the growth or evolution of diversity) have recently been shown to be capable of explaining a number of patterns observed in the fossil record. These models maintain that diversity is held in a steady state for long periods of time by balanced rates of origination (such as speciation) and extinction among taxa. When combined with elementary growth models of diversification, the resulting composite models suggest common patterns in the evolution of diversity over the whole of the Phanerozoic, from prior to the beginning of the Cambrian period up until the present day.

But in addition to being capable of explaining specific phenomena, equilibrial growth models possess several general attributes that are important for paleontological analysis: (1) These models are independent of any specific taxonomic group, ecological setting, or period of time, permitting their use in examining evolutionary relationships among seemingly disparate taxocenes and geologic situations. (2) They can be easily formulated in mathematical terms, permitting rigorous statistical application to numerical data compiled from the fossil record. (3) They can be used as a basis for factorizing the fossil record, that is, for identifying second-order evolutionary processes and events superimposed upon broader underlying patterns of growth and equilibrium. Several of these aspects of equilibrial growth models will be examined in the generalized, qualitative considerations that follow.

Early stages of diversification. Species diversity within an ecosystem increases any time that more species are added than are eliminated by extinction or emigration. Within a closed ecosystem, new species are added through branching from preexisting species. Although the processes that govern speciation are complex, the actual branching is topologically simple. It is similar in many respects to the branching of new individuals from parents in asexual populations and to the branching of daughter cells from mother cells in single individuals. If one wishes to model the effects of this branching upon the growth of diversity, one might begin by assuming that all species share some average probability of speciation (that is, branching) during any given interval of time and that extinction is unimportant, at least during early stages of diversification. Then, the number of new species added during a single interval of time will be expected to approximate the probability of speciation times the existing diversity. If the probability of speciation remains constant, the number of new species generated during the next interval, which again will approximate the product of the probability and the diversity, will be larger as a result of the species added previously. And if the process continues unabated, the quantity of new species added during each successive interval will become larger and larger, causing the total diversity to grow geometrically, or exponentially, through time.

This simple exponential model of the early stages of diversification probably is applicable to a variety of evolutionary radiations, as has been suggested by S. M. Stanley and others. But perhaps most significantly, this model seems to provide an adequate description of the greatest radiation of all time -- the "explosion" of marine invertebrates at the beginning of the Cambrian Period. Figure 1 illustrates a semilogarithmic graph of numbers of metazoan orders known from eight stratigraphic units around the Precambrian-Cambrian Boundary plotted against geologic time. Orders rather than species are shown because the fossil record is far better known for these higher categories; because orders are large, composite entities, their known fossil diversity is far less influenced by poor preservation and uneven paleontologic sampling than that of species. The eight stratigraphic intervals used in the graph represent a rather tentative subdivision of the Vendian (latest Precambrian) and Early Cambrian periods; this subdivision is based primarily on recent work by geologists in the Soviet Union. The first three points in the lower left of Fig. 1 represent the first metazoan burrows and the remarkable soft-bodied Ediacaran fossils known from Australia and other parts of the world. The

remaining five points, spanning the interval from 575 to 540 million years (m.y.) ago, mostly reflect the diversity of the first skeletal fossils, including various mollusks, brachiopods, echinoderms, arthopods, and wormlike animals of uncertain taxonomic affinities.

The data in Fig. 1 appear to be consistent with the simple model of the early stages of diversification. The points fall very closely to a straight line on the semilogarithmic graph, suggesting an exponential growth in ordinal diversity across the Precambrian-Cambrian Boundary. This growth is illustrated more dramatically by the inset linear graph in Fig. 1. This positive result is significant for reasons beyond merely testing the model. For nearly a century, paleontologists have postulated any number of special events or abrupt changes in geological or biological conditions as being responsible for what has been perceived as a sudden appearance of abundant fossils near the beginning of the Cambrian. However, the data presented here suggest that any abruptness in the appearance of marine metazoans is illusionary, reflecting simply a rapid growth phase in a continuous process of diversification. In fact, the exponential increase suggests that the more fundamental parameter of diversification, the probability of ordinal origination and, by inference, of species origination, remained nearly constant for at least 100 m.y. around the Precambrian-Cambrian Boundary. Thus, the real problem in explaining early diversification in the seas appears to reduce to elucidating the processes and events surrounding the initial appearance of metazoans near the beginning of the Vendian Period.

Diversity dependence and equilibrium. The rate of diversification indicated by data in Fig. 1

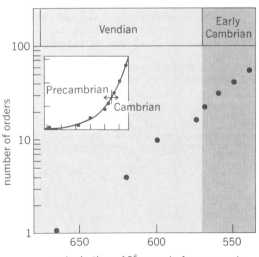

Fig. 1. Numbers of orders known from eight intervals (dots) of geologic time within the Vendian and Early Cambrian periods plotted against the ages of these intervals. The locus of points in the large, semilogarithmic graph is nearly straight, suggesting an exponential diversification of marine metazoans during this time. The data are replotted on linear axes in the inset graph. *(After J. J. Sepkoski, Jr., A kinetic model of Phanerozoic taxonomic diversity: I. Analysis of marine orders, Paleobiology, 4:223–251, 1978)*

is quite high and could not have persisted for much longer than the 100 m.y. illustrated. Extrapolation of the implied curve to the present day provides the prediction that the world's oceans should now be teeming with nearly a billion orders of metazoan animals. This number clearly is excessive, surpassing even the number of species in the modern oceans by more than a thousand times.

Rates of diversification must slow as taxa become more numerous because all ecosystems are ultimately limited; they contain only finite quantities of biomass, nutrients, and habitat space. The role of these limitations in regulating diversity has been pointed out by a number of ecologists, including R. H. MacArthur and E. O. Wilson, who considered theoretical aspects of diversification on biogeographic islands. They argued that the introduction of new species onto an island often causes the average population size of all resident species to decline. This in turn may reduce the chances for any given species to survive unfavorable environmental perturbations such as storms and droughts, or biological calamities such as adverse changes in competition or predation and epidemics of disease or parasites. Together, these factors should cause the average probability of extinction to increase as diversity on an island increases. M. L. Rosenzweig, J. J. Sepkoski, Jr., S. D. Webb, and others

have argued that similar diversity-dependent increases in extinctions should occur on the larger continents and in the oceans, both of which can be considered aggregates of smaller, semi-independent ecosystems.

Shrinking population sizes also should affect speciation. Current hypotheses on the mechanisms controlling speciation maintain that most animal species arise geographically, most commonly from either spatially isolated populations or clinally varying populations arrayed along environmental gradients. Shrinking population sizes and intensified competition associated with increasing diversity should cause species to restrict their ranges to narrower portions of environmental gradients and to experience increased difficulty in traversing geographic barriers. This will reduce clinal variation and decrease opportunities for establishing isolated populations. Furthermore, when local populations do begin to differentiate, their smaller sizes in diverse ecosystems should make extinction more probable, thereby reducing their chances of persisting sufficiently long to become distinct species. The combination of all these factors will cause the probability of speciation to be diversity-dependent, so that it decreases as diversity increases.

If this concept of diversity dependence is combined with the initial model of exponential growth, a more complete, logistic model of the evolution of taxonomic diversity through geologic time can be constructed. As illustrated in Fig. 2a, this model predicts that only the earliest history of diversification will be exponential. Increasing taxonomic diversity causes the rate of diversification to decline as a result of increasing extinction and decreasing origination probabilities. This deceleration generates a sigmoidal curve for the whole of the growth phase of diversity. Eventually, extinctions and originations become balanced, so that the rate of diversification falls to zero and diversity remains indefinitely in a dynamic equilibrium.

Fossil record. The entire known history of ordinal diversity of marine animals is shown in Fig. 2b. As evident, the pattern of diversification is basically consistent with the expectations of the simple equilibrial growth model. The nicks and bumps in the curve, which reflect the peculiarities of the taxa and environments at various points in the geologic record, are minor compared with the overall pattern of early rapid growth and later near-steady-state. The diversity curve increases approximately exponentially into the Early Cambrian, declines somewhat anomalously to about 60 orders in the later Cambrian, but then rapidly increases to about 135 orders during the Ordovician Period. This approximate diversity is then maintained for most of the next nearly half billion years despite various "mass" extinctions, which appear as abrupt but small downward plunges in the curve. Even the great extinctions of the late Permian, which depress the number of orders to about 90, appear to affect the curve only temporarily; by the Cretaceous, diversity reattains approximately the same level as earlier in the Paleozoic. This recovery is particularly important since it indicates that the apparent equilibrium has been a robust feature of the history of life, a level to which

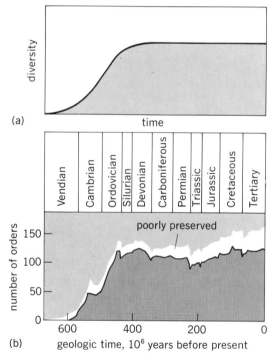

Fig. 2. Expected and observed patterns of diversification among marine metazoan orders. (a) The diversity curve has an initial sigmoidal phase followed by a steady-state phase of indefinite duration. (b) The numbers of marine metazoan orders known from throughout the Phanerozoic are basically consistent with this model. ("Poorly preserved" depicts the ordinal diversity contributed by soft-bodied animals that are rarely preserved in the fossil record.) (Part b after J.J. Sepkoski, Jr., A kinetic model of Phanerozoic taxonomic diversity: I. Analysis of marine orders, Paleobiology, 4:223–251, 1978)

diversity has returned repeatedly after being perturbed.

These data on marine orders provide only a preliminary test of the idea of long-term equilibrium in the evolution of diversity. Orders are large ensembles of species and, as such, tend to smooth out short-term fluctuations in the record of diversity in a manner much like statistical moving averages. Thus, many of the downward fluctuations evident in Fig. 2b probably represent much greater extinction events among species. But, as suggested at the outset of this article, these can be considered simply as superimposed upon a more basic underlying pattern of equilibrium. More complex models might profitably treat extinctions in this context. Indeed, T. J. M. Schopf and D. S. Simberloff have done just that in their treatment of the great Permian extinctions as resulting from a decline in apparent equilibrium caused by shrinking continental seas. Continued analyses and modeling along these lines promise to aid significantly the understanding of the evolution of life.

For background information see ANIMAL EVOLUTION in the McGraw-Hill Encyclopedia of Science and Technology.   [J. JOHN SEPKOSKI, JR.]

Bibliography: R. H. MacArthur and E. O. Wilson, *The Theory of Island Biogeography*, 1967; M. L. Rosenzweig, in M. L. Cody and J. M. Diamond (eds.), *Ecology and Evolution of Communities*, pp. 121–140, 1975; J. J. Sepkoski, Jr., *Paleobiology*, 4: 223–251, 1978; D. S. Simberloff, *J. Geol.*, 82: 267–274, 1974.

Bacteriology, medical

The decade of the 1970s witnessed the recognition of anaerobic bacteria as major causative agents of infectious disease. Prior to the early 1970s, most diagnostic laboratories used isolation techniques that could detect only aerobic (requiring oxygen) or facultative (able to grow with or without oxygen) bacteria. Anaerobic bacteria are microbes that cannot grow in the presence of oxygen, and these organisms were often overlooked in the clinical laboratory.

Characterization of anaerobic bacteria. The introduction of new methods of isolation and identification of anaerobic bacteria, and the documentation of their role in disease, was primarily due to the work of W. E. C. Moore, L. V. Holdeman, and S. M. Finegold. As the result of their work and that of others, it is now recognized that people live in close association with anaerobic bacteria, most of which are primarily innocuous epiphytic members of the normal microbial flora. Anaerobes are present on the skin and mucous membranes, in the genitourinary tract, upper respiratory tract, mouth, and gastrointestinal tract. Anaerobic bacteria outnumber aerobic bacteria by about 10:1 on the skin, 30:1 in the mouth, and 100:1 in the lower gastrointestinal tract.

Anaerobes are usually considered opportunistic pathogens; that is, they cause disease by taking advantage of a break in the normal defense system of the body. These bacteria frequently cause infections after leaving one of their normal ecological niches in the body and penetrating into an otherwise sterile area. Thus, gunshot wounds, injuries caused by automobile accidents, and septic abor-

tions are frequent causes of anaerobic infections. The anaerobe can multiply by taking advantage of the decreased oxygenation of a site caused by impaired blood supply, tissue necrosis, or growth of a facultative organism. Some typical predisposing factors are vascular disease, cold, shock, edema, trauma, surgery, foreign bodies, and malignancy.

Anaerobic bacteria can now be isolated from a majority of clinical specimens. Formerly, about 30% of all wound exudates was dismissed as "sterile pus." Sherwood Gorbach has referred to sterile pus as "the former refuge of bacteriologic ineptitude." Surveys have found anaerobic bacteria in from 40–85% of clinical specimens that are positive for any bacteria. The table presents a typical survey which demonstrated anaerobic bacteria in 58.5% of 689 bacteriologically positive specimens. It is now recognized that, in addition to their involvement in the classical anaerobic infections of gas gangrene, tetanus, and botulism, anaerobic bacteria have a role in many types of infectious disease.

About 70% of anaerobic infections are polymicrobic; that is, a number of different species of bacteria can be isolated from a single clinical specimen. According to Koch's postulates, a single microbe isolated from an infection and inoculated into an experimental animal should reproduce the symptoms of the infection. However, with anaerobic infections, as many as 8–12 different species of bacteria have been isolated from a single clinical specimen, and often only certain combinations of organisms will reproduce the original disease. Thus, anaerobic infections are often synergistic.

Some unique problems arise in the clinical evaluation and management of disease. In the approximately 30% of anaerobic infections in which an anaerobe is isolated in pure culture, the offending organism is readily apparent. However, the physician is often faced with a confusing number of bacterial isolates and can only speculate on the pathogenic significance of many of them. Once the pathogen or pathogens have been identified, the selection of an antimicrobial agent may be difficult. *Bacteroides fragilis*, the most prevalent anaerobe in human infections, is often resistant to many antimicrobial agents used today. To further complicate the situation, until 1979, there was no acceptable uniform method for determining the

Bacterial isolates obtained from 826 specimens cultured aerobically and anaerobically*

Category	Specimens		
	Number	% of total	% of positives
Patients	562		
Total specimens	826	100.0	
Positive specimens	689	83.4	100.0
Positive for aerobes	608	73.6	88.2
Positive for anaerobes	403	48.8	58.5
Mixed aerobes-anaerobes	322	38.9	46.7
Aerobes only	286	34.6	41.5
Anaerobes only	81	9.8	11.8
Negative specimens	137	16.6	

*From J. W. Holland, E. O. Hill, and W. A. Altemeier, Numbers and types of anaerobic bacteria isolated from clinical specimens since 1960, *J. Clin. Microbiol.*, 5(1)20–25, 1977.

antimicrobial susceptibility pattern of anaerobic bacteria.

This article describes some recent investigations in the area of antimicrobial susceptibility testing of anaerobes and also describes previously unrecognized disease syndromes that have recently been associated with anaerobic bacteria.

Antimicrobial susceptibilities of anaerobic bacteria. In 1979 the National Committee for Clinical Laboratory Standards (NCCLS) recommended that an agar-dilution procedure be adopted nationally as a reference method for testing the antimicrobial susceptibilities of anaerobic bacteria. The method entails incorporating various concentrations of antibiotic into a specifically formulated agar medium and observing which concentration of antibiotic inhibits visible bacterial growth. This procedure was tested in various laboratories throughout the United States and was found to be accurate and consistent. The committee is proposing that this agar-dilution procedure be adopted only as a reference method, since it is practical only when large numbers of bacterial isolates are tested at one time. All alternative methods for determining susceptibilities of anaerobic bacteria will, undoubtedly, be standardized in relation to this agar-dilution method.

Transfer of antibiotic resistance. Many pathogenic bacteria have extrachromasomal deoxyribonucleic acid (DNA), called plasmids, and some of these plasmids carry genetic information for antibiotic resistance. These resistant plasmids can frequently be transferred from one bacterial species to another, conferring antibiotic resistance to the recipient species. In the presence of the antibiotic, the resistant species has a competitive advantage over sensitive species. Thus, resistant species tend to proliferate in the wake of increased use of antibiotic therapy. In 1979 a number of antibiotic-resistant plasmids were demonstrated in *B. fragilis*. These plasmids have been shown to code for resistance to the antibiotics erythromycin, tetracycline, and clindamycin, which have been used for the treatment of many anaerobic infections. Plasmid transfer has been shown to occur between different *Bacteroides* species, and some evidence exists for plasmid transfer between different genera of intestinal bacteria. With the continued widespread use of antibiotics, the emergence of highly resistant strains will continue. Indiscriminate use of antibiotics may exhaust the supply of effective antimicrobial agents at a faster rate than scientific research can provide alternative effective compounds for the treatment of infectious disease.

Antibiotic-induced pseudomembranous colitis. Diarrhea is a side effect of treatment with many antimicrobial agents, including penicillins, cephalosporins, erythromycin, tetracycline, chloramphenicol, and metronidazole. This diarrhea is usually a transient phenomenon, but it can evolve into a life-threatening intestinal disease called pseudomembranous colitis. Antibiotic-induced colitis is most often associated with the use of lincomycin and, especially, its derivative clindamycin. Since clindamycin is the current drug of choice for the prophylaxis and treatment of most anaerobic infections, clindamycin-induced colitis has become a

significant problem in clinical medicine. This situation may be due more to the widespread use of this antibiotic than to its relative toxicity. At the present time, no antibiotic has been found to replace clindamycin for the treatment of most serious anaerobic infections, so it is not possible simply to cease using the drug.

Since 1977 researchers have been accumulating evidence on the causative agent of pseudomembranous colitis. It is now generally agreed that clindamycin kills the majority of the anaerobic intestinal flora and that, during repopulation of the colon, *Clostridium difficile* may proliferate. This anaerobic bacterium was not previously considered to be a potential pathogen, but recent work has shown that it secretes a powerful toxin, with as little as 19 ng lethal for a mouse. As *C. difficile* multiplies in the intestine, the colon responds to the toxin by the formation of gray, white, or yellow patches on the intestinal wall known as pseudomembranes. The patient experiences severe abdominal pain with a watery, nonbloody diarrhea. Complications include dilation of the colon and intestinal perforation. The mortality rate is high: death occurs in 27–44% of patients with pseudomembranous colitis. The diagnosis of pseudomembranous colitis is often difficult. Most patients show a nonspecific deterioration over a period of weeks. The toxin of *C. difficile* is present in the feces of pseudomembranous colitis patients, and a specific, high-titered antitoxin will soon be available which should facilitate laboratory diagnosis. *Clostridium difficile* is susceptible in laboratory cultures to vancomycin, metronidazole, and bacitracin, but there are no firm data on their efficacy in treatment.

Infant botulism. A previously unrecognized disease has recently been implicated as one of the mysterious causes of sudden death in infants. This disease, infant botulism, has similarities to the well-known adult form of botulism food poisoning. All forms of botulism are caused by an anaerobic bacterium, *Clostridium botulinum*. In the classical cases of adult food poisoning, *C. botulinum* replicates in improperly preserved food, and the organism secretes a toxin that is the most potent poison known. It is difficult to express the potency of this toxin in units that are meaningful to most people. Ingestion of only 10^{-8} g of this toxin can kill a human—the amount of toxin held on the head of a pin could kill 40 adults! In the adult form of food poisoning, preformed toxin is ingested with contaminated food; if live organisms are ingested, they cannot replicate and produce toxin in the intestinal tract, because of competition from the host's normal bacterial flora. In infant botulism, the infant ingests the spores of *C. botulinum*. These spores germinate, and the bacteria then multiply and produce toxin in the infant intestinal tract. The reason *C. botulinum* can multiply in the infant intestinal tract is probably because the composition of the intestinal flora of infants does not have the comparable barrier effect provided by that of the adult intestinal flora. Also, most cases of infant botulism occur at the time of introduction of solid foods, which causes shifts in the bacterial populations.

Infant botulism usually affects previously healthy infants. Infants from 3 to 26 weeks of age seem to be at the greatest risk. The first symptom is usually constipation, but this may be so slight that it is not detected. The toxin is transported across the intestinal tract and becomes fixed to tissues of the central and peripheral nervous system, causing muscle paralysis. The infant becomes very weak, feeds poorly, and may have difficulty swallowing. Involvement of the respiratory muscles is usually the life-threatening feature of the disease. Death may occur within hours after the first signs of respiratory disease.

To date, less than 100 cases of infant botulism have been reported, but many cases have undoubtedly gone unrecognized. In most investigated cases of infant botulism, the environmental source of *C. botulinum* has not been found. In 6 of 58 investigated cases, *C. botulinum* was found in yard soil, vacuum cleaner dust, and honey. Since 3 cases have been linked to honey, V. R. Dowell has recommended that honey be excluded from the diet of infants under 1 year of age.

Periodontal disease. Considerable evidence exists to support the hypothesis that anaerobic bacteria play an important role in periodontal disease. The term periodontal disease refers to various disease processes that affect the tissues that surround and support the teeth, including the gingiva, periodontal membrane, alveolar bone, and cementum. Obligate anaerobes constitute 77–90% of the cultivable flora of periodontal lesions. Many of these organisms belong to existing species of bacteria, but many belong to species not yet described. Accumulation of dental plaque between the teeth and the gums results in an increase in the concentration of bacteria and a deepening of the space between the tooth and the gum. In active periodontal disease, this periodontal pocket can be described as an open abscess containing a high concentration of a complex microbial flora. Inflammation occurs, possibly in response to enzymes or antigens associated with the bacteria, and extensive bone loss can take place. The exact role of anaerobic bacteria in periodontal disease is still somewhat speculative. Several laboratories are now actively engaged both in identifying the species of bacteria found in the area of the periodontal lesion and in determing the pathogenic potential of individual members of this complex microbial flora. It is hoped that this area of scientific research will provide insight into the prevention and treatment of periodontal diseases.

For background information *see* ANTIBIOTIC; DRUG RESISTANCE; TOXIN, BACTERIAL in the McGraw-Hill Encyclopedia of Science and Technology.   [CAROL L. WELLS; TRACY D. WILKINS]

Bibliography: V. R. Dowell, Jr., *Hospital Practice*, pp. 67–72, October 1978; J. W. Holland, E. O. Hill, and W. A. Altemeier, *J. Clin. Microbiol.*, 5(1):20–25, 1977; G. Privitera, A. Dublanchet, and M. Sebald, *J. Infant. Dis.*, 139(1):97–101, 1979.

Behavioral neuroscience

Animals have an excellent spatial memory, providing them with information about the places they have been and the events that happened in those

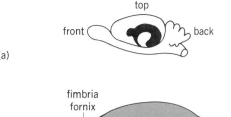
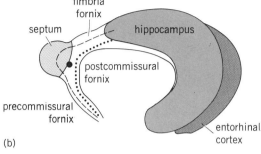

Fig. 1. Left side view of the hippocampal system in the rat brain. (*a*) Outline of the entire rat brain; the hippocampal system is indicated in black. (*b*) Expanded view of the hippocampal system, showing its different components. (*From D. S. Olton, Spatial memory, Sci. Amer., 236:82–98, 1977*)

places. Recent experiments have examined the neuroanatomical mechanisms underlying spatial memory, focusing on a part of the brain called the hippocampus (Fig. 1). This article describes a new experimental procedure for testing spatial memory, summarizes the data indicating that the hippocampus plays a critical role in spatial memory, and discusses the current debate about whether hippocampal function should be described primarily in terms of spatial factors or memory ones.

Test of spatial memory. A new procedure for examining spatial memory uses a radial-arm maze (Fig. 2). Each of the arms extends from a central platform like spokes of a wheel. The entire maze is elevated above the floor. There are no sides on the arms, providing the animals with a clear view of the room in which the maze is placed. At the beginning of each test, one pellet of food is placed at the end of each arm. A hungry animal is placed in the center of the maze and allowed to choose freely among the arms. The optimal strategy for the animal is to run to the end of each arm once and only once to obtain the food. The animal thus can obtain all the food in the minimum number of choices.

A series of experiments by A. Black, B. Maki, D. Olton, and W. Roberts demonstrated that rats and other rodents perform very accurately on this type of maze. For example, on an eight-arm maze, after about 30 tests, rats chose almost perfectly, averaging 7.9 correct responses (different arms) in the first 8 choices. On a 17-arm version, rats averaged about 15 correct responses within the first 17 choices. A number of tests, especially by Black with this type of maze and by J. O'Keefe with other mazes, demonstrated that rats use room cues to identify and remember each of the locations of the arms. Thus the radial-arm maze test requires spatial memory of places which the rat has visited and which should not be repeated during that test.

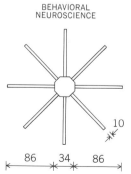

BEHAVIORAL NEUROSCIENCE

Fig. 2. Top view of eight-arm radial maze. (*From D. Olton, in S. H. Hulse et al., eds., Cognitive Processes in Animal Behavior, Lawrence Earlbaum Associates, 1978*)

Lesion experiments. Experiments with lesions have shown that the hippocampus is necessary for rats to perform accurately on the radial-arm maze test. A part of the brain is destroyed by using an appropriate neurosurgical technique. This approach to the structure-function question assumes that if the brain area destroyed by the lesion normally functions in the behavior being tested, then the animal should be impaired in that test. Following destruction of the hippocampus or its connection with other brain areas, rats performed at chance levels in the radial-arm maze test and showed no evidence of recovery of function, even after extended postoperative testing. Lesions of other brain areas, such as the frontal cortex and caudate nucleus, had little if any influence on performance. These results demonstrate that the hippocampus is a critical link in the spatial memory system and that normal performance in the radial-arm maze test requires an intact hippocampal system.

Recording from single cells. A second way to address the structure-function question is to record from nerve cells while the animal is performing a behavioral test. Here the presumption is that if the brain area in question normally functions in the behavior being tested, cells in that brain area ought to be active during the behavior. Furthermore, the activity of these cells ought to have an identifiable behavioral correlate, that is, a behavior which is related to a marked increase or decrease in the "firing rate" of the cell. J. Ranck, who pioneered the technique of recording from single cells in the hippocampus of freely moving rats, used a movable microelectrode which could be lowered gradually into the brain until an appropriate recording was obtained. His experiments demonstrated a variety of behavioral correlates of hippocampal cells. In the context of spatial memory, the most important behavioral correlates are spatial ones, which have been examined by O'Keefe using a three-arm maze. Rats were trained to go to a particular arm, and then run through the rest of the maze, returning to the starting location. Different stimuli were placed in the room and were used to tell the rat which arm should be chosen first. In this procedure, a prominent behavioral correlate was the animal's position in the maze. For example, one hippocampal cell was most active when the rat was at the end of one arm, irrespective of what the rat was doing there. Another cell was active only at the choice point of the maze where the three arms came together. Because the activity of these cells was so closely related to the position of the animal in the maze, O'Keefe called these types of cells the place cells, reflecting their behavioral correlate.

Subsequent experiments by others confirmed and extended O'Keefe's results. For example, one experiment recorded from single cells in the hippocampus while rats ran around on the radial-arm maze. Almost all of the cells with electrophysiological characteristics similar to those of place cells had behavioral correlates reflecting the rat's position in the maze. In addition to cells that showed an increase in the rate of activity in a particular place, there were also cells that showed a decrease in the rate of activity in a particular place. These

"on" and "off" behavioral correlates reflect the same types of activity patterns found in sensory systems, and suggest that the hippocampus may be coding information in a similar fashion.

Space or memory. Although the data reviewed above show that the hippocampus is important for behaviors requiring spatial memory, they do not specify whether the critical factor in these behaviors is their spatial nature or the type of memory required. Data suggesting that the primary factor is spatial come from a variety of experiments conducted by Black, L. Nadel, O'Keefe, H. Mahut, M. Mishkin, and others. These have all been summarized by O'Keefe and Nadel, and suggest that following hippocampal damage animals perform poorly in tasks that require spatial abilities but perform normally in nonspatial tasks. The behavioral correlates of place cells in the hippocampus also support this spatial theory.

An alternative view suggests that the variable of primary importance is not the spatial characteristics of a task, but rather the type of memory the task requires. The hippocampus has been implicated in memory of some form for many years. One example is the work of B. Milner, testing humans who had the hippocampus removed in an effort to control epilepsy. Other data supporting a memory interpretation come from the work of R. Thompson, who showed that the activity of hippocampal cells changes as rabbits learn to associate two stimuli in a Pavlovian conditioning procedure. Although the memory interpretation has been subject to much disagreement, it has been useful in summarizing the functional role of the hippocampus.

Current research is being designed to distinguish between the spatial and memory interpretations of hippocampal function by independently manipulating the spatial and memory characteristics of behavioral tests. For example, L. Weizkrantz and E. Warrrington showed that the amount of interference present in the memory system influences the ability of brain-damaged people to recall material that has been presented to them to remember. L. Jarrard suggested that a similar factor is important in animal experiments. Recent tests of this idea by Jarrard and Olton have held the spatial characteristics of the test constant, but varied the memory requirements. Rats were tested on an elevated radial-arm maze which had two sets of arms. Each arm in the baited set had one pellet of food at the end of it, so that the optimal strategy for these arms was to choose each arm once and only once. Accurate performance on this set of arms required short-term memory. Each arm in the unbaited set never had food on it, so that the optimal strategy for these arms was never to choose them. Accurate performance on this set of arms required long-term memory. Normal rats performed almost perfectly, never choosing an unbaited arm and choosing each baited arm only once. Following hippocampal damage, rats were able to distinguish between the baited and unbaited arms, and rarely made mistakes by choosing an unbaited arm. However, they were unable to remember which of the baited arms had been chosen, and made many errors by returning to baited arms from which they had already taken the single piece of food. These results show that the rats with lesions had a specific impairment in short-term memory but not long-term memory. Because both the baited and the unbaited arms were identified on the basis of their spatial characteristics, the selective impairment on the set of baited arms must have arisen because of the short-term memory requirement for performance on these arms. Data such as these suggest that the memory aspects of the radial-arm maze test are responsible for hippocampal involvement, rather than its spatial aspects.

Summary. Recent experiments have demonstrated that animals have a very good spatial memory, and that the hippocampus is necessary for this memory to be used effectively. At the present time, enough data are not available to determine whether hippocampal involvement is in response to the spatial nature of these tests or to their memory requirement. Ongoing experiments which dissociate the spatial and memory requirements of behavioral tests using experimental designs similar to that described above should provide a resolution of this issue.

For background information *see* NEUROPHYSIOLOGY; PSYCHOLOGY, PHYSIOLOGICAL AND EXPERIMENTAL in the McGraw-Hill Encyclopedia of Science and Technology.

[DAVID S. OLTON]

Bibliography: T. W. Berger and R. F. Thompson, *Proc. Nat. Acad. Sci.*, 75:1572–1576, 1978; J. Horel, *Brain*, 101:403–455, 1978; J. O'Keefe and L. Nadel, *The Hippocampus as a Cognitive Map*, 1978; D. S. Olton and B. C. Papas, *Neuropsychologia*, in press.

Biofeedback

Biofeedback is a procedure to make the individual aware of the ongoing state of various bodily events of which there is usually minimal or no awareness (that is, ability to discriminate). These events include activities of the cardiovascular and gastrointestinal systems, the tension of the striate muscles, and the electrical activity of the central nervous system as expressed by various patterns of the electroencephalogram (EEG). The term "state" must be elaborated. Although one is not usually aware of specific biological activities such as heartbeat or brain waves, biofeedback is aimed at creating awareness of more generalized states of the organism which are associated with one or more of these specific events. For example, biofeedback can facilitate a relaxation state which in turn acts to minimize muscle tension, decrease the heart rate, or change the pattern of brain waves. However, in certain instances biofeedback can act to modify specifically some biological event without requiring a more generalized change in the state of the organism. This is illustrated when one achieves a relaxation of some striate muscle, such as the frontalis muscle of the forehead, without similar changes in other muscles.

Typically, biofeedback procedures use either an auditory or visual signal to indicate whether a given biological process is increasing, decreasing, or not changing from some previous state. For example, if the intent is to lower the blood pressure, the feedback signal can be given on any heartbeat where the blood pressure is below some criterion value. Because the blood pressure, like most other biological events, is continually varying, one can usually detect in any given period of time some values below this criterion value. A shaping procedure is commonly used whereby the criterion value is gradually lowered contingent on the individual's success at keeping the blood pressure below this criterion value.

Basic research. Initial efforts with biofeedback were concerned with demonstrating whether various biological events could be brought under feedback control and then, once this was demonstrated, with issues such as why and how biofeedback works. In the initial demonstration studies, a common procedure was to determine if a given biological event could be trained to both increase and decrease. This procedure, it was believed, has a built-in control to minimize acclimatization effects, that is, the possibility that the given event might decrease or increase over time regardless of the feedback. However, one cannot assess acclimatization with this procedure, because acclimatization might foster decreases more than increases or vice versa. A better procedure, now in use, is a noncontingent control whereby biofeedback is given independently of any given change in biological activity.

This early work focused primarily on biological processes involving heart rate, blood pressure, skin temperature, galvanic skin activity, EEG alpha waves, and striate muscle tension. In general, modest changes in such events were demonstrated. However, these early demonstrations generated a series of issues which are very much of interest today.

One important issue is whether the biofeedback per se is generating these changes or whether just the instruction to either increase or decrease the particular event would suffice, such as "Just relax and try to decrease your heart rate." It has been observed that instructions alone have a significant influence in some instances, with some investigators finding instructions as effective as biofeedback alone and others noting that the combination of instruction and biofeedback produce the maximum effect.

Another important issue is the question of how biofeedback works and what its mechanisms are. Some investigators view the effects of biofeedback like the learning of any kind of sensory motor act, with biofeedback serving a function similar to knowledge of results. That is to say, any type of learning is a function of its consequences which are fed back along any sensory channel or channels.

A particularly appealing variation on this model proposes that the biofeedback enables the individual to utilize the natural afferent feedback from these biological events which are continuously bombarding the central nervous system. For example, the cardiovascular system continuously feeds into the nervous system information about its status because this functions to maintain homeostasis. Whether that same information can be used by the organism to modify any type of biological function remains problematic. There are reports that sensory feedback from the gastrointestinal tract can be used as conditioned stimuli during classical conditioning (interoceptive conditioning). The individual is said not to be aware of the sensory qualities of these stimuli. If so, this would indicate that such visceral afferent information can be used by the brain for purposes other than homeostasis. However, a note of caution needs to be interjected. When dealing with biological processes that are under tight homeostatic control, such as the cardiovascular system, there are inherent restraints on the extent that these processes can be modified without evoking homeostatic resetting mechanisms.

Clinical research. A second area of research which has drawn an intense effort is clinical biofeedback. It represents efforts to treat a variety of medical conditions. These include hypertension, cardiac arrhythmias (abnormal electrical activity of the heart), migraine and tension headaches, intractable pain, epilepsy, and a variety of neuromuscular conditions, such as muscular paralysis resulting from either injury or disease. A particular impetus for this work was the dramatic early demonstration in rats of appreciable learned changes in several types of cardiovascular and gastrointestinal activity. Unfortunately, this work ran into replication problems, and thus cannot be used as a crutch for more clinically oriented work. More recent animal studies, however, are again demonstrating appreciable effects. For example, baboons have been trained to increase their blood pressure in excess of 25 mmHg and heart rate by 50 beats per minute.

The initial clinical efforts were demonstration studies without the necessary controls for such factors as placebo effects and spontaneous remission. More recently these clinical studies have been better designed with appropriate controls. Nonetheless, there remains considerable controversy about the effectiveness of the technique, even though there is a growing clinical application of the procedure. The problem is that the understanding of several of these conditions, particularly with respect to their etiology, remains very primitive. The result is that biofeedback is used to treat a symptom and not necessarily its causes. In such instances, it is somewhat comparable to treating an elevated body temperature with aspirin. The aspirin might temporarily decrease the fever, but does not eliminate the infecting organism.

Hypertension. The treatment of essential hypertension illustrates this point. This type of hypertension, which is the most common, is only vaguely understood. Recent evidence on its etiology suggests a complex process, with the control of the blood pressure possibly changing from the early to later stages, when it is most commonly detected and treated. It is also suggested that the role of neural and humoral factors in affecting blood pressure is of more significance in the early than in the later stages. This is important because the success of any type of behavioral treatment technique such as biofeedback would be critically dependent on such neurogenic mechanisms. To complicate matters even more, blood pressure is known to be quite variable, and can be appreciably reduced just by acclimating a person to having the blood pressure measured. This variability must be controlled in any evaluation of a treatment technique. At present, available evidence on the effectiveness of biofeedback as the sole method of treatment of hypertension at best suggests promise.

Striate muscle activity. In contrast to its application in hypertension, use of biofeedback to modify striate muscle activity (as indicated by electromyography, or EMG) appears to hold considerable promise in the treatment of certain conditions such as retraining paralyzed muscles and eliminating subvocal speech, a condition which interferes with reading speed. This is not too surprising since the striate muscles, in contrast to the cardiovascular system, are amenable to training of rather specific skills which likely do nothing to violate homeostatic mechanisms. Also, the striate muscles provide considerable natural feedback from muscular activity which one continually uses in learning these skills and executing various behaviors. EMG biofeedback has also been used to treat tension headache, while migraine headache has been treated by using biofeedback to train either increases in hand warming or constriction of the extracranial arteries. Both procedures are reported to meet with some degree of success, but one must be cautious with both conditions since their etiology is not well understood. One advantage of treating headache in contrast to something like hypertension is that the headache patient is continuously aware of his symptom. This awareness can help to motivate the individual to continue to use biofeedback and related home-practice therapeutic procedures.

Future applications. Although biofeedback has been used in the treatment of other, less common medical problems, space limitations do not permit a discussion of these. It can be said that the effectiveness of these applications is likely to be similar to that with hypertension and headache. Another potential use of biofeedback is as a tool of preventive medicine. This is illustrated by efforts which seek to minimize the cardiovascular responses to certain important environmental events in the belief that this will prevent the eventual development of more serious pathophysiological conditions such as hypertension. As yet, there has not been much research on this problem, but it holds promise because, in principle, this use will be dealing with etiological events rather than symptoms. Its success, however, will require some understanding of the mechanisms involved in the disease process as well as of the situations which evoke these events.

[PAUL A. OBRIST]

Bibliography: A. H. Black and A. Cott, in J. Beatty and H. Legewie (eds.), *Biofeedback and Behavior,* pp. 7–20, 1977; J. Brener, in J. Beatty and H. Legewie (eds.), *Biofeedback and Behavior,* pp. 235–260, 1977; N. E. Miller, *Annu. Rev. Psychol.,* 29:373–404, 1978; D. Shapiro, in J. Beatty and H. Legewie (eds.), *Biofeedback and Behavior,* pp. 307–322, 1977.

Blood

Platelets are small anucleate cellular fragments which circulate in the blood of mammals and play a major role in the control of hemorrhage ("hemostasis") by their ability to adhere to the transected end of injured vessels, aggregate into a hemostatic plug, secrete several constituents, and become enmeshed in a fibrin clot. In addition to their physiological function of maintaining the integrity of the vascular endothelium, platelets play a significant role in atherosclerosis, thrombosis, local reactions on prosthetic surface, and certain immune and inflammatory processes. Platelets are the end products of the thrombocytic series, a sequence of four cellular compartments derived from the hemopoietic stem cell: the megakaryocyte progenitor compartment; the immature megakaryocyte compartment; the mature megakaryocyte compartment; and the platelet compartment. The cellular flow occurring through these compartments is enormous, since the 420,000,000 megakaryocytes of the adult human release daily about 210,000,000,000 platelets whose average life-span is about 7 days.

Megakaryocyte progenitor compartment. Diploid precursors committed to the thrombocytic series differentiate from the multipotent, self-perpetuating stem cells. Although not morphologically recognizable, these progenitors can be assayed and studied because of their ability to develop in laboratory cultures. Colonies containing from 1 to 100 megakaryocytes have been cloned from mouse or human bone marrow in semisolid gels supplemented with erythropoietin or with media conditioned by stimulated lymphocytes or leukemic monocytes. Megakaryocytes may develop as pure colonies whose founders are termed CFU-M (colony forming unit — megakaryocyte), or they can grow with erythropoietic or granulopoietic precursors in mixed colonies which probably originate from pluripotent stem cells.

Although mouse megakaryocyte progenitors can undergo up to seven mitotic divisions in optimal plasma clot cultures, the average number is 1.3. Assuming equal transit times for megakaryocytes and CFU-M, estimates based on their relative concentration (70 and 50–100 per 100,000 nucleated cells respectively) suggest that no or few amplifying divisions occur in animals as well. In mice, acetylcholine and eserine trigger normally quiescent CFU-M into cycle and increase the number of megakaryocyte colonies. Acute thrombocytopenia induced by the injection of platelet antiserum to normal animals stimulates the production of a humoral factor, thrombopoietin, and within 2 hr results in a three- or fourfold increase in CFU-M. Thrombocytopenia is also known to act on a late diploid precursor, as shown by the 12–18-hr delay between the fall in platelet count and the earliest detectable change in recognizable megakaryocytes.

Because megakaryocytes do not multiply, their production is determined solely by the rate of commitment of stem cells to the megakaryocyte progenitor compartment and by the mitotic amplification which occurs in it. Whereas this production cannot be measured, the steady-state number of megakaryocytes (which is influenced by their maturation time) has been determined by using a radioisotope dilution technique. In normal humans, this number is equal to 6,000,000 cells per kilogram. Platelet hyperdestruction and hypersequestration may multiply this value by a factor of 1

to 5; a higher factor has been found in platelet hyperproduction associated with myeloproliferative diseases, but has not been confirmed in immunologic platelet hyperdestruction. Reduced megakaryocyte number, the major cause of platelet hypoproduction, may be caused by a variety of disorders (Table 1).

Immature megakaryocyte compartment. Following the mitotic divisions of the progenitor stage, thrombocytic cells go through a wave of two to four (in rodents) or two to five (in humans) deoxyribonucleic acid (DNA) endoduplications without cytodieresis. The resulting uninucleated polyploid cell becomes recognizable at a DNA level of 4–8c (chromosome complements). In rodents 70%, and in humans 55%, of megakaryocytes arrest their endoduplications at the 16c level, while the remaining cells do so at the 8 or 32c level in rodents or at the 8, 32, or 64c level in humans. The duration of the ploidization phase is about 9 hr per endoduplication cycle.

Stimulation of megakaryocytopoiesis by platelet depletion tends to increase the number of endomitotic cycles with little prolongation of the total ploidization phase. Clinical thrombocytopenia, whether associated with hypoproduction, hyperdestruction, or hypersequestration, is thought by several researchers to have similar effects on the ploidy histogram, except in the ineffective thrombocytopoiesis which is believed to occur in megaloblastosis and erythroleukemia. Reactive thrombocytosis (but not essential thrombocythemia) is considered to inhibit endomitosis. The most notable abnormalities in the ploidy histogram are found in certain leukemias, where highly abnormal diploid or tetraploid megakaryocytes may be present in both the marrow and blood.

Mature megakaryocyte compartment. Although specific endoplasmic reticulum enzymes (such as peroxidase and, in some species, acetylcholinesterase), granules, demarcation membranes, and ^{35}S-sulfate incorporation are demonstrable in immature diploid or polyploid megakaryocytes, platelet territories are delineated only during the 35–60-hr period (in rodents) which follows the cessation of ploidization. This delineation is achieved by a sequence of territory growth and fragmentation steps in which the percentage of volume variation appears to change randomly from step to step. The combined effects of

Table 1. Etiology and pathogeny of platelet hypoproduction

Pathogeny of thrombocytopenia	Etiology
Megakaryocytopenia	Congenital and acquired megakaryocyte aplasia
	Cyclic thrombocytopenia
	Paroxysmal nocturnal hemoglobinuria
	Marrow infiltration (myelophthisic thrombocytopenia)
Ineffective thrombocytopoiesis	Megaloblastosis
	Erythroleukemia
Reduced demarcation of megakaryocyte territories*	Mediterranean macrothrombocytosis
	May-Hegglin anomaly

*Causes reduction in number of platelets produced, not in volume of platelet material produced.

growth and demarcation determine the number of platelets released by each megakaryocyte, and also account for the inverse correlation between platelet size and number, the gaussian shape of platelet logvolume distributions, and the alterations of their mean and dispersion in several conditions. In platelet hyperdestruction, the latter parameters tend to increase, and several studies suggest that thrombocytopoietic stimulation enhances in a coordinate fashion both territory growth and demarcation. Mediterranean macrothrombocytosis and the May-Hegglin anomaly are characterized by largely increased platelet volume, reduced platelet count, and normal thrombocytocrit, suggesting decreased membrane demarcation in maturing megakaryocytes. In Mediterranean macrothrombocytosis, the number of platelets produced is often decreased (Table 1), whereas the total volume of platelet material produced is elevated. In contrast, the formation of platelet territories is far more anarchical in some leukemias, where significant platelet anisocytosis is often associated with heterogeneous territory demarcation and low ploidy in megakaryocytes.

Platelet compartment. In the last 15 years, labeling with ^{51}Cr-chromate, a radioisotope which binds to platelet cytoplasm and is released only at cell death, has made it possible to calculate in normal humans the mean life expectancy of newborn platelets (6.1–9.3 days), the mean age of circulating platelets (3.8–4.8 days), and the daily production (calculated by dividing the platelet count by the mean life-span and correcting for the splenic pool; reference values range from 46,900 to 89,300/μl/day). Thrombocytopenia may result from platelet hypoproduction (Table 1), hyperdestruction and loss (Table 2), or hypersequestration. Typical examples are shown in the illustration.

While these life-span measurements provide no information on the mechanism of destruction, it has become apparent recently that platelets can be removed from the circulation, either transiently or permanently, by several distinct processes.

Splenic pooling. This is the temporary sequestration of platelets due to their slow transit through the splenic cords. The relative size of the pool, a function of spleen volume, is about 35% of the total platelet mass in normal humans, and may reach values up to 90% in hypersplenism.

Reversible platelet aggregation. This process can be mediated by adenosine diphosphate, serotonin, epinephrin, thrombin, collagen, antigen-antibody complexes, viruses, and bacteria. Unless they are stabilized by fibrin, many aggregates formed in animals are unstable and can dissociate. Recent studies involving the combined labeling of platelets with ^{51}Cr-chromate and ^{125}I-diazotized diidosulfanilic acid (which can be bound to the exposed glycoproteins of the platelet membrane) have shown that platelet survival is not affected by these reversible hemostatic interactions, even though they result in the shedding of unaltered

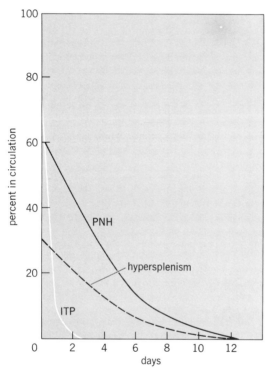

	Plat, per mm³	μ, days	a, days	F, %	P, per mm³ per day
PNH	120,300	5.4	3.9	62	35,900
HS	86,400	6.0	4.0	30	48,000
ITP	90,000	0.83	0.58	61	177,800
Norm	144,000	6.1	3.8	49	47,000
	405,000	9.3	4.8	70	90,000

Typical platelet survival curves obtained in three patients suffering from platelet hypoproduction due to paroxysmal nocturnal hemoglobinuria (PNH), hypersplenism associated with cirrhosis (HS), and platelet hyperdestruction due to idiopathic thrombocytopenic purpura (ITP). The platelet count (Plat), mean life-span (μ), mean age (a), percent recovered at zero time (F), and production (P) are indicated for each case, and for control subjects (Norm).

Table 2. Conditions associated with platelet hyperdestruction and loss

Pathology of thrombocytopenia	Associated disease
Immunological destruction	Drug-induced, posttransfusion, postinfectious, and other thrombocytopenias associated with circulating antigen-antibody complexes
	Idiopathic thrombocytopenic purpura (ITP) and other autoimmune thrombocytopenic purpuras occurring in malignant lymphomas, chronic lymphatic leukemia, systemic lupus erythematosus, and infectious mononucleosis
Platelet consumption	Intravascular coagulation
	Consumption on abnormal vascular surfaces or on synthetic prostheses
Direct platelet damage	Newcastle disease
	Influenza
Platelet loss	Hemorrhage
	Massive blood transfusions
Intrinsic platelet abnormalities	Wiskott-Aldrich syndrome
	Megaloblastosis

pieces of membrane and the release of various intracellular constituents. Sensitive radioimmunoassays are now developed to measure the blood level of specific platelet proteins which are secreted during reversible thrombus formation, such as β-thromboglobulin or platelet factor 4.

Platelet lysis. Lysis can occur when the intensity of the aggregating stimulus is sufficient, for instance, in the clinical disorders of platelet hyperdestruction mediated by isoimmune antibodies, immune complexes, or platelet consumption.

Platelet phagocytosis. Phagocytosis occurs in immunological hyperdestruction, wherein platelets coated with autoantibodies or immune complexes are selectively cleared, without prior aggregation, by the reticuloendothelial organs.

Platelet senescence. Of great theoretical interest is the finding that removal of membrane sialic acid (which can be effected by purified neuraminidase treatment or by the direct action of the influenza and Newcastle disease viruses) causes a drastic reduction in platelet survival. This treatment may provide a model of the mechanisms which determine platelet senescence and death. In animals the immediate destruction of platelets depleted from their membrane sialic acid contrasts with the normal survival after shedding of membrane fragments during reversible aggregation. In fact, the life-span of normal platelets is not prolonged when membrane loss is prevented by use of aspirin and dipyridamole, suggesting that normal senescence is not caused by the "multiple hits" which platelets suffer during reversible hemostatic interactions.

The difficulty in characterizing platelet senescence is compounded by the fact that in several experimental or pathological conditions young stress cells, rather than physiological young platelets, were compared with old or control platelet populations. Furthermore, a significant part of the heterogeneity ascribed to platelet aging may have been generated by the growth and demarcation processes inside maturing megakaryocytes and by the resulting wide variation in platelet territory size and properties. Despite these reservations, it is nevertheless established that when the densest platelets are labeled with ^{51}Cr-chromate and reinfused, they can subsequently be isolated from light-density-gradient fractions, whereas the converse does not occur when the lightest platelets are submitted to the same treatment. Since density is correlated with several metabolic, enzymatic, and functional platelet activities, it can be concluded that senescence, together with thrombocytopoiesis, is a significant factor in the causation of platelet heterogeneity.

For background information *see* BLOOD in the McGraw-Hill Encyclopedia of Science and Technology.

[JEAN-MICHEL PAULUS]

Bibliography: R. H. Aster, in W. J. Williams et al. (eds.), *Hematology*, pp. 1317–1367, 1977; M. Bessis, D. G. Penington, and J. M. Paulus (eds.), Megakaryocyte Differentiation, *Blood Cells*, vol. 5, no. 1, 1979; L. A. Harker, in T. H. Spaet (ed.), *Progress in Hemostasis and Thrombosis*, vol. 4, pp. 321–347, 1978.

Breeding (plant)

Breeding is the procedure of sexually combining desirable genetic traits followed by selection and testing. Conventional methods involving cross-breeding have been used very successfully for several decades (see illustration). In recent times several innovative techniques have been explored to enhance the scope, speed, and efficiency of producing new, superior cultivars.

Advances have been made in extending conventional sexual crossing procedures by laboratory culture of plant organs and tissues and by somatic hybridization through protoplast fusion (see illustration).

New sexual crosses. Several crosses have been made between cereal grains. These include, in addition to *Triticale* (wheat × rye), crosses of wheat × oat, barley × wheat, barley × rye, and corn × sorghum. The fertilization process is facilitated by treatment with growth hormones. The developing embryos, which normally would deteriorate, are removed and cultured in sterile nutrient agar media until the seedling stage. Similar procedures also have been successful in making hybrids between various types of beans. Such crosses have been sought for some years to obtain better resistance to diseases and environmental stresses, as

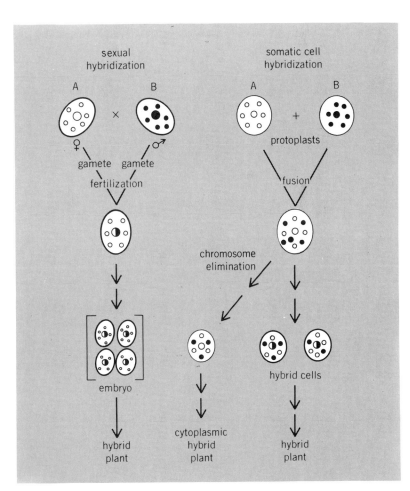

Hybridization procedures in plants.

well as increased yields and improved qualities. Production of these hybrids constitutes major advances in the technology of crossbreeding.

Plants from somatic cells. Mature plants can be regenerated from tissues and single cells cultured in sterile nutrient media. This procedure has permitted development of a rapidly expanding technology in plant breeding. Large numbers of plants can be obtained from leaf, embryo, or shoot tissue. The method has been used to select new disease-resistant cultivars in sugarcane and corn of types which could not be obtained by conventional methods. Corn plants, when regenerated from immature embryos on a nutrient medium containing a toxin isolated from cultures of the pathogen *Helminthosporium maydis*, will survive if they are resistant to the toxin. Such plants are also resistant to attack by the fungal pathogen and are thus disease-resistant. Recent progress in anther culture to obtain haploid plants from pollen has made it possible to utilize the method in the production of superior cultivars of tobacco, wheat, and rice. Haploid plants can be treated to obtain chromosome doubling and the formation of homozygous plants. The anther culture method is being implemented to accelerate selection and shorten the period of crossbreeding because homozygous plants can be produced in one generation.

Somatic hybrids. A rapidly expanding technology using plant protoplasts has been developing. Protoplasts are plant cells with their walls removed by enzymatic treatment. The discovery of polyethylene glycol (PEG) as an agent promoting the fusion of protoplasts, in conjunction with new advances in protoplast isolation and culturing technology, has made somatic hybridization a successful process. Protoplasts from different plant species, genera, and families can be fused, and may subsequently give rise to hybrid cells (see illustration). An outstanding example is the production by K. Kao of *Nicotiana glauca* + soybean cell lines, which are interfamilial hybrids. There is a tendency for deletion of the tobacco genome, but many cells have remained as hybrids during culture for more than 2 years. Hybrid cells have been obtained after fusion of protoplasts from a broad spectrum of plant families, but isolation of the growing hybrid cells was not always achieved.

Selection by genetic complementation. Selection of hybrids has been made possible through the use of albino (chlorophyll-deficient) mutants. Hybrid cells arising after fusion of parental albino protoplasts are recognized as green colonies among nongreen cells on agar plates. Since the species used to date belong to the the genera *Nicotiana*, *Datura*, *Petunia*, or *Daucus*, which have the capacity to regenerate into plants, hybrid cells developed into green plants, and it has thus been possible to obtain several interspecies hybrids. The hybrid nature of the plants is verified by chromosome counts, morphology, isoenzyme patterns, and other characteristics.

A recently developed selection procedure employed two nonallelic mutants of *N. tabacum* which lacked nitrate reductase. The mutant cells required reduced nitrogen and could not grow on nitrate media. The somatic hybrid cells obtained by protoplast fusion regained the ability to grow on media with nitrate as the sole nitrogen source. Auxotrophic mutants requiring essential nutrients are ideal, but have not been available in higher plants until very recently, when a pantothenate-requiring *Datura* mutant was obtained.

Cytoplasmic hybrids. The cell fusion method has enabled the transfer of cytoplasmic-inherited male sterility, which is a desirable quality in plants used for breeding stocks, because the need for removing stamens is eliminated. E. Galun and associates used cells of *N. tabacum* containing *N. suaveolens* cytoplasm which carry genes for male sterility. Protoplasts from these plants were treated by x-irradiation to destroy the nuclei and then were fused with those of *N. sylvestris*. Among the hybrids were male sterile plants with a *N. sylvestris* nucleus.

Intergeneric hybrids. The production of intergeneric plants has been possible in the case of *Daucus carota* + *Aegopodium podograria*, *Solanum tuberosum* + *Lycopersicon esculentum*, and *Arabidopsis thaliana* + *Brassica campestris*. These are model systems, but indicate the feasibility of hybrid plant production. In other examples in which tobacco + tomato and tobacco + potato were combined by fusion, the plants were corrected for the albino trait, but showed unmistakable resemblance to tobacco and did not reflect obvious traits of the tomato or potato. Chromosome elimination appears to have taken place.

The somatic hybridization method would be of particular use in plants where sexual crosses are impossible because the plants are sterile. An example is the commercial banana, which is triploid.

Recombinant DNA and transformation. Considerable research is in progress to determine if exogenous deoxyribonucleic acid (DNA) can be inserted into plant protoplasts and transfer new characteristics to plants regenerated after the transformation process. A potential vector for inserting the DNA is the T_i plasmid of the bacteria (*Agrobacterium tumefaciens*) responsible for crown gall formation in plants. During infection the plasmid, and not the bacteria, enters the plant cells and codes for tumorigenesis. The plasmid DNA becomes integrated with plant DNA, but no further details are available. There are indications that plasmid DNA can be taken up by plant protoplasts. Insertion of DNA and complete transformation with the presence of a stable, new phenotypic expression without the use of plasmids have been reported, but the studies have not been confirmed. The prospect nevertheless exists that in the future it may become feasible to insert into plants specific genes coding for desirable attributes.

Germ plasm preservation. New advances have been made in the methodology of freeze preservation of plant tissues and cells. Shoot meristems of a few species can be frozen and kept in liquid nitrogen at −196°C. After storage the tissues are rapidly thawed and grown to maturity on agar nutrient media. These procedures are important for preservation of plant breeding stocks as well as wild species and progenitors of crop cultivars.

For background information *see* BREEDING (PLANT); GENETICS, SOMATIC CELL in the McGraw-Hill Encyclopedia of Science and Technology.

[OLUF L. GAMBORG]

Bibliography: G. Fedak, *Nature*, 266:529–530, 1977; O. L. Gamborg et al., Genetic modifications in plants, in W. R. Sharp et al. (eds.), *Plant Cell and Tissue Culture*, Ohio State University Biosciences Colloquium IV, 1978; O. L. Gamborg et al., Protoplasts and tissue culture methods in crop plant improvement, in *Proceedings of the Symposium on Plant Tissue Culture*, Peking, 1978; T. Thorpe (ed.), *Frontiers of Plant Tissue Culture*, University of Calgary, 1978.

Bridge

In the field of bridge engineering, the 20th century has seen the coming of age of many structural concepts. This has been made possible by important developments in construction materials, the methods of combining the materials in structures, and the ways of arranging the materials to minimize the dynamic effects of wind. Some primary new materials are high-strength wire, high-strength structural steels and concrete, and welding materials. Among the methods of combining these materials are prestressing and fabrication by welding instead of riveting. Arrangement of the materials to minimize the wind effects refers to the shape of structural members in the wind stream, whether they be individual members of large rigid structures or the overall cross sections of great suspended spans. Such dynamics have been the subject of intensive study and development since the failure of the first Tacoma Narrows Bridge (Washington) in 1940.

Recent years have seen many interesting applications of these basic concepts. This article will be limited to a description of the following examples: (a) the stayed-girder bridge, (b) the prestressed-concrete segmental-box-girder bridge, (c) the inverted-stayed-girder bridge, and (d) the Humber River Bridge in England, as the latest example of the hull-type cross section for the traffic-bearing structure of a major suspension bridge.

Stayed-girder bridge. The stayed-girder bridge has been in use in Europe for several years and has recently been introduced in the United States. It is a suspended structure which looks more like a guyed tower or a maypole than a suspension bridge of the more conventional type with curved cables (Fig. 1). Many people consider it to have a

beauty of its own with its simple straight lines. The straight cables may be anchored at both the floor and the tower, or they may run continuously through the tower, in which case they rest on saddles on the tower structure. The cables are sometimes deployed in a single plane on the centerline of the bridge. Sometimes they lie in two parallel vertical planes, one at each side of the bridge. In other cases, they emanate from the apex of an A-frame tower to anchorages at both edges of the bridge structure. This structure is frequently of steel-box-girder construction, which is basically a large hollow steel box. The steel plate which forms the top surface of this box is sometimes used as the direct support for the traffic in lieu of a typical concrete slab. In this case, the bare plate is normally covered with a bituminous wearing surface. The project illustrated in Fig. 1 does not use the box-girder principle, but instead the suspended structure consists of latticework trusses. One reason for this is that the structure is visualized as a two-level bridge with railroad traffic on the lower level. In come cases, the suspended structure is designed in concrete.

Prestressed-concrete segmental-box-girder bridge. This interesting application of the principles of prestressed concrete also began in Europe and is now gaining popularity in the United States. It makes possible the construction of relatively long spans without the use of falsework or temporary supports. It is most effective for a bridge of many spans. The basic principles are illustrated in Fig. 2. The bridge is made up of a series of building blocks which are held together and made to act as a beam by means of tension members made of high-strength wire. The individual sections, or blocks, may be precast at a central location and moved to the site and then put into place on the bridge, or may be cast in place. The cantilever method of construction is used. In Fig. 2a one can see the manner in which two adjacent spans are started by placing their initial sections at the top of one of the piers. In Fig. 2b a typical scheme for continuing erection is shown. It is quite clear that the loads on either side of the pier must be carefully balanced until the spans between the piers can be completed and closed. Current literature seems to imply that the development of this type of struc-

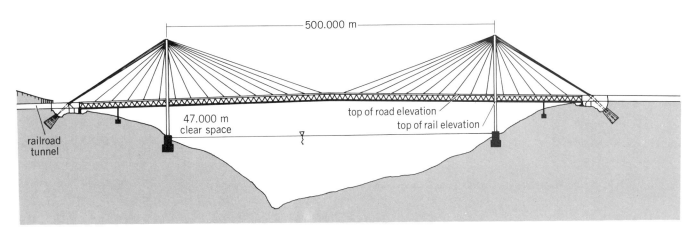

Fig. 1. Proposed Matadi bridge in Zaire. (*From International Engineering Company, Inc.*, and Steinman, Boynton, Gronquist & Birdsall, Final Report: Feasibility Study for Republic of Zaire – Matadi Crossing, December 1974)

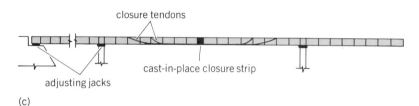

(a)

(b)

(c)

Fig. 2. Schematic diagram of segmental bridge and its construction. (a) Construction of hammerhead. (b) Segmental cantilevering. (c) Closure of span. (From J. E. Breen, *Design of segmental structures, Public Roads, 41(3):101–111, December 1977*)

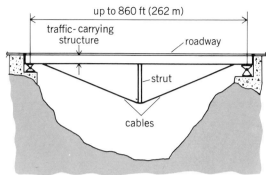

Fig. 4. Schematic diagram of an inverted-stayed-girder bridge showing structural principles.

ture began in the 1960s. However, there were earlier examples, such as a bridge over the Rhine at Worms, which was completed in 1953. One of the more elegant examples of this type of construction is the Fray Bentos Bridge over the Uruguay River between Argentina and Uruguay, which was completed in 1976 (Fig. 3). Its main river span has a length, center to center of piers, of 795 ft (215 m).

Inverted-stayed-girder bridge. The inverted-stayed-girder bridge (Fig. 4) is a reincarnation of an ancient structural concept, the hog-braced beam. In this type of structure, the space above the

bridge deck is uncluttered (or unadorned, depending on the taste of the beholder) by towers or cables. A single tower or strut extends downward from the middle of the span, and cables run from the ends of the span down to the bottom tip of this strut. Here again the basic enabling principle is that of prestressing—made viable by the use of high-strength wire. Whereas the old hog-braced beams might have had span lengths of up to 30 ft (9.2 m), the bridge over the Neckar Valley of West Germany has such inverted spans as long as 864 ft (263 m).

Humber River Bridge. The Humber River Bridge at Hull, England, which will have the longest single span in the world, 4630 ft (1410 m), is the latest example of the box-girder suspension bridge developed by the British and first used over the Severn River in southwestern England. A cross section of the suspended structure looks like a ship's hull. Another interesting feature is that the suspenders, the wire members which connect the main cable with the suspended floor, are not vertical, as on most suspension bridges, but slightly inclined.

The basic principles of the Humber River Bridge are shown in Fig. 5. In silhouette it appears to be no more rigid than the ill-fated Tacoma Bridge, but its stabilizing feature is the shape of its box girder cross section, which represents a full development of aerodynamic theory and practice, plus a minor assist from the nonverticality of the suspenders. The steel-plate on top of the box is the floor of the bridge and will be covered with a bituminous wearing surface.

State of the art. There is no mystery (nor any implied inferiority of creativity on the part of American engineers) about the fact that some developments in this field have arisen first in Europe rather than in the United States. The basic criterion which controls bridge design is economy. Historically, the ratio of material cost to labor cost has always been higher in Europe than in the United States. This has led European engineers on a crusade for reduction of materials, while American engineers have been primarily motivated by the search for mass production. The resulting new developments in one area have not been automatically viable in the other but, with time, have in many cases been made viable by adjustments in

Fig. 3. Fray Bentos Bridge, Uruguay River, under construction.

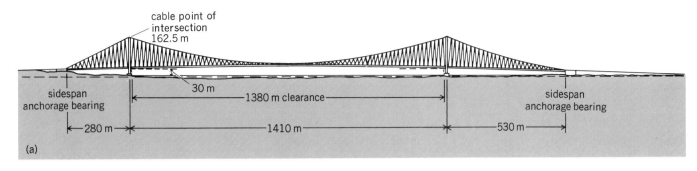

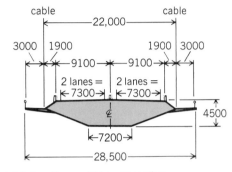

Fig. 5. Humber River Bridge, England: (a) elevation and (b) cross section.

design, manufacturing, or construction procedures.

For background information *see* BRIDGE in the McGraw-Hill Encyclopedia of Science and Technology.   [BLAIR BIRDSALL]

Bibliography: J. E. Breen, *Pub. Roads*, 41(3): 101–111, December 1977; Inverted stayed girder bridges soil problems, *Eng. News. Rec.*, pp. 18–19, Nov. 23, 1978; W. Podolny, Jr., and J. B. Scalzi, *Construction and Design of Cable-Stayed Bridges*, 1976; G. Roberts, in *Proceedings of the Institution of Civil Engineers*, vol. 41, pp. 1–47, September 1968.

Bryophyta

Bryophytes constitute what might be termed a marginal group among archegoniate land plants (embryophytes). Although they share sound similarities with higher vascular plants (tracheophytes), bryophytes have a number of quite distinct characteristics. Two of the most conspicuous are (1) their life cycle, in which the sporophyte almost never becomes entirely independent from the gametophyte, and (2) their lack of a conducting tissue system (xylem and phloem) readily comparable with that of tracheophytes. Also, in terms of species numbers, bryophytes are far less numerous than vascular plants (some 24,000 species versus about 250,000).

In bryological research the recent introduction of such tools as the electron microscope and radioactive tracers has produced much new information and has provided new insights into the structure and life of these plants. Comparative studies on their conducting tissues, in particular, have largely revised concepts in the field. In this respect, bryophytes no longer appear as distinct from tracheophytes as previously thought.

Among the three major classes usually distin-guished within contemporary bryophytes—the hornworts (Anthocerotae), liverworts (Hepaticae), and mosses (Musci)—only the last show the common occurrence of conducting tissues. Such tissues are rarely encountered in the liverworts, and none exist in the hornworts. Whereas only water-conducting cells are seen in the liverworts, both water- and food-conducting cells may be found in mosses (Fig. 1a).

Water-conducting cells of mosses. The water-conducting cells of mosses are called hydroids. They constitute an axial strand in the leafy stem and seta (stalk of the capsule) of many species. Hydroids are also encountered in the leaf nerves of certain mosses, although, oddly enough, true leaf traces linked to the central strand are found only rarely. The relative development of the central strands varies greatly according to the species, and these strands may even disappear entirely from either one or both generations, gametophytic and sporophytic. Although no absolute rule can be made, the leafy gametophytes of acrocarpous mosses frequently possess a well developed water-conducting strand, whereas that of the (more advanced?) pleurocarpous ones usually exhibit only a reduced strand or no strand at all.

The hydroids themselves bear a strong resem-blance to the water-conducting cells (tracheids) of the more primitive vascular plants. This fact has been noticed by G. Haberlandt and J. R. Vaisey toward the end of the 19th century, but has since been somewhat underestimated. Recent ultra-structural investigations have largely confirmed these early observations. Like tracheids, the hy-droids of mosses are elongated cells with more or less inclined end walls, are devoid of a living proto-plast when mature, and may show partial hydroly-sis of their contact walls (Fig. 1b). The functional

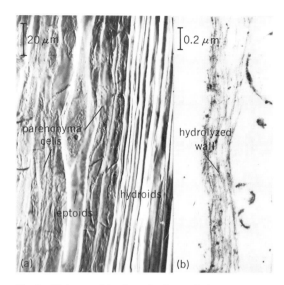

Fig. 1. Water- and food-conducting cells in mosses. (a) Light micrograph showing portion of longitudinal section through the axial conducting system of the leafy stem of *Polytrichum commune* (from C. Hébant, *The Conducting Tissues of Bryophytes*, Cramer, 1977). (b) Electron micrograph showing detail of a hydrolyzed wall of contact between two hydroids.

value of features such as the lack of protoplasm (which would constitute an obstacle to water circulation) and the hydrolysis of the walls (which renders them highly permeable) is easily understood. One important characteristic distinguishes the hydroids from tracheids: the absence of lignified secondary thickenings from the hydroid walls. In fact, some claims concerning the presence of lignin in hydroids have been put forward in the past 10 years. However, the recent chemical studies of G. Miksche and coworkers have confirmed the absence of true lignin from a number of bryophytes, including various "giant" antipodial species *(Dawsonia superba, Dendroligotrichum dendroides)*.

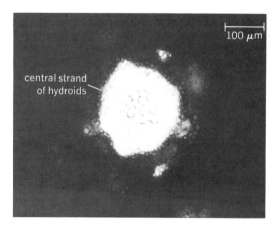

Fig. 2. Transverse section, viewed under ultraviolet light, of a leafy stem of *Polytrichum* fed with a solution of the fluorescent tracer calcofluor white. The conducting role of the central strand of hydroids is demonstrated. *(From C. Hébant, The Conducting Tissues of Bryophytes, Cramer, 1977)*

The effective role of hydroids in water conduction has been demonstrated. Early experiments were made with stain solutions, whereas recent workers use fluorescent dyes or radioactive tracers (Fig. 2). Fairly rapid rates of conduction were measured when leafy stems of mosses with large central strands were investigated—for instance, up to 140 cm/hr or even more for *Polytrichum commune* kept under a relative humidity of 70%. These values fall well within the range known for a number of vascular plants.

Food-conducting cells of mosses. All the living cells in a plant are capable of translocating organic substances. Transport through ordinary cells of conducting parenchyma, however, is slow, and indeed vascular plants have developed a specialized food-conducting tissue, the phloem, which proves particularly efficient when medium-to-long-distance translocation is needed. Most bryophytes, perhaps as a consequence of—or as a possible cause for—their modest size, lack such a tissue. A notable exception, however, is presented by a small number of mosses, nearly all of which belong to the order Polytrichales. In these mosses, a tissue called leptom exists which parallels the phloem of vascular plants (Fig. 1a).

The leptom may be present in both generations, gametophytic and sporophytic, and shows important structural variability. When in its most highly differentiated form, its food-conducting cells, the leptoids, are practically indistinguishable from those food-conducting elements of vascular plants which are the first to develop in young growing organs and are known as protophloem sieve elements. Up to now, such highly differentiated leptoids have been found only in the gametophyte generation of certain Polytrichales (such as *Polytrichum commune* and *Polytrichadelphus magellanicus*), where they are associated with parenchyma cells (Fig. 1a). In the setae of the same mosses, as well as in both generations of other Polytrichales, the leptom tissue as a whole is less specialized.

The similarity of the most highly differentiated leptoids to protophloem sieve elements is so great that they could hardly be separated even at the ultrastructural level. This refers in particular to shared characteristics such as the degeneration of their nucleus, the occurrence of stacked endoplasmic reticulum (sieve element reticulum), or the enlargement of the plasmodesmata (intercellular connections across the walls). It must be pointed out, however, that the connections between moss leptoids remain smaller than those between sieve elements in vascular plants; their diameter does not usually exceed about $0.1-0.2$ μm in leptoids, whereas that of pores between late-formed (metaphloem) sieve elements of tracheophytes is currently $1-2$ μm and may reach 10 μm or even more.

Experiments with radioactive tracers have shown that, in the moss *Polytrichum commune*, labeled organic material is translocated with a velocity of at least 32 cm/hr. Histoautoradiographic studies have also shown that, in the same moss, the leptom of the stem is the effective pathway for transport (Fig. 3). Organic substances synthesized in green portions of gametophytes are thus translo-

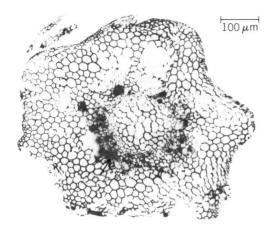

Fig. 3. Histoautoradiograph of transection through a leafy stem of *Polytrichum commune* fed with ¹⁴C. A strong labeling of the food-conducting tissue (leptom) is evident (arrow). (*After W. Eschrich and M. Steiner, Autoradiographische Untersuchungen zum Stofftransport bei Polytrichum commune, Planta, 74:330–349, 1967*)

cated rapidly to such sites (sinks, according to the terminology of phloem workers) as underground growing axes and young developing sporophytes. Here again, the measured rates of conduction fall within the range known for vascular plants.

Leptoids of mosses thus constitute interesting examples of primitive sieve elements, and their study is likely to provide new insights into structure-function relationships in conducting cells of land plants.

Conducting tissues of Hepaticae. Conducting tissues are not of common occurrence in liverworts, and only water-conducting cells are found in these plants. Up to now, true leptoids have not been identified in them.

A water-conducting strand is known from the gametophytes of a few rare leafy forms belonging to the genera *Haplomitrium* and *Takakia*, and of a small number of thalloid ones, all of them belonging to the order Metzgeriales. The sporophytes are always devoid of such conducting tissues.

The water-conducting cells of liverworts are structurally distinct from those of mosses (hydroids), and most certainly from those of primitive vascular plants (tracheids)—those of the Hepaticae (with one known exception, the liverwort *Moerckia*) possess small plasmodesmata-derived pores in their walls. This organization is not found elsewhere, and further points to the distinctiveness of the liverworts within the plant kingdom.

Although experimental work in the field remains scanty, it has been shown that the gametophytic strands of the liverworts constitute preferential pathways for water circulation within leafy stems or thalli.

External conduction in bryophytes. The complete absence of specialized conducting tissues in a number of bryophytes is better understood if reference is made to the phenomenon of "external" conduction by capillarity in these plants. Closely imbricated leaves, aggregated rhizoids, and so forth, provide the necessary capillary channels. Since most bryophytes are of modest size, this external conduction is sufficient in itself to meet the water requirements of many species that otherwise lack a specialized internal conducting system.

Some evolutionary problems. The occurrence in bryophytes, especially in mosses, of conducting tissues strikingly similar to those of higher vascular plants raises fascinating evolutionary problems. Two alternative interpretations have been proposed for these facts. According to the first one, this similarity in organization is the result of a convergent evolution (parallel development). According to the second, the conducting tissues of bryophytes are more or less homologous to those of vascular plants. Indeed, a large body of data from various sources lays stress on the affinity of bryophytes to tracheophytes. It has also been long noted that some of the earliest known vascular plants, which lived in the Late Silurian–Early Devonian, were more or less bryophytelike in organization. A reasonable hypothesis is that both groups (bryophytes and tracheophytes) were derived from a common stock of primitive archegoniate land plants. Bryophytes might thus exhibit relictual types of conducting tissues from these early land invaders. However, new findings in the pre-Devonian fossil record must be awaited before a more definitive picture of the first steps in the evolution of terrestrial plants can be presented.

For background information *see* BRYOPHYTA; PLANT ANATOMY; PLANT EVOLUTION in the McGraw-Hill Encyclopedia of Science and Technology.   [CHARLES HÉBANT]

Bibliography: W. Eschrich and M. Steiner, *Planta*, 74:330–349, 1967; C. Hébant, *The Conducting Tissues of Bryophytes*, 1977; G. E. Miksche and S. Yasuda, *Phytochemistry*, 17:503–504, 1978.

Buildings

The Arab oil embargo of 1973 and the subsequent rapid escalation of fuel costs precipitated a myriad of energy-related studies, ranging from the search for alternate nondepletable energy sources (solar, wind, biomass) to the development of energy-efficient appliances. A number of these studies relate to energy consumption in buildings—commercial, industrial, institutional, and residential.

This article discusses results and conclusions from three of these studies: (1) energy conservation in existing office buildings, (2) energy-conscious design for new office buildings, and (3) energy performance standards for new commercial and residential buildings.

Energy conservation in existing buildings. The main objectives of a study of energy conservation in existing office buildings in New York City—sponsored by the U.S. Energy Research and Development Administration (ERDA), now superseded by the Department of Energy (DOE)—were to: (1) determine physical and operation energy-related chacteristics of office buildings and their energy consumption patterns; (2) establish the value of retrofitting or conservation measures beyond those adopted voluntarily since the 1973 oil embargo; (3) determine the economic, technical, and practical feasibility of reaching for additional energy savings; and (4) develop energy conservation recommendations and evaluate their applicability to other building types, to other geographical areas, and to new buildings.

Table 1. Highlights of the energy conservation study of New York City office buildings

Characteristic	Range	Mean	SI
1975 consumption, normalized,* 10^3 Btu/ft²	65–223	112	$(7.39-25.30 \times 10^8; 12.70 \times 10^8$ J/m²)
1975 consumption, actual, 10^3 Btu/ft²	67–225	115	$(7.62-25.60 \times 10^8; 13.10 \times 10^8$ J/m²)
Age, years	8–82	44	
Total building area, ft²	17,000–1,850,000	401,000	(1580–171,865; 37,253 m²)
Total number of floors	4–51	24	
Computer area, ft²	100–14,000	2700	(9.29–1300; 251 m²)
Total wall area, ft²	7400–503,000	130,000	(687–46,729; 12,077 m²)
Percent glass on wall	13–67	29	
Temperature, winter day, °F	68–75	71	(20.0–23.9; 21.7 °C)
Temperature, winter night, °F	42–68	58	(5.5–20.0; 14.4 °C)
Temperature, summer day, °F	68–78	75	(20.0–25.6; 23.9 °C)
Watts/ft² m – lighting	1.5–5.3	2.8	(16–57; 30 W/m²)
Commercial area (nonoffice), ft²	0–130,000	17,300	(0–12,077; 1607 m²)
Number of persons in building	65–6000	1300	
Average floor area, ft²	1500–46,000	15,000	(139–4273; 1394 m²)
Core area, ft²	0–475,000	65,600	(0–44,127; 6094 m²)

*For weather, percentage of space occupied, and hours used.

The study consisted of three phases. First, 1037 buildings were sampled with a simple questionnaire designed to elicit information on building area, height, window area, lighting, heating and air conditioning, and occupancy. From the 436 responses received, a representative sample of 44 buildings was selected. These buildings were studied in great detail to determine basic physical and operating characteristics, including energy consumption over a 5-year period; the year of the embargo (1973) and 2 years before and after. The results are given in Table 1.

In the second phase, five buildings were selected and studied in even greater depth to establish the potential value and feasibility of retrofitting conservation measures and to determine how best to achieve desired energy savings practically. From cost-benefit analyses of proposed conservation measures, along with computerized simulations of the five "retrofitted" buildings, the results in Table 2 emerged. If these savings are expanded to the entire New York City inventory of office buildings, it is estimated that more than 6000 equivalent barrels (954 m³) of oil per day could be saved at the source. For the entire inventory of United States office buildings, the savings would be over 150,000 bbl (23,847 m³) of oil per day.

Finally, in the third phase, the results of the study were discussed with owners to learn why more had not been done to conserve energy and to explore ways in which their commitment to con-

servation could be strengthened and accelerated. The inteviews revealed that most owenrs: (1) had little appreciation or understanding of energy consumption patterns or their significance; only 10% of the building owners monitored and compared their energy consumption with that of others; (2) had little faith in advertising claims for energy-cutting devices or systems; (3) were seeking advice from consultants as well as developing their own in-house expertise before making energy conservation decisions; (4) waited for feedback from those who have already installed retrofitted products or systems; few owners wanted to be the first; (5) believed that costs of retrofitting would decrease when production of newer and improved energy-saving systems increased; and (6) might be attracted by energy conservation incentives such as tax credits, accelerated depreciation, and elimination of tax assessment on retrofitted energy-saving equipment.

All of these data are now stored and form a priceless source of current state-of-the-art design techniques for all types of buildings in all areas of the country, as well as a source of measures adopted by architects and engineers when given freedom to design for maximum energy conservation.

New office buildings. When New York State was considering the development of an energy code for new buildings which was to be based on the American Society of Heating, Refrigerating and Air Conditioning Engineers (ASHRAE) Stan-

Table 2. Energy savings for a group of New York City office buildings

Age,* years	Gross area, 10^3 ft²; (m²)	1975 Energy index, 10^3 Btu/yr-GSF†; (10^8 J/yr-m²)	Payback period of up to 3 years; 10^3 Btu/yr-GSF†; (10^8 J/yr-m²)‡	Ultimate consumption, 10^3 Btu/yr-GSF†; (10^8 J/yr-m²)
8	1842 (171,000)	164 (18.6)	36.9 (4.19) (23%)	127 (14.4)
16	968 (89,900)	128 (14.5)	17.2 (1.95) (13%)	121 (13.8)
26	392 (36,400)	112 (12.7)	31.5 (3.58) (28%)	80 (9.1)
67	170 (15,800)	69 (7.8)	5.2 (0.59) (8%)	64 (7.3)
48	125 (11,600)	67 (7.6)	7.5 (0.85) (11%)	59 (6.7)

*As of 1977.   †GSF = gross square feet.   ‡Percentage (%) represents annual energy saved as a percentage of the 1975 Energy Index. The savings will pay back the retrofit investment in 3 years or less.

dard 90-75, there was a need to identify the impact it might have on first cost, operating cost, and marketability. A typical New York City speculative office building designed to conform to the Standard was studied and evaluated. The major findings were that the implementation of an Energy Code based on ASHRAE Standard 90-75 would: (1) add little or nothing to the construction cost; (2) reduce energy consumption by 54% in the building, based on net rentable area, when compared with existing buildings which had undergone energy conservation programs; (3) result in a 32% savings in energy cost; (4) reduce the impact of new office building construction on Con Edison electric and steam systems; and (5) increase demand for insulation double glazing and efficient lighting fixtures, while reducing size of mechanical systems.

The New York State Energy Code was implemented and became effective Jan. 1, 1979.

Energy performance standards. The Energy Conservation and Production Act of 1976 directed the U.S. Department of Housing and Urban Development (HUD) to develop energy performance standards for new commercial and residential buildings. Responsibility for this effort was later transferred to the DOE. A program was developed whose first phase called for an assessment of how much energy buildings are currently designed to use. The second phase of the program called for an assessment of how much less energy buildings could be designed to use.

A statistical sample of buildings designed in 1975–1976 was selected, representing a broad cross section of building types in different climatic regions of the nation. The architectural and engineering design teams of these buildings provided data which was used to simulate, by computer model, the annual energy consumption of each building. These same buildings were then redesigned to maximum practical feasible energy performance. Simulations of the energy consumption of the redesigned buildings were made, and the results were compared with those of the original buildings. Results of the simulations indicated 40% average improvement over the original designs.

With the energy performance of the original buildings as a baseline representing current practice, and the performance of the redesigned buildings representing what the profession could do when encouraged and pushed toward the forefront of energy-efficient design, the DOE can select performance standards which will reduce current designed energy consumption to a level that can be practically attained by the design professions.

For background information *see* BUILDINGS in the McGraw-Hill Encyclopedia of Science and Technology.

[IRVING SILVERMAN; CHARLES E. SCHAFFNER]

Bibliography: *Energy Conservation in Existing Office Buildings*: *Phase I*, ERDA, June 1977, *Phase II*, DOE, June 1978, *Phase III*, DOE, July 1979; *Impact Evaluation of New York Energy Code ASHRAE Standard 90-75 on Office Building Construction in New York City*, New York State Energy Resource and Development Administration, December 1977; *Phase I/Base Date for the Development of Energy Performance Standards for New Buildings*, DOE, Jan. 12, 1978.

Cell (biology)

In recent years there has been a substantial increase in understanding the aging processes. Some of this knowledge has been acquired through the study of human cells in tissue culture, as discussed below.

Proliferative capacity of cultured fibroblasts. For many years it was thought that all cells, when taken out of the organism and placed into tissue culture, would have an unlimited capacity to divide. However, in the early 1960s Leonard Hayflick demonstrated that normal human cells from fetal lung tissues (obtained from abortuses) have a limited division potential. When fetal tissue is treated with enzymes, such as trypsin, individual cells are dissociated. These cells can then be placed into flasks with culture medium containing the necessary vitamins, salts, and energy source to permit growth and replication. Within a week or two, spindle-shaped cells (fibroblasts) cover the entire flask. These fetal lung fibroblasts can then be removed with trypsin and transferred to two new flasks. During the first 30 or 40 transfers, these fibroblasts replicate quite rapidly and maintain a relatively uniform spindle-shaped morphology. Then the first signs of "senescence" become apparent. There is a decline in the replication rate of these cells and the appearance of large epithelioidlike (rectangular-shaped) cells. Cell replication rate continues to decline and then ceases. This limited proliferation capacity of normal human diploid fibroblasts is in contrast to the unlimited proliferation of human tumor cells when placed into tissue culture, and has led to the extensive use of this model system for studying human cellular aging.

Numerous laboratories have confirmed Hayflick's observation that a variety of different human cell types have limited replicative capabilities when placed into tissue culture. Investigators have demonstrated that this limitation is due not to the chronologic time in tissue culture, but to the actual number of previous divisions of the cell population.

Theories of cellular aging. Many theories have been proposed to explain this interesting behavior of human cells. One theory, developed by G. Martin, suggests that fibroblasts are undifferentiated cells which after many divisions in tissue culture become differentiated. Since most differentiated cells within the body do not divide, he proposes that differentiation leads to cessation of proliferation. L. Orgel proposed a theory which postulates that errors accumulate as a function of aging in the enzymes involved in protein synthesis. With increasing cell proliferation, errors would become more and more frequent, resulting in the increasing production of altered proteins, and finally in an error catastrophe where cellular replication would no longer be possible. A particularly interesting theory was recently proposed by R. Holliday and T. Kirkwood. These authors suggest that not all cells in a tissue are committed to senescence, and that as fibroblasts are transferred in cell culture, the uncommitted or immortal cells are lost. Despite the abundance of theories to explain cell culture senescence, the mechanism for this decline in replication as a function of serial tissue culture subcultivation remains to be elucidated.

Rejuvenating senescent cells. One approach to rejuvenating senescent cultures was tried by Y. Mitsui and E. L. Schneider. These investigators fractionated a senescent cell population on the basis of cell size by velocity sedimentation. They successfully obtained fractionated senescent cell populations which morphologically resembled early-passage or "young" cell populations. However, these cells quickly reverted to their senescent phenotype when they were returned to tissue culture. Thus the loss of chronic proliferating ability does not appear to correlate with cellular morphology.

Aging in culture and in the organism. Although fetal lung fibroblasts at early and late passage have been extensively employed to study human cellular aging, it was uncertain whether changes observed as a function of serial passage in tissue culture truly reflected changes that occurred with aging in cells within the organism. To examine this question, cell cultures were established from volunteer members of the Baltimore Longitudinal Study of Aging. These healthy men came to the Gerontology Research Center of the National Institute on Aging, located in Baltimore, MD, for a comprehensive series of physiologic and psychologic tests. As part of their visit, a 2-mm piece of skin was removed from the inner aspect of the left upper arm (after informed conset had been obtained). The piece of skin was placed between two coverslips and incubated with culture medium. After 4 or 5 weeks sufficient cells had grown out from the explant to cover the flask surface. These skin fibroblasts were transferred to new flasks each week, with the same techniques used for fetal lung fibroblasts.

Total replicative potential. When the onset of cell culture senescence was examined as a function of the age of the cell culture donor, it was found that skin fibroblast cultures derived from

young subjects had a significantly later onset of senescence than parallel cell cultures derived from older subjects. The total number of cell population replications was also significantly greater in young-subject cell cultures. Thus the total replicative capacity of a skin fibroblast culture appears to correspond to the age of the cell culture donor.

Acute cellular replication. In addition to examining total replicative capacities, it is also important to investigate the acute replication of cell cultures from young and old subjects. Measurements of three separate parameters in early-passage skin fibroblasts indicates acute cell replication is also impaired as a function of aging. The illustration shows one of these measurements. Note the more rapid increase in cell number in the young-donor cell culture. These results indicate that cellular replication, both acute and long-term, are affected by the aging of skin fibroblasts in the organism.

Individual cells. While examination of cell populations can provide valuable information on cellular aging, it is equally important to analyze individual cells. Recently, James Smith, Olivia Pereira Smith, and E. L. Schneider have examined the replicative abilities of cloned cells from skin fibroblast cultures derived from young and old human subjects. These studies revealed that even at the individual cell level, skin fibroblasts from younger donors had significantly greater proliferative abilities.

Other cell types. This decline in cellular replication with aging is not limited to one specific type of cell. Although human lymphocytes do not normally divide in tissue culture, cellular replication can be examined in these cells when they are stimulated to divide by exposure to the antigen phytohemagglutinin. R. Tice, Schneider, D. Kram, and P. Thorne demonstrated that when cellular replication was examined in lymphocytes derived from blood samples taken from young and old members of the Baltimore Longitudinal Study, a significant decline was observed in the replication rate of lymphocytes derived from older subjects.

Significance of studies. These studies indicate that cellular replication may be seriously affected as a function of normal human aging. The next step is to elucidate the mechanisms for this decline. If this can be achieved, scientists may be able to restore the lost proliferative abilities of replicating cell populations and to diminish the effects of many age-related diseases.

For background information *see* CELL DIFFERENTIATION, SENESCENCE, AND DEATH in the McGraw-Hill Encyclopedia of Science and Technology.

[EDWARD L. SCHNEIDER]

Bibliography: C. Finch and L. Hayflick (eds.), *Handbook of the Biology of Aging*, 1978; R. Holliday et al., *Science*, 198:366–372, 1977; G. Martin et al., *Amer. J. Pathol.*, 74:137–154, 1974; Y. Mitsui and E. L. Schneider, *Exp. Cell Res.*, 103: 23–30, 1976; L. Orgel, *Proc. Nat. Acad. Sci.*, 49: 517–521, 1963; E. L. Schneider and Y. Mitsui, *Proc. Nat. Acad. Sci.*, 73:3584–3588, 1976; J. R. Smith, O. M. Pereira Smith, and E. L. Schneider, *Proc. Nat. Acad. Sci.*, 75:1353–1356, 1978; R. Tice et al., *J. Exp. Med.*, 149:1029–1041, 1979.

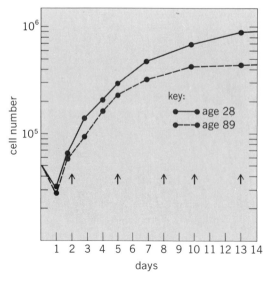

Cell culture growth curves from a typical young-donor and old-donor cell culture.

Cell membranes

The target specificity of drugs and hormones, in that they act only upon certain kinds of cells, has been explained by the hypothesis that these substances are bound chemically to the membranes of the cells on which they act by weak bonds with specific receptor molecules on the cell surface. This hypothesis has been tested in recent years in experiments using drugs or hormones labeled with radioactive isotopes, or with the dye fluorescein, in combination with microautoradiography or fluorescence microscopy.

Earlier, the distinction was made between fat-soluble substances such as the steroid hormones or other substances of small molecular size, presumed able to penetrate the cell membrane, and water-soluble substances such as polypeptide hormones and antibiotics, to which the cell membranes are presumably impermeable. The latter type of agent was presumed to act through effects on the cell surface, but the former type through effects on internal cellular processes directly.

Polypeptide hormone receptors. There is now evidence that polypeptide hormones, including somatotropin or growth hormone, prolactin, and insulin, are bound specifically to cell surfaces and also enter the cells. Receptors for insulin on cell surfaces have been purified and identified as glycoproteins. In addition, there are other receptors in the cytoplasm, on the membranes of the endoplasmic reticulum and the Golgi bodies. The suggestion has been made that, of the many diverse effects of insulin in cells, the immediate or short-term effects—such as increasing penetration of glucose into cells—are mediated through the surface receptors. The delayed or long-term effects—such as increasing glycolysis or changing ionic balance—are then considered to depend on the internal receptors as mediators.

Thyroid hormone receptors. The thyroid gland secretes several related hormones, and one, triiodothyronine (T_3), is bound to receptors in the cell nucleus. This suggests that the developmental actions of this hormone, and possibly its effects in increasing metabolic rate, are mediated through the processes whereby the nucleus controls the synthesis of enzymes and other proteins in cells.

Steroid hormone receptors. The steroid hormones, including the sex hormones and those produced by the adrenal cortex, enter cells without prior binding to the cell surface. In the cytoplasm, they combine with specific receptor molecules and enter the nucleus in this combination. Again, this suggests that these hormones act through influence on nuclear control processes.

Morphine receptors. The search for the receptors for the narcotic drug morphine has led to the discovery of two new classes of substances. These substances are polypeptides known generically as enkephalins and endorphins. Distribution of these substances in the brain suggests that the two types of substance occur in different types of cells. The well-known action of morphine in relieving pain is mimicked by these substances, suggesting that they may function, in the absence of morphine, as "natural" analgesics.

For background information *see* CELL MEMBRANES; ENDOCRINE MECHANISMS; PSYCHOPHARMACOLOGIC DRUGS in the McGraw-Hill Encyclopedia of Science and Technology.

[BRADLEY T. SCHEER]

Bibliography: S. Jacobs, *Science*, 200(4347): 1283, June 16, 1978; C. B. Kolata, *Science*, 201(4359):895, Sept. 8, 1979; B. T. Scheer, *Chem. Zool.*, 11:103, 1979; S. G. Younkin et al., *Science*, 200(4347):1292, June 16, 1978.

Chemotaxonomy

Plants produce many types of natural products in varying amounts, and quite often the biosynthetic pathways responsible for these compounds also differ from one taxonomic group to another. The distribution of these compounds and their biosynthetic pathways correspond well with existing taxonomic arrangements of plants based on more traditional criteria such as morphology. In some cases, chemical data have contradicted existing hypotheses which necessitates a reexamination of the problem or, more positively, have provided decisive information in situations where other forms of data are insufficiently discriminatory. Thus, plant chemistry has become a powerful adjunct tool in plant taxonomy, resulting in a new research field with its own appropriate label, chemotaxonomy.

Historical perspective. Chemotaxonomy in its crudest form was practiced by primitive cultures as they began to label and classify edible plants by taste, smell, color, and whether or not they were poisonous. In the case of poisonous plants, for example, the Greek philosopher Socrates was forced (for political reasons) to take his own life by drinking a broth of poison hemlock, *Conium maculatum*, which contains the toxic alkaloid coniine. In later times, the extensive world explorations by post-Renaissance European adventurers and merchants were fueled by the desire of European societies for exotic spices and condiments. The post-Linnaean taxonomists of the 1800s utilized descriptive chemical characters in their work. Thus, A. P. de Candolle, a Swiss taxonomist, recognized the antifever properties of *Cinchona* species, although not until 1904 was the antimalarial substance, quinine, completely identified. Today, as P. M. Smith has shown, routine chemotaxonomic surveys often uncover many such compounds of practical as well as taxonomic value.

Natural product classification. Modern chemotaxonomists often divide these natural plant products into two major classes: (1) micromolecules, that is, those compounds with a molecular weight of 1000 or less, such as alkaloids, terpenoids, amino acids, fatty acids, flavonoid pigments and other phenolic compounds, mustard oils, and simple carbohydrates; and (2) macromolecules, that is, those compounds (often polymers) with a molecular weight over 1000, including complex polysaccharides, proteins, and the basis of life itself, deoxyribonucleic acid (DNA).

Current technology. A crude extract of a plant can be separated into its individual components, especially in the case of micromolecules, by using one or more techniques of chromatography, including paper, thin-layer, gas, or high-pressure liquid chromatography. The resulting chromatogram provides a visual display or "fingerprint" characteristic of a plant species for the particular class of compounds under study. Figure 1 shows the paper chromatographic pattern (or fingerprint) of flavonoid pigment spots of *Parinari parilis*. The flavonoid spots are normally visible only when observed under long-wave ultraviolet light (360 nm) and have been outlined for clarity.

The individual, separated spots can be further purified and then subjected to one or more types of spectroscopy, such as ultraviolet, infrared, or nuclear magnetic resonance or mass spectroscopy (or both), which may provide information about the structure of the compound. Thus, for taxonomic purposes, both visual patterns and structural knowledge of the compounds can be compared from species to species.

Because of their large, polymeric, and often crystalline nature, macromolecules (for example, proteins, carbohydrates, DNA), can be subjected to x-ray crystallography, which gives some idea of their three-dimensional structure. These large molecules can then be broken down into smaller individual components and analyzed by using techniques employed for micromolecules. In fact, the specific amino acid sequence of portions of a cellular respiratory enzyme, cytochrome *c*, has been elucidated and used for chemotaxonomic comparisons in plants and animals, and with some success.

In the case of proteins, however, it is often not necessary to know the specific amino acid sequence of a protein (a most laborious process) but, rather, to observe how many different proteins, or forms of a single protein, are present in different plant species. The technique of electrophoresis is used to obtain a pattern of protein bands or spots much like the chemical fingerprint of micromolecules. Because each amino acid in a protein carries a positive, negative, or neutral ionic charge, the total sum of charges of the amino acids constituting the protein will give the whole protein a net positive, negative, or neutral charge. A protein extract of seeds, for example, is placed into a slab of a semisolid gel of starch, agar, or polyacrylamide. The gel is placed in contact with electrically conductive salt solutions, and an electric current is passed through the system. Because of their differing charges, the various proteins in the extract migrate at various rates toward the positive (or negative) pole, spreading out into a series of separate bands. A colored marker dye which moves at a specific rate indicates when the electrophoretogram is done. The slab of gel containing the proteins, which are colorless, is then soaked in an appropriate stain or dye which complexes with and visualizes the proteins, giving the desired protein band pattern of fingerprint. Such a gel is shown in Fig. 2. The several columns of band patterns in Fig. 2 represent the variation in forms of a single protein enzyme within individual seeds from a single tree of the red maple, *Acer rubrum*. Thus electrophoresis is sensitive enough to pick out protein (and hence genetic) differences between individual offspring of a single tree.

Specific chemotaxonomic applications. In terms of taxonomic studies, the chemistry of plants has been used extensively, and in some cases it provides the primary method of distinguishing between plant groups. Fungi, for example, are distinguished from plants because they lack the chlorophyll pigments typical of photosynthetic plants. Within the fungi, it has been found that some of them synthesize the amino acid lysine via the precursor aminoadipic acid, whereas others synthesize lysine via the precursor diaminopimelic acid. The presence of one or the other of these biosynthetic pathways correlates with current taxonomic arrangements within the fungi and provides information as to which groups of fungi are closely related to each other.

In contrast, the major divisions of algae are recognized on the basis of the different types of chlorophyll and their starch reserves. In bacteria, beyond a certain point, the external appearance of a cell is no longer sufficiently diagnostic, and bacteria are then classified on the basis of the types of substrates on which they will grow or digest, what type of respiratory by-products they produce, whether they will grow in air (aerobic) or without it (anaerobic) or equally well in both cases (facultative), and most recently, by differences in cell wall chemistry.

In green plants, and particularly in the flowering plants (angiosperms), micromolecular data have been most useful. One such group of compounds, the flavonoid pigments, have been effectively employed at many taxonomic levels. One group of these pigments, the anthocyanins, are responsible for the red-to-orange and the blue range of colors typical of most flowers. The second group, the anthoxanthins, are pale yellow or colorless to the naked eye, but occur in the largest number and variety and are easily separated into distinct chro-

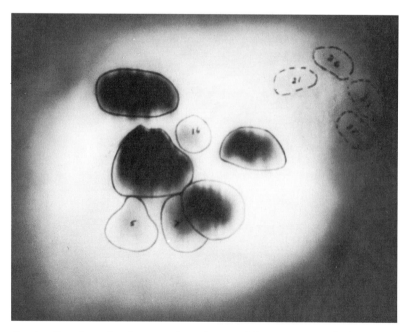

Fig. 1.   Two-dimensional paper chromatogram ("fingerprint") of *Parinari parilis* (Chrysobalanaceae), a Brazilian plant. A portion of the extract was spotted on the paper in the lower right-hand corner and separated into the pattern in solvents flowing from right to left and bottom to top. The pattern was photographed under long-wave ultraviolet light. *(Courtesy of Ralph Rocklin)*

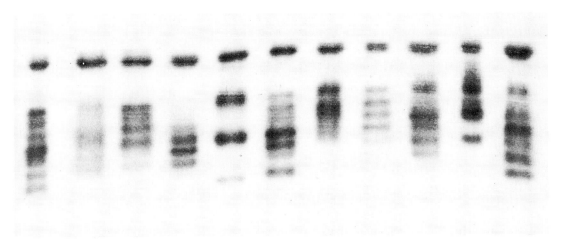

Fig. 2. Electrophoretograms of the protein enzyme phosphoglucoisomerase (PGI). Each column represents the protein band pattern for a single seed from the same tree of red maple, *Acer rubrum*. Note the variation in band patterns between the 11 individual seeds (columns). *(Courtesy of G. M. Iglich, University of Georgia)*

matographic patterns or fingerprints. These patterns are clearly observable under long-wave ultraviolet light (see Fig. 1).

Documentation of hybrids. One of the earliest uses of flavonoid patterns was in the classic studies of chemical documentation of hybridization between species in the legume genus *Baptisia*, by R. E. Alston and B. L. Turner. Occasionally, when two species of *Baptisia* occur together, hybridization occurs. The hybrids resulting from this fortuitous type of cross often look different from both parents. Some early taxonomists even described them as new species of *Baptisia* rather than recognizing them as transient hybrids. Alston and Turner suspected that these "new" species were hybrids. They first examined the flavonoid profiles of the two parents and found them to be different from one another. The profile of the suspected hybrid proved to be a summation of the two contributing parents, a phenomenon termed complementation. If the hybrid had instead been a legitimate species, its flavonoid profile would have been expected to be different from either of the two species present in the area owing to its own morphological and chemical evolution as an individual species. Indeed, so distinct are the flavonoid profiles of each *Baptisia* species that at least one hybrid was shown to possess a complementary profile involving up to four different species.

The same approach has been used employing other classes of compounds such as terpenes to study cases of suspected hybridization in other taxa such as the junipers *(Juniperus)*. In one case, studied by R. Adams and Turner, suspected hybridization between *J. ashei* and *J. virginiana*, based on earlier morphological work, was shown, by terpene chemistry, to be unfounded. In contrast, terpene chemistry indicated that hybridization had occurred between *J. virginiana* and *J. scopulorum* when no such definite conclusions could be drawn from morphological and ecological information alone.

Fossil chemotaxonomy. Recently, K. J. Niklas and D. E. Giannasi found flavonoids to be preserved in fossil angiosperm leaves from the western United States that are about 22,000,000 years old. The flavonoid profiles of the fossils, which could be compared with profiles of extant (living), related species, included elm *(Ulmus)*, hackberry *(Celtis)*, maple *(Acer)*, oak *(Quercus)*, and *Zelkova*. The flavonoid chemistry of the fossil oak *Quercus consimilis* proved especially interesting since paleobotanists considered the fossil oak species to be most closely related, morphologically, to two extant Asian species, *Q. myrsinaefolia* and *Q. stenophylla*. Using combined gas chromatography and mass spectroscopy, the researchers obtained steroid, fatty acid, and alkane profiles of the fossil leaf, demonstrating that it was an oak. The flavonoid profile of the fossil oak leaf, however, was not like that of the previously proposed living relatives, but was in fact nearly identical to that of two other extant Asian oaks, *Q. acutissima* and *Q. chenii*, which, incidentally, also have a leaf morphology similar to that of the fossil. Thus, in the table, the occurrence of flavonoids of group A in the fossil oak (taxon 1) shows it to be chemically most similar to *Q. acutissima* and *Q. chenii* (taxa 2 and 3, respectively) rather than to the putative relatives (see taxon 4) that were chosen on the basis of morphology alone and that lack pigment group A. Other randomly chosen, superficially similar taxa (taxa 5 and 6) also do not match the fossil species.

Although the chemical data confirmed that the fossil was an oak and pointed out its close relationship to two living oak species not previously suspected on the basis of morphology alone, perhaps more important was the fact that Niklas and Giannasi were able: (1) to show that a definite taxonomic and evolutionary link exists between extant Asian oaks and the extinct oaks of western North America, and (2) to compare the chemistry of fossils with that of living relatives and determine how much evolutionary change, if any, had occurred over millions of years.

Potential. Obviously, in studies of existing taxonomic systems, a knowledge of plant chemistry can be a powerful adjunct tool, and in many cases can indicate possible evolutionary relationships between plants more precisely than other evidence

118   CLIMATIC CHANGE

Comparative distribution of flavonoids in fossil and extant species of oak (Quercus) and related taxa*

Taxon	Flavonoid distribution† A						B						C		D	E			
	1	3	5	9	10	11	2	4	12	13	16	17	14	15	7	6	8	18	19
Quercus consimilis	+	+	·	·	·	·	+	+	·	·	·	·			+	+	+	·	+
Q. acutissima	+	+	+	·	·	·	+	+	·	·	·	·			+	+	+	·	+
Q. chenii	+	+	·	+	+	+	+	+	·	·	·	·			+	+	+	·	·
Q. myrsinaefolia	·	·					·	·	+	+	+	·	+	+	+				
Q. stenophylla	·	·					·	·	+	+	+	·	+	+	+				
Q. variabilis							·	+	·	+	+	+	+	+	tr				
Castanea dentata							·	·	·	·	+	+	+	+	tr	·	·	+	·

*From K. J. Niklas and D. E. Giannasi, Angiosperm paleobiochemistry of the Succor Creek flora (Miocene), Oregon, U.S.A., *Amer. J. Bot.*, 65:943–952, 1978.

†A = glycoflavones; B = flavonols; C = flavanonols; D = ellagic acid; E = unknown; + = presence of compound; tr = trace amounts.

alone can. Indeed, the potential of chemistry in plant taxonomy has finally been recognized, resulting in a close synergism between chemist and taxonomist, and in the establishment of the chemotaxonomist, a person trained in both fields.

For background information *see* PLANT TAXONOMY in the McGraw-Hill Encyclopedia of Science and Technology.   [DAVID E. GIANNASI]

Bibliography: R. Adams and B. L. Turner, *Taxon*, 19:728–751, 1970; R. E. Alston and B. L. Turner, *Biochemical Systematics*, 1963; K. J. Niklas and D. E. Giannasi, *Amer. J. Bot.*, 65:943–952, 1978; P. M. Smith, *The Chemotaxonomy of Plants*, 1976.

Climatic change

The entire field of climatic change has been receiving a tremendous amount of attention for the last 5 to 10 years. This article will focus on one rather narrow (and controversial) question: Does solar variability affect the climate? R. H. Olson favors a positive answer, while many other meteorologists have an opinion in the opposite direction. One such is A. B. Pittock, who has been critical of much of the work done on Sun-climate relationships, particularly the statistical methods used in attempting to establish 11-year and 22-year periodicities in weather phenomena. A rule of thumb he suggests is that a weather "periodicity" should not be taken seriously until it has repeated five or six times.

Indirect evidence for Sun-climate effects. Before discussing a few of the results which seem to point to a Sun-climate effect based on analysis of various types of data, some circumstantial evidence that there may well be some effects must be considered. The first of these bits of circumstantial evidence is the ability to make forecasts. H. H. Lamb has recently analyzed 24 long-range (up to several decades) climate forecasts made over the last few decades. Of the ones which can be verified, there are seven which were based on solar periodicities. Of these, one was wrong, five were mainly correct, and one was, in Lamb's words, "completely right." This was a forecast issued in 1951 by H. C. Willett.

Another bit of circumstantial evidence is that increasing numbers of studies lately are turning up evidence that short-term relationships between solar parameters and weather seem to exist. If

such relationships exist over periods of days and weeks, it may not be unreasonable to expect that they also exist on time scales of decades and longer.

That third bit of circumstantial evidence is that periodicities in paleoclimates of 100,000, 41,000, and 23,000 years have been demonstrated by J. Hays and his collaborators. These three periodicities correspond to changes in the orbital parameters of the Earth-Sun system, namely the eccentricity, the obliquity, and the precession of the equinoxes. A sensitivity of the atmosphere to some rather minor changes in solar input is thus demonstrated.

Direct evidence of Sun-climate effects. One of the first relationships sought by early workers was the 11-year sunspot cycle effect. Increased sunspot number has been correlated with both increased and decreased surface temperatures in various parts of the world, but there is general agreement that in rainy tropical areas the effect seems to be cooler weather with more sunspots. R. G. Currie has found what seems to be a definite 11-year periodicity in surface temperatures in the northeastern United States (Fig. 1). However, he does not give the phase relationship, that is, one cannot tell whether the correlation between sunspots and temperature is positive or negative. Others have reported that surface pressures are higher over Alaska and Canada during solar maximum than during minimum, and that the Aleutian Low migrates westward out over the Aleutian chain at times of solar maximum.

22-year solar cycle. Long after the 11-year cycle in the number of sunspots was discovered, it was found that the true solar cycle is 22 years, when the magnetic field of the Sun is considered. At the time of maximum on the 11-year cycle, the dipole field of the Sun reverses, so that it takes 22 years for the complete solar cycle to repeat itself. The 22-year periodicity shows up in the magnetic polarity of the individual sunspots also. It may be significant that research workers, both in the United States and in the Soviet Union, have found that 22-year periodicities in weather seem to be sharper than 11-year effects. However, as Pittock points out, even the stronger correlations, if they are real, are generally so weak as to be of little use in forecasting the weather and climate. Nevertheless, there is so little else that has been useful in climate forecasting that a possible solar influence cannot be ignored.

Precipitation data. One of the parameters that has been used to look for long-term solar influence is worldwide rainfall data. Precipitation data are notoriously hard to correlate with other data because of the spotty geographic coverage and because of the high variance of such data. A few large storms can dominate the statistics. Since it is difficult to find high-quality data sets in conventional meteorological parameters that extend back more than 100 years, a great many studies have been made using proxy data such as tree ring widths. J. M. Mitchell and coworkers found that by examining tree rings in the western United States they could establish a drought area index back to the year 1700. It turns out that the area of the drought, but not its intensity, correlates with the 22-year sunspot cycle rather well, in both wavelength and phase (Fig. 2). Mitchell points out that weather indices based on area rather than on intensity tend to show the solar signal better than others. As an example, the vorticity area index, which is a measure of the size of the major cyclones in the Northern Hemisphere, shows some short-term relationships to variable solar activity. Since the Mitchell study was completed, there has been another drought in the western United States. It started in 1974 in the High Plains and by 1977 shifted to the Far West. Although this was not a drought of the intensity of the famous dust bowl drought of the 1930s, it fits the solar sequence rather well, since 1976 was a year of solar minimum.

These droughts occur near the sunspot minimum of every alternate 11-year sunspot cycle, with an average delay of 2 years past minimum. Some idea of the regularity of the droughts in the western United States can be obtained from the table, which lists the last five droughts along with the date of the appropriate sunspot minimum.

Periodicity in temperature. Another example of a 22-year periodicity in temperature, using proxy data, is the work of S. Epstein and C. J. Yapp (Fig. 3). They examined the ratio of deuterium to hydrogen (D/H) in bristlecone pine trees in California. The D/H ratio is an estimator of the temperature in California and neighboring ocean waters during the time of growth of the trees. They found, over a 1000-year data sample, that the principal period shown was 22 years. Unfortunately, their analysis does not tie the temperature change to the particular phase of the 22-year solar cycle, so it does not prove that the relationship is solar. The suggestion is strong, however, that the warm periods coincided with the drought periods in the Mitchell study, because abnormally warm and dry conditions normally occur together.

Still another 22-year periodicity in temperature was reported by C. J. E. Schuurmans, based on 344 years of record at de Bilt, Netherlands. He was able to tie the temperature periodicity to a change in the circulation in the North Atlantic region. During odd-numbered solar cycles—the same ones which contain the Mitchell droughts—the Icelandic low-pressure cell, which controls the North Atlantic circulation, tends to be found near Ice-

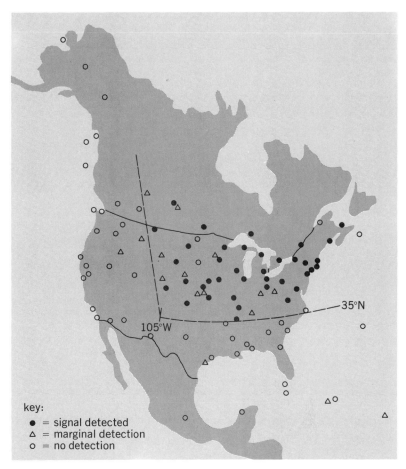

Fig. 1.   Distribution of solar cycle signal in surface air temperatures at 100 stations over North America. Out of 54 stations in the northeast quadrant of the continent, the signal is detected or marginally detected in 51 instances. (*From R. G. Currie, Distribution of solar cycle signal in surface air temperature over North America, J. Geophys. Res., 84:753–761, 1979*)

land. This allows occasional outbreaks of cold Siberian air to sweep across northern Europe. This produces winter temperatures at de Bilt which are cooler and more variable in the odd cycles than in the even cycles. During the even-numbered cycles the low-pressure cell shifts to a position near Scandinavia. This tends to block the Siberian air and brings more maritime air over Europe, so that the temperatures at de Bilt are warmer and less variable. The differences are not great—only about .23°C in average temperature and .12°C in standard deviation, but over such a long period of time they appear to be significant.

Longer-period changes. Several other solar periodicities have been claimed, such as one of 80–90 years and one of 180 years. Also, longer periods, from a hundred to a few thousand years, may turn out to be solar-related. However, the evidence for these relationships does not appear to be as strong as that for the 22-year periodicity.

Mechanisms. The subject of physical mechanisms is of the utmost importance, and the pursuit of empirical studies should always be done with certain mechanisms in mind. There has been so little progress that the best that can be done is to indicate the most popular of the many suggested.

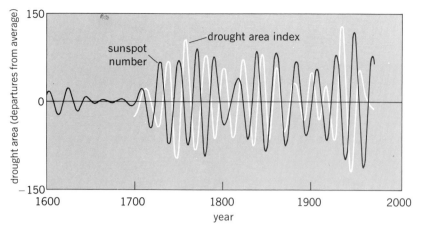

Fig. 2. Comparison of drought area index with sunspot numbers. Note that sunspot numbers are plotted alternately positive and negative to simulate the concept that alternate maxima are of opposite polarity. The time before the year 1700 is referred to as the Maunder minimum, a period when sunspots virtually disappeared. (From C. W. Stockton, D. M. Meko, and J. M. Mitchell, Jr., Tree ring evidence of a 22-year rhythm of drought area in western United States and its relationship to the Hale solar, Meeting of Working Group VIII, U.S./U.S.S.R. Agreement of Protection of the Environment, Crimean Astrophysical and Kislovosk Mountain Observatories, Sept. 13–23, 1978)

Relationship of droughts in the western United States to the solar minimum

Drought dates	Solar minimum
1974, 1977	1976 (start of cycle 21)
1953	1954 (cycle 19)
1934	1933 (cycle 17)
1912	1912 (cycle 15)
1892	1889 (cycle 13)

ted through the atmosphere, both outgoing and incoming.

Wave reflection. Under conditions of high solar activity, the ionosphere becomes unusually disturbed. Wave energy from the troposphere which normally propagates upward until it is lost to space might be reflected back down to the troposphere by the disturbed ionosphere, thus tending to heat the troposphere.

Summary. The subject of Sun-climate relationships is still in its infancy, despite 200 years or more of effort. Intensive efforts are now under way to correct this situation. Initial results allow for an attitude of cautious optimism that significant progress will be made in the coming years.

For background information *see* CLIMATIC CHANGE; SUN; WEATHER FORECASTING AND PREDICTION in the McGraw-Hill Encyclopedia of Science and Technology.

[ROGER H. OLSON]

Bibliography: J. R. Herman and R. A. Goldberg, *Sun, Weather and Climate,* 1978; H. H. Lamb, *Climate, Past, Present, and Future,* vol. 2: *Climate History and the Future,* 1977; T. Seliga (ed.), *Solar-Terrestrial Influences on Weather and Climate,* a symposium at Ohio State University, 1978; O. R. White (ed.), *The Solar Output and Its Variations,* 1977.

Clinical microbiology

Clinical microbiology has continued its rapid growth rate. Many new techniques have been developed to assist in the diagnosis of traditional as well as new diseases. Notable among these are the advances in the elucidation of the agent that causes legionellosis, new techniques for the detection of venereal disease agents, analysis of the toxin that causes antibiotic-associated diarrhea, and methods for detection of the various causes of gastroenteritis, as well as the development and introduction of a number of machines that promote the rapid diagnosis of infectious disease.

Legionnaire's disease. It is interesting that only rarely in the last 2 to 3 decades have new bacterial diseases been discovered. The majority were discovered at the turn of the century. However, in 1976 a group of American Legion conventioneers were struck with a hitherto unknown disease, Legionnaire's disease, or legionellosis as it is now known. After 2 to 3 years of intensive study, a scientific name has been recommended for the agent of legionellosis, *Legionella pneumophila.* This agent is a bacterium and not a virus or fungus. It can grow on artificial media. The bacterium is apparently a new genus and species. Significant also during 1978–1979 was the development of rapid methods for the diagnosis of legionellosis. These included the ability to detect small soluble

Cirrus clouds. It has been suggested that solar particles may cause increased freezing nuclei at the tropopause level, thus encouraging the formation of cirrus clouds. The resulting radiation losses from the top of the cirrus deck, plus the trapping of heat below the clouds, would increase the vertical instability of the atmosphere.

Thunderstorms. The increase in electric conductivity in the air above thunderstorms following solar flares would possibly affect the strength of the thunderstorms.

Ozone. Solar flare particles can cause a dramatic decrease in the concentration of ozone. Also, the slow change of the solar "constant" throughout the solar cycle would affect the concentration of ozone. Magnetic storm particles can cause either a decrease or an increase in ozone. Ozone changes would affect the amount of radiation transmit-

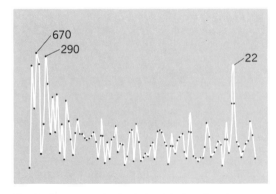

Fig. 3. Power spectrum of simulated surface temperature in California, as estimated from D/H ratio in Bristle cone pines. The numbers labeling the peaks are lengths in years of the resolved periods. The periodicities of 670 and 290 years are probably not real, but the 22-year periodicity appears to be highly significant. (From S. Epstein and C. J. Yapp, Climatic implications of the D/H ratio of hydrogen in C-H groups in tree cellulose, Earth Plan. Sci. Lett., 30:252–261, 1976)

components of *L. pneumophila* by immunochemical techniques. Rapid diagnosis of a disease leads to early initiation of treatment and less patient morbidity and mortality.

Chlamydia. An organism known as *Chlamydia trachomatis* has renewed the interest of many microbiologists since this bacterium is now known to cause many cases of venereal disease, as well as pneumonia in very young children. *Chlamydia* is not a new organism, having been discovered in the 1940s. However, its role in venereal disease and infantile pneumonia has been only recently appreciated. This bacterium is known as an energy parasite because it cannot produce enough energy on its own to live; it needs to use the energy-containing substances of living cells. Because these energy-producing substances are rather large molecules, the outer covering of *Chlamydia* must be different than other bacteria so as to allow the penetration of these phosphorus-containing molecules into the cell for its eventual use. Even though chlamydiae require living cells to grow, they are not viruses.

Pseudomembranous enterocolitis. Important among very recent discoveries has been the cause of antibiotic-associated pseudomembranous enterocolitis (PMC). For a number of years, many patients have developed serious diarrhea following antibiotic therapy. Although many mechanisms were proposed, only recently have a number of investigators discovered the reasons for this severe illness. There are many bacteria in the human large intestine. When a patient is given antibiotics, many of the bacteria in the large intestine are destroyed by the antibiotics. The organisms that are resistant to the antibiotics are given a selective advantage, and they grow in large numbers in the intestine. Even though these organisms when present in small numbers are not harmful, when present in large numbers they may secrete poisonous substances which can cause damage to the host. J. G. Bartlett and coworkers discovered that patients with PMC had a toxin or a poisonous protein in their stools. This toxin was secreted by an organism known as *Clostridium difficile*, which proliferated when competing bacteria were reduced in number by the antibiotic therapy. Tissue culture methods have been proposed for the routine detection of *C. difficile* toxin in the feces of suspected patients. Inclusion of this assay in the clinical laboratory test battery may allow more precise identification of this disease entity.

Gastroenteritis. There are many causes of gastroenteritis. In most clinical microbiology laboratories, the emphasis is on the causative agents of gastrointestinal infection such as *Salmonella* and *Shigella*. They are major causes of such infection, but during the past few years it has been recognized that a number of other organisms cause diarrhea as well. These include toxin-forming *Escherichia coli*, the agent of traveler's diarrhea, which is common to people who vacation in different parts of the world. This organism is present in many different forms, but the form that causes traveler's diarrhea produces a toxin which activates adenyl cyclase, an enzyme in the large intestine, which promotes excessive water loss. During the past year, immunochemical and tissue culture methods

have become available to detect this toxin in *E. coli* isolates from human feces.

Campylobacter, a curved organism, has also been discovered as a major cause of diarrhea. Methods have been published for the routine detection of *Campylobacter* in fecal cultures. These methods rely on the use of antibiotics in media to which *Campylobacter* may be resistant, the heat tolerance of *Campylobacter*, and filtration methods.

Rotavirus, another new agent which has recently been discovered, causes diarrhea in very young children. Rotavirus is related to the virus that causes diarrhea in calves. Methods such as enzyme-linked immunosorbent assay (ELISA) and counterimmunoelectrophoresis have been discovered which aid in the identification of this virus. Although the virus cannot be grown in tissue culture, it can be seen with an electron microscope as very tiny particles in the feces.

Antibiotics and drug resistance. As microorganisms continue to survive antibiotics, in some instances significant differences have been seen in the resistance of certain bacteria to commonly used antibiotics. For many years, the organism that causes pneumonia, *Streptococcus pneumoniae*, was highly susceptible to penicillin. Recently a few have been detected which are resistant to penicillin. Continued resistance of *Haemophilus influenzae*, a common cause of meningitis in children, and *Neisseria gonorrheae* to ampicillin and penicillin have been seen. Many of the very potent antibiotics, such as gentamicin, are less useful in some hospitals because of growing resistance to such antibiotics. Drug companies have met the challenge with the introduction of many new antibiotics, such as amikacin and netilmicin, both aminoglycoside antibiotics. New cephalosporin antibiotics, which resemble penicillin, have been developed. These include cefamandole and cefoxitin. An old antibiotic, erythromycin, appears very active against *Legionella pneumophila*.

Several instruments have been introduced which can detect susceptibility to antibiotics in a span of 3–4 hr. They include the Abbott MS-2; a modification of the Pfizer Autobac which enables the quantitation of antimicrobial susceptibility; and an instrument originally designed for the space program called the Automicrobic system.

Outlook. The 1980s promise new discoveries which will significantly increase the speed at which infectious disease is diagnosed. Such rapidity not only can save money by reducing the length of stay in the hospital, but will save lives through the choice of more appropriate antibiotics.

For background information *see* BACTERIOLOGY, MEDICAL in the McGraw-Hill Encyclopedia of Science and Technology.   [RICHARD C. TILTON]

Bibliography: J. G. Bartlett et al., *New Engl. J. Med.*, 298:531–534, 1978; D. J. Brenner, A. G. Steigerwalt, and J. E. McDade, *Ann. Int. Med.*, 90:656–658, 1978; R. C. Tilton, *Ann. Int. Med.*, 90:697–698, 1978.

Clinical pathology

Prenatal diagnosis of chromosome abnormalities is now routinely offered in many parts of the world to women known to be at special risk. This article summarizes the conditions under which this is undertaken, the risks and scope of the procedures involved, and some of the problems encountered in the laboratory.

It is theoretically possible to identify all pregnancies involving a fetus with a chromosome abnormality at an early enough time for selective abortion to be offered. This would eliminate the 5 out of every 1000 infants born with a chromosome anomaly associated with phenotypic abnormality. However, not only is termination of pregnancy, even for "genetic" reasons, not universally acceptable, but transabdominal amniocentesis, even in practiced and skilled hands, is an invasive procedure and carries a risk. There is the risk of precipitating a sequential miscarriage of as high as 1.5% in addition to perhaps an increase in perinatal morbidity, and possibly mortality. Also, there is the theoretical possibility of needle damage to the fetus. Prenatal diagnosis is currently limited to pregnancies with increased risk of fetal abnormality because of the ethical considerations involved in subjecting a fetus to risk when the benefits are questionable. Moreover, the resources in staff and apparatus needed to carry out amniocentesis with prior ultrasound scanning followed by the difficult amnion cell culture and cytogenetics are not available.

Indications for prenatal diagnosis. As the risk of chromosome abnormality increases with maternal age, this tends to be the most important indication. Up to the age of 35 the total risk is less than 1 in 200; that of Down's syndrome (mongolism) is only 1 in 400. Thereafter, the risk rises rapidly so that by 40 years the total risk becomes 1 in 50, and over 45 years 1 in 10. Indeed, 1.6% of amniocenteses carried out on women between the ages of 35 and 39 led to termination of chromosomally abnormal fetuses; termination followed for 3.9% of those aged 40 and over (see table).

Young women who have had a previous Down's child are the next most frequent category, for the risk of recurrence of a chromosomally abnormal fetus is 1.1% The highest risk of abnormality is in those pregnancies where one parent, though phenotypically normal, carries a balanced translocation. Of the relatively few such pregnancies (219) examined cytogenetically, 9.6% turned out to have an unbalanced translocation or some other chromosome abnormality. All pregnancies with the other "indications" for prenatal diagnostic cytogenetics, such as a previous child with some malformation other than Down's, a family history of Down's (in the absence of a parental translocation), high risk for neural tube malformation, or just parental anxiety result in no greater number of chromosomally abnormal fetuses than would be expected in the general population of women under 35 years. The only exception to this is the "miscellaneous" group of pregnancies exposed to x-rays, immunosuppressants and other drugs, virus infections, hypoxia, anesthesia, and the like.

Amniocentesis followed by amnion cell culture and karyotyping for fetal sexing is offered to women who are carriers for a serious sex-linked inherited disorder, such as Duchenne muscular dystrophy, where it is planned to abort all male fetuses. Social reasons, such as infant sex selection, are not considered adequate reasons for offering the test.

Amniocentesis. This procedure, which is quick and usually easy, should not be carried out before the fifteenth week of gestation (estimated from date of the last menstrual period). Before that stage, not only may it be more difficult and the removal of some fluid more hazardous to the pregnancy, but there is also a danger of fetal damage by the needle point. In addition, amnion cell culture seems to be less successful after this stage. If amniocentesis is carried out after 19 weeks, ethical problems may arise if a late termination has to be carried out.

The amniocentesis should be preceded by an ultrasound scan to confirm the supposed period of gestation, to determine the position of the placenta (which will dictate the precise approach as well as the technique), and to exclude multiple pregnancy (which may make prenatal diagnosis difficult or even impossible).

Amnion cell culture. The cells in the amniotic fluid are generally derived entirely from the fetus or the membranes. Most of the desquamated cells come from the dermis and are dead, but some are viable or potentially viable. It is likely that those which grow in culture are derived from the respiratory, genitourinary, or gastrointestinal tracts and may be macrophages. However, these cells in culture are sometimes derived from the fetal mem-

Results of amniocentesis in Europe and the United States

Indication	United Kingdom (1970-1976)		United States (1970-1978)		West Germany (1973-1977)		Total	
	Number	% abnormalities	Number	% abnormalities	Number	% abnormalities	Number	% abnormalities
Maternal age: 40 years	2,926	3.6	561	4.8	1,490	4.2	4,977	3.9
35-39 years	2,858	1.4	1,843	1.7	2,027	1.7	6,728	1.6
Previous Down's child	1,202	1.2	240	1.2	522	0.7	1,964	1.1
Parental translocation	135	8.9	14	0	70	12.8	219	9.6
Previous other malformation	188	0	17	0	73	1.3	278	0.4
Family history of Down's	557	0.4	—	—	—	—	557	0.4
High risk for neural tube malformations	4,786	0.4	—	—	268	0.8	5,054	0.4
Anxiety alone	471	0.6	—	—	—	—	471	0.6
Miscellaneous	547	3.1	55	9.0	90	3.3	692	3.6
Total	13,670	—	2,730	—	4,540	—	20,940	—

branes, and occasionally maternal cells are grown by mistake. The latter case may cause confusion or even mistaken diagnosis.

Amnion cell culture is difficult and requires meticulous attention to detail. The culture methods vary from laboratory to laboratory, but generally the sample, which should be as free from blood as possible, is centrifuged. The cell deposit is suspended in a culture medium such as TC 199, enriched with 30% fetal bovine serum, and inoculated into two or more culture vessels such as Leighton tubes, which are gassed with CO_2 and sealed. After incubation at 37.5°C for 7 days, they are examined for growth. If growth is satisfactory, the cells are resuspended, centrifuged, and inoculated into fresh tubes. These cells are then reincubated and examined daily. When sufficient growth is seen, preparations are made for karyotyping. A minimum of 10–20 cells is examined. Banding techniques as well as ordinary staining should be employed. Culture times vary from 10 days to 4 weeks with a mean time in most laboratories of 14–20 days. Heavy blood contamination of the fluid will lead not only to a lengthening of the culture time, but also to a reduction of the success rate. Generally, success rates well in excess of 95% are achieved; it should not be less than 95% on satisfactory uncontaminated and bloodless fluids. Cells from more than one primary inoculation should always be examined, as cultural artifacts rarely arise spontaneously in all inoculations.

Mosaicism arising in culture is a common finding. Occasionally, the mosaicism is present in the fetus, in which case it should be found in all cultures. XX,XY mosaicism should arouse suspicion of contamination with maternal cells. Such contamination should always be suspected when a strong culture is established very quickly. Polyploids found in culture suggest that the growth has been from cells shed from the fetal membranes.

Termination decision. Pregnancies of chromosomally normal fetuses are allowed to go to term, but when the fetus is shown to have a chromosome abnormality known to be associated with phenotypic abnormality, immediate termination of the pregnancy should be recommended. After termination, umbilical cord blood, as well as tissues from the fetus, should be submitted for examination for confirmation of the chromosome abnormality. All pregnancies with an apparently chromosomally normal fetus should also be followed up to confirm that a cytogenetic mistake has not been made. Such checks on the work of a prenatal diagnostic service are essential for controlling the quality of the work. Of all aborted fetuses, 50% are Down's, 13% are E trisomy, 20% are sex chromosome anomalies, and 8% are unbalanced translocations. Fewer than 10% are true mosaics.

Difficulties in making a decision may be encountered when a de novo balanced translocation or a chromosome variant is found in the fetus which is not present in either parent. This is a problem because some of these may be associated with phenotypical abnormality, though there should be no cause for concern when the chromosome abnormalities have been inherited from a phenotypically normal parent with the same chromosome finding.

Problems will arise when an XYY chromosome constitution is discovered since only a small minority will grow up to be psychopaths. Some parents may also have difficulties in accepting termination for XXY, Klinefelter's, and XXX females, since a considerable proportion of these will be fairly normal physically and also mentally. De novo balanced translocations and some de novo variants may present problems, as it is not always possible to be certain that they may not be associated with some phenotypical abnormality.

Prenatal diagnosis should not be embarked upon without careful counseling of both parents so that they know the scope and risks of the tests and procedures and that the diagnosis may lead to the recommendation of a termination.

For background information see CLINICAL PATHOLOGY in the McGraw-Hill Encyclopedia of Science and Technology. [K. MICHAEL LAURENCE]

Bibliography: Deutsche Forshungsgemeinshaft, *Prenatale Diagnostik Genetish Bedingter Defecte*, Informations Blatt no. 13, Munich, 1977; M. A. Ferguson-Smith et al., *The Provision of Services for the Prenatal Diagnosis of Abnormality in the United Kingdom*, Bull. Eugenics Soc. London, suppl. no. 3, 1978; M. S. Golbus et al., *N. Engl. J. Med.*, 300:157-163, 1979; K. M. Laurence and P. J. Gregory, in *Pathology Annual*, vol. 8, pp. 155–187.

Coffee

Large increases in green coffee prices caused by a bean shortage in recent years have stimulated research in the development of more efficient coffee extraction technology. Until July 1975 there was very little concern for improved efficiency in extraction from roasted and ground coffee. The green coffee prices, based on years of surplus stocks, were holding about 50–60¢/lb ($1.10–1.32/kg) despite diversification efforts in producing countries, U.S. dollar devaluation, producer support programs, and several natural and political disasters. The United States coffee industry was actively working on campaigns to reverse the gradual trend of declining per capita consumption in the United States. The consumers had grown accustomed to paying $1.25/lb ($2.76/kg) or less for a product from which they could brew 50–60 servings/lb (110–132 servings/kg). They were almost unaware of the good value this beverage offered at less than 3¢ per serving.

Pre-1975 conditions. During the 10-year period prior to July 1975, most technological advancements in the coffee industry centered on either soluble (instant) coffee improvements, with the development of freeze-dried and agglomerated coffee, or specialized grind developments for fresh-brew coffee (with the grind suited specifically for automatic drip or electric perk coffeemakers). In the food service and vending industries, there was already some "stretching" of coffee, that is to say, getting more cups from each pound of ground coffee. This was accomplished mostly by roasting the coffee darker and grinding it finer. There was some sacrifice in flavor quality, but at least the color and body of the brew were acceptable.

On July 17 and 18, 1975, a killer frost swept across the coffee-growing areas of Brazil, not only destroying the following crop but actually killing the trees in many parts of the country. (It takes about 5 years to grow a coffee tree that is productive.) The impact of this frost was indeed severe. Brazil, which produced about one-third of the world's exportable coffee, lost almost an entire crop. For the first time in 20 years there was no surplus of green coffee beans in the world. As the news spread, the importers, roasters, retailers, and consumers began to hoard coffee, and the tight supply situation turned into a shortage. The green coffee market rose with only a few small adjustments from about 55¢/lb ($1.21/kg) in July 1975 to over $3.00/lb ($6.61/kg) by March 1977.

High extraction coffee. In an atmosphere of extreme pressure the coffee industry sought ways to soothe the outcry about the high prices and to curb the dramatic drop in consumption. The early efforts were based on the use of additives to reduce the cost of the unit to the consumer. Grains, garbanzo beans, chicory, soya beans, and almost anything roastable were researched as possible cost cutters. While several products received massive advertising support and wide distribution, none really offered true coffee flavor.

The more successful approach was to find technological ways to increase the consumer's yield (number of servings) and thus cut the cost per serving. Basically two approaches were taken. Both accomplished the objective by increasing the amount of surface area of the coffee particle exposed to the brewing water, thus prompting more efficient extraction.

Flaking. In the flaking process the previously ground coffee particles are run through a set of smooth rollers so as to flatten them. This process not only increases the surface area but also breaks down the cellulose fiber. Both of these changes result in greater extraction when brewing. There are, however, some negatives: much of the aroma-carrying gases is compressed out of the particle (besides loss of aroma, this leads more rapidly to staleness). Also, the breaking down of the cellulose structure causes excessive sediment during brewing, so that such coffee should ideally be restricted to brewing equipment that uses paper filters. This flaking process has been used successfully and, if the product contains high-quality coffee beans, offers the consumer about a 20% increase in the number of servings. It should be pointed out that this process results in a dense product, and the consumer must be instructed to reduce the normal measurements by about 20% to achieve the desired economics.

Puffing. The puffing process uses roasting technology to increase the surface area of the coffee particle. The roasting process expands (opens) the cell structure of the coffee particle by applying heat, using extremely high air velocity. This process requires a delicate balance between feed rate, dwell time, air velocity, and heat. The high velocity of the heat application provides a greater penetration of the coffee beans, thus creating more potential extractable solids, and also opens the cell structure to reduce the density and promote a "flow-through" effect when brewing. The open cell structure also carries a greater amount of aroma-saturated gases than in the flaking process. The processor can increase the number of servings by

about 20% and reduce the density (weight of a given measure) by about 20%. Thus, normal brewing measurements can be used while the expected economic benefits are received. This puffing process also allows production in any desired grind for any type of brewing equipment.

Success of advances. The simultaneous introduction of several new coffee concepts (such as adulteration, flaking, and puffing) was found to be confusing both in terms of measurement and in quality results. The food service and vending industries were already accustomed to dealing with beverage products on a cost-per-serving basis, and have been able to weigh the economy and quality results much faster. There is little doubt that the high-extraction coffee brands have a place in the coffee market. There is reason to believe that consumers will continue to buy, and develop confidence in, the new products. The products that will survive will be the ones that meet quality standards while offering economy.

For background information *see* COFFEE in the McGraw-Hill Encyclopedia of Science and Technology.   [RICHARD L. THOMPSON]

Communications satellite

New techniques are under development which use spot beams to increase both the capacity of satellite communications and the effective power of the signals transmitted by satellites. These techniques involve operation at frequencies higher than those used by present-day satellites, and the use of large-aperture satellite antennas to radiate the relatively narrowly focused spot beams. Special antenna systems are being developed to implement these techniques, in which radiation patterns produced by a small array of antennas are magnified and projected onto a single large aperture. *See* SATELLITE COMMUNICATIONS.

Present-day satellites. Commercial communications satellites share frequency bands with terrestrial microwave radio. A 500-MHz bandwidth, allocated around a carrier of 6 GHz, is dedicated for uplink transmissions, and another 500-MHz bandwidth at 4 GHz is dedicated for downlink transmissions. In all existing commercial communications satellites, the 500-MHz frequency band is divided into 12 equally spaced channels.

Access to most communications satellites is achieved through a single wide-area-coverage antenna, whose beam is shaped to serve best a particular geographic region, such as the United States, Canada, or Indonesia. A single beam per country, however, limits the possibility of frequency reuse, so additional capacity generally must come from another satellite placed in a different orbital slot.

Another method is to use orthogonal-polarization transmission as employed on the AT&T/GTE COMSTAR and RCA SATCOM domestic satellites. In this technique, two signals of the same frequency are transmitted simultaneously at right angles to each other. Thus a conventional 12-channel satellite would use 24 transponders for transmission.

Communication links. A common approach to increasing satellite access is frequency division. However, as the nodes (Earth stations) in a net-

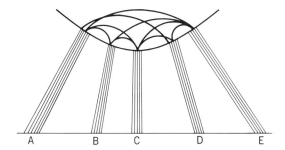

Fig. 1. Schematic diagram of communications system with five nodes, showing how interconnection of N nodes requires $N(N-1)$ paths.

work grow, the number of frequency paths increases as the square of the number of nodes, as shown in Fig. 1, which illustrates a system providing dedicated transponders for only five Earth stations.

One alternative to the rapid proliferation of required channels is to use frequency-division multiple access and operate the satellite's traveling-wave tube amplifier in a "backed-off" (nearly linear) fashion. Earth stations can then transmit in a portion of a frequency spectrum so that several stations can share a transponder. A drawback is that capacity is considerably lower than would occur when only one modulated carrier is used per transponder, because less power is transmitted, and because intermodulation distortion occurs when several modulated carriers are amplified together.

The other alternative is to assign to each Earth station a time slot during which it can transmit to other stations. When the number of stations in the network rises above a very few, this time-division multiple-access (TDMA) approach can provide greater capacity than frequency division can. TDMA uses bit rates that are compatible with an entire transponder, and assigns time slots to stations upon demand.

TDMA allows better load balancing than is possible with frequency-division multiple access. TDMA adjusts more easily to varying traffic demands because it operates with a variable-length time slot rather than a variable bandwidth. In TDMA systems, messages are sent in bursts, necessitating buffering at the Earth terminals. The number of bits that must be stored depends upon the traffic, that is, the data rate and the burst repetition rate.

Frequency reuse. To increase capacity significantly, future satellites will employ several simultaneous uses of the same frequency band. Furthermore, in order to locate Earth stations near sources of traffic, it will be advantageous to operate in one of the higher-frequency bands, where there is less interference. Fortunately, these two requirements go hand in hand, because frequency reuse is obtained by spatial separation of the satellite antenna beams, and beam isolation increases with frequency, for an antenna of the same size. As the size of the antenna aperture increases, the angular spread of energy decreases. This angle may become so small that only a small "spot" of a few hundred miles (1 mi = 1.6 km) diameter is illuminated on the Earth.

Spot beams offer significant advantages in satellite system design. They provide high gain and thus high effective radiated powder. With the use of large-aperture antennas that might be deployed by the space shuttle, antenna gains as high as 50 dB can be realized at 12 GHz. In the United States, particularly on the east coast and in the South, rain attenuation is particularly severe, and large margins of signal over noise are required to ensure that signals can get through all but the heaviest rains, that is, for all but an hour or two per year. Another advantage of spot antenna beams is that the same frequency band can be reused several times within the desired coverage region. Special antenna designs (called offset Cassegrain) make it possible to form several essentially independent beams with only one large main reflector.

Scanning-beam technique. Spot-beam antennas are not without problems, however. It is impossible to reuse the frequency band in contiguous zones, even if orthogonal polarization is employed. Antenna patterns cannot fit together precisely, since they do not have well-defined edges. As a result, more than four independent signal sets may be required, depending upon the degree of interference. To get complete area coverage with spot beams, several trade-offs must be made in terms of available bandwidth or antenna efficiency.

A scanning spot beam can give total coverage to the entire service area, while still providing the high antenna gain of a spot-beam satellite, by independently sweeping both its uplink and downlink in synchronism with a time-division format. The advantages are clear: The high gain of a spot beam is combined with the organization efficiency of TDMA. An attractive technique is to use a phased-element array, in which a digital phase shifter is employed in each element. The phase shift is controlled by high-speed logic, and losses are overcome by low-power transmitters at each element. To make the gain of the array large, either the number of elements or the gain of each element can be increased.

The element gain basically determines the total coverage area in which service is available. The element gain depends directly on the physical size of the element aperture, while the coverage area depends inversely on element size. Usually the element gain can be tailored to conform with the desired coverage area. The Earth, as seen from a geostationary orbit of 3.6×10^4 km (22,300 mi) from the Equator, is a disk about 18° across. If the 3-dB point of the element gain is set at the edge of the Earth, the resultant on-axis antenna gain (that is, the gain in the direction of the center of the Earth) is approximately 20 dB. Combining 100 such elements in an array provides an antenna that has a minimum 37-dB gain at any point on Earth from which the satellite is visible. To cover only the continental United States, a higher-gain element can be employed.

The scanning spot beams must be controlled to move in a certain sequence, with the time spent at each location being proportioned to the traffic needs of that area. Such beam times at a given location can be as little as 1 or 2 μs. Once the scanning sequence was started, it would repeat many times before changing, and changing is required

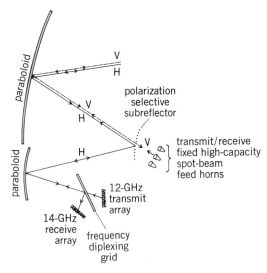

Fig. 2. Imaging satellite antenna with fixed spot beams and a scanning beam.

only as the capacity requirements of individual Earth stations change.

Small-array satellite antennas. There are imaging techniques, similar to classical optics, that allow the physical size of the array elements of a satellite antenna to be greatly reduced. The patterns produced by the elements are magnified through a system of lenses and projected onto a single large aperture.

A large array is unattractive because of its weight, loss, and the complexity of the long interconnections required by the large spacing between elements. A small array combined with several reflectors has therefore been proposed. The reflectors are arranged so that a magnified image of the array is formed over the aperture of the main reflectors. The magnification factor relating to the diameters of the main reflector and the array is chosen to be much greater than unity, so that the array is much smaller than the main reflector.

This approach was applied to the design of an antenna that transmits at 12 GHz and receives at 14 GHz in a 4.2-m-diameter satellite, assuming a field of view of $3 \times 6°$ is required. An arrangement suitable for this purpose, shown in Fig. 2, uses a diplexer between the two arrays, one for transmission and one for reception.

Although the design calls for several reflectors, its advantage is that the array itself is now very small and compact. The other unique factor of this antenna is a polarization-sensitive subreflector. This permits the simultaneous and independent operation of two different antennas, which share a common main reflector. Because only the horizontal polarization H is reflected, the subreflector allows scanning-beam imaging techniques to be performed, as mentioned earlier. Since the vertical polarization V passes through the subreflector, the fixed beams can be placed in the prime focus of the main reflector. By offsetting the feeds slightly from the prime focus, the beams emerge from the main reflector at slightly different angles, thereby allowing beams to be formed on geographically separated metropolitan areas. As long as these beams are separated by 2.5 or more 3-dB bandwidths, interference is minimal.

For background information *see* ANTENNA (ELECTROMAGNETISM); COMMUNICATIONS, ELECTRICAL; COMMUNICATIONS SATELLITE in the McGraw-Hill Encyclopedia of Science and Technology.

[D. O. REUDINK]

Bibliography: M. J. Gans and C. Dragone, *Bell Syst. Tech. J.*, 58(2):501–515, February 1979; D. O. Reudink and Y. S. Yeh, *Bell Syst. Tech. J.*, 56(8): 1549–1561, October 1977.

Control systems

The major force at work in the area of control systems during the past several years has been the continuing revolution in solid-state technology. The development of inexpensive mini- and microcomputers with arithmetic and logical control capability orders of magnitude beyond that obtainable with analog and discrete digital control elements has resulted in the rapid substitution of conventional control systems by computer-based ones. At the same time, the inexpensive nature of microcomputer-based control systems has opened up major new commercial markets; hence, the more traditional applications of large computer systems to the control of satellite launch systems, chemical and petroleum processing plants, and manufacturing facilities have been joined very recently by applications in the laboratory instrumentation and consumer products areas. The use of digital computer systems will become familiar as manufacturers incorporate microcomputer control systems in automobile ignition systems, microwave ovens and other kitchen appliances, sewing machines, stereo music systems, television sets, and the like.

Generalized digital control system. The object that is controlled is usually called a device or, more inclusively, process. A characteristic of any digital control system is the need for a process interface to mate the digital computer and process, to permit them to pass information back and forth (Fig. 1).

Digital control information. Measurements of

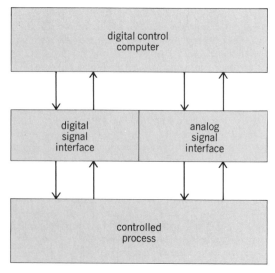

Fig. 1. Generalized digital computer control system.

the state of the process often are obtained naturally as one of two switch states; for example, a part to be machined is in position (or not), or a temperature is above (or below) the desired temperature. Control signals sent to the process often are expressed as one of two states as well; for example, a motor is turned on (or off), or a valve is opened (or closed). Such binary information can be communicated naturally to and from the computer, where it is manipulated in binary form. For this reason the binary or digital computer/process interface usually is quite simple: a set of signal-conditioning circuits for each measured or controlled signal and a set of registers to transfer the bits of digital information in each direction. Each register usually would contain the same number of bits as would be manipulated and stored within the digital computer.

Analog control information. Process information also must be dealt with in analog form; for example, a variable such as temperature can take on any value within its measured range or, looked at conceptually, it can be measured to any number of significant figures by a suitable instrument. Furthermore, analog variables generally change continuously in time. Digital computers are not suited to handle arbitrarily precise or continuously changing information; hence, analog process signals must be reduced to a digital representation (discretized), both in terms of magnitude and in time, to put them into a useful digital form.

The magnitude discretization problem most often is handled by transducing and scaling each measured variable to a common range, then using a single conversion device—the analog-to-digital converter (ADC)—to put the measured value into digital form. An ADC suitable for measurement and control purposes typically will convert signals in the range −10 to +10 volts direct current, yielding an output with 12 bits of accuracy in 10 to 50 μs. A multiplexer often is used to allow a number of analog inputs to be switched into a single ADC for conversion. High-level signals (on the order of volts) can be switched by a solid-state multiplexer; low-level signals (on the order of millivolts, from strain gages or thermocouples) would require mechanical relays followed by an amplifier to boost the signal to an acceptable input level for the ADC. Microcomputers are now available which contain integral analog conversion circuitry for several channels on the processor chip.

Discretization in time requires the computer to sample the signal periodically, storing the results in memory. This sequence of discrete values yields a "staircase" approximation to the original signal (Fig. 2), on which control of the process must be based. Obviously, the accuracy of the representation can be improved by sampling more often, and many digital systems simply have incorporated traditional analog control algorithms along with rapid sampling. However, newer control techniques make fundamental use of the discrete nature of computer input and output signals. Analog outputs from a computer most often are obtained from a digital-to-analog converter (DAC), a device which accepts a digital output from the computer, converts it to a voltage in several microseconds, and latches (holds) the value until the next output

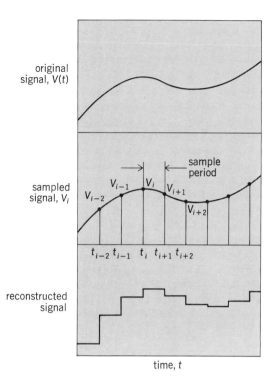

original signal, $V(t)$

sampled signal, V_i

sample period

V_{i-2} V_{i-1} V_i V_{i+1} V_{i+2}

t_{i-2} t_{i-1} t_i t_{i+1} t_{i+2}

reconstructed signal

time, t

Fig. 2. Discretization in time of an analog signal.

is converted. Usually a single DAC is used for each output signal.

Real-time computing. In order to be used as the heart of a control system, a digital computer must be capable of operating in real time. Except for very simple microcomputer applications, this feature implies that the machine must be capable of handling interrupts, that is, inputs to the computer's internal control unit which, on change of state, cause the computer to stop executing some section of program code and begin executing some other section. Using its extraordinary computational abilities (on the order of a million instructions per second), the computer, which basically is a sequential or serial processor, can be made to appear to perform operations in parallel by proper design of its hardware and executive software. By attaching a very accurate oscillator (a so-called real-time clock) to the interrupt line, the computer can be programmed to keep track of the passage of real time and, consequently, to schedule process sampling and control calculations on a periodic basis.

The requirements of real-time computing also imply that the computer must respond to interrupts from the process. Thus a key process variable may be used to trigger an interrupt when it exceeds preset limits. The computer might be programmed to service such an interrupt immediately, taking whatever control action is necessary to bring the variable back within limits. The ability to initiate operations on schedule and to respond to process interrupts in a timely fashion is the very basis of real-time computing; this feature must be available in any digital control system.

Programming considerations. Much of the programming of computer control systems now is done in a high-level language such as BASIC or

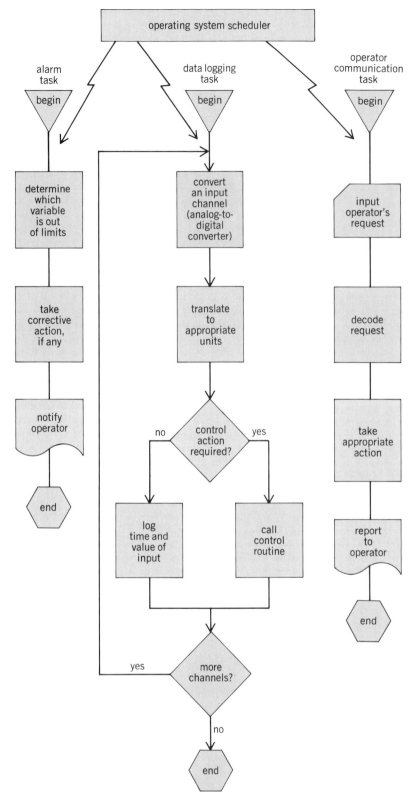

Fig. 3. Example of a multitasked control program.

of a particular digital input register so as to check the status of some process digital element, or write out a digital result to a particular DAC channel. Additionally, the programmer will have the capability to schedule operations periodically or at particular times.

Computer control systems for large or complex processes may involve complicated programs with many thousands of computer instructions. Several routes have been taken to mitigate the difficulty of programming control computers. One approach is to develop a single program which utlizes data supplied by the user to specify both the actions to be performed on the individual process elements and the schedule to be followed. Such an executive program, supplied by a control system vendor, might be utilized by a variety of users; and this approach often is taken for relatively standardized operations such as machining and sequential processing (manufacturing).

Another approach is to develop a rather sophisticated operating system to supervise the execution of user programs, scheduling individual program elements for execution as specified by the user or needed by the process. The multiple program elements, called tasks when used with a multitasking operating system, or called programs with a multiprogramming system, will be scheduled individually for execution by the operating system as specified by the user (programmer) or needed by the process. A simple example is given in Fig. 3 for three tasks—a data logging task which might be scheduled every 30 s, a process alarm task which would be executed whenever a process variable exceeds limits and causes an interrupt, and an operator communication task which would be executed whenever the operator strikes a key on the console. The operating system scheduler would resolve conflicts, caused by two or three tasks needing to execute simultaneously, on the basis of user-supplied priorities; presumably the alarm task would have the highest priority here.

Control algorithms. Many applications, particularly machining, manufacturing, and batch processing, involve large or complex operating schedules. Invariably, these can be broken down into simple logical sequences, for example, in a highspeed bottling operation the bottle must be in position before the filling line is opened. Hence the control program reduces to a set of interlocked sequential operations. Some applications—in the chemical process industries, in power generation, and in aerospace areas—require the use of traditional automatic control algorithms. Such algorithms as the lead-lag network and proportional-integral-derivative controller can be implemented relatively easily, using computers.

Recent developments involving digital control algorithms have moved away from modified analog algorithms toward new ones which exploit the advantages of the digital medium. For example, digital control techniques can account more naturally for process time-delay elements than can analog techniques; hence so-called minimum-settling-time algorithms have been a natural development. The design of digital control systems has prompted the development of discrete techniques which parallel earlier developments for continuous systems. For

FORTRAN; however, many microcomputer applications are carried out in machine (assembly) language. In either case the programmer must utilize program routines which access the devices in the process interface, for example, fetch the contents example, discrete algebraic equations substitute for differential equations used traditionally to describe processes and controllers; and discrete transform techniques (the Z transform) often replace the Laplace transform approach used in analyzing continuous systems.

For background information *see* ANALOG-TO-DIGITAL CONVERTER; CONTROL SYSTEMS; DIGITAL COMPUTER; DIGITAL-TO-ANALOG CONVERTER; INTEGRATED CIRCUITS; PROCESS CONTROL in the McGraw-Hill Encyclopedia of Science and Technology.   [DUNCAN A. MELLICHAMP]

Bibliography: D. M. Auslander et al., *Proc. IEEE*, 66:199–208, 1978; T. J. Harrison (ed.), *Minicomputers in Industrial Control*, Instrument Society of America, 1973; D. A. Mellichamp (ed.), *CACHE Monograph Series in Real-Time Computing*, vols. 1–8, 1977–1979.

Coordination chemistry

Tertiary phosphine ligands are molecular compounds of trivalent phosphorus which form relatively stable metal-phosphorus bonds (coordination compounds) with a wide variety of metal ions, often with a range of formal oxidation states. The metal and tertiary phosphine ligand partners are held together by dipole and induced-dipole interactions between the metal center and the lone pair of electrons on phosphorus. During the past 15 years some controversy has arisen over the relative importance of the σ- and π-bonding contributions; more recent ^{31}P nuclear magnetic resonance data and metal-phosphorus bond distances (from x-ray crystallography) suggest that the importance of π-bonding in determining equilibrium molecular parameters may have been overemphasized in the past.

A large number of tertiary alkylphosphines and arylphosphines (for example, R_3P) can now be synthesized by treating a Grignard or alkyllithium reagent with a phosphorus halide compound. Organic molecules containing more than one phosphine group (for example, bi-, tri-, tetra-, penta-, and hexaphosphines) are becoming more accessible recently as a result of new synthetic methods that utilize R. B. King's base-catalyzed and D. W. Meek's free-radical-promoted addition of a phosphorus-hydrogen bond to a carbon-carbon double bond of various vinylphosphines and allylphosphines.

The number and variety of tertiary phosphine complexes of transition metal ions that have been synthesized and characterized have accelerated rapidly during the past 20 years. Compared with neutral N- and O-donor ligands, the electronic versatility of tertiary phosphine ligands appears to be reflected in kinetically stable complexes with formal oxidation states of the metal ranging from +4 to −1; this feature is often attributed to the metal-phosphorus bond having both σ- and π-components, the relative importance of which may be qualitatively anticipated from the electroneutrality principle.

Polyphosphine ligands. The steric versatility of tertiary phosphines derives from the purposeful variation of the substituent groups on phosphorus and from the synthesis of polydentate phosphines, both of which can provide for a definition of par-

ticular stereochemistries and electronic configurations of complexes so that (among other things) coordinatively unsaturated, reactive species of interest in homogeneous catalytic reactions can be isolated. In recent years, transition-metal complexes of tertiary phosphine ligands have been used in many important and diverse chemical reactions such as: homogeneous hydroformylation, oxidation, hydrogenation, and hydrosilylation (the addition of a Si-H bond to a carbon-carbon double bond of an olefin) of olefins; oxidative addition to the metal center; stabilization of small elemental fragments (for example, S_2, Se_2, P_3, and As_3); and asymmetric synthesis via optically active metal-phosphine catalysts. It is expected that the use of polyphosphine ligands will become increasingly important in such future studies, since the special properties of phosphine ligands can be accentuated by a chelating polyphosphine; also, the steric and electronic properties of the ligands can be altered relatively easily by changing the organic substituents on phosphorus.

Many studies have shown that the structure, stability, and magnetic properties of the transition-metal complexes of such chelating ligands are sensitive functions of both electronic and steric effects. Compared to a monodentate phosphine, a polyphosphine ligand can provide simultaneously: more control on the coordination number, stoichiometry, and stereochemistry of the resulting complex; an increased basicity (or nucleophilicity) at the metal; and detailed structural and bonding information in the form of metal-phosphorus and phosphorus-phosphorus coupling constants. For example, polydentate ligands may be used to change the magnetic states and coordination geometries of complexes by judiciously selecting parameters such as types of donor-atom sets, "chelate bite angle," and sterically demanding substituent groups. Flexible "tripodlike" ligands (for example, structures I–IV), which contain either ethylene or

trimethylene linkages, have been used to form complexes with different coordination numbers (4, 5, and 6) and different structures (trigonal-pyramidal, distorted tetrahedral, planar, square-pyramidal, trigonal-bipyramidal, and octahedral).

Coordination geometries. The coordination stereochemistry for a given donor set (for example, P_3X_2) around a metal ion is sensitive to the length of the connecting chains of a polyphosphine ligand. The flexible ligands $PhP(CH_2CH_2PPh_2)_2$

(etp) and $PhP(CH_2CH_2CH_2PPh_2)_2$ (ttp) differ by only one methylene group in the connecting chains; however, the structures of the five-coordinate cations $\{Co(etp)[P(OMe)_3]_2\}^+$ and $\{Co(ttp)-[P(OMe)_3]CO\}^+$ are trigonal-bipyramidal structure V and square-pyramidal structure VI, respectively.

The change from square-pyramidal coordination geometry in $\{Co(ttp)[P(OMe)_3]CO\}^+$ to trigonal-bipyramidal in $\{Co(etp)[P(OMe)_3]_2\}^+$ can be attributed to the decrease in the chelate chain length of $PhP(CH_2CH_2PPh_2)_2$.

The trigonal symmetry of tripodlike tetradentate ligands (structures I–III) is ideal for synthesis of trigonal-bipyramidal complexes, especially of cobalt(II) and nickel(II). The magnetic ground state in such complexes is a sensitive function of the ligand-donor set. For example, $[Co(np_3)X]Y$ complexes (X = Cl or Br; Y = X, BF_4, PF_6, or BPh_4) are high-spin; the thiocyanate complexes and all iodo derivatives, except for Y = BPh_4, are low-spin. In contrast, all the $[Ni(np_3)X]Y$ complexes are low-spin. If the tripod ligand is changed to the tetraphosphine $P(CH_2CH_2PPh_2)_3$ (structure III), then all of the Co(II) and Ni(II) complexes are low-spin.

Stabilization of unusual metal oxidation states can be achieved relatively easily with phosphine ligands. For example, interesting tetrahedral nickel(I) and coablt(I), complexes of general formula $[M(p_3)X]$, where $p_3 = CH_3C(CH_2PPh_2)_3$ (structure IV) and X = halide, are obtained when the metal salts are reduced with $NaBH_4$ in the presence of $CH_3C(CH_2PPh_2)_3$. In the absence of a coordinating anion, the tripodlike symmetry of this triphosphine ligand leads to binuclear hydride complexes of the type shown in structure VII, where M = Fe(II) or Co(II). Very recently, the triphosphine ligand $CH_3C(CH_2PPh_2)_3$ has been used to stabilize unusual As_3 and P_3 fragments between two metal atoms, that is, the P_3 or As_3 unit is located in between two CoP_3 units like the middle of a sandwich (structure VIII).

Asymmetric synthesis. An obvious growth area for phosphine-metal catalysts is that of asymmetric synthesis. Impressive results have been obtained recently with complexes of *P,P'*-bis-(*o*-methoxyphenyl)-*P,P'*-diphenylethylenediphosphine (structure IX), a ligand which is optically active at both phosphorus atoms. The catalyst is uniquely effective in the homogeneous hydrogenation of α-acylaminoacrylic acids. High optical purities can be obtained also with catalysts which are not chiral at phosphorus but have a chiral substituent or become rigid and chiral as a result of chelation around the metal; examples are rhodium(I) complexes of the diphosphines 4,5-bis(diphenylphosphinomethyl)-2,2-dimethyl-1,3-dioxolan-4,5-diol (DIOP; structure X), and (*S*,*S*)-chiraphos (structure XI), which are useful for asymmetric hydrogenation reactions.

For background information *see* COORDINATION CHEMISTRY in the McGraw-Hill Encyclopedia of Science and Technology.

[DEVON W. MEEK]

Bibliography: M. Di Vaira et al., *Angew. Chem. Int. Ed.*, 17:676–677, 1978; D. L. DuBois, W. H. Myers, and D. W. Meek, *J.C.S. Dalton Trans.*, pp. 1011–1015, 1975; M. D. Fryzuk and B. Bosnick, *J. Amer. Chem. Soc.*, 101:3043–3049, 1979; R. B. King, *Acc. Chem. Res.*, 5:177–185, 1972; R. Mason and D. W. Meek, *Angrew. Chem. Int. Ed.*, 17:183–194, 1978.

Corrosion

The corrosion phenomenon involves an interaction between a metal or alloy and its environment. The interactions are generally undesired chemical reactions which can lead to wastage and can alter the structural integrity of the materials. Corrosion frequently degrades the material to the point where it no longer can meet the functions for which it was selected and designed. Such conditions lead to failure of the part or system. The problem of corrosion was brought into focus in a recent study by the National Bureau of Standards of the U.S. Department of Commerce. The analysis indicated that metallic corrosion costs the United States about $70,000,000,000 each year. The study also suggested that a good share of this corrosion cost could be avoided "by making the most economical use of presently available corrosion technology."

High-temperature corrosion. In recent years, high-temperature corrosion has been recognized as a severe problem in the development of advanced energy systems, especially those designed for the utilization of fossil fuels. This type of corrosion involves a number of fundamental processes such as oxidation, sulfidation, carburization, and hot corrosion. Alloy selection for high-temperature applications is based on the strength of the material as well as its resistance to complex environments which are usually both corrosive and erosive. It has been well established that to be viable in such environments alloys should develop protective oxide scales on the surface upon exposure. For this reason, most high-temperature alloys have been designed to develop chromium or aluminum oxide scales. In general, the effects of alloy chemistry and microstructure on the behavior of corrosion-resistant alloys are reasonably well understood. The minimum level of chromium required for chromium oxide formation ranges from about 18 to 20 wt % for Fe-Cr and Ni-Cr alloys, and from 25 to 30 wt % for Co-Cr alloys. These levels are dependent on the temperature and oxygen partial pressure in the service environment, and are also proportionately reduced by the presence of additional oxide-forming elements, such as aluminum, in the alloy.

Deposit-induced corrosion. Deposit-induced corrosion of materials is a potential problem in environments that contain a high sulfur content and contaminants such as alkali metals, vanadates, and chlorides. Salt deposits on the surfaces of alloys usually increase corrosion and metal wastage. It is generally accepted that the mechanism of hot corrosion involves the formation of a liquid phase that consists mainly of sodium and potassium sulfates. The liquid phase dissolves the protective oxide layer, which allows the base metal to react with sulfate ions to form sulfide ions and nonprotective oxides. The deterioration of the surface layer results in additional penetration of the sulfur into the metal. Control of hot corrosion has been attempted via (a) the development of alloys that form renewable protective scales, (b) the formation of corrosion-product layers that have a melting temperature considerably above the temperature of the service environment, (c) application of thin layers of protective coatings, and (d) use of hot gas cleanup systems to reduce the sulfur and alkali concentrations in the environment.

Erosion. The erosion phenomenon is the result of impact, cutting action, or frictional wear produced by small solid particles, entrained in a fluid medium, moving freely at an angle to the direction of fluid flow and frequently undercutting portions of the material they strike. Although estimates of the annual cost of erosion losses are not available, such costs are expected to be significant. For example, the replacement of erosion-damaged helicopter turbines in southeastern Asia cost the U.S. government $150,000,000 per year. In general, the erosion process and the extent of damage to materials are influenced by a number of factors such as angle of impingement; particle size, shape, strength, velocity, and rotation at impingement; particle loading in the fluid stream; properties of the target surface; and process conditions. Erosion

has been identified as a serious problem in many branches of industry, causing rapid and often life-limiting damage to components such as gas turbines, steam turbines, rocket nozzles, cyclone separators, valves, pumps, transfer lines, and boiler tubes.

A number of investigators have attributed high-velocity particulate erosion of material surfaces to absorption of a significant portion of the kinetic energy of each impacting particle by the target material. Mathematical modeling of erosion behavior suggests that for normal impacts on ductile metals erosion rates are higher for materials with lower ultimate strength and elastic modulus. For impacts at more acute angles to the surface, the frictional force between the impacting particle and the surface plays a role in the erosion behavior. Measures taken to reduce erosive wear have often involved attempts to reduce the particle size and loading, and to decrease the velocity of the erosive stream. In recent years, the emphasis has been on modifications in the design of components and in the flow paths of the fluid stream to minimize the chances of particle impact on the material surfaces. For example, a new valve concept that involves a spiral letdown channel of considerable length for gradual pressure reduction is being developed for coal liquefaction service. The use of blocked-tee bends in the design of piping has resulted in significant reduction in erosive wear loss. Wear-resistant materials compatible with the service environment are also being used to minimize erosion. These materials are generally alloys that contain a fine distribution of hard particles in a tough matrix. Typical examples are tungsten carbide composites, high-chromium irons, and cobalt-

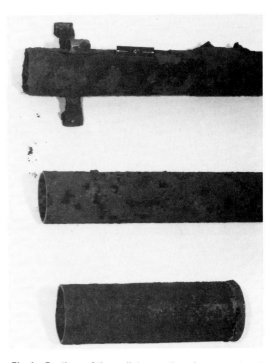

Fig. 1. Portions of three dipleg sections from a coal gasification plant. *(From S. Danyluk and S. Greenberg, Argonne National Laboratory Report, ANL/MSD/FE-77-3, 1976)*

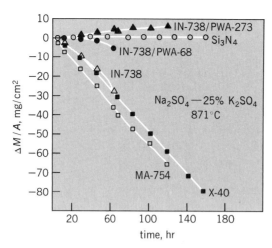

Fig. 2. Weight change versus time data for baseline hot-corrosion test. Approach velocity of gas was about 200 m/s; salt deposition rate was about 0.05 mg/cm²/hr. *(From R. H. Barkalow, J. A. Goebel, and F. S. Pettit, DOE/EPRI/GRI/NBS Conference Proceedings, CONF-781018, p. V-38, 1978)*

and nickel-base alloys. The wear-resistant material is usually used as an overlay by welding or spraying onto an engineering material.

Erosion-corrosion. A combination of corrosion and erosion occurs in a number of energy generation systems, and a priori predictions of the interaction between the two modes of material degradation are difficult. The relative contributions of the mechanical and chemical components vary widely with process conditions such as temperature, pressure, hydrodynamic factors, and chemistry of the environment. For example, the erosion-corrosion process leads to purely mechanical damage in the case of raindrops falling on a supersonic aircraft; on the other hand, the process leads to mechanically accelerated corrosion in the case of particle impingement attack on heat-exchanger tubes,

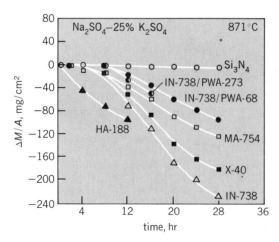

Fig. 3. Weight change versus time data for erosion–hot corrosion with 2-μm Al₂O₃ abrasive. Test conditions were the same as those of Fig. 2 except for the presence of abrasive powder. *(From R. H. Barkalow, J. A. Goebel, and F. S. Pettit, DOE/EPRI/GRI/NBS Conference Proceedings, CONF-781018, p. V-38, 1978)*

where a major effect of erosion is the degradation of protective films on the alloy surface.

The combined effect of corrosion and erosion on material behavior is of considerable concern in the effective use of structural materials in coal utilization schemes. Most parts of coal-processing systems are subject to erosion from crushed coal particles or the ash and insoluble particles in coal-derived slurries or gas mixtures. Erosion is the major contributor to the degradation of materials used for slurry pumps and pipelines. For these applications, corrosion resistance of the alloys is of minor concern since the slurries are normally handled under neutral pH conditions at relatively low temperatures. In such cases, the wear caused by the interaction between abrasive particles and the component material is effectively minimized by the use of hard-matrix alloys such as HC-250 and Ni-Hards. A data bank of erosion information on various materials, reported in a Bureau of Mines document, can be used as a guide for the selection and application of erosion-resistant materials for coal gasifier valves.

At elevated temperatures, both the corrosion and erosion processes play a definite role in the performance of engineering materials. Figure 1 shows an Incoloy 800 dipleg that was exposed to a corrosion-erosion environment in a coal gasification plant. The component, after an approximately 7.2-megasecond exposure at a temperature of about 1010°C, exhibited a thick corrosion-product layer (predominantly sulfides of Fe, Cr, and Ni) and a significant reduction in wall thickness. In this example, the relative contributions of corrosion and erosion to the material degradation are difficult to assess. However, it is fairly well established that the growth rate of the brittle sulfide layer developed by the corrosion process follows a parabolic rate law; the observed wall thinning can be explained only by the erosion of the brittle scale by the impinging particles, thereby exposing base metal which undergoes additional corrosion.

Research programs. Separate and combined effects of aggressive erodent and corrosive-gas atmospheres on material behavior are being evaluated in programs funded by the Department of Energy. The preliminary results have shown that the effects of high-velocity erodent and high-velocity corrosive gas are not additive; rather, their combined effect is synergistic.

Research is in progress to evaluate the behavior of materials simultaneously exposed to hot corrosion and high-velocity particle impact. The erosion–hot-corrosion behavior of advanced turbine materials (such as uncoated IN-738, X-40, MA-754, and HA-188; and Si₃N₄, aluminide, and CoCrAlY coatings on IN-738) is being evaluated under conditions that simulate coal gasification and coal combustion systems. Weight losses of several test materials under hot corrosion and erosion–hot-corrosion test conditions are shown in Figs. 2 and 3, respectively. It is evident that a particle loading of 300 ppm of 2-μm Al₂O₃ produced a significant increase in weight loss when compared with baseline hot-corrosion conditions.

A number of research programs are being conducted to develop a fundamental understanding of the synergistic effect of corrosion and erosion on material behavior. Because of the complexity of the processes and the large number of variables that play a role in material degradation, additional concerted effort by researchers will be required to quantify the effects of these processes on material performance.

For background information *see* ALLOY; CORROSION in the McGraw-Hill Encyclopedia of Science and Technology. [K. NATESAN]

Bibliography: *Bureau of Mines Report of Investigations*, RI 8335, 1979; B. R. Cooper (ed.), *Proceedings of Conference on Scientific Problems of Coal Utilization*, U.S. Department of Energy, 1978; *High Temperature Erosion/Corrosion of Alloys*, EPRI FP-557, April 1978; *Metal Properties Council Report*, FE-1784-48, 1979; *Proceedings of a Conference on Corrosion/Erosion of Coal Conversion System Materials*, NACE, 1979; *Proceedings of the 3d Annual Conference on Material for Coal Conversion and Utilization*, CONF-781018, October 1978.

Cosmic rays

At the present time, cosmic rays are studied primarily in relation to astrophysics. The matter that makes up cosmic rays is in an extraordinary state of motion; the energy per nucleus is typically 10^{10} eV, and in rare instances is more than 10^{20} eV. Cosmic-ray studies are aimed at locating the sources and discovering how the acceleration to

such high energies comes about. Such studies also give information about the interstellar medium, which alters cosmic-ray composition during propagation from the sources to the solar system.

Highest-energy cosmic rays. There is little hope for direct evidence of where the bulk of cosmic rays originate; their paths through interstellar space are too strongly curved by the action of magnetic fields. However, results obtained with giant air shower arrays at Volcano Ranch in New Mexico, Haverah Park in England, and Yakutsk in Siberia have shown that the highest-energy cosmic rays are anisotropic.

An example of the data is given in Fig. 1, which shows the arrival directions of individual primary cosmic rays in celestial coordinates: declination and right ascension. It is believed that cosmic rays in this energy range, above $3 \cdot 10^{19}$ eV, originate in galaxies that are considerably more active than the Milky Way Galaxy. Their magnetic rigidity is so great that the directions shown have hardly been affected by the fields belonging to the Galaxy.

Observations like these are made possible by the fact that cosmic rays entering the Earth's atmosphere generate showers of secondary particles. When the primary energy is greater than about 10^{15} eV, the number of shower particles reaching sea level is great enough that the primaries can be detected with high efficiency over a large area by means of an array of scintillation

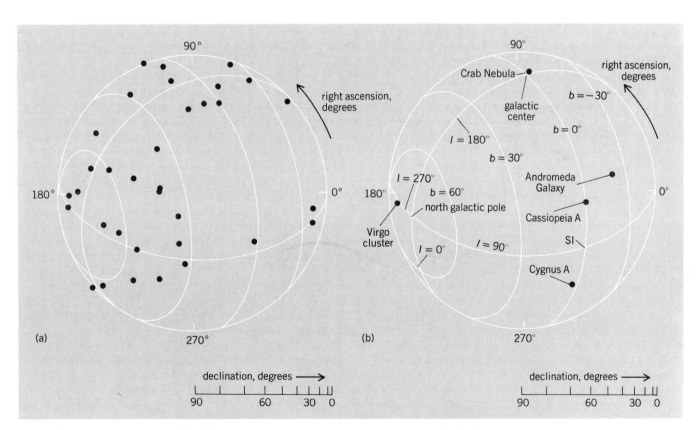

Fig. 1. Data showing anisotropy of the highest-energy cosmic rays. (*a*) Arrival directions (black dots) of individual cosmic rays with energies over 3.2×10^{19} eV, observed at Haverah Park, England, in celestial coordinates (right ascension and declination). The declination scale has been adjusted so that each 10° band represents an equal exposure time. Contour lines correspond to galactic coordinates. (*b*) Key to galactic coordinates (*l* and *b*), together with locations of interesting astronomical objects. Direction along which galactic spiral arm spirals in toward galactic center is designated SI.

counters or the like. The number of shower particles, which is roughly proportional to the energy of the primary cosmic ray, is found from the amplitudes of the counter signals. The direction of the shower, which is the same as the direction of the primary, is found from the time delays between the signals.

The intensity of extragalactic cosmic rays is so small that a counter array for detecting them must cover an area of 10 km² or more. Only four such giant arrays have been built, the three mentioned above and one in the Southern Hemisphere near Sydney, Australia. Even with arrays of that size, gathering information about extragalactic cosmic rays is painfully slow work. It required almost a decade to collect the 32 events shown in Fig. 1.

Less energetic cosmic rays. The giant arrays are also used to study less energetic (and correspondingly more abundant) cosmic rays, down to 10^{16}–10^{17} eV. These cosmic rays are thought to be produced by sources belonging to the Galaxy, probably the same sources that produce the typical 10^{10}-eV cosmic rays referred to earlier. It is found that the highest-energy galactic cosmic rays are also anisotropic. The anisotropy has been detected by using muons deep underground produced by 10^{11}–10^{12}-eV primaries, and has been tracked upward in energy, through a region dealt with by means of small air shower arrays, to the range covered by using giant arrays. At the lower energies the anisotropy is quite small, less than 0.1%. It increases with increasing energy, and there is a progressive change in the direction of maximum intensity. These features are understood by supposing that cosmic rays tend to stream away from the denser portions of the Galaxy along the spiral arms, and they also tend to diffuse radially outward toward the surface of the arms. There is evidence that the intensity pattern becomes complex above 10^{17} eV, and that the pattern then changes rapidly with increasing energy.

Fig. 2. One element of the Fly's Eye air shower detector, showing mirror and cluster of photomultiplier tubes in front of it.

However, the numerous features that appear to be present cannot yet be made out distinctly because the cosmic-ray intensity is so low in relation to the collecting power of the receivers. Furthermore, interpretation of the evidence is hampered by a lack of information about the composition of the primaries. While it is reasonably certain that they are atomic nuclei, as at lower energies, it is not yet possible to tell whether a given air shower was initiated by a light nucleus, such as a proton, or a heavier one with a correspondingly greater charge. Consequently, there is a large uncertainty in regard to magnetic rigidity, which is the crucial parameter for interpreting arrival directions.

Fly's Eye. It is anticipated that a new device being developed at the University of Utah will have 10 times the collecting power of all the giant arrays combined. Called the Fly's Eye, it is also expected to play an important role in settling the question of cosmic-ray composition at the highest energies.

The Fly's Eye will consist of sixty-seven 1.5-m-diameter mirrors pointed in different directions so as to view the entire sky. Mounted in the focal plane of each mirror will be a cluster of photomultiplier tubes (Fig. 2), each one viewing a different narrow cone. A preliminary test carried out at the Volcano Ranch array, using only three mirror units, proved that large air showers are "visible" with such a system, at night, by virtue of the scintillation light they produce in the atmosphere. Calculations based on that test indicate that 10^{20}-eV air showers will be detectable at ranges up to 50 km. The complete Fly's Eye is being set up at an isolated location at Little Granite Mountain on the Utah salt flats some distance from Salt Lake City. Tests made there early in 1979, when only sixteen mirror units had been installed, show that the time relations between signals from the line of phototubes activated by a given shower allow determination of the distance and orientation of the shower trajectory. The amplitudes of the signals provide a profile of the shower development, showing the growth in number of particles until a maximum is attained, followed by a decline. According to theory, showers produced by heavy primaries will begin their development at higher elevations and will develop more rapidly than proton-initiated showers of the same energy. Shower profiles registered by the Fly's Eye will provide a favorable basis for applying that theory in order to distinguish between highest-energy cosmic rays according to their mass.

For background information *see* COSMIC RAYS; GALAXY, EXTERNAL in the McGraw-Hill Encyclopedia of Science and Technology.

[JOHN LINSLEY]

Bibliography: D. M. Edge et al., *J. Phys., G: Nucl. Phys.*, 4:133–155, 1977; J. Linsley, *Sci. Amer.*, 239(1):60–70, 1978; H. E. Bergeson et al., *Phys. Rev. Lett.*, 39:847–849, 1977.

Cosmology

Two of the most important questions in cosmology are how the structure of the universe arose and what the eventual fate of the universe will be. One way of investigating these questions is through the study of the cosmic microwave radiation. Recent observations have confirmed the blackbody shape of the radiation spectrum, while isotropy measurements of the radiation have given important information about both the large-scale structure of the universe and the velocity of the Milky Way Galaxy. A quite different way of studying the evolution and structure of the universe, which has recently been developed, is through computer simulations of the motions of large numbers of galaxies. In such simulations the galaxies form clusters and superclusters remarkably like those actually observed.

Cosmic microwave radiation. As of 1979, it was 13 years since the discovery of cosmic microwave radiation by A. Penzias and R. Wilson of Bell Laboratories, but it has been only recently that the origin of this radiation as the remnant fireball from the "big bang" has become nearly unanimously accepted by astrophysicists. The clear observation of the peak of the blackbody spectrum and its high-frequency tail has answered the doubts of many skeptics. The correct prediction of the entire spectrum by P. J. E. Peebles, R. H. Dicke, R. G. Roll, and D.T. Wilkinson at Princeton, after the observation of the intensity at only one frequency, must certainly be considered one of the great triumphs of cosmological theory. Not only is cosmic microwave radiation the strongest evidence available for a hot, compressed early universe, but is provides one of the most direct means for studying the large-scale structure of that universe. One of the surprising results from the careful observation of this radiation is the detection of an unexpectedly high velocity for the Milky Way Galaxy through the universe.

Origin of radiation. In the big-bang model of the universe, as originally detailed by G. Gamow, H. Alpher, and R. Hermann, the universe began with an explosion about 15×10^9 years ago. A half million years after its beginning, the universe was still filled with a hot plasma of photons, electrons, protons, and heavier atoms. But at about this time the expansion of the universe cooled the plasma to the point that the electrons and protons combined

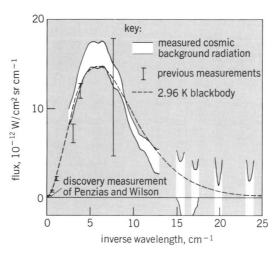

Fig. 2. Measured spectrum of cosmic background radiation. One-standard-deviation error limits are shown for the measured cosmic background and for previous measurements. Frequency resolution of the measurements = 1 cm⁻¹.

to form neutral hydrogen. The previously opaque universe suddenly became transparent, since photons are much less strongly absorbed by neutral hydrogen than by plasma. From that period until now, the cosmic blackbody photons have been traveling virtually unscattered, carrying information about the nature of the universe at the time of the decoupling (Fig. 1). To an observer moving with the plasma, the photons have a blackbody spectrum with a characteristic temperature of about 4500 K; however, due to the Hubble expansion the plasma has a recessional velocity with respect to the Earth large enough to redshift the radiation by a factor of 10^3 into the microwave region, with a characteristic temperature of 3 K. When one measures the radiation, one is really "looking at" the shell of matter which last scattered the radiation.

This microwave radiation is coming from the most distant region of space ever observed, and was emitted earlier in time than any other cosmological signal that has been received. Although "background" radiation was originally given that name because of its potential interference with satellite communications, the word has taken on a new and vivid meaning: the radiating shell of matter forms the spatial background in front of which all other astrophysical objects, such as quasars, lie. Until methods are devised to detect the neutrinos or gravitons which decoupled earlier, there will be no direct means of viewing beyond this background.

Spectrum measurements. There have been more than 20 articles published on measurements of the spectrum of the radiation. However, only in the last few years have accurate measurements truly verified the predicted high-frequency fall-off, or "tail," of the spectrum. Particularly important are the recent results of D. Woody and P. L. Richards at Berkeley. These are shown in Fig. 2, where one-standard-deviation error limits on their measurements and on previous observations are indicated. The gaps in the data are at frequencies at which strong atmospheric emission lines cause the errors

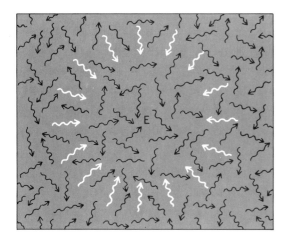

Fig. 1. Origin of cosmic background radiation. Arrows represent photons at a time when the universe became transparent, about a half million years after the big bang. Those photons which reached Earth in 1978 are indicated by the white arrows; the future position of Earth is indicated by letter E.

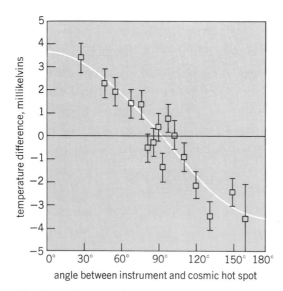

temperature difference, millikelvins

angle between instrument and cosmic hot spot

Fig. 3. Anisotropy of cosmic background radiation. Points and bars represent measured data; the curve gives the cosine fit.

to become very large. The spectrum of a blackbody with a temperature of 2.96 K is shown for comparison. As can be seen in the figure, most of the energy of the background radiation is in the range of frequencies measured in the Woody-Richards experiment, a region which had few previous measurements because of interference by intense atmospheric emission. Woody and Richards sent their experiment above 99.8% of the atmosphere, using a spectrometer flown in a balloon at an altitidue of about 140,000 ft (43 km).

The striking confirmation of the characteristic blackbody shape of the spectrum is the most important result of the Woody-Richards experiment. However, these researchers do see a statistically significant deviation from that shape: the observed spectrum is slightly high in the region of 3 to 7 cm^{-1} and low from 9 to 12 cm^{-1}. The probability of a random fluctuation accounting for this effect is about 10^{-6}. Great care was taken by the experimenters to determine if the deviation could be false and due to a defect in their apparatus; such care is important since an overall reduction in signal by 22% brings the data into good agreement with a 2.8-K blackbody curve. They found no explanation other than the cosmological one. If the deviation from the blackbody curve *is* cosmological in origin, its interpretation should significantly affect understanding of the universe.

Isotropy measurements. The large-angular-scale isotropy of the cosmic background radiation (that is, the uniformity of its intensity in different directions in the sky) is the most sensitive available probe of several phenomena of cosmological interest. These include the uniformity of the Hubble expansion, possible rotation of the universe, and the existence of very-long-wavelength gravitational radiation. Anisotropy can also arise from nonuniform distribution of matter at the time of the decoupling.

Anisotropy has now been clearly observed by groups at Princeton and Berkeley; however, its smooth "cosine" behavior suggests that it should be attributed to a local rather than a cosmological

effect: the motion of the Earth relative to the radiation. Except for the cosine term, there is no statistically significant anisotropy. The most sensitive experiment so far is that by M. Gorenstein, R. Muller, and G. Smoot at Berkeley using a 33-GHz twin-horn Dicke radiometer carried in a series of flights aboard the NASA-Ames Earth Survey Aircraft (U-2). In this experiment the measured upper limit to the existence of a noncosine component to the background anisotrophy is about 4 parts in 10,000 of a 2.8-K signal. This limit puts strong constraints on the cosmological phenomena mentioned above.

Perhaps the most interesting result, however, is the value for the local velocity of the Earth obtained from the cosine component. From this value one can also infer the velocity of the Milky Way Galaxy and the nearby galaxies (the "local group") which accompany it through space. The data for seven U-2 flights between April 1977 and May 1978 are shown in Fig. 3, along with the cosine fit. Each data point is the average of approximately five 20-min "legs" of the U-2 aircraft at an altitude of 65,000 ft (20 km). Anisotropy, expressed as a temperature difference in millikelvins from the average value, is plotted against the angle between the instrument axis and the observed direction of maximum temperature. The amplitude of the cosine is $3.61 \pm 0.54 \times 10^{-3}$ K, in a direction of R.A. (right ascension) $= 11.23 \pm 0.46$ hr, declination $19 \pm 7.5°$. For a cosmic background temperature of T, this implies a velocity for the Earth of $361 \times (3 \text{ K}/T)$ km/s. For $T = 2.8$ K, one finds $v = 390$ km/s. When this is folded with the known velocity of the Sun relative to the Milky Way Galaxy (from galactic rotation), one finds a net velocity for the local group of galaxies of about 550 ± 75 km/s or about 1,200,000 mph in the direction R.A. $= 10.5$ hr, declination $= -11°$, just south of the constellation Leo.

The large magnitude of this velocity is particularly surprising, since velocities for nearby galaxies relative to the Milky Way Galaxy are small. For example, the relative velocity of the Milky Way and Andromeda galaxies is only 80 km/s, and Peebles has concluded that the peculiar (non-Hubble) velocity of the Virgo cluster is similarly small. Thus all these galaxies must be moving with the Milky Way at its high cosmological velocity.

Similarly large peculiar velocities have been observed recently in diverse areas of cosmology. V. C. Rubin and W. K. Ford reported a high velocity for the Milky Way measured with respect to distant galaxies with recessional velocities between 3500 and 6500 km/s, but they found a different direction than that found in the microwave experiments. Peebles recently reported measurements made at Princeton which show high-velocity differences in pairs of (presumably) bound galaxies. It is difficult to reconcile the high kinetic energy of such galaxies with the belief that they are gravitationally bound to each other unless one supposes that the mass of each galaxy is much greater than the visible mass. The missing mass must be comparable to that required to close the universe, that is, to curve space enough so that the total volume of the universe is finite. However, these results are so new and unexpected that their interpretation is still unclear.

Future measurements. The deviations observed in the spectrum of the background radiation are tantalizing, and an order-of-magnitude improvement in sensitivity will be necessary in order to study them in detail. In addition, one would like to study whether these deviations are the same in different directions in the sky; such a study requires a combined spectrum and isotropy experiment. The recent isotropy measurements have already yielded an unexpected discovery in the high peculiar velocity of the local group of galaxies, but as yet no anisotropy of cosmological origin has been observed. Again one would like to improve over present experiments by a factor of 10.

Fortunately just such an improvement in both the spectrum and isotropy should be possible with COBE (Cosmic Background Explorer), a satellite which may be flown by NASA as early as 1983 (Fig. 4). The tentative technical specifications for the experiments are as follows: (1) Spectrum measurement from 3.3 to 0.33 mm, based on differential comparison with an adjustable-temperature blackbody, with a sensitivity of 2×10^{-14} W/cm^2 sr cm^{-1} for each 7°-diameter field of view. This instrument should be capable of measuring the spectrum to 0.1% of the peak flux for each field of view on the sky. (2) Anisotropy measurements at four frequencies: 23.5, 31.4, 53, and 90 GHz, again with a 7° field of view. The differential microwave radiometers should be able to detect deviations from isotropy as small as 10^{-4} of the 2.8-K radiation. The use of four frequencies should allow some compensation for radio interference from the Milky Way. (3) Diffuse infrared background photometer to measure the infrared region 8–300 μm in six octave bands; sensitivity of 10^{-13} W/cm^2 sr for each 1° field of view.

Without a satellite, a factor-of-10 improvement over the existing data would be nearly impossible to obtain; with COBE one can expect to have it in the next few years. The advantages of COBE come not only from its sensitivity, but also from the complete sky coverage afforded by its polar orbit. The COBE sensitivity is such that the uncertainty in the galactic synchrotron emission should set the limit to the accuracy of its measurements. If this is the case, COBE may provide the ultimate limit to the knowledge that can be derived from measurements of the cosmic background.

[RICHARD A. MULLER]

Computer simulations of universe. Most galaxies, like the Milky Way system, are members of small groups, clusters, or superclusters. There is evidence that at earlier epochs the distribution of matter was much more uniform that it is today. The growth of structure in the universe can be simulated on a computer by following the dynamics of a system of masses which are allowed to interact through their mutual gravitational attraction. As the simulated universe expands, the distribution of masses becomes more clumpy, finally producing a hierarchical pattern of clustering quite similar to that observed.

Cosmological theory. In 1929 Hubble discovered that the universe is expanding. This expansion is similar to that of raisin bread baking in the oven, with the galaxies as raisins. As the bread expands, the distance between any two raisins increases by the same factor. As discussed above, the micro-

wave radiation left over from the big-bang explosion from which the universe originated was last scattered at an epoch when the universe was a thousand times smaller than it is today. The fact that this radiation is the same intensity in different directions to at least 1 part in 1000 shows that at that epoch the universe was much more uniform that at present. Protogalaxies began as regions where the density was perhaps a few percent higher than the mean density of the universe. Because of gravitational attraction, the excess density causes the expansion in the region to be slowed, and the density contrast between the region and the universe as a whole grows. When the universe has expanded to approximately one-thirtieth of its present size, the protogalaxies have densities about twice the mean density and can begin to be treated as separate masses. Computer simulations can model the evolution from this point to the present and trace the formation of groups and clusters of galaxies.

In the simplest general relativity models, the fate of the universe is determined by the value of $\Omega = \rho/\rho_{CRIT}$, where ρ is the current mean density in the universe and ρ_{CRIT}, which equals approximately 5×10^{-30} g cm^{-3}, is the critical density for the observed expansion rate. If $\Omega > 1$, the expansion will slow down and the universe will eventually begin to recontract, finally reaching a big crunch at the end. If $\Omega \leq 1$, the universe will continue to expand forever. In such a low-density universe, the mutual gravitational attraction of the galaxies is insufficient to ever overcome the kinetic energy of the expansion. Simulations with different values of Ω may be compared with the observations.

Observed clustering of galaxies. The Milky Way Galaxy is a member of the local group of galaxies which also includes M31, M32, M33, NGC205, the

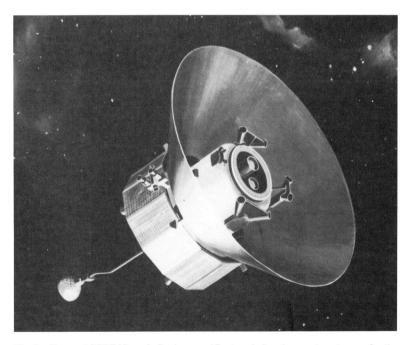

Fig. 4. Planned COBE (Cosmic Background Explorer). Apertures at center are for the 3.3- to 0.33-mm spectrum radiometer and for the diffuse infrared photometer; the four pairs of microwave horns are for the isotropy radiometers. Radiometers are shielded from the Sun and Earth by the large conical shade. *(NASA)*

Fig. 5. A 4000-body simulation with $\Omega = 0.1$ seen at an epoch 60×10^6 years after the big bang. *(From S. J. Aarseth, J. R. Gott, and E. L. Turner, N-body simulations of* galaxy clustering, I: Initial conditions and galaxy collapse times, Astrophys. J., 228:664–683, 1979)

Large and Small Magellanic Clouds, and other smaller galaxies. The mean density in this group is about a thousand times as great as the mean density in the universe. The group's diameter is approximately 0.7 megaparsec (Mpc). (For comparison, the Sun is at a distance of about 0.01 Mpc from the center of the Galaxy; 1 Mpc $\cong 3.1 \times 10^{19}$ km $\cong$ 3,200,000 light-years.) The local group appears to have a sufficient density inside it to have halted its original expansion, and it now appears to be collapsing. The local group is one of many similar groups within the local supercluster. The supercluster, centered approximately on the Virgo cluster, has a diameter of approximately 40 Mpc and is about four times as dense as the mean density in the universe. It is expanding at a rate similar to that of the universe as a whole. The mass of the local supercluster is about 1000 times that of the local group. The richest bound clusters, like the Coma cluster, are several hundred times as massive as the local group. Clusters like Coma were originally density enhancements participat-

ing in the expansion which have slowed down, reached a point of maximum expansion, collapsed, and are approaching equilibrium.

Galaxy clustering can be studied statistically by measuring the covariance function $\xi(r)$. This function measures the excess probability of finding a galaxy at a distance r from a random galaxy. Peebles found that $\xi(r) \sim 70 \, (1 \text{ Mpc}/r)^{1.8}$. This means that at a distance of 1 Mpc from a random galaxy, one finds an excess density of galaxies that is 70 times the mean density of galaxies in the universe. This formula applies from galactic sizes to the sizes of superclusters ($0.01 \text{ Mpc} < r < 10 \text{ Mpc}$).

Simulations of galaxy clustering. A typical simulation will be described with $\Omega = 0.1$ by using 4000 masses representing galaxies. The simulation starts at an epoch when the universe was a factor of 32 smaller than it is today. The masses are placed randomly in a spherical volume of radius 1.65 Mpc and started with a uniform expansion. This simulates a typical spherical volume in the universe. By a theorem in general relativity, if the

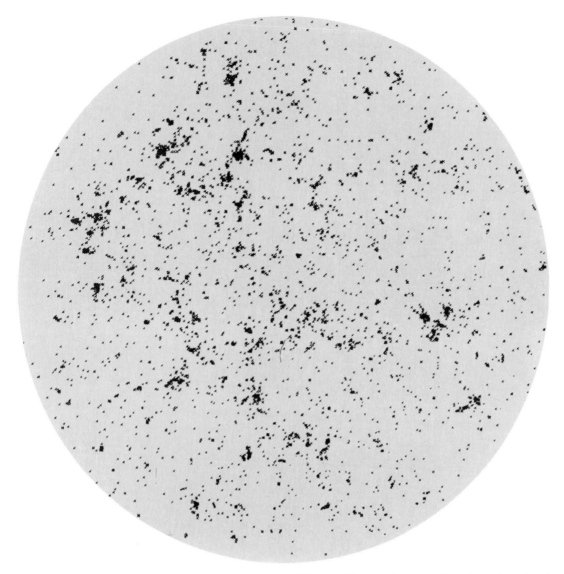

Fig. 6.   Same simulation as in Fig. 5 at an epoch 4×10^9 years after the big bang. *(From S. J. Aarseth, J. R. Gott, and E. L. Turner, N-body simulations of galaxy clustering, I: Initial conditions and galaxy collapse times, Astrophys. J., 228:664–683, 1979)*

matter outside the sphere is uniformly distributed, it will not affect the mass inside the sphere. For each mass, the gravitational acceleration due to each of the other 3999 masses is computed. After the model has expanded by a factor of approximately 2, the distribution of the masses is as indicated in Fig. 5. The universe has an age of 60×10^6 years at this point, and the sphere has a radius of 3.3 Mpc. In Fig. 5 a number of regions with slightly higher than average density have formed. Figure 6 corresponds to an age of the universe of $t = 4 \times 10^9$ years, where the sphere has expanded to a radius of 14 Mpc. Many small groups and clusters have formed. Comparison with Fig. 5 shows that the groups and clusters have formed from the regions that were the largest density enhancements.

Figure 7 shows the end point of the calculations corresponding to the present epoch with $t = 17.6 \times 10^9$ years and a radius of 50 Mpc. A number of tight groups and rich clusters have formed. The richest clusters in the simulations are similar in size to large clusters like Virgo or Coma. The clusters are

not distributed at random, but form superclusters such as the group of four large clusters in the upper-left portion of the figure. There are large regions, approximately 40 Mpc in diameter, which have average densities perhaps a factor of 2 or so over the background (for example, the supercluster in the upper-left and the region immediately below the center). These regions are apparent with similar locations and sizes in Fig. 6. This shows that these regions of small density enhancement are expanding at a rate similar to that of the universe as a whole, just as the Virgo supercluster is observed to do. The cluster to the right in Fig. 7 was formed out of three clusters visible in Fig. 6, which have fallen together and coalesced. This is a region so dense that it is gravitationally bound and has undergone a collapse and relaxation toward equilibrium.

The covariance function for this simulation is $\xi(r) \sim 90 \ (1 \ \mathrm{Mpc}/r)^{1.9}$, in remarkable agreement with the observed relation $\xi(r) \sim 70 \ (1 \ \mathrm{Mpc}/r)^{1.8}$. Both $\Omega = 0.1$ and $\Omega = 1$ models can be constructed

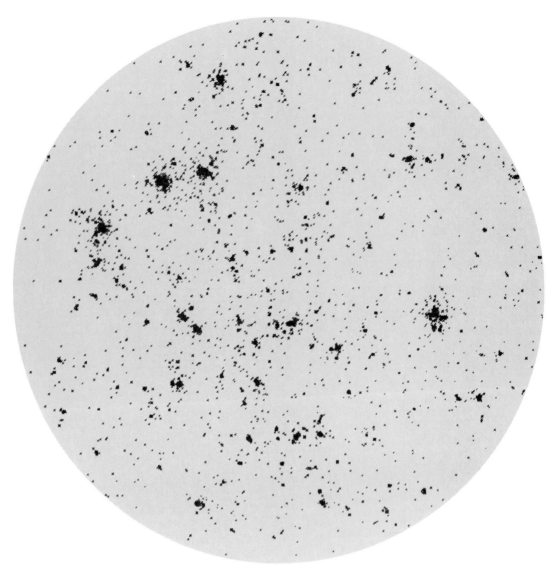

Fig. 7. Same simulation as Figs. 5 and 6 seen at the present epoch, 17.6×10^9 years after the big bang. (From S. J. Aarseth, J. R. Gott, and E. L. Turner, N-body simulations of galaxy clustering, I: Initial conditions and galaxy collapse times, Astrophys. J., 228:644–683, 1979)

which give good covariance functions at the present epoch. Prior to these simulations, it was not known whether $\Omega = 0.1$ models could produce a sufficient degree of clustering. Thus the covariance function data alone do not give a unique value of Ω. However, the velocities of galaxies orbiting within groups and clusters in the $\Omega = 1$ simulations are three times higher than those in the $\Omega = 0.1$ simulations. The velocities observed in the universe are low, favoring $\Omega = 0.1$ models. The $\Omega = 0.1$ models also correctly predict the observed abundance of isotope deuterium in the universe, as well as being consistent with the available data on the age of the elements and the oldest stars as discussed by J. R. Gott, J. E. Gunn, D. N. Schramm, and B. M. Tinsley.

These simulations vindicate the simple gravitational interaction theory of how clusters and groups of galaxies form. Cosmology, unlike biology or chemistry, is not an experimental science in that one can observe the universe only as it is and cannot perform experiments. Computer simulations change this. Simulated universes can be constructed whose properties are known exactly, which can be compared directly with actual observations. Simulations such as these hold great promise for improving understanding of the universe.

For background information see COSMOGONY; COSMOLOGY; GALAXY, EXTERNAL in the McGraw-Hill Encyclopedia of Science and Technology.

[J. RICHARD GOTT III]

Bibliography: S. J. Aarseth, J. R. Gott, and E. L. Turner, *Astrophys. J.*, 228:664–683, 1979; J. R. Gott et al., *Sci. Amer.*, 234(3):62–79, 1976; R. A. Muller, *Sci. Amer.*, 238(5):64–74, 1978; P. J. E. Peebles, *Astrophys. J. Lett.*, 189:L51–L53, 1974; P. J. E. Peebles, *Physical Cosmology*, 1971; G. F. Smoot, M. V. Gorenstein, and R. A. Muller, *Phys. Rev. Lett.*, 39:898–901, 1977; S. Weinberg, *The First Three Minutes*, 1977; D. Woody and P. L. Richards, *Phys. Rev. Lett.*, 42:925–929, 1979.

Crown gall

Crown gall disease, formerly a minor topic in plant pathology, has become of major interest in molecular genetics during the last few years. This change has come about with the realization that the disease represents a unique genetic interaction between a bacterium and a higher plant, in which bacterial deoxyribonucleic acid (DNA) becomes permanently incorporated into the plant genome.

Crown gall disease affects nearly all dicotyledonous plants. It is caused by a soil bacterium, *Agrobacterium tumefaciens*, which enters the stems or leaves of the plants at wound sites and stimulates the plant tissue to proliferate into a tumor, or gall (Fig. 1). Often the galls are located at the crown of the plant, hence the name. The properties of the cells in the galls are permanently changed, just as they are in animal cancers. If they are grown in culture, they retain indefinitely the capacity to divide in the absence of added plant hormones (auxins and cytokinins), whereas cultures of normal tissue from the same plants need such hormones added to the growth medium.

Plasmids and crown gall. Strains of *Agrobacterium* capable of causing crown gall tumors always possess at least one of a class of genetic elements called plasmids. Plasmids are DNA elements carried by bacteria of many different kinds; they bear genes conferring particular characteristics on the organisms that possess them. For example, many clinically important bacteria owe their newly acquired resistance against antibiotics to plasmids, while members of the genus *Pseudomonas* can sometimes grow on unusual compounds, such as hydrocarbons or camphor, because they carry plasmids bearing genes which code for the enzymes needed to metabolize these compounds. Other plasmids code for antibiotic production by *Streptomyces*. Even the mating processes of bacteria are controlled by conjugative plasmids, which code for the cellular appendages necessary for copulation; when a bacterium carrying a plasmid mates with one lacking it, a copy of the plasmid is usually the first genetic material to pass from cell to cell.

The plasmids carried by virulent strains of *A. tumefaciens* are among the largest known, with molecular weights of 100,000,000 – 200,000,000: enough DNA for 100 – 200 genes. Some of these genes control the replication and maintenance of the plasmids in their bacterial hosts — properties possessed by all plasmids — and others determine mating functions comparable with those conferred on bacteria by other conjugative plasmids. But the characteristic features of crown gall disease are controlled by genes specific to these Ti (tumor-inducing) plasmids.

Unusual metabolites in crown galls. The cells of crown gall tissue, as well as being altered in the normal controls of cell division so that they grow as tumors, have a characteristically changed metabolism. It was these metabolic peculiarities which provided the first clues to the intricate "molecular parasitism" represented by this disease. In 1964 A. Menagé and G. Morel, working at Versailles, France, identified an unusual amino acid, octopine, as a characteristic component of crown gall

tissue but not of normal plant cells. Octopine can be considered as a condensation product of the basic amino acid arginine with pyruvic acid (Fig. 2). Tumors induced by the same strain of *Agrobacterium* contain, in addition to octopine, other compounds, each representing condensation products of pyruvic acid with other basic amino acids: ornithine (to give octopinic acid), lysine (lysopine), or histidine (histopine). Other tumors, induced on the same host plants by a different strain of *Agrobacterium*, contained, instead of this series of opines, another series of condensation products of α-ketoglutaric acid with arginine (nopaline) or ornithine (nopalinic acid). It then emerged that the difference between *Agrobacterium* strains inducing tumors containing the octopine family of compounds and those of the nopaline family depended solely on the plasmids they carried.

How do the Ti plasmids determine production of these characteristic crown gall metabolites? Are they produced by plant genes activated by the process of infection, or is part or all of a Ti plasmid actually transferred to a plant cell to become a permanent part of its genetic complement? Recent developments in techniques for handling DNA have led to definitive answers.

Plasmid DNA transfer to plant cells. Two groups of workers, one in Seattle, WA, and the other in Ghent, Belgium, have recently identified DNA sequences in crown gall tissue, but not in normal tissue from the same plants, which corre-

Fig. 1. Crown gall tumors induced on a tomato plant by inoculation with *Agrobacterium tumefaciens*. (*Courtesy of J. L. Firmin*)

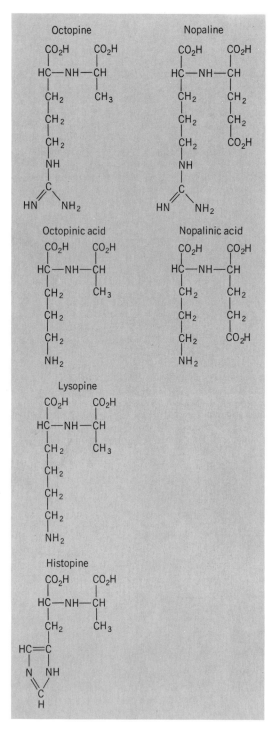

Fig. 2. Chemical structures of the opines. The octopine series represent condensation products of a basic amino acid (arginine, ornithine, lysine, or histidine, respectively) with pyruvic acid; in the nopaline series, arginine or ornithine is condensed with α-ketoglutaric acid.

spond to specific short regions of Ti plasmids. The experiments were made possible by the use of restriction enzymes to cut isolated plasmid DNA into a number of specific segments. These were radioactively labeled, and separated into single strands (denatured) for use as probes to seek corresponding sequences in denatured DNA from plant tissue. It was found that a specific segment

of Ti plasmid DNA would hybridize with DNA from crown gall tissue, even tissue that for years had been in artificial culture entirely free from viable bacteria. No hybridization occurred with DNA from normal tissues. The conclusion from these experiments is that when a virulent strain of *Agrobacterium* interacts with plant cells, a certain region of its Ti plasmid, called the T-DNA, is transferred to the plant cells and becomes stably incorporated into them.

Genes on the T-DNA presumably are responsible for the special characteristics of crown gall cells. One or more genes code for the enzymes necessary for opine synthesis. Whether or not there are also specific "oncogenes" for initiation and maintenance of the cancerous state of gall cells is not yet known, but presence of the T-DNA is evidently required. It is not certain how many copies of the T-DNA reside in each transformed plant cell, or where the T-DNA is located: Is it integrated into the chromosomes, or into the DNA of the mitochondria or chloroplasts?

The T-DNA represents only a small proportion of each Ti plasmid; the rest of the plasmid does not become incorporated into plant cells, but it carries some other important genes. These include genes coding for enzymes which enable agrobacteria on the outside of the crown galls, and in their interstices, to use the opines as sources of carbon and nitrogen. Only strains stimulating production of tumors containing the octopine series of compounds can use these compounds for their own growth. Strains leading to nopaline synthesis in tumors, on the other hand, can use the nopaline series of compounds. Other identified genes on Ti plasmids are those which cause conjugation of bacteria carrying the plasmid with those which do not, and less well-characterized genes responsible for the tranfer of the T-DNA into plant cells.

Genetic colonization by Agrobacterium. The relationship between *A. tumefaciens* and higher plants is clearly a highly evolved and intricate one. By transferring specific genes from its Ti plasmid into plant cells, the bacterium subverts the plant's metabolism to produce considerable quantities of materials, the opines, which can be metabolized by the bacterium but appear to be of no use to the plant, as well as to most other microorganisms. Not only do the opines reach high concentrations in the cells of the tumor, but precisely because the cells divide in an unregulated fashion, large masses of tumor tissue are produced. Thus the quantities of opines in the immediate vicinity of the galls may (perhaps when the galls begin to rot) be significant and may be capable of supporting large populations of agrobacteria.

As well as being metabolized by bacteria carrying Ti plasmids, the opines specifically stimulate expression of the conjugative functions coded by the Ti plasmids. Thus nopaline promotes the spread of the "nopaline" class of Ti plasmids within a population of agrobacteria, while octopine does the same for "octopine" plasmids, thus giving selective advantage to the recipients of the plasmids. The whole system appears to have evolved in a highly specific way to ensure the survival and spread of the Ti plasmids and their bacterial hosts by "genetically colonizing" plant tissues.

Genetic engineering of plants. In the last few years it has become possible to insert genes, potentially from any source, into bacteria and, more recently, other microorganisms, in such a way that they become a permanent part of the new host's genetic machinery. To do this, a vector is required: a plasmid or virus genome capable of replicating in the host and into which the foreign genes can be spliced artificially before reintroducing the vector into the host cell. Such genetic engineering techniques, applied to crop plants, could potentially lead to the development of valuable new varieties by insertion of foreign genes determining, for example, disease resistance, increased nutritional value of plant proteins, or, conceivably, the ability to fix atmospheric nitrogen. The Ti plasmids are obvious potential vectors for this purpose, since they are naturally capable of transferring their T-DNA segments into plant cells. Thus foreign DNA could be inserted into the T-DNA of isolated Ti plasmids; transfer into plant tissue could be achieved, either via *Agrobacterium*, or possibly by exposing plant protoplasts themselves to the isolated plasmid DNA.

The first steps in developing such a system have already been taken, but much more remains to be done. Apart from ensuring the satisfactory expression of the foreign genes, major unknowns concern the fate of the T-DNA when crown gall tissue regenerates, as it sometimes does, into normal plants which undergo sexual reproduction. However, progress in the understanding of crown gall disease has been so startling over the last few years that further major advances can be confidently predicted.

For background information *see* PLANT DISEASE in the McGraw-Hill Encyclopedia of Science and Technology.   [D A. HOPWOOD]

Bibliography: M.-D. Chilton et al., *Cell*, 11: 263–271, 1977; J. L. Firmin and G. R. Fenwick, *Nature*, 276:842–844, 1978; C. I. Kado, *Ann. Rev. Phytopathol.*, 14:265–305, 1976; J. Schell et al., *Proceedings of the Royal Society*, ser. B, 204: 251–266, 1979.

Cryptography

As society becomes increasingly dependent upon computers in both the business and private sectors, vast amounts of data have to be communicated, processed, and stored within the computer systems and networks. Such data often have to be protected. Cryptography, which embraces the various methods for writing in secret code or cipher, is one means of achieving protection.

Cryptography is the only known practical method for protecting information transmitted through accessible communication networks, for example, telephone lines, satellites, or microwave systems. However, cryptography has much broader uses. It can be the most economical way to protect stored data in some instances. Cryptographic procedures can be used for message authentication, personal identification, and signature verification authorizing electronic funds transfer and credit card transactions.

Cryptographic algorithms. An obvious requirement of cryptography is that it resist decoding or deciphering by unauthorized personnel. Messages (plaintext) must be transformed into cryptograms (codetext or ciphertext) that can withstand intense cryptanalysis. This can be done by code systems or by cipher systems.

Code systems rely on a code book that transforms the plaintext words, phrases, and sentences into code groups. To prevent cryptanalysis, there must be a great number of plaintext passages in the code book and the code-group equivalents must be kept secret, making it difficult to utilize code books in electronic data-processing systems.

Cipher systems are more versatile. Messages are transformed through the use of two basic elements: a set of unchanging rules or steps called a cryptographic algorithm, and a set of variable cryptographic keys.

The algorithm is composed of enciphering (E) and deciphering (D) procedures which usually are identical or simply consist of the same steps performed in reverse order, but which can be dissimilar. The user selects the keys, which consist of a sequence of numbers or characters. An enciphering key (Ke) is used to encipher plaintext (X) into ciphertext (Y) as in Eq. (1), and a deciphering key (Kd) is used to decipher ciphertext (Y) into plaintext (X) as in Eq. (2).

$$E_{Ke}(X) = Y \qquad (1)$$

$$D_{Kd}[E_{Ke}(X)] = D_{Kd}(Y) = X \qquad (2)$$

Algorithms are of two types—conventional and public-key. The enciphering and deciphering keys in a conventional algorithm either may be identical or are easily computed each from the other (Ke = Kd = K, denoting $E_K(X) = Y$ for encipherment and $D_K(Y) = X$ for decipherment). In a public-key algorithm, the public key is used for enciphering and is not secret; the private key is used for deciphering and is secret; the private key cannot be deduced from the public key.

When an algorithm is made public, for example, as a published encryption standard, cryptographic security completely depends on protecting those cryptographic keys specified as secret.

Privacy and authentication. Anyone can encipher data in a public-key cryptographic system

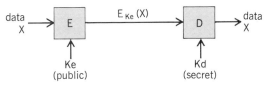

Fig. 1. Public-key cryptographic system used for privacy only.

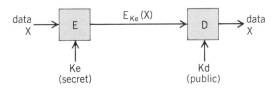

Fig. 2. Public-key cryptographic system used for message authentication only.

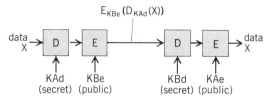

Fig. 3. Public-key cryptographic system used for both message authentication and privacy. KAe and KAd are enciphering and deciphering keys of the sender (A). KBe and KBd are enciphering and deciphering keys of the receiver (B).

(Fig. 1) by using the public enciphering key, but only the authorized user can decipher the data through possession of the secret deciphering key. Since anyone can encipher data, the identity of a message's sender cannot be known with certainty, which is necessary for message authentication. A procedure for message authentication can be devised (Fig. 2)—provided the enciphering key cannot be obtained from the deciphering key—by keeping the enciphering key secret and making the deciphering key public. This makes it impossible for nondesignated personnel to encipher messages, that is, produce $E_{Ke}(X)$.

By insertion of prearranged information in all messages, such as originator identification, recipient identification, and message sequence number, the messages can be checked to determine if they are genuine. Since the contents of the messages are available to anyone having the public deciphering key, privacy is unavailable.

Figure 3 shows a public-key algorithm that provides privacy as well as authentication, since encipherment followed by decipherment, and decipherment followed by encipherment, produce the original plaintext, as in Eq. (3). A message to be

$$D_{Kd}[E_{Ke}(X)] = E_{Ke}[D_{Kd}(X)] = X \qquad (3)$$

authenticated is first deciphered by the sender (A) with a secret deciphering key (KAd). Privacy is ensured by enciphering the result with the receiver's (B) public enciphering key (KBe).

A conventional cryptographic system protects data effectively only if the sender and receiver of the message share a common secret key. Such a system automatically provides both privacy and authentication (Fig. 4).

Digital signatures. Digital signatures can authenticate messages by ensuring that (1) the sender of the message cannot later disown it, (2) the message receiver cannot forge messages or signatures, and (3) the message receiver can prove to

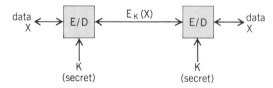

Fig. 4. Conventional cryptographic system in which message authentication and message privacy are provided simultaneously. K represents a common secret key.

others that the contents of a message are genuine and that the message originated with that particular sender.

Digital signatures can be obtained with both public-key and conventional cryptographic algorithms. The digital signature is a function of the message, a secret key or keys possessed by the sender of the message, and sometimes other data that are not secret or that may become so as part of the procedure, such as a secret key that is later made public.

Digital signatures are more easily obtained with public-key algorithms. When a message is enciphered with a private key known only to the originator of the message, anyone deciphering the message with the public key can identify the originator of the message.

Since enciphering and deciphering keys are identical in a conventional algorithm, digital signatures must be obtained differently. One method is to use a set of keys to produce the signature. Some of the keys are known to the message's receiver to permit signature verification, and the rest of the keys are known only by the message's originator in order to prevent forgery.

Strong algorithms. Unbreakable ciphers are possible. To design one, the key must be randomly selected and used only once, and its length must be equal to or greater than that of the plaintext to be enciphered. However, such long keys, called one-time tapes, are not practical in data-processing applications.

To work well, a key must be of fixed length, relatively short, and capable of being repeatedly used without hazarding security. In theory, any algorithm that uses such a finite key can be analyzed; in practice, the effort and resources necessary to break the algorithm would be unjustified.

Construction of an unbreakable algorithm is not necessary to achieve effective data security, provided that the work factor (a measure, under a given set of assumptions, of the requirements necessary for a specific analysis or attack against a cryptographic algorithm) required to break the algorithm is sufficiently great. The set of assumptions include the type of information perceived to be available for cryptanalysis. Such types of information may be ciphertext only; plaintext (not chosen) and corresponding ciphertext; chosen plaintext and corresponding ciphertext; or chosen ciphertext and corresponding recovered plaintext.

A conservative approach should be taken in the construction of a strong algorithm when an algorithm's life expectancy is relatively long and the algorithm is to be utilized in various applications and environments.

A strong cryptographic algorithm must satisfy the following conditions: (1) The complexity of the mathematical equations describing the algorithm's operation prevents, for all practical purposes, their solution through analytical methods. (2) The cost or time necessary to unravel the message or key is too great when mathematically less complicated methods are used, because too many computational steps are involved, for example, in trying one key after another, or because too much storage space is required, for example, in an analysis requiring data accumulations such as dictionaries and statistical tables.

A strong algorithm must satisfy these conditions, even when the analyst has the following advantages: (1) Relatively large amounts of plaintext (specified by the analyst, if so desired) and corresponding ciphertext are available. (2) Relatively large amounts of ciphertext (specified by the analyst, if so desired) and corresponding recovered plaintext are available. (3) All details of the algorithm are available to the analyst, that is, cryptographic strength cannot depend on the algorithm remaining secret. (4) Large high-speed computers are available for cryptanalysis.

However, an unbreakable algorithm implies that even with an unlimited amount of computational power, data storage, and calendar time, the message or key cannot be obtained through cryptanalysis. Thus, in theory, a strong algorithm may be breakable, even though in practice it is not. And so, the term "strong" is a variable, and the term "unbreakable" is that variable's maximum value.

Therefore a cryptographic system must be based on a cryptographic algorithm of validated strength if it is to be acceptable. The Data Encryption Standard (DES) is such a validated conventional algorithm already in the public domain. Since public-key algorithms are relatively recent, their strength has yet to be validated.

Data Encryption Standard. A cryptographic procedure was developed during 1968–1975 at IBM that is composed of alternate steps (rounds) of key-controlled substitution and fixed permutation. The National Bureau of Standards accepted this algorithm as a standard and it became effective on July 15, 1977.

DES enciphers a 64-bit block of ciphertext under the control of a 56-bit key. The encryption procedure consists of 16 separate rounds of encipherment, where at each round a cipher function (f) is employed together with a 48-bit key. The interaction of data, cryptographic key (K), and f are shown in Fig. 5. The externally supplied key consists of 64 bits (56 bits are used by the algorithm, and up to 8 bits may be used for parity checking). A special shifting scheme on the original 56-bit key is utilized so that a different subset of 48 key bits is used in each round. These subsets of key bits are labeled $K_1, K_2, \ldots, K_{16}$. During decipherment the rounds are performed in reverse order (K_{16} is used in round one, K_{15} in round two, and so on).

Block and stream ciphers. These ciphers constitute two different cryptographic approaches to data security. A block cipher (Fig. 6) transforms a string of input bits of fixed length (termed an input block) into a string of output bits of fixed length (termed an output block). In a strong block cipher, the enciphering and deciphering functions are such that every bit in the output block jointly depends on every bit in the input block and on every bit in the key.

A stream cipher (Fig. 7) enciphers (or deciphers) a bit stream of plaintext (or ciphertext) of arbitrary length into a bit stream of ciphertext (or recovered plaintext) with a bit stream—defined here as the cryptographic bit stream—of equal length, which is generated (or regenerated) by a ciphering algo-

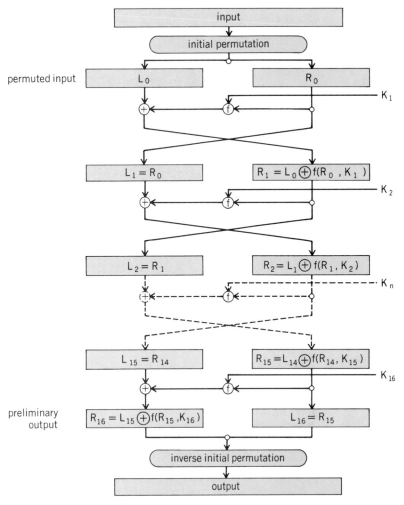

Fig. 5. Enciphering computation in the Data Encryption Standard. (*From Data Encryption Standard, FIPS Publ. no. 46, National Bureau of Standards, 1977*)

rithm. The cryptographic bit stream may be produced on a bit-by-bit or a block-by-block basis. (The term "key stream" normally denotes the bit stream produced by the ciphering algorithm; the term "cryptographic bit stream" is used here to avoid confusion between "key stream" and the fixed-length key used by the ciphering algorithm.)

Since the message's sender and receiver in a stream cipher must produce cryptographic bit

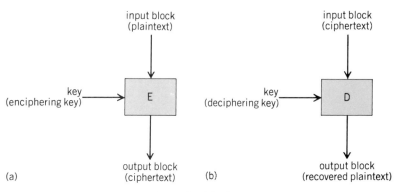

Fig. 6. Block cipher. (*a*) Enciphering. (*b*) Deciphering.

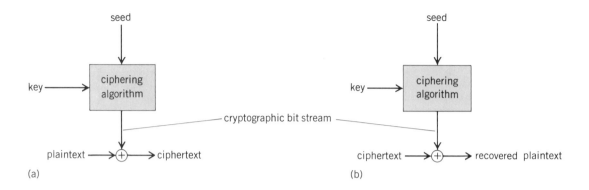

$\oplus$ plaintext/ciphertext: 0 1 0 1
     cryptographic bit stream: 0 0 1 1
     ciphertext/recovered plaintext: 0 1 1 0

Fig. 7. Stream cipher. (a) Enciphering. (b) Deciphering.

streams that are both equal and secret, their keys must also be equal and secret. Public-key algorithms would obviously not be used in a stream-cipher mode.

For security purposes, a stream cipher must never start from the same initial condition in a predictable way, thereby producing the same cryptographic bit stream. This is avoided by making the cryptographic bit stream dependent on a non-secret quantity known as seed, initializing vector, or fill, which is used as an input parameter to the ciphering algorithm.

For background information *see* INFORMATION THEORY in the McGraw-Hill Encyclopedia of Science and Technology.

[CARL H. MEYER; STEPHAN M. MATYAS]

Bibliography: *Data Encryption Standard*, FIPS Publ. no. 46, National Bureau of Standards, January 1977; W. Diffie and M. Hellman, *IEEE Trans. Inform. Theory*, IT-22:644–654, November 1976; H. Feistel, *Sci. Amer.*, 228(5):15–23, May 1973; C. H. Meyer and S. M. Matyas, *Cryptography: A New Dimension in Computer Data Security*, 1980; R. L. Rivest, A. Shamir, and L. Adleman, *Commun. ACM*, 21(2): 120–126, February 1978.

Crystal growth

Techniques for growing crystals have recently been developed which permit creation of minute three-dimensional structures within crystal films. Since some of the structures which have been created contain dimensions as small as interatomic distances, they demonstrate the elements needed for the controlled architecture of crystals at the atomic scale where the characteristic properties of matter are defined. The key to the new architecture is the use of molecular and atomic beams for crystal growth.

Thin films of single crystals can be formed by directing streams of atoms or molecules in ultrahigh vacuum directly onto a suitable substrate. This method of crystal growth, known as molecular beam epitaxy (MBE), provides the finest dimensional control of any growth technique presently available. In contrast to other types of crystal growth, in which larger numbers of atoms are being arranged in regular three-dimensional arrays in a more nearly equilibrium process, the molecular beam technique allows layer-by-layer crystal formation. The control of the thickness of the deposited layers derives from the fact that the layers can be grown with very smooth surfaces and little interdiffusion between surface and subsurface layers. Layer definition can be controlled because the beams of atoms and molecules can be interrupted in times shorter than needed to deposit one atomic thickness of material. The possibility thus exists to define the structure of the crystal at the atomic layer thickness level and to control the properties of the resultant crystal by design of its layer structure.

Alternate monolayer crystals. The material for which MBE has been most highly developed and from which the first alternate monolayer crystals were grown is the semiconducting compound gallium arsenide, GaAs. Many other materials are capable of growth by MBE, but interest in the crystalline, electrical, and optical properties of GaAs has made the development of a precision means for its growth especially desirable. GaAs may also be grown by chemical vapor deposition, liquid-phase epitaxy, and metal-organic vapor deposition. However, the direct synthesis in vacuum from atomic and molecular beams by MBE is a preferred technique for smoothness and control of the crystal layers.

Through use of the MBE technique, alternate ultrathin layers of GaAs and the isomorphic mixed compound with fraction x of Ga sites occupied by Al, $Al_xGa_{1-x}As$, have been deposited to build up synthetic layered structures. The depositions were made by alternately shuttering open and shut heated evaporation ovens from which gallium and aluminum beams effused by evaporation of elemental metallic charges. A third oven contained arsenic. The beams impinged directly on a heated GaAs substrate in ultrahigh vacuum.

The actual deposited structures were tested by transmission electron microscopy and by x-ray diffraction. These tests demonstrated that, for substrate temperatures of 500–600°C, layered crystals resulted with exactly the compositional period expected from the deposited quantities of materials. An example of a transmission electron microscopy image of a cross section of a sample built up by alternately depositing two monolayers (0.6 nm) each of GaAs and AlAs on a GaAs (111) crystal face is shown in Fig. 1. The space between adjacent dark lines on the figure is thus about 1.3 nm, and contains two GaAs monolayers and two AlAs monolayers. Samples consisting of alternate single monolayers have also been fabricated and, although they could not be clearly imaged with the electron microscopy, they showed electron and x-ray diffraction patterns with definite layering at the alternate monolayer repeat distance. Evidence for layers as thin as 0.28 nm was found, corresponding to the shortest artificially synthesized compositional period created to date. The layers at this small scale were not completely ordered, however, indicating that some mixing occurred between adjacent monolayers of AlAs and GaAs.

The primary importance of the alternate-monolayer crystals is that they demonstrate that substances which are essentially new materials can be built up by variation of chemical composition in layers at nearly arbitrarily fine scale. The properties of the crystals thus formed were tested after growth and showed electron energy levels and phonon spectra that were distinctly different from GaAs, AlAs, or $Al_xGa_{1-x}As$. The crystals showed tetragonally anisotropic optical properties, distinct from the cubic properties of their constituent materials. Band structure calculations have been made, corroborating the electronic structure imposed on the monolayer materials by their new structures.

Alternately evaporated ultrathin layers of metals have also been studied recently and have shown unusual new properties. Although they do not grow as single crystals and they interdiffuse more readily than the semiconductors, the metal crystallites can be oriented and their layer structure can be analyzed by x-ray diffraction. In layered copper-gold structures, an interesting composition-dependent interdiffusion leading toward an ordered Cu_3Au structure was observed in x-ray diffraction studies of sufficiently thin layers. In the range of compositional periods below 3 nm, gold-nickel and copper-palladium layered foils have shown unusual increases in elastic modulus. Copper-nickel layered films have been reported to have increased magnetization densities relative to alloys or thicker films.

Quantum wells. The thin layers which can be assembled with the epitaxial techniques described above offer the opportunity to experimentally produce structures in which quantum-mechanical particle confinement and tunneling can be observed. Since GaAs offers a lower potential energy to conduction electrons than does an immediately adjoining layer of $Al_xGa_{1-x}As$, an electron in a layer of GaAs enclosed between two layers of $Al_xGa_{1-x}As$ experiences a confining potential. If the layer in which the electron in confined is less than a few tens of nanometers in thickness, its available energy levels will be discretely quantized. Barrier thicknesses of $Al_xGa_{1-x}As$ of tens of monolayers

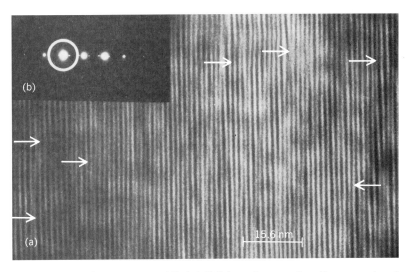

Fig. 1. Alternate bilayer crystal of $(GaAs)_2(AlAs)_2$. (a) Cross-sectional image produced by transmission electron microscope. (b) Electron diffraction pattern produced by the crystal. The cross-sectional image is produced from electrons within the circle in the diffraction pattern.

or more will confine electrons, while thicknesses of one or a few monolayers can be penetrated by quantum-mechanical tunneling of electrons. Quantum states of confined electron waves are observed in the 5- to 50-nm thickness range of GaAs for sufficiently smooth and pure layers. This constitutes a clear realization of the theoretical quantum-mechanical particle-in-a-box familiar in quantum mechanics texts. Since holes in the normally filled valence bands are also bound in the GaAs by the confining potential, quantum effects for the valence band hole states are also observed. The energies of the electrons and holes in their quantum-mechanical GaAs "boxes" have been most clearly observed in optical absorption measurements in which absorbed photons excite electrons from the filled valence band to the empty conduction band.

Optical absorption spectra for stacks of quantum wells consisting of GaAs layers of 5.0-, 11.6-, 19.2-, and 400-nm thickness, bounded by $Al_{0.3}Ga_{0.7}As$, are shown in Fig. 2. The spectrum of the 400-nm-thick sample corresponds to bulklike absorption, which occurs above 1.51-eV photon energy, where electrons are excited from filled valence bands up to the unoccupied conduction band. The spectra for thinner samples show discrete band-edge peaks labeled $n = 1, 2, 3, \ldots$, for excitation of electrons into quantized energy levels, where n corresponds to the number of electron half-wavelengths confined in the GaAs layers in the particular energy level. The fit of the energies to the predictions of quantum mechanics for one-dimensional potential wells is excellent.

Several applications of these quantum wells have been achieved. Optical absorption of circularly polarized light by GaAs quantum wells provides a new means to achieve spin-polarized electrons in the conduction band of GaAs. Since two types of normally energetically equal GaAs valence electrons have separated energies in quantum wells of GaAs, it becomes possible to selectively pump only the desired type of valence band

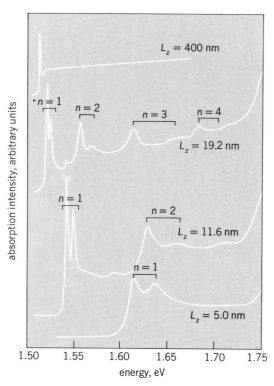

Fig. 2. Optical absorption spectra for stacks of quantum wells consisting of GaAs layers having thickness L_z of 5.0, 11.6, 19.2, and 400 nm. Layers are bounded by $Al_{0.3}Ga_{0.7}As$.

electron into the conduction band. This produces a greater spin polarization of conduction electrons than in bulk GaAs (which is currently the most favorable available source of spin-polarized free electrons).

In other work with quantum well structures, laser emission has been produced between the quantum well states both by optical pumping and by electrical injection, for quantum layers created both by MBE and by metal-organic chemical vapor deposition.

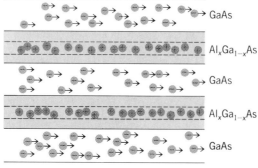

key:

⊕→ mobile electron

⊕ silicon dopant atom

Fig. 3. Modulation-doped structure for enhanced mobility of conduction electrons.

Quantum layers and barriers composed of alternating layers of the semiconductors GaSb and InAs have also been grown by MBE. Because the band gap of InAs lies entirely below the band gap of GaSb, conduction electrons can be localized in InAs layers and valence band holes can be localized in GaSb layers, producing a separation of electrons and holes. In some instances, tunneling in these layers has led to semimetal electrical conductivity characteristics.

Modulation-doped structures. Epitaxial growth of crystals allows the successive growth of semiconductor layers with different types of electrical dopants. In layers with thickness near 1 μm, this is used in production of various conventional electrical and optical devices. In thinner layers, MBE crystal growth has provided access to a new regime of charge distribution where charge carriers may be separated from the dopant atoms which contributed them to the crystal. This permits charge carriers to propagate in selected layers where there is a reduced probability for collision with impurities or imperfections. In turn, this can produce higher conductivities, mobilities, and lifetimes between scattering events.

The technique used for separating carriers from impurities is known as modulation doping and involves doping electron donor atoms such as silicon only into the electron barrier layers ($Al_xGa_{1-x}As$) in a multilayer structure (Fig. 3). The donor atoms contribute electrons which are attracted to the adjacent GaAs layers, due to the greater electron affinity of GaAs. The electrons propagate freely in the GaAs with reduced chance of collision, since the donor impurity atoms are not in the conduction paths. Electrons are contributed from distances up to several tens of nanometers.

Modulation-doped structures made in the above fashion have shown electron mobilities and conductivities higher than GaAs samples prepared by any other means at similar doping level. The mobility enhancement is most marked at low temperatures, where impurities scatter the most strongly in conventional uniformly doped semiconductors. The electrons in the layered structure approximate very closely a stack of two-dimensional electron gases.

Other new structural elements. A number of other new elements have been devised with MBE in the development of epitaxial crystal architecture.

One area under investigation involves the definition of structures within the planes of the deposited layers. Two general approaches have been followed. One involves deposition through a mask near the crystal surface. Photolithographically defined masks on the surface of GaAs permit micrometer-width stripes to be grown and even permit "writing" with the evaporants by means of moving the mask. Masks slightly above the surface have allowed layers with tapered edges to be grown due to the penumbral region of the evaporant shadows. The other approach to lateral control involves growth over photolithographically defined oxide patterns. This method results in polycrystalline growth over oxide surfaces and single-crystal growth over exposed crystal substrate surfaces. Definition within the plane of the layers by these techniques has not yet approached either the monolayer or the quantum-well dimensions, but progress in that direction may be anticipated.

Other recent advances in molecular-beam crystal growth have involved in-place contacting and isolating of layers within the crystal growth vacuum chamber and the use of ion beams for in-place doping with otherwise nonincorporable atoms. These techniques hold promise for future flexibility in interfacing molecular-beam crystal architecture with the macroscopic outside world.

For background information *see* CRYSTAL GROWTH; MOLECULAR BEAMS; QUANTUM THEORY, NONRELATIVISTIC; SEMICONDUCTOR HETEROSTRUCTURES in the McGraw-Hill Encyclopedia of Science and Technology.

[ARTHUR C. GOSSARD]

Bibliography: A. Y. Cho, *J. Vac. Sci. Technol.*, 16:275–284, 1979; R. Dingle et al., *Appl. Phys. Lett.*, 33:665–667, 1978; A. C. Gossard, *Thin Solid Films*, 57:3–13, 1979; G. A. Sai-Halasz, R. Tsu, and L. Esaki, *Appl. Phys. Lett.*, 30:651–653, 1977.

Ctenophora

Tentaculate ctenophores feed on planktonic animals such as copepods, crustacean larvae, and fish eggs. They trap these prey in their tentacular apparatus, which is widely spread out in the sea like a fishing net. When a prey organism collides with a lateral branch (tentillum) of a tentacle, it is retained by the sticky properties of the tentillum, like a fly in a spider's web. The struggling movements of the prey lead to a sequential stereotyped feeding response of the ctenophore which conveys the food to the mouth for swallowing.

The gluing function of the tentillar apparatus was known to depend upon adhesive cells, or colloblasts, which constitute the most abundant cell type of the tentacular and tentillar epithelium. With the light microscope, each colloblast appears to consist of a bell-shaped structure limited by eosinophilic granules and bearing two cords designated as straight and spiral filaments. Brilliant vesicles are often observed outside and up against the eosinophilic granules. However, before electron microscopy, many points concerning colloblast structure and the gluing function remained uncertain. Successive studies by R. Hovasse and P. de Puytorac, W. Bargmann and coworkers, and J. M. Franc have now proved that each colloblast is a single cell, containing only one filament, helical in shape, and connected to the eosinophilic granules, which are responsible for the gluing function of the cell. Evidence now exists that the capture phenomenon is controlled by the nervous system of the animal and requires the disruption of the colloblast, which appears to be a disposable capture cell.

Colloblast features. Seen by scanning electron microscopy (Fig. 1), a colloblast looks like a mushroom with a helicoidal twisted foot. The head of the colloblast is embossed by numerous inner spherical granules and may often bear depressed outer vesicles.

In transmission electron microscopy, thin sections of tentacular or buccal tentilla have demonstrated that colloblasts are single cells (Fig. 2). The size of these cells varies with the species. The heads of the colloblasts are 4 to 10μm wide, whereas the length of the whole cell depends on its degree of extension. The basal slender end of the cell is anchored in the dense fibrillar jelly surrounding the neuromuscular axis of the tentillum. This basal end forms a cone-shaped root bristling with numerous thick filaments, giving it the appearance of a bottlebrush. From this root arises a helical thread which appears as a more or less empty tube with a circular cross section. It corresponds to the spiral filament of light microscope observations. This intracellular structure twists toward the expanded head of the colloblast, lifting up the cell membrane until it reaches into the head. At the center of the colloblast head, the distal end of the helical thread differentiates into a star-shaped body with fibrous radii. Each radius reaches one of the numerous mucous granules which are situated at the inner periphery of the colloblast head. The granules together constitute a dome capped by the cytoplasmic membrane. These granules correspond to the "eosinophilic" granules of the light microscope description. The external "brilliant" vesicles, previously described and eventually seen with the scanning electron microscope on the outer surface of the colloblast head, correspond to the remnants of accessory cells which disappear during the last stages of differentiation of the colloblast. They must be considered as foreign to the colloblast, and do not seem to play any part in the sticking phenomenon.

The second filament, or "straight filament," does not exist as such. Its description from light microscopy resulted from a confusion with the cell

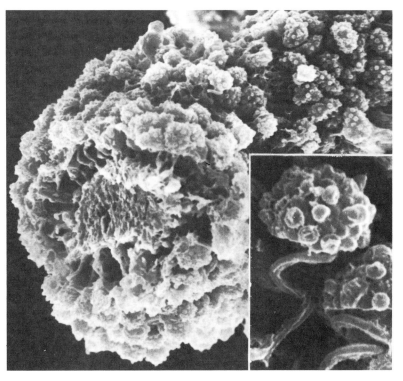

Fig. 1. Scanning electron micrograph of ctenophore tentillum. The surface is covered with numerous colloblasts, whereas the neuromuscular axis is visible on the broken section. The inset shows a magnified view of two partly extended colloblasts.

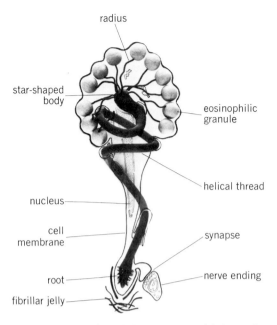

Fig. 2. Diagram of a colloblast as seen with transmission electron microscope.

nucleus. The latter closely joins and partially surrounds the star-shaped body. The nucleus elongates toward the basal end of the cell, but has no relation with the anchoring structure.

The most recent finding is a nerve connection for each colloblast. This connection always appears just above the root where a well-differentiated synapse occurs. This synapse is of the polarized type and is always directed toward the colloblast. This indicates that the colloblast acts as a motor effector, and cannot be considered as having a sensory function. This role is assumed by clusters of sensory cells situated between the colloblasts all along the tentacular apparatus.

Capture. Ultrastructural examination of captured prey shows that the eosinophilic granules are responsible for the gluing action of the colloblast. The collision between a prey and a tentillum causes the disruption of the cytoplasmic membranes of the struck colloblasts, their eosinophilic granules burst, and a gluing mucus is liberated onto the surface of the prey, which is thus stuck to the fibrous radii of the helical thread. The gluing mucus is not toxic to the captured prey, which can live until it is engulfed by the mouth, but its struggling movements make it more firmly attached as more neighboring colloblasts become involved.

Some authors have attributed elastic properties to the so-called straight filament, and a spring-propeller function to the helical thread. Since in fact there is no straight filament, it cannot be elastic. As to the helical thread, its function remains unknown. However, because of its elaboration in a spiral shape, it probably does not project the colloblast head toward the prey, but rather absorbs the shocks of the movements as the captured prey struggles to be free.

The connection of each colloblast to the nervous system of the ctenophore seems to permit a regulation of the gluing phenomenon since this disappears when the animal has its pharynx full of prey. However, the existence and the mechanism of this inhibiting or activating regulation are not yet resolved.

As described above, the capture mechanism of the colloblast is self-destructive, and new colloblasts must continually be formed. This occurs at the base of the tentacle, where young cells continually proliferate and differentiate into tentacle and tentilla. This growing of the tentacular apparatus occurs at a slow rate, but it can eventually accelerate, regenerating a whole tentacle in half a day in case of accidental breakage.

For background information *see* CTENOPHORA in the McGraw-Hill Encyclopedia of Science and Technology.   [JEAN-MARIE FRANC]

Bibliography: W. Bargmann et al., *Z. Zellforsch. Mikrosk. Anat.*, 123:121–152, 1972; J.-M. Franc, *Biol. Bull.*, 155:527–541, 1978; R. Hovasse and P. de Puytorac, *C.R. Hebd. Séances Acad. Sci.*, 255: 3223–3225, 1962.

Deep-ocean mixing processes

Mixing in the ocean is clearly important for pollutant dispersal, but it also affects the climate, through its influence on sea-surface temperature, and biological productivity. Knowledge of mixing rates comes from fitting models to observations of large-scale patterns of temperature, salinity, and other tracers, as well as from various direct measurements and studies of particular stirring and mixing processes (such as ocean eddies, internal wave breaking, double-diffusive convection, and turbulent boundary layers).

Work in recent years has led to a fuller appreciation of the variety and complexity of oceanic processes than existed hitherto, and a number of general results or problems seem to be emerging: New data on the distribution of natural and anthropogenic radiochemical tracers (such as tritium) provide new clues to ocean circulation and mixing, but are still hard to interpret unambiguously. Direct measurements of mixing in the ocean interior show it to be less intense than expected. The apparent vertical mixing, deduced from large-scale patterns, may be due to mixing at ocean boundaries (the sea surface and sea floor) followed by advection into the interior, or it may be due to horizontal mixing across sloping density surfaces followed by advection along these surfaces. Internal wave breaking is probably unimportant on a global basis. Double-diffusive convection occurs in the ocean and is particularly important in boundary zones between different water masses.

A schematic and highly distorted view of the oceans is shown in Fig. 1. The left-hand boundary is the Antarctic, with the right-hand side representing a mid-latitude boundary. The top few tens of meters are thoroughly mixed by convection due to surface cooling and by turbulence associated with breaking waves and shear instabilities at the base of the layer. In polar regions in winter, surface cooling generates water of sufficient density to reach the sea floor. This bottom water spreads equatorward, slowly upwelling as it is replaced by more bottom water. Some vertical mixing is required to maintain the observed vertical temperature profile $T(z)$ against this upwelling.

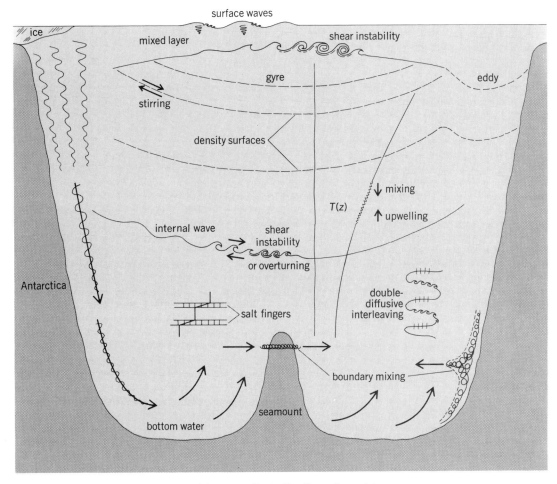

Fig. 1. A schematic and distorted view of the ocean, illustrating the various mixing processes.

Inference of mixing rates. A simple model assumes that the upwelling of cold bottom water occurs uniformly over the ocean basin with speed w, and that the vertical eddy diffusivity K_v is constant. The temperature profile is then governed by Eq. (1), with solution $T = T_0 + T_1 \exp (wz/K_v)$.

$$K_v \frac{d^2T}{dz^2} - w \frac{dT}{dz} = 0 \qquad (1)$$

Matching this to observed, smoothed profiles and using $w = 4$ meters per year (corresponding to an estimated 0.1% of the oceans per year becoming bottom water) implies $K_v \simeq 10^{-4}$ m² s⁻¹. However, analysis of time scales shows that the mean temperature profile could be maintained by intense mixing in a few locations followed by horizontal advection, rather than by a spatially uniform K_v.

The values of w, K_v can be checked by comparing measured profiles of radiochemical tracers to solutions of Eq. (1) with a decay term added. However, for isotopes originating in the bottom sediments, it again seems that horizontal advection from the sloping sides of the ocean is more important than local vertical mixing.

The balance discussed above does not apply, even in an average sense, to the ocean gyres whose density surfaces outcrop at the surface away from regions of bottom water formation. In these gyres (such as the Sargasso Sea) profiles of tritium, introduced at the sea surface as a by-product of the atmospheric testing of nuclear weapons, give useful quantitative results on apparent downward diffusion, but again with some evidence suggesting the dominance of advection along sloping density surfaces.

Large-scale horizontal mixing rates may be deduced by relating models to major features such as the tongue of warm salty water that spreads into the Atlantic from the Mediterranean, yielding a horizontal diffusivity K_h of order 10^3 m² s⁻¹. Much of what is termed horizontal mixing is in fact mixing along density surfaces, but truly horizontal mixing, across the sloping average density surfaces, can occur through the formation of detached rings of fluid. As these mix into their new environment, they produce an effect equivalent to that of vertical mixing. The process has not yet been adequately quantified.

Direct measurement. The turbulent diffusive flux of temperature T may be written as the eddy correlations $\overline{u'T'}$ horizontally and $\overline{w'T'}$ vertically, where u', w' are the fluctuations in horizontal and vertical velocity, T' is the temperature fluctuation, and the overbar denotes a time average.

Data from moored instruments in the deep ocean give fluxes equivalent to a horizontal eddy diffusivity (flux/mean gradient) of order 10^3 m² s⁻¹, but with very little statistical reliability because of the long time scale (many weeks) of the dominant fluctuations. It is worth remarking that in some areas the horizontal eddy flux of momentum acts to intensify mean velocity gradients, that is, an eddy viscosity is negative!

Direct estimates of horizontal mixing in the Sargasso Sea have been made from observations of the dispersal of neutrally buoyant, acoustically tracked floats in the sofar channel at a depth of about 1500 m (Fig. 2). Implied mixing rates are of order 10^3 m² s⁻¹ for patches more than 100 km or so (less for smaller patches). The values vary horizontally, but on the whole are compatible with deductions from large-scale balances.

In the vertical the dominant fluctuations are due to internal waves, with a time scale of a few hours, but w', T' are almost totally uncorrelated and a useful measurement of $\overline{w'T'}$ from moored sensors is not possible.

The best direct measurements of vertical mixing rates in the ocean interior have come from measurements of vertical profiles of temperature with resolution of order 10^{-3}°C in temperature and 1 mm in depth. The idea is that the persistence of small-scale fluctuations in the face of molecular diffusion requires some stirring. A quantitative theory leads to formula (2), where k is the thermal

$$\overline{K_v} = \overline{w'T'}/(d\overline{T}/dz) \simeq 2k\overline{(dT'/dz)^2}/(d\overline{T}/dz)^2 \qquad (2)$$

diffusivity (1.4×10^{-7} m² s⁻¹), T' is the fluctuation away from a smooth profile $\overline{T}(z)$ and an overbar indicates a vertical average. Measurements in many locations give an average K_v in the range 10^{-6} to 10^{-5} m² s⁻¹, much less than the 10^{-4} m² s⁻¹ expected.

Direct measurements of diffusion through use of injected dye have been successful in shallow seas, but are largely untried, and difficult, in the deep ocean interior.

Knowledge of present diffusion rates from inference or direct measurements may be adequate for some problems, such as pollutant dispersal, but greater confidence and predictability for studies of ocean circulation and climate can come only from an understanding and parameterization of the mixing processes themselves.

Mixing processes. Horizontal mixing by mesoscale eddies is being explored through the use of numerical models, with a fine (20-km) grid scale, that can resolve the individual eddies and relate their statistics to the atmospheric driving and the mean flow. Such studies will resolve the horizontal stirring, but final mixing at the molecular level probably depends on intermediate-scale processes, which are not yet understood but which seem to occur largely at fronts (where sharp horizontal gradients exist) and to be accompanied by vertical mixing.

Vertical mixing in the ocean interior may be discussed in terms of a number of possible mechanisms.

Internal waves. Water particles in the ocean interior are constantly moving to and fro on a time scale of a few hours, with horizontal excursions of a kilometer or so and vertical excursions of order 10 m. This is mainly a result of internal waves, which occasionally cause enough shear, or become steep enough, to break down into turbulence and cause vertical mixing. Until recently it seemed quite likely that energy entered the internal waves through meteorological, tidal, or other forcing and was transferred by internal interactions to small-scale waves with sufficient shear to cause turbulent mixing. However, there is now increasing suspicion that internal waves actually cause rather little mixing in general (indeed, they cannot if the direct measurements of ocean turbulence are to be believed). A vertical diffusivity of order 10^{-6} m² s⁻¹ is compatible with some ideas on the expected scale of mixing events caused by internal waves, but considerable uncertainty still exists about these ideas and about the overall energy balance of the waves.

Double-diffusive convection. For much of the ocean, the mean vertical density gradient is stable (density decreasing upward), but the density gradient associated with either the salinity or temperature gradient on its own would be unstable. Double-diffusive instabilities, which depend on the different diffusion rates of heat and salt, are then possible. One case is with hot salty water above colder, fresher water. Any blob of the hot salty water displaced downward rapidly loses its heat by thermal diffusion, but not much of its salt. It is then heavier than its surroundings and continues to sink. Such salt fingers (typically 1 mm or less across in laboratory experiments) have a width of about 10 mm in the ocean. There is a tendency for multiple convecting layers to form, separated by thin interfaces across which salt fingers transport heat and salt. There is increasing evidence that such processes occur in the ocean. The equivalent eddy diffusivity for heat and salt, using formulas based on laboratory experiments, may be of order 10^{-5} m² s⁻¹, but the flux of density is downward (equivalent to a negative eddy diffusivity) and cannot maintain the deep density structure.

Small-scale horizontal mixing. There are many locations in the ocean where water masses of different temperature and salinity, but comparable density, come into contact. This seems to lead to interleaving on a vertical scale of order 10 m, with subsequent double diffusive convection (Fig. 3).

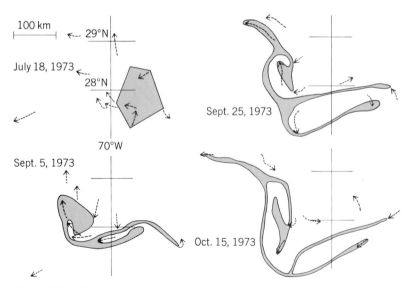

Fig. 2. Distortion, over a 3-month period, of a polygon connecting five sofar floats. The solid line is a sketch of the smoothest connecting line enclosing a fixed area, and illustrates the extent to which the ocean is stirred horizontally. (From P. B. Rhines, *Geostrophic turbulence, Annu. Rev. Fluid Mech., 11:401–441, 1979*)

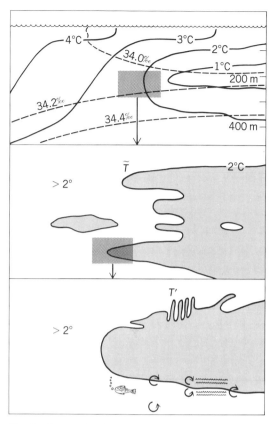

Fig. 3. Schematic of different scales of mixing near a water mass boundary. The large-scale front breaks down into medium-scale intrusions and small-scale mixing. *(From T. M. Joyce, A note on the lateral mixing of water masses, J. Phys. Oceanogr., 7:626–629, 1977)*

Theory and careful laboratory simulations are helping oceanographers understand this process, but there is not yet a precise parameterization of its role in ocean mixing.

Boundary mixing. The discrepancy between direct measurements of vertical mixing in the ocean interior and the apparent requirement of large-scale balances can perhaps be reconciled if the mixing occurs at the sloping sides of ocean basins and seamounts, followed by advection into the interior. Direct estimates of the efficiency of this process are hampered by lack of knowledge of bottom currents, which cause the mixing, and of the way in which mixed fluid escapes from the boundaries. Estimates based on energy requirements make boundary mixing seem plausible for the bottom kilometer or two of the ocean, but less likely at mid-depth.

For the top kilometer or so of the mid-latitude oceans, density surfaces are connected to the sea surface, not the sea floor. A quantitative study of boundary mixing associated with the sea surface is only just beginning.

The future. Oceanographers now have a substantial body of precise data on tracer distributions and small-scale fluctuations, and an increasing understanding of individual mixing processes, together with the tools to investigate them further. The next decade should bring considerable progress toward a detailed understanding of ocean mix-

ing, and a correct parameterization of it in terms of large-scale external forcing.

For background information *see* OCEANOGRAPHY in the McGraw-Hill Encyclopedia of Science and Technology.

[C. J. R. GARRETT]

Bibliography: W. S. Broecker and H. Göte Östlund, *J. Geophys. Res.*, 84:1145–1154, 1979; K. N. Fedorov, *The Thermohaline Fine Structure of the Ocean*, 1978; C. J. R. Garrett, in *Dynamics of Atmospheres and Oceans*, pp. 239–265, 1979; M. C. Gregg and M. G. Briscoe, *Rev. Geophys. Space Phys.*, in press.

Dielectrics

The resonance mechanism of dielectric polarization based on the harmonic oscillator model represents a good approximation to the response of materials at frequencies above 10^{12} Hz, that is, in excess of the infrared region of the spectrum. In these circumstances, the inertial processes dominate over viscous effects and the response may be treated as arising from simple summation of one-particle effects. At lower frequencies, however, and this means in the entire technologically significant radio-, audio-, and power-frequency ranges, the response is dominated by viscous forces and yet the classical Debye model is wholly inadequate to describe the behavior. Significant developments have taken place recently in the understanding of the dielectric response of solids in the low-frequency range extending from 10^{10} Hz to as low as 10^{-5} Hz—30 hr per cycle—using many-body interactions.

The commonly considered polarizing species are permanent dipoles found typically in molecular solids such as polymers and arising from inherent charge separation in molecules, for example, C^+-Cl^-. The two charges of opposite sign are

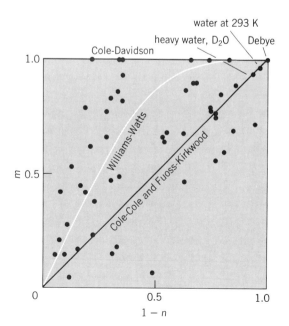

Fig. 1. Plot of the exponents m and $1-n$ determined for 50 different loss peaks corresponding to a wide range of polymeric and nonpolymeric dipolar solids and liquids. *(From R. M. Hill, Characterization of dielectric loss in solids and liquids, Nature, 275:96–99, 1978)*

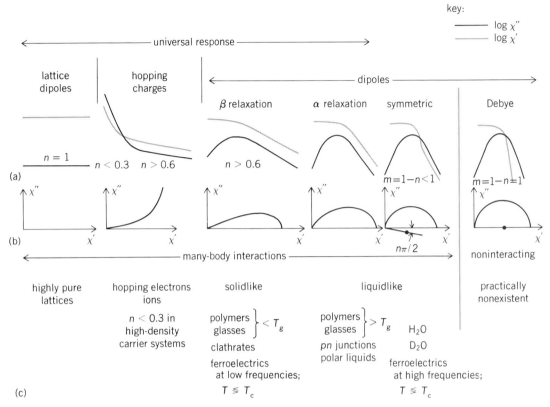

Fig. 2. Complete classification of the various observed types of dielectric response of all the known solids in the "low"-frequency region. (a) Schematic plots of $\log \chi'$ and $\log \chi''$ against $\log \omega$. (b) Corresponding schematic plots of χ' against χ''. (c) Typical materials following the various types of response, together with the suggested mechanisms responsible. T = temperature of dielectric; T_c = Curie temperature, T_g = glass transition temperature. (From K. L. Ngai, A. K. Jonscher, and C. T. White, On the origin of the universal dielectric response in condensed matter, Nature, 277:185–189, 1979)

kept at a fixed distance, but their orientation is not rigidly specified and they may align themselves in an external field, thus giving rise to polarization. However, the same effect may be achieved if one has oppositely charged carriers such as ions or localized electrons which are not held rigidly at a fixed distance apart but may move within certain limits. They are in many ways indistinguishable from dipoles, except that in the limit of long times and high electric fields they may become completely separated from one another, giving rise to conduction of steady current.

Universal response. The actual frequency response of the great majority of dielectric materials has long been known to depart drastically from the ideal Debye model for which the complex dielectric susceptibility χ, that is, the contribution of the material medium to the total dielectric permittivity $\epsilon = \epsilon_0(1 + \chi)$, follows the law given by Eq. (1),

$$\chi(\omega) = \chi'(\omega) - i\chi''(\omega) = (1 + i\omega\tau)^{-1} \qquad (1)$$

where ω is the angular frequency, χ' and χ'' are the real and negative imaginary parts of χ, and τ is the relaxation time. The departures are often so drastic that there is no similarity with Eq. (1), and the explanations for these departures are as many and varied as the different materials to which they purport to apply: polymers, glasses, ceramics, crystalline solids like mica, amorphous semicon-

ductors, and so on. However, a detailed survey by A. K. Jonscher has revealed the surprising experimental fact that the susceptibility follows an empirical "universal law" given by relation (2), where

$$\chi(\omega) \propto (i\omega)^{n-1} \qquad (2)$$

the exponent n falls in the range $0 < n < 1$. The universal response is followed normally over many decades of frequency above any loss peak frequency ω_p that may be observable in dipolar materials, and the complete description of the loss peaks is given by empirical relation (3), where the exponent

$$\chi''(\omega) \propto \left\{ (\omega/\omega_p)^{-m} + (\omega/\omega_p)^{1-n} \right\}^{-1} \qquad (3)$$

m falls in the same range as $1 - n$. Relation (3) gives limiting form (2) for $\omega \gg \omega_p$, while for $\omega \ll \omega_p$ the loss follows the law $\chi''(\omega) \propto \omega^m$.

Classification of observed responses. The classical Debye response corresponds to $m = 1 - n = 1$. The actual response of a wide range of solid and liquid dipolar materials is shown in Fig. 1, which gives for every material one point with coordinates $(m, 1 - n)$. Superimposed on the plot are the regions of validity of the empirical functions due to K. S. Cole and R. H. Cole, R. H. Cole and D. W. Davidson, R. M. Fuoss and J. G. Kirkwood, and G. Williams and D. C. Watts. Figure 1 shows clearly that the pure Debye response is not seen in any of the materials plotted, even though water and heavy water, D_2O, fall close to it. It is concluded that the exponents m and n in relation (3) are not correlated, suggesting that they correspond to different physical mechanisms.

On the other hand, most dielectrics in which the polarization is dominated by displacements of hopping electrons or of ions show at low frequencies a second universal law with a smaller value of the exponent n than the high-frequency response, giving a strongly dispersive behavior. This may be regarded as an example of the widely known Maxwell-Wagner dispersion, for which no plausible model appears to have been proposed so far either. A complete classification of all types of observed dielectric response is given in Fig. 2, which shows the plots of log χ' and log χ'' against log ω and also the corresponding plots of χ' versus χ''. The figure also lists typical materials and indicates specific mechanisms.

Many-body interactions. The wide applicability of universal relation (2), regardless of the detailed chemical structure, chemical properties, or the prevailing type of polarizing species, has led Jonscher to propose that the dominant physical processes are not the classical "viscous" forces acting on individual dipoles or charges, as in the traditional approach, but that a much more fundamental and all-embracing mechanism is at play in the form of many-body interactions which are to be expected, anyway, in all condensed matter. It is also significant that all three polarizing species in solids have the common property of moving by discontinuous jumps between preferred orientations or positions.

The Fourier transform of universal law (2), which represents the time dependence of the polarizing current under step-function excitation, is of the form of relation (4), and it was noted by K. L. Ngai that this is the same as the type of many-body response known under the name of infrared divergence. Ngai, Jonscher, and C. T. White have described many-body interactions which are present in most dielectrics and which would be expected to give rise to observed universal response (2) and (4) by a process analogous to infrared divergence. The exponent n can take on any value in the empirically determined range (0,1).

Very recently this argument has been extended further by L. A. Dissado and R. M. Hill to cover the general form of response of interacting dipolar systems, accounting fully for empirically determined relation (3). Their model of interacting dipoles having a net dipole moment is represented by the double potential well of Fig. 3. The surprising conclusion is that simultaneous transitions by many particles require much lower energies than would be required had only one particle been moving at a time. This explains, in particular, the remarkable fact that the dielectric loss, unlike almost any other transport property, retains high values down to temperatures near absolute zero. The exponents m and n are explained in terms of the detailed nature of specified many-body transitions.

This many-body interpretation of the dielectric response constitutes therefore a completely fresh start, and it achieves at once two separate and important objectives. It explains for the first time in a coherent manner the entire range of the observed dielectric responses in all types of solids at "low" frequencies, thus laying the foundation for a proper interpretation of a wide range of other dielectric phenomena, such as transport of charges and the nature of dielectric breakdown. It also shows that the dielectric response provides a uniquely sensitive tool for the study of many-body interactions in a much broader context where they would not be so readily accessible by other experimental methods.

For background information *see* DIELECTRIC CONSTANT: DIELECTRICS in the McGraw-Hill Encyclopedia of Science and Technology.

[ANDREW K. JONSCHER]

Bibliography: L. A. Dissado and R. M. Hill, *Nature*, 279:685–689, 1979; R. M. Hill, *Nature*, 275:96–99, 1978; A. K. Jonscher, *Nature*, 267:673–679, 1977; A. K. Jonscher, The universal dielectric response: A review of data and their new interpretation, *Phys. Thin Films*, in press; K. L. Ngai, A. K. Jonscher, and C. T. White, *Nature*, 277:185–189, 1979.

Echinodermata

Recent paleobiological studies of fossil echinoderms by B. N. Haugh have revealed anatomical details about the visceral organization of several extinct classes. Among the organs preserved three-dimensionally in these remarkable fossils are the digestive system, several coelomic systems (including the hydrocoel), and several nervous systems. The amazing fidelity of preservation is due primarily to the presence of a fairly rigid, biologically secreted, spicular framework that pervaded much of the soft internal tissues of the once-living organism. As might be expected, some of the preserved viscera generally resemble visceral components of living forms. Minor discrepancies in relative size or shape of a given organ are to be expected in the light of the hundreds of millions of

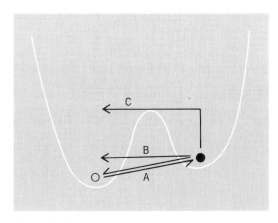

Fig. 3. Double potential well representing a system of interacting dipoles. A and B represent tunneling transitions with very little thermal energy, and C represents thermally activated transitions over the potential barrier. Many-body tunneling transitions are possible not only for very light particles, such as electrons or protons, but also for relatively much heavier ions or dipoles. (From L. A. Dissado and R. M. Hill, Non-exponential decay in dielectrics as a consequence of the dynamics of correlated systems, Nature, 279:685–689, 1979)

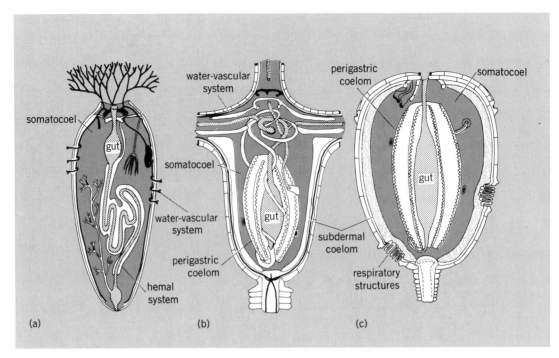

Fig. 1. Diagrams of the visceral anatomy of three echi-noderms representing informal subphyla based on visceral anatomy: (a) a holothurian (subphylum III), (b) a camerate crinoid (subphylum I), and (c) a rhombiferan cystoid (subphylum II). Corresponding graphic patterns indicate possible homology. (*Courtesy of Bruce N. Haugh*)

years that separate fossil and living forms. In other instances, fossilized organs are truly bizarre in that they have no readily distinguishable close modern analogs. Among these bizarre fossilized organs are a convoluted perigastric coelom which coils around the gut, and a perivisceral or subdermal coelom that occurs as a mineralized lining internally adjacent to the plated thecal wall of the organism (Fig. 1b,c). Because of these discrepancies in

visceral anatomy, a comprehensive reclassification of echinoderms, at high taxonomic rank, is probably warranted. Major anatomical differences such as those described also indicate the possibility of fundamentally different embryological development, which might specifically characterize high-ranking taxa of extinct echinoderms.

Other recent paleobiological studies by B. M. Bell, who investigated growth sequences in the extinct class Edrioasteroidea, also support this assertion. The small individuals (0.5–2 mm diameter) of such growth sequences demonstrate that triradiate ambulacral symmetry characterized the early stages of ontogeny. This primary triradiate symmetry is modified into pentaradiate symmetry in larger adult forms. However, remnants of the triradiate symmetry plan are retained in the adults (Fig. 2d). Significantly, comparable cases of triradiate symmetry are known in adult forms of several other classes of extinct echinoderms. Onefold and twofold symmetry, as well as asymmetry, characterize certain other fossil taxa. Hence, not all echinoderm taxa were primarily pentameral. The diverse ambulacral symmetry patterns among fossil forms, as opposed to strictly pentameral symmetry in living forms, indicate greater fundamental morphological diversity within extinct members of the phylum. According to studies conducted jointly by Haugh and Bell, this diversity should be, but has not yet been, taken into account in the overall classification of echinoderms. As a generalization, one is compelled to conclude, based on fossilized visceral anatomy and ambulacral symmetry, that the four living classes of echinoderms do not provide a morphologically representative sample; nor do they likely provide a physiologically or ontogenetically representative

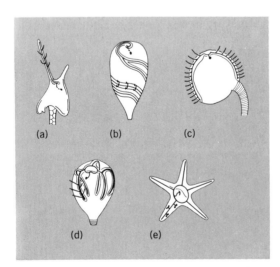

Fig. 2. Diversity of primary ambulacral symmetry patterns in echinoderms: (a) asymmetry; (b) onefold symmetry; (c) twofold symmetry; (d) threefold (2-1-2) symmetry; and (e) fivefold symmetry. Patterns a, b, and d appear in the Early Cambrian; c and e appear in the Middle Ordovician. Only the regular pentameral symmetry of e is known to be extant. (*Courtesy of Bruce N. Haugh*)

sample of the phylum which, in the Paleozoic Period, featured 20 classes of organisms.

Visceral paleobiology. Anatomical investigations by Haugh have revealed details of the visceral morphology of the camerate crinoids, a prominent group of extinct Paleozoic echinoderms (Fig. 1b). Their visceral anatomy differs significantly from that of extant classes (Fig. 1a) by virtue of the presence of a convoluted perigastric coelom, which wraps around the gut, and the presence of a subdermal coelom that shrouds all the remaining viscera. The perigastric coelom may be functionally comparable to the hemal system in living echinoderms (Fig. 1a). The subdermal coelom of camerates has no apparent analog in living echinoderms. Points of similarity among extinct camerates and living echinoderms include the somatocoel and the hydraulic water-vascular system. Certain nervous systems may also be homologous.

Until quite recently, details of the visceral paleobiology of extinct taxa, other than camerates, have been largely conjectural. However, investigations being conducted jointly by Haugh and Bell have revealed the presence of paleoviscera in several other classes of extinct echinoderms: edrioasteroids, blastoids, diporite cystoids, and rhombiferan cystoids. There exists, therefore, the rare opportunity to compare the visceral organization of several extinct and living classes of echinoderms.

Excellent three-dimensional organ preservation has been found in specimens of rhombiferan cystoids (Fig. 1c). Analysis of rhombiferans from the Rochester Shale of New York confirmed the presence of a coiled fusiform structure that is almost certainly homologous to the perigastric coelom in camerate crinoids. Also preserved are the remains of a subdermal coelom which is associated with thecal respiratory structures, termed respiratory rhombs. Thus, the subdermal coelom probably had a respiratory function. Major visceral differences among camerates and rhombiferans are also evident (Fig. 1b,c). The rhombiferans lack radially disposed openings through the body wall that could accommodate tubular extensions of body coeloms (such as a water-vascular coelom) and

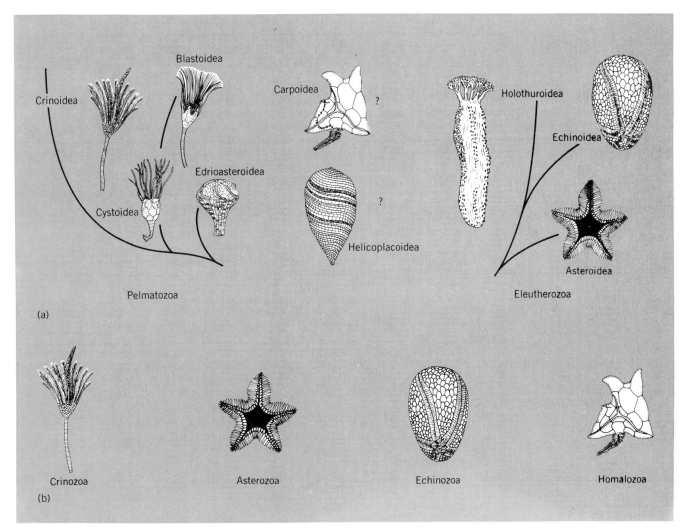

Fig. 3. Formally established, high-rank classification schemes: (a) subphyla Pelmatozoa and Eleutherozoa appear to represent ecological convergence among forms with an attached versus a free life-mode; (b) subphyla Crinozoa, Asterozoa, Echinozoa, and Homalozoa appear to represent groupings of taxa with general shape convergence. (*Courtesy of Bruce N. Haugh*)

nerve trunks (such as the aboral motor nervous system). The tubular extensions and nerve trunks characterize true exothecal ambulacral structures, euambulacra, in all living and in many fossil taxa (Fig. 1*a,b*). Cystoids likely possessed simple, acoelomate parambulacra, in the terminology of Haugh and Bell. Parambulacra appear to have been ciliate food-gathering tracts that, in all but a few cases, are recumbent on the theca. At least one of the two nervous systems preserved in parambulacra does not appear to be homologous with any nervous system in euambulacra.

The visceral anatomy revealed by Haugh and Bell's studies, therefore, establishes three morphological groups recognizable by the presence of specific coelomic structures and nervous systems, and delimited as follows (Fig. 1): A hydraulic water-vascular system is present in living forms and in camerate crinoids, but is lacking in rhombiferan cystoids. A subdermal respiratory coelom is present in camerate crinoids and rhombiferan cystoids, but is lacking in living forms. Two, and perhaps three, of the nervous systems in camerate crinoids and living forms appear to be homologous, and are distinct from the known nervous systems in rhombiferan cystoids. The convoluted perigastric coeloms associated with the gut in rhombiferans and camerates are probably homologous, and may be only functionally analogous to the hemal system in extant classes.

Yet-unpublished data permit the assignment of most of the other extinct classes of echinoderms to one or the other of the three, currently informal, subphyla (Fig. 1). The resulting high-rank classification is at odds with two formally established, alternative schemes (Fig. 3), which are currently favored by various neontologists and paleontologists.

Primary ambulacral symmetry. A controversy has raged for many years regarding the primary ambulacral symmetry of primitive echinoderms. Many paleontologists have suggested that the earliest echinoderms were bilaterally symmetrical in terms of coelomic anatomy, like the hypothetical "dipleurula larva," and that early in the history of the phylum, ontogenetic evolution produced a coelomically degenerate adult with threefold adult ambulacral symmetry. The earliest potential echinoderm fossil, *Tribrachidium*, from the late Precambrian, may be an example of this stage of evolution. The scenario suggests that subsequent evolution produced a branching of the two lateral rays of the triad, resulting in a fivefold branching pattern with remnants of the triad near the mouth—commonly termed a 2-1-2 pattern (Fig. 2*d*). Later on, degeneration of the 2-1-2 pattern yielded derived twofold and onefold symmetries, as well as asymmetry (Fig. 2*a–c*); whereas perfection of the 2-1-2 pattern produced the regular pentameral symmetry manifest in extant classes and their fossil ancestors (Fig. 2*e*). Conversely, many neontologists, on the basis of ontogeny in living forms, believe that it is unlikely that there were ever any adult echinoderms with a bilateral complement of coeloms; and further, that pentameral symmetry is primitive and other patterns are derived.

Bell's study of edrioasteroid growth series strongly supports the hypothesis that triradiate symmetry is indeed a primitive echinoderm character. A subsequent investigation of other extinct echinoderms revealed that at least seven extinct classes are characterized by 2-1-2 adult symmetry. Therefore, these taxa may have evolved from primitive ancestors with adult triradiate symmetry; and their early ontogenetic stages were almost certainly triradiate. Based on fossil evidence, Haugh and Bell concluded: (1) No one primary ambulacral symmetry pattern was fixed in primitive echinoderms during the ontogeny of the early bilateral larva. Onefold symmetry and threefold (2-1-2) symmetry are manifest earliest in the geologic record, and characterize the majority of extinct taxa, with fivefold (regular pentameral) and a few twofold taxa appearing later on (Fig. 2). (2) These diverse symmetry patterns can serve as a basis to delimit several superclasses because they likely represent early ontogentic derivations.

A classification based on visceral anatomy and primary ambulacral symmetry is to be preferred because it likely reflects primitively established ontogenetic patterns, and thereby delimits phylogenetically related high-rank taxa. If this is the case, then the "two-subphylum system" (Fig. 3*a*) merely expresses gross ecological convergence; and the "four-subphylum system" merely expresses gross shape convergence (Fig. 3*b*).

For background information *see* ECHINODERMATA; PALEONTOLOGY in the McGraw-Hill Encyclopedia of Science and Technology.

[BRUCE N. HAUGH]

Bibliography: B. M. Bell, *J. Paleontol.*, 50: 1001–1019, 1976; B. N. Haugh, *J. Paleontol.*, 49: 472–493, 1975; B. N. Haugh and B. M. Bell, *Geol. Soc. Amer. Abstr. Prog.*, 10:417–418, 1978; R. C. Moore (ed.), *Treatise on Invertebrate Paleontology*, pts. S and U: *Echinodermata*, 1966–1967.

Egg (fowl)

The developing avian embryo is enclosed in a rigid, calcite shell which only the respiratory gases (O_2 and CO_2) and water vapor move across. Much recent work with avian embryos has centered on analyzing the gas exchange of the egg and has led to insight into the physiological requirements of embryonic development and the way in which these requirements interact with the physical and biological environment of the egg. The eggshell, in a functional sense, satisfies time and energy constraints placed on the embryo by the need to maintain water balance and exchange respiratory gases in a variety of ecological situations. Most bird eggs appear to be required to lose a fixed fraction of their initial mass in the form of water and to produce similar respiratory environments for the embryo despite large variation in egg mass and incubation time. At the same time all bird eggs seem to require similar quantities of O_2 per unit of egg mass during incubation.

Shell pore geometry and conductance. The avian egg shell is traversed by many tiny tubular pores through which gases move by random molecular motion from an area of high concentration to one of low concentration. Movement by diffusion means that a concentration gradient (often expressed in terms of partial pressure) must be established across the shell if a net exchange of gas is to occur. Conversely, if a gradient is present, net gas exchange will occur. The diffusion process can be described quantitatively by relating the flux of the gas across the shell (mass or volume per unit time) to the shell conductance (mass or volume per unit time per unit pressure gradient) multiplied by the gas pressure gradient across the shell (partial pressure). The conductance of the eggshell (which is the inverse of the resistance of the shell to gas movement) is directly proportional to the ratio of total pore area to pore length (the pore geometry) and to the diffusion constant of the gas, and inversely proportional to the gas constant and the absolute temperature of the system.

Since the pore geometry is determined as the shell is deposited on the egg in the maternal oviduct, and the other factors may be treated as constants (in a given environment), the shell conductance should not change during incubation unless the integrity of the shell is challenged. However, the pore geometry and therefore the eggshell conductance may vary from egg to egg, and certainly vary from bird species to bird species. Furthermore, the physical environment can also influence the eggshell conductance since the diffusion constant term in the conductance is inversely proportional to pressure (altitude) and directly proportional to temperature. The temperature effect on the shell conductance is relatively small in the range of biological interest, but the pressure effect can be substantial: a 4000-m increase in altitude nearly doubles the diffusion constant and the eggshell conductance.

Egg water balance. The water vapor content of the environment is usually less than that found inside eggs, so water vapor is lost continually throughout incubation. Research early in the 1970s suggested that the daily water loss from the egg (measured as the change in egg mass) was a relatively constant function of egg mass, increasing as egg mass increased. Subsequent work showed that eggshell conductance was directly related to egg mass, and inversely related to incubation time. These alterations in shell conductance were due to species-specific changes in pore geometry. The relationship between eggshell conductance and daily water loss, egg mass, and incubation time predicts that all bird eggs should lose about 15% of their initial mass as water vapor during incubation and that the vapor pressure gradient across the shell is similar for all bird species and relatively constant throughout incubation. The fractional water loss is apparently obligatory since eggs prevented from losing water show increased embryonic mortality and deformity. The obligatory nature of egg water loss may be related to the observation that the embryo inflates its lung just prior to hatching with the gas in the air space of the egg. The volume of the air space is related to the quantity of water loss since the egg shell is rigid. Measurements of egg mass change on a variety of species (Fig. 1) ranging in egg mass from 20 to 600 g and in incubation time from 20 to 50 days appear to verify the predicted mass loss.

The measurements of change in egg mass during incubation would also seem to verify the prediction that the vapor pressure gradient is constant for all birds. This is interesting because birds

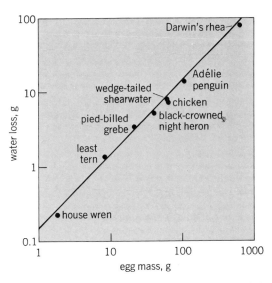

Fig. 1. Mass of water lost by eggs of different mass plotted on logarithmic axis. The line represents the mass of water lost during incubation equal to 15% of the initial egg mass.

breed in a variety of environments where average vapor pressure may vary from 4 torrs (0.5 kilopascals; Antarctic and Arctic) to over 30 torrs (4.0 kPa; marshy, wet nests). Furthermore, the vapor pressure gradient predicted exists between the inside of the egg and the gas in the nest. If the ambient vapor pressure is lower than expected (as may be the usual case), the incubating adult should play a role in elevating the vapor pressure of the nest to an appropriate level. Recent work with pheasant eggs suggests that this may true. Eggs incubated in very moist environments or at high altitude, where the diffusion of water vapor through the shell is increased, appear to meet water loss constraints in another way. Since parental modification of nest

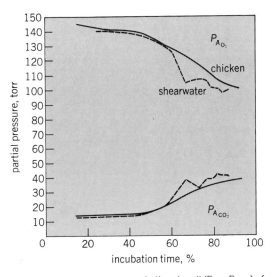

Fig. 2. Gas partial pressures in the air cell ($P_{A_{O_2}}$, $P_{A_{CO_2}}$) of chicken and wedge-tailed shearwater eggs measured throughout incubation. The incubation period of the chicken egg is 21 days, and 52 days for the shearwater egg. Measurements of air-cell gas partial pressures terminate just prior to eggshell pipping.

vapor pressure may be of little use in these environments, the water vapor conductance of the shell is altered to accommodate the new environment. In the case of wet environments, shell water vapor conductance is increased, while at altitude, the shell water vapor conductance is decreased.

Respiratory gas exchange. While the daily water loss from avian eggs during incubation appears to be relatively constant, such is not the case for the daily oxygen consumption (O_2 flux) and carbon dioxide production (CO_2 flux). The metabolic activity increases throughout incubation as the embryo develops and increases in mass, so both the O_2 and CO_2 exchange must similarly increase. Since the gas conductance of the shell is constant (all gases diffuse through the same pores), the pressure of O_2 inside the egg must fall and the pressure of CO_2 must rise if the embryo is to gain O_2 and lose metabolically produced CO_2. This appears to be so, as just prior to hatching the P_{O_2} and P_{CO_2} (as measured in the air space of the egg) reach about 100 and 40 torrs (13.3 and 5.3 kPa) respectively (Fig. 2). The egg O_2 exchange of a variety of bird species measured just prior to hatching appears to increase in proportion to egg mass in much the same way that the eggshell gas conductance (as measured with water vapor) increases with increase in egg mass. These relationships allow one to predict that P_{O_2} and P_{CO_2} (in the air space) are the same for all bird embryos just prior to hatching and are in fact similar to values one might expect to see in adult bird lungs. It would appear that one function of the eggshell is to prepare the embryo for transition to lung breathing at birth.

The O_2 flux just prior to hatching is a function of incubation time as well as egg mass. As incubation

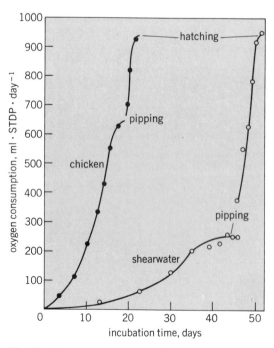

Fig. 3. Oxygen consumption (M_{O_2}) of chicken and wedge-tailed shearwater eggs measured throughout incubation.

time increases, the O_2 flux tends to decrease. This is clearly shown in Fig. 3, where the lines represent the O_2 fluxes throughout incubation of chicken and wedge-tailed shearwater *(Puffinus pacificus)* eggs. The eggs of both species weigh about 60 g, eliminating the effect of change in egg mass. As hatching begins, the oxygen uptakes of both eggs increase but that of the shearwater increases more so that hatchlings of both species consume similar quantities of O_2. The lower O_2 consumption of the shearwater egg during prehatching incubation is probably related to the observation that the embryo is growing much more slowly than the chicken embryo. The area under the curve described by plotting O_2 consumption against time prior to the initiation of hatching is similar in both species and yields a total quantity of O_2 consumed equal to 90 ml O_2 standard temperature and pressure, dry (STPD) per gram of egg. This is consistent with an earlier prediction that all bird eggs ought to consume the same quantity of O_2 per gram of egg prior to the initiation of hatching.

For background information *see* EGG (FOWL) in the McGraw-Hill Encyclopedia of Science and Technology.   [RALPH A. ACKERMAN]

Bibliography: C. V. Paganelli, R. A. Ackerman, and H. Rahn, in *Respiratory Function in Birds, Adult and Embryonic*, pp. 212–218, 1978; H. Rahn, R. A. Ackerman, and C. V. Paganelli, *Physiol. Zool.*, 50:269–283, 1977; H. Rahn, A. Ar, and C. V.. Paganelli, *Sci. Amer.*, 239:46–55, 1979; H. Rahn, C. V. Paganelli, and A. Ar, *Resp. Physiol.*, 22:297–309, 1974.

Electric power generation

Cogeneration, an old technology with a new name, began in 1977–1978 to challenge the underlying concepts of appropriate power generation. Defined as the simultaneous generation of electric energy and useful low-grade heat from the same source, cogeneration is reemerging as a viable alternative to central generation of electric power. This 60-year-old technology of combined heat and electric power generation was renamed on Apr. 19, 1977, when President Jimmy Carter, in unveiling the National Energy Plan, introduced cogeneration to the electric power generation lexicon.

This article explores several features of cogeneration, including on-site and mass-produced plants, district heating, energy efficiencies, past economic deterrents, other remaining obstacles, and the challenge to the underlying concepts of electric generation.

To cogenerate, or recapture and use the low-grade heat remaining after generating electric power, two approaches are possible: moving electrical generation to the user's facility, and moving the waste heat to the user.

Cogeneration at user's site. This first approach has been commonly practiced with large industries, but has declined as electric rates fell. In 1950, 15% of United States electricity was cogenerated in large on-site industrial plants, but the percentage fell to 4% by 1970. These plants typically have 15–150 MW of capacity and make high-pressure steam, creating electricity via a back-pressure turbine which drives an electric alternator, and then feeds the 0.21–0.28-kilopascal

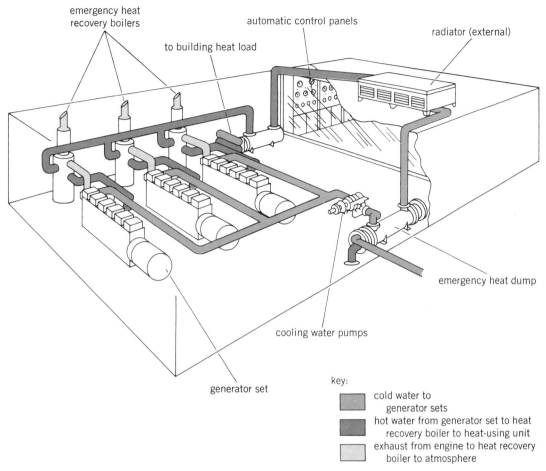

key:
cold water to generator sets
hot water from generator set to heat recovery boiler to heat-using unit
exhaust from engine to heat recovery boiler to atmosphere

Fig. 1. Typical three-diesel-engine cogeneration plant.

(150–200-psig) exhaust steam to plant processes — using the industrial plant as a condenser.

Mass-produced plants. Since 1960, roughly 600 on-site cogenerating plants using diesels or combustion engines have been constructed in the 200-kW to 6-MW range, using prime movers or engines that were mass-produced for mobile power markets. To these prime movers — diesels, natural gas engines, and gas turbines — are attached alternators to generate electricity, heat recovery devices, and suitable controls (Fig. 1).

Any two of the diesel generator sets in Fig. 1 can meet peak facility electrical load using 11,000–11,600 kilojoules per kilowatt-hour (10,500–11,000 Btu/kWh) generated. A control panel monitors key parameters and automatically matches input fuel to facility electric loads, holding frequency constant. The internal logic automatically shuts down engines which are unneeded or malfunctioning, starts up engines, and electrically parallels generators as loads rise.

Cooling water circulates through the engine block, gaining roughly 2530 kJ (2400 Btu) of power per kilowatt-hour of electricity produced. Heated block water then passes through exhaust recovery boilers, capturing heat from the 480–530°C exhaust gases.

The heated engine coolant passes through heat exchangers, and warms facility-process or space-heating water. Any remaining heat is removed in a cooling tower before the coolant is pumped back to the engine circuit. A variation in plumbing produces up to 0.22-kPa (150-psig) steam from the exhaust gases or low-pressure steam from the engine and exhaust.

When the prime mover is a turbine, higher-input fuel per kilowatt-hour is required, but all waste heat is in exhaust gases, allowing recapture of high temperature hot water or steam up to 200°C.

Cogeneration by district heating. District heating plants that centrally generate electricity and simultaneously supply steam or hot water via a network of underground pipes represent the other approach to cogeneration. Such plants have operated in the United States since 1877, typically in dense center-city areas. About 250 utilities operate district heating plants in the United States today, over 100 cogeneration plants feed district heating systems in Germany, and over 30% of all power generated in the Soviet Union comes from district heating cogeneration plants. The older systems supplying 0.22-kPa (150-psig) steam with no condensate return span a maximum radius of 3.5 mi (5.7 km). The efficiency of these older systems is hampered by the relatively low thermal efficiency of electric power generation, escaping steam vapor, losses to surrounding earth, and loss of condensate. The present United States systems nearly all suffer

Relative fuel efficiencies of various generating technologies

Type of energy converter	Gross kJ/kWh (Btu/kWh)	Recoverable kJ/kWh (Btu/kWh)	Net electrical heat rate, kJ/kWh* (Btu/kWh)	Net electrical heat rate as a percent of central stations
Central fossil boiler, nondistrict heating	12,500 (11,800)	–	12,500 (11,800)	100
Central fossil boiler, district heating	13,200–17,950 (12,500–17,000)	6300–9000 (5950–8500)	4800–5950 (4550–5650)	39–48
Low-speed diesel (6–30 MW each)	8650–9080 (8200–8600)	1270–1310 (1200–1240)	6950–7350 (6600–6950)	56–59
High-speed diesel (0.2–1-MW capacity)	11,100–12,100 (10,500–11,500)	4650–4850 (4400–4600)	4900–5650 (4650–5350)	39–45
Combustion turbines (0.8–2.5 MW)	18,500–21,100 (17,500–20,000)	8500–11,100 (8000–10,500)	6300–7150 (5950–6750)	50–58

*Assumes a 75% boiler efficiency in producing equivalent recoverable heat with direct-heat fossil-fired boiler.

from age and are generally considered unprofitable. The inherent energy inefficiencies and the cost of maintaining older systems have recently combined to make the cost of central steam greater than the cost of noncogenerated steam from on-site boilers. This form of cogeneration has thus been losing market penetration and declining in both absolute and relative terms.

Newer district heating systems use high-temperature water, return cooled water to the generating stations, and can operate at 65–70% overall efficiency. Their use is economically attractive, for example, on campuses where single owners control all land. However, given pipe costs of $400 to $1200 per foot ($1300 to $3900 per meter) and many legal and physical barriers to installing new systems in builtup areas, district heating as a form of cogeneration faces difficult expansion prospect.

Fuel efficiencies of generating technology. Current central generation of electricity without heat recapture delivers on average only 29% of the input heat energy to the end user. By contrast, cogeneration has theoretical efficiencies of up to 75–80% and commonly delivers 60–65% of input energy. Some currently utilized energy converters and their efficiencies are given in the table. In each case, the waste energy can be utilized for low-temperature-process heat, space heating, or cooling with absorption chillers. However, low-grade byproduct energy does not transport over long distances efficiently and requires large expenditures in underground insulated piping if the heat user is any distance from the generation site.

Past economic deterrents. Oil-fired cogeneration, replacing conventional oil-fired systems of central generation and on-site boilers, saves on average 111–127 m³ (700–800 barrels) of oil per MM (million) kWh, and can save 160 m³ (1000 barrels) of oil per MM kWh. Nonetheless, cogeneration declined in use between 1950 and 1970. Several reasons explain why larger-scale industrial cogeneration has declined and why cogeneration in smaller sizes is only now emerging.

Between 1950 and 1969, the average cost per kilowatt of central generating capacity fell in real terms, due to economies of scale and a stable design climate, while real energy prices fell by 42%. Since the late 1960s, new central generating stations have encountered new environmental requirements, inability to further increase plant size,

litigation delays, and requirements for highly specific designs causing quintupling of their installed cost per kilowatt. Figure 2 shows the change in price of central generating capacity between 1967 and 1978. On-site cogeneration plants' average cost per kilowatt were about $275 in 1967, or 2.2 times the cost of new central capacity. By 1978, on-site plants at $600 per kilowatt of capacity were costing under 50% of the cost of new central capacity.

Equally significant to cogeneration's economic viability has been the trend toward increasing the value of low-grade heat. Between 1950 and 1972, real energy prices fell from an index of 100 to 58, lowering the value of recovered heat every year. In the past 5 years, the cost of heat energy has increased 5–18 times. In 1970, interstate pipeline gas sold for 10¢ per MM kJ (11¢ per MM Btu), and diesel oil at 10¢ per gallon ($26 per cubic meter) cost 68¢ per MM kJ (72¢ per MM Btu). By 1978, pipeline gas, when available, cost $1.89 per MM kJ ($2.00 per MM Btu), an increase of 1800%. Diesel oil has risen 500%. The impact has been to increase the recovered heat value of cogeneration.

Finally, prior to the mid 1970s, smaller cogeneration plants were largely uneconomical since they required daytime or even full-time operators, costing $40,000 to $175,000 per year. This effectively eliminated all but larger facilities from on-site cogeneration prior to 1975–1976.

Unattended cogeneration plants in the 200-kW to 6-MW range have been made possible by improvements in control technology and general increases in product reliability. One major manufacturer of diesel engines now offers unattended, automatically operated cogeneration plants with guarantees of fuel efficiency and reliability. These new options remove a major hurdle for on-site cogeneration at smaller plants.

Other obstacles to cogeneration. Several problems which bar widespread moves to cogeneration at this time include shortage of experienced well-financed vendors of complete cogeneration systems, current central utility overcapacity, environmental concerns, and national fuel use policies.

System development. Cogeneration systems combine several distinctly different pieces of equipment, typically of diverse manufacture, which then must operate together, such as engines or boilers, generators, control equipment, heat

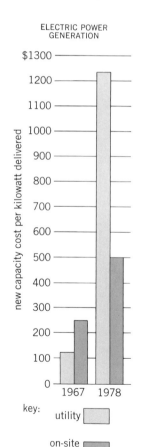

ELECTRIC POWER GENERATION

Fig. 2. Relative costs for electric generating capacity in 1967 and 1978.

exchangers, pumps, sensors. Many past attempts to cogenerate experienced extensive debugging and unplanned system startup problems. Vendors of complete cogeneration systems who have credible backing are only beginning to emerge, and their absence has slowed commericial development of cogeneration.

Central utility overcapacity. A second major problem is the current central utility overcapacity. Since 1973, the growth of peak electrical demand has been sharply curtailed. Central utilities have been attempting to slow completion of new capacity under construction, but in each year since 1973, have added more new capacity then required to cover that year's growth in peak loads. National generation capacity was 25% over peak demand at the end of 1973, and rose to 33% over peak in 1978. Reserve capacity of 20% is considered adequate for central system reliability, with the excess capacity thought to be unproductive use of capital. Given this large and growing overcapacity, managements have focused on delaying construction of planned capacity and have not been in the market for new small capacity, regardless of its energy efficiency. See ELECTRICAL UTILITY INDUSTRY.

Rate structures. Existing rate structures provide a third block to widespread moves to cogeneration. Present utility standby rates are typically significantly higher than the cost of amortizing on-site redundant capacity. This forces most potential cogenerators to isolate themselves from the utility grid in order to achieve acceptable economic returns. The National Energy Act of 1978 mandates state public utility commissions to review standby rates by 1981, ensuring that the rates are equitable and based on statistical probability of outage.

Environmental impact. A fourth concern is impact on environment of any large-scale move to cogeneration. In moving generation to the site of heat use, cogeneration also moves some extra emissions to the site, often a builtup area. Although cogenerators burn 40–60% less fuel than conventional energy conversion systems, environmental impacts are in dispute. Total emission of SO_2, hydrocarbons, and particulates are all lower from cogeneration than from equivalent conventional energy conversion, CO production appears to be roughly equal, and oxides of nitrogen can be 30–150% higher from cogeneration. Cogeneration reduces both CO_2 and thermal emissions proportional to the oil saved. Street-level effects are theoretically predicted by dispersion equations, but only to accuracies of a factor of 3 in urban settings. Ultimate judgment of on-site cogeneration's environmental effect awaits empirical testing.

Challenge to underlying concepts. Cogeneration provides a case study in the problems any technology faces in displacing older, less efficient approaches. Existing regulations were not drawn with cogeneration in mind and do not recognize its inherent fuel efficiency. No automatic emission offsets are allowed under the Federal Clean Air Act Amendments of 1977, creating the paradox that an existing facility can be prevented from cogenerating even though to do so will lower most local emissions. Users with long histories of reliable service from central utilities are reluctant to self-generate. State and local governments often use central utilities as tax collectors, and cogeneration typically escapes some local taxes, while paying more Federal income taxes on users' savings.

The substitution of efficient on-site cogeneration capacity for less efficient but existing central generating stations is perhaps the most difficult policy conundrum. The central utilities enjoy a near monopoly of power generation in their franchise areas, and while on-site generation is often competitive, it may or may not include utility involvement. To the extent on-site, privately owned cogeneration replaces central-utility-owned stations, those stations become economically less viable. Regulated utilities, unlike competitive business, do not have a history of removing operating capacity, however inefficient, from their rate base. New cogeneration, either utility-owned or privately owned, that displaces present and planned load will confront regulatory bodies and utility managements with a challenging period of new thinking.

For background information *see* ELECTRIC POWER GENERATION in the McGraw-Hill Encyclopedia of Science and Technology. [THOMAS R. CASTEN]

Bibliography: *Cogeneration: Its Benefits to New England—Final Report of the Governor's Commission on Cogeneration*, October 1978; International District Heating Association (Pittsburgh), *District Heating Mag.*, all issues; U.S. Department of Energy, Office of Conservation and Solar Applications, *The Potential for Cogeneration Development in Six Major Industries by 1985*, Resource Planning Associates, Inc., Cambridge, MA, December 1977.

Electrical utility industry

Three events during 1979 portend major changes in the electrical utility industry in the United States. A major accident at the Three Mile Island nuclear plant of Metropolitan Edison Company near Harrisburg, PA, was the first occurrence in a commercial nuclear station to affect the public; President Jimmy Carter requested that Congress make mandatory a 50% reduction by 1990 in the use of oil burned by utilities; and, for a second year, annual peak growth was below 3%, touching only 0.5% for 1979.

Three Mile Island incident. The Three Mile Island incident occurred when a combination of equipment malfunctions and operator misjudgments resulted in a partial meltdown of the unit's nuclear fuel and a subsequent major release of radioactive gases into the atmosphere. Though Federal authorities insist that the exposure of the public to radiation was within permissible limits, this has become a major area of contention. The interior of the containment vessel was so severely contaminated by the presence of 500,000 gal (1.9×10^6 liters) of highly radioactive water that entry might not be possible for a year or more afterward.

The effects on the industry are manifold. All reactors of this design are now subject to shutdown for modifications designed to prevent recurrence of this kind of accident. National legislation imposing a moratorium on nuclear construction in any state that does not have an evacuation plan for nuclear disasters has been passed. And, most importantly, the financial costs attributed directly or indirectly to the accident must heavily influence other utilities' policy on nuclear capacity. Replacement power that the Three Mile Island station owners have to buy from neighboring utilities costs as much as $500,000 a day, and the $1,200,000,000 installation has been removed from the rate base, pushing the utility to the brink of bankruptcy. *See* NUCLEAR POWER.

Mandate for reduced oil use. The proposed mandate for a 50% reduction in use of oil could have an even more profound effect than the nuclear accident, however. Utilities now depend on oil for about 19% of their total electrical generation. This cannot be displaced by nuclear units because such units ordered now could not be in service until after 1992. A lead time of 12–14 years is now normal. The industry has little confidence that enough coal could be mined and shipped to displace oil as fuel; and in any event, environmental constraints would make it impossible to burn that much coal and meet air-quality standards. The policy, therefore, poses a major question whose answer is not readily apparent at this time.

Growth rate. The low growth rate, compared to consensus forecasts of 4.5–5.0% or higher, may presage a new plateau of growth for the future. Many utilities feel that the summer of 1979 did not display the pattern of hot periods that cause high demands. It was, however, a hot summer. Should this 2-year succession of low peak growth represent a new plateau, utility finances would be greatly affected, since construction programs could be substantially lower.

Ownership. Ownership of electric utility facilities in the United States is pluralistic, being shared by private investors, customer-owned cooperatives, and public bodies on city, district, state, and Federal levels. Investor-owned companies constitute by far the major portion of the industry. They serve 67,923,000 customers, representing a 77.5% share of all electric customers, and own 78.3% of the installed generating capacity. Cooperatives serve 9.7% of the total, but only own about 2% of the generating capacity. Public bodies at all levels serve 12.8% of the total electric customers, and own 19.7% of the installed generating capacity. Of this amount, Federal agencies hold 9.5%, and all others 10.2%.

The small amount of generating capacity owned by the cooperatives reflects the fact that most such organizations are distribution companies which buy their power either from investor-owned utilities at wholesale rates, from special generation and transmission cooperatives, or from publicly owned utilities.

There is a growing tendency for cooperatives and, to some extent, municipal utilities to purchase shares in large generating units built by investor-owned utilities. This arrangement permits small utilities to share in the economies of scale of very large units and in the lower costs of nuclear units. For the investor-owned builder, the arrangement eases the financial drain, since cooperatives and public entities have access to lower cost financing and avoid antitrust requirements. Typical is the joint ownership of the Black Fox nuclear plant now under construction in Oklahoma. Public Service Company of Oklahoma owns a 700-MW share, Associated Electric Cooperative owns 250

MW, and Western Farmers Electric Cooperative owns the remaining 200 MW.

Capacity additions. Utilities had a total generating capability of 574,365 MW at the end of 1978, having added 23,935 MW during that year (see table). By the end of 1979, industry capability had increased to 602,549 MW. The annual 1979 summer peak demand for the entire United States was 416,400 MW which, with the substantial capacity additions in 1979, gave a new reserve margin of 34.3%. This is a slight increase from the 33.7% recorded in 1978. The normal target is taken by most utilities to be 25%, though the norm can vary regionally from 13 to 28%.

A major concern in 1979 was the timing of the Environmental Protection Agency's imposition of fines against utilities which have coal-fired plants that are not in compliance with the latest air quality standards for point sources of particulates and of sulfur dioxides. These noncompliance fines will be levied against some 70 plants, of an aggregate capacity of about 25 GW, representing about 10% of all coal-fired plants in the United States. The magnitude of the proposed fines would render continued operation of these plants financially unfeasible in many cases. Should utilities elect to shut any major portion of their plants down rather than

United States electric power industry statistics for 1979*

Parameter	Amount	Increase compared with 1978, %
Generating capability, ×10³ kW		
Conventional hydro	65,207	8.7
Pumped-storage hydro	12,795	12.0
Fossil-fueled steam	408,655	2.9
Nuclear steam	61,053	16.1
Combustion turbine and internal combustion	54,839	1.2
Total	602,549	4.9
Energy production, ×10⁶ kWh	2,301,000	4.3
Energy sales, ×10⁶ kWh		
Residential	702,500	3.4
Commercial	493,300	2.6
Industrial	813,800	4.0
Miscellaneous	78,400	5.2
Total	2,100,000	3.5
Revenues, total; ×10⁶ dollars	77,238	10.6
Capital expenditures, total; ×10⁶ dollars	32,866	8.9
Customers, ×10³		
Residential	79,997	2.9
Total	89,670	2.3
Residential usage, kWh	8,904	0.3
Residential bill, ¢/kWh (average)	4.26	5.9

From 30th annual electrical industry forecast, *Elec. World*, 192(6)69–84, Sept. 15, 1979, and extrapolations from monthly data of the Edison Electric Institute—The Association of Electric Companies.

pay the fines, conditions of severe power inadequacy could result in several regions of the country. The application of these fines was protested by the utility industry, but unsuccessfully, and the expectation was that they would probably be imposed in early 1980.

The 28,184 MW of capacity added during 1979 consisted of 11,708 MW of fossil-fuel capacity, 8486 MW of nuclear units, 5196 MW of conventional hydroelectric, 2155 MW of pumped-storage hydroelectric, 527 MW of combustion turbines, and only 112 MW of diesels.

Added capacity by type of ownership was 14,839 MW by investor-owned utilities, 2644 MW by cooperatives, 7711 MW by Federal agencies, and 2990 MW by public bodies.

The total plant types as of the end of 1979 were 67.9% of fossil-fueled (of this capacity, 59% was coal, 27% was oil-fired, and 14% was gas-fired), 10% nuclear, 10.8% conventional hydroelectric, 2.1% pumped-storage hydroelectric, 8.2% combustion turbines, and 1% internal combustion engines such as diesels (see illustration).

Fossil-fueled capacity. Fossil-fuel units constitute 41.6% of the total new capacity added in 1979. A total of 32 individual units went into service, of which 24 were coal-fired, 3 were oil-fired, and 1 was fueled by gas; 2 small (16 MW each) units that burn waste came into service, and 2 geothermal plants with a total of 190 MW were started up.

The $10,679,336,000 expended in 1979 for fossil-fired construction, though up from the $9,705,817,000 expended in 1978, was an increase of only 1.6%.

Nuclear power. Utilities added 7 more nuclear units in 1979 to bring the total of reactors operating in the United States to 79. The total capacity of the plants brought into service during 1979 was 8486 MW, raising the total operating to 61,043 MW. Of the units added, 2 were boiling-water reactors (BWR) and 5 were pressurized-water reactors (PWR). There are now 46 PWRs and 28 BWRs operating in the United States: the remaining 5 units use other technologies. Nuclear units planned or in construction have a total capacity of 186,998 MW which, if current plans hold, will bring nuclear capacity to about 22% of all installed capacity by 1995. During 1978 nuclear units generated a total of 2.8×10^{11} kWh, accounting for 13% of that year's total.

Combustion turbines. Combustion turbines have served admirably as peak-load units because of their quick-start capabilities, that is, the ability to go from cold to full load in 2–3 min, and their low initial capital cost of less than $200/kW. Utilities keep an average of about 9% of their peak demand in gas turbine capacity, using them for several hundred hours a year to meet annual peak loads, or to go on line quickly to supply load in emergency situations. However, because of the uncertainty of the future supply of the distillate oil or gas that these machines burn, and the uncertainty of national policies concerning the permissible use of petroleum fuels and the permissible levels of the nitrogen oxides that these machines produce in their exhaust gases, this percentage will undoubtedly decline substantially.

Units installed in 1979 ranged from 20 to 79 MW.

Utilities brought 2172 and 527 MW into service in 1978 and 1979, respectively. The 527 MW in 1979 was made up of 12 individual units. The total installed combustion turbine capacity for the entire industry is now 48,683 MW, or 8.5% of total capacity. Of the total, however, 1314 MW represent the combustion turbine portion of combined cycle installations, in which the 900–1000° (482–538°C) exhaust gas of the turbines is used to generate steam for a conventional steam turbine. Utilities spent $263,000,000 on combustion turbine construction in 1979.

Hydroelectric installations. Installation of conventional hydroelectric capacity continued, with 5196 MW coming on line in 1979, and a total of about 17,207 MW additional capacity planned for the future. Hydroelectric capacity, amounting to 60,011 MW, provided 10.4% of total industry ca-

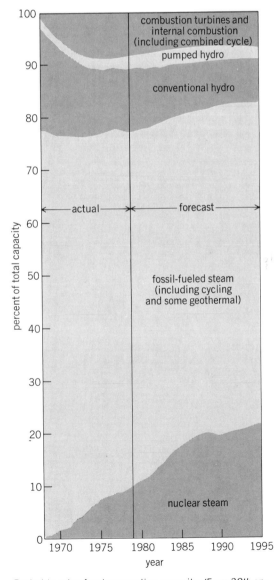

Probable mix of net generating capacity. (*From 30th annual electric industry forecast, Elec. World, 192(6):69–84, Sept. 15, 1979; used with permission of the Edison Electric Institute—The Association of Electric Companies*)

pacity. However, as sites become increasingly more difficult to find and develop, this percentage is expected to decline to 9% by 1985. During 1979 the industry spent $602,695,000 on hydroelectric projects, of which $341,330,000 was by Federal agencies and $218,810,000 by investor-owned utilities. The future of hydroelectric power will depend on economical development of low-head, or run-of-river, turbines, since environmentally acceptable high dam sites are almost nonexistent. During 1979 the smallest unit brought on line was rated 2 MW. The largest, at Grand Coulee Dam in Washington, was rated at 700 MW.

Pumped storage. Pumped storage represents one of the few methods by which electrical energy can be stored. In the mechanical analog, water is pumped to an elevated reservoir during off-peak periods and is released through hydraulic turbines during the subsequent peak period, recovering 65% of the original fuel energy. Although the 12,795 MW installed at the end of 1979 was only 2% of the industry's total capacity, another 12,000 MW are planned for the future, and utilities continue to spend about $321,000,000 annually on such projects. The world's largest pumped-storage installation is Ludington Station on Lake Michigan, rated at 970 MW.

Rate of growth. Electric demand rose appreciably less than had been anticipated for 1979, inching up only 0.5%. This was the second year of low growth—although 1978 grew somewhat more at 2.3%—and may signal a basic change in growth pattern. It is clear that the price of electricity has had a substantial suppressant effect on peak demand. It is still not entirely clear, however, whether this is a permanent long-range change or reflects the short-term psychology of current sensitivity to the energy situation, emphasized by the gasoline shortage.

The low growth rate of peak demand will have the effect of further shaving expectation for long-range growth, which had been projected in 1978 at 5.0%. The annualized compound growth rate of the last 5 years since the OPEC oil embargo of late 1973, was 3.5%. These 5 years were marked by the embargo itself, the radical upward movement of the price of electricity (the first real period of rise in history), introduction of double-digit inflation, and the longest and deepest recession since the 1930s. Longer-term growth in the future, in the face of such strength during such adversity, should be considerably higher, or about 4.7% over the next decade. The pattern will be one of higher early growth, tapering off as energy-efficient appliances, equipment, and processes phase in and as conservation cuts use in existing installations.

The pattern of peak growth generally follows that of growth in total usage. Though it is not possible to separate out the residential, commercial, and industrial components of peak demand, it appears that in 1979 industrial growth helped sustain what little growth there was, with residential and commerical sectors lagging. This is the opposite pattern from 1978. Industry is operating at close to full capacity because of the lack of capital expansion over the last few years, and may have found little opportunity to reduce peak demand as a result. Residential and commercial load, on the other hand, is considerably more flexible and has been able to effect conservation.

Regionally, only the Pacific Northwest and the Arizona–New Mexico areas attained substantial growth. The Pacific Northwest, rebounding from a low-growth year in 1978, rose 4.8%, while the Arizona–New Mexico area rose 6%. The entire northern tier of industrialized states had negative growth in 1979, however, and the southeastern and south-central regions, normally regions of above-average growth, barely attained the peak levels of 1978. In the latter regions, the weather did not reach forecast basis—that is, there were lower temperatures, and high temperatures were sustained for shorter periods than had been forecast. Rainy weather throughout the Gulf Coast states reduced both air-conditioning and irrigation loads. But the fact remains that a new low peak growth in peacetime was achieved by the total electric utility industry.

Usage. Sales of electricity rose at a higher rate than peak demand, reaching 3.5% for 1979. Total usage for 1979 was 2.1×10^{12} kWh. Energy sales for the first half of the year were pushed higher than in 1978 by cold weather, recovery from the coal strike, and a sharp increase in industrial sales. Total industrial sales for the year were 8.138×10^{11} kWh, compared with 7.821×10^{11} kWh in 1978, for an increase of 4%.

Residential and industrial growth were much stronger than that of the commercial sector. Residential usage rose to 7.025×10^{11} kWh for 1979, and commercial use to 4.933×10^{11} kWh—increases of 3.4% and 2.6%, respectively, over 1978. Electric heating continued to contribute strongly to total industry sales, accounting for 1.38×10^{11} kWh. In the residential sector it accounted for over 20% of sales. Residential revenues also rose from $29,800,000,000 in 1978 to $29,900,000,000 in 1979, a 0.3% rise in constant 1979 dollars. The 1979 sales reflect an average annual use per residential customer of 8904 kWh and an average annual bill of $379.31.

In the industrial sector generation by industrial plants contributed 6.98×10^{10} kWh, down from the 7.23×10^{10} kWh of 1978. This cut industrial generation's share of the total industrial kilowatt-hour use about 0.5%, to 7.9%. The long-term trend of industrial generation as percent of total usage is consistently downward, however, and by 1989 is expected to decline to only 5.2% of the total.

Fuels. The 1973 oil embargo brought tremendous Federal pressures on utilities to convert from petroleum fuels to coal, and the cutoff of Iranian oil greatly increased this pressure. President Carter called for a 50% cut in use of oil as boiler fuel by 1990. As a result, all new planned stations have been designed to fire coal. The use of coal in 1978 rose from 4.770×10^8 tons (4.327×10^8 metric tons) in 1977 to 4.824×10^8 tons (4.376×10^8 metric tons) in 1978, a rise of only 1.1%. Oil also rose from 6.242×10^8 bbl (9.925×10^7 m³) in 1977 to 6.333×10^8 bbl (10.069×10^7 m³) in 1978, a rise of 1.5%. Gas used rose from 3.1912×10^{12} ft³ (9.037×10^{10} m³) in 1977 to 3.224×10^{12} ft³ (9.129×10^{10} m³) in 1978.

Energy generated by each of these major fuels and their percentage share of total generation were as follows for 1978: coal, 9.802×10^{13} kWh (50.7%); oil, 3.621×10^{11} kWh (18.8%); gas, 3.089×10^{11} kWh (16.0%); and nuclear, 2.800×10^{11} kWh (14.5%). Converting these quantities to coal equivalent gives an equivalent total fuel consumption of 9.519×10^8 tons (8.635×10^8 metric tons) of coal for the entire industry in 1978. During that year, it took only 0.984 lb (446 g) of coal to produce 1 kWh.

Transmission. Utilities spent $3,590,000,000 on transmission construction in 1979. This included $754,000,000 for overhead lines below 345 kV, $835,000,000 for overhead lines at 345 kV and above, $44,500,000 for underground construction, and $896,000 for substations. These costs bought 5121 km of overhead lines at 345 kV and above, and 13,216 km at lower voltages. Only 217 mi (349 km) of underground cables were installed, primarily because of the 8:1 ratio of underground to overhead costs. Utilities also installed a total of 81.13 GVA of substation capacity during the year. Maintaining existing lines cost utilities $420,000,000 in 1979.

Distribution. Distribution facilities required the expenditure of $4,833,000,000 in 1979. Of this, $1,258,000,000 was spent to build 36,342 km of three-phase equivalent overhead primary lines ranging from 5 to 69 kV, with the majority at 15 kV. Of all overhead lines constructed in 1979, 3% was rated at 4 kV, 80% at 15 kV, 10% at 25 kV, and 7% at 34.5 kV. Expenditures for underground primary distribution lines amounted to $646,670,000 in 1979. In underground construction of the 14,138 km of three-phase equivalent lines built, 4.6% was rated at 5 kV, 76.3% at 15 kV, 12.2% at 25 kV, and 6.9% at 34.5 kV. Utilities energized 20,738 MVA of substation distribution capacity in 1979 at a total cost of $588,352,000. Maintenance costs for distribution were $1,555,000,000 in 1979.

Capital expenditures. Utilities increased their capital expenditures in 1979 to $34,527,000,000, up 3.9% in constant 1979 dollars from 1978. Of this total, $24,441,000,000 was for generation, $3,635,000,000 for transmission, $4,833,000,000 for distribution, and $1,618,000,000 for miscellaneous uses, such as headquarters buildings, services, and vehicles, which cannot be directly posted to the other categories. Total assets for the investor-owned segment of the industry rose from $171,093,000,000 in 1977 to $221,000,000,000 in 1978, the last year for which figures are available.

For background information *see* ELECTRIC POWER GENERATION; ELECTRIC POWER SYSTEMS; ENERGY SOURCES; TRANSMISSION LINES in the McGraw-Hill Encyclopedia of Science and Technology.

[WILLIAM C. HAYES]

Bibliography: Edison Electric Institute, *Statistical Yearbook of Electric Utility Industry*, 1979; 1979 annual statistical report, *Elec. World*, 191(6): 51–82, Mar. 15, 1979; 30th annual electric industry forecast, *Elec. World*, 192(6):69–84, Sept. 15, 1979; 20th steam station cost survey, *Elec. World*, 188(10):43–58, Nov. 15, 1977.

Electron beam channeling

Electron beam channeling is the technique of transporting high-energy, high-current electron beams from an accelerator to a target through a region of high-pressure gas by creating a "channel" through the gas. Along this channel the gas density may be temporarily reduced, the gas may be ionized (conducting), or the channel may carry a current whose magnetic field "focuses" the electron beam on the target.

The primary need to transport high-energy, high-current electron (or ion) beams arises in schemes to produce fusion power by the rapid implosion of a fuel pellet irradiated by such beams. Within the accelerator, even a very-high-current beam (on the order of 10^6 A) can be contained and transported. But between the output of the accelerator and the pellet, a standoff distance of several meters is required. To propagate the particle beam across this distance and to allow many beams to overlap on the same target, channels must be created that will guide the beams to the target.

Beams with self-fields. The passage of a high-current electron beam through vacuum differs from that of a single electron because of the interaction of the many electrons in the beam with each

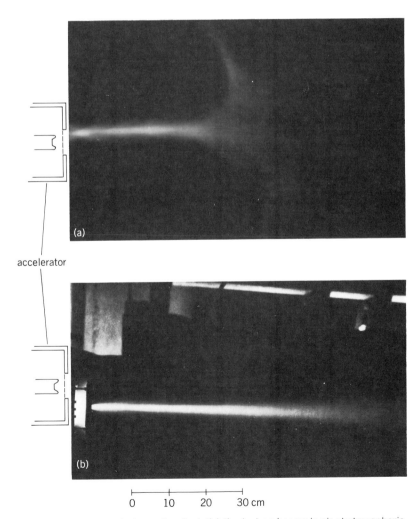

accelerator

Fig. 1. Open-shutter photographs of relativistic electron beams in air at atmospheric pressure. (a) Net current $\sim 40 \times 10^3$ A, accelerating voltage $\sim 10^6$ V, giving rise to oscillation of the beam after it leaves the accelerator and subsequent disruption. (b) Net current $\sim 4 \times 10^3$ A, accelerating voltage $\sim 2 \times 10^6$ V, giving rise to stable propagation of the beam with gradual expansion due to scattering.

other. For a uniform, paraxial, monoenergetic beam, it is easy to see that in the rest frame of the electrons their interaction is just the mutual repulsion of like charges. Reverting to the laboratory rest frame and including the interaction of the beam with its own magnetic field (self-pinching), the net radial acceleration of a beam electron at radius r may be written as Eq. (1). Here γ is

$$\gamma m_0 \ddot{r} = \frac{Ne^2 r}{2\pi a^2 \epsilon_0}(1 - \beta^2) \qquad (1)$$

$1/(1 - \beta^2)^{1/2}$, $\beta = v/c$ where v is the electron velocity and c the velocity of light, m_0 is the electron rest mass, N is the number of electrons per unit length of the beam, e is the electron charge, and a is the beam radius. According to Eq. (1), electron beams always expand, but as the electron energy increases (β approaches 1) this expansion becomes less important. Unfortunately, the electron beam constitutes a significant negative charge; therefore a large electrostatic field develops and prevents further propagation. Even when a return current path is provided, if it is external to the beam, the beam is hydromagnetically unstable.

Beams with self-fields and scattering. When a high-energy, high-current electron beam is launched into a gaseous atmosphere, the beam ionizes the gas molecules. Slow secondary electrons are expelled from the beam path, leaving a channel which is nearly charge-neutral. At low gas pressure, the radial acceleration of a beam electron is modified to Eq. (2), where f is the fraction of the

$$\gamma m_0 \ddot{r} = \frac{Ne^2}{2\pi a^2 \epsilon_0} r(1 - f - \beta^2) \qquad (2)$$

beam charge density that is neutralized. If $(1 - f) \lesssim \beta^2$, the beam does not expand, but because the gas particle density is comparable to the beam electron density the two-stream instability couples beam energy to the slow electrons, producing a hot plasma and stopping the beam. At higher gas densities, collisions of electrons with neutral atoms quench this instability.

For a charge-neutralized ($f = 1$) beam, Eq. (2) shows that the beam electrons oscillate about the beam axis. The wavelength of this oscillation is the betatron wavelength λ_β and is given by Eq. (3). In

$$\lambda_\beta \sim 2\pi a \sqrt{\frac{I_A}{I_N}} \qquad (3)$$

this equation, I_N is the net current (that is, beam current plus plasma current) and I_A is the Alfvén current. For electrons, $I_A \sim 17{,}000\,\beta\gamma$ amperes and is the current at which the self-magnetic field of the beam is so strong that the beam stops itself. The Alfvén current therefore represents the maximum possible net current that can propagate without externally applied fields. In Fig. 1a, betatron oscillations can be seen as the electron beam leaves the accelerator. After a few half-wavelengths, these oscillations become blurred, probably by collisional phase mixing, and within a few λ_β the beam is disrupted by large-amplitude hose oscillation.

To include the effect of gas scattering, the transverse pressure of the electron beam must be added to Eq. (2). Then for a charge-neutralized beam, stable propagation occurs when the magnetic pressure balances the transverse beam pressure. This is known as a "matched" beam; an example is shown in Fig. 1b. Because of continuing collisions, the radius of a matched beam increases with time according to the Nordsieck equation, which may be written as Eq. (4), where ϵ is the rate of increase of transverse kinetic energy due to scattering.

$$\frac{d}{dt}(\ln a^2) = \epsilon \frac{8\pi}{e\beta c\mu_0 I_N} \qquad (4)$$

Propagation of beams in channels. The above discussion indicates that high-energy, high-current electron beams do not propagate stably through either vacuum or a gaseous atmosphere without the aid of externally applied fields. There is a narrow "window" at relatively low pressure (around a few torrs) in which a matched beam will propagate modest distances without significant Nordsieck expansion and with no appreciable growth of hydromagnetic instabilities. But for pellet fusion, both higher currents and higher gas pressures are needed. Several groups have shown that electron beams of the desired energy and current propagate stably along the axis of a cylindrical plasma column while the plasma is carrying a current I_P which is approximately equal to, or less than, I_A. This situation is shown schematically in Fig. 2. Because the electron beam is by nature short-lived (total beam duration of 10^{-7} s or less) and is traveling through highly conducting plasma, the magnetic field produced by the beam cannot penetrate the plasma in the lifetime of the beam: that is, the plasma "freezes" the total net current (beam current plus all plasma currents) to its initial value I_P. Consequently, the electron beam is "pinched" not by its own magnetic field but by the current I_P flowing through the plasma column, which produces the toroidal magnetic field B_θ.

Laser-produced channels. In the experiments done so far, the current-carrying plasma column has been established by using either a z-pinched gas discharge or a "wire-guided" discharge. Neither of the systems is suited to reactor use, and in present experiments they are being replaced by laser-guided discharges as indicated in Fig. 2.

To guide an electric discharge through the atmosphere, the laser beam must change the atmosphere so that prebreakdown electrical phenomena travel more rapidly along the desired path than

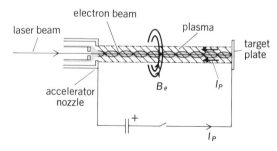

Fig. 2. Schematic diagram of a relativistic electron beam guided along a laser-initiated, electrical-discharge-heated channel.

along any other. A fast high-voltage source (with a rise time of 10^{-7} s or less, and a voltage of approximately 300,000 V) is required to effect electrical breakdown and may also be used to heat the gas in the channel. Heating could be achieved in a few microseconds, even by the laser alone, but for a channel of diameter on the order of 1 cm no appreciable expansion can occur in that time. Thus the plasma particle density equals the original gas density, and an electron beam propagating through the plasma column suffers large collisional losses. Over the next 10 μs or so, the hot plasma channel expands to pressure equilibrium with the surrounding atmosphere. An electron beam propagating through this rarefied channel suffers considerably less collisional losses.

For background information *see* ELECTRON MOTION IN VACUUM; FUSION, NUCLEAR; PINCH EFFECT; WAVES AND INSTABILITIES IN PLASMAS in the McGraw-Hill Encyclopedia of Science and Technology.   [R. F. FERNSLER; J. R. GREIG]

Bibliography: J. Benford, *J. Appl. Phys.*, 48(6): 2320–2323, 1977; J. R. Greig et al., *Phys. Rev. Lett.*, 41(3):174–177, 1978; P. A. Miller et al., *Phys. Rev. Lett.*, 39(2):92–94, 1977; G. Yonas, *Sci. Amer.*, 239(5):50–61, 1978.

Endocrine mechanisms

The ability of nerve cells to form and set free, or secrete, specific organic products has been known for many years. Recent work, which earned the Nobel Prize in medicine or physiology for Rosalyn Yalow, Andrew Schally, and Roger Guillemin in 1977, demonstrated a complex set of mechanisms whereby neurosecretory cells in the brain regulate the secretion of hormones by the endocrine glands.

Neurosecretion. In the chemical transmission of nerve impulses from one nerve cell to another at a synapse or junction, or from a nerve cell to a muscle, gland, or other effector, ordinary motor nerve cells release transmitter substances from their endings. These are chemical compounds such as acetylcholine, noradrenalin, or serotonin which cause a state of excitation, or alternatively, a state of inhibition, in the postsynaptic unit.

In 1928 Ernst Scharrer first described nerve cells in the midbrain of a minnow which have the granular cytoplasmic structure characteristic otherwise of gland cells. Similar cells had previously been noticed in the nervous systems of invertebrates, and C. C. Speidel had emphasized their glandular nature in 1922. Since then, the secretory function of these cells has been demonstrated conclusively, and they have been found in the central nervous systems of all major groups of animals having such systems. These cells produce a characteristic product which is then transported in the axon or nerve fiber to an ending on or near a blood capillary or sinus, into which the product is discharged. Neurosecretory cells are, in a word, endocrine cells.

Pituitary hormones. The neurosecretory cells of the midbrains of vertebrates are concentrated in the hypothalamus, on the ventral surface of the brain. The axons of these cells pass along a ridge, the median eminence, and through the pituitary stalk into the neurohypophysis, or posterior lobe of the pituitary gland. The posterior lobe has long

been known to be the source of two hormones, oxytocin and vasopressin, which have been shown to originate in the hypothalamic neurosecretory cells. Oxytocin functions especially in childbirth, stimulating the contractions of the muscle of the uterus which, during parturition or labor, expel the fetus through the birth canal. Vasopressin is the antidiuretic hormone of mammals which, by promoting the reabsorption of water from the kidney tubules, concentrates the urine and decreases the volume formed. Large intake of fluid, resulting in an increase in volume of the blood, inhibits the secretion of vasopressin, with the result that the excess fluid is eliminated in the urine. Alcohol and caffeine also inhibit the secretion of vasopressin, and this accounts for the diuretic effects of beverages containing those substances, beyond the effect of the volume consumed. Vasopressin also has the effect of increasing the blood pressure.

Blood circulates in the pituitary through a portal system. The capillary bed of the posterior lobe drains into a vessel which leads into another capillary bed in the anterior lobe (Fig. 1). This arrangement suggested the possibility that neurosecretory hormones, liberated into the capillaries of the posterior lobe, then have some function in the control of secretion of hormones by the anterior lobe. The six hormones of the anterior lobe act on other endocrine glands and have a trophic function, promoting and supporting the growth and secretion of hormones by their target organs: the gonads or sex glands, the thyroid gland, and the cortex of the adrenal gland. In addition, the growth hormone, somatotropin, which stimulates growth and the synthesis of proteins in all parts of the body, is a product of the anterior lobe.

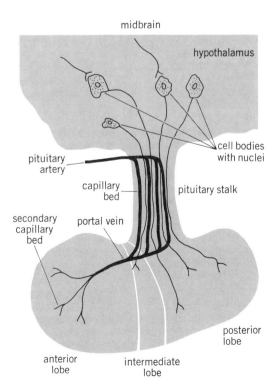

Fig. 1. Arrangement of capillaries and neurons in the pituitary.

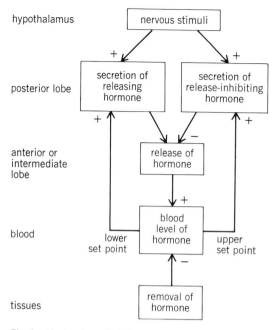

Fig. 2. Mechanism of pituitary hormone regulation.

Releasing factors. Schally, in his doctoral research at McGill University, demonstrated that the posterior pituitary contains an agent or factor which stimulates secretion of adrenocorticotropic hormone (ACTH) by the anterior lobe. Schally and Guillemin, in Houston, then set about the task of obtaining this and other, similar factors in pure form. All the evidence suggested that it was chemically a peptide, a compound of several amino acids joined by peptide linkages. Finally, between 1966 and 1968, a few milligrams of a peptide which stimulated the release of thyroxine from the thyroid gland were isolated by Guillemin and associates in Houston, and by Schally and collaborators in New Orleans. Since then, eight different factors have been purified, and the structure or amino acid sequence of some of them has been determined and confirmed by demonstrating activity of the synthetic product. In all of this work, the very sensitive and specific technique of radioimmune assay, developed by Yalow with Solomon Berson in 1957 and extended later, was essential.

Releasing factors, and a parallel set of release-inhibiting factors, have now been demonstrated for all the hormones of the anterior lobe and for the hormone intermedin from the intermediate lobe of the pituitary. The result is that a complete picture of the mechanisms which ensure a constant normal level of the pituitary hormones in the blood can now be formulated, and that adjustment of these levels according to circumstances can be made (Fig. 2.). This is the central mechanism of homeostasis, whereby the cellular processes of the body are integrated in maintaining that constancy of the internal environment which, in the famous phrase of Claude Bernard, is the condition of the free and independent life.

For background information *see* ENDOCRINE MECHANISMS in the McGraw-Hill Encyclopedia of Science and Technology.   [BRADLEY T. SCHEER]

Bibliography: R. Guillemin, *Science*, 390(4366): 390–402, Oct. 27, 1978; A. V. Schally, *Science*, 202(4363):18–28, Oct. 6, 1978; B. T. Scheer, *Chem. Zool.*, 11:103–158, 1978; R. S. Yalow, *Science*, 200(4347):1236–1245, June 16, 1978.

Environment

The international organization for identifying global and regional environmental concerns, and for coordinating appropriate actions, is the United Nations Environment Program (UNEP). The 58-nation Governing Council, which held its seventh session from April 18 to May 4, 1979, provides policy guidance to the UNEP Secretariat, headquartered in Nairobi, Kenya. A voluntary fund, maintained by member nations, is the source of support for UNEP that enables it, in a catalytic way, to bring much larger resources of participating nations and international bodies upon the environmental problems of common concern. UNEP consists of three major parts: environmental assessment (monitoring, research, evaluation and review, information exchange), environmental management (goals, criteria, conventions), and supporting measures (training, education, public information, technical assistance). The designation Earthwatch has been given to the environmental assessment part of the program. This article addresses the activities under the Earthwatch program, as reported to the Governing Council in the course of its seventh session.

Earthwatch. The executive director of UNEP defines Earthwatch as "a dynamic process of integrated environmental assessment by which relevant environmental issues are identified and necessary data are gathered and evaluated to provide a basis of information and understanding for effective environmental management."

As part of its program of monitoring, Earthwatch has devoted special attention to environmental variables that describe the state of soil and vegetation cover, contribute to or determine climatic changes or fluctuations, affect oceans and their living resources, and make possible the assessment of human exposure to pollutants that affect health.

The Earthwatch program improved access to the world's environmental information sources through regional seminars, visits to various countries, workshops, and the establishment of national focal points. During 1979 the number of countries participating in the International Register of Potentially Toxic Chemicals almost doubled.

The Earthwatch program continues to suffer, however, from the lack of a clearly defined research extent that is uniquely supportive of the goals of Earthwatch. Research efforts need to be organized and directed at a better understanding and interpretation of environmental processes and data. In particular, research is needed to improve sampling and intercalibration methods, and to develop more effective means for the remote sensing of natural resources from satellite platforms.

Progress is being made in the Earthwatch program, however, toward understanding and solving environmental problems relevant to the atmosphere, the oceans and regional seas, the Earth's land surfaces, and health.

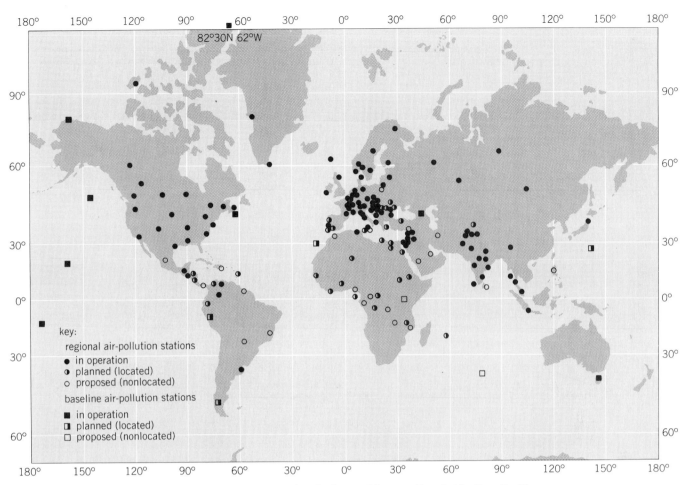

Fig. 1.   Network of 100 regional and baseline stations for measuring background air pollution. (*From United Nations Environment Program: Report of the Executive Director, Governing Council, 7th Session, p. 7, 1979*)

Atmosphere. The present network of about 100 regional and baseline air-pollution monitoring stations is shown in Fig. 1. The regional stations, reasonably distant from pollution sources — those little affected by short-term fluctuations in pollution levels — measure turbidity under clear sky conditions, and collect and analyze suspended particulate matter and precipitation samples. The baseline stations in Fig. 1 are located in even more remote areas, and are the sites where samples for the study of long-term changes in atmospheric carbon dioxide concentrations are collected. Additional measurements include those for chlorofluorocarbons, carbon monoxide, and ozone profiles. A new baseline station is being planned for location on Mount Kenya.

A large number of data on glacier volume and mass balances from 45 countries now provide a reference against which changes due to climatic fluctuations can be assessed in the future.

Measurements of solar radiation flux, albedo, ocean surface temperature, and extent of snow cover will be embodied in a comprehensive project on climate-related monitoring that began operation at the end of 1979 as part of the World Climate Program.

Acid rains have caused problems in Europe for many years. A network of 42 stations (see Fig. 2) in 12 countries collects and analyzes samples of air, rain, and airborne particulates over Europe. Arrangements have also been made to collect and synthesize meteorological information in western Europe that is required to assess the transport of pollutants across borders and to test transport prediction models. Similar arrangements are planned for eastern Europe.

Oceans and regional seas. A pilot project on monitoring the pollution of the oceans by petroleum hydrocarbons is soon to be completed. Data will provide the extent, frequency, and location of oil slicks and tar balls along some of the main shipping lanes. The first phase of the World Register of Rivers Discharging to the Oceans documented the hydrological characteristics of 260 major rivers.

Agreement has been reached by the First International Monk Seal Conference, Rhodes, Greece, on a plan of action for the Mediterranean monk seal. This plan will provide an important element of the global plan of action on marine mammals, including a proposed revision of the 1946 International Convention for the Regulation of Whaling.

Governments have become increasingly aware of the need for collective action to prevent marine pollution at the regional level so as to achieve efficient management of coastal and marine resources on a sustainable basis. Action plans within the regional seas program have been substantially approved for the Mediterranean, the Red Sea, and the Persian Gulf. Other action plans are in preparation for the Caribbean, the Gulf of Guinea, east Asian waters, the southeast Pacific, and the southwest Pacific.

Eleven Mediterranean coastal states have ratified the 1976 Convention for the Protection of the Mediterranean Sea against Pollution. Sixteen states have agreed to participate in the Coordinated Mediterranean Pollution Monitoring and Research Program, a progress report which has been prepared for the period 1975–1978.

Land. The United Nations Conference on Desertification, held in August 1977, highlighted this worldwide problem of desertification, affecting about 36% of the Earth's land surface. Desertification has long been associated with the misuse of land by such practices as overgrazing, slash and burn, and removal of tree cover for fuel and fodder. In an insidious feedback mechanism, the destruction of vegetation affects the climate adversely, thus accelerating the process of desertification.

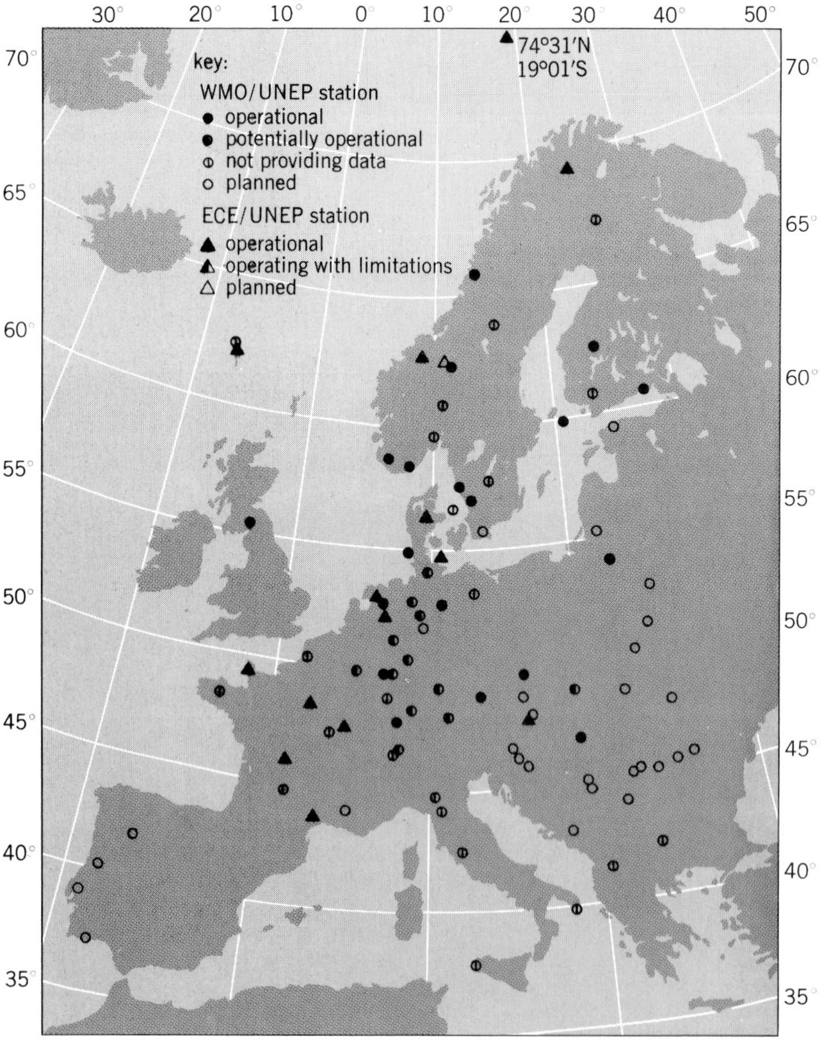

Fig. 2. Network of 42 stations in 12 countries for sampling air, rain, and airborne particulates. *(From United Nations Environment Program: Report of the Executive Director, Governing Council, 7th Session, p. 12, 1979)*

A pilot project on monitoring of tropical forest in west Africa has been completed, and results (reports and maps) will be published soon. A fact-finding mission to seven west African countries in mid-1978 recommended the initiation (due in mid-1979) of a pilot project on rangelands monitoring. This project is expected to combine the use of satellite imagery, aircraft reconnaissance flights at very low altitudes, and ground surveys.

A framework methodology for the assessment of soil degradation is being devised, and will be published soon. The methodology should provide guidelines for field testing before it is applied on a large scale in selected areas throughout the world. The first maps to be printed will cover soil degradation rates and hazards in Africa north of the Equator, and in the Middle East.

A worldwide assessment of tropical forest resources is under way, and will be completed by mid-1980. This assessment will make use of both conventional information and information obtained through remote sensing.

Health. A network of stations monitoring air pollution in urban areas includes some 60 cities in 42 countries. In most cities in the network, there are three stations, located respectively in residential, commercial, and industrial areas, that measure suspended particulate matter, smoke, and sulfur dioxide. Beginning soon, analyses of nitrogen dioxide, carbon monoxide, ozone, and airborne lead will be added, where appropriate.

Food contamination by chlorinated hydrocarbons and selected metals has been monitored by 19 countries, covering the period 1971–1977. These data are being published. Beginning in 1979, the data will be collected under strict quality control to ensure their effective use in comparative analyses.

A Government Expert Group met in 1977, and proposed the monitoring of human exposure to air pollutants through the collection and analysis of specimens of human tissues and body fluids. A pilot project with a limited number of countries will be initiated in 1979.

Comprehensive assessment of the impact of pollutants on humans and their environment is a prerequisite to appropriate environmental management measures. The recent enactment by the United States of the Toxic Substance Control Act to protect human health and the environment from potentially toxic chemicals is an example of management measures that have been enacted in similar form by Sweden, Japan, Canada, and other countries. The total assessment of the impact of pollutants, and its relationship with Earthwatch, is shown schematically in Fig. 3.

Outer limits. The UNEP Governing Council has identified five fields under the heading of "outer limits" for special attention in the Earthwatch program. These elements are climatic changes, weather modification, risks to the ozone layer, bioproductivity, and social aspects.

A draft has been prepared for a World Climate Program that will focus on climate monitoring, natural climate change and variability, and human impact on climate and, conversely, the impact of climate changes and fluctuations on human activities.

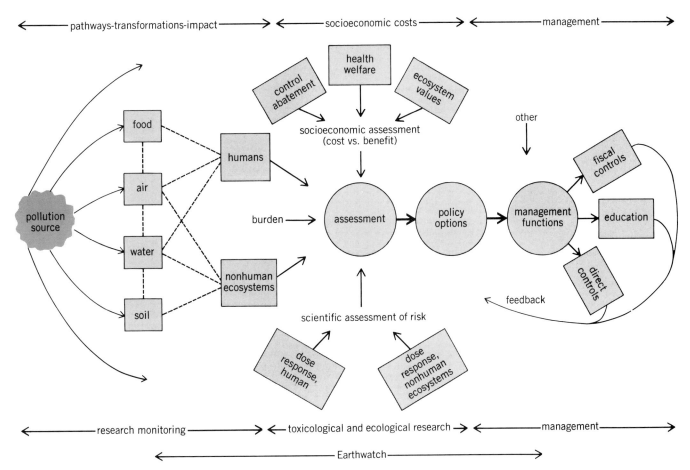

Fig. 3. Schematic of the assessment of pollution within the Earthwatch program. (*From United Nations Environment Program: Report of the Executive Director, Governing Council, 5th Session, p. 48, 1977*)

The report of the joint World Meteorological Organization and UNEP informal meeting of experts on legal aspects of weather modification that recommended nine draft principles of conduct for the guidance of states concerning weather modification was submitted to governments for comment. These comments were to be considered by a meeting of experts scheduled in 1979.

An intergovernmental meeting on chlorofluoromethanes was convened in Munich from December 6 to 8, 1978. In considering the harmonization of national regulatory policies on chlorofluoromethanes, the participants called upon all governments to work toward achieving a significant reduction in the global release of chlorofluoromethanes during the next few years. The United States is participating in this reduction program.

With respect to bioproductivity, an international nitrogen unit has been set up in the Royal Swedish Academy of Sciences to serve as a focal point for nitrogen cycling studies. A regional workshop on nitrogen cycling was held in west Africa in December 1978 to evaluate gaps in knowledge of the cycling of the element in humid tropical areas of Africa.

Three field surveys were undertaken in India, Egypt, and Malta, on the social aspects of outer limits, to determine attitudes, values, and perceptions of people with respect to environment and development programs. Completion of two studies

on social response to energy conservation and options were planned for 1979.

For background information *see* ECOLOGY; ENVIRONMENT in the McGraw-Hill Encyclopedia of Science and Technology. [CLAYTON E. JENSEN]

Bibliography: C. E. Jensen, D. W. Brown, and J. A. Mirabito, *Science*, 190:432–438, 1975; *United Nations Environment Program: Report of the Executive Director, Governing Council, 7th Session,* 1979.

Enzyme

Multiple forms of proteins, distinguishable by properties other than their shared substrate specificities and catalytic activities, have been useful probes of a number of biological phenomena ever since their discovery several decades ago. Such protein systems have provided valuable insights into a wide variety of questions, ranging from analysis of the structure and function of subcellular organelles to the ecology and genetics of natural populations.

Formation and adaptation. Multiple-enzymatic forms arise by several mechanisms, however, and the nature of the information they provide is dependent upon the mechanism of their formation. In some cases, multiple forms occur because the peptide products of several different gene loci participate in the functioning of a particular protein or series of proteins, with a resultant array of molecules representing various combinations and permutations of gene products. In other instances, some molecules in a population of molecules may be secondarily modified either by the action of other enzymes or by binding to various ligands or substrates. These and other factors may generate an isozymal series of proteins with different electrophoretic or chromatographic properties as well as somewhat different functional properties. Isozyme series which arise from multiple-gene expression are useful probes of developmental genetic phenomena, and if the gene activation is dependent upon environmental conditions, such enzyme systems may be powerful tools for examining physiological and biochemical mechanisms of compensation to environmental perturbation. Another class of multiple-protein forms arises from charge alternatives that result from genetic polymorphisms at the loci encoding the protein subunits of particular enzymes.

These genetically determined allozymes, which represent the expression of variant alleles at a gene locus, are a major focus of microevolutionary studies on the nature of molecular adaptation to environmental variation. A number of studies over the last decade have indicated that the expression of certain isozymes and the frequency of occurrence of particular allozymes are correlated with environmental conditions. By loose analogy, the hope arose that the adaptive nature of these two mechanisms of generating multiple-molecular proteins, although not necessarily two sides of the same coin, might nevertheless be understood in common biochemical terms. Subsequent work, particularly in the last year or so, has led to a realization that this initial view may have

been overly optimistic. The adaptiveness of both isozymal induction and allozyme frequency change in response to environmental conditions is largely unresolved and is an area of active intellectual ferment.

Isozyme changes and temperature acclimation. The observation that certain isozymal forms appear possessing molecular properties seemingly adaptive to the conditions under which they appear was made some years ago in studies primarily on the physiology and biochemistry of fishes. This observation led to the suggestion that such isozymal changes represented a qualitative strategy of temperature adaptation, and as such might represent a primary molecular mechanism of thermal compensation. The notion was that an organism might have at its disposal a variety of gene loci, each encoding products capable of catalyzing a particular reaction, but each optimally suited to different environmental conditions. Thus, those gene products most appropriate to a given set of conditions would be expressed, and a qualitative change in the environment would be met with a qualitative change in gene expression. Problems with this view have surfaced. First, in the original studies, sample sizes may have been inadequate to screen out the confounding effects of allozymic variation. Second, the noteworthy examples of such inductive effects are largely confined to studies on fishes of polyploid ancestry, such as salmonids and carp.

The relative importance of isozyme induction in temperature acclimation has recently been investigated in a more prosaic diploid fish, the green sunfish, by James Shaklee and his colleagues. These investigators examined a variety of enzymatic proteins in several tissues of fishes variously acclimated to different temperatures and oxygen concentrations. Enzymatic activities and electrophoretic patterns were determined to evaluate the relative roles of quantitative changes in enzyme activity versus qualitative changes in gene expression. In short, although many of the enzymes showed changes in activity levels, no enzyme involved in central intermediary metabolism responded to acclimation by the elaboration of new electrophoretically detectable isozymal forms. In only two instances, both involving tissue-specific esterases of unknown function, was there any evidence of a change in isozymal pattern. Shaklee and coworkers concluded that major changes in isozyme complements are not a necessary concomitant of thermal acclimation, and that, indeed, quantitative changes in activity levels may play a more important role at the molecular level. Only further study of additional experimental systems will determine if this view represents a better generalization.

Temperature dependence of substrate affinities. The molecular property of the fish isozymes which generated the most interest in their potential role in temperature acclimation, and which attracted the attention of many allozyme-oriented biochemical geneticists, was the temperature dependence of their apparent substrate binding affinities. This affinity, measured reciprocally by the apparent Michaelis constant, K_m, was purported to be strongly temperature-dependent such that when environmental temperatures fell (or rose)

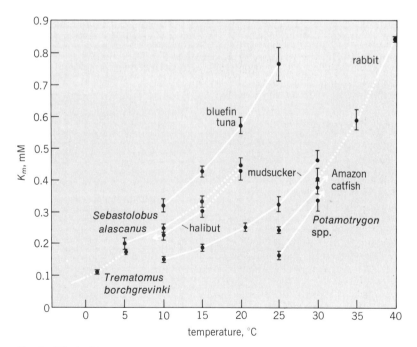

Fig. 1. Effect of temperature on apparent Michaelis constants (K_m) for pyruvate of M_4-LDH of several vertebrate species. Assays were conducted in phosphate buffer, pH 7.4. (From P. H. Yancey and G. N. Somero, Temperature dependence of intracellular pH: its role in the conservation of pyruvate apparent K_m values of vertebrate lactate dehydrogenases, J. Compar. Physiol. B, 125:129–134, 1978)

from some optimal temperature, substrate binding ability was drastically reduced. Furthermore, isozymes elaborated at a particular temperature seemed to have optimal binding affinities at that temperature.

This temperature dependence of K_m was developed into a model of thermal compensation called positive thermal modulation. Under this model, the effect of increased temperature on enzymatic reaction rate would be ameliorated by the antagonism of increased catalytic turnover rate and decreased substrate affinity. A given protein might provide compensation of reaction rate over a narrow range of environmental temperatures, but the elaboration of different isozymes at different temperatures could provide a wide range of thermal insensitivity. This model of thermal compensation was easily transposable to the analysis of allozymes. It was particularly appealing in that it provided a readily intuitive molecular mechanism for heterozygote superiority. In a variable environment, a homozygous organism, capable of synthesizing only a single allozymic form of a particular protein (presumably capable of thermal compensation over a relatively narrow range of conditions), might well be at a disadvantage in comparison with a heterozygous organism possessing the genetic information to encode two slightly different enzymatic forms (with different optima). Such a pattern of heterozygote superiority would be a strong force in preserving genetic variability in natural populations.

The model of positive thermal modulation presupposes two conditions. The first is that subcellular metabolite concentrations in the immediate vicinity of the enzyme are low enough that binding affinities are important constraints on enzyme activity, and that these concentrations remain relatively constant at different temperatures. The second condition is that K_m is, in fact, robustly and strongly dependent upon temperature in a manner sufficient to counteract the change in catalytic rate. The first condition has generally not been investigated with sufficient precision to falsify the positive thermal modulation model. Detailed investigations of subcellular concentrations of metabolites are difficult enough in isolated tissues, let alone intact organisms. The second condition, however, that the binding affinities themselves are temperature-dependent, has been recently evaluated in a comparative study of the muscle lactate dehydrogenases (M_4-LDH) of various fish species.

Paul Yancey and George Somero have reexamined the thermal dependences of K_m in a homologous series of enzymes purified from fishes from a wide range of thermal conditions. They have found that K_m values determined on these various proteins, rather than being strongly dependent on temperature, are in fact quite temperature-independent (Fig. 1). Furthermore, fishes originating in very different thermal environments nevertheless have proteins with rather similar binding affinities, suggesting that the K_ms of these enzymes are conserved or constained rather than being variable. Even more striking is their observation that when the test-tube assay conditions are chosen so as to mimic the interaction between pH and temperature that would be expected intracellularly, the

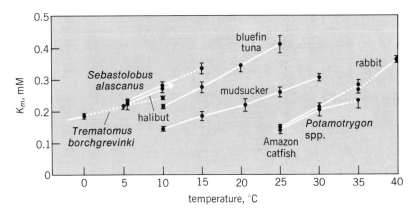

Fig. 2. Effect of temperature on apparent Michaelis constants (K_m) for pyruvate of M_4-LDH of several vertebrate species. Assays were conducted in imidazole–HCl buffer. (From P. H. Yancey and G. N. Somero, Temperature dependence of intracellular pH: its role in the conservation of pyruvate apparent K_m values of vertebrate lactate dehydrogenases, J. Compar. Physiol. B, 125:129–134, 1978)

substrate binding affinities are even more constrained and temperature-independent (Fig. 2).

These observations, if subsequently found to be general for other protein systems, create serious problems for the positive thermal modulation model. They also have implications for studies on the adaptive nature of allozyme variation. If organisms in disparate evolutionary lines, each with a long history of occupancy of different thermal environments, nevertheless have strikingly similar substrate binding affinities, it is doubtful that allozymic forms within a species (differing by only a few amino acid residues) would be divergent for this property. If K_ms are tightly constrained in response to the demands of the cellular metabolic machinery, other aspects of enzyme structure and function might be more promising for further analysis. Indeed, studies in progress in a number of laboratories on the regulation of enzyme levels, on the genetics of elements controlling posttranslational modifications of proteins, and on membrane-enzyme interactions may lead to much-needed insight into the adaptive significance of both allozymic and isozymic variation.

For background information see ENZYME in the McGraw-Hill Encyclopedia of Science and Technology.   [CLAY SASSAMAN]

Bibliography: J. B. Shaklee et al., J. Exp. Zool., 201:1–20, 1977; P. H. Yancey and G. N. Somero, J. Compar. Physiol. B, 125:129–134, 1978.

Epidermis (plant)

Specialized openings can be found in the epidermis of most aerial parts of plants. These openings are called stomata, and each stoma consists of a pore or opening surrounded by two or more highly differentiated guard cells. The guard cells may be surrounded by undifferentiated epidermal cells, as in the onion, or by specialized cells called subsidiary cells, as in the grasses, in which case the whole unit is referred to as a stomatal complex.

Stomata have an important role in the physiology of plants, since changes in their aperture regulate the gas exchange between the interior of the leaves and the environment. The stomata maintain efficient levels of photosynthesis by regulating the pore aperture to permit sufficient diffusion of CO_2 into the leaf while avoiding excessive transpiration of water vapor. They usually close at night, or in the absence of light, when photosynthesis cannot take place.

The control of stomatal opening and closing depends on the morphological and physiological properties of the guard cells. The walls of the guard cells are unusual on two accounts: they are greatly thickened in the area surrounding the pore, and their cellulose microfibrils are oriented radially, like a reinforced radial tire. These special properties allow the pores to open as the guard cells swell and to close as they become flaccid. To swell, the guard cells must increase their water content; but since water diffuses very rapidly across cell membranes, a water content higher than that of the neighboring epidermal cells cannot be maintained by the accumulation of water alone. Rather, salts are accumulated as well, mainly KCl and K^+ salts of organic acids. A higher solute content raises the osmotic concentration of the guard cells; water is taken in and the cells swell. During closing, the sequence is reversed. Thus guard cells convert chemiosmotic energy into mechanical work which opens the pore.

Much remains to be learned about the cellular processes underlying ion transport, energy transduction, and the water balance of stomata. Also, from a developmental point of view, very little is known about how the guard cells differentiate and acquire the properties that allow them to perform their specialized work.

Isolation of guard cell protoplasts. A new approach to studying the biology of guard cells is by isolating guard cell protoplasts. The most convenient way to obtain protoplasts is to treat a tissue with a mixture of fungal enzymes to hydrolyze the wall components. Several enzyme mixtures are now commercially available; one of these is Cellulysin which has been used for guard cell protoplast isolation. The absence of a wall requires that the osmotic concentration of the solution containing the protoplasts be adjusted to prevent them from bursting. This can be accomplished by including 0.5 M mannitol, and adding 0.5 mM $CaCl_2$ to stabilize the cell membranes.

Two methods to prepare guard cell protoplasts are currently available. One involves the use of microchambers. A piece of leaf tissue is mounted in a small, sealed chamber. The digestion, washing, and any experimental procedures are conducted by perfusing the microchambers with the required solutions through the O-ring seal of the chamber with hypodermic needles and small perfusion pumps. This method yields a small number of guard cell protoplasts (100 to 200) but allows microscopic examination and minimizes manipulation. The other method involves larger amounts of tissue in test tubes and requires centrifugation of the released protoplasts for purification. The latter is the method of choice for biochemical analysis because of the higher yield.

Light-induced swelling of protoplasts. Epidermal peels are easily obtained from onion leaves by tearing a small portion of the tissue with a pair of fine forceps and then pulling upward. The epidermal layer separates from the rest of the leaf, and

the peel is mounted in the microchamber. After the chamber is sealed, the peel is treated overnight with the enzyme solution. The next morning both epidermal cell and guard cell protoplasts are found in the chamber. The enzyme solution is then replaced by 0.5 M mannitol and 0.5 mM $CaCl_2$. In the onion, epidermal and guard cell protoplasts are morphologically distinct. Epidermal protoplasts are highly vacuolate and average 45 μm in diameter, while the guard cell protoplasts are 18 to 25 μm in diameter and have dense, granular cytoplasm (see illustration).

The preparation is then ready to test the swelling properties of the protoplasts. A population of guard cell protoplasts is kept in the dark, and their average diameters are measured in a compound microscope (dim, green light is used during the short time required to make the measurements). They are then illuminated with white light for 60 min, and their diameters measured again. The volumes of the spherical protoplasts can be calculated from their diameters. The results show that light causes swelling of the guard cell protoplasts. In the presence of light and 30 mM KCl, the guard cell protoplasts show up to a 60% increase in volume. This swelling is not seen in guard cell protoplasts kept in the dark or in illuminated epidermal cell protoplasts.

It seems likely that the swelling is a manifestation of a normal physiological response of the guard cells. As in the intact tissue, light causes K^+ uptake, which increases the osmotic concentration inside the cells. In the absence of a cell wall, the water uptake that follows K^+ entry should lead to swelling.

A study on the wavelength dependence of the light-induced swelling also showed that protoplasts exposed to blue, green, and red light of equal in-

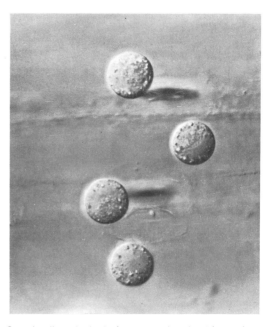

Guard cell protoplasts from an onion, kept in a microchamber. The protoplasts are about 20 μm in diameter. (From E. Zeiger and P. K. Hepler, Production of guard cell protoplasts from onion and tobacco, Plant Physiol., 58: 492–498, 1976)

tensities swell in the presence of blue light only. Presumably a specific blue light photoreceptor is present in the guard cell protoplasts and is involved in the light-induced ion uptake.

Other experimental approaches. In general, it has been found that conditions normally leading to stomatal opening, such as light, the presence of fusicoccin, or high concentrations of KCl, cause swelling of the protoplasts, whereas treatments conducive to stomatal closing, such as abscisic acid, cause shrinking of the protoplasts. Hence studies with protoplasts provide a unique opportunity to investigate the physiological properties of guard cells isolated from the rest of the leaf tissue.

Two other uses of this novel experimental system may be noted. One is the isolation of subcellular organelles. The protoplasts can be lysed by osmotic shock, leading to the release of the intracellular contents. This technique has been used to isolate the guard cell vacuoles and, as a result, the green fluorescing substance previously discovered in the guard cells of the onion was localized in the vacuoles. Similarly, guard cell chloroplasts can be isolated and their properties studied.

Another experimental approach offered by this system is the culture of guard cell protoplasts in synthetic media. The study of cultured guard cell protoplasts could provide answers to some interesting questions regarding the development and differentiation of these specialized cells. Would guard cell protoplasts regenerate a guard cell–like wall? Would they dedifferentiate? Would they maintain the ability to divide in culture as mesophyll cell protoplasts do, or does the process of differentiation imply a concomitant loss of their division properties? The answers to these and other questions are likely to enrich understanding of the complex biology of the stomatal guard cells in the near future.

For background information *see* EPIDERMIS (PLANT) in the McGraw-Hill Encyclopedia of Science and Technology.   [EDUARDO ZEIGER]

Bibliography: H. Schnabl, *Planta*, 144:95–100, 1978; E. Zeiger and P. K. Hepler, *J. Cell Sci.*, 37: 1–10, 1979; E. Zeiger and P. K. Hepler, *Plant Physiol.*, 58:492–498, 1976; E. Zeiger and P. K. Hepler, *Science*, 196:887–889, 1977.

Epithelium

In several groups of animals, epithelia conduct electrical impulses. Mechanical stimulation such as stroking or prodding, or electrical shocks, cause depolarizations which propagate from cell to cell in the epithelium by direct-current flow through low-resistance junctions. As pathways mediating behaviorally meaningful signals, conducting epithelia function much like nerves, or like muscles such as the heart which are capable of myoid conduction. Conduction velocities are lower than in nerves, and rarely exceed 30 cm/s. Repetitive firing at frequencies up to about 40 Hz is described in some conducting epithelia, and conduction obeys the all-or-nothing principle. In a few cases, electrical coupling has been specifically demonstrated between the excitable cells, and gap junctions are known to be present. These junctions almost certainly provide the pathways serving for communication between the coupled cells. The ionic basis of epithelial impulses is poorly understood; in some cases at least, an influx of sodium ions is responsible for the initial depolarization, but calcium may enter too, and in conducting myoepithelia and ciliated and glandular epithelia, activation of the effector component generally requires calcium ions.

Occurrence of epithelial conduction. In hydrozoan coelenterates, pelagic tunicates, and amphibian larvae, epithelial conduction plays important behavioral roles.

Hydrozoan medusae. Hydrozoan medusae are particularly well endowed with excitable epithelia, and may serve as an example (see illustration). The medusa responds to strong tactile or electrical stimulation of the outer surface by involution, or folding inward, of the margin of the bell-shaped body (illustration *a*). The tentacles, gonads, and other delicate components are thus brought into a protected space within the interior of the bell. Involution is chiefly due to contraction of the four radial muscles lying on the inside of the bell (illustration *b*). These muscles are excited during

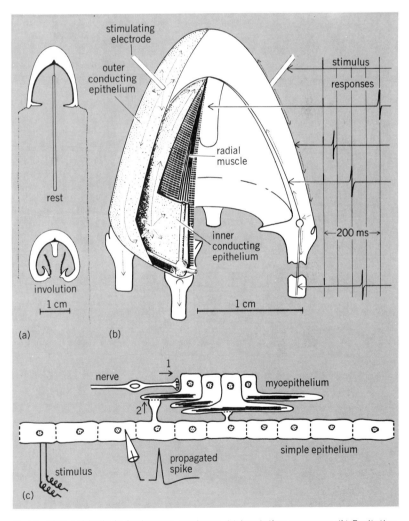

Conducting epithelia in hydrozoan medusae. (*a*) Involution response. (*b*) Excitation pathway from outer epithelium to muscles on the inside (*after G. O. Mackie et al., Physiologie du comportement de l'Hydroméduse Sarsia tubulosa Sars.: Les systèmes à conduction aneurale, C.R. Acad. Sci. Paris, 264:466–469, 1967*). (*c*) Model of dual excitatory control of the radial muscles; the muscle (myoepithelium) can be excited either by nerves through synapses (pathway 1) or by processes from an adjacent excitable epithelium through coupling junctions (pathway 2).

involution by impulses conducted to them via epithelial pathways. The epithelium covering the outside of the bell acts both as a receptor zone and as a conduction pathway for impulses generated by stimulation. Nerves are absent in these regions. The impulses propagate in all directions from the site of initiation. On reaching the lower margin of the bell, they pass inward and enter an inner conducting epithelium, whence they spread to the muscles. The four radial muscles contract symmetrically as they are excited simultaneously by the diffusely conducted epithelial signals.

In some variations of the protective involution response, epithelial impulses evoke activity in nerves located near the margin (by some unknown mechanism), and this can lead to secondary, neurally mediated muscle responses which are superimposed on and intensify the primary, epithelially mediated response. A simplified version of how the radial muscles are excited is shown in illustration c. The inner and outer conducting epithelia are composed of single layers of unspecialized, squamous cells (simple epithelia), while the radial muscle is a myoepithelium. Impulses travel directly from the simple epithelium to the myoepithelium through coupling pathways. Chemically transmitting junctions are not known to occur between the cells in conducting epithelia, in contrast to the neuromuscular junctions in the same animals, which are conventional synapses.

This example illustrates a commonly encountered feature of epithelially mediated responses: effectors over a wide area are simultaneously excited in a simple direct way. Subtler, or more local, responses are generally mediated by nerves, which can activate effector units more selectively.

Siphonophores. Epithelial conduction in coelenterates is not confined to hydrozoan medusae, but occurs in some hydroid polyps, where it is implicated in the spread of protective contractions, and in the siphonophores, a group of pelagic hydrozoan colonies. In some siphonophores the central axis of the colony, or stem, is lined with a simple conducting epithelium which excites the adjacent myoepithelium, causing slow, or tonic, contractions. The same muscles are innervated by nerves, and respond to nervous input by fast, facilitating, twitch contractions. The layout is similar to that shown in illustration c, but a novel feature of this system is that when the myoepithelium is sufficiently depolarized by nervous input in the twitch response, impulses are generated secondarily (backfiring) in the simple conducting epithelium. Thus tonic contractions always follow strong twitches. The two epithelia are evidently coupled by junctions which can transmit in either direction. Such two-way interactions between diversely specialized conducting epithelia, and between epithelia and nerves, are the subject of active investigation at the present time.

Tunicates. In the tunicates, conducting epithelia function either as extensions of the sensory system or for the spread of effector responses. The sensory role is best exemplified by forms such as *Oikopleura* (Larvacea), where the entire skin of the tail is a simple conducting epithelium. Impulses generated in this epithelium enter nerves leading to the caudal ganglion and modify the locomotory rhythm. A similar arrangement has been found in ascidian tadpoles, salps, and doliolids. An example of a conducting effector epithelium is the ascidian gill. The gill epithelium is ciliated, and the cilia normally beat continuously, showing a metachronal rhythm. Ciliary beating is, however, subject to neurally mediated intermittent arrests. The innervation is sparse, and very few ciliated cells actually receive synapses. Spread of the depolarizations responsible for arrest apparently takes place from cell to cell within the epithelium itself.

Amphibia. Several amphibian larvae go through a stage in their early development when the epidermis is excitable and conducts impulses. The impulses enter the nervous system and cause an escape swimming response. This precocious excitability is lost later in development, when the definitive sensory innervation of the skin becomes established.

Types of conducting epithelia. A complete list of conducting epithelia would include examples of: Simple epithelia, with no known effector capability; these serve for reception and transmission of stimuli to the nervous system or to epithelial effectors. Ciliated epithelia, as, for example, in tunicates. Epithelial glands, where the secretomotor response is triggered by impulses conducted from cell to cell, for instance, in some mollusks. Bioluminescent epithelia, as in some polynoid worms. Conducting myoepithelia, such as the swimming muscles in hydrozoan medusae. In one siphonophore, *Hippopodius*, four of these five types coexist.

Not all epithelia conduct; electrical coupling and the presence of gap junctions do not in themselves necessitate the assumption that conduction occurs. The cell membranes must also be capable of generating action potentials, and this is far from being a universal phenomenon. Most ciliated epithelia are richly innervated, and impulses are not propagated between the epithelial cells themselves. This is also true of epithelial glands and myoepithelia in vertebrates. Some workers have suggested that epithelial conduction represents a survival of, or reversion to, a primitive, possibly pre-nervous type of signaling. The existing examples are scattered through the animal kingdom without any hint of a phylogenetic sequence. Regardless of the evolution of the adaptation, it endows the epithelia in question with the ability for rapid transmission of simple signals, and thus reduces the organism's dependency on nerves, and so lessens the heavy metabolic investment that this entails.

For background information *see* COELENTERATA; EPITHELIUM; TUNICATA in the McGraw-Hill Encyclopedia of Science and Technology.

[GEORGE O. MACKIE]

Bibliography: J.-M. Bassot et al., *Biol. Bull.,* 155:473–479; Q. Bone and G. O. Mackie, *Biol. Bull.,* 149:267–286, 1975; R. K. Josephson and W. E. Schwab, *J. Gen. Physiol.,* in press; A. N. Spencer, *Amer. Zool.,* 14:917–929, 1974.

Exciton

One of the most interesting developments to emerge from semiconductor physics research in the 1970s was the discovery that within a single crystal of germanium, Ge, a "gas" of certain elementary entities called excitons can condense at the temperatures of liquid helium (below 4 K) into a liquid state—a liquid which is, in fact, metallic. This dramatic condensation effect has suggested new directions of research into the nature of the metal-insulator transition, the interaction between liquid and other kinds of excitations in the crystal, and the unusual properties of the condensate proper. In some respects the liquid may be viewed as almost a new state of matter; it is the most metallic of metals and the most quantum of liquids; it is by its simple composition an especially elementary kind of matter; also unprecedented for a conventional metal, the liquid emits radiation (fluoresces) at a characteristic energy (in the near-infrared) due to its internal decay processes. All of these aspects are appealing to physicists seeking new and at the same time simple and basic systems from which to learn the laws of nature.

Recently the main attention has been focused on a phenomenon called the phonon wind. For some time it has been recognized that the liquid condensate in Ge possesses a high mobility, that is, it is able to move with relative ease through the Ge crystal when pushed by certain kinds of forces. The latest chapter is the discovery that a flux of nonthermal phonons (quantized high-frequency acoustic modes in the solid) is effective in propelling drops of the liquid phase; hence the term phonon wind, coined by the Soviet theorist L. V. Keldysh. This concept has inspired a series of investigations involving phonons and the condensed phase.

Excitons and electron-hole drops. A pure semiconductor such as Ge is practically devoid of free charge carriers at the temperatures of liquid helium (1–4 K); in other words, the solid behaves as an insulator. However, carriers can be introduced by optical excitation, for example, by a laser, a process which promotes an electron from the filled valence band into the empty conduction band, leaving behind a free hole, the positively charged counterpart to the electron. (The condensate decays by the inverse process, that is, the recombination of a hole and electron with the emission of an infrared light quantum.) Holes and electrons at low temperatures can bind pairwise through their mutual Coulomb attraction to form excitons.

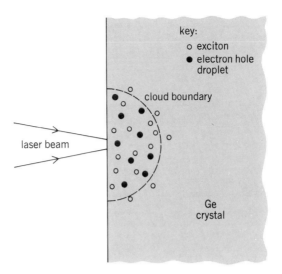

Fig. 2. Schematic representation of the cloud of electron-hole drops. Droplets are excited by a laser beam focused on the sample surface.

These electrically neutral entities, analogous to hydrogen atoms, move freely in the crystalline lattice with thermal kinetic energy. Collectively the excitons behave as a gas in the crystal, this gas being generated by optical pumping.

With increasing excitation intensity, the exciton gas becomes denser; when the density reaches a certain prescribed level (at a given low temperature), a new, liquid phase begins to condense in the form of fine but macroscopic droplets. Their diameter in Ge is typically of the order of 5 μm. This phenomenon is analogous to the condensation of water droplets from the vapor phase. The two-phase system, liquid drops and exciton gas, can be represented by the simple phase diagram shown in Fig. 1. To the right (low average density), there exists an exciton gas only; at intermediate densities, under the curve, is the two-phase coexistence region; and at the left, at high densities, only liquid is present.

By nature the condensate is a metallic, electron-hole liquid, the electrons and holes both being itinerant, as opposed to an exciton liquid where electrons and holes would remain bound together as atoms. In this respect the condensate is more closely analogous to a liquid alkali metal such as liquid Li or Na, and unlike a molecular liquid such as liquid hydrogen. Observations show that electron-hole drops are dispersed in the form of a cloud or fog extending into the crystal some distance, typically 1 mm, from the laser-pumped crystal face as schematically pictured in Fig. 2.

Phonon wind. The cloud concept as represented in Fig. 2 seems straightforward enough; yet its understanding has for a long time remained elusive. The question is simply this: Since the optical excitation used to produce excitons does not penetrate much beyond the crystal surface, how does it happen that the drop cloud extends up to 1 mm or so in depth into the sample? A number of explanations were advanced, all of which upon closer scrutiny proved untenable. Yet, out of these exercises one fact soon became clear, namely, that there

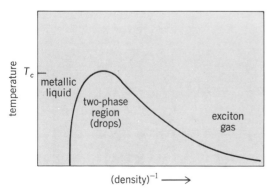

Fig. 1. Schematic diagram of the electron-hole liquid and exciton gas system. The critical temperature T_c in Ge is 6.7 K.

must exist a force of some description acting to propel the drops into the interior of the crystal. The breakthrough came when Keldysh announced that the answer might be streams of nonthermal phonons, that is, a phonon wind. As envisioned by Keldysh, phonons are generated copiously in the vicinity of the exciting laser spot, a by-product as it were of photoexcitation and decay processes. As these phonons radiate ballistically in a more or less isotropic pattern, they are absorbed by drops, resulting in a net force tending to push the drops radially outward from the origin. Ultimately, at a certain radial distance, both because the force becomes ineffectual (due to its falling off according to the inverse-square law) and because the drops decay during the time of flight, a cloud boundary develops. Its size is controlled largely by the laser excitation intensity.

Physicists in the United States and Soviet Union quickly seized upon these ideas and devised experiments to test their validity. They included V. S. Bagaev (in collaboration with Keldysh) at the Lebedev Institute in Moscow; J. C. Hensel and R. C. Dynes, Bell Labs, Murray Hill, NJ; J. Doehler and J. M. Worlock, Bell Labs, Holmdel, NJ; and M. Greenstein and J. Wolfe, University of Illinois. The experiments of the four groups, each taking a very different approach, have verified the phonon wind concept; moreover, the investigations brought to light unanticipated but important results regarding the structure of the droplet cloud, the interaction of phonons and electron-hole drops, and the propagation of phonons in crystals. A synopsis of this work follows.

Bagaev and coworkers reasoned that the phonon wind produced by intense laser excitation at one end of a Ge bar would set in motion a "test" packet of drops injected by weak excitation near the midpoint of the bar. A drift of drops approximately 1 mm "downwind" was observed, the first direct evidence for the phonon wind.

The approach of Hensel and Dynes was more quantitative, utilizing time-resolved spectroscopy and submicrosecond bursts of acoustic phonons called heat pulses. Basically the experiment is a phonon transmission experiment. Heat-pulse phonons propagating ballistically in the Ge crystal from source to detector pass through an interposed cloud of electron-hole drops. A decrease in transmission due to phonon absorption measures the interaction between drops and phonons, which underlies the phonon-wind force. When the heat pulse intensity was raised to very high levels, the drops were observed to move, as much as 0.5 cm under certain conditions, and their time of flight indicated a velocity of 4×10^4 cm/s, 10% of the sound velocity.

These studies yielded two important facts: (1) the absorption is strongly directional with respect to the crystallographic axes; (2) phonons are selectively "channeled" along certain directions in the crystal (an effect called phonon focusing). These facts led to the startling conclusion that the cloud is structured in shape, and not isotropic as originally thought.

This set the stage for an experiment by Greenstein and Wolfe. The result is shown in Fig. 3. These workers, using some ingenious instrumentation, succeeded in making a photograph of the cloud by means of the infrared radiation emitted from the droplets in their decay. The sharp flares along the <001> and <010> axes of the crystal represent the phonon focusing effect; the broader lobes in between arise from maxima in the phonon absorption anisotropy.

The movement of drops in the cloud has been "seen" by Doehler and Worlock in an experiment based on the Doppler effect. Radiation (in the infrared, where Ge is transparent) is scattered from droplets; the frequency shift of the scattered radiation then is a direct probe of the drops' velocity.

At this juncture the phonon-wind model seems firmly established and the controversy about droplet clouds essentially settled. Some questions remain, but by and large the experiments outlined above confirm the essential ideas and complete one of the most interesting chapters in the story of electron-hole drops.

For background information *see* EXCITON in the McGraw-Hill Encyclopedia of Science and Technology.

[J. C. HENSEL]

Bibliography: V. S. Bagaev et al., *Pis'ma Zh. Eksp. Teor. Fiz.*, 70:702–716, 1976 (transl., *Sov. Phys. JETP*, 43:362–370, 1976); J. Doehler, J. C. V. Mattos, and J. M. Worlock, *Phys. Rev. Lett.*, 38: 726–729, 1977; M. Greenstein and J. P. Wolfe, *Phys. Rev. Lett.*, 41:715–719, 1978; J. C. Hensel and R. C. Dynes, *Phys. Rev. Lett.*, 39:969–972, 1977; J. C. Hensel, T. G. Phillips, and G. A. Thomas, *Solid State Phys.*, 32:87–314, 1977; L. V. Keldysh, *Pis'ma Zh. Eksp. Teor. Fiz.*, 23:100–103, 1976 (transl., *JETP Lett.*, 23:86–89, 1976); T. M. Rice, *Solid State Phys.*, 32:1–86, 1977.

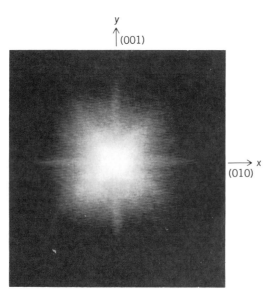

Fig. 3. Cloud of electron-hole droplets in Ge at 1.7 K. Droplets are produced by a laser focused on the rear surface of the crystal. The crystal shown is approximately 5 mm square. *(From M. Greenstein and J. P. Wolfe, Anisotropy in the shape of the electron-hole-droplet cloud in germanium, Phys. Rev. Lett., 41:715–719, 1978)*

Extinction (biology)

At the end of Mesozoic time, some 65,000,000 years ago, a very large segment of the biota of the oceans disappeared in one massive wave of extinctions. Among the organisms that died out were most of the coccolithophores, the dominant phytoplankton in the oceans. Evidence recently uncovered indicates that the coccolithophores that repopulated the oceans had evolved well before the catastrophe, and were actually thriving, while the diverse Cretaceous phytoplankton, which had been singularly successful for more than 50,000,000 years, died abruptly. Some of the phytoplankton that thrived during the catastrophe favored brackish waters, and this suggests that the catastrophe was caused by the presence of a low-salinity surface layer over the entire oceans. The source for this vast amount of brackish water was the Arctic Ocean, which had become isolated and flushed of salt water during latest Cretaceous time. When Greenland and Norway separated 65,000,000 years ago, the lower-density Arctic Ocean waters spilled out onto the surface of the entire world oceans, causing extinction of the stenohaline surface-dwelling plankton (those unable to tolerate the salinity change). Many deeper-dwelling organisms died when their food chain collapsed or their oxygen supply became exhausted as a result of the intense density stratification. The low-density Arctic Ocean water was also substantially cooler than the ocean surface in general, and this probably caused modification of the Earth's climate, which ultimately led to the extinction of large reptiles, the dinosaurs, on land.

Danian coccolithophores. The crucial evidence for this model comes from the North Sea, where oil is produced from chalk beds which were deposited continuously during latest Cretaceous and earliest Tertiary time. A major constituent of these and all chalk beds is coccoliths, the fossil remains of the planktonic coccolithophores. The table gives the coccolith succession as it appears in one of several wells drilled and cored through these chalk beds. In this well, and in others nearby, early Tertiary (Danian) coccoliths first appear in the sediment nearly 200 ft (about 60 m) below the level where Cretaceous coccoliths disappear. Surprisingly, the previously common Cretaceous coccoliths are virtually absent at the level where the Tertiary species occur, and the few specimens that are present can probably be accounted for by sediment mixing.

Following this first appearance of Tertiary coccoliths, the Cretaceous species returned, and the Tertiary species, which had been quite successful for a very brief period, were again completely excluded. When the Tertiary coccolithophores again returned, after a period represented by deposition of about 200 ft (60 m) of chalk, the Cretaceous coccolithophores disappeared forever. They became extinct. Significantly, during both their first and second appearance the Danian coccolithophores produced large numbers of very small coccoliths, a condition generally indicative of rapid rates of reproduction or blooms.

Clearly the Danian coccolithophores did not evolve from a few stragglers that survived the great extinction at the end of Mesozoic time; they already existed in latest Cretaceous time, although

Checklist of calcareous nannofossils near the Cretaceous-Tertiary boundary for Phillips Ekofisk Well 2/4-A8*

Depth (m)	Depth (ft)	Arkhangelskiella cymbiformis	Cribrosphaerella ehrenbergii	Nephrolithus frequens	Prediscosphaera cretacea	Kamptnerius magnificus	Biscutum sp.	Cretarhabdus surirellus	Micula mura	Eiffellithus turriseiffeli	Lithraphidites quadratus	Chiastozygus sp.	Zygodiscus sp.	Watznaueria barnesae	Microrhabdulus decoratus	Ahmuellerella octoradiata	Prediscosphaera spinosa	Markalius astroporus	Chiasmolithus danicus	Biantolithus sparsus	Thoracosphaera sp.	Neochiastozygus spp.	Thoracosphaera operculata	Braarudosphaera bigelowi	Cruciplacolithus tenuis	Coccolithus cavus	Zygodiscus sigmoides	Prinsius bisulcus	Ericsonia? brotzenii	Coccospheres	% small placoliths of total nannofossils	% nannofossils	% detrital carbonate (diagenetic?)
3093.7	10,150	X	X	X														R	A	R	C	F	F	R	F	C	R	C	R	C	65	80	20
	10,190																	A	R	C		R			C	C	F	C	R	C	80	<10	>90
	10,250																				R					R							99
3139.4	10,300																	R	A		C	R	F		F	C	R			F	40	80	20
	10,330																	R	A		C				A	C	C				50	80	20
	10,340	C	R	R	C	C	R	F						R				R	R						F	R					0	20	80
3169.9	10,400	A	F	C	F	R	F	F		R	R	F			R																0	20	80
	10,430	C	R		F	F	R	R		F					R	R	R	R													0	15	85
	10,460	A			C	R	R	R	F					R		R	R	R													0	25	75
	10,490	A	R		C	R	R	F		R				R				R													0	35	65
3206.5	10,520	R												R				C	F						C	R	R		R	F	25	20	80
	10,524	R						R				R						R		R					F	R	R		R	R	80	<10	>90
	10,570	C			R	R		F	R	R					R	R		R													0	10	90
3249.2	10,660	A			R	F	F			R	F			R	R	R		R													0	10	90

*R = rare; F = few; C = common; A = abundant; X = present.
SOURCE: From S. Gartner and J. Keany, The terminal Cretaceous event: A geologic problem with an oceanographic solution, *Geology*, 6:708–712, December 1978.

seemingly they were unable to compete with the Cretaceous coccolithophores in the normal oceanic environment. How then could they appear and flourish in the North Sea for a brief period when Cretaceous species were excluded, then completely disappear again from the record when the normal Cretaceous phytoplankton returned, only to reappear all over the world, simultaneously with the extinction of all Cretaceous coccolithophores and planktonic foraminifers, as well as many metazoan species?

The answer is that while the Danian coccolithophores were unable to compete successfully in the stable Late Cretaceous seas, they had an abnormal salinity tolerance and may, in fact, have evolved in a low-salinity brackish water body. A likely candidate for such a brackish water body is the Arctic Ocean, which became isolated from the world oceans in latest Cretaceous time. The passage between the Arctic and the Pacific oceans was closed by compressional tectonics—mountain building—in the Bering Sea area; similarly, mountain building, extending from northeastern Greenland across the Canadian Arctic Archipelago, probably closed the seaway between Greenland and Canada; the seaway between Norway and Greenland did not yet exist; and the shallow seaways over North America and Eurasia had emerged or been blocked by sedimentation sufficiently to prevent exchange of significant amounts of water.

Arctic Ocean salinity. This isolated Arctic Ocean received, then as now, an excess amount of precipitation and runoff from the surrounding continents. This excess precipitation diluted and flushed most, possibly all, the salt water out of the Arctic Ocean, depending on the length of the period of isolation, and thereby made that basin perhaps the largest fresh-water body ever to exist on Earth.

Then, approximately 65,000,000 years ago, rifting occurred between Greenland and Norway. The initial opening probably was temporary and allowed for only limited exchange of water between the Arctic Ocean and the North Sea. This would correspond to the level of first occurrence of Danian coccoliths. When the permanent connection to the North Atlantic was established, it was probably no larger than a midocean rift valley (that is, a few tens of kilometers wide and probably several hundred meters deep). Through this opening the much denser normal-salinity waters of the North Atlantic intruded into the Arctic Ocean, filling the basin from the bottom up, while the much lighter brackish or fresh waters in the Arctic Ocean spilled out on the surface of the North Atlantic and from there into the South Atlantic, through the Tethyan and Caribbean seaways out over the vast Pacific Ocean.

During the dispersal the Arctic Ocean water was mixed with the normal-salinity water of the oceans, and the final product of this mixing probably was a water mass of intermediate salinity. But even with a very substantial amount of mixing the resulting water mass was less saline and, therefore, significantly less dense (lighter) than normal sea water. Moreover, the volume of the lighter water—and, hence, the thickness of the low-density surface

layer—was increased by mixing. This is important because the Arctic Ocean probably did not contain more than 1–1.5% of all of the water in the oceans (that is how much water is in the Arctic Ocean today), which would be enough to form a layer on the surface only 38–57 m thick. In clear, tropical waters, light can penetrate to a greater depth than this, and if the low-salinity layer were only this thick, some of the stenohaline phytoplankton might have survived beneath it and repopulated the oceans after the surface layer had been thoroughly mixed. If the Arctic water was mixed with an equal volume or even two or three times as much normal-salinity ocean water, the density contrast between the resulting water mass and normal sea water would still be sufficient to maintain intense stratification, while the thickness of the low-density surface layer would be increased two-, three- or even fourfold. Such a thick layer of low-salinity water would ensure that all of the phytoplankton unable to tolerate the lower salinity would become extinct.

A low-density surface layer 70 m or more in thickness would also be an effective barrier to mixing by surface winds, because even the largest storm waves do not cause significant mixing at such depth. Consequently, mixing and the eventual breakdown of this stable stratification could take a long time—certainly decades and possibly centuries—long enough to severely deplete the oxygen supply in the deeper waters. So long as these waters were isolated from the atmosphere and below the photic zone, there would be no opportunity to replenish the oxygen, and many deep-dwelling organisms, most especially organisms that were very active, would suffocate. Prominent among these would be squids and their relatives.

The low-salinity surface layer itself would kill not only the phytoplankton that had to live near the surface within the photic zone, but also such organisms as planktonic foraminifers, some of which may have obtained their food from symbiotic algae, and all of which probably had to spend at least part of their life cycle close to the surface. Bottom-dwelling organisms also were not immune if they lived in shallow water and were unable to tolerate low salinities. Rudists and many echinoderms probably were among the latter.

Other consequences of the catastrophe. Finally, some organisms probably died because the organisms that they fed on disappeared. Among them were no doubt many marine predators, including certain seagoing reptiles such as the plesiosaurs and mosasaurs.

The marine organisms best equipped to survive the catastrophe probably were those whose natural habitat was the brackish water of bays, estuaries, and lagoons, as well as those deep-dwelling species that were able to get on with only a minimum supply of oxygen.

The potential effects of this catastrophe on land are highly speculative. However, it is reasonable to think that the surface temperature of the oceans dropped appreciably everywhere, at least for a short time, as a consequence of the dispersal of the Arctic Ocean water. This drop in ocean surface temperature may have been sufficient to cause a radical change in the climate of the Earth, and this climatic change, either catastrophically or gradually, led to the decline and extinction of the large land reptiles, the dinosaurs.

For background information *see* EXTINCTION (BIOLOGY) in the McGraw-Hill Encyclopedia of Science and Technology.   [STEFAN GARTNER]

Bibliography: M. M. Bramlette, *Science*, 148: 1696–1699, 1965; S. Gartner and J. Keany, *Geology*, 6:708–712, December 1978.

Fertilization

The objectives of fertilization are twofold: (1) to incorporate a spermatozoon into the egg and then to prevent any additional sperm from fusing with the egg, and (2) to activate the egg toward development. Research during the last few years, primarily on sea urchin fertilization, has provided important new insights into how both processes occur. The findings obtained with these relatively simple invertebrate systems appear to be universal and have shed much light on similar mechanisms in higher organisms, such as the mammals.

Program of fertilization. An important realization is that the sperm initiates a sequence of events which are remarkably similar in the various eggs that have been studied (see illustration). The generalization that may now be made is that the early sequence of events emanates solely from changes in the ionic composition and ion permeability of the egg. The initial changes can be measured with microelectrodes. These measurements, which are evidenced as changes in the membrane potential, reveal a very rapid depolarization of the egg plasma membrane, which goes from about −70 mV to +10–20 mV in a few seconds after sperm addition. This change results from an influx of sodium. The next major change is a rise in intracellular free calcium, which can be detected in eggs previously injected with the light-emitting protein aequorin. This protein emits light only in the presence of calcium and is thus an excellent indicator of changes in the intracellular concentration of this ion. The rise in calcium is only transient, and lasts but a few minutes. Coincident with or shortly after the beginning of the rise in calcium, there begins a release of hydrogen ion from the egg. This is an extremely large release which is mediated by an exchange process in which extracellular sodium is exchanged with intracellular protons. The result is an increase in the intracellular pH of the egg cytoplasm. Also beginning about the time of the calcium rise is an interconversion of pyridine nucleotides from nicotinamide adenine dinucleotide (NAD) into nicotinamide adenine dinucleotide phosphate (NADP). There is also a large increase in the respiratory activity of the cell and a massive secretion of the contents of numerous granules from the egg surface (cortical granules). This secretion results in the elevation of a new egg envelope referred to as the fertilization membrane.

About 5 min after fertilization there begin changes centering on the permeability and the synthetic activity of the cell. The permeability of the membrane for potassium, phosphate, amino acids, and nucleosides begins to increase dramatically, but these changes in transport are not fully

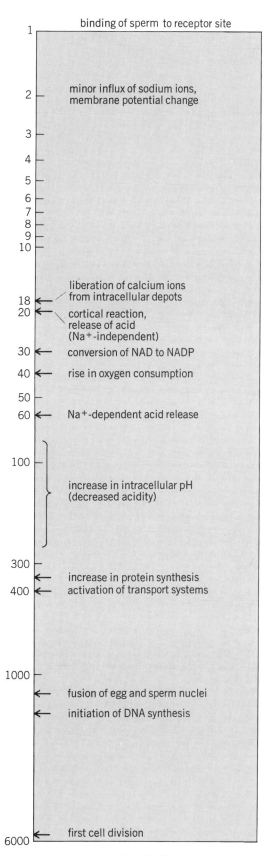

Program of events that follows fertilization of eggs of the sea urchin (*Strongylocentrotus purpuratus*). Time is indicated in seconds on a logarithmic scale. A similar sequence of events is seen for other sea urchins.

established for an additional 20 min to 2 hr. Also beginning at this time is a large increase in the rate of protein synthesis, carried out on templates of messenger RNA that are present in the egg cytoplasm; about 20 min after fertilization the first cycle of the DNA synthesis is initiated.

The above paragraphs summarize the major changes that occur in the sea urchin egg. Do similar sorts of changes occur in eggs of other species? Unfortunately knowledge is very meager about the events in these other eggs, but the rise in calcium has also been detected in eggs from starfish and fish. As discussed below, the increase in this ion is critical for development and is probably a universal concomitant of fertilization. The cortical granule secretion, although not universal, is very common. All organisms do not raise a fertilization membrane, but it appears that fertilization generally results in alterations of the extracellular coats that surround the egg and that a cortical granule secretion is involved in these surface changes.

Entry of sperm into egg. Although the sperm has a flagellum which is involved in getting it to the egg surface, the activity of the flagellum does not appear necessary for incorporation of the sperm into the egg. For example, sperm which have had their flagellum removed can fertilize eggs. Also, motion picture analyses indicate that tail motility ceases before sperm entry. Finally, the drug cytochalasin B, which prevents microfilament function, inhibits the incorporation of the sperm into the egg; the effect appears to be on the egg and not on the sperm and suggests that the actin-containing microfilaments of the egg are involved in bringing the sperm into the egg. Indeed, scanning electron micrographs indicate that the microvilli of the egg, which contain microfilaments, elongate and surround the sperm and probably function to bring it into the egg; this appears to be true in both the sea urchin and mammalian egg.

Precluding extra sperm. In order for normal development to ensue, it is imperative that only one sperm fuse with the egg cytoplasm; if more than one enters, the constancy of the genome would be lost and species could not be conserved. For these reasons, eggs possess potent mechanisms to ensure that only one sperm will fuse with the egg.

The primary mechanism for precluding extra sperm appears to be the aforementioned change in the membrane potential that occurs at fertilization. The major line of evidence supporting this is that if one injects current into the egg so that the membrane potential is raised to that level which normally occurs after fertilization, sperm will attach to such eggs but will not fuse with them. If the membrane potential is now dropped, the attached sperm will fuse. This indicates that the depolarization, by itself, can prevent sperm-egg fusion; the mechanism by which this occurs is not yet known.

A secondary mechanism to preclude supernumerary sperm from entering the egg ensues from the aforementioned cortical granule secretion. Part of the secretory products include two proteases which destroy sperm receptors and result in the elevation of the fertilization membrane. If the activity of these proteases is interfered with, supernumerary sperm will fuse with the egg at sites where the fertilization membrane has remained attached to the egg plasma membrane. Similar mechanisms may also be involved in precluding supernumerary sperm entry in other organisms.

Activation of the egg: The transient increase in calcium appears to be the important change resulting in egg activation. A major part of this evidence comes from experiments in which the calcium level of the egg cytoplasm is experimentally increased. One means of doing this is with the ionophorus antibiotic A23187. This compound abolishes the selective permeability of the plasma membrane for calcium, with a resultant increase in the calcium level of the cell. A prediction would be that if the calcium rise is critical, application of this drug to the egg should result in its activation in the absence of sperm (artificial parthenogenesis). This prediction was fulfilled; eggs of animals from all phyla so far tested have been activated, including those of mollusks, annelids, echinoderms, and vertebrates (including fish, frogs, and hamsters). Similar results have also been obtained by directly injecting calcium into eggs.

A major question is how this rise in calcium can activate egg metabolism. A recent insight has come about from the analysis of the enzyme NAD kinase. The activity of this enzyme, which increases after fertilization, is responsible for the large increase in the coenzyme, NADP. This enzyme is activated during the period when the calcium level is high, suggesting that calcium is somehow involved. Recent work indicates that the involvement of calcium is through the binding of calcium ion to an enzyme regulator protein known as calmodulin. Calmodulin is a ubiquitous protein in cells which, when it binds with calcium, can subsequently attach to target enzymes and regulate their activity. NAD kinase, the enzyme responsible for the NADP formation, is calmodulin-activated.

These findings suggest that the sequence of events after fertilization is a sperm-induced rise in calcium ion, the calcium then binding to calmodulin and the calmodulin-calcium complex then activating NAD kinase. Calcium also somehow triggers the cortical granule secretion and undoubtedly causes many other yet undiscovered reactions.

In the sea urchin egg the rise in calcium, by itself, does not appear to be an adequate stimulus to trigger development. A second change is required, which is an increase in the intracellular pH, rising from approximately pH 6.8 to about 7.2–7.3. This large change in pH is catalyzed by an exchange reaction in which extracellular sodium is exchanged for intracellular hydrogen. The pH rise appears to directly initiate the large increase in protein synthesis that occurs after fertilization, and may also be involved in a polymerization of actin in the egg surface. The pH increase might also indirectly initiate DNA synthesis and other synthetic changes of the cell.

A similar sequence of ionic changes may be involved in the activation of the sperm, which also undergoes a sequence of changes necessary for fertilization. As it approaches the egg, there occurs a secretion of enzymes necessary to allow it to pass through the outer egg coats so that it can fuse with the egg surface. This secretion requires calcium ion. In the sea urchin there is in addition a polymerization of actin, resulting in the formation of a filament which is necessary to attach the sperm to the egg. This actin polymerization, as in the egg, may be pH-mediated.

Summary. The initiation of development thus involves complex changes in both sperm and egg. The changes in the sperm are initiated as the sperm comes near the egg, probably triggered by substances released by the egg. The changes in the egg are triggered by sperm-egg contact. The nature of these important changes is unclear, but it is apparent that the initial changes involve ions; release of calcium appears to be universal, and in the sea urchin egg an increase in intracellular pH, mediated by a sodium-hydrogen exchange, is also critical. One target of calcium is calmodulin. Important problems to be answered in the coming years will concern how these ion changes are mediated by sperm-egg contact and how these are transduced into the initiation of new metabolic pathways necessary for development.

For background information *see* FERTILIZATION in the McGraw-Hill Encyclopedia of Science and Technology.

[DAVID EPEL]

Bibliography: D. Epel, Mechanisms of Activations of Sperm and Egg during Fertilization of Sea Urchin Gametes, in A. A. Moscona and A. Monroy (eds.), *Current Topics in Developmental Biology*, 1978; L. J. Jaffe, *Nature*, 261:68–71, 1976; F. J. Longo, *Develop. Biol.*, 67:249–265, 1978; R. A. Steinhardt, R. Zucker, and G. Schatten, *Develop. Biol.*, 58:185–196, 1977.

Flash floods

In the United States more people are dying each year from flash floods than ever before. But there are important opportunities for science and technology to reverse this trend.

Causes. Flash floods most often result from heavy rainfall. Some result from damming of a stream by mud slides, avalanches, or ice jams; release of water from a glacier-dammed lake (Icelandic, *jokulhlaup*); rainfall combined with snowmelt; or failure of a dam (either natural or erected structures). Dam failures often are the result of intense rainfall producing reservoir inflow that exceeds the damming and spilling capabilities. Flooding resulting from the sudden release of impounded water by dam failure has been responsible for a major portion of the loss of lives and property in many flash floods. Major dam failures occured at Buffalo Creek, WV, in February 1972; Teton Dam, ID, in June 1976; Johnstown, PA, in July 1977; and Toccoa, GA, in November 1977 (see table). In these four cases alone there were 251 killed and nearly $139 million in damages.

Forecasting. In the United States, success in decreasing the tragic loss of life (averaging nearly 200 deaths per year) and property (over $1 billion in damages each year) depends on the ability to understand and forecast the physical processes which produce flash floods, and on the ability to satisfactorily cope with the human aspects associated with protective or preventive measures when flood warnings are issued.

Dam failure. Satisfactory techniques have recently been developed for predicting the flood flows resulting from dam failures, based on assumptions as to the type of failure, reservoir contents, and streamflow into the reservoir. Under actual conditions, specific information on these three factors can be used to improve the predictive accuracy and timing of the downstream rapid flooding.

Technical limitations. There are substantial gaps in knowledge of how extremely heavy rainfalls of short duration are produced. In many cases, storms that produce flash floods are neither forecast nor even detected by the present river and rainfall reporting networks.

New approaches to observing the occurrence and magnitude of intense precipitation are providing forecasters with increased information for issuance of flash flood warnings. Automation of many rain and stream gages in drainages subject to flash flooding have helped to provide information more rapidly, but less than 10% of the reporting networks are automated. Many means of communication are presently in use, including AM and FM radio, radio via meteor bursts, satellites, and microwave links, as well as the telephone. Many gages are programmed to report periodically (as at hourly intervals). An event-reporting rain gage is now being used to report directly to computers each time a specified amount of rainfall occurs. In many areas of the United States large numbers of people have rain gages and serve as spotters to provide telephone or citizen-band radio reports on heavy storms, as is done for reporting of tornadoes.

However, rainfall gages can measure only those intense storm cells that occur immediately over the gage. Radar scopes monitored by meteorologists of the National Weather Service (NWS)—a component of the National Oceanic and Atmospheric Administration (NOAA)—provide surveillance of precipitation events. The intensity of the reflected microwave energy, from heavy precipitation, does provide a measure of the water flux in the cloud and thus an index to the precipitation rate. Some radars are equipped with automatic digitizers that allow the data to be processed by minicomputers for quickly providing maps of the areal distribution and amount of rainfall to the forecaster. Photographs and infrared data from NOAA meteorological satellites are used to derive estimates of the amount of rainfall. The satellite infrared data provide a measure of the temperatures of the cloud tops. For very large storm centers the cloud tops are generally very high and thus have colder temperatures than warmer and lower clouds that generally produce only moderate or average precipitation. Papers describing the equipment and techniques for observing and measuring heavy rainfall were presented at the 1st National Conference on Flash Floods held in Los Angeles on May 2–3, 1978.

Watch-warning forecasts. The NWS has a watch-warning forecast service to alert people to the possible occurrences of a flash flood. The issuance of a flash flood watch for a particular area indicates that heavy rains may result in flash flooding and people in the area should be prepared for the possibility of a flash flood warning signifying an emergency which will require immediate action. A warning is issued when flash flooding is occurring or is imminent in the area. Persons in the area should move immediately to safe ground.

The flash flood watches and warnings are based on the prediction or observance of heavy rainfall. In addition to the use of various observing and measurement methods, the NWS offices predict the occurrence of heavy rains or other causative factors. Basic weather guidance is prepared by the National Meteorological Center of the NWS, located in Camp Springs, MD. These forecasts are based on observations and numerical, diagnostic, and predictive models of the atmosphere, and include quantitative precipitation forecasts. The large-scale numerical models do not provide adequate forecasts for the extreme amounts of rainfall generally associated with flash floods.

Research. Studies by C. F. Maddox and R. A. Chappell of the Environmental Research Laboratories, NOAA, and others are beginning to develop a knowledge of the characteristics of flash flood storms. A cooperative experiment is being funded by NOAA, the National Science Foundation, and other government agencies, to improve the scientific understanding of mesoscale storms and rainfall processes. SESAME (Severe Environmental Storms and Mesoscale Experiment) was conducted in the spring and summer of 1979 to provide a massive measurement of mesoscale storms in "tornado alley" centered near Norman, OK. The SESAME data will be of great value for testing and development of mesoscale weather models needed to predict the convective activity that often results in flash floods. Of 20 significant flash floods in the

United States between 1972 and 1977, striking similarities have been observed. For example, in 75% of the cases studied, the floods began during the night hours of 6 P.M. to 6 A.M.

Many flash flood–producing storms are associated with extratropical storms that degenerate from hurricanes or tropical storms. The movement of these storms, such as occurred from hurricanes Camille in 1969 and Agnes in 1972, can be predicted with some degree of success. However, it has been almost impossible to predict accurately the timing, intensity, and location of flash flood–magnitude rainfalls that occur with severe local storms.

A review of the major flash floods during the past decade (see table) shows that four were associated with storms derived from tropical storms, and five were associated with one or more dam failures that contributed to the severity of the flooding. The table indicates whether telemetered rainfall reports, radar, and satellite data were used for issuing flash flood forecasts prior to the flooding.

Human factors. If there is inadequate public response to warnings, even a perfect forecast of impending disaster may be of little value. Case studies have shown that people do not always recognize the direct relationship between heavy precipitation and flash flooding. Therefore, they are often unwilling to evacuate or take other protective measures even when warned by national, state, or community officials. In many situations, the potential victims have not been trained as to protective or preventive measures. An example is the false

Major flash floods in the United States, 1969–1978

	James River Basin, VA, August 1969	Arizona, September 1970	Buffalo Creek, WV, February 1972	Rapid City, SD, June 1972	Hurricane Agnes, northeast U.S., June 1972	Teton Dam, ID, June 1976	Big Thompson, CO, July 31 – Aug. 1, 1976	Appalachia, April 1977	Johnstown, PA, July 1977	Kansas City, MO, September 1977	Toccoa, GA, November 1977	Southern California, February 1978	Texas, August 1978
Observations (used for issuing forecasts prior to flooding)													
Precipitation, telemetered	No	No	No	No	Yes	*	No	Yes	No	No (?)	No	No	No
Radar (from area of heavy precipitation)	No	No	No	No	Yes	*	No (?)	No	Yes	Yes	Yes	No	Yes
Satellite (from area of heavy precipitation)	No	No	No	No	Yes	*	No	Yes	No	Yes	No	Yes	No
Forecast issued (for principal flood area prior to flooding)													
Flash flood watch	No	—	Yes	—	Yes	*	No	Yes	No	Yes	Yes	Yes	Yes
Flash flood warning	No	Yes	No	Yes	Yes	*	No	Yes	No	Yes	No	No	Yes
Highest amount of precipitation, mm	810+	290	142	381	478	*	304	394	305	406	178	120	1067
Previous tropical storm	Yes	Yes	No	No	Yes	No	No	No	No	No	No	No	Yes
Dam failure involved	No	No	Yes	Yes	No	Yes	No	No	Yes	No	Yes	No	No
Report available (NOAA or USGS-NOAA)	Yes	Yes	No	Yes	Yes	No	Yes	No	Yes	Yes	No	Yes	Yes
Lives lost†	153	23	125	237	105	11	139	22	76	25	39	20	33
Damages, millions of dollars†	116	5	10	165	4020	500	30	424	330	90	2.5	83	100+

*Dam failure — not weather related.   †For some cases, statistics are for all flooding.

security that people often place in their cars; a major percentage of deaths is related to automobiles. An increased and continuous education program is essential for success in reducing the annual loss of life and damages from flash floods.

Flash floods occur in all parts of the United States. Some areas, such as the recreational areas of the Southwest, can quickly turn into death traps for the unobserving traveler. In all areas, simple precautions such as never driving into inundated areas could save hundreds of lives.

Community involvement. The key to an effective preparedness program lies with local communities. Training of individuals in how to protect themselves and their property and how to respond when warned by local officials is best accomplished at the local community level.

The local community should also establish an appropriate organization which can cooperate with Federal and state officials responsible for issuing warnings of impending disasters. This must be a 24-hr operation. It must also have sufficient authority to properly coordinate with the police, fire, and other preparedness organizations, and have the necessary communications for collection of data, receipt of information, and contact with all emergency personnel, as well as with the public news media.

In addition to the flash flood watch and warning service, the NWS assists communities on the development of local warning systems. Such systems include the installation of reporting precipitation and stream gages, some of which may automatically sound an alarm when a specified amount of precipitation occurs in a watershed or when a stream reaches a specified level. The NWS forecast offices in many cases can provide daily information on the amount of rainfall needed to cause flooding. Local officials can then predict the streamflow which would occur on small, fast-rising streams, using this information and the amount of precipitation from their gages. The U.S. Army Corps of Engineers and other government agencies provide extensive help and guidance for both structural and nonstructural alternatives for flash flood protection and preventive actions.

Needed action. The American Meteorological Society has issued a statement of concern, "Flash Floods—A National Problem." The following actions were recommended to help alleviate the problem: (1) increase regulation of the use of areas subject to flash flooding; (2) certify and monitor the safety of dams; (3) improve information on frequency of maximum precipitation and associated runoff for design and planning; (4) plan and carry out an extensive and continuous public awareness program; (5) strengthen ties among meteorologists, hydrologists, engineers, social scientists, and action agencies in communities; (6) improve the ability to monitor and detect flash flood conditions, partly by increased use of automated ground measurements, radar, and weather satellites; (7) increase the capability to forecast the location and magnitude of rainfall; (8) improve the capability to forecast intense, small-scale phenomena; (9) improve community warning programs, with emphasis on encouraging individual response to warnings.

For background information see DAM; STORM DETECTION; WEATHER FORECASTING AND PREDICTION in the McGraw-Hill Encyclopedia of Science and Technology.   [EUGENE L. PECK]

Bibliography: Conference on Flash Floods: Hydrometeorological Aspects, Amer. Meteorol. Soc. Preprint, 1978; Flash Floods—A National Problem, Bull. Amer. Meteorol. Soc., 59:585–586, 1978; D. L. Fread, in Proceedings of the Dam-Break Flood Routing Model Workshop, Bethesda, Oct. 18–20, 1977, U.S. Water Resources Council, pp. 164–197, 1977.

Flight controls

Developments in a number of technologies related to flight controls are leading to major changes in aircraft flight control systems. Examples of these recent developments are advances in digital technology, fiber optics, and electric motors that use rare-earth magnetic materials. These developments will make possible reduced volume, weight, power requirements, and cost and improved reliability of conventional flight control systems, as well as providing additional options for the aircraft designer for improving aircraft efficiency. One option is to incorporate active control technology (ACT) in the design of the aircraft, with resulting greater dependence upon the flight control system and improved aircraft efficiency. The fuel efficiency of commercial transports is of growing importance due to the rising value of liquid petroleum and its various derivatives, including jet fuel.

Primary flight controls. Flight control system design is in a state of change as a result of the improvements in electrical and electronic component reliability. These improvements have led to flight control designs for military aircraft with electrical control paths from the pilot's controls to the surface actuators. In these designs, motions or forces applied to the cockpit controls are transmitted by electrical means to the aircraft control surfaces instead of by cables or push-pull rods. As flight control systems become more complex to meet requirements for improved civil aircraft performance, savings in aircraft weight and cost can result from this fly-by-wire (FBW) approach.

Auxiliary flight controls such as flaps, commonly called secondary flight controls on commercial transport aircraft, have conventionally been used for improving lift capability on fixed-wing aircraft. In addition to this function, they now are being used for aircraft maneuvering and regulation of aerodynamic loads. Applications such as these are finding increased usage on short takeoff and landing (STOL) aircraft and highly maneuverable aircraft, such as military fighters.

Hydraulic actuators are used to power primary flight control surfaces for most high-performance military and civil aircraft. However, there is increased interest in the development of high-torque, low-inertia electric motors using rare-earth samarium-cobalt magnetic materials. For certain low-power applications this form of actuation may result in lower actuation weight and cost.

Automatic flight controls. Automatic flight control systems, such as autopilots and autothrottles, are used to control aircraft position and speed. These systems are installed on most aircraft, from single-engine private aircraft to the large swept-wing commercial transport. Characteristically, these systems have employed conventional analog electronic circuitry, but rapid advances in digital computer technology have resulted in widespread acceptance of the use of digital computers for these and many other functions related to aircraft systems, such as air data, navigation, inertial reference, and radio altimeters. In addition to these uses of digital computation, information transfer between such system elements occurs on digital data buses.

Digital implementation of automatic flight control systems provides the advantage of flexible system design. System changes may be made in software, thereby reducing costly modification of electronic circuitry. Improvements in the ability to accomplish system tests, both on the ground and during flight, also provide maintenance cost savings.

Augmentation systems. This is a class of electronic servomechanisms or feedback control systems which provide improvements in aircraft performance or pilot handling characteristics over that of the basic unaugmented aircraft. Augmentation system designs are similar to automatic flight control systems which utilize sensing devices, computers, and servomotors.

Stability augmentation systems are generally used to improve aircraft handling characteristics so that pilots may accomplish their task with greater ease and precision. These systems utilize feedback information from motion sensors and command control surface position in such a manner as to improve aircraft stability. A typical example of an augmentation system is a yaw damper, commonly used on swept-wing commercial transport aircraft. This system improves dutch-roll damping, a lightly damped lateral-directional oscillatory mode, by commands introduced into the rudder actuators. Several classes of aircraft employ extensive use of augmentation systems, such as military aircraft, supersonic transports, STOL aircraft, and vertical-takeoff-and-landing (VTOL) aircraft. Almost all modern high-performance transport aircraft employ stability augmentation in some form. Command augmentation systems employ feed-forward paths, from the pilot's controls,

in addition to feedback elements as in stability augmentation systems. Figure 1 shows how the command augmentation system is implemented in the longitudinal (pitch) axis of an aircraft.

Aircraft designs employing FBW primary flight control systems may integrate all or portions of the augmentation system as part of the primary flight control function. This is particularly true for aircraft configurations that are purposely designed for improved mission performance capability regardless of the aircraft's unaugmented stability characteristics. Increased emphasis on the use of augmentation systems to improve performance is evidenced by NASA- and Air Force-sponsored programs in the area of ACT and control configured vehicles (CCV), respectively.

Characteristics of ACT. The terminology ACT, in commercial airplane design, and CCV, in military design, is used to describe a special form of augmentation. These augmentation systems are used to stabilize otherwise unstable airplane configurations, and to limit, or tailor, the design loads that the airplane structure must support. Traditional airplane design practice produces airplanes that are stable as a result of the relative placement of the center of gravity and sizing of the aerodynamic surfaces, such as wings and empennage. An ACT airplane is stabilized with feedback control systems (augmentation) included in the design. Such design typically yields an airplane with a smaller empennage, resulting in less drag and therefore better performance.

The pilot maneuvers an airplane by either increasing or decreasing the lift required to sustain level flight and by banking as required to produce the desired flight path. Traditionally, the airplane structure is designed to accommodate the additional lift with wing-trailing-edge controls fixed. Maneuver load control systems (one form of ACT load relief) are designed to produce much of the incremental lift necessary for maneuvering through deflection of wing-trailing-edge controls. Thus, most of the additional lift is inboard (Fig. 2), and less wing structural material is necessary. Such design practice allows the structural weight to be reduced, compared with conventional wing design, and results in a lighter, more efficient airplane.

Other ACT functions can be designed to stabilize wing flutter, improve the ride qualities for crew and passenger, reduce the loads that the structure must accommodate when the airplane encounters gusts, and improve the airplane's fatigue life.

Motivation for ACT. The motivation for incorporation of ACT functions in airplane design stems from the desire to produce lighter and smaller airplanes to accomplish a specific mission. This reduces initial costs as well as cost of operation. These apparently feasible fuel savings are especially important to commercial airplane manufacturers and operators because of the dwindling petroleum resources. Maturing airborne digital computer technology has reduced the weight, cost, and volume of computing elements and increased their reliability. Further, these technology advances make feasible the widespread application of ACT in commercial aircraft.

Application of ACT. Current military airplanes

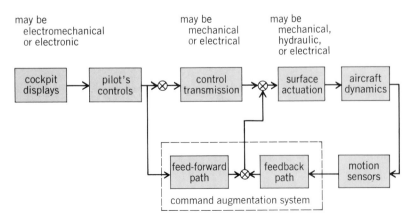

Fig. 1. Block diagram of flight control system incorporating a command augmentation system.

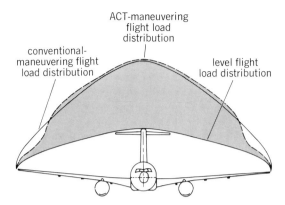

Fig. 2. Effect of ACT on the lift distribution of a maneuvering aircraft.

incorporate many of these concepts. In contrast, commercial transports include only limited applications of ACT, such as yaw dampers and ride improvement systems. Airframe manufacturers are using ACT concepts in the design of major derivatives of existing transports. These manufacturers have flight-tested experimental versions of wing-load-relief systems on such jet transport aircraft as the L-1011 and 747. New airplanes, designed for commercial use beginning in 1983, may also incorporate some of these functions.

Concerns about the continuing availability and cost of jet fuel for commercial transportation have led NASA to sponsor a number of programs designed to yield technology that would allow the production of airplanes which require less fuel to accomplish a specified mission. One of these programs, Energy Efficient Transports (EET), includes work focused on the commercial application of ACT.

For background information *see* FLIGHT CONTROLS in the McGraw-Hill Encyclopedia of Science and Technology.

[RICHARD L. SCHOENMAN; HENRY A. SHOMBER]

Bibliography: R. W. Howard, Automatic flight controls in fixed wing aircraft: The first one-hundred years, *Aeronaut. J.* 77(755):533–562, November 1973; J. McWha and L. R. Smith, *The Coming of Age of Digital Electronics in Commercial Transport*, AIAA Pap. no. 79-0686, presented at AIAA/RAeS/CASI/AAAF Atlantic Aeronautical Conference, Mar. 27–28, 1979; R. L. Schoenman and H. A. Shomber, *Impact of Active Controls on Future Transport Design Performance, and Operation*, SAE Pap. no. 751051, presented at Society of Automotive Engineers National Aerospace Engineering and Manufacturing Meeting, Nov. 17–20, 1975; D. M. Urie, *L-1011 Active Control Design Philosophy and Experience*, presented at AGARD Flight Mechanics Panel, Symposium on Stability and Control, Sept. 25–28, 1978.

Flow of solids

Particulate solids comprise ores, coal, grain, flour, and chemicals in the form of separate particles. Gravity flow of these solids with particles in contact and the voids between them filled with gas, usually air, will be discussed.

Particulate solids are a two-phase, solid-gas system. They are compressible; their bulk density

changes during flow. Since the volume of the particles changes little during this process, it is the size of the voids that is mostly affected. Changes in the size of the voids cause changes in gas pressure and result in gas-pressure gradients across a flowing solid which tend to reduce the rate of gravity discharge. When the solid is made up of large particles, that is, its permeability is high, and when the required flow rates are low, the gas-pressure gradients are not significant; the gaseous phase can be ignored and the solid treated as a one-phase, solid-only system. The theory of flow of such a system dates to the early 1960s and has since found general acceptance. Its application to storage vessels will now be outlined.

Bins. A bin (silo, bunker) generally consists of a vertical cylinder and a converging hopper. From the standpoint of flow, there are three types of bins: mass-flow, funnel-flow, and expanded-flow.

Mass-flow bins. Mass flow occurs when the hopper walls are sufficiently steep and smooth to cause flow of all the solid, without stagnant regions, whenever any solid is withdrawn. Valleys are not permitted, nor are ledges or protrusions into the hopper. In addition, the outlet must be fully effective, that is, if the hopper is equipped with a shut-off gate, the gate must by fully open; if it is equipped with a feeder, the feeder must draw material across the full outlet area. The range of hopper slope and friction angles leading to mass flow is shown in Fig. 1. Funnel flow will occur unless both the conditions for mass flow shown in this figure (the condition on θ_c in Fig. 1b and the condition on θ_p in Fig. 1c) are satisfied.

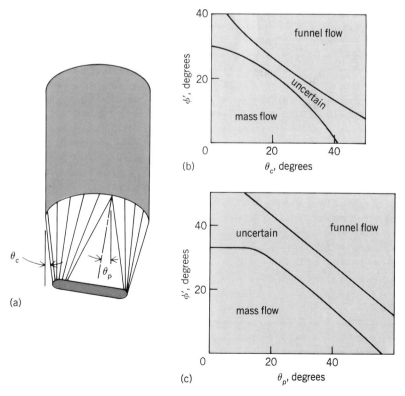

Fig. 1. Bounds on mass flow and funnel flow. (a) Geometry of hopper, showing slope angles θ_c and θ_p. (b) Range of kinematic angle of friction (ϕ') and θ_c leading to mass flow and funnel flow. (c) Range of ϕ' and θ_p leading to mass flow and funnel flow.

Mass-flow bins have advantages. Flow is uniform, and feed density is practically independent of the head of solid in the bin; this frequently permits the use of volumetric feeders for feed rate control. Indicators of low material level work reliably. Segregation is minimized because, while a solid may segregate at the point of charge into the bin, continuity of flow enforces remixing of the fractions within the hopper. Special bin design can be used for in-bin blending by circulation of the stored solid.

Mass-flow bins have a first-in-first-out flow sequence, thus ensuring uniform residence time and deaeration of the stored solid. Airlocks can often be dispensed with, provided a critical flow rate is not exceeded; otherwise channeling, that is, the development of a fast-flowing column of material within the bed of material, may occur; material will remain fluidized within this column and flush (pour out like a liquid) on exiting the bin.

The mass-flow type of bin is recommended for cohesive materials, for materials which degrade with time, for powders, and when segregation needs to be minimized.

Funnel-flow bins. Funnel flow occurs when the hopper walls are not sufficiently steep and smooth to force material to slide along the walls or when the outlet of a mass-flow bin is not fully effective.

In a funnel-flow bin, solid flows toward the outlet through a channel that forms within stagnant material. The diameter of that channel approximates the largest dimension of the effective outlet. When the outlet is fully effective, this dimension is the diameter of a circular outlet, the diagonal of a square outlet, or the length of an oblong outlet. As the level of solid within the channel drops, layers slough off the top of the stagnant mass and fall into the channel. This spasmodic behavior is detrimental with cohesive solids since the falling material packs on impact, thereby increasing the chance of material developing a stable arch across the hopper so that a complete stoppage of flow results. A channel, especially a narrow, high-velocity channel, may empty out completely, forming what is known as a rathole, and powder charged into the bin then flushes through. Powders flowing at a high rate in a funnel-flow bin may remain fluidized due to the short residence time in the bin and flush on exiting the bin. A rotary valve is often used under these conditions to contain the material, but a uniform flow rate cannot be ensured because of the spasmodic flow to the valve.

Funnel-flow bins are more prone to cause arching of cohesive solids than mass-flow bins, and they therefore require larger outlets for dependable flow. These bins also cause segregation of solids (Fig. 2) and are unsuitable for solids which degrade with time in the stagnant regions. Cleanout of a funnel-flow bin is often uncertain because solid in the stagnant regions may pack and cake.

The funnel-flow type of bin is suitable for coarse, free-flowing, or slightly cohesive, nondegrading solids when segregation is unimportant.

Expanded-flow bins. These are formed by attaching a mass-flow hopper to the bottom of a funnel-flow bin. The outlet usually requires a smaller feeder than would be the case for a funnel-flow bin. The mass-flow hopper should expand the flow channel to a dimension sufficient to prevent ratholing.

These bins are recommended for the storage of large quantities of nondegrading solids. This design is also useful as a modification of existing funnel-flow bins to correct erratic flow caused by arching, ratholing, or flushing. The concept can be used with single or multiple outlets.

Feeders. The hopper outlet must be fully effective. If flow is controlled by a feeder, the feeder must be designed to draw uniformly through the entire cross section of the outlet. It is also essential that the feeder be either suspended from the bin itself or supported on a flexible frame so as to readily deflect with the bin.

Vibrators are suitable for materials which are free-flowing under conditions of continuous flow, cake when stored at rest, but break up into separate particles when vibrated—for example, granular sugar. Fine powders and wet materials tend to pack when vibrated; hence vibration is not recommended. A mass-flow bin with a screw, belt, or similar feeder is best suited to those materials.

Spouts. These are acceptable for free-flowing, granular materials but cause flow problems with cohesive powders, unless the powder passes the spout in free fall. This usually requires a feeder

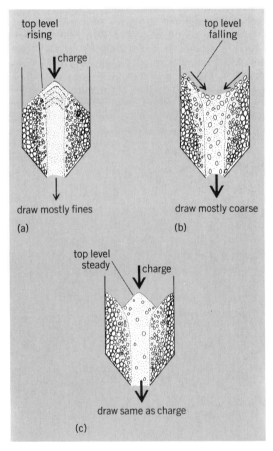

Fig. 2. Flow patterns in a funnel-flow bin, showing segregation of solids. (*a*) Charge coming in is greater than draw coming out, resulting in draw with a preponderance of fine particles. (*b*) Draw is greater than charge, resulting in draw with a preponderance of coarse particles. (*c*) Draw and charge are of same magnitude, resulting in draw with same consistency as charge. (*From A. W. Jenike, Why bins don't flow, Mech. Eng., May 1964*)

above the spout to control the flow rate, and sufficient surge capacity below to prevent settling out within the spout. Vertical spouts are preferred for powders.

Design for flow. The outlet dimensions of a hopper outlet must be sufficient to assure unobstructed flow. Flow may be obstructed by interlocking of large particles and by cohesive doming and ratholing across or above the outlet. Prevention of interlocking requires that the outlet be larger than several particle sizes. Prevention of cohesive obstructions requires that the outlet be sufficiently large to ensure the failure of potential cohesive obstructions.

Under the head of solid in a bin, pressure p develops within the solid. The solid consolidates and gains compressive strength f. This strength permits the development of stable domes and of stable ratholes in funnel-flow bins.

Consider a bin with a mass-flow conical hopper (Fig. 3). The solid is permitted to flow into it. The solid is unconsolidated when deposited at the top but, as an element of the solid flows down, it becomes consolidated under the pressure p acting on it in the bin. This pressure is shown by the p-line. At first, p increases with depth, then levels off in the vertical part of the channel. At the transition to the hopper, there occurs an abrupt change of pressure which may increase or decrease, depending on the frictional properties of the walls; then pressure decreases approximately linearly toward zero at the imaginary vertex of the hopper. To each value of consolidating pressure p, there corresponds a strength f of the solid, so that the strength f, generated by pressure p, similarly increases and decreases as indicated by the f-line. Design for flow is based on the flow–no flow criterion postulated as follows: a solid will flow provided the strength f which the solid develops is less than the stress s which would act in a stable obstruction to flow, $f < s$.

In a mass-flow bin, a dome across the hopper (Fig. 3) is the potential obstruction. The stress s in a dome is proportional to the span of the bin, as shown by line s. It is now observed that lines s and f intersect. Above the point of intersection, $f < s$, the flow criterion is satisfied and the solid will flow. Below that point, $f > s$, the solid has enough strength to support a dome and will not flow. The point $f = s$ determines the critical diameter B of the hopper which has to be exceeded for dependable flow.

Flowability of hoppers. As shown in Fig. 3, both the stress s in a stable dome and the consolidating pressure p are linear functions of the width of the hopper B. Since s and p are zero at the vertex of the hopper, their ratio is constant for a given hopper. This ratio is referred to as the flow factor, $ff = p/s$. It measures the flowability of a hopper. Values of ff have been computed for a wide range of hoppers.

Flowability of solids. The relation between the consolidating pressure p and the generated compressive strength f is called the flow function FF of a solid and is a measure of its flowability. The flowability of solids depends on several parameters.

Surface moisture. Solids are least free-flowing when moisture content is in the range of 70–90% of saturation. A saturated solid will usually drain in

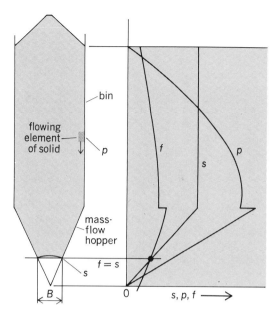

Fig. 3. Section of bin with mass-flow conical hopper, together with graphs showing relative values of pressure p, compressive strength f, and stress s, and illustrating criterion ($f = s$) for minimum outlet diameter B. (From A. W. Jenike, *Quantitative design of mass-flow bins, Powder Technol., 1:237–244, 1967*)

storage to the range of minimum flowability. Flowability tests may be carried out for a range of moisture content to determine the maximum moisture at which a solid can be handled by gravity.

Temperature. Many solids, especially those with a low softening point, are affected by temperature. The critical conditions usually occur when a solid is placed in storage at an elevated temperature and allowed to cool while at rest. Flowability tests indicate the maximum temperature at which a solid can be stored and whether or not it can be cooled without excessive caking.

Time of storage at rest. Many solids are free-flowing if they are kept in motion, but cake severely if stored at rest for a period of time. Flowability tests predict the maximum time that a solid can remain in storage at rest.

Effect of gaseous phase. Incidental effects of the gaseous phase are flushing of powders and limitation of discharge flow rate. In the latter, a speed-up of the bin outlet feeder produces no increase in feed rate, which is limited by the hopper outflow rate. Gas can also be introduced purposely into a bin to promote flow. Analytical information in this field is scant and, by and large, designs are developed empirically. This is due to the higher order of difficulty in the mathematical formulation of two-phase systems, the prevalence of nonsteady flow requiring transient treatment, and the shortcomings of testing methods used to measure the necessary additional material parameters. However, work in this field is in progress, with many potential applications in fixed- and moving-bed reactors, lock hoppers, and standpipes, but these results were proprietary in mid-1979.

For background information *see* BULK-HANDLING MACHINES; FRICTION in the McGraw-Hill Encyclopedia of Science and Technology.

[ANDREW W. JENIKE]

Bibliography: A. W. Jenike, *Trans. Inst. Chem. Eng.*, 40(5):264–271, October 1962; A. W. Jenike, P. J. Elsey, and R. H. Woolley, *Proc. ASTM*, 60: 1168–1181, 1960; A. W. Jenike and T. Leser, *Proc. 4th Intern. Congr. Rheol.*, pt. 3, pp. 125–142, August 1963; J. R. Johanson, *Chem. Eng.*, 85(11): 183–188, May 8, 1978; J. R. Johanson, *Chem. Eng.*, 86(1):77–86, Jan. 1, 1979; J. R. Johanson, *Chem. Eng. Progr.*, 66(6):50–55, June 1970: J. R. Johanson and A. W. Jenike, *ASME J. Appl. Mech.*, 39E(4):863–868, December 1972.

Fluvial erosion landforms

Landforms result from forces acting on resistant rocks and soils at the interface between the land surface and the atmosphere. Rivers long inspired the human imagination because of their capability of imposing forces on the Earth's surface. However, most scientists have observed that rivers do their work very slowly. Valley walls are weakened by the prolonged action of rain splash, frost action, soil formation, and other processes. Rivers are usually thought of as transport agents that remove debris during their occasional floods. The process generally takes place over millions of years, but there are a few spectacular exceptions. In a brilliant study of the Channeled Scabland region of Washington State (Fig. 1), J Harlen Bretz showed that great channelways had been scoured in rock by a few cataclysmic outbursts of Glacial Lake Missoula in Montana. The idea of valleys up to 250 m deep being carved by a few catastrophic floods was considered outrageous by many of his contemporaries when Bretz posed it 50 years ago. However, the recent discovery of probable catastrophic flood channels on Mars has given new relevance to Bretz's insights.

Channeled Scabland. The Channeled Scabland consists of a complex of anastomosing channels, abandoned cataracts, streamlined hills, and immense gravel bars created by the catastrophic fluvial erosion of the basalt and sedimentary rocks of the Columbia Plateau in eastern Washington. The last of the great outbursts from Glacial Lake Missoula occurred about 13,000 years ago during the wastage of the great ice sheets that advanced southward in the Pleistocene, or Ice Age.

The Missoula flooding of the Columbia Plateau region entered a network of preexisting stream valleys. These valleys did not have the capacity to convey the immense flood flows. The flooding spilled over the valley divides, eroding the loose, wind-deposited soil, called loess, that mantled the divides. Thus, the flooding created a complex pattern of dividing and rejoining channels. At least 100 such channelways were formed over a region

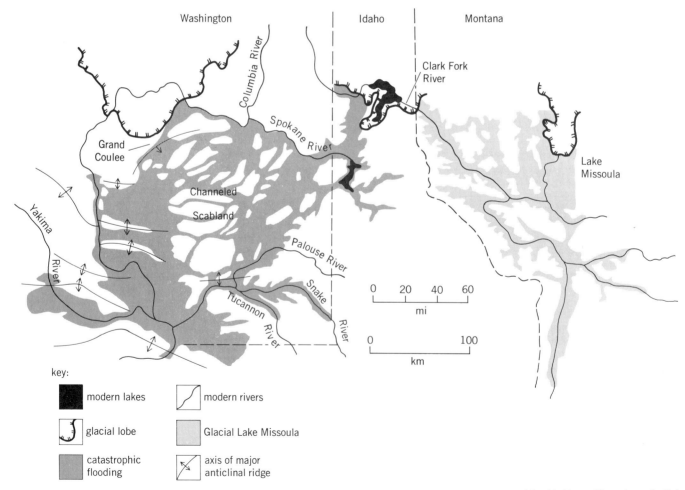

key:

modern lakes   modern rivers

glacial lobe   Glacial Lake Missoula

catastrophic   axis of major
flooding   anticlinal ridge

Fig. 1. Relationship between Glacial Lake Missoula and the Channeled Scabland. (*From V. R. Baker, The Spokane Flood controversy and the Martian outflow channels, Science, 202:1249–1256, 1978*)

Fig. 2. Dry Falls cataract in the Channeled Scabland. This former flood cataract is 120 m high and 5.5 km wide. Note the grooves on the relatively flat rock surface above the cataract lip. The large lake above the cataract is a part of the Columbia Basin irrigation project. (*Photograph by V. R. Baker*)

150 × 200 km. Most of the channels, called coulees, are now streamless because of the region's modern arid to semiarid climate.

Within the scabland channelways the Missoula flooding produced a distinctive assemblage of landforms. Residual areas of the loess which mantled the preflood divides were scoured to form shiplike prows that point upstream. Some loess hills were completely streamlined to a teardrop shape. Deposition of the boulders and gravel transported by the flood also formed great streamlined mounds. Some of these gravel bars are over 30 m in height and several kilometers in length.

The most spectacular fluvial erosion occurred on the bare basalt rock surfaces. Because of its fractured nature, the basalt was easily attacked by a plucking-type erosion beneath deep, high-velocity flood water. On broad basalt surfaces the floods scoured a linear pattern of grooves, aligned with the direction of current flow. At points where the basalt layers could be undercut, the flood eroded great steps into the basalt. As these steps, or cataracts, receded headward, they formed deep inner channels. Dry Falls in the Grand Coulee is an excellent illustration of these processes (Fig. 2).

Reconstructing an ancient flood. Although the erosional features of the Channeled Scabland have been well described by Bretz, the physics of the flood flows were not studied until recently by V. R. Baker. In 1973 Baker described evidence of the high-water marks left by the largest of the Lake Missoula outbursts. Using engineering hydraulic calculation procedures, he estimated the flood discharges and flow velocities.

At maximum outflow, near the end of the last major Pleistocene glaciation, Lake Missoula yielded as much as 20,000,000 m³ of water per second in the vicinity of its ruptured glacial ice dam in northern Idaho. Flow velocities in the scabland channelways ranged from 10 to 30 m/s. A very important discovery was that these paleohydraulic conditions are consistent with the scale and the variety of

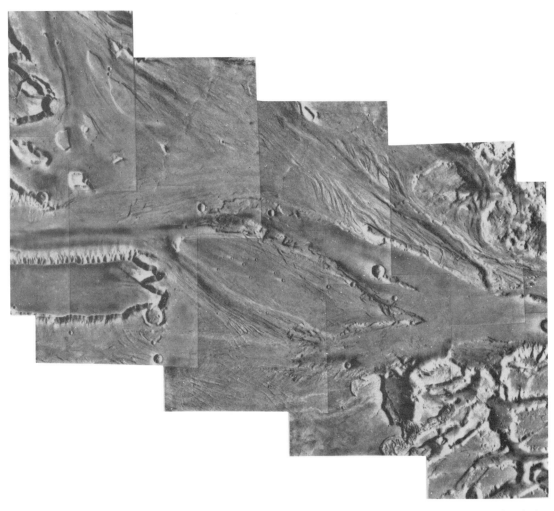

Fig. 3. Viking photomosaic of a portion of Kasei Vallis channel on Mars. The scene measures 300 km across and shows inner channels with recessional headcuts, grooved rock surfaces, and streamlined uplands.

erosional landforms in the Channeled Scabland. After 50 years Bretz's hypothesis for catastrophic flood erosion is no longer considered "outrageous."

Channels on Mars. Orbital photographs of Mars obtained during the Viking space mission (Fig. 3) reveal that the planet possesses a remarkable variety of channeled terrains. Some channels form relatively small networks up to 100 km in length that resemble the dendritic valley systems of the Earth. The most spectacular channels, however, comprise great anastomosing complexes up to 100 km wide that can be traced 2000 km or more across the planet's surface. These zones are called outflow channels because they appear full-born at localized source regions, usually collapse zones. Baker and others suggested that immense floods of water were released from within the planet, emanating from troughs that receded headward to form the channels. The collapse zones at the heads of the channels mark the last points of fluid release.

Much of Mars is heavily cratered, similar in some respects to the Moon. Closer scrutiny reveals that the Martian craters are very different than those on the Moon. Instead of being surrounded by rays of ballistically ejected debris, the Martian cra-

ters appear as though they were formed by meteors that impacted a soupy mud. They are bounded by lobes of debris that flowed from the impact sites. Abundant photographic evidence suggests that Mars was once underlain by a thick layer of permafrost which probably contained a high percentage of ice. The great outflow channels probably developed from a variety of processes related to this permafrost. Local zones of volcanic heating may have built up pressure beneath the permafrost cap until the liquid water burst forth as great floods. Another idea is that liquid water was trapped beneath the frozen surface layers until crustal warping caused pressure differences to develop. When the pressure in a local zone reached a critical point the water burst forth, draining the subsurface reservoir in much the same way that relatively small springs sometimes develop on the Earth.

Whatever the mechanism of water release onto the Martian surface, the outbursts produced an assemblage of erosional landforms that includes streamlined hills, anastomosing patterns, inner channels with recessional headcuts, barlike forms, and grooves aligned parallel to the indicated fluid flow direction. The scour features all occur below a definite trim line, indicating that the responsible fluid had an upper flow boundary. The only terrestrial landscape that contains analogs to all the Martian outflow channel landforms is the Channeled Scabland of eastern Washington.

The ages of the Martian outflow channels are very difficult to estimate from photographic evidence alone. One method is to look at the number of impact craters that occur on the channeled surface, and to reason that more craters per unit area will occur on older surfaces. Although this method provides relative ages, the absolute age in years requires an assumption concerning the rate of cratering. Current thinking relies on assumptions that suggest very old ages for the outflow channels, perhaps 1 to 3×10^9 years. If these ages prove correct, then the channels may have formed during an ancient episode of Martian history, perhaps when the climate was warmer and the atmosphere denser than at present.

For background information *see* FLUVIAL EROSION LANDFORMS; MARS; RIVER in the McGraw-Hill Encyclopedia of Science and Technology.

[VICTOR R. BAKER]

Bibliography: V. R. Baker, *Paleohydrology and Sedimentology of Lake Missoula Flooding in Eastern Washington*, Geol. Soc. Amer. Spec. Pap. no. 144, 1973; V. R. Baker, in *Proceedings of the 9th Lunar and Planetary Sci. Conference*, pp. 3205–3223, 1978; V. R. Baker, *Science*, 202: 1249–1256, 1978; J H. Bretz, *Geogr. Rev.*, 18: 446–477, 1928.

Foraminifera

In the last few years scientists from the United States, Denmark, Switzerland, Germany, and Israel have made some surprisingly rapid progress on a problem that had been mystifying protozoologists and micropaleontologists for over half a century. Certain families of foraminifera (ameboid protozoans with calcium carbonate shells) produced almost unbelievable giants hundreds of times the linear dimensions of their ancestors during the course of evolution. For example, ancestral fusuline foraminifera were 0.5 mm in length, while some descendants 40,000,000 to 50,000,000 years later reached 60 mm in length. *Nummulites gizehensis*, a member of a different family of larger foraminifera, reached over 120 mm in diameter! Although among the animals there are many examples of the evolutionary trend for the increase in size of descendant forms (Cope's law), foraminifera are only single-celled animals. As such, they should have been limited by the boundaries which usually affect cellular function, or should have some special adaptations to escape those limitations. Since there was a strong evolutionary trend toward giantism in four separate families of foraminifera (Orbitoididae, Nummulitidae, Discocyclinidae, and Miogypsinidae), particularly near the close of the Cretaceous and early Tertiary, the question of the factors underlying the adaptation for cellular giantism is particularly intriguing.

Endosymbionts. There were clues in the fossil record to answer the question, but they did not fall into place until after the completion of recent biological investigations of living representatives of some of these families. It is now apparent that there has been a long and continuous association between symbiotic algae and these animals which probably was the driving force in their evolution.

Evidence. There are now five lines of evidence (morphological, paleoecological, phycological, cytological, and physiological) which lend support to this hypothesis.

The shells of larger foraminifera are well adapted for symbiosis. Most are disk- or coin-shaped, exposing a large surface area relative to volume. In addition, large portions of their surfaces are thin and windowlike. In order to give strength to the shells, the windowlike portions are buttressed by various patterns and types of wall thickenings, which give many shells an almost miniature greenhouse appearance (Fig. 1). Some species have fingerlike projections from the inner surfaces of the shell which apparently serve to surround and hold algae like eggs in a grocery store container.

Paleoecological data and contemporary distribution data suggest that these animals characteristically grow, or grew, in the photic zones of warm tropical seas along with hermatypic corals. As is well known, the gastrodermis of all species of hermatypic corals is packed with endosymbiotic dinoflagellates (a particular group of unicellular olive-brown algae).

Very little is known of the cytology of larger foraminifera. Only one organism, *Sorites marginalis*, has been studied in detail. This animal had undergone extreme organellar replication, and is regionally highly differentiated. It has hundreds of generative nuclei, scores of larger somatic nuclei, and a generous supply of mitochondria and golgi. The generative nuclei, which give rise to the nuclei of the next generation, are located in the center of this disklike animal. They are surrounded by a doughnut-shaped ring of photosynthetically active symbiotic algae and large somatic nuclei. The outer third of the cell, at the periphery of the disk, is filled with digestive vacuoles. Radionuclide tracer field and laboratory experiments suggest that primary production by the endosymbionts of *Sorites marginalis* contributes to approximately 10% of their total carbon budgets. Quite the opposite has emerged from various experiments with *Heterostegina depressa*, another modern giant foraminifer. The data suggest that 90% of the carbon budget of this animal may be supplied by the photosynthetic activities of its endosymbionts.

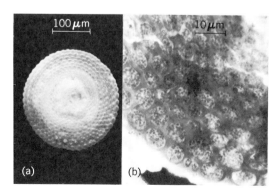

Fig. 1. *Sorites marginalis.* (a) Overall view. (b) Section showing symbionts within chamberlets. (*Courtesy of J. J. Lee and M. E. McEnery*)

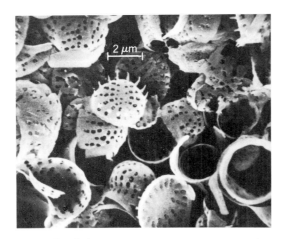

Fig. 2. *Fragilaria* sp., the algal endosymbiont from *Amphistegina* spp. in the Red Sea. (*Courtesy of J. J. Lee and M. E. McEnery*)

Identity. Investigations in the last 3 or 4 years into the identity of the endosymbionts inhabiting larger foraminifera have provided some of the best evidence that these animals are functionally well adapted for algal symbiosis. Unlike hermatypic corals, which all play host to a single species of endosymbiotic dinoflagellate, *Symbiodinium microadriaticum*, a great variety of algal types belonging to four different algal classes—Dinophyceae (dinoflagellates), Chlorophyceae (green algae), Bacillariophyceae (diatoms), and Rhodophyceae (red algae)—have been found in various larger foraminiferal species. This is true even among species belonging to the same family. In the family Soritidae, for instance, a dinoflagellate, *Symbiodinium microadriaticum*, is found in species of *Sorites* and *Amphisorus hemprichii*; a green alga, *Chlamydomonas hedleyi*, is found in *Archaias angulatus*; a different species of *Chlamydomonas*, *C. provasolii*, is found in *Cyclorbiculina compressa*; and a red alga, *Porphyridium* sp., has been found in *Peneroplis pertusus*. Certainly no one would argue against the idea that members of this family are generally adapted for symbiotic relationships. It was suspected that diatoms and chlorophytes (green algae) are the endosymbionts in two other families of giant foraminifera, Nummulitidae and Amphisteginidae, but proof was lacking. The critical morphological features needed to identify the algae as particular species of diatoms do not develop when they are endosymbiotic within their hosts. The diatom symbionts are not unique in this respect, since cell wall and thecal formation of the chlorophyte and dinoflagellate symbionts is also repressed within foraminiferal hosts but not in culture.

Isolation and culture. It became important therefore to isolate and cultivate the suspected diatom endosymbionts in order to identify them. In 1979 that goal was accomplished. With the aid of scuba, the animals were collected at Eilat in the Red Sea. After carefully scrubbing the outsides of several larger foraminiferal species suspected of harboring diatom endosymbionts, the animals were gently crushed with sterile glass rods to release the algae. Since their nutritional requirements were unknown, the algae were inoculated into a variety of media known to support the growth of diatoms. Fortunately the diatoms grew in some of the media and formed frustules (shells). All of them were extremely small (3–14 μm) pennate diatoms. Several were new species belonging to the genus *Fragilaria* (Fig. 2) and were endosymbiotic in all three species of *Amphistegina* which were collected in the Red Sea. Double infections were found in about 20–30% of the animals. A new species of *Navicula* was isolated from two specimens of *Amphistegina papillosa*. *Heterostegina depressa*, another larger foraminifer, harbored a very small variety of *Nitzschia panduriformis* (Fig. 3).

With the techniques for isolation of the diatom endosymbionts worked out, two of the same species of larger foraminifera were collected a few months later from a tide pool in Hawaii. Both harbored different but closely related species of diatoms. A diminutive form of *Nitzschia laevis* was found in *Amphistegina lessonii*, and a small variety of *Nitzschia valdestriata* came from all specimens of *Heterostegina depressa*.

While solving one problem, the new research raises many new questions relating to host range, depth distribution, mechanisms of transmission, and host-symbiont interactions.   [JOHN J. LEE]

Significance of symbionts. The nutritional and physiological significance of algal symbiosis in invertebrates has been investigated for nearly 200 years. Among foraminifera, this significance is apparently twofold: Algal endosymbionts provide a reliable source of reduced carbon that can be utilized for growth and reproduction. They may also stimulate the precipitation of calcium carbonate used in skeletal-formation. Theories of how algal symbionts enhance skeletogenesis vary, and the actual mechanism is poorly understood. In general, it is believed that symbionts can enhance the precipitation of calcium carbonate (1) through the removal of CO_2 in photosynthesis, (2) by providing substrates for the synthesis of an organic matrix required for skeletogenesis, (3) by providing

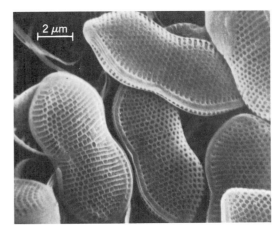

Fig. 3. *Nitzschia panduriformis*, the algal endosymbiont from *Heterostegina depressa* in the Red Sea. (*Courtesy of J. J. Lee and M. E. McEnery*)

energy for the transport of Ca^{++} and CO_3^{--}, and (4) through the removal of crystal poisons such as phosphate ions.

These various hypotheses were tested by L. E. Duguay and D. L. Taylor in 1978, in the symbiosis involving *Archaias angulatus* and its symbiont *Chlamydomonas hedleyi* (Chlorophyceae). They have not been examined rigorously in associations involving diatoms. However, present information does suggest that diatom symbiosis is similar, and provides the same nutritional-physiological benefits. This assumption is supported by the work of R. Röttger (1972) and R. Röttger and W. H. Berger (1972) which clearly shows the enhancement of growth, reproduction, and calcium carbonate deposition in the foraminifer *Heterostegina depressa*. Assuming that the fundamental attributes of associations involving diatoms are similar to those found in *Archaias angulatus*, it is expected that calcium carbonate production in these symbioses will be nearly directly proportional to the photosynthetic rate of the diatom endosymbiont.

Symbiosis with diatoms and other unicellular algae has allowed a variety of larger foraminifera to successfully exploit shallow, nutrient-poor marine environments. They achieve high population densities and substantial rates of calcium carbonate production. As a result, they are frequently the dominant component of marine sediments in these areas, and are thus of considerable geological significance.

For background information *see* DIATOM; ECOLOGICAL INTERACTIONS; FORAMINIFERIDA in the McGraw-Hill Encyclopedia of Science and Technology.   [DENNIS L. TAYLOR; LINDA E. DUGUAY]

Bibliography: L. E. Duguay and D. L. Taylor, *J. Protozool.*, 25:356–361, 1978; J. J. Lee, Towards Understanding the Niche of Foraminifera, in R. N. Hedley and C. Adams (eds.), *Foraminifera*, vol. 1, 1974; J. J. Lee et al., Symbiosis and the evolution of larger foraminifera, *Micropaleontology*, 25:118–140; J. J. Lee and W. D. Bock, *Bull. Mar. Sci.*, 26(4):530–537, 1976; S. Leutenegger, *Cahiers Micropaleontol.*, 3:5–53, 1977; R. Röttger, *Verh. Deutsch. Zool. Ges.*, 65:42–47, 1972; R. Röttger and W. H. Berger, *Mar. Biol.*, 15:89–94, 1972.

Forest and forestry

In 1964 Congress passed the Wilderness Act, establishing a legislatively protected National Wilderness Preservation System (NWPS). This legislation replaced administrative procedures for wilderness preservation first implemented in 1924. The principal objectives of the Wilderness Act were (1) to protect selected areas so that ecological processes would be, to the maximum extent possible, uninterrupted by human influence, and (2) to provide outstanding opportunities for solitude or a primitive kind of recreation. But rapid growth in the recreational use of wilderness poses serious threats to these objectives. To assure adequate protection, use is being managed to some degree on nearly all wildernesses, and this management will probably become more widespread and stringent in the future.

Wilderness management. The term wilderness management seems contradictory. Wilderness is generally conceived as beyond human control. Certainly, at one time, use levels were so low that there was little reason for concern about management. But recent growth in wilderness use no longer permits this unconcern. For example, in California, wilderness use grew an average 16% annually between 1970 and 1975. With such growth, the problems—damage to vegetation and soil, water pollution, wildlife disturbance, and crowding—that arise threaten the values that prompted the establishment of wildernesses in the first place. Thus, the real question is not whether to manage or not to manage, but rather how to manage.

There are two principal means for controlling impacts. First, managers could "harden" wilderness sites. That is, engineering-type solutions could be employed to offset or prevent undesired impacts. Trails could be paved, campsites provided with facilities, and visitor activity channeled through the use of barriers, signs, and facilities.

But the kind of opportunity wilderness is intended to provide, with emphasis on naturalness and the lack of human influence, makes such a solution inappropriate. Both legal constraints and the historical spirit of wilderness decry engineering solutions. The alternative is to manage use to control impacts through regulation of visitor behavior.

G. Gilbert and colleagues outlined a continuum of management actions (see table). Direct management imposes regulation over individual behavior, with managers holding a high degree of control; an example is the imposition of use rationing. At the opposite end of this continuum, indirect management features efforts to subtly influence or modify behavior, with individual freedom of choice largely retained; providing visitors with information is an example of this approach.

Research by G. Stankey and others indicates a preference among visitors for indirect management techniques. Such an approach is also more consistent with the motives commonly associated with wilderness users. However, both types of visitor management are appropriate in certain circumstances.

Visitor management. Three specific examples of visitor management are regulations, use redistribution, and rationing.

Regulations on use. Regulations on use are intended to reduce the per capita impact of use. That is, if the impacts of an individual's use can be ameliorated or prevented by regulation, more users can be accommodated in the area without adverse impact. An example of such a regulation would be limitations on visitor party size. Research by Stankey suggests that large groups have a disproportionate impact upon the experience of other groups. There is also concern that large groups have excessive impacts upon vegetation and soils. In the 1970s permissible party size in most areas dropped from about 30 people to about 15.

Such regulations can be extremely useful in reducing or eliminating visitor impacts without imposing more drastic controls on behavior. However, effectiveness depends upon visitors' knowing and complying with regulations. Studies have revealed that the levels of visitor knowledge and compliance are low. For example, current Forest

Service policy calls for visitors to carry out all non-combustible trash. One investigation revealed that only about one-quarter of surveyed visitors knew that this was the required method of disposal. Most visitors burned and buried such materials, a technique no longer considered appropriate since buried materials can be dug up by animals or unearthed by frost heaving.

The usefulness of these regulations also depends upon knowing that they are, in fact, effective at controlling the impact. For instance, many wildernesses have imposed regulations that prohibit camping within some minimum distance of lakes and streams. These regulations are intended to reduce water pollution, to protect shoreline vegetation, and to prevent bank erosion. Studies by L. C. Merriam and associates, however, reveal little water pollution danger from camps located in proximity to water. Moreover, such setbacks may greatly reduce the area available for overnight camping. In the Spanish Peaks Primitive Area in Montana, a 200-ft (60-m) no-camping restriction around lakes would result in a loss of 95% of all existing overnight sites.

Regulations can easily proliferate to the point at which they seriously conflict with the kind of recreational experience visitors seek. Excessive numbers of regulations become counterproductive, with visitors overwhelmed by their sheer volume. At some point, it is more satisfactory to impose some other form of use control (for example, rationing), leaving those who are admitted with a minimum of control. Proliferation of regulations also leads to an increased need for enforcement, a task that is often not welcome.

Use redistribution. The distribution of visitors both between and within wildernesses is typically very uneven. On a visitor-day-per-acre basis, the Galiuro Wilderness in Arizona received only 0.02 visitor-day per hectare in 1975; at the other end of

Direct and indirect techniques for managing the character and intensity of wilderness use*

Type of management	Method	Specific techniques
Indirect (Emphasis on influencing or modifying behavior. Individual retains freedom to choose. Control less complete, more variation in use possible.)	Physical alterations	Improve, maintain, or neglect access roads. Improve, maintain, or neglect campsites. Make trails more or less difficult. Build trails or leave areas trailless. Improve fish or wildlife populations or take no action (stock, or allow depletion or elimination).
	Information dispersal	Advertise specific attributes of the wilderness. Identify range of recreation opportunities in surrounding area. Educate users to basic concepts of ecology and care of ecosystems. Advertise underused areas and general patterns of use.
	Eligibility requirements	Charge constant entrance fee. Charge differential fees by trail zones, season, and so on. Require proof of camping and ecological knowledge or skills.
Direct (Emphasis on regulation of behavior. Individual choice restricted. High degree of control.)	Increased enforcement	Impose fines. Increase surveillance of area.
	Zoning	Separate incompatible uses (hiker-only zones in areas with horse use). Prohibit uses at times of high damage potential (no horse use in high meadows until soil moisture declines, say, July 1). Limit camping in some campsites to one night, or some other limit.
	Rationing use intensity	Rotate use (open or close access points, trails, campsites). Require reservations. Assign campsites or travel routes to each camper group. Limit usage via access point. Limit size of groups, number of horses. Limit camping to designated campsites only. Limit length of stay in area (max/min).
	Restrictions on activities	Restrict building campfires. Restrict horse use, hunting, or fishing.

*Modified from C. G. Gilbert, G. L. Peterson, and D. W. Lime, Toward a model of travel behavior in the Boundary Waters Canoe Area, *Environ. Behav.*, 4(2):131–157, 1972; used in J. C. Hendee, G. H. Stankey, and R. C. Lucas, *Wilderness Management*, USDA Forest Serv. Handb. no. 1365, 1978.

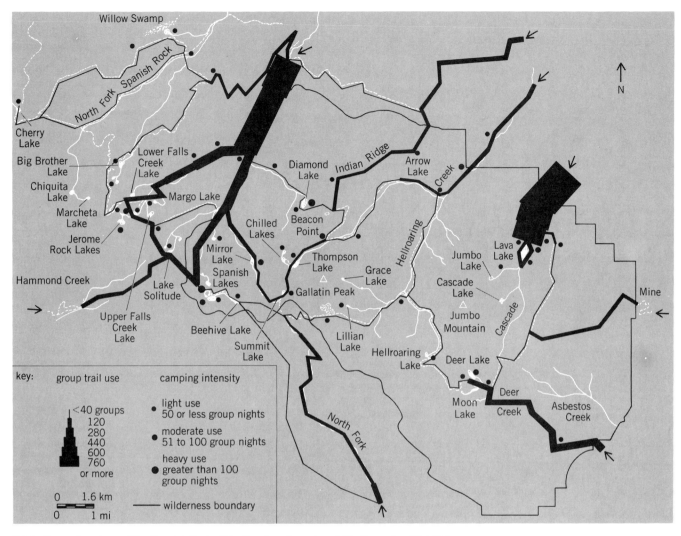

Fig. 1. Recreational use of the Spanish Peaks Primitive Area from June 14 to November 13, 1970.

the scale, the San Gorgonio Wilderness in California received over 12. Reasons for this variable drawing power are poorly understood. Proximity to population is one major factor, and the presence of attractions is another. But the visitor-day-per-hectare index is a crude measure. For example, it does not account for length of season that ranges nationally from 2 months in the northern Rocky Mountains to year-round in other areas. Nor does it reflect variations in the proportion of land actually available for use. In steep, rough country, use is concentrated in a few lake basins and stream bottoms, whereas in areas of more gentle topography, almost the entire area may be available for visitors.

Two measures that more realistically describe distribution potential are the number of entry points per 100 hectares and trail distance per 1000 hectares. The number of entry points per 1000 hectares provides an index of the potential for distribution around the boundary of the area. For the nation as a whole, there is an 80-fold difference among areas ranked on this variable. Trail distance per 1000 hectares provides an index of the potential for distribution within the area. Again, there is a great range: the area with the densest trail network has nearly 20 times the potential for

distribution of the lowest-ranked area.

Use distributions are similarly skewed. For example, in the Boundary Waters Canoe Area in Minnesota, nearly 70% of the user groups entered on 10% of the available entry points. In the Mission Mountains Primitive Area in Montana, however, the respective figures were 90% of the use on 10% of the entry points.

Wilderness travel is typically concentrated along a few trails. Figure 1 is a use distribution map of the Spanish Peaks Primitive Area. Most use is located along certain rather short segments of the trail system. This concentration can be illustrated by graphing cumulative visitor travel (total distance traveled by all visitors) against cumulative trail distance. Figure 2 shows this concentration index (CI) for the map in Fig. 1. For example, only 10% of the total trail distance accounted for 50% of the total visitor distance traveled. This typical pattern of spatial concentration is matched by a temporal concentration, with intense use on weekends. These factors combine to create serious crowding in nearly all areas.

Achieving more even use would bring important benefits. Much of the crowding would be relieved by better distribution of visitors in space or time,

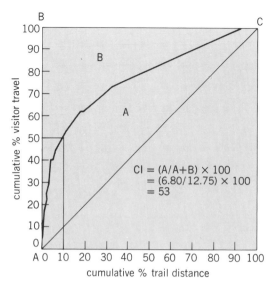

Fig. 2. Spanish Peaks use concentration index (CI) for all trail use during 1970.

or both. Some of the crowding has been reduced without directly assigning routes or visiting dates. Instead, visitors have been provided with information about use levels, attractions, alternative locations, and so forth. With such information, visitors themselves decide when or where to go. For example, about 75% of the visitors sampled in the Boundary Waters Canoe Area reported that they used information contained in a simple brochure, especially to find out when and where crowding occurred. The effect of these programs may take time to be fully realized, but it is now known that use can be shifted without direct control.

Shifting use is a two-edged sword, however. Raising use levels in areas once lightly used can mean the loss of "outstanding opportunities for solitude," as prescribed by the Wilderness Act. It can also result in use pressures on wildlife that may have relied upon these lightly used areas for sanctuary. Nevertheless, redistribution of visitors remains a useful management tool.

Use rationing. Rationing was first instituted in 1972 in the backcountry portions of several national parks and is found today in a number of other areas as well. Rationing involves the direct regulation of the number of persons permitted in an area at any one time. The specific number permitted is based upon a calculation of the area's capacity to withstand use without losing its wilderness qualities.

Stankey and J. Baden described five general rationing systems: reservations, lottery, queuing, pricing, and merit. Each technique has advantages and disadvantages for visitors and administrators alike. Currently, most areas with a rationing program use reservations or a queue (first-come, first-served). Studies of visitors to areas currently rationed indicate that an overwhelming percentage of visitors, including those not selected, support such controls. J. Fazio and D. Gilbert report that 80% of the unsuccessful applicants for backcountry use permits in Rocky Mountain National Park felt that rationing was necessary, a percentage

similar to that reported by Stankey for two southern California wildernesses.

One critical problem with rationing through reservation is the "no-show" with which airlines must also contend. Some areas report that more than 50% of reservations made are not picked up. Failure to allocate these unused openings would result in an underutilization of an area's capacity. By combining types of rationing schemes, as has been done in some areas, particularly with a queuing system, unused reservations can be taken by the next in line, so that optimal use of the area is achieved.

However, rationing is clearly an extreme step. It should be used only after other, less authoritarian and less direct measures have proved to be ineffective, or when existing conditions of resource impact or crowding are so severe that extreme action is warranted. That is, by using only that level of management action needed to achieve a management objective, the potential value of rationing will not be lost through premature use, nor will visitors' experiences be unnecessarily interfered with by unneeded controls. This principle of minimum regimentation will help ensure that excessively restrictive measures are not inappropriately applied. Judicious use of regulation, use redistribution, and rationing will help ensure the long-term preservation of an "enduring resource of wilderness."

For background information *see* FOREST AND FORESTRY; WILDLIFE CONSERVATION in the Mc-Graw-Hill Encyclopedia of Science and Technology.   [GEORGE H. STANKEY]

Bibliography: J. C. Hendee, G. H. Stankey, and R. C. Lucas, *Wilderness Management*, USDA Forest Serv. Handb. no 1365, 1978.

Forest soil

Detailed qualitative studies on fungi in forest soils have been published regularly since the early 1950s and, through these, ideas on successional patterns of fungi on forest litter (particularly leaf litter) have been developed. A. Hayes has reviewed these data and reemphasized the concept of a three-stage successional pattern on plant litter. He stated that despite the qualitative data which have been obtained, there is considerable ignorance of the ecological functions of the fungi which have been isolated. This ignorance is likely to continue until the decomposer processes are studied by biochemists and enzymologists.

Recent interest has centered on quantitative determinations of fungal standing crop and fungal biomass in the organic and mineral horizons of forest soils. This interest developed initially from the realization that fungi produce considerable biomass in both organic and mineral soil horizons (that is, tying up considerable amounts of nutrients) and that they are probably the predominant microorganisms in the early stages of decay of plant debris, thus playing a vital role in nutrient cycling. The possibility of competition between actively growing fungal hyphae and higher-plant roots for nutrients present in soil in limiting concentrations has been suggested. Additional attention has been directed to the importance of fungal-faunal interactions in the processes of organic matter decomposition and consequent nutrient cycling within the organic horizon.

Assessing fungal biomass. The absence of suitable methods to determine fungal biomass in samples of forest soils rapidly and accurately has impaired the development of such studies. Several new approaches have been developed to assess fungal biomass in forest litter and soil samples.

Fluorescein diacetate staining. Observations of soil smears, soil-agar films, or membrane filter preparations allow an estimate of the fungal standing crop in soil or litter samples, but clear distinction between live and dead hyphae is frequently difficult (that is, determination of fungal biomass is rarely made). B. Söderström has given data on the amounts of metabolically active fungal biomass in a pine forest soil (central Sweden) derived from the application of staining soil samples with fluorescein diacetate. Using this method, he calculated that in the A_{01}/A_{02} horizons the mean value for active fungal biomass was 680 mg m^{-2}, dry weight, as compared with a fungal standing crop (that is, live + dead) in similar samples of 42,000 mg m^{-2} (the latter figure being obtained by the use of conventional soil-agar films). This method holds promise, but it is difficult to apply for samples high in organic content, and the soil preparations stained with fluorescein diacetate must be observed very quickly because fluorescence fades rapidly (about 30 s) under high-intensity ultraviolet illumination.

Chitin determination. J. Frankland and coworkers studied the relative merits of direct-observation measurements (using soil or litter-agar films) and a chemical method of chitin determination to assess fungal biomass in decomposing forest litter. Apparently, while chitin determinations may be valuable for specific studies, such as the biomass of a single fungus growing on a wood substrate, the direct-observation methodology is more effective for studies on complex fungal communities with associated microfauna (which may also contain chitinous exoskeletons) in decomposing forest litter. This is particularly the case if direct observation also attempts to distinguish between live and dead hyphae, either by staining soil preparations with phenolic aniline blue or by observing under phase-contrast microscopy.

Agar film. It has been accepted that the agar film method for direct-observation measures of fungal biomass is likely to provide underestimates. For use in forest litter, it is necessary to optimize the maceration of litter prior to preparation of the agar films. Furthermore, factors used in the conversion of hyphal length measurements to the biomass (relative density of fungal hyphae and moisture content of fungi) must be as accurate as possible.

Soil respiration rates. Another approach to both total microbial biomass and fungal biomass determinations was taken by J. Anderson and K. Domsch. They measured initial rates of soil respiration following glucose amendment (optimal concentration for this purpose being predetermined) to determine total microbial biomass in soil samples. For most soils this method is very speedy in application, replicate determinations being made in a maximum of 4–5 hr. D. Parkinson and coworkers tested applicability of this method in studies of forest soil and found it suitable for total microbial biomass determinations in composite organic horizons as well as in individual layers of the organic horizon. In a study of a *Picea abies* forest (Solling, Germany), an average yearly total microbial biomass of 9.6 g microbial carbon per square meter was calculated using this method (maximum biomass in June: 14.1 g C m^{-2}; minimum biomass in March: 8.8 g C m^{-2}).

Bacterial-fungal biomass relations. The application of selective inhibitors to these litter samples indicated the relationship of bacterial to fungal biomass in this total microbial biomass to be in the ratio 25:75. There are very few studies on microbial biomass in the organic horizon of coniferous forest soils with which these data can be compared. Work on F and H layers of a Soviet *Picea abies* forest indicated 1.9–37.9 mg fungal carbon per 100 g dry weight of F layer and 1.6–31.4 mg fungal carbon per 100 g of H layer. Similar work on the H layer of a *Pinus contorta* forest indicated fungal biomass of 5.7–114.4 mg fungal carbon per 100 g.

Energy and nutrients. An accurate knowledge of fungal biomass in soil and organic debris allows estimates of tie-up of energy and nutrients by that biomass. It has been shown that fungal biomass can concentrate elements such as nitrogen and calcium. Until more is known about the death and decay rates of fungal tissue in forest systems, the quantification of their role in nutrient cycling must remain ill-defined.

Little progress has been made in measuring the rates of production of fungi in forest systems. Absence of critical methods is the reason for this. Certainly changes in fungal biomass over short time periods can be made by the techniques described above, but such measurements give only a crude indication of the turnover of fungal tissue in soil.

The substantial microbial biomass in forest litter has implications for the carbon economy of forests. The study in *Picea abies* forest indicated that when the carbon requirements for growth and maintenance of bacteria and fungi in that system were considered, the input of carbon in litter fall and so forth was sufficient for the maintenance of the measured biomass. Such calculations are very crude, primarily because they take no account of the recycling of microbial carbon and other nutrients during the turnover of microorganisms (as mentioned previously).

Ecological efficiency. Studies of ecological efficiency on dominant fungi in forest litter require an accurate measure of fungal biomass. Frankland and coworkers described this requirement with respect to their studies on *Mycena galopus* growing on *Corylus* leaf litter and *Quercus* leaf litter, where the calculated efficiencies were 0.07 and 0.04, respectively.

Undoubtedly the application of the fluorescent-antibody method to studies of the distribution and development of specific fungal taxa in forest soils will provide important data. This may well be of particular importance for studies on the growth and competitive abilities of ectomycorrhizal fungi in forest soils.

For background information *see* Forest soil;

SOIL MICROBIOLOGY in the McGraw-Hill Encyclopedia of Science and Technology.

[DENNIS PARKINSON]

Bibliography: J. P. E. Anderson and K. H. Domsch, *Soil Biol. Biochem.*, 10:215–221, 1978; J. C. Frankland, D. K. Lindley, and M. J. Swift, *Soil Biol. Biochem.*, 10:323–334, 1978; A. J. Hayes. *Sci. Prog. Oxf.*, 66:25–42, 1979; D. Parkinson, K. H. Domsch, and J. P. E. Anderson, *Oecol. Plant.*, 13:355–366, 1978; B. Söderström, *Soil Biol. Biochem.*, 11:149–154, 1979.

Fossil

Much recent progress in paleontology concerns the early history of several groups of animals and plants, highlighted by discoveries of the earliest known fossils of several life-forms. These fascinating new discoveries provide evidence both to test old hypotheses and to propose new ones concerning the origins of these life-forms.

For animals, recent advances include discovery of the oldest known vertebrates and bryozoans, a new synthesis of the origin and early evolution of the pelecypods, and the description of a well-preserved fossil worm fauna of Middle Cambrian age. Paleobotanical advances include the recent discovery of the oldest known land plants, a report on specialized shapes in some of the oldest known algal cells, and the recognition of fossilized organelles in some Miocene leaves.

Oldest known fish. In 1882 C. D. Walcott reported Ordovician vertebrate fossils among the rich faunas of the Harding Sandstone in Colorado. Until very recently, those Middle Ordovician fish remained the oldest known undisputed vertebrates. Other Ordovician fish fragments have been discovered since then in Wyoming, Montana, Oklahoma, Ontario, and Quebec. All these remains were assigned to Walcott's two genera *Astraspis* and *Eriptychius*, and all are also Middle Ordovician (about 450,000,000 years old).

Spitzbergen specimens. In the past few years, however, several paleontologists, most of whom were looking for other fossils, pushed back the age of the oldest known vertebrates. In 1976 the Europeans T. Bockelie and R. Fortey reported fish remains from the upper part of the Lower Ordovician in Spitsbergen, an arctic island north of Scandinavia. The Spitsbergen specimens are tiny plates bearing even tinier scales. Thin sections of these plates show an internal structure quite in accord with the histology of the well-studied Harding Sandstone fish. The plates consist of outer and inner lamellar layers enclosing a cavernous, or spongy, middle layer. These fossils were described as a new fish genus, *Anatolepis*. This first known occurrence of fish in the Lower Ordovician pushed back the age of earliest known vertebrates by about 20,000,000 years.

Anatolepis fragments. In 1978 J. E. Repetski reported an occurrence of *Anatolepis* in the Upper Cambrian part of the Deadwood Formation of northeastern Wyoming. While examining a sample of limy siltstone containing Late Cambrian trilobites, hoping to recover some conodonts (phosphatic tooth-shaped microfossils that are useful in dating marine sedimentary rocks), Repetski

found about two dozen specimens of *Anatolepis*. The fragments are flat to strongly curved plates with inclined scales, most of which have a rhomboidal shape (illustration *a, b*). The scales are 0.05 to 0.15 mm long, and the largest plate fragment is several millimeters long. X-ray analysis confirms that *Anatolepis* fragments consist of the calcium phosphate mineral apatite, the substance that makes up vertebrate bony tissue. The ultrastructure of *Anatolepis* plates is seen in scanning electron microscope photographs. An edge-on view (illustration *c*) clearly shows the inner and outer layers that display an indistinct lamination parallel to the plate surfaces. Between these layers is a cavernous zone. In thin section (illustration *d*), the Wyoming specimens closely resemble *Anatolepis* from Spitsbergen. The basal lamellar layer is penetrated by pores (one of which is shown just above center in illustration *d*) that lead into the middle spongy layer of bony tissue called aspidin. Dermal scales (one is shown as the lower right part of illustration *d*) are made of dentine which is constructed of lamellae parallel to the scale surface; tiny dentine tubules pass through these lamellae perpendicular to the scale surface.

Ordovician fish fossils. Although this evidence that *Anatolepis* was a vertebrate was convincing, one more problem remained. The dermal armor of Middle Ordovician fish (class Agnatha, order Heterostraci) consists of many discrete plates, called tesserae, embedded in the skin. This condition was considered primitive and ancestral to later fish that had fused tesserae making up a rigid dermal shield. Recently published findings from Australia seem to settle this issue. A. Ritchie and J. Gilbert-Tomlinson described some excellently preserved natural molds of articulated fish fossils of earliest Middle Ordovician age. There are at least two different species, both considered similar to *Anatolepis*. These fossils, though not preserving any of the actual bone, are good enough to permit a partial reconstruction. Ritchie and Gilbert-Tomlinson concluded that their fish, and *Anatolepis*, had large dorsal and ventral anterior shields that were nontesserate but that consisted of large, very thin, scale-bearing rigid plates. Behind the shields, the fish probably were covered with smaller plates, articulated to allow flexibility necessary for locomotion. Ritchie and Gilbert-Tomlinson estimated a total body length of 12 to 14 cm for the Australian fish. *Anatolepis* probably was no longer than 2 to 6 cm.

Age and environment of Anatolepis. The North American occurrences of *Anatolepis* (illustration *e, f*) document an age range for this fish of about 40,000,000 years (from about 510,000,000 to about 470,000,000 years before present). *Anatolepis* was a very successful animal! Perhaps more significantly, *Anatolepis* is known now from more than a dozen localities, always in rocks of undisputed marine origin. The earlier, widely held theory that the vertebrates originated in fresh water is thus dealt a serious blow. In addition, all these *Anatolepis*-bearing sedimentary rocks were within about 25° of the Equator at the time they were deposited, indicating that *Anatolepis* favored tropical to subtropical marine environments.

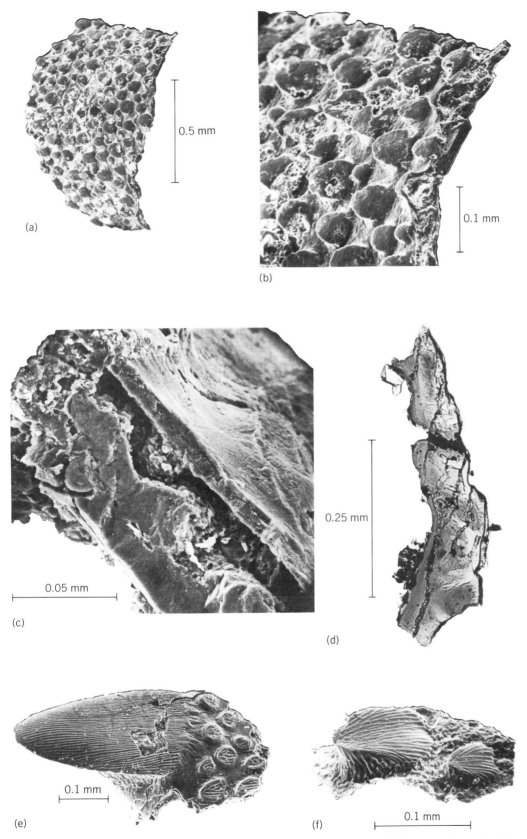

Fragments of *Anatolepis*, the oldest known vertebrate.
(a) Curved plate fragment from the Upper Cambrian of
Wyoming. (b) Closeup of (a), showing scales and broken
edge. (c) Closeup of broken edge, showing plate layers;
(d) thin section of plate, showing internal structure (*both
from J. E. Repetski, A fish from the Upper Cambrian of
North America, Science, 200:529–531, 1978*). (e) *Anatolepis* fragments from the Lower Ordovician of Utah and (f)
of Texas.

Other fossil animals. J. D. McLeod reported recently that the oldest known fossils of the colonial marine animals known as Bryozoa are specimens found in Arkansas and Missouri. These specimens are the first undoubted bryozoans of Early Ordovician age.

After studying more than 40,000 specimens, J. Pojeta, Jr., documented a synthesis concerning the origin and early evolution of the class Pelecypoda. Pojeta traced the history of pelecypods from their oldest known representative *(Fordilla troyensis)* in the late Early Cambrian through their first major radiation and diversification in Ordovician time.

S. Conway Morris described a very rare priapulid worm fauna from British Columbia, Canada. Marine worms, almost never preserved as fossils, were found in the black mudstones of the Middle Cambrian Burgess Shale. Conway Morris's painstaking examination of these fossils revealed both the external morphology and internal anatomy of these burrowing worms and suggested that some of them were predators.

Paleobotanical advances. The oldest known land plants, reported in 1978 by L. Pratt, T. Phillips, and J. Dennison, occur in fluvial (river-deposited) sediments of the Massanutten Sandstone in Virginia. These plants may be as old as Early Silurian (about 425,000,000 to 430,000,000 years old). They are certainly older than the previously known oldest land plant fossils, which are of Late Silurian age (about 395,000,000 to 400,000,000 years old).

L. A. Nagy, using sophisticated microscopical techniques, studied cysting cells in blue-green algal microfossils from Transvaal, South Africa, and indicated that plant cells were diversifying morphologically, and possibly functionally, as early as about 2,300,000,000 years ago. K. J. Niklas and his colleagues reported that Miocene leaves (about 16,700,000 to 25,000,000 years old) from Oregon have well-preserved cell walls, chloroplasts with grana stacks and starch bodies, and nuclei with "chromatinlike material." These leaves were exceptionally well preserved by rapid burial in volcanic ash.

For background information *see* FOSSIL; PALEOBOTANY; PALEONTOLOGY in the McGraw-Hill Encyclopedia of Science and Technology.

[JOHN E. REPETSKI]

Bibliography: T. Bockelie and R. A. Fortey, *Nature (London)*, 260(5546):36–38, 1976; S. Conway Morris, *Spec. Pap. Palaeontol.*, 20:iv–155, 1977; J. D. McLeod, *Science*, 200:771–773, 1978; L. A. Nagy, *J. Paleontol.*, 52(1):141–154, 1978; K. J. Niklas et al., *Proc. Nat. Acad. Sci. USA*, 75(7): 3263–3267, 1978; J. Pojeta, Jr., *Phil. Trans. Roy. Soc. London, B*, 284:225–246, 1978; L. M. Pratt, T. L. Phillips, and J. M. Dennison, *Rev. Paleobot. Palynol.*, 25:121–149, 1978; J. E. Repetski, *Science*, 200:529–531, 1978; A. Ritchie and J. Gilbert-Tomlinson, *Alcheringa*, 1:351–368, 1977.

Fusion, nuclear

The prospect of supplying economically a large fraction of the world's energy requirements from the thermonuclear fuel deuterium, a heavy isotope of hydrogen which can be obtained from water, has prompted research programs to explore the various techniques by which this desirable objective might be achieved. The interest in using lasers to initiate nuclear burning of a thermonuclear fuel, or fusion, arises from the laser's ability to produce the enormously high power densities required. The energy stored in the lasing medium can be released in a very short time, a nanosecond (10^{-9} s) or less, and focused into a very small volume (10^{-1} to 10^{-4} cm^3). Although this power density is sufficient for the process, other conditions described later must also be met. Because both the generation of laser energy and its coupling to the fuel in a useful manner are inefficient processes, it is necessary to capture in the fuel part of the nuclear reaction energy released and to contain and insulate the fuel against heat conduction losses until a sufficient fraction is consumed to give the net energy return required for a power plant cycle.

Inertial containment. In the laser-fusion concept, the fuel is contained by its own inertial mass while the burning takes place, hence the term inertial containment. Since the energy which the laser can supply and the energy release which can be contained are relatively limited, the fuel mass must be small and the burning time correspondingly short (less than 1 ns). This is possible only when the fuel density is very high. The criterion for satisfactory fuel burnup is that the product of density times radius of the fuel lies between 2 and 3 g/cm^2 for the optimum fuel, a mixture of deuterium and tritium (d-t). Therefore, the fuel must be strongly compressed (10^3–10^4 times) and reach ignition temperature only at the center during maximum compression. This can be achieved most efficiently by an ablatively driven isentropic compression followed by shock heating of the central region to produce the central temperature spike needed for ignition. Preheat of the fuel must be avoided.

Laser-fusion programs in the past have concentrated on developing the tools required for this process, including lasers, target design codes, fabrication techniques, and diagnostic instrumentation, and on understanding laser-matter interaction physics.

Laser capabilities. During the past year, two kinds of lasers, the Nd glass laser operating at 1.06 μm at the Lawrence Livermore Laboratory (LLL), and the CO$_2$ laser operating at 10.6 μm at the Los Alamos Scientific Laboratory (LASL), have achieved over 10 terawatts peak power on target, while a number of other Nd glass lasers in the United States and other countries are now doing target work at the terawatt level. Laser systems designed to operate at 200 TW and 100 kilojoules or more for both CO$_2$ and Nd glass at LASL and LLL, respectively, are under construction.

Absorption of laser light. Light is absorbed by scattering of electrons by photons (inverse bremsstrahlung), by ion acoustic waves (Brillouin scattering), and by driving electrons resonantly in a density gradient by the electric field of the coherent optical wave (resonant absorption). Very fast "hot" electrons escape, dragging ions with them, while slower electrons thermalize with the ions in the plasma. There is now general agreement that in the intensity regions of interest, 10^{14}–10^{16} W/cm^2, the fraction of light absorbed by a spherical target is 25–30% for both 1.06-μm and 10.6-μm irradiation; that the hot electron temperatures are about 15–20 keV; and that fast ion losses are small (less than 10%). This result is in agreement with calculations when the pondermotive force terms (radiation pressure) are included so that a steep density gradient arises in the region of critical density where absorption takes place, particularly for 10.6 μm where a large density change occurs in a small fraction of a wavelength. In this region the hot electron temperature is a function of laser intensity I and wavelength λ and is given by the relation Te_{hot} = constant $(I\lambda^2)^\delta$, where δ goes from 1 at low intensities to $^1/_3$ at useful intensities. Hot electrons tend to preheat the fuel and thus limit the amount of compression to unacceptable values.

The laser absorption is almost entirely due to resonant absorption at 10.6 μm, while both inverse bremsstrahlung and resonant absorption appear to be important at 1.06 μm. For pulses of a nanosecond and longer, a large amount of stimulated Brillouin backscatter is observed at 1.06 μm, while less than 5% occurs at 10.6 μm. Recent theoretical predictions give an $I\lambda$ dependence for Brillouin backscatter at low effective intensities, and a $1/\lambda$ dependence at high effective intensities, in agreement with these results. This is due to density profile steepening so that insufficient length exists in the plasma between $^1/_4$ and 1 times critical density for the Brillouin instability to develop for either for the Brillouin instability to develop for either 10.6 μm or subnanosecond pulses at 1.06 μm.

Shorter-wavelength lasers are predicted to give better coupling and lower hot electron temperatures than existing lasers because of the shift toward inverse bremsstrahlung absorption (which does not produce hot electrons). However, a note of caution seems advisable since no hard data exist at these wavelengths. If the stimulated Brillouin scattering does not saturate, it could invalidate the prediction. Also, Compton scattering processes could drive up the electron temperature at the higher critical densities which vary as $1/\lambda^2$. More measurements are needed before a major effort to develop such lasers is justified.

Energy transport. Calculation of energy transport from the critical surface where absorption takes place to the ablation surface where it is used still presents difficulties. At 10.6 μm, the transport appears to be classical and good agreement is obtained between theory and experiments. At 1.06 μm, transport seems to be impeded and a flux limit is included as an empirical constant in the calculations to obtain agreement. Transport inhibition might be due to self-generated electrostatic or magnetic fields or perhaps a calculational difficulty associated with two-dimensional effects.

Exploding pusher mode. Pellet experiments have been primarily concerned with the exploding pusher mode until recently. A thin-walled glass or metal shell is filled with high-pressure d-t gas and irradiated with a laser pulse which burns through the shell and continues to add energy to the shell material. This causes the material to explode both outward and inward, heating and compressing the central gas. This mode gives the largest neutron yield for presently available laser energy, as it increases the temperature at the expense of density, and the yield increases with the fifth power of the

temperature and only linearly with density before ignition is achieved. The higher neutron flux is easier to measure and thus permits calibration of design codes. Neutron fluxes in the range 10^9–10^{10} have been achieved for both 1.06-μm and 10.6-μm wavelengths in agreement with calculations.

Current experiments. With the introduction of the LLL and LASL higher-power lasers in 1979, experiments on pellets are now directed at achieving the maximum density in an ablatively driven isentropic compression mode. This is done by adding a thick plastic ablator to the outside of the glass or metal shell. Both laboratories appear to have exceeded 10 times liquid density, and each may be able to achieve several hundred times liquid density when the full potential of their respective on-line lasers is realized.

The important experiments remaining to be done with existing lasers are to understand the transport inhibition and stimulated Brillouin scattering losses, to obtain good wavelength scaling of the hot electron temperature for wavelengths shorter than 1 μm, and to develop diagnostics and optimize target design for the generation of lasers now under construction.

Future lasers. The next generation of lasers will be needed to accomplish ignition and to obtain experimental data on the stability of thin shells which tend to break up under acceleration forces. These lasers may permit scientific breakeven to be reached by the mid-1980s. The amount of laser energy needed to give sufficient gain for a power plant cycle is strongly dependent on this stability and on whether shorter wavelengths or other drivers can provide higher overall efficiencies.

For background information *see* FUSION, NUCLEAR; LASER in the McGraw-Hill Encyclopedia of Science and Technology.   [KEITH BOYER]

Galaxy, external

Recent work has clarified the origin of the spiral structure in disk galaxies. In 1962 A. Sandage noted that the sequence of galactic morphological types developed by E. P. Hubble comprises in reality two (or more) types: one consisting of galaxies that display a large-scale spiral pattern with a high degree of symmetry and whose arms are coherent structures that extend from the nucleus to the periphery of the galaxy; and another type consisting of galaxies that appear axisymmetric in a general way, showing spiral structure, but where no single arm can be traced from the center to the periphery of the galaxy. Examples of the two types are shown in Figs. 1 and 2. The recent advances have come from both theoretical and observational work and seem to indicate that the spiral structures of the two types are produced by different physical mechanisms.

Spiral density waves. The large-scale coherence in the first type has been taken to imply that a large-scale interaction is responsible for it. In 1964 C. C. Lin and F. Shu developed the theory of spiral density waves, where the spiral arms are seen as excited modes of the stellar disk under its own gravity. This theory has been very successful, particularly when applied to the galaxies with a well-developed global pattern. There are difficulties, however, in explaining the persistence of such fea-

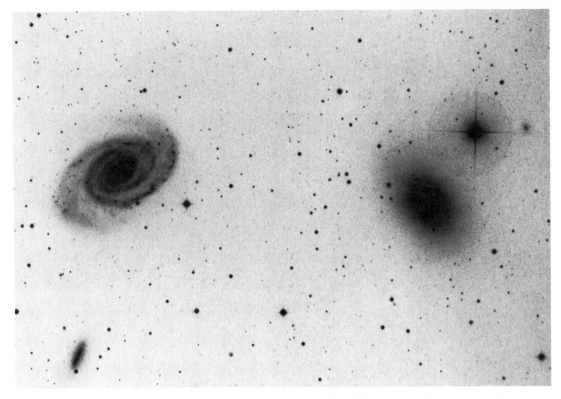

Fig. 1. NGC 5364, with coherent spiral arms, and companion galaxy. The photograph was made with the Palomar 48-in. (1.22-m) Schmidt telescope. (*Courtesy of J. Kormendy*)

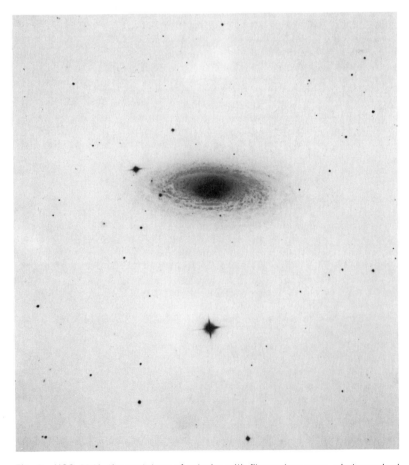

Fig. 2. NGC 2841, the prototype of galaxies with filamentary arms, photographed with the 100-in. (2.54-m) Mount Wilson telescope. (*Courtesy of J. Kormendy*)

tures because the spiral density waves decay in a few rotations of the galaxy. A mechanism to reexcite such waves is then required.

Recently J. Kormendy and C. Norman have shown that all known galaxies of the first type, those with a global pattern, either are barred spirals (galaxies whose central regions are not axisymmetric) or have in the neighborhood a companion galaxy. Both events are able, through gravitational interaction, to excite natural frequencies of the disk galaxy. These excitations appear as spiral density waves. In all the other cases where the galaxy is isolated and rotates in a highly differential form, the structure is of a multiple-arm type. This filamentary type of spiral structure has been postulated to have a completely different origin.

Self-propagating star formation. Observations in the Milky Way Galaxy have shown that the formation of clusters of stars is a self-propagating phenomenon. A typical young cluster contains several massive stars. During their short lifetime (a few million years) these stars generate, through stellar winds (that is, emission of streams of particles) and ultraviolet radiation, shock waves in the gas surrounding the cluster. At the end of their life cycle these same stars explode as supernovae, creating additional shocks in the surrounding gas. These shock waves compress the surrounding gas to densities high enough to trigger the formation of new clusters. These new clusters themselves contain massive stars, and the process can now be repeated. Thus there is a chain reaction that propagates through the galaxy and creates groups or strings of young clusters. The differential rotation of the galactic disk then shears these groups into segments of spiral forms.

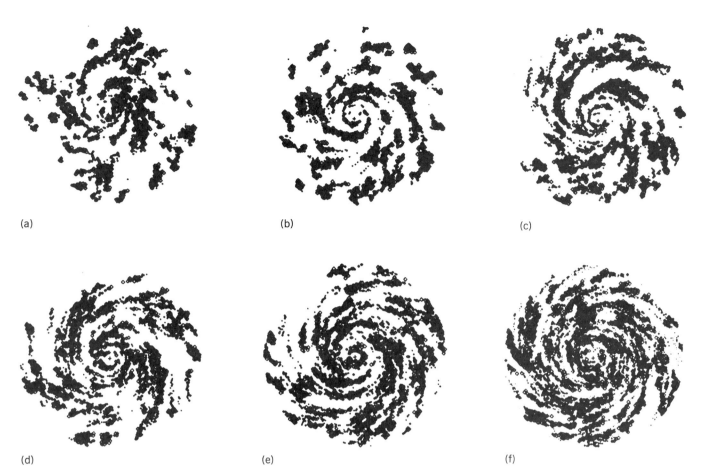

(a)   (b)   (c)

(d)   (e)   (f)

Fig. 3. Simulations of galaxy structure resulting from stochastic self-propagating star formation. Galactic disk is assumed to have tangential velocity, constant with radius, of (a) 50 km/s, (b) 100 km/s, (c) 150 km/s, (d) 200 km/s, (e) 250 km/s, and (f) 300 km/s.

Numerical simulation. H. Gerola and P. E. Seiden have made a simple numerical calculation which simulates such a chain reaction, and have found that this process generates patterns having the observed filamentary spiral structure. An initially random distribution of stars, representing young clusters, is established in a differentially rotating disk. After a certain time, the stars in a cluster explode. When they do, there is a certain finite probability, $P<stim>$, that a new cluster is created in its vicinity. These new clusters will then age and produce new explosions, repeating the cycle. This mechanism has been dubbed stochastic self-propagating star formation (SSPSF). The results of this simulation are shown in Fig. 3. In this figure, the galactic disk was assumed to have a tangential velocity of rotation constant with radius. Each frame displays the appearance at one time of a galaxy with a particular tangential velocity. The model galaxies display the ragged structure of the filamentary-type galaxies. Moreover, as the velocity of rotation increases, there is a regular progression from the irregular structure of the 50 km/s model to the very tightly wound structure at 300 km/s. This correlation between the velocity of rotation and the appearance of the galaxy, that is, its morphological type, is well known from observations.

Chain reaction. The SSPSF mechanism has the characteristics of a chain reaction, that is, there is

a narrow range of values of the probability $P<stim>$ above which the galaxy undergoes an explosion of star formation, while below this range the galaxy dies, ceasing all activity of star formation. When one imposes the condition that the galaxy neither die nor explode, spiral structure is the stable configuration of the system. The arms are the sites of star formation activity, and as time passes individual arms are dispersed while new ones are regenerated, maintaining the appearance of the spiral structure.

Dependence on galactic rotational velocity. A basic property of this model is that the rate of star formation in disk galaxies is determined by the velocity of rotation: as the rotational velocity increases, the star formation rate increases. That this is the case is not hard to understand. The larger the differential rotation, the faster new regions will be brought into contact with active star-forming regions, thereby enlarging the area of the galaxy available for stochastic star propagation. This dependence of the rate of star formation on the rate of rotation is an intrinsic property of the model, independent of the exact details of the physics of the propagation of star formation. This property of the theory predicts a well-defined set of correlations among properties of spiral galaxies. If all galaxies are coeval, that is, of equal age, those that rotate faster will have converted more of their original gas into stars, and hence at present have a smaller fractional gas content than spiral galaxies with a lower velocity of rotation. By the same token, spiral galaxies with higher velocities of rotation appear redder than those with lower velocities because of the integrated effect of the larger number of old red stars. What is gratifying for this theory is that these correlations are the observed ones.

It appears, then, that a second major mechanism responsible for the production of the shapes of galaxies has been identified. Future work will clarify the interaction between this process and the gravitational mechanisms such as the spiral density waves of the Lin-Shu theory.

For background information *see* GALAXY, EXTERNAL; STAR CLUSTERS; SUPERNOVA in the McGraw-Hill Encyclopedia of Science and Technology.   [HUMBERTO GEROLA]

Bibliography: B. G. Elmegreen and C. J. Lada, *Astrophys. J.*, 214:725, 1977; H. Gerola and P. E. Seiden, *Astrophys. J.*, 223:129, 1978; W. Herbst and G. E. Assousa, *Astrophys. J.*, 217:473, 1977; J. Kormendy and C. Norman, *Astrophys. J.*, 233:539, 1979.

Gastropoda

The ability to learn enhances an organism's chances of adapting successfully to the demands of the environment. Although it seems at first glance that this ability occurs widely in the animal kingdom, there is considerable debate as to which animals have the capacity to learn and whether the same behavioral and neural principles hold for all cases. This problem has arisen in part from the fact that most theories of learning have been developed from studies on higher animals and humans, and partly from the lack of a workable consensus of what actually constitutes learning. A systematic comparative analysis of the available theories may yield a better understanding of the fundamental behavioral laws that govern learning. The simpler behaviors of some of the invertebrate animals provide an excellent opportunity for this endeavor. Animals such as the gastropod mollusks (the snails and slugs), which have relatively simple nervous systems with large, visually identifiable nerve cells, offer the additional opportunity to study the physiological basis of learning, about which little is known.

This article will begin by discussing the central feature of the definition of learning that all studies must address. It will then describe some of the recent evidence for learning in the gastropod mollusks, and discuss progress made in physiological investigations.

Definition of learning. In the broadest sense, learning may be defined as a behavioral change that comes about from the experiences of an animal. This definition emcompasses many forms of behavior. Generally, these may be grouped into either nonassociative or associative catagories. Examples of the simpler nonassociative behaviors are habituation, sensitization, and pseudoconditioning. The distinguishing feature that sets associative learning apart from all simpler forms is the property of association itself.

The property of association can be illustrated best by the Pavlovian training procedure. In this procedure an initially ineffective stimulus is pre-

sented to the animal in close temporal pairing with another stimulus that normally elicits a strong response from the animal; these are called, respectively, the conditioned stimulus (CS) and the unconditioned stimulus (UCS). Although the initial responses are quite different, training causes the animal to respond to the CS and UCS in a similar manner. It is the acquisition of this similarity that constitutes associative learning.

The pairing of the CS with the UCS is called the experimental procedure. There are actually several different types of procedures. They are all defined by the operations performed by the experimenter and the responses made by the animal. The purpose of each experimental procedure is to change the behavior of the animal, but in itself it is only part of the definition of associative learning. The above discussion assumes that the behavioral changes are a direct result of pairing of the CS and UCS. It is possible, however, that some effect of the stimuli other than pairing may produce the behavioral changes. In order to demonstrate the property of association sufficiently, it is necessary to compare the results obtained by using the experimental procedure in one group of animals with the results obtained by using control procedures in other groups of animals. The control procedures consist of presentations of either the CS or UCS alone or in a temporally unpaired manner. The value of these controls is to assure that the behavioral changes produced by the experimental training procedure are critically dependent on the close temporal relationship between the CS and UCS. If the experimental-control differences are not significantly different from one another, then the changes produced by the experimental training procedure may be attributed to sensitization, pseudoconditioning, or some other putatively nonassociative phenomenon.

Behavior. Firm evidence for associative learning has been obtained in several subclasses of the gastropod mollusks. In the opisthobranchs, learning has been obtained in the sea slugs, *Pleurobranchaea* and *Hermissenda*; and in pulmonates, the evidence is from the garden slug, *Limax*, and the fresh-water snail, *Physa*. The work on *Pleurobranchaea* will be stressed here because it provides some of the most extensive evidence for learning in the invertebrates.

Observations. Pleurobranchaea (Fig. 1a) is a voracious carnivore that exhibits a characteristic feeding response when presented with a suitable food stimulus. The feeding response consists of two phases, extension of a proboscis and a bite-strike movement in which the proboscis is thrust forward to grasp the food between the jaws (Fig. 1b). Initially, it was shown that *Pleurobranchaea* could be trained by using an operant conditioning procedure (unlike Pavlovian techniques, in which the experimental animal has no control of the stimulus presentations, operant techniques require the animal to make an appropriate response before the UCS is presented—this is sometimes called trial-and-error learning). Experimental animals were trained by pairing a solution of squid homogenate (the CS) with electrical shocks (the UCS) to the anterior region of the animal (Fig. 1d). The experimental animals could avoid the shocks if they with-

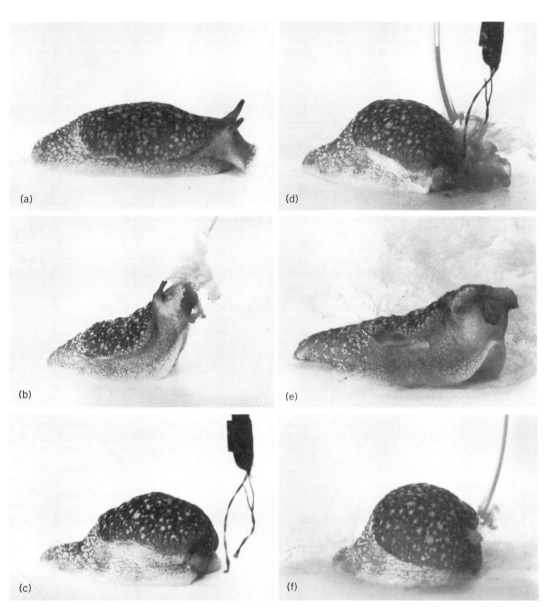

Fig. 1. Behavioral observations in *Pleurobranchaea*. (a) Specimen at rest. (b) Bite-strike response of naive animal to food. (c) Unconditioned withdrawal response to shocks applied via spanning electrodes. (d) Shock-elicited withdrawal response during conditioning when food was present. (e) Approach-avoidance response. (f) Fully conditioned withdrawal response elicited by food (compare with b). (From G. J. Mpitsos and S. D. Collins, Learning: Rapid aversive conditioning in the gastropod mollusc(s.s.) Pleurobranchaea, Science, 188:954–957, © 1979 by the American Association for the Advancement of Science)

drew from the squid food stimulus. Control animals received the same amount of stimulation, but in an unpaired control procedure in which the squid was separated temporally from shocks in each trial. As a result of training, the presentation of the food CS to the experimental animals caused both a suppression of feeding in the form of an approach-avoidance response (Fig. 1e), and then a strong, fully conditioned withdrawal response (Fig. 1f) which was similar to the response produced by electrical shocks (Fig. 1c). In contrast, the control animals continued to feed to squid despite the fact that they too were shocked. Since pairing of squid and shocks was the major difference between the experimental and control procedures, it can be concluded that the behavior of the experimental

animals was produced by the association between the two stimuli.

Complex associations. Recent studies have been designed to determine whether *Pleurobranchaea* is capable of more complex associations. Two sets of experiments will be described here in which discriminative test and discriminative conditioning procedures were used. Both procedures involve the use of two different food stimuli to test the animals before and after training. In discriminative test procedures, only one of the food stimuli is presented during the training. In discriminative conditioning, both stimuli are presented to the animals; only one of them is paired with the UCS, while the second stimulus is presented by a control procedure. The discriminative tests are used to determine whether animals can discriminate between stimuli presented after training. Discriminative conditioning is used to determine whether selective association can be made in the more difficult situation when there are two similar stimuli present during training.

To conduct these experiments, therefore, two different food stimuli are necessary. In addition, the animal must be capable of forming associations to each. The conditioning experiments described above provide one of the stimuli, namely, squid. The discriminative test experiments will show that *Pleurobranchaea* can be trained to avoid a second food stimulus, and furthermore, that it can discriminate between the two stimuli after training. The ability to discriminate indicates that

Pleurobranchaea perceives the two stimuli as being different.

Discriminative experiments. The behaviors here are qualitatively similar to those described above and shown in Fig. 1. The following discussion will illustrate the more quantitative aspects of the experimental analysis. The results of the discriminative test experiments are illustrated in Fig. 2. The two food stimuli consisted of specially prepared solutions of squid and beer. Before training, during preconditioning tests, the experimental and control animals had roughly equal feeding thresholds to both stimuli. Experimental animals were then given five trials, spaced 2 hr apart, of beer and electrical shocks presented together; control animals were given the beer and shocks separately 1 hr apart in each trial. After training, the experimental animals exhibited greatly increased thresholds only to the beer stimulus; their responses to squid remained unchanged (Fig. 2a). The control animals, however, retained low feeding thresholds to both stimuli (Fig. 2b).

The results of the discriminative conditioning experiments are illustrated in Fig. 3. During training, the beer stimulus was paired with the electrical shocks, whereas the squid stimulus was presented to the same animals in the unpaired fashion. All other methods were as described above. After training, the animals exhibited increased thresholds only to the beer stimulus. Nonetheless, they continued to respond with low thresholds to the squid stimulus.

Significance of experiments. The results of the above experiments show that *Pleurobranchaea* can discriminate between similar stimuli based on the temporal sequence of events during training. These rather simple animals are capable, therefore, of complex associative behaviors. Many of the traits attributed to seemingly more intelligent animals may indeed be part of the behavioral repertoire of simpler life-forms as well. Clarification of both the numbers and types of behavioral strategies which evolved in different groups of animals to enable them to deal more effectively with their environment awaits further work.

Physiology. Behaviors consist of coordinated sequences of muscular movements which are controlled by the nervous system. Thus, the basic mechanisms of associative learning must be sought in the nervous system. There are a variety of criteria that experimental preparations must fulfill in order to facilitate neurophysiological research. The most important of these is that the learned response should persist when the animal is dissected, and that the neurocircuitry which produces the response should be analyzable in detail from sensory neuron to motoneuron. Progress has been made in this direction in several gastropod preparations.

Nervous system analysis. T. Crow and D. Alkon have obtained evidence in *Hermissenda* for an associative behavioral change which they have begun to analyze in the nervous system. Although they have not presented their training procedures in the context of Pavlovian conditioning, their experiments will be phrased in these terms for comparison with the preceding discussion. In their studies the CS consisted of a light stimulus toward

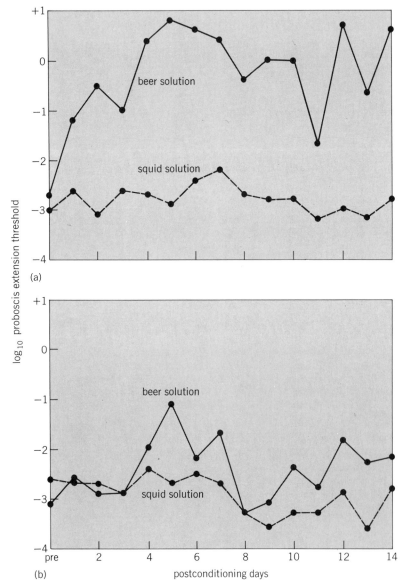

Fig. 2. Pavlovian conditioning and discriminative tests. Threshold of proboscis response to specially prepared stimuli for (a) experimental and (b) control animals. Preconditioning tests are shown at pre on the abscissa. The ordinate indicates the concentration of test solutions (based on serial 10-fold dilutions of standard solutions indicated by zero) just necessary to elicit the response. (*Courtesy of C. Cohan and G. J. Mpitsos*)

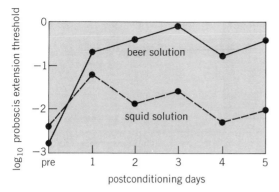

Fig. 3. Pavlovian discriminative conditioning. Thresholds for proboscis response to beer and squid stimuli. (*Courtesy of C. Cohan and G. J. Mpitsos*)

which the animals readily moved. For the UCS, the animals were rotated by means of a turntable. By comparison with controls, pairing of these two stimuli produced a significant decrease in the movement of the experimental animals toward the light. These same animals were then prepared for neurophysiological recording. It was found that the spontaneous neural activity of the photoreceptor cells in the eye of *Hermissenda* was significantly changed.

A. Gelperin has obtained behavioral evidence showing that *Limax* can be rapidly trained to avoid a natural food substance. Unlike the experiments on *Hermissenda*, the physiological analysis of this behavior has not been possible. However, in analogous experiments on *Limax*, J. Chang and Gelperin have shown that nervous systems taken from untrained animals can be conditioned in laboratory cultures by methods similar to the ones used in the behavioral studies. These preparations consisted only of the lips and attached central nervous system. Food substances, such as potato, when placed on the lips, produced rhythmic electrical activity in the nerves that innervate the lips. By pairing one of these foods (CS) with a noxious substance, such as quinine (UCS), the rhythmic activity produced by the food was suppressed.

W. Davis and R. Gillette have attempted to obtain neurophysiological evidence in *Pleurobranchaea* for the squid avoidance conditioning described above. Previously trained animals were prepared for neurophysiological study with a minimum of dissection. By their methods it was possible to observe the behavioral responses of the animal while simultaneously recording the electrical activity of the nervous system. It was found that presentation of the squid stimulus to experimental animals suppressed their feeding behavior and at the same time produced physiological changes in the activity of nerve cells in the brain. By comparison, the responses of the control animals were similar to those obtained from untrained animals.

Significance of physiological findings. Whether the findings of any of the studies described here are critically related to the mechanisms which produce the learned behaviors or whether they represent changes which occur in parallel with the actual mechanisms awaits further elaboration of

the neural circuitry in these animals. In any event, these changes should be recognized as significant in their own right as changes which occur in nervous systems undergoing learning and, therefore, they are worthy of continued study.

Conclusions. Learning is a powerful adaptive behavior that enables animals to interact more effectively with their environment. In its more complex forms it allows selective associations to be made between events which occur closely together in time. Considerable progress has been made in showing that simple animals, such as the gastropod mollusks, have the capability for discriminative learning.

The neurophysiological mechanisms that underlie associative learning are largely unknown at this time. The advantages offered by the gastropod nervous systems, though, have permitted some initial observations to be made. It is now clear that differences in neuronal activity can be demonstrated between experimental and control groups. New techniques will be needed to obtain a more complete analysis of the nervous system. Although ways of identifying the input and output (sensory and motor) nerve cells are readily available, there is no consistent way to identify the other cells that link these two levels of the nervous system. A thorough analysis of learning demands that the complete neurocircuitry be known.

Finally, the behavioral and neurophysiological properties of learning may not be constant throughout the animal kingdom. Although basic biological mechanisms tend to be phylogenetically conserved, it is quite possible that different mechanisms have evolved to meet similar adaptive requirements. Nonetheless, it is hoped that a comparative approach to learning will provide the base of information from which experimental hypotheses may be formulated to guide further research.

For background information *see* LEARNING, NEURAL MECHANISMS OF; REFLEX, CONDITIONAL; REFLEX, UNCONDITIONED in the McGraw-Hill Encyclopedia of Science and Technology.

[CHRISTOPHER S. COHAN; GEORGE J. MPITSOS]

Bibliography: J. J. Chang and A. Gelperin, *Soc. Neurosci. Abstr.*, 4:189, 1978; T. Crow and D Alkon, *Soc. Neurosci. Abstr.*, 4:191, 1978; W. J. Davis and R. Gillette, *Science*, 199:801–804, 1978; G. J. Mpitsos et al., *Science*, 199:497–506, 1978.

Geothermal energy

This novel energy source has received a great deal of attention in recent years as its position as a proved, economical, relatively clean energy source has been established in certain geographical locations. It is one of the alternative energy sources cited by the Department of Energy to receive preferential treatment so that its development and utilization will be expedited. In spite of all these favorable happenings, the growth of geothermal has been quite modest.

By definition, geothermal means earth-heat. The heat is generally considered to originate in the Earth's core largely from radioactivity. Near-surface manifestations are generally related to magmatic formations which, due to volcanism, tectonic uplift, erosion, or a combination thereof, are relatively close to the surface (within 10,000 ft or 3.0

km). When this magma is in contact with permeable rock which is saturated with groundwater or through which groundwater is circulating, the water will be heated. Depending upon the temperature of the rocks, rates of circulation, and limiting pressure, the water may be partially or completely converted to steam. The steam may constitute a commercial geothermal deposit if a well can be drilled into the reservoir, and if economically useful production rates can be secured (Fig. 1).

There are basically four types of geothermal deposits: hot dry rock, geopressured formations, dry steam, and hot water. At this time dry steam and hot water attract the most commercial interest.

Hot dry rock. In the United States, the major activity is primarily research, and is being carried out by a Federal agency at Los Alamos, NM. On an experimental basis, these workers have been successful in proving that two wells can be drilled into hot dry rock, a fracture can be artificially created (hydraulically induced) between the two wells, and cold water can be pumped down one well and produced through the adjacent well after absorbing sufficient heat to raise its temperature more than 100°F (56°C) while traveling from one well bore to the other through the fractures.

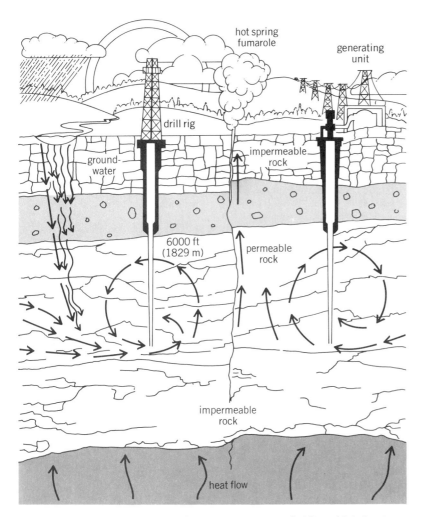

Fig. 1. Formation and penetration of geothermal deposits. (*R. J. Reynolds Industries*)

Unfortunately, the cost of this type of geothermal operation is so high that its commercial applicability is not anticipated for at least 10 years.

Geopressured formations. The major geopressured geothermal activity to date has taken place in the Gulf Coast area of Texas and Louisiana. The Department of Energy has sponsored a program to drill test wells into deep permeable rock formations which contain hot water under abnormally high pressures. The water produced from these formations is reasonably hot (300°F or 149°C) and contains limited amounts of methane gas (maximum solubility is 40 SCF [standard cubic feet] per barrel or 7.1 m³/m³). The cost of drilling these deep and difficult wells is extremely high (up to $6,000,000 each), and for an ongoing operation, disposal wells (also expensive) would be required and the probable subsidence problem (surface sinking due to fluid withdrawal) would have to be solved. Even with the concurrent capture of the methane to supplement the economic value of the hot water, the economics of a geopressured operation are somewhat questionable. Very little commercial development, if any, is anticipated during the next 10 or more years.

Dry steam. The world's largest geothermal-powered electric generating plants, and the only such plants operating commercially in the United States, are located in the Geysers area of northern California, approximately 75 mi (121 km) north of San Francisco (Fig. 2). From a reservoir viewpoint, it is unique since it is basically a dry stream reservoir in contrast to the more frequently encountered geothermal reservoir which contains hot water. There are 12 electric generating power plants in the Geysers, varying in size from 12.5 to 110 MW, with a total electric power–generating capacity of 608 MW. By the end of 1982, this is anticipated to reach at least 1344 MW, or sufficient electric generating capacity to supply the requirement of two cities the size of San Francisco. The ultimate capacity of the Geysers has been estimated to be anywhere from 2000 to 10,000 MW, with perhaps 3000 MW being generally considered a reasonable forecast for dry steam power–generating capabilities. The remaining 7000 MW have to be developed from hot water energy sources.

The Geysers operation is very economical and can supply cheaper electric power at the bus-bar (transmission-line distribution point) than any other alternative new power-generating system, except possibly hydropower. Lower bus-bar costs are a result of relatively low fuel costs plus lower capital investment costs, since no boilers are required and environmental abatement requirements are considerably less. At the Geysers, the cost of electricity at the bus-bar is no more than 60% of what the cost is for electricity from an oil-fired generating plant.

It is easy to see why power derived from geothermal, based on the experience at the Geysers, is considered a preferred energy source: it is both cheaper and cleaner than almost all competitive energy sources.

Hot water. The future growth of geothermal development will depend upon the ability to locate and economically develop hot water reservoirs,

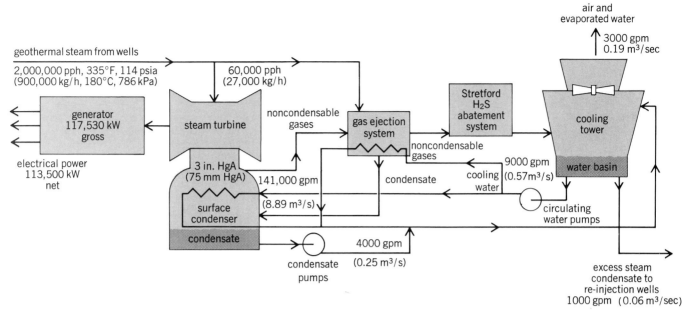

Fig. 2. Typical Geysers (California) power plant.

since the most common geothermal reservoir is expected to be hot water rather than dry steam. There are no commercially operating electric generating plants in the United States fueled by hot water at this time. Throughout the world, however, there are approximately 500 MW of hot water generation plants in operation and another 1000 MW planned by the year 1982. In the United States a few pilot plants are federally subsidized demonstration plants are in operation or under construction, but thus far, hot water development has been very restricted for a number of reasons.

Foremost among these reasons has been the concern about reservoir life and the site-specific nature of the geothermal reservoir. As distinguished from other power plants — coal, oil, or nuclear — there are, in theory at least, supplies available from a number of sources, and fuels can be stockpiled. Unfortunately, for geothermal power plants such is not the case since it is impractical to transport geothermal energy much more than 4000 ft (1.2 km) without having excessive heat and pressure losses. This site-specific limitation has resulted in many utilities having a rather limited interest in geothermal.

Another, and perhaps dominating, restraint in hot water development is the technological problems associated with both producing a reservoir and handling the hot water in the power plant when the water is highly saline. In terms of high temperatures and large resources, the most attractive geothermal area in the United States is the Imperial Valley of California. Ultimate geothermal resources have been estimated to be on the order of 20,000 MW or more. However, the highly saline nature of the fluid (up to 300,000 ppm) has caused such significant corrosion and scaling problems that it is not anticipated that there will be many, if any, unsubsidized commercial-size plants (greater than 50 MW) placed in operation within the next 5 to 10 years.

In other areas of the country such as Roosevelt Hot Springs, UT, or Valles Caldera, Baca, NM, exceptionally high-quality (low-salinity, high-temperature) hot water reserves have been discovered, but the development of these reservoirs into commercial-size operating electric generating plants has been slow, because of the utilities' concern about reservoir life and the high front-end costs for exploring and developing the hot water reservoirs.

Nonelectric uses. Most of the foregoing alludes to the commercial aspects of geothermal energy for the generation of electric power from high-temperature reservoirs. There are also multiple nonelectric uses, including space heating, industrial processing, and fish farming, which can be achieved through the use of the more frequently encountered moderate-temperature (less than 350°F or 177°C) geothermal reservoirs. However, this type of reservoir is not of sufficient economic interest to most geothermal operators to warrant high-cost lease acquisition and exploration activities, and therefore most nonelectric projects will evolve from the development of a site-specific economic method of utilizing the geothermal energy from a known (minimum exploration risk), readily accessible, moderate-temperature reservoir.

Summary. A recent U.S. Geological Survey report has concluded that in the United States the total amount of heat energy in the upper 6 mi (10 km) of the earth is about 32×10^{21} Btu (34×10^{24} joules), of which possibly 6.4×10^{18} Btu (6.8×10^{21}J; equivalent to 1.2×10^{12} barrels or 1.9×10^{11} m^3 of oil) could be utilized. However, this assessment does not take into account the many technological and economic problems that are involved in converting this energy into an economic reserve. It is therefore not generally predicted that the amount of energy supplied by geothermal will ever exceed 5% of the United States' energy requirements on a yearly basis. Where geothermal energy is found to be an alternative, economically viable, and clean energy source, its development will be expedited to cope with the burgeoning energy problems of the world.

For background information *see* GEOTHERMAL POWER in the McGraw-Hill Encyclopedia of Science and Technology.   [CLAUDE B. JENKINS]

Bibliography: Department of the Interior, *USGS New Release*, INT 3686–79, pp. 1–3, May 10, 1979; M. H. Dofman and R. W. Deller, *Electric Power Research Institute Proceedings*, EPRI WS-78–97, pp. 44–45, October 1978; T. M. Doscher et al., *Oil Gas J.*, 77(17):178–183, 19   ; Stanford Research Institute, *Economic Analyses of Geothermal Energy Development in California*, vol. 1, p. 74, May 1977.

Glaciation

Earth scientists have struggled with the enigma of glacial causes for more than 100 years. Proffered explanations have run the gamut from extremely simplistic climatic hypotheses (colder and wetter then, warmer and drier now) to highly complex geophysical models utilizing the most sophisticated statistical methodologies. As the range of evidence bearing upon the problem has widened, especially in the decades following World War II, it has become apparent to most investigators that glaciation on a massive scale has been the result of a multiplicity of causes, some of which are reasonably well understood today, some of which remain only poorly known. Any proposed model of glaciation must thus take into account a number of possible causal factors; it must also be founded upon the fact that most of the available evidence of past glaciations is geological, not meteorological, in nature.

Causal factors. With general acceptance in the 19th century of the reality of former large-scale glaciations, attempts to identify possible causes proceeded apace. Probably the most significant factors would include at least the following, not necessarily arranged in order of relative importance : (1) mountain building and other diastrophic activity; (2) changes in oceanic characteristics; (3) changes in the Sun's radiational output; (4) explosive volcanic activity, with an associated "dust veil"; (5) variations in the Earth's surface albedo; (6) changes in the Earth's orbital geometry; (7) changes in the location of continents (continental drift). Other factors, such as fluctuations in cosmic or galactic parameters, may also have played a role, but a usable model of the causes of glaciation can be assembled from these seven, with the latter five of particular relevance.

High-latitude continental location. Although the evidence is obviously subject to differing interpretations, analysis of the geological record of past glaciations leads to the conclusion that the fundamental requirement has been high-latitude continental location. Assuming a relatively fixed position of the Earth's rotational axis, it is reasonable to suppose that climate on this planet has been arranged in broad latitudinal belts during most of Earth history, with cold conditions prevailing at the poles and a warm zone centering on the Equator. Much of the paleomagnetic data suggests that glaciation on a continental scale has occurred only when large land masses have been in the appropriate latitudes to catch and hold a lot of snow, as they are today and were at the end of Paleozoic time. Attainment of such locations is presumed to have been by continental drift via the contemporary sea-floor spreading–plate tectonics model. Under these assumptions, climatic change as such is not required to initiate glaciation; rather, from time to time the continents have been moved into latitudes in which the climatic potential for glaciation has always existed.

Initiation of glaciation. Given the requisite high-latitude positioning of continental land masses, how may an episode of glaciation be started? A feasible and probable "trigger" is a comparatively rapid increase in the Earth's surface albedo, leading to a decrease in atmospheric temperatures and a related increase in snowfall in a number of critical high-latitude areas. An increase in the Earth's snow and ice cover must induce a further drop in tropospheric temperatures. This brings into existence a self-sustaining feedback mechanism: increasing surface albedo favors lowered atmospheric temperature, which in turn leads to the occurrence of still more precipitation in the form of snow. Continuation of the pattern for a few tens or hundreds of years launches the Earth into an ice age.

Little Ice Age. Crucial to this postulated developmental sequence is achievement of the initial albedo increase. This, it is believed, can be engendered by the effects of a coincidence in time of a slight decrease in total solar radiational output with an increase in explosive volcanic activity, leading inevitably to atmospheric cooling and the attendant increase in surface snow and ice.

That such a seemingly fortuitous coincidence can indeed take place is indicated by the events of the so-called Little Ice Age of a few hundred years ago, for which reliable historical documentation is available. Sunspot activity declined precipitously (with an apparent decrease in solar radiation) from about 1645 to 1715, explosive volcanism was so prolific that the Moon's luminosity was perceptibly dimmed during total eclipses, and atmospheric temperatures fell on a worldwide basis. The growing season was dramatically shortened in much of the Northern Hemisphere, and an increase in regional snow cover and lowering of the snowline took place in sizable parts of northeastern Canada (and probably elsewhere as well). The Earth seemed to be on the verge of another ice age. Yet, atmospheric temperatures generally rose after about 1730 and, with occasional departures, have fluctuated around long-time mean values since then.

Effect of orbital changes. Evidently the influence of one or more additional factors is needed to prolong a surface albedo increase for a period sufficient to nourish a major glaciation, and the effect on atmospheric temperatures of the changing orbital geometry of the Earth is a probable cause. Small but systematic variations in the orientation and inclination of the rotational axis combined with changes in the eccentricity of the Earth's orbit conspire to produce fluctuations in seasonal radiational receipt at any given latitude on a regular, recurring basis. Such changes occur independently of any deviations in absolute solar radiational output and should produce discernible variations in average seasonal temperatures.

The geometry of the Earth-Sun system is presently such that much of the Northern Hemisphere, where the major land masses are located, is receiving relatively high wintertime insolation: 10,000–15,000 years ago, wintertime radiational receipt was significantly lower. The warming effect of higher wintertime radiation is apparently adequate to override short-time cooling induced by an increase in atmospheric volcanic dust coupled with a decrease in total solar radiation. Little ice ages of relatively limited duration may develop under these circumstances, but big ones, lasting tens of thousands of years, cannot.

Major glacial episodes are generated during periods of comparatively low wintertime insolation in the appropriate hemisphere, and that will be the situation on this planet in another 8000–10,000 years. At that time, Northern Hemispheric summer insolation receipt will also be low, and the glaciation that should then ensue may well be much more extensive than any comparable advance of the last few hundred thousand years.

Mountain building and oceanic changes. The effects of mountain building and changing oceanic characteristics are debatable. In all probability, orogenic activity of sufficient magnitude at the right place and time could intensify preexistent major glaciation and almost certainly would lead to localized alpine glaciers. However, there is little geological evidence persuasively demonstrating an unequivocal cause-and-effect relationship between most past orogenies and the beginnings of major ice ages.

By the same token, whereas the deep-sea core record has expanded remarkably in the last couple of decades and stimulated construction of numerous paleoclimatic scenarios, what is most likely represented by the bulk of that record is evidence of the oceanic consequences of glaciation and related atmospheric variability, not evidence of causes. Long-lasting changes in oceanic circulation patterns directly attributable to continental drift would seem not to have occurred rapidly enough to have played a significant role as initiators of glaciation. On the glacial-interglacial time scale, the atmosphere appears to have led and the oceans to have followed.

Synthesis. A coherent model of the causes of glaciation can accordingly be formulated, one which does not demand unreasonable atmospheric behavior and meshes reasonably well with the available geological evidence. Drifting of continental land masses into high latitudes (and thus into climatic belts in which glaciation may be initiated and sustained) creates the geographical condition essential for the development of ice ages. What then happens is dependent upon the intensity of seasonal insolation receipt in the hemisphere with most of the land. Random fluctuations in explosive volcanic activity and probable secular variations in total solar radiational output act on occasion to reinforce one another, thereby bringing about a lowering of atmospheric temperatures, an increase in the Earth's surface albedo, and the triggering of an incipient ice age. If a little ice age evolves during a period of low wintertime insolation in the appropriate hemisphere, a major glacial episode will be generated. Little ice ages that come into being during periods of high insolation receipt are fated for relatively rapid extermination.

Once initiated, glaciation becomes, up to a point, a self-sustaining, snowballing process, with lessened seasonal insolation receipt favoring continued volume growth and areal expansion. But ultimately, climatic limitations are imposed as advancing ice masses penetrate into lower latitudes and altitudes. Deflection of storm tracks away from centers of accumulation by growing continental glaciers leads to a temporary reduction of moisture supply and a condition of starvation. The outer parts of glaciers will thus tend to experience short-time advances and retreats, with constant fluctuation in the location of ice margins.

Major glacial episodes come to an end, apparently very rapidly, when seasonal insolation receipt rises to a high and a decisive change in the Earth's heat balance is experienced. The overall pattern that unfolds, one supported by considerable geological evidence, is oscillation of relatively long glacial intervals (100,000 years ±) with rather brief periods (20,000 years ±) of interglacial amelioration.

Fundamental to the model of causes outlined here is the thesis that major climatic change is not required for the initiation of glaciation, only the drift of continents (by whatever mechanism) into latitudes in which a climatic potential is present. It follows that ice ages are not preceded by a sudden worldwide drop in atmospheric temperatures; they produce such a drop. And the "normal" global climate need not necessarily be warmer, but rather simply nonglacial.

Finally, taking into consideration the inherent controlling factors, the Earth would seem at present to be locked into an ice age condition, with orbital geometry changes dictating repetition of the glacial-interglacial cycle so long as parts of North America and Eurasia are situated in comparatively high latitudes.

For background information *see* GLACIAL EPOCH in the McGraw-Hill Encyclopedia of Science and Technology.   [CHESTER B. BEATY]

Bibliography: C. B. Beaty, *Amer. Sci.*, 66: 452–459, 1978; CLIMAP project members, *Science*, 191:1131–1137, 1976; C. Emiliani and J. Geiss, *Geol. Rundschau*, 46:576–601, 1955; J. D. Hays, J. Imbrie, and N. J. Schackleton, *Science*, 194:1121–1132, 1976.

Gravitation

One of the long-standing predictions of the general theory of relativity is that an accelerating mass should radiate energy in the form of gravitational waves, in much the same way as an accelerated electric charge radiates electromagnetic waves. Both kinds of waves travel at the speed of light and carry energy, momentum, and information, but whereas electromagnetic waves interact only with electric charges and currents, gravitational waves should interact with all matter and energy. The amount of energy radiated as gravitational waves by an oscillating mass is extremely small under most conditions, and laboratory experiments to detect the waves directly are not yet feasible. However, recent observations of a distant pulsar have provided indirect evidence that strongly supports the existence of gravitational radiation. The new results are difficult to reconcile with existing theories of gravitation other than general relativity, and represent the first test of relativity beyond its first-order corrections to Newtonian theory.

Binary pulsar PSR 1913+16. The pulsar in question, PSR 1913+16, was discovered in 1974 by R. A. Hulse and J. H. Taylor of the University of Massachusetts, using the 1000-ft-diameter (305-m) radio telescope at Arecibo, Puerto Rico. This pulsar is unique among the more than 320 known pulsars in that it is a member of a binary system, orbiting another object of comparable mass every 7 hr 45 min. The orbital motion is observed as Doppler shifts in the period of the pulsar emission, the inferred orbital velocity being approximately 0.001 times the speed of light. According to general relativity, such a system should emit gravitational waves that carry energy away from the system and cause the pulsar and its companion to spiral slowly closer together.

Change in orbital period. Pulsars are excellent timekeeping devices; typically, pulses from a given pulsar arrive at the Earth at times that remain stable to within less than 0.001 s over several years. This excellent timekeeping ability is characteristic of binary pulsar PSR 1913+16 as well, and makes it possible to observe small changes that may occur in the period of its orbit. The change predicted by general relativity on account of gravitational radiation is a decrease of about 10^{-7} s per orbit, which is far too small to be detected directly. However, the effect is a cumulative one, like a clock that is running fast at an ever-increasing rate. After 4 years, the accumulated shift of the "zero point" of the elliptical orbit—the time when the pulsar and its companion are closest together—should amount to about 1 s, as shown by the curve in the illustration. This curve corresponds to the predicted rate of accumulation of excess orbit phase if the pulsar and companion star are both

about 1.4 times the mass of the Sun and if the system radiates gravitational waves at the rate predicted by general relativity.

Observational results. The University of Massachusetts group (including L. A. Fowler, P. M. McCulloch, and J. M. Weisberg, as well as Hulse and Taylor) by mid-1979 had been observing the binary pulsar for over 4 years, with results summarized by the data points in the illustration. The measured points are not consistent with a constant orbital period, which is represented by a straight line in this diagram; however, the data fit very well on the parabolic curve representing the general relativistic prediction. This agreement is taken as very strong evidence that the speedup of the orbit is in fact the result of gravitational radiation.

Conclusions. A number of possible alternative explanations of the change of orbital period have been considered, but all of them appear to be either implausible or of negligible magnitude. For example, differential galactic rotation and mass loss from the system can contribute at most about 1% of the observed rate of period change. Tidal interactions between the pulsar and the companion star (the nature of which is not yet known) could possibly be important if the companion is larger than a white dwarf. However, the absence of any eclipses of the pulsar or of any evidence of gas inside the orbit places strong constraints on the size of the companion star, and the most probable companion appears to be another neutron star. In this case, it is quite certain that gravitational radiation is the dominating dissipative effect, and that general relativity has passed one of its most probing tests.

For background information *see* GRAVITATION; PULSAR; RELATIVITY in the McGraw-Hill Encyclopedia of Science and Technology.

[JOSEPH H. TAYLOR, JR.]

Bibliography: P. M. McCulloch, J. H. Taylor, and J. M. Weisberg, *Astrophys. J. Lett.*, 227: L133–137, 1979; J. H. Taylor, L. A. Fowler, and P. M. McCulloch, *Nature*, 277:437–440, 1979; J. H. Taylor and P. M. McCulloch, *Ann. N.Y. Acad. Sci.*, in press.

Hail

Hail suppression activities in the United States changed dramatically during the 1977–1978 period. Research concerning hail suppression had been the focus of the American, as well as Canadian, weather modification research from the mid 1960s until 1977. More than $30,000,000 had been spent on research, with additional millions on privately supported operational programs. Much of the total research costs were associated with the National Hail Research Experiment (NHRE) in the United States. A large part of the United States hail suppression experimental effort effectively ceased by 1978. A short review of the past efforts in hail suppression provides understanding for the 1977–1978 activities.

Operational hail suppression programs. Efforts to suppress hail began in the United States in 1950. Privately supported cloud-seeding projects over small areas took place in high-hail-loss crop areas in Nebraska and West Virginia, well before scientific experimentation had established proof of the

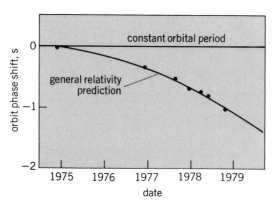

Accumulated orbit phase shift of the binary pulsar PSR 1913+16. Curvature of line gives prediction of general relativity, and data points give measured values. Uncertainties are comparable to the sizes of the points themselves. (*From J. H. Taylor, L. A. Fowler, and P. M. McCulloch, Measurements of general relativistic effects in the binary pulsar PSR 1913+16, Nature, 277:437–440, 1979*)

effectiveness of cloud seeding. The efforts then, as now, were based largely on the hypothesis that additional ice nuclei in the hail formation zone of a thunderstorm increase competition for the available supercooled moisture, producing many small hailstones that either melt by the time they reach the surface or are sufficiently small to be harmless. In these early years, there was no way to directly inject the materials for cloud modification into the hailstone formation zones at high levels in storms, so the materials were released either from the ground or from airplanes circling below the storm clouds. The first major experimentation with hail suppression occurred in northeastern Colorado in 1959, but results were inconclusive.

Locally sponsored operational programs to suppress hail continued intermittently into the 1960s and 1970s in Colorado, Kansas, Texas, North Dakota, Utah, Idaho, and South Dakota, where weather modification companies were employed to suppress hail. About 70,000 mi² (180,000 km²) of the United States were seeded in 14 different operational projects for hail suppression in 1974. The first major operational hail suppression program began in South Dakota in 1972. State and county funding led to hail suppression activities over more than half the state for 4 years. Funding ended in 1976 as grassroots-organized opposition halted the program. A similar large state program developed in North Dakota in 1975.

In this same general time period, a series of events in the Soviet Union had a considerable bearing on the eventual hail suppression activities in the United States. The Soviet Union, along its southern boundary, has major crop areas that suffer greatly from hail. The Soviets first experimented with hail suppression by using a systematic engineering and empirical approach and then began operational (nonexperimental) hail suppression projects in their high-hail-loss regions during the early 1960s. Their published claims in 1964 indicated 50–80% reductions in hail losses of crops. Their claims had a considerable impact on both the American scientific community and the Federal agencies concerned with weather and its modification. It helped cause a national hail research effort in the United States.

National experiment on suppression. A committee of atmospheric scientists from several Federal agencies called for a national experiment with the aim of checking the 1964–1965 Soviet claims of success. A Federal plan for hail suppression research was developed, with the National Science Foundation (NSF) as the lead agency to provide most of the support.

Some atmospheric scientists accepted the difficult challenge of hail suppression. Presumably, many saw in this new national goal the promise for substantial increases in research funding, and many also believed a major modification breakthrough could be achieved with hailstorms of the Great Plains.

Hail suppression research developed at several universities, at the weather service laboratories, and at the National Center for Atmospheric Research (NCAR) in the late 1960s. Experiments involving hail suppression and rain increase were conducted in the Dakotas in 1966–1972. Ultimate-

ly most of these research groups and efforts, and their related hail research support from NSF, were redirected into the NHRE, which was directed by NCAR.

The NHRE had two complex goals: (1) to verify whether hail could be suppressed experimentally (with overtones of testing the Soviet hail suppression hypotheses), and (2) to study all facets of hailstorms so as to understand storms and explain the modification results. Facility installations and testing in northeast Colorado began in 1971 in a 600-mi² (1550-km²) experimental area. Throughout this area there were numerous sites with surface instruments to measure rain and hail.

Large sophisticated weather radars designed to detect hail also were developed and employed in the NHRE, along with an armored jet aircraft that could safely penetrate storms. The full seeding experiment began in 1972 and was conducted in 1973 and 1974.

A series of problems related to the hard-to-achieve goals, governmental shifts in project policies, and lack of timely analysis caused the seeding experiment to be stopped after 1974. The studies suggested that the frozen-drop-embryo assumption, on which the seeding strategy was based, was not valid in the Colorado area, and statistics revealed that even a high rate of suppression could not be achieved in 2 or 3 more years of experimentation. Research and key field measurements were pursued in 1975–1978, but no experimentation.

Other Federal and state-sponsored research programs concerning hail suppression were conducted in 1973–1977. These addressed subjects such as public attitudes toward hail suppression, environmental consequences of silver from the seeding with AgI, economic impacts, legal consequences, and effects on rain and hail in areas downwind of seeded areas.

Another major research area concerned basic hail research and the subsequent development of a design for a future hail suppression experiment for the Midwest. This Illinois-centered research was sustained to provide the experimental background in an area with a hailstorm climatology quite different from that in Colorado.

A major hail suppression research effort was technology assessment of hail suppression, conducted in 1976–1977. Its results furnished useful guidance for future hail suppression research in the United States, and for the first time dimensionalized the hail problem in the United States and the impacts of varying potential hail suppression capabilities.

This history of events up to 1977 reveals development of disillusionment about hail suppression on the scientific front, and limited usage by the public. In 1978 only three operational projects existed in the United States.

Advances due to research. However, the NHRE and the other extensive research projects had brought about major advances. Many new observational techniques had been developed, including dual-wavelength weather radars to detect hail, surface hail sensors, and armored aircraft. These new devices, their data, and the resulting information have led to major advances in understanding hailstorms. A wide variety of different types of hailstorms have now been identified. Numerical modeling has been heavily employed, but at this stage, modeling still cannot handle the complexities of hailstorm research.

Research into analysis of hailstone structures, with the belief that the hailstone is a sensor of storm history, has reached a point of understanding. Researchers have learned about all that can be derived readily about storm structure.

The tremendous attention to hail through the late 1960s and 1970s led to the development of definitive work on the climate of hail in North America. Information on the local, regional, and national scales of hail was evolved, revealing the extreme time and space variability of hail. The major hail zone of the United States is in the lee of the Rocky Mountains, and much is now known about the frequency of hailstones and detailed surface distribution patterns of hail. After considerable attention to forecasting of hail, scientists have still not achieved a desirable level of skill, being accurate in forecasting hailstone size only about 60% of the time.

Certain major accomplishments achieved by 1977-1978 have resulted from intensive research on the social, legal, economic, and environmental aspects of hail suppression and other forms of weather modification. A major technology assessment pointed to the fact that a highly developed capability, greater than 60%, in hail suppression would lead to wide adoption of hail suppression in the Great Plains. The primary winner, economically, on a national scale would be the United States consumer. Farmers in the United States would both win and lose, depending on the type of farming in their locale. The research also showed that a high capability was needed for hail suppression to be really useful and that if hail suppression in any way decreased rainfall it would be a disservice. The approach to public adoption and wise use depends on careful public education and development of means to compensate the losers when hail is suppressed. The insurance industry would benefit from a well-developed hail suppression capability, but national policy for development and management of hail suppression was found to be poor.

In hail research the concepts of hail suppression are still not well understood. Effectiveness of hail suppression varies according to variations in the storm updraft speeds and the temperature at the level of maximum updrafts. Future field experiments failing to allow for meteorological and in-storm differences, when choosing how to seed the storm, would fail to produce convincing results.

The uncertainty about actual mechanisms of hail suppression that remains after so many field projects is very disconcerting, Basically, there has not been much progress in the past 15 years in hail suppression concepts and seeding approaches. An approach using hygroscopic seeding to deplete the liquid water in the lower cloud, and to capture natural hail embryos, appears to offer some value.

Of importance is an ongoing test in Switzerland of the Soviet seeding rocket system. This randomized experiment is being performed by Swiss, French, and Italian scientists. The experiment, coupled with a better understanding of how to de-

sign and evaluate experiments, is the major recent advance in hail suppression.

Future progress in suppression. There is now a considerable understanding of the impacts, both social and environmental, of hail suppression, and many of the key instruments that would be needed in future research have been developed. However, the concepts for hail suppression are still not well delineated, and modeling lags. Forecasting and the conduct of experiments still appear to achieve less than desired. Although current beliefs concerning the status of hail suppression vary widely, research will likely continue. Discussions on the best scientific course of action to follow during the next several years have been marked with controversy between those who favor large statistical experiments and those who favor greater emphasis on more fundamental research on cloud physics. Basically, the lack of definitive information about the physics of the modification of hailstorms, and severe convective storms in general, is still so great that it is unlikely the complexities can be totally resolved in the next 20 years. The rate of future progress will depend on the amount of national attention to the hail problem.

For background information see HAIL; WEATHER MODIFICATION in the McGraw-Hill Encyclopedia of Science and Technology.

[STANLEY A CHANGNON, JR.]

Bibliography: S. A., Changnon, *Hail Suppression*: *Impacts and Issues*, Illinois State Water Survey, 1977; S. A., Changnon, Jr., B. C. Farhar, and E. R. Swanson, *Science*, 200:387–394, Apr. 28, 1978; G. B. Foote and C. A. Knight, *Hail*: *A Review of Hail Science and Hail Supression*, Amer. Meteorol. Soc. Monogr. no. 38, 1978.

Heat treatment (metallurgy)

The processing of engineering materials has two major objectives: shape change and property control. Traditionally, these separate procedures were performed by a mechanical working operation designed to change shape in a rapid, efficient, and economical fashion, and by a heat treatment operation to modify internal structure and control physical, mechanical, or chemical properties. In some cases the procedures were combined to reduce manufacturing costs.

Thermomechanical treatment. The mechanical deformation and heat treatment operations are normally considered separately because the heat treatment usually removes the worked structure. However, it has been known for many years that these operations interact and can have an appreciable influence on each other. But only recently, after metallurgists began to understand the processing-structure-property relationships, were attempts made to design processing procedures combining mechanical working and heat treatment operations for microstructure-property control. The combined procedures which produce structures and properties not obtainable by simple mechanical working and heat treatments alone are referred to as thermomechanical treatment (TMT).

Much of TMT technology has been developed for the processing of steels, although it has also been applied to nonferrous alloys, including those of aluminum, titanium, and nickel. That the topic has been the subject of over a thousand papers and reports illustrates its scope and importance. This article describes briefly the physical metallurgy important in the TMT of aluminum alloys and reviews recent applications of this technology.

Background of TMT. Most TMTs are directed at improving strength without the accompanying degradation of other properties that normally occurs when using conventional methods. TMT procedures involve optimizing grain shape and size, introducing a homogeneous distribution of dislocations, or achieveing a better distribution of precipitates. For convenience, TMTs are divided into two classifications: intermediate thermomechanical treatments (ITMTs) and final thermomechanical treatments (FTMTs). The first are used primarily to control grain structure and dispersoid size and distribution in order to improve fabricability, ductility, and toughness (and strength in non-heat-treatable alloys). The second are used to modify the hardening precipitates and dislocation structure in age-hardenable alloys for strength improvements. The ITMT is normally applied to the ingot before final heat treatment, while the FTMT is applied to the wrought product, after the final hot-working operation.

Although most of the methodology of TMT has evolved since the first recognized work on the subject by E. M. H. Lips and H. Van Zuilen in 1954, the type of processing now referred to as FTMT has been used by the aluminum industry since 1944. In that year E. H. Dix published a paper which described a procedure for improving the strength of aluminum alloys by inserting a mechanical working step after solution heat treatment and prior to warm aging (precipitation step). This TMT, designated T8 by the Aluminum Association, is used extensively for strengthening 2XXX series alloys. However, most attempts to produce stronger materials in a T8 temper with 7XXX alloys were unsuccessful.

The mixed results obtained in the two alloy systems with TMTs are now understandable. In some cases the mechanical working step has an accelerating influence on the precipitation reaction by increasing the vacancy concentration (which enhances the diffusion necessary for precipitation). In addition, heterogeneous nucleation of particles occurs on the dislocations, pinning them and producing higher strengths than obtained for undeformed alloys. In 2XXX alloys the precipitates are responsible for large interfacial strains which are minimized by heterogeneous precipitation on dislocations. In the 7XXX alloys, however, the precipitates do not create the large interfacial strains, thus heterogeneous nucleation on dislocations does not occur. As a result, the dislocations are not pinned in the same way as in 2XXX alloys and no improvement in strength is obtained by this particular type of TMT. These results illustrate why identical TMTs are not transferable from alloy system to alloy system, and underline the importance of understanding the specific processing-structure-property relationships.

Conventional processing. Figure 1 illustrates the conventional processing sequence for 7XXX aluminum alloys. Beginning with the as-cast ingot, a homogenization treatment (step 1) is given, fol-

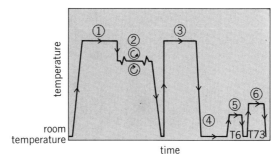

Fig. 1. Schematic of conventional-type processing used for 7XXX alloy plate. Steps 1–6 are described in the text.

lowed by hot rolling (step 2) for ingot breakdown and shape change. Step 2 is normally followed by a solutionizing treatment (step 3) and a coldwater quench (step 4). Depending on the application, the alloy is given either a warm aging treatment (step 5) to produce a fine dispersion of precipitates for maximum strength, or an overaging treatment (step 6) to produce a high resistance to stress corrosion cracking. Unfortunately, step 6, as compared to the results obtained by step 5, also results in considerable sacrifice in strength. These last two steps are designated as T6 and T73, respectively, by the Aluminum Association.

After step 2 the grain structure consists of partially recrystallized "pancake"-shaped grains which are stable with respect to subsequent processing. This structure forms during hot working because of a fine dispersion of precipitates that aids in the development of a stable subgrain structure and the suppression of recrystallization. The pancake-shaped grain structure has been considered responsible for the poor fracture resistance in the short transverse direction of 7XXX thick plate material.

ITMT processing. Recently E. DiRusso and coworkers in Italy and J. Waldman, H. Sulinski, and H. Markus in the United States developed ITMT processes designed to produce a much finer controlled grain structure in 7XXX alloy plate than is obtained with conventional processing. The Waldman process, shown schematically in Fig. 2, establishes a coarse precipitate structure which aids recrystallization. This is accomplished by applying an appropriate thermal treatment (step 2A) prior to working at lower than conventional hot-working temperatures (step 2B). This treatment introduces a relatively high degree of strain

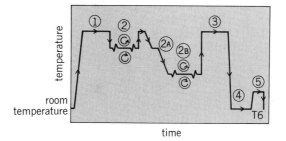

Fig. 2. Schematic of ITMT processing used for 7XXX alloy plate. Steps 1–5 are described in the text.

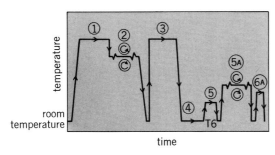

Fig. 3. Schematic of FTMT processing used for 7XXX alloy plate. Steps 1–6 are described in the text.

hardening which promotes recrystallization to a fine, relatively equiaxed grain structure during the solutionizing treatment. Significant increases in elongation and fatigue crack–initiation resistance are obtained in ITMT-processed material when compared with conventionally processed counterparts. However, a degradation in crack-propagation resistance is associated with this fine-grained structure. Although the T6 strengths of ITMT and conventional processing are about the same, the stress corrosion resistance of ITMT material is much better, approaching that obtained by the T73 treatment.

FTMT processing. This relatively new procedure offers a way of recovering the strength decrease that accompanies the T73 treatment. Numerous investigators have shown that 7XXX aluminum alloys will respond favorably to the FTMT treatments. The most successful procedures are variations of that illustrated schematically in Fig. 3. N. E. Paton and A. W. Sommer have demonstrated that to be effective the initial deformation step in FTMT processing of 7XXX alloys must produce a homogeneous distribution of dislocations. This can be accomplished either by room temperature deformation in the solutionized condition after step 4 or at elevated temperature after step 5. Unfortunately, the dislocation structure introduced at room temperature is unstable and rapid recovery occurs on subsequent aging (step 5). This accounts for the poor response of 7XXX alloys

to the T8 treatment mentioned previously. However, the dislocation structure resulting from deformation after the T6 treatment (step 5A) is stabilized by the strengthening precipitates. The optimum temperature for the deformation treatment has been found to be at or just above the critical temperature for homogeneous precipitation, where a transition from planar to random slip occurs.

Although FTMT processing results in significant improvements in strength over the T6 condition, the stress corrosion resistance is inadequate for many applications. Consequently, an additional aging step is normally employed to produce a precipitate size comparable with that in the T73 condition. Because of the high dislocation density, this requires less time than the normal T73 treatment. The relationship between strength and processing is shown in Fig. 4. The curves show the normal aging response of 7XXX alloys and the response due to FTMT processing. Note the large strengthening increment due to step 5A, and the increased substructure strengthening between the conventional T73 condition and the FTMT (step 6A) condition. Thus, with proper FTMT a combination of properties superior to those obtained by conventional processing can be realized. It is intriguing to speculate on the properties that might be achieved in aluminum alloys by appropriate combinations of ITMT and FTMT.

Recent applications. The technology of TMT has contributed to the development of new aluminum conductor alloys now widely used for communication cables and building, magnet, appliance, automotive, and aircraft wiring. These applications require a balance of good electrical conductivity and mechanical properties—properties which are not jointly optimized in most conventional strengthening methods. The desired properties were obtained from rigidly controlled TMT that yielded fine subgrains stabilized by a homogeneous distribution of small intermetallic precipitates. Moreover, during the past several years, metallurgists have developed new aluminum alloys for the packaging industry with higher strengths and more formability than previously available. Most of the new materials were developed through minor compositional variations plus new TMTs. Processing variables are controlled so as to optimize deep-drawing properties needed for a given part without altering other material properties in undesirable ways. For example, use is made of the baking process, required during coating, to polygonize the microstructure, and polygonization leads to more favorable strain hardening and homogeneous deformation during subsequent forming.

TMTs have frequently resulted in significant improvements in such properties as strength, stress corrosion resistance, and fatigue crack initiation resistance. Simultaneous improvements in several of these properties are sometimes observed, but often they are obtained at the expense of other properties, such as fatigue crack–propagation resistance or toughness. It is interesting to note that in some cases the desirable properties obtained by complex TMTs have been duplicated by simple modifications in chemical composition and heat treatment. Even so, some structures and

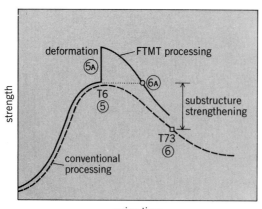

Fig. 4. Schematic comparison of aging treatments of FTMT and conventional processing. Steps 5, 5A, 6, and 6A are described in the text.

properties are not economically feasible by other than TMT manufacturing methods.

TMTs have some limitations associated with product size, shape, and fabrication. Large sections cannot be cold-worked easily, and it is difficult to achieve uniform deformation in complex shapes. Moreover, welding and machining of hard TMT materials may also limit applications. Manufacturing problems arise from the logistics involved with these complex processing schedules. However, the normal processing of some alloy systems routinely incorporates the same controls required by TMT processes.

During the past 10 years the development of new aluminum alloys has been based on a better understanding of processing-structure and structure-property relationships. The success of the new high-strength aluminum alloys, aluminum conductor wires, and packaging alloys is attributable to the unique advantages of TMTs as well as close control of alloy chemistry.

For background information *see* ALUMINUM; HEAT TREATMENT (METALLURGY); METAL, MECHANICAL PROPERTIES OF in the McGraw-Hill Encyclopedia of Science and Technology.

[EDGAR A. STARKE, JR.]

Bibliography: M. Azrin, *Soviet Progress in Thermomechanical Treatment of Metals*, AMMRC TR 76-36, November 1976; E. H. Chia and E. A. Starke, Jr., *Met. Trans. A.*, 8A:825–832, 1977; J. W. Morris (ed.), *Proceedings of the Symposium on Thermomechanical Processing of Aluminum Alloys*, AIME, St. Louis, September 1978; E. A. Starke, Jr., *Mater. Sci. Eng.*, 29:99–115, 1977.

Hormone

One area of continued interest in endocrinology has been the search for how polypeptide and other types of hormones regulate intracellular events in target cells. The smaller lipophilic hormones, such as the steroid and thyroid hormones, readily enter target cells, interact with intracellular structures, and then directly influence cellular functions. In the case of the large hydrophilic polypeptide hormones, however, in many instances it is not known how intracellular events are regulated. Polypeptide hormones first bind to a receptor protein (or glycoprotein) on the surface of target cells. With some hormones, after binding, the hormone-receptor complex generates a signal at the cell surface, and this signal then enters the cell interior and carries out the hormone's actions. For other hormones, after binding, the hormone-receptor complex is internalized, and in the cell interior the hormone interacts with intracellular organelles. Recent progress in understanding both of these mechanisms will be reviewed.

Polypeptide hormones and cyclic AMP. In the liver, the polypeptide hormone glucagon and the peptide hormone epinephrine, after binding to receptors on the cell surface, increase glucose output by activating several glycogenolytic and gluconeogenic enzymes. For a number of years, it has been known these hormones stimulate the production of the intracellular messenger cyclic adenosinemonophosphate (AMP), also known as cyclic adenylic acid. In the liver cell, cyclic AMP triggers a cascade of enzymes leading to glucose

release. Cyclic AMP is formed from the nucleotide adenosinetriphosphate (ATP) through the action of adenylate cyclase (AC), an enzyme that is present in the plasma membrane. The mechanism whereby the hormone-receptor complex activates adenylate cyclase, however, has only recently been understood. It appears that another nucleotide, guanosinetriphosphate (GTP), plays two critical roles in this process. GTP converts the hormone receptor to a state which, in the presence of hormone, can activate adenylate cyclase. As a separate action, GTP converts adenylate cyclase to a form where it can be activated by the hormone-receptor complex (Fig. 1).

In addition to glucagon and epinephrine, several other hormones also activate adenylate cyclase. In their respective target tissues they generate cyclic AMP, which in turn activates specific enzymes and carries out hormone action. These other hormones include corticotrophin (ACTH), vasopressin (ADH), calcitonin, follicle-stimulating hormone (FSH), luteinizing hormone (LH), thyrotrophin (TSH), parathormone (PTH)), and secretin. For many of these hormones, GTP is also needed to generate cyclic AMP.

Polypeptide hormones and calcium. In addition to cyclic AMP, it has recently been learned that Ca^{++} mediates the intracellular actions of certain hormones (Fig. 1). The best example of how Ca^{++} functions as an intracellular messenger is seen with the hormone cholecystokinin (also known as pancreozymin). In acini of the exocrine pancreas, cholecystokinin stimulates the release of stored intracellular Ca^{++} from sites in either the plasma membrane, mitochondria, or endoplasmic reticulum. Ca^{++} in turn interacts with zymogen granules, leading to the release, by exocytosis, of

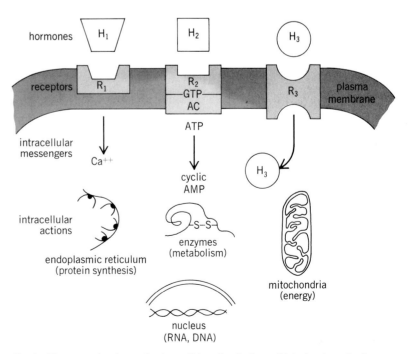

Fig. 1. Three mechanisms of polypeptide action in the cell interior. In order to carry out intracellular actions, after binding to their surface receptors, hormones may either generate the second messengers Ca^{++} and cyclic AMP or serve as their own second messenger.

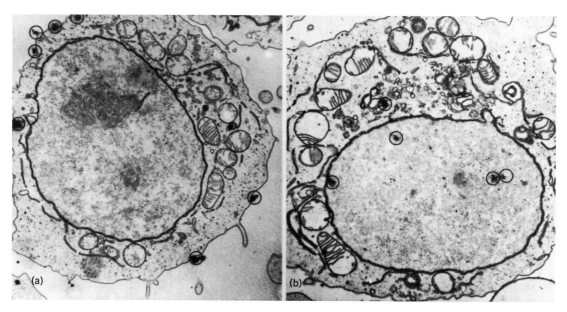

Fig. 2. Demonstration of time-dependent insulin uptake into the interior of intact cells. Human lymphocytes were incubated with radio-labeled insulin, and electron microscopic autoradiographs prepared. (a) At early incubation time (30 s), insulin is seen on the cell surface. (b) At later time (30 min), insulin is seen in the cell interior. (From I. D. Goldfine et al., Entry of insulin into human cultured lymphocytes: Electron microscope autoradiographic analysis, Science, 202:760–763, 1978)

digestive enzymes from the cell into the pancreatic duct and duodenum. Ca^{++} appears to mediate other cellular effects of cholecystokinin in pancreatic acini, such as the regulation of glucose and amino acid transport and cyclic GMP generation. Other hormones, such as oxytocin, angiotensin, and epinephrine (in addition to its actions via cyclic AMP), may act through Ca^{++} as a second messenger. In liver, for example, angiotensin elevates Ca^{++}, leading to glycogenolysis.

Absence of known second messenger. A number of polypeptide hormones, however, have no known second messenger. These hormones include insulin, insulinlike growth factors (for example, somatomedin), growth hormone (GH), prolactin (PRL), and somatostatin. Because of its profound effects on cellular metabolism, insulin has been studied most extensively.

One major difficulty in understanding the action of insulin is that it regulates a wide variety of functions in target cells. This regulation ranges from rapid effects on membrane transport of glucose, amino acids, ions, and nucleotide precursors that take place within seconds to minutes; through intermediate effects on cytoplasmic functions such as protein synthesis and enzyme activation that take place within minutes to hours; to delayed effects on nuclear functions such as ribonucleic acid (RNA) and deoxyribonucleic acid (DNA) synthesis that take place within hours to days. Specific receptors for insulin have been demonstrated on the plasma membrane of target cells, and the activation of surface functions such as transport are believed to be a result of the direct interaction of insulin with these receptors. How insulin mediates intracellular effects is unknown. It has been speculated that insulin acts through a second messenger. Candidates for a second messenger for insulin have included ions such as Ca^{++}, K^+, and

Mg^{++} and nucleotides such as cyclic GMP. No second messenger analogous to cyclic AMP has been discovered for insulin.

The concept of polypeptide hormones entering intact cells and then functioning as their own messengers has long been considered, but until recently this type of action mechanism was not thought possible because it was believed that polypeptide hormones such as insulin did not enter intact target cells. In the 1970s a number of even larger proteins, however, were shown to enter intact cells and, in certain instances, regulate cellular functions. These proteins include plant toxins and lectins, serum lipoproteins, and immunoglobulins.

An example of how a large protein can directly influence intracellular functions is seen with abrin, a potentially lethal toxin which comes from castor beans. Recent studies indicate that this toxin first binds to a specific receptor on the surface of cells, and the toxin-receptor complex is then endocytosed into the cell via small vesicles. In the cell interior, the toxin escapes from vesicles and binds to specific sites on ribosomes, where it inhibits protein synthesis, leading to cell death. Similar models of intracellular translocation followed by direct intracellular actions, either stimulatory or inhibitory, can be developed for a variety of polypeptide hormones.

Recent studies, employing the technique of electron microscopic autoradiography, have demonstrated that insulin and other hormones, after binding to their surface receptors, enter the interior of target cells (Fig. 2). Further, after internalization, insulin and other hormones interact with specific binding sites within the cell interior, including the nuclear membranes, Golgi apparatus, and smooth and rough endoplasmic reticulum (Fig. 1). It is possible, therefore, that insulin or other polypeptide hormones, after binding to surface receptors, may enter the interior of target cells, bind to intracellular structures, and directly regulate intracellular events such as protein and RNA synthesis (Fig. 1).

Conclusion. Polypeptide hormones bind to specific receptors on the cell surface. In certain instances, the hormone-receptor complex generates a second messenger, such as cyclic AMP or Ca^{++}, that regulates intracellular events. In other instances, the hormone-receptor complex is internalized, raising the possibility that the hormone itself may interact with intracellular structures and then regulate intracellular events. It thus appears that there are multiple mechanisms whereby polypeptide hormones regulate target cells.

For background information *see* ADENYLIC ACID, CYCLIC; HORMONE in the McGraw-Hill Encyclopedia of Science and Technology.

[IRA D. GOLDFINE]

Bibliography: I. D. Goldfine, *Life Sci.*, 23: 2639–2648, 1978; M. E. Maguire et al., *Adv. Cyclic Nucleotide Res.*, 8:1–184, 1977; J. A. Williams et al., *Amer. J. Physiol.*, 235:517–525, 1978.

Hydroponics

Over the last few years, technological and practical developments have resulted in an enhanced feasibility of hydroponics—the method of growing of plants without soil—as a method for professional and amateur growers. These developments are the following: (1) availability, at relatively low prices, of plastics used in the form of noncorrodible containers (or as a material to make growing beds waterproof); (2) perfecting of a range of growing methods based on a very efficient and flexible hydroponic system called the nutrient flow technique; (3) development of some simple methods for estimating the level of various nutrients in the nutrient solution.

In this age of an increased desire for self-reliance, evident in, among other things, attempts by many people to grow some of their own food, these developments may be especially important, since potential hydroponic methods would enable far more people to succeed in these attempts than would be possible with conventional growing methods. Because of the elimination of the need for soil, areas with horticulturally unsuitable soils can be made productive. Another advantage is that some of these techniques use space so effectively that even small urban backyards can be turned into highly productive units as long as sufficient light is available.

Nutrient flow technique. Allen J. Cooper developed a practical hydroponic method with a recirculating nutrient solution which he called the nutrient film technique. The uniqueness of this method is that the roots of the plants are bathed in a continuously flowing nutrient solution without the presence of a growing medium. Originally, it was thought that the solution in the growing channel should be as thin as a film of only a few millimeters, so that the solution would be kept sufficiently aerated. However, P. A. Schippers found that the layer of solution can be much thicker (such as 2 or 3 cm, or about an inch)—in fact, a film cannot be maintained when the roots fill the bottom of the channel—and that a continuous flow of the solution is much more pertinent for this method than the thickness. He renamed it, therefore, nutrient flow technique (often abbreviated, as was the case with nutrient film technique, as NFT).

An important feature of this system is that, irrespective of the size of the unit, it can be built with readily available and relatively inexpensive materials.

The actual version of the method, of which the principle is shown in Fig. 1, is determined by the growing bed, but the other elements are essentially the same for any version. The only distinctions which can be made here are those caused by the size of the unit. If the growing area is not larger than, say, 1.2 m × 4.8 m (4 ft × 16 ft), the various containers can be plastic refuse cans, wastepaper baskets, dishwashing basins, and the like. In systems of a more commercial size, special provisions are made such as a receiving basin consisting of a wooden frame built on—or a trench dug into—the ground and made waterproof by a polyethylene lining. In those systems, manifolds are constructed which consist of plastic pipe with outlets in each of a large number of growing channels, all served by one nutrient container. A continuous flow of 100 to 200 ml (approximately ¼ to ½ pint) of nutrient solution per minute out of each outlet is maintained.

Commercial application. For commercial enterprises the advantages of having no substrate usually outweigh the disadvantages for crops such as tomatoes, cucumbers, and others of similar size (possibly also for lettuce), because there is no growing medium to be disinfected after each crop. The need for sterilization of growing bed and receiving basin is also eliminated because the polyethylene lining can be discarded and replaced by fresh material. Disinfection of the nutrient container and pipeline is very simple to carry out.

Seeds of the crops mentioned above are usually sown in a growing block made from peat or from some synthetic material in which the plants can stand by themselves in the channels for all of their growing period (such as lettuce) or until they are supported with the help of suspended twine (such as tomatoes or cucumbers). For most vegetable crops, however, it would not be economical to use growing blocks. A substrate in the growing bed is then needed to give the roots the necessary anchorage. Perlite, in which the seeds can be sown directly, is the preferred medium, because it ab-

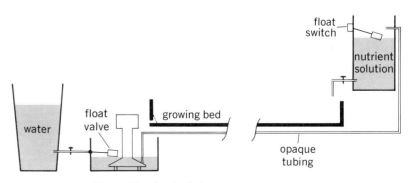

Fig. 1. Diagram of the nutrient flow technique.

sorbs the nutrient solution from the bottom of the bed without becoming soggy, and the seeds do not need any further watering. A layer of 3 to 5 cm (1 to 2 in.) thickness is sufficient except for root crops such as carrots and potatoes which need a depth of at least 15 cm (6 in.). These crops, however, would not be of interest to commercial growers.

At the moment, none of the vegetable crops presently grown commercially in American greenhouses needs any substrate. However, if the European trend of increased greenhouse production of crops such as radish, kohlrabi, and Chinese cabbage is an indication of what could be expected in the United States, a substrate would be needed if these crops were to be grown hydroponically. This might cause some problems with keeping the substrate free from diseases in the long run. Research on this subject is needed.

Growing bed. With regard to the growing bed, several variations are possible.

Sloping troughs. The most general type for commercial growers will probably be slightly sloping (1:50 to 1:100) troughs of a width of approximately 15 cm (6 in.), consisting entirely of wood or of a wood bottom and fiberboard sides and lined with 4-mil (0.1-mm) polyethylene, which are in use for larger vegetables such as tomatoes and cucumbers. Troughs as long as 33 m (100 ft) have been reported in England, but shorter channels offer some advantages. They are moved around more easily, they are more readily adjusted in slope, the nutrient solution will not back up to such an extent at the high end, and aeration of the solution may be better.

For crops which are seeded directly, growing beds of plywood or fiberboard lined with polyethylene and filled with perlite are very suitable.

Vertical plastic pipes. In order to make better use of the vertical dimension—outside or in a greenhouse—plants can be grown in vertical plastic pipes with a diameter of about 5 cm (2 in.) with holes in which plants such as lettuce, beans, peas, or strawberries are placed (Fig. 2). The plants are fed by nutrient solution dripping into the top of the pipes and falling either into the receiving basin itself or, in larger systems, into channels leading to the receiving basin. This system is more complicated to construct, since it requires a frame for support of feeding lines and hanging pipes. As many as 25 heads of lettuce have been grown on pipes of 1.5 m (5 ft) length. Although designed for lettuce, the system was commercially not completely satisfactory for this crop, since differences in light intensities between top and bottom of the pipes caused too uneven a growth. It may, therefore, be more suitable for crops which are harvested repeatedly. However, lettuce would be a good crop for amateur growers, since the uneven rate of growth, preventing the heads from maturing all at the same time, would be an advantage rather than a disadvantage.

Cascade system. In a third system, vertical space is also used very effectively by suspending four, six, or eight slightly sloping beds above each other (Fig. 2). In this cascade system, the nutrient solution runs successively through these beds, which usually consist of polyvinylchloride pipes with a diameter of 7.5 cm (3 in.) filled with perlite, starting at the high end of the highest bed and running out at the low end of the bottom bed into the re-

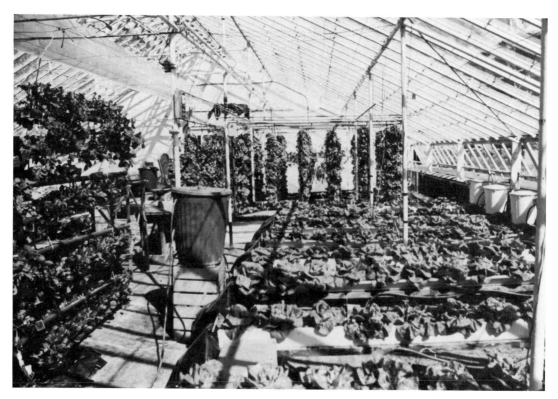

Fig. 2.   Horizontal system (utilizing roof gutters) with lettuce is shown on the right, cascade system with various vegetables on the left, and vertical pipes with lettuce in the background.

ceiving basin or a channel leading to it. From there the solution is pumped up again to the nutrient container which, as is the case with the vertical system, is placed 2.5 to 3 m (8 to 10 ft) high.

Maintenance of nutrient solution. Although it is often recommended that the nutrient solution should be discarded and replaced periodically by a fresh one, this is not in line with the present trend toward conservation and away from possible groundwater pollution. It would be preferable to add those nutrients to the solution which have been taken out by the plants. This has not seemed feasible, because in the first place it was thought that the range of the nutrients in the solution was very limited, and in the second place no practical chemical methods were known which could be applied by growers without any chemical background. In both these respects, considerable progress has been made. Recent research results indicate that the range in nutrient levels from which the plants can derive nutrition is much larger than was thought before and that the plants can be without any nutrients in the solution for several days before starting to show deficiency symptoms. This means that accurate chemical tests are not needed, but that a rough indication of the nutrient status would suffice.

Lately, several simple spot tests have been devised, in addition to the well-known test for acidity, in which color reactions indicate the approximate levels of phosphorus, iron, and nitrate-nitrogen. With some turbidimetric reactions an idea can also be obtained about the levels of potassium, calcium, and chloride in the nutrient solution. These tests will be of great help in deciding not only when to fertilize, but also what fertilizers to administer. They can also be used in deciding whether the whole nutrient solution should be replaced because of possible ionic imbalances which can be caused by accumulation of salts added with the water from the water container (to compensate for the water lost through transpiration by the plants), but not used by the plant.

For background information *see* HYDROPONICS in the McGraw-Hill Encyclopedia of Science and Technology.   [P. A. SCHIPPERS]

Bibliography: A. J. Cooper, *Nutrient Film Technique of Growing Crops*, 1976; Hydroponic Growing Systems (Calverton, NY), *The Nutrient Flow Technique for Growing Plants*, 1979; P. A. Schippers, *The Nutrient Flow Technique: Three Years of Hydroponic Research at the Long Island Horticultural Research Laboratory of Cornell University*, Vegetable Crops Department, Cornell University, V.C. Mimeo 212, 1979.

Immunology, cellular

The immune system of higher vertebrates is a very complex multicellular system that has evolved to detect and eliminate threats to the health and life of the individual animal. A key property of the system is its ability to respond to the unpredictable, because the antigenic spectrum of harmful infectious agents is potentially infinite. This property is achieved by generating a very large number of cells (lymphocytes), each of which expresses on its surface antigen-recognition sites of a single specificity, with the total repertoire of

specificities covering a virtually infinite spectrum of antigens (but as will be seen below, this statement must be qualified). A given antigen stimulates appropriate lymphocytes to proliferate and give rise to an expanded number of progeny cells of the same specificity (clonal selection and expansion). These cells mediate various effector mechanisms leading to elimination of the infection, or of other threats such as aberrant self cells ("immunological surveillance").

It is equally important, however, that the destructive forces of the immune system are not unleashed against normal self components. The achievement and maintenance of this self tolerance is not a simple problem, because some self antigens are also diverse, for example, blood group and histocompatibility (transplantation) antigens of cell surfaces. The antigen-sensitive cells in the developing immune system of each individual must therefore undergo processes of selection and control so that the system "learns" not to respond to inherited self antigens, but retains the ability to respond to all foreign antigens.

In this context, the recent discovery that a certain class of antigen-sensitive lymphocytes termed T cells actually recognizes self histocompatibility antigens together with foreign antigens has aroused considerable interest among immunologists. This dual specificity is exhibited by several subclasses of T cells, including the cytotoxic T cells. The latter are capable of lysing other cells which display appropriate antigens, a capability that is readily measured by simple assays with laboratory cultures. Thus, cytotoxic T cells generated in response to a given viral infection will lyse self target cells infected with the same virus used for immunization, but will not lyse virus-infected nonself cells, or uninfected self cells, or self cells infected by a different virus.

H-2 restriction. This phenomenon was first shown by using inbred mouse strains. The use of congenic mice clearly demonstrated that the self antigens recognized by T cells were coded in the major histocompatibility gene complex, the *H-2* gene complex. Other self antigens were found to be irrelevant to T cell recognition. Hence, the term H-2 restriction is generally used to describe the phenomenon in mice. Mouse strains with recombinant or mutant *H-2* haplotypes, and F1 hybrids, were then used to show that each of the T cells (or clones of identical progeny derived from a single precursor) recognizes a single H-2 molecule together with a viral antigen, and thus far three loci *(K, D, L)* in the *H-2* complex have been identified as relevant to cytotoxic T cell recognition. Various loci in the *I* region of the *H-2* complex are required for other subclasses of T cells. Comparable though less extensive and detailed findings have been reported for chicken, rat, and human, so that the phenomenon seems generally applicable to the immune systems of all higher vertebrates.

The major histocompatibility antigens are highly polymorphic. Except for members of inbred strains, each individual within the species has a different set of cell-surface protein antigens (coded by allelic genes) that provoke the immune responses causing rejection of transplanted tissue from another individual, and the genes coding for these antigens are inherited and expressed in mendelian dominant fashion. Some sections of the amino acid sequences of the major histocompatibility antigens of mouse (H-2) and human (HLA) are strikingly similar, despite differences related to their antigenic variation. Thus, it is clear that modern H-2 and HLA antigens have evolved from a common ancestral protein, and parts of these molecules have been conserved for over 100,000,000 years of vertebrate evolution.

Elimination of viral infection. Evaluation of the role of histocompatibility antigens in immune responses to viral infections in modern vertebrates requires background knowledge of the mechanisms involved in recovery from these infections. Over the last decade it has become apparent that recovery from several major groups of viral infections depends upon a T cell–mediated immune response rather than antibody production. T cells were so named because a major part of their differentiation from multipotential stem cells into antigen-sensitive lymphocytes takes place in the thymus. After leaving the thymus, T cells recirculate via blood and lymph through most tissues, but concentrated populations are present in certain thymus-dependent areas of the spleen and lymph nodes. It is in these latter sites that antigen-driven proliferation of T cells takes place and gives rise to enlarged populations of antigen-specific effector T cells that migrate via blood to sites of infection in other organs. The effector functions relevant to elimination of viral infection are: first, the ability of the cytotoxic subclass of T cells to recognize and lyse any virus-infected cells that display virus-specific antigens on their surfaces before the virus has completed its intracellular replication cycle; second, an influx of blood-borne phagocytes, triggered by the T cell activity, that ingest and largely destroy remaining infectious virus particles; and third, the local production of various soluble factors, such as interferon, that inhibit viral replication within infected cells, and may also augment and complement the functions of cytotoxic T cells and mononuclear phagocytes.

The mechanisms outlined above that lead to elimination of a viral pathogen from its natural murine hosts can be induced in nonimmune infected mice by the injection of virus-immune T cells, but only if the H-2 type of the T cell donors matches that of the recipients. The clearance of infection from the recipients' tissues is dramatically efficient, with the numbers of infectious viral particles being reduced 10,000-fold or more in 24 hr. This type of experiment clearly demonstrates both the importance of T cell–mediated events in viral clearance and the fact that recognition of self H-2 antigen, as well as viral antigen, is an essential part of T cell triggering in this process of recovery from potentially lethal infection.

It is obvious that the T cell cytotoxic function cannot be exercised against free viral antigen molecules. Indeed, T cell recognition sites that bind free antigen molecules would be inhibited in binding similar antigens on an infected cell surface. Thus, the T cells would compromise their viral clearance function. The desirability of selectively binding antigen only on cell surface may have led to the evolution of the observed T cell property of

simultaneous recognition of a self cell-surface antigen together with foreign viral antigen. However, it must be stressed that self antigen recognition must be achieved in such a manner that T cells do not bind to uninfected self cells and that potentially destructive anti-self T cell responses do not generally occur, or are quickly suppressed.

Models of T cell recognition. The precise mechanisms by which T cells recognize antigens with dual specificity, or the nature of the molecules employed for this purpose, are not yet defined, but current speculation centers on two types of model. One type, often designated as altered self, states that an individual T cell expresses on its surface only one type of antigen binding site that is specific for an antigenic pattern produced by interaction or complex between a self histocompatibility antigen and a foreign antigen. The T binding site must bind only to the complex but not to either of the separate components, or it falls into the second category of model generally designated as dual recognition. The latter model states that a T cell possesses two different antigen binding sites, one specific for self histocompatibility antigen and the other specific for a foreign antigen. But this statement must be qualified by saying that only a connected pair of these sites, upon binding to the appropriate antigens, can trigger T cell functions. The experimental data clearly show that if the two antigens are on two different cells, both of which are in contact with a T cell, then no triggering occurs. Thus, both models, though they differ in principle, require the foreign and self antigen molecules to be close to one another in the same cell membrane.

The key problem is that the T cell pool apparently has an almost infinite repertoire of antigen binding sites to cover the spectrum of foreign antigens, but the recognition of foreign antigen is always accompanied by self recognition, and the self antigen is always a major histocompatibility antigen. Thus, great diversity is exhibited by one facet of T cell recognition, while strict control is imposed on the specificity of the other facet. This implies that the genetic units coding for the two different binding capabilities are subject to separate processes of regulation, regardless of whether they are ultimately expressed on a single molecule or on two molecules at the T cell surface.

Experimental evidence supporting this idea has been obtained by examining the expression of self recognition capability in T cells derived from multipotential stem cells taken from F1 hybrid mice. F1 cells express H-2 antigens of both inbred parental strains, and the F1 T cell pool that develops in the thymus of an F1 animal can generate two separate but equally numerous virus-immune populations of T cells with specificity for each of the two parental types of H-2 antigen. However, if the same F1 stem cells develop into T cells in a thymic environment derived from only one parental strain, then the virus-immune response is heavily biased toward recognition of that parental H-2 type but not the other. Thus, only one half of the self-recognition potential of the F1 genotype is fully expressed in F1 T cells leaving the thymic environment of one parental strain. This result indicates that a rigorous selection process takes place in the thymus during T cell differentiation. Subsequently, the pool of differentiated T cells predominantly recognizes histocompatibility antigens of thymic type. The development of this restricted capability contrasts with the very much broader capability of the same cells to recognize foreign antigens. Therefore, the two types of capability for antigen recognition (self and foreign) may develop in the thymus independently of one another.

A recent series of experiments has shown that with certain viruses not all alleles of the K, D, and L loci of the H-2 gene complex are associated with strong T cell responses. Some combinations of a given virus and a given H-2 allele show little or no response. However, if stem cells of such low responder genotype are allowed to differentiate in a thymic environment of high responder H-2 type, they give rise to T cells of high responder capability. This observation emphasizes the importance of the thymic environment in regulating the ultimate capacity of the T cell pool to respond to a particular viral antigen. In turn, the ability to respond affects the potential survival of the animal and the occurrence of histocompatibility antigen polymorphism. Much current effort is being devoted to further investigation of the basis of these phenomena.

For background information see IMMUNOLOGY, CELLULAR in the McGraw-Hill Encyclopedia of Science and Technology.   [R. V. BLANDEN]

Bibliography: G. Moller (ed.), Immunological Reviews, vol. 35, 1977, and vol. 42, 1978; G. Moller (ed.)., Transplantation Reviews, vol. 19, 1974, and vol. 29, 1976.

Indian Ocean

Recent physical oceanographic studies in the Indian Ocean are providing new insight into both its deep and upper circulation.

Deep circulation. Prior to the International Indian Ocean Expedition (IIOE) of 1962–1965, deep measurements of water properties were so sparse and widely spaced in the Indian Ocean that only very general remarks could be made about its deep circulation, and those dealt mainly with origins of features in property distributions. The data coverage of the IIOE, however, combined with results from special-purpose cruises stimulated by simple theoretical models of circulation systems, has since generated some fairly specific ideas about patterns and rates of deep flow in many basins of the Indian Ocean.

Among the accomplishments of the IIOE was the first comprehensive mapping of the bottom topography of the Indian Ocean. The existence of the Ninetyeast Ridge, for example, had not even been suspected before then. The geometry of the Indian Ocean (Fig. 1) has emerged as the most complicated among all the oceans, for its several ridges divide it into a multiplicity of separate basins: the north-south series of small basins in the west, the Central Indian Basin in the middle, and the great West Australian Basin in the east. One might anticipate that the pattern of deep flow, constrained by this geometry, is correspondingly more complicated than in other oceans.

Since there is no sinking of water to great depth in the North Indian Ocean, all of the cold deep water in the Indian Ocean must enter it from the Antarctic. It does not do so in a broad, ocean-wide manner, however, because slow, large-scale motions on the rotating Earth take very different forms from those of everyday experience in which the constraint of rotation and the curvature of the Earth's surface are unimportant. Rather, idealized models of deep circulation systems suggest that equatorward flow from the Antarctic should take place in narrow, relatively swift currents along the western boundaries of basins. These currents then are thought to feed the deep water into the interiors of the basins, which subsequently moves slowly upward and poleward.

Western basins. In the west, the deep Indian Ocean is open to the Antarctic through the entrances to the Mozambique and Crozet basins (Fig. 1). Because of the Madagascar Ridge and the sill joining Madagascar to Africa, however, the Mozambique Basin appears to be isolated from the rest of the deep Indian Ocean. Water from the Antarctic therefore flows northward into the Crozet Basin and passes through fractures in the Southwest Indian Ridge into the Madagascar Basin. There and in the Mascarene Basin this northward flow occurs as predicted by theory as a narrow current adjacent to Madagascar, the effective western boundary for the deep southwestern Indian Ocean. The current has been identified at latitudes 12, 18, and 23°S by cold water pressed up against Madagascar at depths greater than about 3500 m; it is some 300–400 km wide, has speeds of a few centimeters per second and a volume transport of about 4×10^6 m^3 s^{-1}.

Some fraction of this current has been found to flow out of the Mascarene Basin into the Somali Basin, and marginal evidence has been obtained for deep northward flow close to Somalia, but the existence of a deep western boundary current in the Somali Basin has not been established with surety, and the pattern of deep flow there must still be considered quite uncertain. For the Arabian Basin to the north, there is no evidence yet available concerning the character of the deep flow, except that the concentration of dissolved oxygen there is lower than anywhere else in the deep Indian Ocean. Dissolved oxygen is introduced to the ocean at the sea surface and consumed at depth by the oxidation of dead organisms; thus its concentration to some extent measures the "age" of sea water, and the low values found in the deep Arabian Sea suggest a relatively small rate of supply of water from the south.

Above the bottom water in the southwestern Indian Ocean there is a layer at depths of 2500–3500 m of relatively saline water, which is brought in from the South Atlantic by the Antarctic Circumpolar Current. The associated salinity maximum can be followed roughly to latitude 20°S, but farther northward, flow at these levels (if any) is obscured by salinity enrichment of the deep water to the north by outflow from the Red Sea.

West Australian Basin. Across the breadth of the Indian Ocean the temperature at depths greater than 2500 m generally decreases from west to east, a fact that would not be consistent with the deep water in the east being supplied from the western part of the ocean. It suggests in-

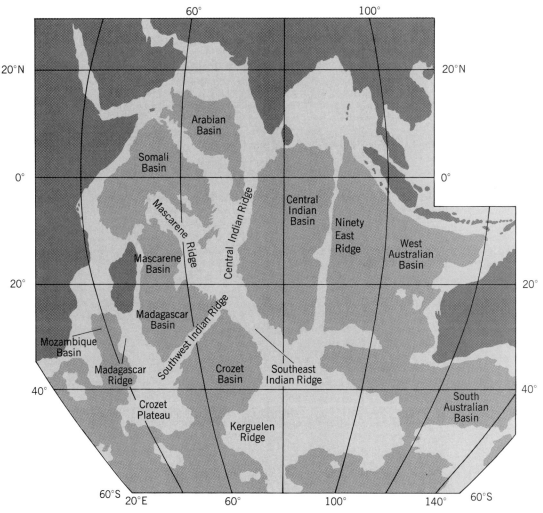

Fig. 1. Index map identifying names of basins and ridges in the Indian Ocean, including an approximate representation of the 4-km isobath, after K. Wyrtki (1971). *(From B. Warren, Bottom water transport through the Southwest Indian Ridge, Deep-Sea Res., 25:315–321, 1978)*

stead that the deep eastern Indian Ocean has sources different from the passage through the Crozet Basin, and in fact the West Australian Basin is open to the Antarctic just southwest of Australia. If deep water does enter the Indian Ocean through this passage, the dynamical theory above requires that it should flow northward in a narrow current adjacent to the Ninetyeast Ridge, which is the western boundary of the West Australian Basin. That current has been identified at latitude 18°S at depths greater than 3000 m; it is about 600 km wide and has a likely volume transport of around 4×10^6 m³ s⁻¹. Suitable observations for detecting the current at other latitudes have not yet been made, but the geometry of the West Australian Basin is simple enough for one to suppose that this current flows along nearly the entire length of the Ninetyeast Ridge, discharging water eastward into the interior of the basin.

Central Indian Basin. The density of deep observation in the Central Indian Basin is much lower than in the basins to the east and west. Nevertheless, lines of observations across it along latitudes 18 and 12°S, made in 1976 and 1979 respectively, show a pattern of density variation indicative of

meridional flow at depths of 2000–3500 m on the eastern flank of the Central Indian Ridge. This zone is also characterized by slightly high concentrations of dissolved oxygen and low concentrations of dissolved silica, which are properties of water from the Antarctic. It seems likely that this is yet a third deep western boundary current, carrying water at mid-depths northward from the Antarctic into the Central Indian Basin. No estimate of its volume transport has been made yet.

The water deeper than 3500 m in the Central Basin is not supplied by this boundary current, and in fact the ridge bounding the basin to the south is probably too shallow to allow direct inflow from the Antarctic of water so deep. The properties of this very deep water, however, are much like those observed at depths of 3500–4000 m in the boundary current on the eastern side of the Ninetyeast Ridge, and there are known sills of such depth on the ridge at latitudes 10 and 3°S, across which water from the West Australian Basin could pass into the Central Basin. In a line of observations made just west of the Ninetyeast Ridge in 1979, between latitude 12°S and the Equator, such an overflow was in fact found at 10°S; none was occurring at 3°S. No estimate is possible yet of the mean rate of this overflow, but the available evidence makes it highly probable that inflow from the Ninetyeast Ridge current is the source of the deepest water in the Central Indian Basin.

For background information *see* INDIAN OCEAN in the McGraw-Hill Encyclopedia of Science and Technology.   [BRUCE A. WARREN]

Upper circulation. The equatorial regions of the world's oceans have received special attention oceanographically in recent years because of their probable role in the global heat balance and the variability on climatic scales. Global atmospheric models have revealed that there is a special sensitivity of the global climate to the equatorial oceanic conditions at and near the surface. Conversely, oceanographic studies have shown that the equatorial oceans respond more rapidly and effectively to atmospheric conditions on the seasonal and longer time scales than at mid-latitudes.

The western Indian Ocean is subject to the most dramatic atmospheric forcing at the seasonal and longer time scales. The atmospheric circulation is dominated by the monsoon: during part of the year, starting in May, the winds are from the southwest; these winds weaken in September, and beginning in October, the winds are from the northeast. During both periods, the winds are strong and steady, although the winds are strongest during the southwest monsoon. The entire Indian Ocean, and part of the western Pacific Ocean, are dominated by this reversing monsoonal circulation, so that the scale of the meteorological disturbance is very large, of the order of the physical dimensions of the Indian Ocean.

The oceanographic response to these large-scale variations in the meteorological forcing is equally dramatic. Although only the equatorial response is discussed here, there are significant changes in the Arabian Sea and in the Somali Current. The equatorial response is largely confined to the upper layers of the Indian Ocean, although there is recent evidence from moored-current-meter data of significant variations in the currents at long time scales in the deep water, as well as in the upper layers.

The Somali Current is an intense narrow current flowing toward the north along the Kenyan and Somali coasts during the southwest monsoon. This current system and its relationship to the equatorial circulation discussed below is presently the subject of a large international observational program, entitled INDEX (Indian Ocean Experiment), whose United States component is sponsored jointly by the National Science Foundation and the Office of Naval Research, Code 480. INDEX is an oceanographic component of the Global Atmospheric Research Program (GARP), and the data will be integrated with the extensive meteorological data from the GARP Global Experiment. The scientific results from this experiment should become available during 1980.

The undercurrent. The equatorial undercurrent is a narrow eastward jet in the thermocline, trapped to within a few degrees of the Equator. In the Pacific and Atlantic oceans, it is a permanent feature of the circulation, and is generally believed to be related to the persistent trade winds from the east. This westward stress at the surface and resulting difference in the sea surface elevation is balanced by a zonal pressure gradient between the boundaries. There is a compensating pressure gradient in the thermocline which drives a flow down the gradient. In the Indian Ocean, a favorable pressure gradient can be established only by the northeast monsoon, when there is a westward component of the wind at the Equator. The observations have shown an undercurrent in the western Indian Ocean only during the latter part of the northeast monsoon, although there are marked interannual variations in its vigor, possibly related directly to variations in the strength of the wind stress from the east.

The basic understanding of both the kinematics and dynamics of the undercurrent is still rather sketchy in the Indian Ocean—for example, it is not known whether the undercurrent exists as a continuous filament. The actual process by which the undercurrent is set up and decays must involve the long-period equatorial planetary and Kelvin waves which propagate rapidly along the equatorial waveguide. One of the objectives of INDEX-79 was to observe the propagation of these equatorially trapped waves with periods of the order of a month or longer. Because of the efficiency with which the local winds can generate currents near the Equator, where the Coriolis accelerations are weak, it becomes difficult to distinguish local currents from the propagating equatorial waves.

The surface jet. During the transitions between the monsoons, in May and October, there is a period of eastward wind stress, during which a narrow intense eastward surface jet is formed along the Equator. J. C. Swallow's observations, in 1967, along 56°E and 62°E showed the surface jet narrow and well developed at the end of April, embedded in an equatorially convergent flow—consistent with an extraequatorial Ekman transport. The eastward flow was confined to the upper 50 m, below which there was a strong flow to the west.

K. Wyrtki, in 1973, using ship drift data, showed that this jet is trapped to the Equator and extends across the breadth of the Indian Ocean. Wyrtki speculates that this flow is associated with a large-scale redistribution of mass—uplifting the thermocline along the east African coast and depressing it along Sumatra.

There is a profound cooling of the surface waters in the equatorial region and Arabian Sea soon after the southwest monsoon begins. It is not clear at present whether this cooling results from the uplifting of the thermocline in response to the eastward flux of the surface jet, or to the onset of the Somali Current and its upwelling, or a combination of the two.

The equatorial trapping and the time scales of the setup of the surface jet are consistent with the simple theory of a near-surface flow accelerating until a sufficient pressure head is developed against the eastern boundary to balance the wind stress, although numerical calculations indicate that the jet rapidly becomes nonlinear. In the absence of Coriolis accelerations at the Equator, this current would persist until dissipated or replaced by the response to the changing monsoon winds.

Multiple equatorial jets. INDEX provided the

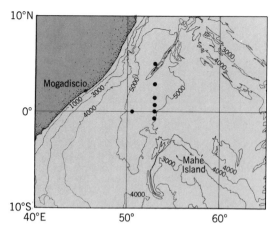

Fig. 2. Bottom topography and spatial configuration of the velocity profile sites dump—the INDEX-76 program.

opportunity in 1976 to obtain time series of vertical profiles of horizontal current, in a section along 53°E between 0° 45′S and 5°N, as shown in Fig. 2. The observations were made in May and June 1976, and spanned the onset of the southwest monsoon. These data showed that the equatorial region was dominated by zonal jets of narrow vertical extent, 100–300 m, and amplitudes of 20–40 cm/s. The entire section was occupied twice during the 45-day period, with more frequent stations in the equatorial band. The time series at the Equator extended over 31 days and showed little variation in the zonal component of velocity, and suggests that the time scale of the zonal jets is of the order of several months. The meridional component of velocity is largely uncorrelated with the zonal component and exhibits significant variations over the time period.

The vertical structure of the velocity changes dramatically across the section, as shown in Fig. 3, from being dominated by short vertical scales in the equatorial region, to the large vertical scales

typical of the mid-latitude low-frequency variability.

The most energetic westward jets occured at the 200-m and 750-m levels. The jet structures extend coherently into the deeper water to 2000 m.

There is ample historical evidence for westward flow below the equatorial undercurrent in all oceans and particularly in the Indian Ocean, where it is consistently identified during the southwest monsoon.

Although these observations provide the first convincing demonstration of the energetic dominance of the velocity field by the small vertical scales in the equatorial regions, one can look in the historical data and find similar but isolated examples.

The fate of these relatively strong westward jets as they encounter the western boundary is largely unknown as yet, although the discussion in the next section speculates on their role.

The meridional equatorial trapping and small vertical scale of the multiple jets is consistent with the equatorial waves of planetary scales whose periods range from the order of a month to a year and longer. The corresponding planetary scale waves at mid-latitudes all have periods longer than 1 year, and are therefore poorly matched to the spectrum of atmospheric forcing.

Upper ocean equatorial circulation. The distribution of properties, particularly temperature and salinity, is the traditional source of information on the long-term or general circulation. The inference is based upon the assumption that these properties are conserved along the fluid's trajectory. This assumption is probably a poor one in the equatorial regions, particularly in the near-surface layers, because of the strong and variable air-sea interaction. In the Indian Ocean, as discussed earlier, the currents are dominated by energetic processes of relatively small spatial scales. These are reflected in the very large degree of variability in the potential temperature-salinity characteristics in the equatorial region. B. Warren, H. Stommel, and

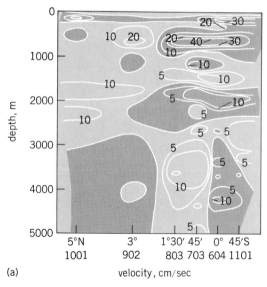

(a)

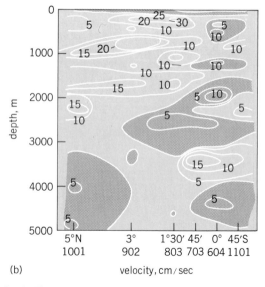

(b)

Fig. 3. Contours of horizontal current velocity in the (a) zonal and (b) meridional directions plotted against depth. These data are from a meridional section along 53°E extending from 0°45′S to 5°N.

Swallow have given an excellent and detailed discussion of the water mass characteristics of the Somali Basin to which little can be added until the analysis of INDEX is completed.

The western Indian Ocean contains several water masses of distinct origin which abut one another in the equatorial region, giving rise to large variations, both in the vertical and horizontal, in the potential temperature-salinity characteristics. In the near-surface equatorial layers, warm, extremely saline water produced by evaporation in the northern Arabian Sea impinges upon cooler, fresh water brought by the Somali Current from the South Equatorial Current. This water is transported by the South Equatorial Current from the region east of the Indian subcontinent. The temperature-salinity characteristics change rapidly both in time and space as these water masses intermingle. The equatorial surface jet appears to play a role here, transporting a large amount of this water toward the east, leaving the western region significantly cooler, although there may be large local air-sea interactions responsible for part of this cooling as well.

At intermediate depths, the structure of the salinity field is complex, with a low-salinity layer overlying a layer of high salinity. There are large variations in the structure of the potential temperature-salinity relationship from north to south across the Equator. The low-salinity layer appears to come from the southern Indian Ocean, while the salinity maximum layer has its origin in the Red Sea. Again, the equatorial region is a region of intermingling between these water masses, resulting in a great deal of interleaving and enhanced fine-structure variability (vertical scales of $10-100$ m) as seen in recent studies with a micro-profiling CTD (salinity-temperature-depth recorder) during INDEX.

The intermediate-depth region, in which the distribution of properties is so complex, coincides with the region in which the velocity field is dominated by energetic low-frequency variability in the current field. The multiple jet structures in the equatorial region with vertical scales of 100 m and time scales of several months are consistent with the intense variability and disorder in the hydrography discussed above.

For background information *see* CORIOLIS ACCELERATION AND FORCE; INDIAN OCEAN; MONSOON; SEA WATER; SOUTHEAST ASIAN WATERS; THERMOCLINE in the McGraw-Hill Encyclopedia of Science and Technology.

[JAMES R. LUYTEN]

Bibliography: J. R. Luyten and J. C. Swallow, *Deep-Sea Res.*, 23:999–1001, 1976; D. W. Moore and S. G. Philander, Modeling of the equatorial oceanic circulation, *The Sea*, vol. 6, 1976; H. Stommel and A. Arons, *Deep-Sea Res.*, 6:217–233, 1960; J. C. Swallow, The equatorial undercurrent in the Western Indian Ocean in 1964, *Study of Tropical Oceanography*, vol. 5, 1967; B. Warren, *Deep-Sea Res.*, 21:1–21, 1974; B. Warren, *Science*, 196:53–54, 1977; B. Warren, H. Stommel, and J. C. Swallow, *Deep-Sea Res.*, 13:825–860, 1966; K. Wyrtki, *Oceanographic Atlas of the International Indian Ocean Expedition*, 1971; K. Wyrtki, *Science*, 181:262–264, 1973.

Industrial engineering

Application of the computer in combining design and manufacturing functions continues to gain momentum in diminishing the time between concept and finished product. Specifically, computer-aided design (CAD) refers to the use of computers to perform design calculations for determining an optimum shape and size for a variety of applications ranging from mechanical structures and tiny integrated circuits to maps of huge areas. Computer-aided manufacturing (CAM) employs computers to communicate the work instructions for automatic machinery in the handling and processing technology used to produce a workpiece. Computer-based automation is drastically changing the way things are made and profoundly changing the jobs of people who make them. Among the benefits of an integrated CAD/CAM system are increased productivity, a significant reduction in nonproductive time, improved product quality, and a payback potential obtained by lowering the cost per piece. This modern concept of manufacturing management has led to important advances in the design and production of components used in aerospace, automotive, electronics, and other industries throughout the world

CAD. This first component of the totally integrated CAD/CAM system yields an impressive number of benefits. One benefit is that the designer is equipped with vastly more efficient computer equivalents of the many drafting tools needed in the performance of the job. Computer programs can be written to generate hard copy of drawings at speeds and accuracies far beyond human skills. The computer has liberated the designer from countless, tedious drafting details. Designers are free from the repetitive task of drawing various lines and, often, even from tasks such as calculating workpiece sizes. In fact, a CAD system not only eliminates the need for drawing dimensions on the views of a part, but also saves the monotonous labor involved in making the arrowheads.

Another time-saving feature of considerable importance is that sectional views as well as auxiliary views can be automatically conceived and drawn by computer methods. The unique system accepts commands like "erase line," "move circle," or "insert dotted or crosshatch lines." CAD systems can be programmed to generate and display a variety of symbols, characters, and points, lines, arcs, and circles—in virtually any form required for the construction of a geometric image. Once developed, part drawings can be stored in computer memory as dynamic, three-dimensional forms or as conventional, shop-type multiview projections.

Another significant benefit of CAD is that a computer-created design can be instantly recalled, either partially or totally. The graphic image can be displayed on a CRT (cathod-ray tube) screen to permit an analysis of the workability of the designer's ideas. So versatile is a computer-graphics system that mirror images of mating parts may be quickly produced and displayed. Computer-refreshed views on a CRT may be manipulated, reduced, enlarged, or viewed from different angles for possible design modifications. With this technique, a great number of design alternatives may be examined within an astonishingly short period of time. Also, when required, a CAD system can automatically generate a control tape that in some manufacturing systems can be employed to drive a numerical control metal cutting or forming machine.

When all of the elements of the design concept are in final and acceptable form, an assembly drawing can be readily generated by recalling each part from computer memory and placing the parts in their appropriate assembled graphical position. A hard copy bill of material may be automatically produced with lines ruled off, correctly spaced, and with all of the components of the design accurately recorded in the proper position.

CAM. The second component of CAD/CAM enables the utilization of processes that allow the machines to perform productive chip removal operations over a much larger percentage of time than heretofore. Using a concept which some machine tool firms call total processing, the manufacturing engineer identifies each processing requirement that interfaces with the computer data base corresponding to the original design. As an example, functional tool design data (that is, proper tool path generation ensured by selected datum plane locations on the workpiece) is integrated with program data relating to the final part design. CAM programs are written that automatically command an optimum machining sequence of processing operations, control the cutter type and size, turn coolant on or off, select appropriate feeds and speeds, and regulate a number of other machining parameters. Correlation of the design phase with the part-processing phase has significantly affected the production cycle. CAM permits a part to progress at a more rapid rate from raw material to finished product.

Programming languages. Computer programs, called software, are the principal form of communication between the programmer and the computer. Compatability of a computer-integrated design with the machine and control begins at the design stage with each aspect taking full advantage of the relative capabilities of the other. While there are many languages available for this purpose, most CAD/CAM software programs use either FORTRAN, COBOL, BASIC, PASCAL PL/1, or NUFORM.

Of all the programs, FORTRAN is currently considered by most to be the easiest programming language to work with in engineering applications. It is the program ordinarily used for solving complex numerical calculations and is a particularly useful language for solving engineering analysis problems. COBOL is mainly a business language. Its principal application is commercial data processing. Despite a somewhat limited arithmetic capability, COBOL is an effective programming language for certain kinds of engineering applications. BASIC, an acronym for Beginners All-purpose Symbolic Instruction Code, was originally developed as an educational tool to be used in teaching the use of remote-control time-sharing systems. The original language has recently undergone significant improvements and standardization. PASCAL was also originally introduced as a

teaching tool for computer programming. The principal advantage of PASCAL is in its effectiveness in structured programming. It is currently being evaluated by several agencies as a possible standard. PASCAL is used extensively in universities and colleges and is gaining wide acceptance in industry and government installations. The PL/1 language in some ways resembles FORTRAN. There are also similarities in block structures to ALGOL, while the data types suggest the influence of COBOL. PL/1 is a large general-purpose language well adapted to engineering and scientific applications. NUFORM features a fixed format with numeric input. This versatile system significantly reduces the syntax and vocabulary requirements associated with an alphanumeric input. NUFORM was designed to simplify the programmer's task, thus saving time and promoting accuracy.

Vendor computer programs. Most CAD/CAM programs that have been developed by industry are proprietary and thus are not available for general use. Often the applications of such programs are limited because of their specialized and restricted nature. Commercially available preprogrammed software in the area of mechanism design has been found to be helpful in certain specific problem-solving applications. There are a number of software firms which can supply a full range of computer services. These range from a surprisingly large assortment of comparatively simple, straightforward software programs to a complete data-base manufacturing management system that, in addition to an engineering and manufacturing system, includes software for inventory control bill of material processing, material requirements planning, work in progress, and production costing modules.

Examples of software that can be purchased from software vendors are Automatic Dynamic Analysis of Mechanical Systems (ADAMS), Dynamic Response of Articulated Machinery (DRAM), Integrated Mechanisms Program (IMP), and Kinematic Synthesis (KINSYN). Examples of other software programs currently available are NASTRAN and CSMP. These programs deal with aspects of engineering analysis. There are also two computer programs in extrusion die engineering currently available at no cost. Developed by Battelle-Columbus under the sponsorship of the U.S. Army, they are ALEXTR and EXTCAM. Both programs have been written in FORTRAN language and operate on a minicomputer. Also available are programs such as STRUDL, ANSYS, and SAP which relate to a newly developed analytical technique called finite element analysis.

Numerical control. Before the advent of numerical control (NC) the regulation of product accuracy and repeatability was directly related to the skill of the operator. The maintenance of precision is no longer considered a serious problem. The limits of accuracy on a workpiece are now controlled entirely by NC machines. The result is that the operator is no longer responsible for positioning and repeating the operation of the tool. Instead of manual controls, electric signals on NC machines precisely guide the movements of the tools and control the position of the workpiece.

NC machines may be operated by manually dialing the machine setting at a console and letting the electronic signals execute the operation. As might be expected, manual operation not only is a very slow production method but is very inefficient for large-volume production of parts. Improved versatility for NC machines is possible by using a punched paper or Mylar tape. In this method the tool and machine instructions are punched onto the tape according to an alphanumeric code. In operation, signals from the punched tape are sent to a data storage unit. As the tape unwinds past the tape reader, electric impulses are sent to the drive mechanisms which automatically control the machine functions. Information on tape can be used to establish the feed rate, spindle speed, machine table positions, traverse rates, tool selection, depth of cut, stops, and dwell intervals for selected periods of time. Unfortunately, there are some inherent problems associated with punched tape. Some of these include difficulties in tape preparation, tape breakage, limitations in adapting programs for a sufficiently wide range of workpiece conditions, and problems associated with editing and with making program changes.

The NC machine was first introduced in the 1950s. It was found that NC machines could be depended upon to operate more hours a day than was possible with traditional machine tools. Also, NC machines were more accurate. Finally, of considerable advantage was the fact that tools and workpieces could be installed and positioned automatically. As programming requirements increased in volume and complexity, however, the tape control machine proved impractical—if not impossible—for many potential jobs. Computer-assisted programming for machine tools appeared to offer a solution. In 1964 graphic computer display terminals were introduced, thus giving the system engineering staff a unique opportunity to visually verify each step in the production sequence of a workpiece.

Direct numerical control. The first industrial direct numerical control (DNC) system became operational in 1968. A DNC system, the lifeline of the CAM concept, includes both the hardware and the software required to drive one or more NC machines simultaneously while connected to a common memory in a computer. The computer may be a minicomputer, several minicomputers linked together, a minicomputer connected to a large computer, or a single large computer. Unlike an NC system, a punched tape is not used. An important advantage of the DNC system is that more than one machine can be operated at a time because the computer can be time-shared. The early DNC systems were produced out of the need to ensure optimum utilization of NC machines.

Some DNC systems consist of combinations with capabilities that encompass CRT display, part program storage, part program edit, and maintenance diagnostics. On some machines, as the program runs, the CRT provides visible part-program information, operating data, current-status messages, error messages, and diagnostic instructions. A typewriterlike keyboard facility located at the NC machine permits the operator to input, delete, or correct data. After the editing and optimizing phases have been completed, the pro-gram changes are stored in memory. The sequence of processing operations in CAM systems begins immediately after the program is recalled from storage in the computer memory. DNC programs are commanded entirely by electronic signals sent from the computer memory.

In addition to the primary function, that of controlling the manufacturing sequence of parts, DNC computers can perform a wide range of other useful data transmission functions: parts program development and verification and job scheduling. DNC systems may be applied to an almost unlimited range of product management activities.

Computer numerical control. The newest CAM innovation is called computer numerical control (CNC). This sophisticated manufacturing concept first appeared in 1970. Unlike DNC, each machine tool has its own computer. The principal advantage of CNC over DNC is that the computer software systems are less expensive. While one standard control can be used for a range of different types of machine tools, the software can be written in such a way as to adapt the control to each particular machine requirement.

Today, extremely large and versatile, totally integrated CAM machining centers are available from a number of well-established machine tool firms. Advantages cited for this trend toward complete manufacturing centers include shorter design lead time and shorter manufacturing make-ready time. In combination, these aspects result in a dramatic reduction of production throughput time. The economic pressures for computer-managed parts technology have never been more apparent. The proof is currently reflected in substantial investments in CAD/CAM research and development activities by machine tool builders both in the United States and in many other countries. The field is changing at an ever-increasing rate, and most forward-thinking parts processers are changing with it.

For background information *see* COMPUTER GRAPHICS; DIGITAL COMPUTER PROGRAMMING; TOOLING in the McGraw-Hill Encyclopedia of Science and Technology.   [HERBERT W. YANKEE]

Bibliography: Happy marriage of CAD and CAM, *Machine Design*, 51(2):36–42, Jan. 25, 1979; J. Harrington, *Computer Integrated Manufacturing*, 1974, reprinted 1979; G. L. Petoff, The look of modern CAM standards, *Manufact. Eng.*, 82(4): 78–80, April 1979; Society of Manufacturing Engineers, *Numerical Control in Manufacturing*, 1975.

Infrared spectroscopy

Infrared lasers are being used to an increasing extent as sources for high-resolution spectroscopy in the gas phase. In some cases, infrared lasers with narrow line width can be tuned easily over wide spectral regions without the need for any monochromator. In other cases, where the laser emits only at certain discrete frequencies, high-resolution spectroscopy can still be accomplished by tuning molecular absorption frequencies into coincidence with those of the laser. Laser spectroscopy offers increases in both resolution and sensitivity compared with conventional (nonlaser) techniques.

Tunable infrared laser spectroscopy. After many years of development, three tunable laser sources, capable of producing high-resolution infrared spectra, are commercially available. Two of these, the spin-flip Raman laser and the color center laser, have yet to prove themselves as routine high-resolution instruments. However, the third device, the diode laser, is being used in an ever-increasing number of applications, from high-resolution spectroscopy to remote detection of atmospheric molecules. Another tunable infrared source, the frequency-difference laser, is currently used in several laboratories for high-resolution spectroscopy.

Diode laser. Diode lasers are now available over the 400- to 5000-cm^{-1} infrared region, although a single diode crystal will operate only over about a 100-cm^{-1} portion of this region. The device has an output power of about 0.25 mW and a line width (typically) of 15 MHz. This power level of 0.5 W per cm^{-1} frequency interval should be compared with that of a hot filament at 1800 K of 5×10^{-7} W/cm^{-1} at 10 μm. The wide tunability of diode lasers makes them ideal for high-resolution infrared spectroscopy. Infrared spectra of gas-phase samples obtained with diode lasers are limited in resolution only by the width imposed on the absorption by the thermal velocity spread of the molecules (that is, the Doppler effect).

Already a number of laboratories are using diode lasers for routine, high-resolution infrared spectroscopy (resolution of ≤ 0.002 cm^{-1} or ≤ 60 MHz). Doppler-limited vibration-rotational spectra of such species as CS, NO, ClO, CH_3F, HNO_3 and NH_3 and a pure rotational transition in H_2 have been analyzed to yield a wealth of molecular information at a precision impossible with a conventional infrared spectrometer (resolution of ≤ 0.03 cm^{-1}).

Perhaps the greatest spectroscopic achievement with a diode laser has been the work of R. S. McDowell, H. Flicker, B. J. Krohn, and others at the Los Alamos Scientific Laboratory. They have used diode lasers to resolve the extremely complex structure in some vibrational bands of OsO_4, CF_4, SF_6, and UF_6. These complex spectra from such heavy molecules are unresolvable with even the best conventional spectrometers. An example of a 0.05-cm^{-1} portion of the ν_3 band of $^{192}OsO_4$ obtained by McDowell and colleagues with a diode laser is shown in Fig. 1. In this 0.05-cm^{-1} region near the band center at 960.705 cm^{-1}, 34 transitions are predicted. Clearly in a conventional infrared spectrometer of relatively high resolution, say 0.05 cm^{-1}, all this information would be entirely unresolved. Even the combined effects of the small OsO_4 Doppler width and the slight laser instability mean that some of the transitions are still unresolved in Fig. 1.

The impetus for the high-resolution study of the infrared spectra of such species as OsO_4 and SF_6 comes from the fact that they undergo dissociation when subjected to high-intensity radiation from a pulsed CO_2 laser (at about the 1 GW/cm^2 level). In fact, under certain conditions, one isotopic form of the molecule will undergo dissociation faster than another molecule containing a different isotope. Thus the relative abundance within the reactants and products of the dissociation is different. The technique, proved to enrich S isotopes in SF_6 and Os isotopes in OsO_4, could have application to the enrichment of ^{235}U in UF_6. Since OsO_4, SF_6, and UF_6 have similar rotational structure in their vibrational spectra, OsO_4 and SF_6 have been extensively studied as models for the UF_6 system. Diode laser spectra provide information on the factors determining the efficiency of, and hence the economic viability of, isotope enrichment via laser techniques.

Frequency-difference laser. The frequency difference laser, developed by A. Pine at the Massachusetts Institute of Technology, operates from 2.2 to 4.2 μm at a power level of about 1 μW. Pine and others have used this device to obtain Doppler-limited spectra over this wide spectral region. The torsional splittings in the vibrational spectrum of ethane, caused by the rotation of the two methyl groups relative to each other, have been observed for the first time through using the frequency-difference laser spectrometer. Such studies will greatly improve the understanding of nonrigid effects in molecules. Other applications have included the study of small perturbations in vibration-rotational spectra of allene and phosphine and the observation of the very weak, electric quadrupole vibration-rotational spectrum of D_2.

Applications. Many studies with the diode and frequency-difference lasers have involved the measurement of molecular parameters that influence the propagation of infrared laser radiation through the atmosphere. Absorption coefficients and pressure broadening coefficients (with N_2 and O_2) for such atmospheric constituents as CO_2, H_2O and N_2O have been accurately measured in the 8-

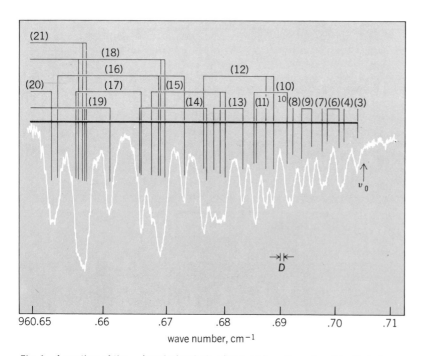

Fig. 1. A portion of the ν_3 band of $^{192}OsO_4$ observed by using a tunable diode laser. The rotational structure of part of the Q branch near the band center ν_0 is shown. D represents the Doppler width for $^{192}OsO_4$ at 245 K. The calculated spectrum is shown as a stick diagram above the observed lines. (*From R. S. McDowell et al., OsO_4 stretching fundamental, J. Chem. Phys., 69:1513–1521, Aug. 15, 1978*)

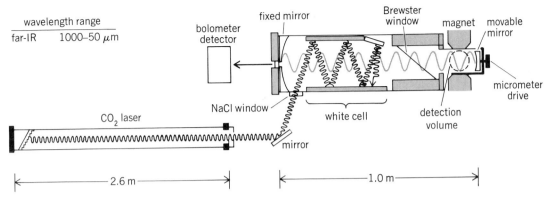

far-infrared laser

wavelength range
far-IR   1000–50 μm

Fig. 2. Diagram of a laser magnetic resonance spectrometer. The CO_2 laser pumps the far-infrared gas laser. By suitable choice of gas, the far-infrared laser will operate from 1000 to 50μm (10 to 200 cm⁻¹). (*From J. T. Hougen et al., Optically pumped laser magnetic resonance spectrometer, J. Mol. Spectrosc., 72:463–483,1978*)

to 12-μm region with diode lasers. Such data are required for quantitative interpretation of remote-sensing experiments on polluted atmospheres. Tunable infrared lasers have been used to study many molecules that pollute urban atmospheres, such as CH_4, H_2CO, H_2SO_4, SO_3, SO_2, O_3 and HNO_3. Sensitivities of 0.1 to 1.0 part per billion have been achieved under atmospheric conditions.

Stratospheric pollutants, responsible for destroying the protective ozone layer, such as CF_2Cl_2, ClO, and NO, have been studied in the laboratory by using tunable infrared lasers as a prelude for possible remote detection via laser heterodyne techniques. Remote heterodyne studies of stratospheric ozone have been made by using a diode laser as local oscillator and the Sun as the continuum source. Heterodyne detection in 200-MHz segments of a 0.8-cm⁻¹ (24-GHz) total range was possible because of (unlike the more normal CO_2 laser local oscillator) the tunability of the diode laser.

Spectroscopy with nontunable lasers. Most gas lasers, particularly in the far infrared, oscillate at only discrete, nontunable frequencies. To use these sources for high-resolution spectroscopy, molecular transitions must be tuned into coincidence with the laser frequency. Three methods of such tuning have been successful, namely, application to the molecules of electric or magnetic fields and, for ions, velocity tuning.

Laser magnetic resonance. The method, by which rotational or vibration-rotational transitions are magnetically tuned into coincidence with fixed-frequency far-infrared or infrared lasers, respectively, is called laser magnetic resonance (LMR) spectroscopy. It is similar to nuclear magnetic resonance (NMR) and electron spin resonance (ESR) techniques but operates at much higher fixed frequencies. Studies have shown that the limiting sensitivity is about 10^7 to 10^8 free radicals per cubic centimeter for detection via LMR. Free radicals studied by LMR in the gas phase include NH_2, PH_2, ND_2, CH_2, HO_2, DO_2, HCO, CH_3O, CH_2F, CCH, CF, CH, SH, SeH, NH, and OH.

A schematic diagram of the far-infrared LMR spectrometer used by J. T. Hougen, J. A. Mucha, D. A. Jennings, and K. M. Evenson for the study of the CH radical is shown in Fig. 2. A continuous-wave CO_2 laser pumps a gas contained in the far-infrared laser cavity and produces a population inversion between rotational levels of an excited vibrational state of the gas. Far-infrared laser action occurs between the rotational levels of the gas that have inverted populations, and is detected by the bolometer. CH is produced by the reaction of F atoms with CH_4 in a cell between the pole faces of the magnet and within the far-infrared

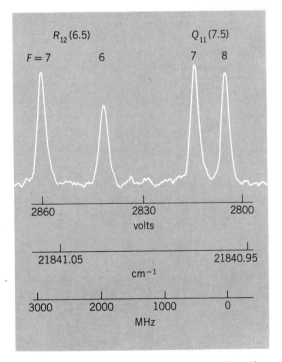

Fig. 3. Observed spectrum of ¹³C¹⁶O⁺ obtained by using the 457.9-nm argon laser line. The top scale shows the accelerating voltage required to bring the ion absorption frequencies into coincidence with the laser line at 21831.01 cm⁻¹. The middle scale is the unshifted frequency. (*From A. Carrington, D. R. J. Milverton, and P. J. Sarre, Electronic absorption spectrum of CO⁺ in an ion beam, Mol. Phys., 35:1505–1521,1978*)

laser cavity. The improved molecular parameters determined in this work for CH should be of use for further detection and study of CH in interstellar molecular clouds via radio astronomy. By using the LMR technqiue, many rate constants for reactions involving HO_2, HO, and C10 radicals of great importance in the stratospheric ozone cycle have been measured.

Gas-phase ions. Recently, further improvements in the high-resolution spectroscopy of gas-phase ions have been reported. Ions in a collimated beam are accelerated so that their Doppler-shifted, transitional frequencies are tuned into coincidence with a fixed frequency from an infrared or visible laser, whose direction is along the ion beam. Several ions have been studied by this technique, HD^+, H_2O^+, O_2^+, and CO^+. A. Carrington and P. J. Sarre have applied it to study of the $A\ ^2\Pi \leftarrow X^2\Sigma^+$ system of CO^+, using the 457.9-nm and 488.0-nm argon laser lines. They detected the absorption process by allowing the CO^+, after interaction with the laser, to charge-exchange with a thermal, CO gas sample. The cross sections for exchange when CO^+ is in the X or A state are substantially different. The resultant attenuation of the CO^+ in the ion beam is then studied with a mass spectrometer. In Fig. 3, an example of the attainable resolution is shown for $^{13}C^{16}O^+$. Two transitions, R_{12} (6.5) and Q_{11} (7.5), in which simultaneous changes occur in the rotational, vibrational, and electronic states of CO^+, are shown in this figure; each transition is split into a doublet by the interaction between the unpaired electronic spin and the ^{13}C nuclear spin. The observed line width of 160 MHz results primarily from the residual velocity spread in the ion beam. Such studies allow the determination of hyperfine coupling parameters free from the effects of perturbations present when ESR studies are undertaken on ions in the solid phase.

Carrington and Sarre have also studied the predissociation of electronically excited O_2^+ into O^+ as a function of vibration-rotational states of the O_2^+ by detecting the appearance of the O^+ with a mass spectrometer. The observed line width in the O_2^+ absorption process is inversely related to the lifetime of the state with respect to the predissociation process.

For background information *see* INFRARED SPECTROSCOPY in the McGraw-Hill Encyclopedia of Science and Technology.   [GRAHAM W. HILLS]

Bibliography: A. Carrington et al., *Mol. Phys.*, 35:1505–1521, 1523–1535, 1978; J. T. Hougen et al., *J. Mol. Spectrosc.*, 72:463–483, 1978; R. S. McDowell et al., *J. Chem. Phys.*, 69:1513–1521, Aug. 15, 1978; A. S. Pine et al., *J. Mol. Spectrosc.*, 74:43–51, 52–69, 1979.

Instrument landing system

The instrument landing system (ILS) has been the international standard system for final approach and landing at airports since about 1947. During the 1960s the large increase in air traffic movements—particularly of jet aircraft—led to the requirement for improved standards of performance.

The ILS comprises a localizer, which provides the azimuth guidance, and the glidepath, which provides the elevation guidance. A marker beacon is often included to give the pilot an indication of the distance to touchdown.

Localizer. The localizer is sited on centerline typically 300 m beyond the stop end of the runway. Good localizer performance can be assured by maintaining a segment of about 5° about the centerline clear of interfering obstacles. Buildings located within 5–10° sectors off centerline must be carefully orientated so that signal is not scattered into the approach path. Localizer antenna design has improved considerably over the years, and it is a relatively straightforward matter to achieve satisfactory performance on a typical site.

Glidepath. The glidepath is sited typically 300 m in from the start end of the runway and 150 m to the side. This system uses the ground in front of the antennas as a mirror to form "image" antennas. The basic system requires typically 600 m of flat ground directly in front of the antennas, and to ensure good performance there should not be any aboveground obstructions beyond this region. At major air terminals, ideal sites are seldom available, and considerable research effort has been conducted worldwide to develop glidepath systems which would operate with short ground planes or in the presence of obstructions.

Far-field techniques. The approach which has been used to design these image systems until very recently is summarized in Fig. 1. An individual antenna plus its image produces a simple repetitive interference pattern in the far field, as shown in Fig. 1a. The angular frequency of this pattern is governed only by the height h of the aerial. Design

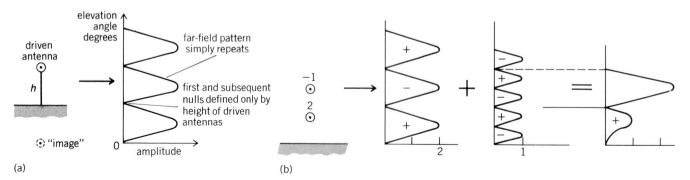

Fig. 1. Far-field design methods for image antenna systems. (a) System with single horizontal antenna, producing simple repetitive interference pattern. (b) System with two antennas, used to synthesize combination of interference patterns.

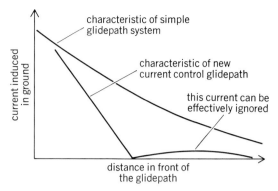

Fig. 2. Control of far ground currents.

techniques have previously synthesized combinations of these interference patterns, as shown, for example, in Fig. 1b, or have used other approaches based on far-field techniques. However, the 600 m of image ground is essentially a part of the radiating system, and if this ground requirement is to be reduced, near-field techniques must be used.

Near-field technique. A near-field technique has been developed by realizing that it is the currents induced in the ground plane by the driven antennas which reradiate to produce the effect of an image. When these ground currents are controlled by reducing the far currents as shown in Fig. 2, it is found that the resulting glidepath needs very little ground for satisfactory operation. This new glidepath uses the same hardware as the common M-array glidepath (also known as the capture-effect glidepath) with a simple rearrangement of the drives, and requires less than 300 m of image ground. This array, called the 3G, has been installed on a severe short-ground-plane site at Sydney (Australia) Airport.

Systems with obstructions. The second major problem is caused by aboveground obstructions in the forward region. To date, the M-array glidepath with its "scooped" pattern just above ground (Fig. 3) has been used on such sites; however, it has not been generally realized that the scooping in the pattern is achieved only in the far field of the system at distances beyond 5 mi (8 km) from the glidepath. Closer than this, the M-array scooping deteriorates, and close-in obstructions are then quite strongly illuminated, causing reflections which give rise to bends or even changes in the glidepath angle.

To circumvent this problem, it is possible to make use of focusing techniques which allow the far-field (perfectly scooped) pattern to be obtained in the near field. At the present time, lateral focusing is used to improve the glidepath quality all along the runway centerline approach and over the threshold of the runway. This is achieved by setting the individual antennas on a lateral arc with the runway centerline as center (Fig. 4). However, in directions away from the centerline the illumination is greater than if lateral focusing had not been used.

The simplest glidepath system (the null reference) has two driven antennas and does not defocus. It is possible to design a four-element glidepath which, like the null reference, does not need

lateral focusing. This new glidepath also has the desirable scooped pattern shown in Fig. 3. It is now possible to set the elements of this so-called 4F antenna on a forward arc so that the array is focused on one dominant aboveground obstacle, thus minimizing the illumination of that obstacle and its subsequent reradiation.

Implementation. One of the most difficult tasks of all is the implementation of glidepath arrays in the field. Following many hundreds of setups on a model glidepath system, a simple logical setting-up technique has been developed. The very simple errors which are so easy to overlook are summarized in Fig. 5: (1) The glidepath must be designed relative to the ground which may be sloping, rather than trying to force a result relative to an arbitrary horizontal reference (Fig. 5a). (2) It is common practice to mount glidepath poles vertically, which clearly results in offset images on a sloping ground plane (Fig. 5b). (3) An array must always be driven and treated as a complete unit; nothing is gained by setting up elements of an array one at a time when the site demands the performance of the complete array (Fig. 5c). (4) If the glidepath is mounted on a local high, the whole array must be lowered to maintain the necessary equispacing (Fig. 5d).

When these simple precautions are attended to, the best possible performance for the particular site will be achieved.

Conclusions. The improvements which are discussed here do not allow operation on the rare

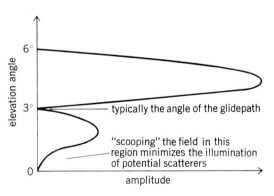

Fig. 3. Scooped interference pattern.

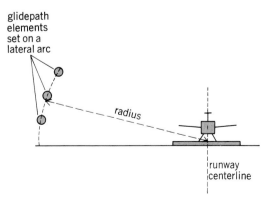

Fig. 4. Cross-sectional view of runway using lateral focusing.

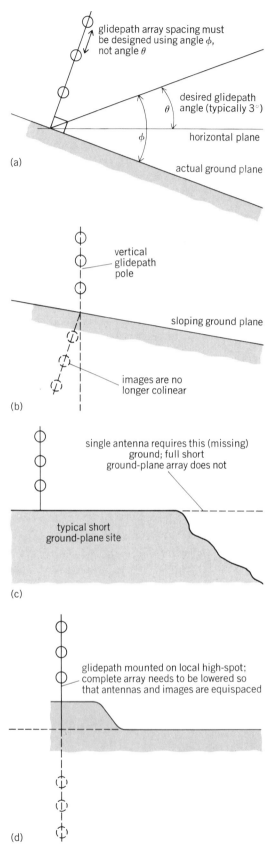

Fig. 5. Precautions in implementation of glidepath arrays. (a) Glidepath must be designed relative to actual ground plane rather than horizontal. (b) Glidepath poles must not be mounted vertically on sloping ground. (c) Array must be treated as complete unit. (d) Complete array must be lowered if mounted on local high.

sites where the ground up to the runway threshold cannot be used for imaging (as in a tidal situation with water directly in front of the glidepath), and they do not overcome the broader criticism that the ILS produces only a single line of guidance which insists that all aircraft queue one behind the other along the approach. The microwave landing system (MLS), which is presently being developed to replace the ILS, will provide positive guidance over a broad azimuth coverage (greater than ± 30°) with a highly accurate central region. This will allow the approach flexibility which is not provided by the ILS.

However, the present glidepath developments will, in the meantime, allow existing ILS operations to be improved, and will also allow the ILS to be implemented in many situations where this was previously impossible.

For background information *see* INSTRUMENT LANDING SYSTEM (ILS) in the McGraw-Hill Encyclopedia of Science and Technology. [J. G. LUCAS]

Bibliography: F. W. Iden, *IRE Trans.*, ANE-6: 100–111, 1959; J. G. Lucas, *Proc. Inst. Electr. Eng. (London)*, 119(5):529–536, 1972; J. G. Lucas and A. C. Young, *IEEE Trans.*, AES-14(6): 873–883, 1978; P. Sandretto, *Electronic Avigation Engineering*, 1958.

Interferon

Interferon is a protein, or more accurately a group of proteins, since several types are known which differ from one another physically, chemically, and biologically, produced by animal cells in response to virus infection. As illustrated in Fig. 1, during the course of viral replication (1), the initially infected cell is stimulated to produce interferon (2). Interferon is secreted by this cell (3), and is free to interact with other as yet uninfected cells (4). Interferon binds at the cell surface and induces the production of a new protein or polypeptide that establishes an intracellular state of antiviral resistance that inhibits replication of a wide range of viruses (5). In addition, it has recently been found that interferon interacts with cells that serve various host defense and immune functions. Such cells are macrophages (6), which are phagocytic cells responsible for clearing foreign substances and debris from the body; T lymphocytes (7), which can be cytotoxic to invading foreign cells and are involved in immunologic processes; and B lymphocytes (8), which are responsible for producing antibodies against foreign substances. These cells interact with one another and regulate the host's immune systems. It is now evident that interferon can affect each type of cell. For example, interferon can activate macrophages to attack and specifically kill tumor cells. It can also increase or decrease the ability of lymphocytes to make antibodies, and the ability of T lymphocytes to respond to foreign substances such as tumor cells. In brief, interferon produced by the originally infected cell can (a) protect other cells from virus infection, thereby limiting the spread and severity of virus infection, and (b) manipulate the host's immune surveillance cells to have an enhanced capacity to destroy foreign substances or cells such as tumor cells.

Classification and properties. Following stimulation of animal cells by a virus or other suitable substances such as mitogens, antigens, bacterial cell wall extracts, double-stranded ribonucleic acids (RNAs), or various low-molecular-weight molecules, various types of interferons can be induced. They vary from one to another, depending upon the cell type induced and the nature of the inducing agent. For example, it is now believed that basically two broad classes of interferons exist. Type I interferons are induced by stimuli such as viruses and are very stable, not losing biologic activity at very high or low pH. Interferons induced by mitogens or antigens, however, are different in that they are not stable at extreme pH, and are called type II or immune-induced interferons. In each class, subtypes of interferons exist. In type I interferons, both lymphocyte and fibroblast interferons are found that differ from one another antigenically and biologically. Basically, however, interferons have common properties that are used in the classification of a substance as an interferon. They are proteins with a molecular weight of 15,000–40,000 and are generally stable at temperatures of up to 56°C for an hour. Interferons are broad in spectrum and inhibit a wide range of viruses, not just the virus that induced their production. They are also species-specific. That is, interferons produced by mouse cells are most active in mouse or rat cells but not human cells; and vice versa, human interferons are active on human or monkey cells but not mouse cells. Therefore the treatment of humans requires human interferon. Interferons are remarkably nontoxic, though some effects on cellular macromolecular synthesis have been observed, but generally, at higher concentra-

tions than those required for establishment of, for example, the antiviral state. The antiviral or immune-modulating action of interferon is only short-lived, lasting up to several days.

Interferons have a broad spectrum of effects on cells. They initiate development of an intracellular antiviral state, modulate the immune response, can affect rates of cellular division and metabolic synthesis rates, and can affect cellular membranes, inhibiting the release of certain types of virus particles and inducing the production of prostaglandins, which are natural cellular products involved in various regulatory functions. It is possible that many of the non-antiviral properties attributed to interferon may in fact be aided or mediated by prostaglandins.

Potential. The very nature of interferons make them ideal broad-spectrum chemotherapeutic agents for the treatment of viral diseases. Such diseases as the common cold, influenza, herpesvirus infections, hepatitis, and shingles may be amenable to interferon therapy. Since interferon can influence the immune response, studies are presently under way to determine their value in organ transplantation, since the longevity of grafts may be prolonged and recurrent viral infections such as cytomegalovirus and herpesvirus, which are significant problems in transplantation, may be effectively suppressed by interferon. Its potential in the treatment of various types of cancer is also under clinical investigation.

Exogenous interferon. There are essentially two approaches being taken in the development of interferon. The first is the mass production of large quantities of human interferon for passive transfer, much the same as gamma globulin might be used. Essentially, three types of human cells are available for the production of this exogenous interferon: human fibroblasts, leukocytes, and lymphoblasts; each is currently being used. The major emphasis during the past several years has been the production of human leukocyte interferon, which has been going on in Finland under the direction of Kari Cantell. The buffy coats from all the blood donated to the Finish Red Cross are collected, and the white blood cells (primarily lymphocytes) are induced to make interferon with a virus. Presently the cost of producing this interferon is $40–50 per million units. Human fibroblast cells have also been used to produce human interferon, and at the present time the cost of producing interferon in this system is also about $50 per million units. The third source of human interferon has been the use of human lymphoblastoid cells (virus-transformed lymphocytes) grown in large fermentors. Through the use of 1000-liter fermentors, the cost of producing human lymphoblastoid interferon has been reduced to $10–20 per million units. Mass-producing human interferon for exogenous transfer is a costly process, and the amount of interferon being produced at present is barely enough for clinical evaluation, and not for routine clinical availability. A number of avenues are being explored to solve these problems, including: production of the interferon proteins by recombinant deoxyribonucleic acid (DNA) research; purification and sequencing the interferon molecule, thereby opening the way for synthesis of selected parts of the molecule; and development of addi-

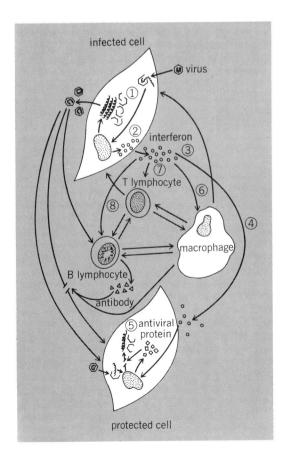

Fig. 1. Interferon induction and action.

Success of interferon and interferon inducers in humans

Disease treated	Dose	Route*	Result†
Interferon			
Influenza B virus	8×10^5 units	I.N.	±
Rhinovirus (common cold)	1.4×10^7 U	I.N.	+
Hepatitis B virus	6×10^3 to 1.7×10^4 U/kg/day (for months)	I.M.	+
Herpes zoster	5×10^5 U/kg/day (for 2–5 days)	I.M.	+
Herpes keratitis	3×10^6 U/ml (2 times daily)	Eye drops	+
Osteogenic sarcoma	2.5×10^6 U/day (3 times weekly for 1 year)	I.M.	+
Hodgkin's disease	5×10^6 U/day for 45 days and 7×10^6 U/day for 7 months (total $= 2 \times 10^9$ U per patient)	I.M.	+
Interferon inducers			
Rhinovirus	Poly I:C	I.N.	±
Influenza A$_2$	Poly I:C	I.N.	±
Rhinovirus	Propanediamine (CP20961)	I.N.	+

*I.N. = intranasal, I.M. = intramuscular.

†Here + = significant effect on disease state, ± = equivocal effect.

tional cells, for example, lymphoblastoid cells, as a source of producing human interferon in large fermentor-type operations.

Interferon inducers. The second approach to interferon production is the development of nontoxic compounds that can be given to animals to induce their own cells to produce interferon, thereby circumventing the requirement of the very costly process of mass-producing interferons. A number of agents are now known to be effective inducers, and four of these are illustrated in Fig. 2.

A wide spectrum of double-stranded RNA molecules of synthetic or natural origin are known to be active interferon inducers. One of these, polyriboinosinic: polyribocytidylic acid (abbreviated poly I:C) has been extensively studied. Although it induces high levels of interferon in various experi-

mental animals, poly I:C is not as active or as well tolerated in humans. Therefore, at the dosage levels given, only limited amounts of interferon have been induced in various clinical studies. Recently, development of this agent has branched in two diverse directions. A group at the Johns Hopkins University has developed more labile analogs of poly I:C that are less toxic but, it is hoped, have maintained their interferon-inducing activity. Going in the opposite direction, a group at the National Institutes of Health headed by Hilton Levy has developed a more stable analog of poly I:C (poly I:C lysine) that shows improved interferon-inducing activity. So far, this analog has induced high levels of interferon in volunteers and mediated antiviral resistance in monkeys infected with yellow fever virus.

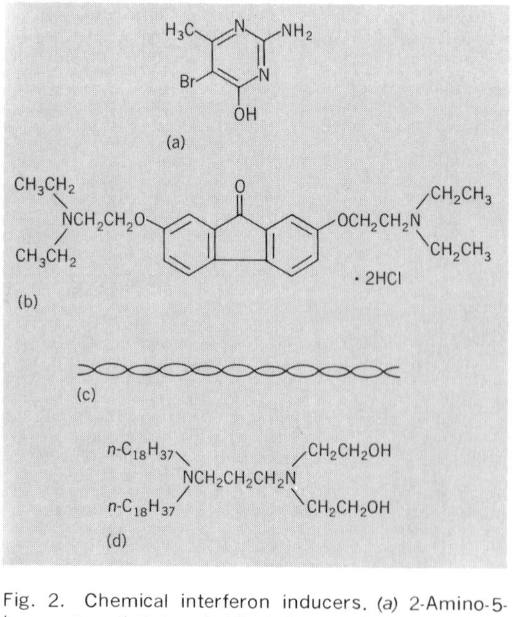

The other compounds illustrated in Fig. 2 have not been developed to the same extent as poly I:C. Tilorone hydrochloride is a compound that induced high levels of interferon in mice but was inactive in monkeys, humans, or various domestic animals. It was an important development, nevertheless, since it was the first inducing agent found to be effective when administered orally. CP20961 is a propanediamine that has induced high levels of local interferon in human volunteers who were given the drug intranasally, and that has, to a limited degree, been effective in humans infected with rhinoviruses (common cold viruses). U25166 is a pyrimidine molecule that is a very active interferon inducer in a variety of animal species, including domestic animals, but has not yet been tested in humans.

Clinical utilization. Both interferon and interferon inducers are presently under clinical investigation for the treatment of viral infections and various types of neoplasias as summarized in the table. Exogenous interferon has been effective in preventing or treating human rhinovirus infections. Likewise, high doses of interferon given systemically (intramuscularly) to patients with chronic hepatitis B virus infections have decreased the virus carrier rate, and in Sweden patients with osteogenic sarcoma have had a longer life-span and less metastasis than patients not treated with interferon. These data indicate a potential for inter-

Fig. 2. Chemical interferon inducers. (a) 2-Amino-5-bromo-6-methyl-4-pyrimidinol (U25166). (b) Bis-diethyl amino ethyoxy-fluorenone (tilorone hydrochloride). (c) Polyriboinosinic: polyribocytidylic acid (poly I:C). (d) N,N-dioctadecyl-N',N'-bis (2-hydroxyethyl) propanediamine (CP20961).

feron. However, in each study, comparatively large doses of interferon were required to be effective.

Conclusions. In summary, interferons are an intriguing group of molecules that appear to have a number of effects on animal cells. They modulate an intracellular antiviral state that effectively blocks virus replication. They can activate macrophages to remove foreign particles and destroy tumor cells and can modulate the immune response. By their very nature, they appear to have a future as effective chemotherapeutic agents in the treatment of viral infections, specific types of neoplasias, and modulation of the immune response. However, practical application of the interferon system has been difficult to achieve. During the past 20 years a number of approaches have been taken to development of the interferon system. Yet, at present, none have resulted in routine clinical availability of interferon. Recent progress in developing methods of producing exogenous interferon and in discovering new interferon inducers may solve many of the problems encountered in the past. Ongoing clinical trials should answer questions as to whether interferon will be effective in the treatment of human diseases.

For background information see INTERFERON in the McGraw-Hill Encyclopedia of Science and Technology.   [DALE A. STRINGFELLOW]

Bibliography: E. A. Havell, T. G. Hayes, and J. Vilcek, *Virology*, 89:330–334, 1978; H. M. Johnson, G. J. Stanton, and S. Baron, *Proc. Soc. Exp. Biol. Med.*, 154:138–141, 1977; R. B. Pollard and T. C. Merigan, *Pharmaceut. Ther.*, 2:783–811, 1978; R. M. Schultz et al., *Science*, 202:320–321, 1978.

Interstellar matter

During the 1970s, radio astronomers and chemists discovered a new form of matter in space, the interstellar molecules. Though only a trace component of the interstellar matter by mass, these molecules have stimulated great advances in knowledge of the birth and death of stars, the motion and structure of the Milky Way, the chemical evolution of life on Earth, and the search for life elsewhere.

Light-years of space separate stars in the Galaxy. That nearly empty space is filled with a very dilute gas of hydrogen atoms typically a centimeter apart. Even more rare than the gas are tiny specks of stardust, a fine smoke of sooty and sandy particles blown out from dying stars, with only one dust grain for every 10^{12} hydrogen atoms. Most of the mass of the Galaxy glows as its 10^{11} stars. The interstellar matter, representing only 5–10% of the galactic mass, is almost entirely gaseous hydrogen and helium. The interstellar dust, containing most of the elements heavier than helium, represents only 1% of the interstellar mass. The gas and the dust clump together in ragged patches, called clouds, extending over light-years.

Interstellar molecules. Since the detection in 1969 of the unique radio signature of gaseous formaldehyde in interstellar clouds, the new field of astrochemistry has grown rapidly. Currently, more than four dozen different interstellar chemical species have been identified. The list of known molecules (see table) has been growing by more than four new species each year over a decade.

Note that most of these molecules are organic (that is, contain carbon) and that oxides of nitrogen are quite uncommon. The list includes molecules containing hydrogen, carbon, nitrogen, oxygen, sulfur, and silicon. Searches are under way for species containing the elements phosphorus, chlorine, iron, and magnesium.

Some of the observed interstellar species are familiar, terrestrial chemicals: carbon monoxide, water vapor, ammonia, methyl and ethyl alcohol, and formic acid. But many on the list are truly exotic, nonterrestrial species previously unknown to chemists, reflecting the unusual conditions in interstellar clouds. One focus of current interest is on the series of long, carbon-chain species detected in a dark cloud in the constellation Taurus: CN, HCN, HC_3N, HC_5N, HC_7N, HC_9N. The longest member of this sequence, cyanotetraacetylene, can be represented as

$$H—C≡C—C≡C—C≡C—C≡C—C≡N$$

With 11 atoms, HC_9N is currently the most complex interstellar molecule known. How such a

Interstellar molecules

Molecule	Chemical symbol	Year of discovery	Part of spectrum
Methylidyne	CH	1937	Visible
Cyanogen radical	CN	1940	Visible
Methylidyne ion	CH^+	1941	Visible
Hydroxyl radical	OH	1963	Radio
Ammonia	NH_3	1968	Radio
Water	H_2O	1968	Radio
Formaldehyde	H_2CO	1969	Radio
Carbon monoxide	CO	1970	Radio
Hydrogen cyanide	HCN	1970	Radio
Cyanoacetylene	HC_3N	1970	Radio
Hydrogen	H_2	1970	Ultraviolet
Methyl alcohol	CH_3OH	1970	Radio
Formic acid	HCOOH	1970	Radio
Formyl ion	HCO^+	1970	Radio
Formamide	$HCONH_2$	1971	Radio
Carbon monosulfide	CS	1971	Radio
Silicon monoxide	SiO	1971	Radio
Carbonyl sulfide	OCS	1971	Radio
Acetonitrile	CH_3CN	1971	Radio
Isocyanic acid	HNCO	1971	Radio
Methylacetylene	CH_3C_2H	1971	Radio
Acetaldehyde	CH_3CHO	1971	Radio
Thioformaldehyde	H_2CS	1971	Radio
Hydrogen isocyanide	HNC	1971	Radio
Hydrogen sulfide	H_2S	1972	Radio
Methanimine	H_2CNH	1972	Radio
Sulfur monoxide	SO	1973	Radio
Diazenylium	N_2H^+	1974	Radio
Ethynyl radical	C_2H	1974	Radio
Methylamine	CH_3NH_2	1974	Radio
Dimethyl ether	$(CH_3)_2O$	1974	Radio
Ethyl alcohol	CH_3CH_2OH	1974	Radio
Sulfur dioxide	SO_2	1975	Radio
Silicon sulfide	SiS	1975	Radio
Vinyl cyanide	H_2CCHCN	1975	Radio
Methyl formate	$HCOOCH_3$	1975	Radio
Nitrogen sulfide	NS	1975	Radio
Cyanamide	NH_2CN	1975	Radio
Cyanodiacetylene	HC_5N	1976	Radio
Formyl radical	HCO	1976	Radio
Acetylene	C_2H_2	1976	Infrared
Carbon	C_2	1977	Infrared
Nitroxyl radical	HNO	1977	Radio
Cyanotriacetylene	HC_7N	1977	Radio
Cyanotetracetylene	HC_9N	1978	Radio
Butadiynyl radical	C_4H	1978	Radio
Nitric oxide	NO	1978	Radio
Methane	CH_4	1978	Infrared

complicated molecule forms in space is still unknown to astrochemists.

The interstellar molecules discovered in recent years constitute a very dilute organic soup in the dust clouds where they are found. Relative to the molecular hydrogen gas in these clouds, the organic gases are just a trace, ranging in abundance from carbon monoxide, at about 30 parts per million (3×10^{-5}), down to cyanotetraacetylene (HC_9N), just detectable at a few parts per trillion (2×10^{-12}). In general, more complicated species are less abundant, which means that future searches with more sensitive radio telescopes will detect even more complicated species.

There is an intriguing similarity between the kinds of molecules found in space and those found in comets and in laboratory simulations of origin-of-life conditions on the primitive Earth. Biologists, chemists, and astronomers are working together to measure the radio signature of the simpler amino acid and deoxyribonucleic acid (DNA) base molecules and then to seek these species in space. The interstellar search for the simplest amino acid, glycine, has already begun.

Star formation. Interstellar molecules send radio messages from the centers of collapsing dust clouds, messages which radio astronomers are reading to learn how new stars are born. Since interstellar dust blocks starlight, optical astronomers can follow the process of stellar evolution only once an infant star has blown away its surrounding placental dust cloud and shown its true color. Before this time, when the star "turns on," all is hidden behind a dusty veil which only radio and infrared radiation can penetrate.

Both infrared and radio astronomy have been applied to the mystery of the early stages of star formation. Infrared measurements tell more about interstellar dust grains, while radio astronomy is most useful for gaseous molecules. Each different molecule requires different conditions of temperature and gas density to produce its characteristic radio signal. By carefully mapping the signals from several different molecules over the same area of a contracting dust cloud, radio astronomers can determine the variation of temperature, gas density, and internal motion throughout these regions of star birth. For example, unusually intense radio signals from steam (hot water vapor) generally mark the dense core of the collapsing cloud, while carbon monoxide signals outline the cooler, extended borders of the cloud.

Current theoretical models link the collapse of clouds with their chemical evolution over times measured in millions of years. Molecules formed in clouds likely foster the formation of new stars by sending the collapse energy out of the cloud in the form of radio waves.

Galactic structure. Since stars occur in groups, star formation is not an isolated, solitary process. Radio mapping of interstellar carbon monoxide is revealing a complex process in which clouds can collide, contract, break into smaller cloudlets, or evaporate. Waves of star birth seem to sweep through dust clouds as shock waves from each generation of stars trigger the cloud contractions leading to the next generation.

On an even larger scale, radio astronomers are tracing out galactic patterns of star formation in dark clouds which seem to be associated with the overall spiral structure of the Milky Way. In particular, galactic observation of dust clouds by radio emission from carbon monoxide has revealed a unique and unexpected picture of the Galaxy. Unlike the pancake-shaped model that emerged from galactic surveys of radio signals from atomic hydrogen, the molecular hydrogen picture of the Milky Way, derived from the CO observations, can best be viewed as a giant doughnut, a great ring of star-bearing molecular clouds with the Sun on the outer edge (see illustration). Most of the interstellar mass is concentrated in these giant molecular hydrogen clouds, each typically 100,000 times more massive than the Sun. Many thousands of such giant clouds make up the ring of star birth, which can also be seen in the galactic distribution of other tracers of stellar activity: x-rays, cosmic rays, supernova remnants, and pulsars. It is now known that from the location of the Sun inward toward the center of the Galaxy, the interstellar hydrogen gas is mostly in molecular form, while going away from the center, the hydrogen is primarily in atomic form.

Elements and isotopes. The blockage of starlight by the interstellar dust grains limits the view to less than a few thousand light-years in the plane of the Milky Way. The Galaxy, as defined in hydrogen atoms, is shaped like a great spinning phonograph record 100,000 light-years across, with the Sun located in the plane of the record, a little more than halfway out from the center.

When the Milky Way formed some 1.2×10^{10} years ago out of a cloud of primordial hydrogen and helium, there was no interstellar dust. The elements frozen in the dust grains were produced by nuclear fusion reactions in earlier generations of massive stars and were recycled to interstellar space to await the next generation of stars. The carbon, nitrogen, and oxygen comprising the human body had a similar origin; humans are, in fact, stardust.

In the process of burning lighter elements to

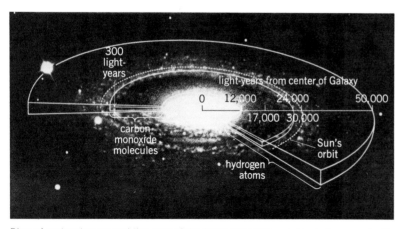

Ring of molecules around the core of the Milky Way Galaxy extends from a galactic radius of about 12,000 light-years (1.1×10^{20} m) to one of 24,000 light-years (2.3×10^{20} m), attaining its highest density at approximately 17,000 light-years (1.6×10^{20} m). (From M. Gordon and B. Burton, Carbon monoxide in the galaxy, Sci. Amer., 240(5):54–67; copyright 1979 by Scientific American, Inc.; all rights reserved)

heavier elements, some chemical elements are produced with more than one stable form or isotope, for example, ^{12}C and ^{13}C. The relative amounts of different isotopes of elements in the interstellar matter can be accurately determined by carefully measuring the different radio signals that isotopically substituted interstellar molecules emit, for example, ^{12}CO and ^{13}CO. In just this way, astrochemists have recently been able to determine for the first time the relative abundances of isotopes of hydrogen, carbon, nitrogen, oxygen, silicon, and sulfur at different locations around the Galaxy. The completely unexpected result is that, with the exception of the unusual region near the center of the Milky Way, the isotope ratios are nearly constant throughout the Galaxy and quite similar to those on Earth, or in the human body.

The one important exception to this general conclusion is heavy hydrogen, or deuterium, which becomes concentrated in certain interstellar molecules 100 to 10,000 times over the natural or cosmic ratio, $(D/H) \sim 10^{-5}$. This natural isotopic enrichment process offers important clues to understanding the chemical pathways by which interstellar molecules are made and broken.

Extragalactic chemistry. Very recently the same radio telescopes used to discover and map interstellar molecules throughout the Galaxy have also detected faint radio signals from these same chemical species in other galaxies, signals emitted more than 10^7 years ago. Such species include OH, H_2O, H_2CO, CO, HCN, and HCO^+. So far at least, it seems that interstellar chemistry in other galaxies proceeds much as it does in the Milky Way. This raises the question as to whether life elsewhere, if it exists, chemically resembles life on Earth.

For background information *see* GALAXY; INTERSTELLAR MATTER; RADIO ASTRONOMY in the McGraw-Hill Encyclopedia of Science and Technology.   [RICHARD H. GAMMON]

Bibliography: R. Gammon, *Chem. Eng. News*, 56:20–33, 1978; M. Gordon and B. Burton, *Sci. Amer.*, 240(5):54–67, 1979; F. Hoyle, *Mercury*, p. 2, January 1978; B. E. Turner, *Sci. Amer.*, 228(31): 51–69, 1973; M. Zeilick, *Sci. Amer.*, 238(4): 110–118, 1978.

Invertebrate architecture

Evidence of the architectural skill and activity of invertebrate animals is of commonplace occurrence. The webs of spiders, hives of bees, nests of wasps and hornets, and mounds or gallery systems of ants and termites are well known—largely because of their cosmopolitan distribution and proximity to human habitation. Such external constructions, as they are termed, are the focus of interest of investigators in a variety of disciplines, ranging from animal behavior and evolution to polymer chemistry and materials science. Although arthropod external constructions—particularly those of the social insects—have attracted considerable attention, knowledge of invertebrate architecture in the lower phyla is much more fragmentary. An exception to this generalization is the marine annelid worm *Chaetopterus variopedatus*. Because of its worldwide distribution along sand and mud regions of the intertidal zone, its relatively large size and bizarre appearance, and its distinctive U-shaped tube house, *Chaetopterus* has been the object of study by researchers from a number of countries for nearly a century. Recent reinvestigation of the behavior of the animal, combined with new data on the structural, chemical, and physical

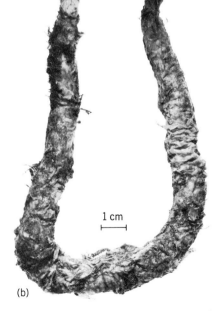

Fig. 1. Tube house of *Chaetopterus variopedatus*. (a) Cutaway view of worm inside its U-shaped tube house in the intertidal mud flat habitat. (*American Museum of Natural History*). (b) External view of complete tube house.

properties of its tube house, reveals for the first time a relatively complete picture of a marine invertebrate "architect" and its external construction.

Chaetopterus life history and habits. *Chaetopterus* spends its entire postlarval life within the confines of its cylindrical U-shaped tube house which, except for the two protruding terminal sections or "chimneys," lies buried within the sand and mud substratum (Fig. 1a). Unlike the majority of tubicolous marine worms, *Chaetopterus* lacks any kind of protrusible tentacular apparatus for obtaining food and oxygen from the surrounding waters. Instead, it possesses three highly modified segments in the middle of its body which act to pump a current of nutrient-laden water through the tube. The worm ingests the small food particles which it filters from this water current by means of a porous mucous sheet held across the lumen of the tube. The house of *Chaetopterus* thus functions not only as a protective enclosure but also as a hydraulic conduit essential for food gathering and respiration. The maintenance of its tube house free of any obstacle which would impede this flow of water is clearly of major importance to the animal, and recent investigations by S. C. Brown and J. S. Rosen have shown that the worm utilizes no less than four complex behavioral sequences specifically to cleanse the tube of potentially clogging suspended material.

Tube house construction and remodeling. The basic mechanism of tube formation has been deduced from observations of the behavior of worms in natural and experimental situations, as well as analyses of the structure of completed and partially completed tube houses. The wall of a tube house of an adult *Chaetopterus* is several millimeters thick and is composed of many concentric layers of a thin, parchmentlike material. At the outer surface, the layers are partially separated and torn in places, and are typically infiltrated with sand and mud from the surrounding region. In addition, the outer layers harbor a microfauna and microflora which, as shown by J. Kohlmeyer, are responsible for continually degrading the tube material, thus producing the shaggy, "old-tree-bark" external appearance of the tube (Fig. 1b). In contrast, the inner wall of the tube is invariably tightly layered, smooth, and intact, because of the continuous secretion of a thin layer of tube material from the anterior ventral region of the worm. The animal thus continually lines the inner surface of its tube as it moves back and forth, turns around, and rotates within its cylindrical home.

The general sequence of activity leading to a "steady-state" tube is shown in Fig. 2. Provided that the rate of application of new tube material along the inner surface equals the rate of degradation along the outer surface, the tube wall will remain of relatively constant thickness. However, over time this process results in a steady decrease in the internal diameter of the tube—a problem compounded by the steady increase in size of the worm. The animal's solution to this progressively intolerable situation is as remarkably simple as it is effective. On either side of its fourth body segment are located sets of (six to eight) heavily thickened, spear-shaped setae. With these, the animal periodically makes a longitudinal incision through the tube wall. The worm then forces the walls apart slightly by expanding and flattening its body. It finally lays down several layers of new tube material, thus sealing the rent and stabilizing the wall in its expanded condition.

The task of remodeling the tube house to accommodate the increasing length of the worm is handled in a somewhat similar manner, but with certain additional features. The tube-lengthening sequence begins with an incision through the tube wall, usually at a point of major inflection where one of the upper "arms" of the tube originates from the horizontal section. The worm then protrudes its anterior region through the newly formed opening and excavates a small cavity into the surrounding sand and mud substratum, secreting a layer of tube material as it progresses. The mud and sand particles loosened by the excavation process are passed backward into the tube and ultimately ejected through the excurrent opening. The worm continues the process until it reaches the mud-water interface, whereupon it constructs a terminal chimney. At this stage the tube has

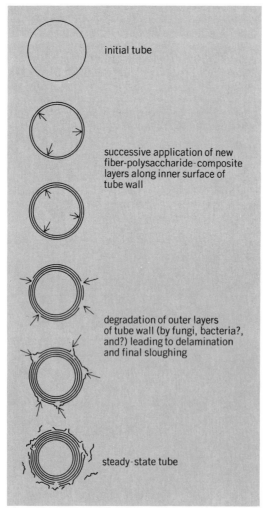

initial tube

successive application of new fiber-polysaccharide-composite layers along inner surface of tube wall

degradation of outer layers of tube wall (by fungi, bacteria?, and?) leading to delamination and final sloughing

steady-state tube

Fig. 2. Basic sequence of tube construction.

three openings, but the worm rapidly seals off the extra arm where it joins the main section. This isolated arm, now lacking the cleansing and renewal activities of the worm, rapidly fills with mud and is ultimately degraded completely.

Tube house microstructure and chemical composition. Studies by Brown and S. M. McGee-Russell have shown that each macroscopic layer of the tube has a complex and highly organized substructure. The tube material is a two-component system consisting of alternating plies of fibers embedded within an unstructured matrix (Fig. $3a-c$). Although the fibers within a given ply run strictly parallel to one another, the fiber orientation in successive plies may vary from parallel to orthogonal. In addition, there appears to be considerable variation in fiber diameter, ply thickness, and relative proportion of fibers to matrix.

The fibers are a highly ordered protein, as indicated by their banded appearance in electron micrographs (Fig. $3d$). Moreover, amino acid analysis of the tube fibers shows them to belong to the keratin family of proteins, with nearly 33% of the residues as acidic or basic amino acids, and 13% as the sulfur-containing amino acid, cysteine. Chemical separation of the fibers from the matrix has been achieved by the use of a chelating agent which removes ionically bound calcium and magnesium. Analysis of the released matrix material by H. Zola has shown it to be a phosphorylated carbohydrate polymer containing mannose, glucose, xylose, and an unidentified hexose phosphate.

From these structural and chemical data, certain inferences can be drawn concerning the macromolecular organization of the tube material. For instance, it is clear that both fiber and matrix components bear a strong negative charge under natural conditions in sea water. This fact, together with the known presence of relatively large amounts of bound divalent cations, and the demonstrated efficacy of chelating agents to separate matrix from fibers, suggests that calcium and magnesium act to form ionic bridges between the phosphate groups of the matrix polysaccharides, as well as between these polysaccharide phosphates and the carboxyl groups of the fiber proteins. The multiple layers of oriented fibers are thus bound together and stabilized by the polysaccharide phosphate –divalent cation adhesive or binder. At present, the cellular secretory activities leading to the production of the fiber and polysaccharide layers are not known in detail, although it appears certain that the orientation of the fiber plies is determined by the direction of the worm's locomotion as it secretes the tube material.

Chaetopterus tube – a "space age" material. Both in composition and fabrication, the tube house of *Chaetopterus* bears an uncanny resemblance to certain of the synthetic nonwoven fabrics and fiber-reinforced composite materials developed by industry in recent years. The combination of low-elasticity fibers well bonded to each other through an amorphous polymer network is similar to a number of epoxy resin–glass filament systems, of which fiber glass (in its many forms) is perhaps the best-known example. (In this connec-

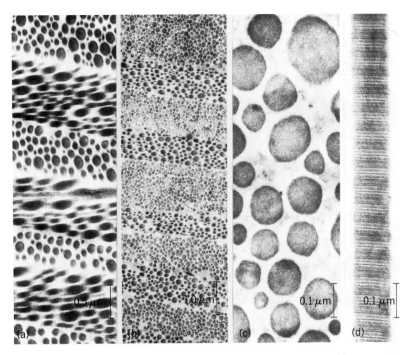

Fig. 3. Ultrastructure of wall of tube house showing two-component (fiber and matrix) composition. (*a*) Adjacent plies oriented at about a 45° angle; (*b*) adjacent plies with parallel orientation; (*c*) matrix slightly stained, range of fiber diameters evident; (*d*) fiber in longitudinal section, showing regular periodic banding.

tion it is of interest that the complex Young's modulus for *Chaetopterus* tube material has been found to be 3×10^9 dynes/cm², a value equivalent to that of a 60% resin–glass composite.) In addition, the layering of thin sheets of fiber-containing material in multiaxial orientation is a now-common industrial procedure utilized to produce laminated structures of great strength. Even the basic process of tube house formation – the application of fiber and matrix strips onto a preexistent "mold" (the inner surface of the tube) – is identical in principle to the fabrication of semirigid objects of complex curvature (such as boat hulls and automobile bodies) by successive applications (so-called "layups") of fiber reinforced composite mats or tapes.

Detailed analysis of external constructions is of major importance in furthering understanding of the biology of invertebrate "architects." Such research has also revealed that the basic sequence of the papermaking process (wasps), the high strength-weight advantage of honeycomb construction (bees), and, now, the construction potential of fiber-reinforced laminated sheets (*Chaetopterus*) have antedated each industrial "rediscovery" by millions of years. More such surprises await future researchers.

For background information *see* ANNELIDA; POLYCHAETA in the McGraw-Hill Encyclopedia of Science and Technology.   [S. C. BROWN]

Bibliography: S. C. Brown and S. M. McGee-Russell, *Tissue and Cell*, 3:65–70, 1971; S. C. Brown and J. S. Rosen, *Anim. Behav.*, 26:160–166, 1978; J. Kohlmeyer, *Mar. Biol.*, 12:277–284, 1972; H. Zola, *Compar. Biochem. Physiol.*, 21:179–183, 1967.

Irrigation of crops

Recent developments in crop irrigation have been concerned with harnessing the energy of the wind to power irrigation pumps, and the use of trickle-drip irrigation for soil water managment.

Irrigating with wind power. Windmills have been used for centuries to grind grain and pump water. Many are still used today for pumping livestock and domestic water supplies. These units normally produce a maximum power of 1 kW and pump less than 3 m³/hr. However, new wind machines are being developed that produce 10–200 kW. These new machines are capable of powering large pumps used for irrigation pumping.

In the United States about 27% of agricultural products grown today, on about 12% of the cropped lands, are irrigated. However, irrigation requires large amounts of energy to pump adequate water. An estimated 87×10^9 kWh of energy were used during 1978 for pumping irrigation water, or enough energy for 4,000,000 homes. Only energy used for manufacturing fertilizer and for tractor and truck fuel exceeds that used in irrigation, making irrigation the largest on-farm, nonvehicular use of energy in agriculture.

Most irrigation pumping is done with electricity, natural gas, or diesel fuel. All of these energy supplies are rapidly increasing in price and are decreasing in supply, thus creating interest in new or alternate energy sources for irrigation pumping.

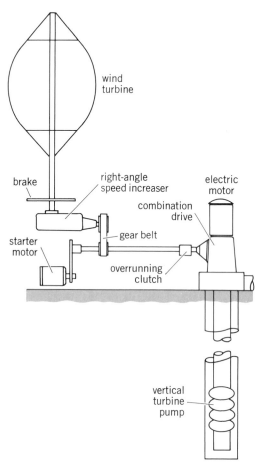

Fig. 1. Schematic of the wind-assisted irrigation pumping system.

Wind is being studied as an alternate power source because most irrigated areas are in windy regions.

Irrigation requires large amounts of water, often supplied by deep wells. Flow rates usually range between 100 and 500 m³/hr, and water is often pumped from wells ranging from 20 to 200 m deep. These wells require power units between 20 and 200 kW to lift the water to the surface and distribute it to the land.

Experimental pumping system. The USDA Science and Education Administration, in cooperation with the U.S. Department of Energy, has been testing an experimental wind-powered irrigation system since early 1977. This pumping system (called a wind-assist system) uses both a wind turbine and an electric motor to power a conventional vertical-turbine irrigation pump (Fig. 1). The electric motor is large enough to operate the pump by itself and runs continuously. The wind turbine is coupled to the pumping system through an overrunning clutch so it will furnish power to the pump only when the wind speed exceeds 6 m/s. When the wind turbine operates, the electric motor is not being replaced, but the electric load is being reduced. This power arrangement allows water to be pumped at the desired flow rate, regardless of the wind power level, and a conventional irrigation well and vertical-turbine pump can be used without modification.

The wind turbine is a Darrieus, or vertical-axis, type that was designed to produce 40 kW in a 15-m/s wind (Fig. 2). The rotor height is 17 m, and the maximum rotor diameter is 11.3 m. The rotor is set on top of a 9-m steel tower and is supported at the top by four 22.2-mm steel cables. When the turbine is producing power, the rotor turns at a steady speed of 90 rpm. A speed increaser and right-angle-gear drive increase the shaft speed from 90 to 1780 rpm (Fig. 1). An overrunning clutch in the wind turbine drive shaft transmits power to the pump without transferring power from the electric motor back to the wind turbine at low wind speeds. Between the electric motor and the pump discharge head is a combination gear drive that permits power to be supplied to the pump from two independent power sources (Fig. 1). A 750-mm-diameter disk brake, with three double-action calipers, is used to stop the rotor in normal or emergency shutdown.

The entire pumping system was assembled from commercially available equipment, and much of the system was installed in an existing irrigation well. The 200-mm vertical-turbine pump was installed in 1964 and is a type commonly used in irrigation pumping. The well supplies 104 m³/hr of water and has a 105-m total dynamic head. Neither the pump nor the well was modified for use here.

Performance. The vertical-axis wind turbine begins to produce power at 6 m/s wind speed, and the power output continually increases until the system is stopped at 20 m/s wind speed. In the Texas Panhandle, where the experimental turbine is located, 90% of the available wind power is within this wind speed range. The power harvested by a wind turbine increases proportionally to the cube of the wind speed. Thus, as wind speed increases, the wind power produced increases rapidly (Fig. 3). With the wind-assist pumping system, as wind turbine power increases, the electric power con-

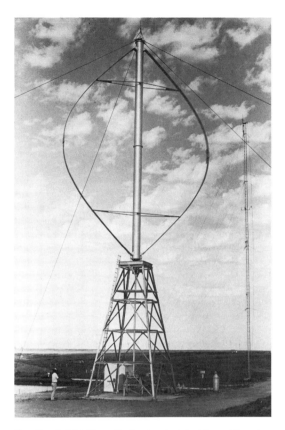

Fig. 2. A 40-kW vertical-axis wind turbine with rotor height of 17 m and maximum diameter of 11.3 m, installed at USDA Southwestern Great Plains Research Center, Bushland, TX.

sumed decreases, until the electric motor is almost completely unloaded. The sum of the wind power and electric power represents the steady power needed to operate present irrigation pumps. For the experimental system shown in Fig. 3, the pump requires a constant 51 kW to lift the water to the surface. At low wind speeds water is pumped primarily with electricity, whereas at high wind speeds water is pumped primarily by wind power.

A normal irrigation season usually requires about 2000 hr of pumping during the spring and summer months. In the Southern Great Plains, between March 1 and October 1, wind speed exceeds 6 m/s at least 3000 hr—time that could be used for irrigation pumping with wind energy. During this period, the experimental turbine could supply 40,000 kWh of power, or about 40% of that required by the irrigation pump. Presently, all the water needed for irrigation cannot be supplied with wind energy, simply because the wind does not blow all of the time during peak crop water-use periods. However, temporary storage of water in surface reservoirs can increase the percentage of irrigation water supplied by wind power.

The wind-assist pumping system has potential for providing an alternate energy source for irrigated agriculture in areas with average wind speeds exceeding 7 m/s. New wind turbines which are being developed should be economically priced and should provide farmers with a good power unit for their pumps.   [R. NOLAN CLARK]

Trickle-drip irrigation. The use of trickling or dripping as a method of irrigating large fields has

become common practice in agricultural production all over the world. A trickle irrigation system consists of emitters which distribute the water. The trickle emitter is essentially a water energy reducer acting as a resistor which dissipates the energy of the flowing water and thus reduces the flow rate to a given discharge.

The surface approach in which the irrigation emitter is placed directly on the soil surface, as opposed to underground application of trickling (subsurface irrigation), has been successfully and widely used. Optimal irrigation criteria are achieved by adjusting the trickling system to the soil hydraulic properties and to the water and nutritional requirements of the specific crop.

Potential advantages. The principal advantages of using trickle-drip irrigation are:

1. Improving soil water regime for greater crop yield. Trickle irrigation systems are capable of delivering matter to the soil in small quantities as often as desired with no additional cost. As the frequency of irrigation increases, the time-average soil water potential increases and is restricted to a narrow range, hence eliminating low average soil water content and high water fluctuations as factors restricting plant growth and crop yields.

2. Minimizing the salinity hazard to plants. This is related to the displacement of salts beyond the main efficient root zone and to the lowering of salt concentration by maintaining a relatively high soil water content due to the high-frequency irrigation.

3. Partial wetting of the soil volume. Water supply is restricted to those parts of the soil where the activity of the root system (with respect to water and nutrients) is greatest. Local application of water to less than the total root volume does not affect the ability of the root to take up sufficient water and nitrate-nitrogen. Roots in the wetted root zone increase in their ability to take up water.

Selective wetting of the soil surface has additional benefits, such as reducing water evaporation (by preventing evaporation of water from outside the wetted surface zone). The partial wetting also

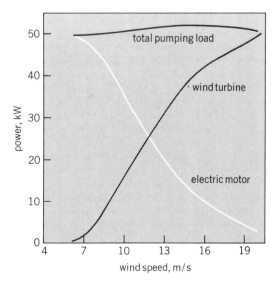

Fig. 3. Wind turbine power, electric motor power, and total pumping load versus wind speed for the experimental wind turbine.

restricts the growth of weeds to the wetted region, reducing the cost of weed control, and of application of herbicides through the drip system.

4. Maintaining dry foliage. Dry foliage retards the development of leaf diseases that require humidity, obviating the necessity for removing plant-protecting chemicals from the leaves by washing, and preventing leaf burns due to the lack of direct contact of the leaves with saline irrigation water.

5. Flexibility in fertilization. Fertilizers can easily be applied along with irrigation water. Optimizing the nutritional balance of the root zone is possible by frequently supplying nutrients directly to the most efficient part of the root zone.

6. Possible water saving. Loss of water due to runoff in low-permeability or crusted soils is reduced. Destruction of the surface-soil structure and the development of surface crust is avoided, and water infiltration into the soil is improved. Water saving may be achieved by restricting the water supply to the region of the most efficient root zone, and by not wetting the entire interrow or intertree space (especially in young crops or trees). On steep hills or under strong wind conditions the use of trickle irrigation prevents water loss beyond the irrigated field by wind convection or runoff.

7. Technical-economical features. Automation is a reliable tool which can easily be used in trickle irrigation for accurate soil water control, for the supply of water as needed, and for a large reduction in worker-power. The trickle irrigation method may rely on a relatively low operational pressure,

as long as the energy losses in the control system unit are not too large.

Additional important technical-economical features are the use of a small pipe diameter and the possibility of operating the system 24 hr a day, including windy hours, and the possibility of using the soil fungicides in the irrigation system.

8. Exploitation of marginal soils. Hardpans of various types, various sands and sand dunes, desert pavements of many kinds, and saline and alkali soils are very common marginal soils in arid zones of the world. In many areas of the humid tropics, soils are leached and become very acidic, which limits the rooting depth. Most of such marginal soils are characterized by low values of available water or water-holding capacity, which are often accompanied by limited rooting systems. Since under trickle irrigation the irrigation cycle is dominated by infiltration, rather than by the extraction stage, water-holding capacity or water availability properties of the marginal soils become relatively unimportant, because the soil water regime is continuously maintained at a relatively high water-content level. Water is supplied to the crop as it is needed, and there is no need to store water within the limited soil root zone.

An additional problem common to many marginal soils is that they may have a low fertility status (as in sands or sand dunes), or a high capacity to fix phosphorus. This fertility problem may be overcome by applying fertilizers simultaneously with irrigation through the trickling system.

Problems in practical use. While the trickle-drip method offers a number of advantages, there are certain operational and design problems.

1. Clogging. Operational difficulties with the trickling method arise from clogging of the drippers. Clogging results in nonuniform water distribution and requires frequent replacement of emitters, which is quite an expensive maintenance procedure. Clogging is caused by several factors, such as root penetration; blockage of the orifice by sand, rust, leaves, small soil animals, and microorganisms; and precipitation of salts. Smaller particles can pass through the filtering system and act as crystallization nuclei in the emitter.

2. Salinity problems. In arid regions where saline water must be used, there is a tendency for salt to accumulate at the periphery of the wetted soil volume. This accumulated salt may be washed by rain into the main effective root zone and causes osmotic shock to plants. A serious problem for seasonal crops is caused by regions of high salt concentration from the previous crop.

3. System design problems. In trickle irrigation three-dimensional flow occurs. The main design problem is to select the proper combination of emitter spacing and discharge for a given set of soil and water characteristics, and for the specific crop. Determination of the proper emitter spacing and discharge requires knowledge of soil hydraulic properties together with crop water and salt regime in the wetted soil volume.

Water and salt distribution. To understand water and salt distribution in a trickle-irrigated field, consider a field that is irrigated by a set of emitters spaced at regular intervals, $2X$ and $2Y$. Due to the symmetry of the pattern, one can subdivide the

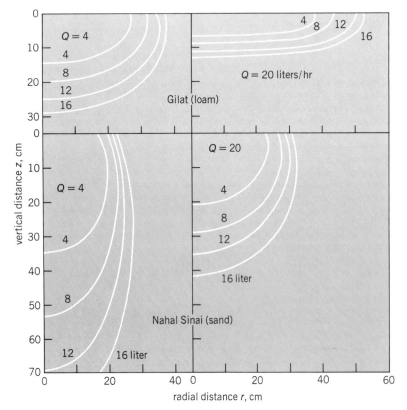

Fig. 4. Wetting front position as a function of infiltration time for two soils, two trickling rates (Q), and four values of cumulative infiltration water (in liters). *(From E. Bresler, Analysis of trickle irrigation with application to design problems, Irrigation Sci., 1: 3–17, 1978)*

radial distance r, cm

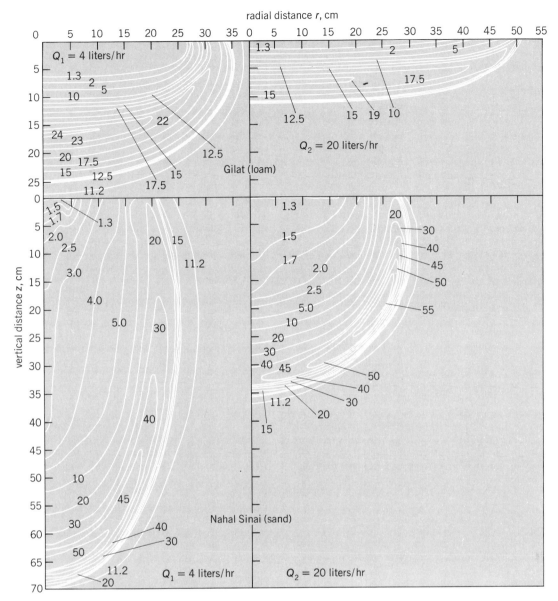

vertical distance z, cm

Fig. 5. Computed volumetric salt content field for two trickle discharges and two soils. The numbers labeling the curves indicate volumetric salt content in milliequivalents per liter of bulk soil. (From E. Bresler, Two-dimen-*sional transport of solutes during nonsteady infiltration from a trickle source, Soil Sci. Soc. Amer. Proc., 39(4): 604–613, 1975)*

entire field into identical volume elements W of length X, width Y, and depth Z, where the last always remains below the wetting front. Each volume element acts therefore as an independent unit in the sense that there is no flow from one element to another. At the soil surface the horizontal area across which water infiltrates into each soil element is a radial area of ponded water which develops in the vicinity of the trickle source. This area is initially very small, but its radius becomes larger (at a decreasing rate) as infiltration time t increases. Mathematical expressions have been developed for relating this radius and flow regimes of water and salts to the rate of trickle discharge Q and to the saturated and unsaturated hydraulic conductivity K of the soil. At large infiltration time when steady-state flow can be assumed, the ultimate size of the water entry zone is approximately

calculated as a function of Q and K. The calculations indicate that the radius of the saturated water entry zone will always be larger as K decreases and Q increases. This fact has a very important implication since the whole wetted zone also depends on Q and K. Examples of trickle irrigation wetting fronts as a function of the space coordinates (r, z) and the total amount of infiltrated water (in liters) for two different trickle discharges and two soils differing in their hydraulic characteristics are given in Fig. 4. The location of the wetting front is an important factor because it indicates the boundaries of the irrigated soil volume. The data in Fig. 4 demonstrate that the rate of trickle discharge and the hydraulic properties of the soil have a remarkable effect on the shape of the wetted soil zone. Increasing the rate of discharge and decreasing the saturated conductivity result in an increase in the zontal component (and decrease in the vertical component) of the wetted soil depth. Practical application of the mathematical treatment (the results of which are given in Fig. 4) for designing field irrigation systems is the possibility of controlling the wetted volume by regulating the trickle discharge according to the hydraulic properties of the soil.

Because of salinity problems, control of the salinity regime is essential to the permanent operation of a trickle irrigation system. Salt distribution in the soil under trickle irrigation is demonstrated in Fig. 5 by an example involving the case of low concentrated inflow solution that miscibly displaces a highly concentrated solution originally present in the soil. Two soil media with different hydraulic properties and salt dispersion characteristics are shown. The figure shows the position of the dissolved salt field in the cylindrical flow pattern. The results of the two soils are compared for an identical amount of cumulative infiltration. It is clear that the salt distribution pattern is soil- and discharge-dependent. The completely leached zone remains at the water-saturated zone in the sandy soil but penetrates deeper in the loam soil. The leached part of the soil in the vertical component of the wetted zone is deeper, and in the radial component it is narrower, as the soil becomes coarser and as the trickle discharge becomes slower. The salt quantities from the leached part of the soil are accumulated and reach a maximum at a certain distance from the source. The location of this maximum salt accumulation zone is also dependent on both soil hydraulic properties and the discharge rate of the trickler. This pattern of salt distribution and its soil and discharge dependence is of practical interest for the creation of a leached zone sufficiently large for most of the roots to concentrate and function without disturbance.

Spacing between emitters. Applications of mathematical treatments to determine the spacing between emitters require a knowledge of soil water pressure and soil hydraulic conductivity as functions of soil water content. The effects of these soil properties on the spacing-discharge relationships for a given soil water tension h_c between emitters are demonstrated in Fig. 6. Larger spacing between emitters is permitted in soils with lower values of K (loam as compared to sand) and also when the crop grown is not sensitive to water stresses (lower h_c values are permitted). Closer spacing is required for soils having higher water conductance when a sensitive crop is being grown. For any given soil, emitter spacing can be increased as the discharge rate Q becomes higher. The proper choice of Q chiefly depends on some optimization criteria in the engineering design of the field irrigation system.

The data as given in Fig. 6 can be combined with principles of hydraulics to obtain diameter and length of the lateral system for design purposes. For a given spacing-discharge combination and uniformity criteria, the length and diameter of the laterals can be calculated from any conventional head loss formula.

For background information *see* IRRIGATION OF CROPS; WIND POWER in the McGraw-Hill Encyclopedia of Science and Technology.

[ESHEL BRESLER]

Bibliography: E. Bresler, *Adv. Agron.*, 29: 343–393, 1977; E. Bresler, *Irrigat. Sci.*, 1:3–17, 1978; E. Bresler, *Soil Sci. Soc. Amer. Proc.*, 39(4): 604–613, 1975; R. N. Clark and A. D. Schneider, *Irrigation Pumping with Wind Energy*, ASAE Pap. no. 78-2549, American Society of Agricultural Engineers, 1978; D. L. Elliott, *Synthesis of National Wind Energy Assessments*, BNWL/Wind-5, U.S. Department of Energy, July 1977; New energy saver harnesses wind, in H. G. Hass (ed.), *Agricultural Research*, USDA, January 1979; *Proceedings of the 2d International Drip Irrigation Congress*, San Diego, 1974; USDA Economic Research Service, *Energy and U.S. Agriculture: 1974 Data Base*, USDA Bull. FEA/D-77/140, 1977.

Jet propulsion

The mechanical definition of an aircraft variable-cycle gas-turbine engine has been the objective of recent research programs. The term variable-cycle engine is, in fact, a misnomer because the Brayton thermodynamic cycle is common to all gas-turbine engines. The "variable" in the term refers to control of the cycle parameters such as pressure ratio, temperature, gas flow paths, and air-handling characteristics during flight. Variability permits the aircraft thrust requirements to be more closely matched with good fuel economy at two or more differing flight speeds than is possible with conventional turbofan or turboject engines.

A typical application of the variable-cycle engine would be a supersonic commercial transport which must be designed to fly most efficiently at supersonic speeds but is also constrained to fly over populated areas at subsonic speeds to avoid causing disturbances on the ground due to sonic

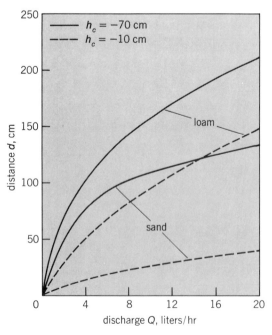

Fig. 6. Distance between emitters *d* as a function of trickle discharge *Q* for two soils and two values of h_c. (From E. Bresler, Analysis of trickle irrigation with application to design problems, Irrigation Sci., 1:3–17, 1978)

booms. For such aircraft, fuel reserves, which may weigh as much as half the revenue payload, must be carried in case the destination airport is closed and the aircraft is diverted to an alternate airport. Even small gains in subsonic fuel economy will have a significant impact on the number of passengers carried or the maximum range. Another application of the variable-cycle engine would be a long-range military aircraft which cruises subsonically but must penetrate enemy skies at supersonic velocity. The radius of action for such an aircraft could possibly be extended if variable-cycle engines were available. At present, there are no variable-cycle engines in existence, although many conceptual designs have been proposed and the operation of unique critical elements have been verified.

Turbojet and turbofan engines. Before describing several concepts of variable-cycle engines, it is appropriate first to understand the operational characteristics of conventional turbojet and turbofan engines and how they interact thermodynamically, aerodynamically, physically, and environmentally with aircraft designed for cruise at speeds twice that of sound or greater. Rather broad generalizations will be made, recognizing that gray areas exist in the distinctions between turbojet and turbofans.

Advantages of turbojet. Of the two types of engines, each with an equal degree of technological development, the turbojet is superior for cruise at the required supersonic speeds. This is because it exhibits better propulsion efficiency and hence better fuel economy, and also because it provides greater thrust per pound of airflow through the engine, which results in smaller frontal area for a given thrust and hence less external drag of the podded engines. The turbofan engine, on the other hand, exhibits a rapid thrust decay with increasing speed in the supersonic range and therefore requires energy addition in the form of fuel burning in the bypass air duct to provide the thrust required to cruise supersonically. (Bypass air is that portion of the engine intake airflow which is pressurized by the fan but flows around the core, which consists of the compressor, combustor, and turbines. The ratio of bypass airflow rate to core flow rate is referred to as the bypass ratio).

Advantages of turbofan. At subsonic flight speeds, a moderate bypass-ratio turbofan can be as much as 20% more fuel-efficient than a turbojet engine because of its lower average jet exhaust velocity. This can be understood by considering that the thrust of a jet engine is the product of its air mass flow rate and the difference between jet exhaust and flight velocity. For equal thrust, the turbofan must have greater airflow because of its lower jet exhaust velocity relative to a turbojet. Thus, the lower jet exhaust velocity of the turbofan produces less dissipative kinetic energy loss in the exhaust stream, resulting in higher propulsive efficiency and a consequent fuel savings. The lower jet exhaust velocity that is characteristic of the turbofan is also advantageous in the vicinity of the airport because jet noise is a very strong function of exhaust velocity. Another factor favoring the use of turbofan engines is that it weighs less than the turbojet for engines sized to

produce equal takeoff thrust. Weight is important, particularly for supersonic transports, since each additional pound of engine weight results in 5–7 lb of additional design takeoff gross weight to fly the required distance.

Choice of engine. The final choice of a type of gas-turbine engine for use in a supersonic transport depends upon many conflicting requirements, such as the need to minimize supersonic and subsonic fuel consumption rates, engine weight, and airport noise. Adequate acceleration margin and takeoff thrust to meet restrictive airport field lengths must also be provided. The engine type providing the best compromise will result in the lowest-gross-weight airplane capable of carrying the required number of passengers the required distance. Designers of the British-French Concorde and the canceled American supersonic transport (SST), whose projects were begun before Federal noise regulations were established, selected turbojet engines. The Soviets selected the turbofan for the TU-144. The fact that both types of gas-turbine engines were selected for supersonic transports, plus the fact that the turbofan studied for the American SST produced almost identical results as the selected turbojet, indicate that the advantages and disadvantages of each type of engine tend to balance each other.

Variable-cycle engines. After the demise of the American SST, the National Aeronautics and Space Administration (NASA) undertook a research program to provide the technology for an advanced environmentally acceptable supersonic transport; one element of the program was to explore the possibility of defining a gas-turbine engine which could be controlled in flight to exhibit the desirable features of both turbojets and turbofans. Through the use of unconventional arrangements of engine components, variable-flow fans and compressors, variable internal-flow-passage area, and diverter, inverter, and closure valves, used either singly or in combination, many concepts of variable-cycle engines have evolved, ranging from minor modification of bypass ratio to true conversion from moderate bypass-ratio turbofans to pure turbojets. As a result of the inherent flexibility created by gas flow path management in variable-cycle engines, the intake airflow rate may be scheduled to minimize aerodynamic drag of the

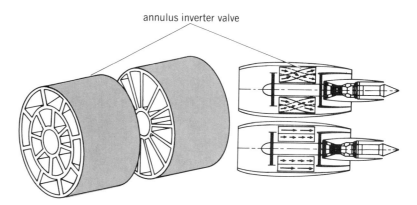

annulus inverter valve

Fig. 1. Series-parallel-fan, variable-cycle engine. *(From G. W. Klees and A. D. Welliver, Variable-Cycle Engines for the Second Generation SST, SAE Pap. no. 750630, Society of Automotive Engineers, May 1975)*

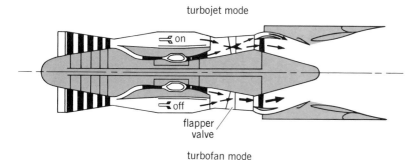

turbojet mode

on

flapper
valve

off

turbofan mode

Fig. 2. Multicycle engine. *(From G. W. Klees and A. D. Welliver, Variable-Cycle Engines for the Second Generation SST, SAE Pap. no. 750630, Society of Automotive Engineers, May 1975)*

propulsion system. Typically, the thrust required at subsonic speeds is one-third to one-half of that available from the engine. The airflow of the conventional turbofan and turbojet engine falls as power is reduced. This causes added inlet drag as the excess air is spilled around the sharp lip inlet required for supersonic flight, and it causes added nozzle external drag as the nozzle exit diameter closes about a smaller exhaust stream.

Satellite engine. Perhaps the most easily visualized example of a variable-cycle engine is the satellite engine concept. This engine can operate supersonically as a true turbojet and provide the low-speed performance benefits of a moderate bypass-ratio turbofan. In the turbofan mode, the fan bypass airflow is ducted directly to the fan exhaust nozzle. The conversion to the turbojet mode of operation is accomplished by valving, which diverts the fan discharge air to supercharge separate turbojet engines surrounding the core engine before delivery to the fan exhaust nozzle. Special control logic is employed to properly sequence the starting of the satellite engine with diverter valve actuation and exhaust nozzle area variation to avoid exceeding the bounds of stable engine operation. The obvious deficiency of the satellite engine is its greater weight and frontal area relative to convential power plants.

Variable-stream control engine. The variable-stream control engine proposed by the Pratt and

Whitney Company is the least complicated of the variable-cycle engines, but it is capable of only minimal variation of bypass ratio. It is a duct-burning low-bypass-ratio turbofan which, through scheduled variation of the minimum exhaust flow area of the core and bypass exhaust streams, can modulate engine flow and bypass ratio within restrictive limits to provide improvements in performance over conventional turbofans. This is achieved without significant weight penalty.

Series-parallel fan engine. The series-parallel fan concept utilizes a unique annulus inverter valve located between elements of the fan to vary the bypass ratio from low to moderate values. The two-element annulus inverter valve shown in Fig. 1 consists of flow-passage segments which, when positioned for series operation, guide the front fan element flow directly to the second fan element. In this position, the engine operates in the low-bypass mode with relatively high fan duct exit velocity. Positioning the annulus inverter valve for parallel operation causes the front fan element airflow to be discharged directly to an auxiliary exhaust nozzle. For this mode of operation, the rear fan element induces flow which bypasses the front fan and delivers lower-pressure air to the core engine and primary fan duct exhaust than takes place in series operation. The annulus inverter valve was successfully demonstrated in a ground stand test facility by the Boeing Company in a modified Pratt and Whitney JT8D turbofan engine. The normal (series operation) bypass ratio of 1 was increased to a value of 3.5 for the parallel mode, with a concurrent increase in total airflow of 70%. Although the series-parallel fan engine appeared to be attractive for use with the SST, the increased engine weight and the necessity for an auxiliary inlet and nozzle caused reduced performance relative to a conventional power plant.

Multicycle engine. The multicycle engine, proposed by the Boeing Company and illustrated in Fig. 2, utilizes an adaptation of the annulus inverter valve, called a flapper valve, which either inverts the core and bypass streams or merges the two streams before the last turbine element. In the high-thrust mode, with combustion of the airflow in the bypass duct, each stream acts as an independent turbojet, exhausting through a common jet nozzle. Without combustion in the outerstream and with the inverter valve set to merge the two streams, performance which is very nearly equal to that of a 2.5-bypass-ratio conventional turbofan is achieved.

Double-bypass engine. The double-bypass engine proposed by the General Electric Company, shown schematically in Fig. 3, utilizes a fan that is separated into two elements and has the capability of bypassing air between elements as well as after the second element. With the valve between fan elements in the closed position, the engine operates as a very-low-bypass-ratio turbofan with good specific fuel consumption at supersonic speeds. With the valve open, the bypass ratio increases by a factor of 2.5, and the two bypass streams of differing energy are mixed efficiently in the bypass duct through the use of a variable-area bypass injector valve. For either mode of operation, the bypass flow and core flow are mixed through the

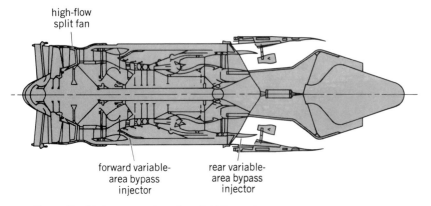

high-flow
split fan

forward variable-
area bypass
injector

rear variable-
area bypass
injector

Fig. 3. Double-bypass engine. *(From R. H. Brown, Integration of a Variable Cycle Engine Concept in a Supersonic Cruise Aircraft, AIAA/SAE Pap. no. 78-1049, American Institute of Aeronautics and Astronautics, July 1978)* use of a similar variable-area ejector valve prior to flowing through the jet exhaust nozzle. Although the engine operates as a variable low-bypass-ratio turbofan in either mode, it provides a better compromise solution to the conflicting engine performance requirement than a conventional turbofan or turbojet does.

For background information *see* TURBINE PROPULSION; TURBOFAN; TURBOJET in the McGraw-Hill Encyclopedia of Science and Technology.

[EMANUEL BOXER]

Bibliography: E. Boxer, S. J. Morris, Jr., and W. E. Foss, Jr., Assessment of Variable-Cycle Engines for Supersonic Transport, in *Variable Geometry and Multicycle Engines*, AGARD Conf. Proc. no. 205, pp. 9–1 to 9–20, September 1976; R. Brown, *Integration of a Variable Cycle Engine Concept in a Supersonic Cruise Aircraft*, AIAA/SAE Pap. no. 78-1049, July 1978; G. W. Klees and A. D. Welliver, *Variable-Cycle Engines for the Second Generation SST*, SAE Pap. no. 750630, May 1975.

Leaf (botany)

The shoots of higher plants are longitudinally asymmetric and exhibit progressive changes in form along their length which are distinctive for an individual species. These changes can be expressed in differences in leaf size or complexity, internode length, and shoot diameter, as well as in differences in leaf arrangement (phyllotaxis). Because serial changes in leaf form can be the most striking, they have received the greatest attention from plant morphologists and have been described for a number of plants. Nevertheless, the more significant question is how such leaf diversity arises developmentally. Although the developmental basis for heterophylly has been the subject of investigation for over a century, only in the past 10 years has new, more detailed comparative developmental research led to a questioning of past morphogenetic interpretations and helped to place knowledge of this subject on a sounder basis. But before turning to a description of these recent studies, it is important to review the nature of the interpretation that has predominated until now.

Metamorphosis theory. The dominant interpretation of the ontogenetic change in leaf form along the length of a shoot has been the metamorphosis theory propounded by the German plant morphologist Karl von Goebel in the latter part of the 19th century. Goebel's theory was based principally on selected observations of developing primordia of bud scales and foliage leaves of woody plants, and also upon a series of defoliation experiments which purported to have transformed bud scale primordia into foliage leaves.

Appendage initiation. According to the metamorphosis concept, all appendages formed along the length of a shoot actually are initiated as foliage leaf primordia which, depending upon the nutritional status of the plant (that is, the general environment in which the leaf is growing), undergo a developmental transformation or metamorphosis from one leaf type to another simply by a shift in the distribution of growth. For example, in the case of bud scale production, the earliest differentiation into a distal precursor of the lamina and petiole

and proximal progenitor of the leaf base would be identical to that of the foliage leaf. The distinctive vaginate form of the scale, however, would be due to an emphasis on the expansion of the leaf base and an arrest in growth of the distal lamina-petiole zone in contrast to the foliage leaf, where elaboration of the lamina would be at the expense of the base. Furthermore, Goebel conceived of all variant appendage forms produced by an individual plant during its life (cotyledons, scale leaves, primary leaves, foliage leaves, bracts, sepals, petals, stamens, and carpels) as representing metamorphosed foliage leaf primordia whose structural specialization was simply the result of an earlier or later arrest in the course of a common developmental pathway.

Criticism. Although elements of Goebel's theory have received general acceptance from those concerned with the physiological control of plant form, the idea has been strongly criticized by plant morphologists interested in development. The principal problem with the metamorphosis concept as originally proposed is that it is too simplistic and fails to take into account qualitative developmental differences that are manifest in differing leaf types from inception, and that become amplified during subsequent morphogenesis. Furthermore, this theory never considers the relationship of leaf morphogenesis to overall shoot development in attempting to account for the foliar heteromorphism produced.

A prime reason for reexamining Goebel's theory with greater rigor is the need to have a more detailed understanding of how morphological change is expressed ontogenetically before sensible investigations of the physiological basis of developmental control can be made. Two recent studies of plant species exhibiting marked heterophylly were carried out by D. R. Kaplan and associates at the University of California, Berkeley, and have cast considerable doubt on the metamorphosis concept, especially as it relates to the interpretation of morphogenetic divergence.

Heterophyllic development in Muehlenbeckia. *Muehlenbeckia platyclados* is a xeromorphic, shrubby member of the buckwheat family (Polygonaceae) and is endemic to the Solomon Islands. Its distinctive stems are conspicuously flattened, green, and photosynthetic. *Muehlenbeckia* also shows a marked difference in leaf structure between the juvenile and the adult phases. The juvenile (nonflowering) phase produces conspicuous laminate foliage leaves that are regionally differentiated into a sagittate blade, a petiole, and a stem-encircling base or ochrea (Fig. 1a). The adult (flowering) phase bears minute, barely visible nonphotosynthetic scale leaves with only a small, protruding, membranous tip and axis-encircling base (Fig. 1e). The transition region between these two phases produces leaves that are intermediate in form, exhibiting a reduced lamina and stalk (Fig. 1a).

Scale and foliage leaf primordia. Merely observing the differences in size and proportion of these successive leaves would lead one to suggest, in agreement with Goebel's theory, that the scale leaf simply represents an arrested developmental derivative of the foliage leaf. However, recent ontogenetic comparisons between these two leaf types by D. Bruck and Kaplan have shown that this is not the case, and that the primordia of these two leaf types are distinctive from inception. Hence, they do not show the developmental parallelism predicted by the metamorphosis concept.

Early in development, the primordium of the foliage leaf shows a regional differentiation into a freely projecting distal sector (the future lamina) and an axis-encircling proximal zone (the future

Fig. 1. Comparative leaf development in *Muehlenbeckia platyclados* (Polygonaceae). (a) Juvenile shoot. (b) Young stages in foliage leaf development. (c) Young foliage leaf showing initiation of ochrea tube from adaxial surface of the lamina, and (d) a distinctly saccate ochrea and a reflexed revolute lamina. (e) Adult plant. (f) Stages of initiation and early growth of scale leaves. (g) More developed scale leaf. (h) Primordium of a transition leaf showing the hooded-type development of the scale leaf, but a distinct, abaxially inserted lamina.

ochrea) (Fig. 1b). Subsequent development involves elongation and broadening of the primordial blade as well as upward growth of the saccate leaf base so that it completely enshrouds the apical meristem and younger leaf primordia composing the terminal bud (Fig 1c and d). Later, a petiole is intercalated between the blade and base. In marked contrast, the scale leaf arises as a broadly inserted, collarlike primordium lacking the differentiation of a distally projecting blade region (Fig. 1f). It grows up and around the shoot apex as a hood which eventually overarches the shoot apex and resembles a helmet (Fig. 1f). There is an axis-encircling portion to the scale leaf, but it never develops the saccate shape of the foliage leaf ochrea (Fig. 1g). Ultimately, as the shoot axis broadens, the overarching portion of the scale leaf is displaced laterally and is evident as the aforementioned membranous projection inserted at the stem periphery (Fig. 1e).

Hood portion of scale. If just the developmental idiosyncracies of these contrasting leaf types were observed, the temptation would be to interpret the overarching hood portion of the scale as the homolog of the lamina region of the foliage leaf. The simplified form of this region could be attributed to an early developmental arrest, per the metamorphosis concept. However, observations of the primordia of intermediate leaf types which have a reduced lamina have shown that this is not the case. Interestingly, young primordia of the transition leaves exhibit a combination of scale and foliage leaf characteristics (that is, the hood development of the scale, but with a distinct lamina attached to the abaxial side — Fig. 1h). Thus it can be concluded that the hooded portion of the primordial scale is equivalent to the leaf base sector of the foliage leaf, and not to the lamina. Since the scale never differentiates a blade, it is obvious that there is no developmental parallelism between these two organ types. Rather, each is a product of the particular developmental status of the shoot as a whole, and as such each manifests its unique properties from the time it arises.

Apical meristem relationships. Beyond its usefulness in testing the metamorphosis theory, the shoot system of *Muehlenbeckia* has also been of interest in evaluating the relationship of shoot apical meristem size to leaf complexity. In the majority of plant species that exhibit changes in leaf form, increasing leaf size and structural complexity are correlated with an increase in diameter of the parent shoot apex, to the extent that some investigators have viewed the two as being causally related. However, *Muehlenbeckia* is unique among investigated species because the largest, most complicated appendage type (the foliage leaf) is initiated from the smaller apical meristem, whereas the simpler, reduced scale arises from a shoot apex which is more than twice the diameter of that of the juvenile phase. This is the converse of the situation described for most species. Although it is possible that these differences in leaf morphology are related in part to the plastochron interval and degree of packing of primordia in the bud (loose vertical packing, in the case of foliage leaf, associated with slower shoot growth and hence longer plastochrons versus tight packing, in the scale leaf phase, associated with rapid shoot growth and shorter plastochrons), such hypotheses not only need to be tested by the appropriate experimentation, but also suggest that the relationships between meristem size and leaf complexity may not be causal but simply an expression of the integration of shoot development.

Heterophyllic development in Acacia. The classic change in leaf morphology is exhibited by the Australian xeromorphic species of the leguminous genus *Acacia*. Here, the seedling produces dissected (once-pinnate and then twice-pinnate) leaves in contrast to the simple (undissected) appendages of the adult region of the shoot (Fig. 2d and l). Typically, the change from dissected juvenile to undissected adult organs occurs within the first few nodes of seedling growth and results in a visually striking leaf dimorphism (compare Fig. 2d with Fig. 2l). Usually, the first leaf after the cotyledons is a once-pinnate leaf with three or more pairs of leaflets (termed pinnae). There then follows a series of intermediate leaves bearing only a single leaflet pair at their apices. In some of these bipinnate transitions, the leaf axis below the leaflet pair may be dilated into a bladelike structure (Fig. 2h). The so-called phyllode or simple leaf type characteristic of adult shoots is then differentiated, and can be distinguished from the dissected leaves by its lack of pinnae (Fig. 2l).

Dissected growth. Since it is possible to observe what appear to be gradual changes in form from mature pinnate to simple leaves along the length of an individual seedling, most plant morphologists of the 19th century deduced the developmental changes responsible for this heterophylly rather than from an actual comparative study of development. As a result, they arrived at developmental deductions that were erroneous. For example, the majority of previous investigators interpreted the simple leaf type in *Acacia* as being equivalent to the petiole or stalk region of the dissected leaf, in which expansion of the pinnate blade region has been suppressed in favor of elongation of the petiole, which therefore has assumed the major assimilative role (Fig. 2l). The bipinnate transitions were seen by these morphologists as intermediates in this process of laminar suppression and petiolar dilation (Fig. 2h).

Implicit in this traditional view of the petiolar homology of the *Acacia* phyllode was an interpretation of the developmental basis for the change in leaf form which was strongly influenced by the metamorphosis theory. Without any tangible basis, Goebel assumed that the phyllode was initiated as a pinnatifid leaf which in the course of its development underwent a metamorphosis into the phyllodic organ by the aforementioned suppression of blade and leaflet elongation and correlative early extension of the petiole (compare Fig. 2i–l with Fig. 2e–h and 2a–d). Thus, as in other expressions of the metamorphosis concept, the central assumption was that there was an exact developmental parallelism between phyllode and pinnate leaf up to a certain stage, and then the divergence occurred (compare Fig. 2i–l with Fig. 2a–d). However, as in *Muehlenbeckia*, actual developmental investigations have not only demonstrated

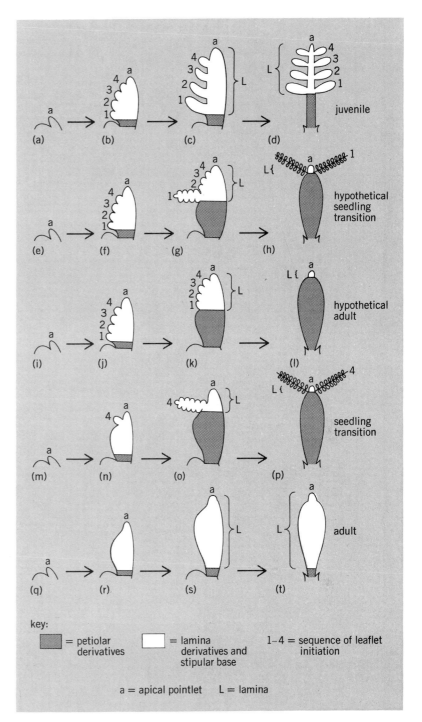

Fig. 2. Developmental relationships between a pinnatifid leaf, transition form, and phyllode in *Acacia*. (a-d) Developmental stages of a once-pinnate juvenile leaf; (e-h) hypothetical developmental stages of a bipinnate transition leaf with a petiolar origin; (i-l) hypothetical developmental stages of a phyllode with a petiolar derivation; (m-p) actual developmental stages of a bipinnate transition at node 2; (q-t) actual developmental stages of a phyllode. Since it is not possible to indicate the vertically extended nature of the phyllode in an adaxial view, h, l, p, and t show the phyllodic blade region expanded from an adaxial view, even though the plane of enlargement is actually at right angles to the plane shown.

a lack of developmental equivalence, but have also drastically altered the interpretation of the homology between these organs and the ontogenetic basis for the heteromorphism.

Unified growth. In an extensive investigation which made a rigorous comparison of the development of each succeeding leaf type between pinnatifid and phyllodic leaves in four species of *Acacia*, Kaplan demonstrated that the phyllode blade is the positional equivalent of the blade region of the pinnate leaf at all stages of development. Except for the early seedling transitions, in which the dilated leaf axis is predominantly petiolar, at no stage in phyllode development is there a developmental displacement of the blade by petiolar elongation (Fig. 2*l*). Instead of the embryonic lamina becoming subdivided into individual subunits (the leaflets), as occurs in the pinnate leaf, from inception the blade of the phyllode grows as a unified structure (Fig. 2*q − t*). Expansion in both the phyllodial and pinate leaves is in an adaxial, radial direction, but since the leaflets of the pinnate organ become oriented secondarily to a more conventional, dorsiventral plane, the two leaf types look radically different at maturity (compare Fig. 2*d* with Fig. 2*t*).

Evidence was also provided that successive transition and phyllodic leaves actually have progressively longer blades and that the increase in blade length is correlated with a progressive increase in shoot apical meristem diameter (compare Fig. 2*m − p* with Fig. 2*q − t*). Thus, in contrast to the conventional view, which indicated a progressive shortening of the blade, there actually is a progressive lengthening.

Conclusions. The marked change in blade morphology (dissected versus unified growth) obscures the limits between petiole and blade, and it generated the superficial developmental evaluation which has predominated. The results of the *Acacia* research, in particular, have demonstrated dramatically the shortcomings of Goebel's theory and have emphasized the need for rigorous developmental investigations in place of its gross abstractions.   [DONALD R. KAPLAN; DAVID K. BRUCK]

Bibliography: D. K. Bruck and D. R. Kaplan, *Amer. J. Bot.*, in press; K. Goebel, *Bot. Z.*, 38: 753 − 760, 769 − 778, 785 − 795, 801 − 815, 817 − 826, 833 − 845, 1880; D. R. Kaplan, *La Cellule*, in press.

Life, origin of

In 1953 Stanley Miller demonstrated that amino acids, the building blocks of proteins, could be produced by passing a spark through a flask containing a mixture of methane, ammonia, and water vapor. Since that time, numerous studies have confirmed that amino acids and other biomolecules can readily be formed under plausibly prebiotic conditions from the simple molecules which are thought to have existed on the primitive Earth. The formation of organic compounds in these experiments is considered to represent the first step in the prebiotic formation of the biomacromolecules necessary for the evolution of the first living organism. Recent experiments have demonstrated that hydrogen cyanide (HCN) may have been an important intermediate in these prebiotic reactions, and it has been shown that a large variety of biomonomers can be obtained from HCN in laboratory experiments. Another area of recent study has been the subsequent steps of chemical evolution resulting in the conversion of biomonomers to

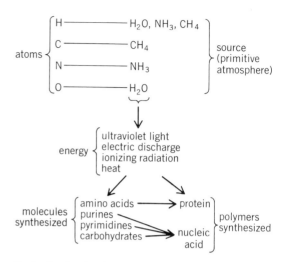

Fig. 1. Synthesis of organic compounds on the primitive Earth.

biopolymers (Fig. 1). The possible role of clays and metal ions as catalysts in prebiotic chemistry has been explored in this context, with some intriguing results.

Role of hydrogen cyanide. In some of the early chemical evolution experiments, it was discovered that HCN and simple aldehydes were intermediates in the formation of amino acids and other compounds from precursors such as methane, ammonia, and water. The discovery of HCN and aldehydes in these reaction mixtures suggested that amino acids were being formed by the Strecker synthesis [reaction (1)], a well-known laboratory reaction for the preparation of amino acids.

$$
RCHO + NH_3 \longrightarrow R{-}\underset{\underset{OH}{}}{\overset{\overset{NH_2}{|}}{CH}} \xrightarrow{HCN} R{-}\underset{}{\overset{\overset{NH_2}{|}}{CH}}{-}CN
$$

Aldehyde   Ammonia   Aminonitrile

$$\big\downarrow H_2O \qquad (1)$$

$$R{-}\underset{}{\overset{\overset{NH_2}{|}}{CH}}{-}CO_2H$$

α-Amino acid

Thus, a scenario developed wherein HCN and aldehydes were formed from the simple compounds in the primitive atmosphere and the products then reacted further to yield amino acids or other compounds necessary for the evolution of life.

Since HCN is quite soluble in water, this prompted other workers to suggest an alternative scenario: that HCN was formed in the primitive atmosphere but eventually dissolved in primitive oceans and other bodies of water to form dilute HCN solutions, and it was there, in aqueous solution, that the chemical reactions leading from HCN to biomolecules took place. A large number of experiments have since been performed which appear to support this hypothesis. This work has been largely the result of the efforts of the research groups of Leslie Orgel and James Ferris, and has dealt primarily with the solution chemistry of HCN.

An attractive feature of this hypothesis is that amino acids and other biomolecules can be formed directly from HCN and water without the need for any other chemical reagents or unusual conditions. Thus, this reaction system appears to be a reasonable simulation of the primitive environment. Typically, experiments have involved the preparation of dilute HCN solutions which are then adjusted to a pH of approximately 9 (close to the probable pH of the primitive ocean). On standing at room temperature, these solutions turn orange-brown and form a dark precipitate which, collectively, has come to be known as HCN polymer or HCN oligomer. The HCN oligomer has proved to be a chemically complex material that has, so far, defied precise chemical characterization. Rather than being a single compound, it appears to be composed of a mixture of species, probably in the molecular weight range of 500–1000; however, it has generated much interest since further chemical or chromatographic separation followed by hydrolysis yields a large suite of amino acids. The mechanisms of these reactions are not completely understood; however, it seems clear that the reaction proceeds via a stepwise oligomerization to yield a tetramer of HCN which then reacts further to yield HCN oligomer, as shown in reaction (2). Although

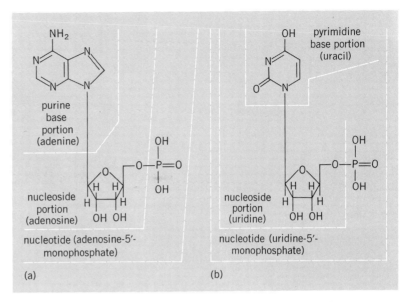

Fig. 2.   Structures of (a) purine and (b) pyrimidine nucleotides and their components.

the mechanistic aspects of the oligomerization of HCN have been the subject of a great deal of scientific research and debate, of equal importance are the products obtainable from HCN oligomers and their significance for the origin of life.

The basic components of contemporary living systems responsible for their ability to reproduce and evolve are proteins and nucleic acids (DNA and RNA). It is believed that these molecules, at least in a primitive form, would also have been necessary for life to originate in the primitive environment. As discussed above, amino acids, the building blocks of proteins, could readily have arisen on the primitive Earth from HCN. A recent report has also shown that both purine and pyrimidine bases can be isolated from HCN oligomers under prebiotic conditions. These bases are the precursors of nucleotides, the building blocks of nucleic acids (Fig. 2). The purine known as adenine and the pyrimidines 5-hydroxyuracil and 4,5-dihydroxypyrimidine were obtained from HCN oligomers. Of further interest, the compounds 4-aminoimidazole-5-carboxamide and orotic acid

were reported. These compounds are significant in that they are intermediates in the contemporary biosynthetic pathways for the formation of purines and pyrimidines, respectively. These results suggest that the contemporary biosynthetic pathways of nucleotides may have evolved from the compounds released from HCN oligomers on the primitive Earth.

As a partial test of this hypothesis, J. P. Ferris and coworkers subjected solutions of orotic acid and orotidine (the nucleoside of orotic acid) at pH 7–8.5 to ultraviolet irradiation. They observed that the orotic acid was converted to uracil and orotidine to uridine in yields of up to 45%. They performed the analogous experiment with the nucleotide (orotidine-5′-monophosphate), and it was similarly converted to uridine-5′-monophosphate in

Fig. 3.   Nucleotide isomers: (a) 2′-AMP, (b) 3′-AMP, and (c) 5′-AMP.

primitive atmosphere

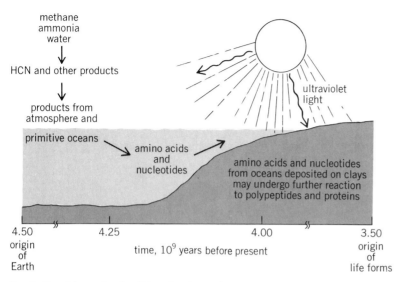

Fig. 4. Possible synthesis of biomolecules under primitive-Earth conditions.

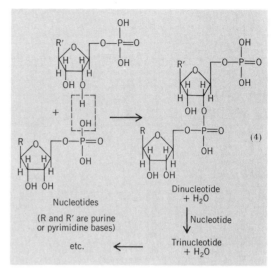

23% yield. Thus, it appears that the hypothesis is supported by experiment, and that contemporary biosynthetic pathways could readily have evolved by reactions of HCN oligomers present in primitive bodies of water in the presence of ultraviolet radiation. Although other types of chemical reactions certainly occurred on the primitive Earth, the above results clearly demonstrate that all three classes of nitrogen-containing biomolecules could have formed exclusively from HCN and water under relatively mild primitive-Earth conditions.

Role of clays and metal ions. Given a mechanism for the formation of biomonomers in the primitive environment, such as production from HCN oligomers, the question remains as to how such a dilute solution, containing a complex mixture of organic compounds, could have resulted in the formation of the types of biomacromolecules known to be important in contemporary biochemistry. These reactions, that is, the formation of polypeptides from amino acids [reaction (3)] and the

formation of polynucleotides from nucleotides [reaction (4)], are dehydration reactions which, in aqueous solution, are thermodynamically unfavorable. Therefore, in order for these reactions to proceed, some driving force would have been necessary. Furthermore, some means would have been necessary to select from the complex "soup" the types of compounds which later became important in biology, and to reject others. J. D. Bernal was the first to suggest that clays and other minerals might have performed both of these functions.

Experimental evidence consistent with this hypothesis has recently been obtained in the research group of James Lawless. They have investigated the effect of different clays on the adsorption and oligomerization of amino acids and nucleotides. In these studies, bentonite clays were employed since they are fairly ubiquitous and are known to be catalytically active. The clay particles are negatively charged and, thus, always have positive counterions associated with them, typically Na^+, Ca^{2+}, or Mg^{2+}. In the reported studies, the clays were first equilibrated with concentrated salt solutions to produce homoionic clays—clays containing only a single type of counterion. When the transition metal–substituted clays (Cu^{2+}, Ni^{2+}) were added to amino acid solutions, it was observed that the amino acids were readily adsorbed from solution. Furthermore, the Ni^{2+}-bentonite, in particular, preferentially adsorbed those amino acids that are found in proteins today over those that are not. Thus, a mechanism was demonstrated for the selection of the biological subset of amino acids.

In another experiment, it was demonstrated that amino acids could be polymerized in the presence of transition metal–substituted clays. In order to simulate a prebiotic tidal environment, the amino acid solutions were mixed with the clay and then subjected to repeated cycles of wetting, drying, and heating. On analysis, it was found that, using Cu^{2+}-, Ni^{2+}-, and Zn^{2+}-bentonites, up to 6.2% polymerization was obtained, with the largest oligomer detected being the pentamer. Thus, the interaction of clays, metal ions, and amino acids could readily have resulted in the formation of polypeptides on the primitive Earth.

Equally intriguing are the results of a series of experiments with nucleotides. The adsorption of 5'-adenosinemonophosphate (5'-AMP) from solution by a series of homoionic bentonites was investigated, and it was found that the nucleotide was readily adsorbed by the Mg^{2+}-, Ni^{2+}-, Cu^{2+}-, Co^{2+}-, and Zn^{2+}-bentonites, but not by the Na^+-, Li^+-, or K^+-bentonites. The strongest interaction was with the Zn^{2+}-bentonite, which adsorbed 98% of the 5'-AMP from solution. When the interaction between the Zn^{2+}-bentonite and five other nucleotides was observed, it was found that the Zn^{2+}-bentonite adsorbed all the nucleotides to some extent, with the purine nucleotides being more strongly adsorbed than the pyrimidine nucleotides. The strongest interaction was between 5'-AMP and Zn^{2+}-bentonite. The nucleotide 5'-AMP and its derivatives are the most ubiquitous of nucleotides found in all living systems today. It is tempting to speculate that this may be the result of interactions with metal ions and clays on the primitive Earth.

Nucleotides can exist in three possible isomeric forms: the 2'-, 3'-, or 5'-isomers (Fig. 3). Contemporary biochemistry utilizes only the 5'-isomer. In previous experiments attempting to synthesize nucleotides under prebiotic conditions by phosphorylation reactions, all three isomers were obtained, in addition to a cyclic isomer. Lawless's group recognized that, because of its strong affinity for 5'-AMP, the Zn^{2+}-bentonite might be capable of selecting this isomer in competition with the others. When Zn^{2+}-bentonite was added to an equal mixture of 2'-, 3'-, and 5'-AMP, they observed as much as a 10-fold preference for the 5'-isomer.

Implications. The research results discussed here together suggest a possible mechanism (Fig. 4) for the origin of life wherein HCN in the primitive atmosphere dissolved in primitive bodies of water and, through the influence of thermal energy and ultraviolet radiation, reacted to form the biomonomers necessary as the first step in chemical evolution. These biomonomers, through the influence of tidal forces, interacted with metal ions and clays in sediments and on shorelines to form the first biopolymers. Equally significant is the implication that the state of contemporary biochemistry is a result of the chemical reactions that occurred on the primitive Earth, manifested, for example, by the crucial role of zinc in contemporary DNA and RNA biosynthesis and the apparent role of zinc in prebiotic chemistry.   [EDWARD EDELSON]

Bibliography: J. D. Bernal, *The Physical Basis of Life*, 1951; E. H. Edelson, L. E. Manring, F. Seidl, and J. G. Lawless, *Science*, in press; J. P. Ferris et al., *J. Mol. Evol.*, 11:293–311, 1978; J. P. Ferris and E. H. Edelson, *J. Org. Chem.*, 43: 3989–3995, 1978; J. P. Ferris and P. C. Joshi, *Science*, 201:361–362, 1978; S. L. Miller and L. E. Orgel, *The Origins of Life on Earth*, 1974.

Light

The speed of light c is one of the most interesting and important of the fundamental physical constants. It is used to convert light travel times to distance, as in the laser measurements of the distance to the Moon. It relates mass m to energy E in Einstein's famous equation, $E = mc^2$. The mea-

surement of c has challenged physicists for over 300 years, and recent measurements of the speed of light have increased the accuracy of its value a hundredfold. Highly precise values of c have been obtained by extending absolute frequency measurements into a region of the electromagnetic spectrum where wavelengths can be most accurately measured. These advances have been facilitated by the use of the laser and a high-speed tungsten-nickel diode which has been used to measure the laser's frequency. The speed of light had been one of the least accurately known fundamental constants, but now it is one of the most accurately known constants. The measurements of the speed of light and of the frequency of lasers are now opening the door for a redefinition of the meter in such a way that the value of the speed of light could assume a fixed value by defining the meter as the distance that light travels in a fraction of a second.

History. Prior to the observations of the Danish astronomer O. Roemer, the speed of light was thought to be infinite. In 1675 Roemer noted a variation of the orbiting periods of the moons of Jupiter that depended on the annual variation in the distance between Earth and Jupiter. He correctly ascribed the variation to the time it takes light to travel the varying distance between the two planets. The accuracy of Roemer's value of c was limited by a 30% error in the knowledge of the Earth's orbit at that time.

The first terrestrial measurement of c was performed by the French physicist H. L. Fizeau in 1849. His measurement of the time it took light to travel to a distant mirror and return resulted in a value accurate to 15%.

J. C. Maxwell's theory of electromagnetic radiation showed that both light and radio waves were electromagnetic and hence traveled at the same speed in vacuum. This discovery soon led to another method of measuring c: c was the product of the frequency and wavelength of an electromagnetic wave. In 1891 a French physicist, René Blondlot, first used this method to determine a value of c by measuring both λ and ν of a radio-frequency wave. His measurement demonstrated that c was the same for radio and light waves. It is this method which now exhibits the greatest accuracy, and it is used in the recent, most accurate measurements of c using a laser's frequency and wavelength.

In 1958 an English physicist, K. D. Froome, reported the speed of light to be 299,792,500 m/s, with an uncertainty of plus or minus 100 m/s. He measured both the frequency and the wavelength of millimeter waves from klystron oscillators to obtain this result. His major uncertainty lay in the difficulty of accurately measuring the wavelength of the radiation. Since short wavelengths can be measured much more accurately than long wavelengths, a shorter-wavelength source was needed; the laser soon provided such a source. However, a means of measuring its incredibly high frequency was needed. This problem, too, was soon overcome with the discovery of the tungsten-nickel point-contact diode.

Stabilized lasers. Before the advent of the laser, the most spectrally pure light came from the emission of radiation by atoms in electric discharges. The spectral purity of such radiation was about 1 part per million. Lasers, in contrast, have exhibited short-period spectral purities some hundred million times greater than this. However, the frequency of this laser radiation was free to wander over the entire emission line, and a means of stabilizing the frequency was necessary before it could be used in a measurement of c. Fortunately, the technique of sub-Doppler saturated absorption spectroscopy was soon discovered. This permitted the "locking" of the frequency of the radiation to very narrow spectral features so that the frequency (and, of course, the wavelength) remained fixed. Three different lasers at different wavelengths have been stabilized, and they serve as precise frequency and wavelength sources: the helium-neon laser at a wavelength of 3.39 μm stabilized with a saturated absorption in methane, the 10-μm CO_2 laser stabilized to a saturated fluorescence in CO_2, and the common red helium-neon laser stabilized to an iodine-saturated absorption. Both the 3.39-μm He-Ne laser and the CO_2 lasers have been used in accurate speed-of-light measurements, but the frequency of the red laser has not yet been measured.

Measurement of wavelength. The most precise measurements of wavelengths are made in Fabry-Perot interferometers, in which two wavelengths are compared by the observation of interference fringes of waves reflecting between two mirrors. A bright fringe occurs when the optical path length between the high-reflectivity mirrors is a multiple of a half-wavelength.

With the use of special frequency-controlled Fabry-Perot interferometers, wavelength measurements more accurate than ever were made, and a new limitation to the accuracy appeared: that of the length standard itself. The 605.8-nm orange

radiation from the krypton atom did not allow a measurement as precise as was possible with laser sources. This limitation now affects all speed-of-light measurements, with a resulting uncertainty of about 3 parts per 10^9.

In 1972 the comparison of the wavelength of the 3.39-μm laser with that of the krypton standard was made using such a Fabry-Perot interferometer. In the same year, the wavelength of the CO_2 laser was measured in a slightly different manner. A CO_2 laser sideband was added to a visible laser, and the wavelength difference was measured in the visible. The accuracies in the measurement of each wavelength were comparable.

Measurement of frequency. The frequency of an electromagnetic wave is the number of oscillations of the electric field in an electromagnetic wave occurring in 1 s. The measurement of frequency (in cycles per second, or hertz) is a counting technique in which the number of these oscillations is counted for a period of time determined by a time standard. Frequency and time are directly related; that is, frequency is the reciprocal of the period of a single oscillation. Electronic counters are available which directly count frequencies up to about 500,000,000 Hz. At higher frequencies, harmonic generation and mixing techniques must be used. Harmonics of a known frequency ν_0 are generated by illuminating a nonlinear device which distorts the electromagnetic wave, creating higher frequencies which are whole-number multiples n of the fundamental frequency ν_0. The resulting radiation at a frequency $n\nu_0$ is then subtracted from an unknown, higher frequency near $n\nu_0$ in a mixer to produce a difference or beat frequency ν_B, which is usually at a directly countable radio frequency. At infrared frequencies a tungsten-nickel point-contact diode serves to perform both of these steps in a single operation. Thus $\nu = n\nu_0 + \nu_B$; and where ν_0 is the known frequency, ν_B is directly counted, and ν is the unknown frequency.

In this way, one laser frequency is measured with respect to another, and a whole series of measurements must be made. In these tungsten-nickel diodes, the harmonic number is typically less than 12 because the signal to noise decreases with increasing harmonic number. The result is an entire chain of frequency measurements linking the fundamental frequency standard (the cesium standard, at 9.3×10^9 Hz) with frequencies of the various lasers used.

The accuracy of frequency measurement is limited only by the stabilities of the oscillators themselves, and is generally greater than that of the wavelength measurements. This was the case in the measurements of c; the uncertainties from frequency measurements were about 10 times smaller than those from the wavelength measurements.

Results. The values of c obtained from the frequency and wavelength measurements of various lasers are shown in the illustration. Because of the stability and reproducibility of the stabilized lasers, separate frequency and wavelength measurements are sometimes combined to give independent values of c, as shown on the graph. The first 1972 measurement did not involve an absolute counting of the laser frequency and was somewhat

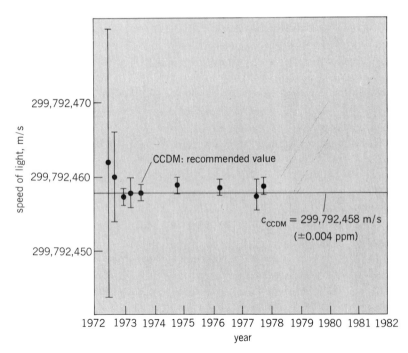

Values of the speed of light obtained since 1972 with the use of lasers. (From K. M. Evenson et al., Laser frequency measurements: A review, limitations, extension to 197 THz (1.5 μm), in J. L. Hall and J. L. Carlsten, eds., Laser Spectroscopy III, pp. 56–58, Springer, 1977)

less accurate. One rather amazing fact shown on the graph illustrates that, for the first time in history, the various values of the speed of light agree. Prior to 1958 the value of c often varied outside the limits of error quoted by the experimenters, and even prompted some observers to think that c might be changing with time. The 1974 meeting of the Consultative Committee for the Definition of the Meter (CCDM) recommended that 299,792,458 m/s be the value of c to be used for converting wavelength to frequency and vice versa, and in all other precise applications involving c. This number was arrived at by the consideration of the first four values of c on the figure. The subsequent measurements of c have shown that this was a good choice.

Future redefinition of the meter. The present definition of the meter is obviously archaic, and the meter will probably be redefined so that a laser can be utilized to measure length. One of the most promising ways of achieving this would be to define the meter as the distance that light travels in $1/c_0$ second, so that the speed of light would be fixed at c_0 meters/second. With such a definition, any suitably stabilized laser whose frequency had been measured could be used to realize the meter. Its vacuum wavelength, to be used to realize the meter, would simply be the fixed value of c divided by its frequency. This scheme would be workable because frequencies can be measured some hundred to a thousand times more accurately than wavelengths can, and the determination of a standard laser's wavelength via a frequency measurement would not affect the uncertainty in its applications as a standard of length. These stabilized lasers are sufficiently stable so that their frequency need be measured only once; then, any other similarly constructed laser would also emit this standard wavelength and could be used as a standard wavelength source without further frequency measurement.

With the possibility of this new definition, the era of speed-of-light measurements may be nearly at an end.

For background information *see* ELECTROMAGNETIC RADIATION; LASER; LIGHT in the McGraw-Hill Encyclopedia of Science and Technology.

[KENNETH M. EVENSON]

Bibliography: K. M. Evenson et al., in J. L. Hall and J. L. Carlsten (eds.), *Laser Spectroscopy III*, pp. 56–68, 1977; K. D. Froome and L. Essen, *The Velocity of Light and Radio Waves*, 1969; D. J. E. Knight and W. R. C. Rowley, *Survey Review XXIV*, 185:131–134, 1977; J. H. Sanders, *The Velocity of Light*, 1965.

Lithosphere

The outer shell of the Earth is composed of a mosaic of rigid plates that move about over the Earth's surface. In current geoscience usage, these plates are synonymous with the lithosphere.

The term "lithosphere," however, has had other meanings which have evolved as interest in, and knowledge of, the Earth's interior and its dynamic processes have changed. The lithosphere was defined in 1914 by Joseph Barrell as the outermost portion of the Earth, including the crust and part of the upper mantle, of sufficient strength to sustain stresses imposed by the Earth's topography. Below the lithosphere, a contrasting weaker region

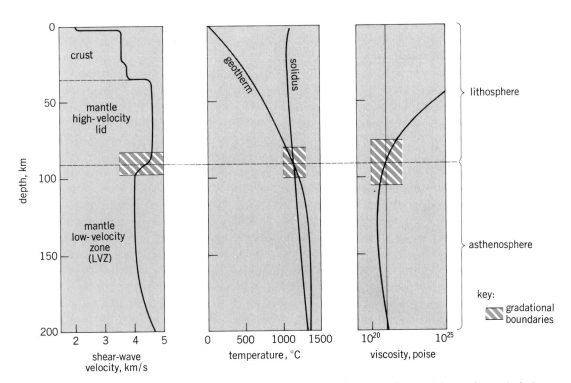

Fig. 1. Schematic patterns of some important physical properties of the crust and upper mantle of the Earth. The depth to the lithosphere-asthenosphere boundary generally increases slowly with increasing geological age but has strong present-day lateral variations.

marked by a capacity to yield readily to small but long-enduring shear stresses was called by Barrell the sphere of weakness or asthenosphere. This mechanical subdivision of the outermost few hundred kilometers of the Earth into the lithosphere and asthenosphere has long been a useful concept, particularly to scientists analyzing loading of the Earth's surface as occurs during glaciation, building of large river deltas, and generation of volcanic islands on the sea floor.

With the acceptance of plate tectonics in the 1960s, the lithosphere became associated with the rigid plates, and the asthenosphere with a mechanically weak zone that serves to decouple rigid lithospheric plates from deeper mantle regions. The lithosphere in this context is considerably thicker than that envisaged in the earlier view of Barrell. Still more recently, T. H. Jordan has attempted to reconcile both views by introducing the term "tectosphere" to describe that outer portion of the Earth composed of plates moving as coherent entities. The current plate definition of the lithosphere will be used in this article.

Thickness. Measurements of lithospheric and asthenospheric thickness have rested principally upon seismological evidence that identifies the asthenosphere with the seismic-shear-wave low-velocity zone in the upper mantle. This zone is postulated to be a region of partially melted ultrabasic rock. The lithosphere includes an overlying lid of upper mantle with higher seismic velocity and the crust (Fig. 1). Early studies had suggested a well-developed asthenosphere under oceanic regions with a uniform lithospheric thickness of about 80 km; in contrast, the asthenosphere under conti-

nental Precambrian shields was not as evident or well developed, and also deeper, between 100 and 150 km.

The results from more detailed studies based on the propagation of seismic surface waves across distinct geologic provinces and innovative analyses of shear waves reflected at the Earth's core-mantle boundary now require a reappraisal of the simplifed early picture. It is now believed that the lithospheric thickness varies from as little as 20 km, near actively spreading oceanic ridges, to as much as 300 km under old shields that have been tectonically stable for more than 1,000,000,000 years. Lithospheric thickness seems to bₑ controlled by the thermal and petrologic properties of the upper mantle.

Thermal model for thickness. Physical properties of the upper mantle, such as seismic velocities and strength, are more or less dependent upon mantle thermal conditions (Fig. 1), most commonly characterized by the ratio of the ambient temperature to the melting temperature (solidus). Thermal conditions appropriate for the present-day Earth are shown in Fig. 2 as a set of temperature-depth curves (geotherms) for oceanic and continental regions.

Continental geotherms depend on the surface heat flow, the vertical distribution of radiogenic heat sources estimated from petrologic arguments, and the experimentally determined temperature dependence of thermal conductivity. All continental geotherms are characterized by near-surface curvature due to crustal heat production and, for those geotherms not terminated by the solidus, a deeper region of curvature due in part to enhanced thermal conductivity at higher temperature and in part to heat production of undifferentiated or metasomatically enriched mantle rock. The oceanic thermal regime is characteristic of transient cooling of the oceanic lithosphere following formation at a spreading ridge, and is consistent with heat flow and bathymetry measurements made over the world's oceans. Each of the oceanic geotherms intersects the mixed-volatile solidus, implying that a partial melt zone may exist beneath the entire oceanic region. Beneath the continents, geotherms corresponding to a heat flow of more than 45 $mW \cdot m^{-2}$ also intersect the mixed-volatile solidus, but the lower heat flow geotherms do not. When examined in detail, the depth at which individual geotherms intersect the mixed-volatile mantle solidus agrees remarkably well with the depth to the top of the seismic low-velocity zone (or base of the lithosphere) determined by independent methods. Thus a careful analysis of regional heat flow can be used to infer the depth to the top of the low-velocity layer, corresponding to the base of the lithosphere.

A global heat flow map for the Earth was developed by D. S. Chapman and H. N. Pollack. This map is contoured from a twelfth-degree spherical harmonic expansion, which smooths local anomalies but accurately represents thermal features more important to lithospheric studies. For each 5° × 5° grid element on the globe, the surface heat flow value was associated with a geotherm, and the depth at which that geotherm intersects the mantle solidus was identified as the local value for lith-

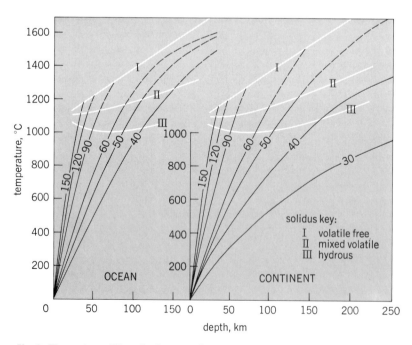

Fig. 2. Thermal conditions for the outer few hundred kilometers of the Earth. Individual curves of temperature versus depth (geotherms) characterize specific oceanic or continental regions, and are identified by their surface heat flow values. The mixed-volatile solidus is likely the best approximation to conditions which will produce partial melting of the mantle. (*From D. S. Chapman and H. N. Pollack, Regional geotherms and lithospheric thickness, Geology, 5:265–268, 1977*)

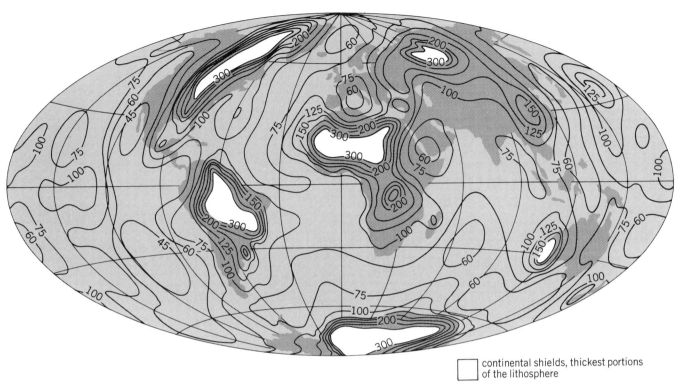

continental shields, thickest portions of the lithosphere

Fig. 3. Thickness of the lithosphere determined from a correlation with a twelfth-degree spherical harmonic representation of global heat flow. Contours are in kilometers with variable interval. The lithosphere is thinnest at oceanic spreading ridges and thickest under continental shields. *(From D. S. Chapman and H. N. Pollack, Regional geotherms and lithospheric thickness, Geology, 5:265–268, 1977)*

ospheric thickness. Figure 3 displays the variations of lithospheric thickness so derived over the entire Earth. The lithosphere is thinnest at active oceanic ridges and thickest under continental shields, with a range of about 300 km over the entire Earth. As previously noted, the shield geotherms probably do not intersect the solidus, and thus in a strict sense the asthenosphere should not exist beneath shields. More realistically it is inferred from independent data on plate motions that plates containing shields do decouple from the deep interior, probably at a depth at which the geotherm makes its closest approach to the melting curve of the mantle. In a sense, the shields may act as viscous keels for the plates.

Deep continental roots. Thermal, seismological, and petrologic evidence supports the hypothesis of thick root zones beneath old continental cratons. The development of a thick root zone, as argued by T. H. Jordan, comes about by basaltic depletion of the upper mantle in continental rifting and subduction processes. The depleted mantle is stabilized at lower temperatures because its solidus is elevated and therefore its viscosity is greater than the surrounding undepleted mantle. This evolution of a thick continental lithosphere probably began in the Archean and, aided by major continent-continent plate collisions, has proceeded until the present, producing the stable cratons of the world. The greater viscosity of thick continental roots beneath cratons may play an important role in retarding plate motions; several investigations have shown that predominantly continental

plates move on the average about one-fifth the velocity of oceanic plates.

Although the precise definition of the lithosphere continues to provoke discussion, the Earth sciences community is now accepting the concept of plates whose thickness varies from a few tens to a few hundreds of kilometers. Understanding the evolution of the thickest portions of the plates underlying cratons and the manner in which these regions affect plate motions and tectonic processes promises to be a fruitful area of future research.

For background information *see* LITHOSPHERE; PLATE TECTONICS in the McGraw-Hill Encyclopedia of Science and Technology.

[DAVID S. CHAPMAN]

Bibliography: D. S. Chapman and H. N. Pollack, *Earth Planet. Sci. Lett.*, 23:23–32, 1975; D. S. Chapman and H. N. Pollack, *Geology*, 5:265–268, 1977; T. H. Jordan, *Sci. Amer.*, 240:92–107, 1979; H. N. Pollack and D. S. Chapman, *Sci. Amer.*, 237:60–76, 1977.

Magnetism

One-dimensional (1D) magnetism has been the subject of extensive research activity in recent years, and it has been demonstrated that various materials are well described by simple 1D magnetic models. The following three compounds constitute especially good model systems and have therefore attracted a great deal of attention — $(CH_3)_4NMnCl_3$ (TMMC): antiferromagnetic Heisenberg system, spin = 5/2; $CuCl_2 \cdot 2NC_5D_5$ (CPC): antiferromagnetic Heisenberg system,

chain direction

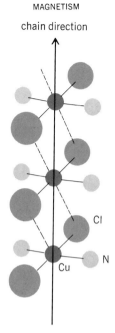

Fig. 1. Linear chain of Cu⁺⁺ ions in CPC.

spin $= 1/2$; and $CsNiF_3$: ferromagnetic easy plane system, spin $= 1$. Very recently, important new results have been achieved from experimental studies of these three compounds, as discussed below.

Properties of 1D magnetic materials. The fundamental structural feature of quasi-1D magnetic materials is the existence of linear chains of magnetic atoms, the chains being well separated from each other. Since the spins of the magnetic atoms couple through short-range forces, essentially only neighboring magnetic atoms within the same chain interact with each other, thus giving rise to the 1D magnetic properties. A sufficiently large separation between chains can be achieved with intervening organic groups like the pyridine group $[NC_5D_5]$ in CPC or the tetramethyl ammonium complex $[N(CH_3)_4]^+$ in TMMC, thus leading to fairly complex chemical formulas for these compounds.

TMMC contains chains of the magnetic ion Mn^{++}, which has a spin $S = 5/2$. Extensive experimental studies of both static and dynamical properties of TMMC conform closely to predictions based on a 1D classical antiferromagnetic Heisenberg model. CPC is a unique realization of the 1D $S = 1/2$ antiferromagnetic Heisenberg model. The magnetic ions are Cu^{++}, which form linear chains linked together by Cl^- ions, as shown in Fig. 1. Due to the quantum limit of the spin value, nonclassical behavior is predominant, as has been beautifully demonstrated by inelastic neutron scattering experiments: the energies of the low-lying magnetic excitations differ by a factor of $\pi/2$ from the values yielded by a classical spin-wave approach.

In $CsNiF_3$ the magnetism arises from the chains of Ni atoms, each carrying a spin $S = 1$. The exchange interaction is ferromagnetic, and causes the neighboring spins within a chain to line up parallel to each other, contrary to the case for the two former compounds. Strong crystal field effects favor alignment of the spins in the plane perpendicular to the chain (the xy plane), so that $CsNiF_3$ is the best available example of the 1D ferromagnetic xy model.

In order to illustrate some very recent developments in the field, three new experimental studies will be discussed. In the first of these, the asymmetrical line shapes of the inelastic neutron peaks demonstrate the quantum nature of CPC and confirm recent theories. In the second, the experimental results found for TMMC doped with Cu impurities are in agreement with theoretical results obtained from an exactly soluble model with disorder. Finally, there is evidence which indicates that $CsNiF_3$ in a magnetic field is the most promising system in the quest for experimental observation of soliton excitations. All three of these studies have been carried out by means of neutron scattering.

Quantum effects in CPC. Figure 2a shows the results of exact quantum-mechanical calculations of low-lying excited states of the antiferromagnetic $S = 1/2$ Heisenberg hamiltonian given by the equation below. The index i numbers the spins along

$$H = 2J \sum_i \vec{S}_i \cdot \vec{S}_{i+1}$$

the chain, and a denotes the nearest-neighbor spacing. The horizontal axis in Fig. 2a shows the wave vector q (in units of $1/a$) of the eigenstates,

while the vertical axis denotes the energy E (in units of J) of the eigenstates, relative to the ground-state energy. Thus, these excitations form a continuum with a lower and upper boundary.

Neutron scattering is a unique experimental technique for directly probing magnetic excitations like those of Fig. 2a for given values of momentum transfer q. Figure 2b and 2c shows energy distributions of neutrons scattered from a CPC crystal with $qa = 1.2\pi$ and 1.65π, respectively. One of the curves in Fig. 2b is calculated for the case where only the lower boundary of excited states is assumed to contribute to the neutron scattering. A second curve is based on a theory by H. J. Mikeska, which takes into account the continuum of higher states. The theory predicts that only for qa values in the vicinity of π is the neutron scattering sensitive to these higher excited states, so that in Fig. 2c the two curves coincide. It is seen that the experimental results conform closely to this

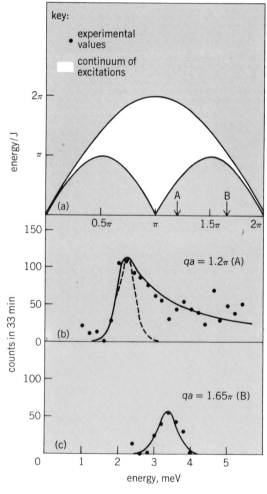

Fig. 2. Quantum effects in CPC. (a) Spectrum of low-lying excitations (eigenstates) of the antiferromagnetic $S=1/2$ Heisenberg hamiltonian. (b) Energy distribution of scattered neutrons probing excitations for wave vector q such that $qa = 1.2\pi$. Broken line indicates values calculated from contribution of lower boundary of excited states only. Solid line indicates values calculated from theory, taking contributions of higher states into account. (c) Energy distribution for $qa = 1.65\pi$.

prediction. The exact theoretical results shown in Fig. 2a clearly demonstrate the quantum-mechanical behavior of this system, and it is worth noting that approximate theories, like classical spin-wave theory, would yield distinctly different results. Thus such an approach would not be able to explain the observed asymmetric line shape, and moreover would yield a spectrum of lowest excited states with a maximum energy of $2J$ instead of πJ, as observed in earlier studies.

Impurity effects of TMMC. The properties of solids are markedly affected by the presence of defects such as impurity atoms, and a proper understanding of such phenomena is therefore important. Impurity atoms randomly substituted for the magnetic atoms are expected to affect the magnetic properties of a quasi-1D system much more drastically than those of a "normal" 3D system. The reason is obvious: if a bond between two atoms is broken, the 1D lattice is divided into two parts, while the 3D lattice is hardly affected. For magnetic materials this phenomenon is conveniently described through the spin-spin correlation length λ, which can be determined directly by means of quasi-elastic neutron scattering. There is no correlation in the relative orientation of two atomic spins separated by much more than the distance λ. In practice, the inverse correlation length $\kappa = 1/\lambda$ is used in analyzing experimental results.

Figure 3 shows recent neutron scattering studies of κ at various temperatures for Cu-doped TMMC, $(Cd_3)_4NMn_cCu_{1-c}Cl_3$. Measurements were made on samples with 15% ($c = 0.85$) and 7% ($c = 0.93$) Cu^{++} ions randomly substituted for the Mn^{++} ions. Also shown are results of earlier measurements of κ on pure TMMC ($c = 1$). Since the Cu^{++} ions themselves are magnetic, there are magnetic interactions J' between a Cu^{++} ion and a neighboring Mn^{++} ion in addition to the interaction J between neighboring Mn^{++} ions. J' is much smaller than J. The solid curves in Fig. 3 show theoretical results for the two impurity concentrations and also for the pure system ($c = 1$). The calculations are based on a classical Heisenberg model, properly accounting for the statistical distribution of the exchange parameters. These calculations show several characteristic features: (1) At a given temperature, κ increases markedly with impurity concentration. (2) At higher temperatures, κ depends linearly on the temperature T. This is the region where J' is unimportant. (3) At lower temperatures, κ bends down, reflecting the onset of correlations being transmitted through J' ("the weak link"). If all the interactions are antiferromagnetic, κ goes to zero ($\lambda \to \infty$) as the temperature approaches zero, indicating a transition to a long-range ordered state at $T_c = 0$. The temperature dependence of κ as shown in Fig. 3 is characteristic of a 1D magnet. In a 3D system, κ increases much more rapidly when the temperature exceeds the transition temperature T_c. Exact theoretical calculations on 3D systems for critical fluctuations corresponding to those in Fig. 3 for 1D systems do not exist.

Solitons in CsNiF₃. Quasi-1D magnets have recently attracted considerable interest in connection with the experimental evidence for solitons. The term soliton has been applied to a special kind of excitation predicted in systems with strongly anharmonic interactions. One of the few exactly

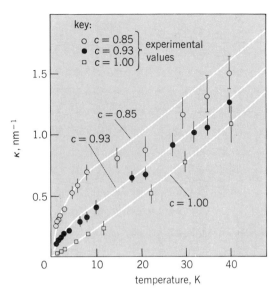

Fig. 3. Inverse correlation length versus temperature for Cu-doped TMMC, $(Cd_3)_4NMn_cCu_{1-c}Cl_3$. Data points give experimental values. Curves correspond to the theory as described in the text.

soluble models exhibiting soliton excitations is the classical 1D chain of pendula oscillating in a plane perpendicular to the chain and coupled via harmonic springs. The equation of motion for this system is the classical "sine-Gordon" equation. In the ground state all pendula hang vertically down, and low-energy excitations consist of propagating waves with small (harmonic) displacement amplitudes. Qualitatively different and of much higher energy are the states where the chain is "twisted" one full rotation. These are the solitons.

There is a close analogy between this model and the 1D ferromagnetic xy model with a magnetic field applied parallel to the xy plane. This model system is realized by the quasi-1D magnet CsNiF₃ in a magnetic field applied perpendicular to the chain direction. The analogy to the model described above is obvious: spins correspond to the pendula, the applied magnetic field corresponds to gravity, and the exchange coupling corresponds to the harmonic springs. The low-lying excitations are the usual spin waves. A soliton excitation is shown in Fig. 4 and requires much more energy to be created than a spin-wave excitation (magnon). J. K. Kjems and M. Steiner have recently carried out neutron scattering measurements on this system for various values of the magnetic field and temperatures. They showed that the results can be interpreted by assuming the presence of a "gas" of

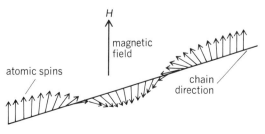

Fig. 4. Soliton in the ferromagnetic xy chain in a magnetic field.

thermally excited solitons, giving rise to an additional quasi-elastic neutron scattering cross section. Experimental confirmation of these ideas is being looked for in further studies of this compound, and other materials with similar properties are being sought.

The recent studies described here, as well as a variety of earlier studies on 1D magnetic systems, all demonstrate the extent to which the experimental results are quantitatively accounted for by theoretical predictions, perhaps the most gratifying feature of 1D magnetism, and of 1D physics in general. Several model systems still remain to be found, and continuing research activity in 1D magnetism can be anticipated in the future.

For background information *see* ANTIFERROMAGNETISM; FERROMAGNETISM in the McGraw-Hill Encyclopedia of Science and Technology.

[I. U. HEILMANN; G. SHIRANE]

Bibliography: R. J. Birgeneau and G. Shirane, *Phys. Today*, 31(12):32–43, 1978; I. U. Heilmann et al., *Phys. Rev.*, B18:3530–3536, 1978; J. K. Kjems and M. Steiner, *Phys. Rev. Lett.*, 41(16): 1137–1140, 1978; M. Steiner, J. Villain, and C. G. Windsor, *Advan. Phys.*, 25:87–209, 1976.

Maser

Laboratory masers were first made a generation ago and now have numerous uses in science and industry. In 1965, however, it was learned that nature had anticipated human efforts throughout the Milky Way Galaxy. Naturally occurring masers are now known to exist near regions where massive, hot stars are being formed, and in the circumstellar atmospheres of certain types of old, dying stars.

The first celestial masers were discovered by Harold F. Weaver and his group at the University of California at Berkeley. They were hydroxyl masers near the Great Nebula in Orion and in the radio source W49. Both regions are known to be sites of recently formed massive stars. Three years later, OH was found in the distended atmospheres

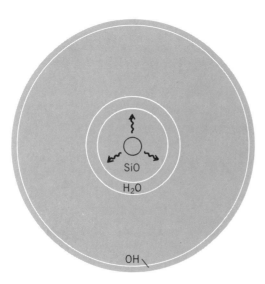

Fig. 1. Schematic view of a red variable star showing the outflow of material into the circumstellar shell and the region where each molecular maser is found.

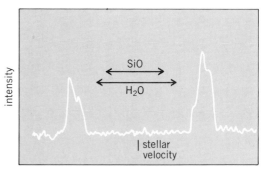

Fig. 2. Representative OH maser spectrum showing the two groups of features produced by gas flowing from the star toward and directly away from the observer. Corresponding Doppler velocities of H_2O and SiO masers are indicated by arrows.

of old red variable stars. Despite the timetable of discovery, the stellar case is the better understood of the two—although the knowledge which has been obtained is still modest.

Three molecules are now known to be astronomical masers: the hydroxyl radical (OH), water (H_2O), and silicon monoxide (SiO). There are other molecules which exhibit slight deviations from equilibrium (nonmaser) conditions, the most notable being the 1.3-cm transitions of methyl alcohol observed in Orion; but only OH, H_2O, and SiO show the very intense emission which warrants the title "maser."

Stellar masers. The OH molecule has three transitions near 18 cm which are often observed. The strongest (in the laboratory, or equilibrium, case) is at a frequency of 1667 MHz, the weakest at 1612 MHz. The latter is normally one-ninth the intensity of the former. In the red variable stars (Mira variables or semiregular variables), however, the 1612-MHz transition usually dominates by a substantial margin. This reversal was one of the original clues that the upper energy level of this transition was overpopulated (inverted) and that the molecule was a natural maser at this frequency.

Pumping process. It was also a clue to the pumping process by which this inversion is achieved. The red variable stars, as their name implies, grow alternately brighter and dimmer in cycles typically about 1 year in length. The OH maser has been observed to follow these changes with a delay of not more than 2 weeks. (This behavior is not, in general, true for H_2O and SiO masers.) Because the OH is observed to come from the outer regions of the atmosphere, some 10^{10} km distant from the star itself, the maser pump could not be a mechanical process like shock waves or collisions. Such processes propagate at speeds of order 10 km/s; thus, if the intensity changes in the star were coupled to the OH in this way, it would take $10^{10} \div 10 = 10^9$ s (about 30 years) to communicate this to the outer atmosphere. Because the OH mimics the stellar changes within about 2 weeks, the excitation or pumping of the maser must be strongly coupled to the radiation of the star which, of course, travels at the speed of light. The strength of the 1612-MHz transition confirms this, because a detailed knowledge of the quantum mechanics of the OH molecule reveals that this transition is readily inverted when the molecule is pumped with infrared radiation. Because red variable stars are known to be rich in infrared radiation, the picture has compelling consistency. And, not only does the 1612 MHz maser confirm the radiative pump, but it tells which frequency domain of the star's radiation is most important.

Observation of gas outflow. The circumstellar shells of these stars are composed of gas and dust being blown out with each pulsation cycle of the star (Fig. 1). The OH maser provides kinematic information which aids in understanding this process. On account of the Doppler effect, the 1612-MHz maser is shifted slightly in frequency because of the OH molecule's motion in the circumstellar gas and because of the star's motion, itself, through the Galaxy. Because the maser radiation is very directional, emission is not observed from the entire atmospheric shell surrounding the star, but only from gas which is moving from the star directly toward the Earth, or from the star directly away from the Earth. This motion Doppler-shifts the 1612-MHz transition into two distinct frequency groups. This characteristic double-peaked spectrum is shown in Fig. 2, a somewhat idealized "average" spectrum. Because the frequency shift is proportional to the velocity, the velocity of outflow of the gas from the star, and, at the midpoint of the two features, the velocity of the star itself along the Earth's line of sight, are immediately obtained.

Dependence on distance from star. As the gas and dust flow away from the central star, they are accelerated by the radiation pressure — the "push" provided by the radiated energy intercepted by the gas molecules and the grains of dust. Hence, the greater the distance from the stellar surface, the greater the velocity of outflow. Also, of course, the greater the distance from the star, the cooler the temperature. Thus, gas near the star is hotter and has a smaller velocity of outflow; gas far from the star is cooler and has a greater velocity of outflow.

This situation is reflected in the H_2O and SiO maser emission. The radiation from the OH molecule comes from its ground state (lowest energy level). Water masers originate from a more excited state whose energy corresponds (in temperature terms) to several hundred degrees Kelvin above its ground state. SiO masers originate in the most excited levels of all, having energies which correspond to approximately 1800, 3600, and 5400 K — depending on which of three transitions is observed. One might expect the masing conditions for OH to be best in cooler outer regions, where the velocity is highest; those for water in more intermediate temperature regions; and those for SiO in regions closest to the star where temperatures are higher and gas velocities lower (Fig. 1). Thus, OH would have the broadest velocity spread with H_2O emission nesting within its extremes, and, finally, SiO would have the narrowest velocity spread of all. The observational evidence indicates that this is the case (Fig. 2).

Irregular variations. Additionally, many H_2O

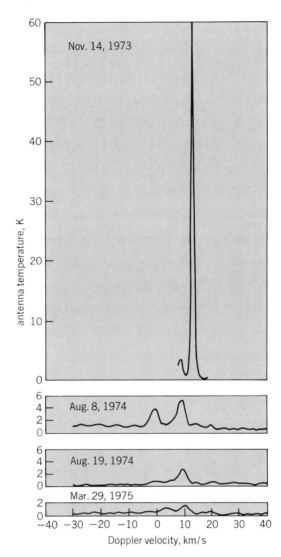

Fig. 3. Group of spectra from the water maser in the bright nebula Messier 17, showing dramatic changes in time. Antenna temperature is a measure of the intensity of radiation received by the radio telescope. *(From D. F. Dickinson, Cosmic masers, Sci. Amer., 238(6):90–105, 1978)*

and SiO masers show irregular variations in time, uncorrelated with the star's intensity changes. This is unlike the OH and immediately suggests that these masers come from a different spatial domain than the OH and could have a different pumping mechanism as well.

Thus, not only is the maser phenomenon interesting in itself, but it provides many insights into the red variable stars, objects of considerable interest in the scheme of stellar evolution.

Interstellar masers. In the neighborhood of newborn, massive stars, OH and H_2O are both seen as strong masers. Water masers, in fact, have come to be regarded as a signature of such regions. Here, variations in time are also seen, but the regular periodicity of the stellar OH is absent. An example of the dramatic changes which sometimes occur is shown in Fig. 3. There are no firm clues to the pumping mechanism, and the velocity structure often provides a bewildering choice of Doppler frequencies. As a further piece to the puzzle, SiO is never seen as a maser in this context, although it is known to be present from observations of SiO transitions in thermal equilibrium (nonmaser emission).

Recent high-resolution observations of some H_2O masers near new stars suggest that their spatial distribution may be orderly. Some, for example, appear to lie along an extended arc. This may indicate they are associated with regions of higher density brought on by shock waves created by the new stars when they "turn on." This is by no means clear, as yet, but observations are being carried out to look for the telltale signs that shock waves would leave.

This idea is doubly important because recent ideas in star formation suggest that shock waves propagating through the interstellar gas and dust "sweep up" material into regions of higher density—dense enough, in fact, that they may then collapse from their own gravitational pull. When the heat and pressure generated by the collapse become high enough, nuclear burning begins and these regions become new stars.

Thus, the case of the nonstellar maser is not well understood, only described, at present. A wealth of good observational data exists, and substantial theoretical inroads may be made in the early 1980s.

For background information *see* DOPPLER EFFECT; MASER; RADIO ASTRONOMY; VARIABLE STAR in the McGraw-Hill Encyclopedia of Science and Technology. [DALE F. DICKINSON]

Bibliography: D. F. Dickinson, *Sci. Amer.*, 238(6):90–105, June 1978; D. F. Dickinson, B. Zuckerman, and M. M. Litvak, *Sky Telesc.*, 39(1):4–7, January 1970.; K. J. Johnston, S. H. Knowles, and P. R. Schwartz, *Sky Telesc.*, 44(2):88–90, August 1972; J. M. Pasachoff, *Contemporary Astronomy*, 1977.

Materials analysis

New ion-beam techniques for measuring the elemental composition of materials are revolutionizing materials analysis. These include: proton-induced x-ray emission analysis, which is capable of determining all the elements with atomic number $Z \gtrsim 10$ in a single few-minute analysis; Rutherford backscattering spectrometry, which is capable of determining quantitative profiles of concentration versus depth for all heavy elements; nuclear reaction analysis, which is capable of quantitatively measuring concentration profiles of light elements such as hydrogen; and ultrasensitive secondary ion mass spectrometry, which is capable of measuring isotope ratios or trace element concentrations as small as 10^{-15}. These new analytic tools are having impact in many fields, including such various ones as medicine, microelectronics, the study of surfaces and surface reactions, art history and archeology, physical chemistry, and geophysics and geochemistry.

Proton-induced x-ray emission. One of the new analytic tools which has become widely used to analyze air and water pollution samples is proton-induced x-ray emission (PIXE) analysis. A typical experimental arrangement is shown in Fig. 1. A small Van de Graaff accelerator provides a proton beam (typically at 2 MeV) which bombards the sample. This bombardment knocks electrons out of low-lying atomic orbitals, thus creating vacancies. These vacancies may be filled by electrons cascading down from higher orbitals, emitting characteristic x-rays in the process. The energy of these characteristic x-rays depends on the atomic number of the element and, as a consequence, can be used to unambiguously identify the elements present in the sample.

PIXE is a powerful technique because it can assay an unknown for, essentially, all elements with $Z \gtrsim 10$ in a single few-minute proton bombardment. It is this feature which makes PIXE so attractive in the broad surveys common in studies of environmental air and water pollution where very large numbers of samples need to be analyzed.

One of the important capabilities of ion-beam techniques, such as PIXE, is the possibility of bombarding with microbeams (having a diameter of about a micrometer or less) in order to analyze a very small part of the sample. By sweeping such small beams over the surface of a sample, it is possible to obtain concentration maps of the various elements present. This microbeam capacity is important in applications as various as medicine biology, metallurgy, microelectronics, and the study of particulate air pollution. *See* SCANNING PROTON MICROPROBE.

Another feature of PIXE is the ability to analyze samples in an air environment by bringing the proton beam from the vacuum necessary for an accelerator through a thin foil into air. The ability to do analysis in air both is convenient and can be important for samples which might change (for example, dehydrate) if placed in vacuum. Analysis can even be made on living organisms.

Rutherford backscattering spectrometry. The study of surface structure and surface reactions, thin films, ion implantation, and diffusion is becoming technologically very important, but presents a different type of analytic problem from those discussed above. In such studies, the elements present are usually known; what is needed is a quantitative determination of the concentration of each element versus depth into the surface. For example, in the microelectronics industry, ion implantation is used to introduce dopants into the semiconductor. In order to properly predict the properties of such a device, it is essential to know the dopant distribution, which depends both on the

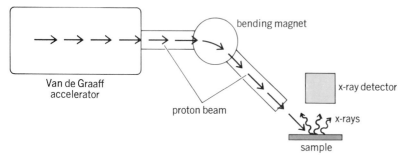

Fig. 1. Schematic representation of an experimental setup for proton-induced x-ray analysis. A small accelerator provides a low-energy proton beam which bombards the sample, and the characteristic x-rays emitted are recorded with an x-ray detector.

range of the implanted ions and on any subsequent diffusion.

The technique most widely used to study such problems is Rutherford backscattering spectrometry. This method relies on the fact that the energy of an ion backscattered from a sample depends only on (1) the mass of the atomic nucleus from which it scattered and (2) the depth at which the scattering occurred. A backscattered ion has a larger fraction of its incident energy when it scatters from a heavy atom than when it scatters from a light atom. The deeper in the sample the scattering occurs, the lower the observed energy of the backscattered ion, because ions lose energy as they traverse matter.

By measuring the energy spectrum of backscattered particles from a sample, the concentration versus depth of atoms of different masses can be quantitatively determined. Such analyses are particularly easy when concentration profiles of heavy atoms in a light-element host are needed, as is often the case when considering dopant distributions in silicon. An illustration of the use of backscattering is shown in Fig. 2a. Here, 2-MeV He ions bombard a silicon sample which had been ion-implanted with arsenic (As). Arsenic (mass = 74.9 atomic mass units) is much heavier than Si (mass = 28.1 amu), and as a consequence, He backscattered from As retains a much larger fraction of its energy (82%) than it does when it backscatters from Si (57%). The high-energy part of the spectrum at energies near 1.5 MeV (giving rise to the peak at the right of Fig. 2a) results only from backscattering from the As; and from this measured energy distribution, using the known rate of energy loss of He in Si, the concentration versus depth of the As can be quantitatively determined. In Fig. 2a the displacement ΔE_{As} of the scattering peak from the initial energy of ions backscattered from arsenic (denoted by the arrow labeled As) is used to determine the depth R_p of the arsenic layer.

In order to understand the effects of dopants, or impurity atoms in general, it is usually necessary to know where they reside in the crystal lattice, that is, whether they reside in interstitial or substitution sites. Such determinations are now routinely made by measuring backscattering (or nuclear reactions, discussed below) in a channeling geometry. Because atoms in a crystal lattice are in a regular array with channels between the rows of atoms, it is possible to direct the bombarding ion beam down a channel where it will preferentially backscatter from interstitial atoms in the center of the channel.

Shown in Fig. 2b are backscattering results illustrating this principle. This figure shows the backscattering yield from a single crystal of silicon implanted with ytterbium (Yb) as a function of the angle between the bombarding beam and the direction of the <110> channel in silicon. The yields from both Si and Yb are shown. In precisely the <110> direction, the yield from the Yb sharply peaks because the Yb atoms are located in the center of the <110> channel. Under the same conditions, the backscatter yield from the Si decreases because of blocking. This blocking occurs because most of the silicon atoms are aligned precisely behind the first silicon atom in a lattice row;

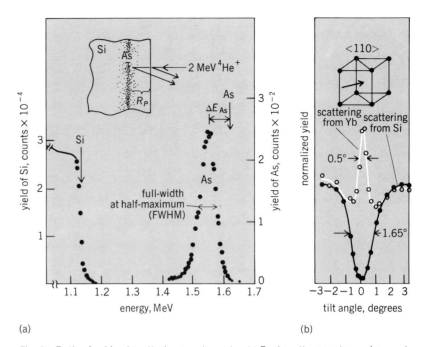

(a) (b)

Fig. 2. Rutherford backscattering spectrometry. (a) Backscatter spectrum observed when a 2-MeV He beam is backscattered from silicon ions implanted with arsenic (As). (b) Normalized backscattering yield from single-crystal silicon implanted with ytterbium, plotted as a function of the tilt angle between the He beam and the <110> direction. (From W. K. Chu, S. M. Mayer, and M. A. Nicolet, Backscattering Spectrometry, Academic Press, 1978)

as a consequence, there is a much smaller chance for backscattering to occur from these "blocked" Si atoms than if they were randomly arranged in the solid (or if the beam were randomly oriented). Such blocking dips occur whenever the incident beam is along a channel direction, but peaking, such as in Fig. 2b, occurs only when the impurity

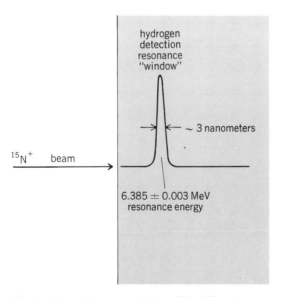

Fig. 3. Schematic representation of the ¹⁵N nuclear resonance method of measuring hydrogen concentration versus depth in any solid. (From W. A. Lanford et al., A new precision technique for determining the concentration versus depth of hydrogen in solids, Appl. Phys. Lett., 28: 566–568, 1976)

atoms are in the channel (as opposed to being in substitution sites which would also be blocked).

The use of channeling and blocking in conjunction with Rutherford backscattering spectrometry

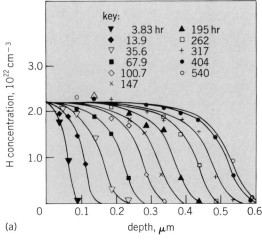

(a)

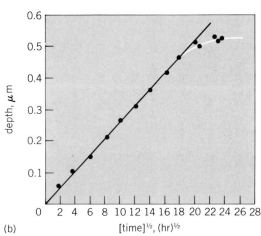

(b)

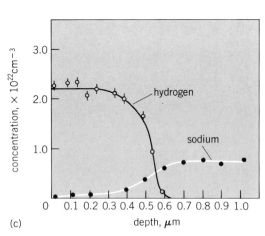

(c)

Fig. 4. Illustration of the use of the ^{15}N hydrogen-profiling technique to study the kinetics of a water-based chemical reaction. The samples are a series of soda-lime glasses exposed to water at 90°C for varying lengths of time. (a) Resulting hydrogen concentration profiles. (b) Depth of penetration versus the square root of the exposure time. (c) Superposition of hydrogen and sodium profiles, showing that the hydrogen (from the water) and the sodium (in the original glass) are anticorrelated. (From W. A. Lanford et al., Hydration of soda-lime glass, J. Non-Crystal. Sol., 33:249–266, 1979)

or nuclear reaction spectrometry is often the only viable method of determining the lattice location of impurity atoms.

Nuclear reaction spectrometry. Light elements present special analytical problems because even the rather general methods discussed above lose sensitivity when applied to light elements. The lightest element, hydrogen, is an extreme example; its atomic number is so low that it has no x-ray transitions useful in PIXE, and its mass is so small that it cannot backscatter even a hydrogen ion. For similar reasons, hydrogen is also invisible to other general analytical methods such as Auger spectroscopy and neutron activation analysis. Recently, however, very successful nuclear reaction techniques for measuring hydrogen concentration versus depth in any solid have been developed. Because hydrogen analysis provides some of the most interesting applications of nuclear reaction spectroscopy, it will be discussed in detail, but there are similar analyses for other light elements.

The presence of hydrogen can have dramatic effect on the physical, chemical, and electrical properties of many materials. The hydrogen embrittlement of steel is a well-known (and poorly understood) example, but hydrogen also plays a central role in the mechanical properties of minerals, giving rise to hydrolytic weakening and hydrofracture. In thin films, hydrogen is important in superconductors (where it can influence the superconducting transition temperature), in semiconductors (where it is an important dopant), and in insulators (where it can affect both chemical durability and electrical properties). Another class of problems where hydrogen analysis can be important is in the study of water-based chemical reactions (for example, in corrosion or weathering studies) where hydrogen can be used as a marker to study the reaction kinetics and reaction mechanisms.

As part of the space program, T. A. Tombrello at the California Institute of Techology pointed out that a ^{19}F-induced nuclear reaction could be used to study the hydrogen distribution in solids, and he applied this new method to the analysis of Moon rocks. W. A. Lanford at Yale developed a more generally useful method based on a ^{15}N-induced nuclear reaction, and he has applied this method in a wide variety of areas, including "neutron bottle" physics, energy, art history and archeology, high-technology thin films, and glass science. *See* NEUTRON.

The ^{15}N hydrogen profiling technique makes use of a narrow isolated resonance in the $^{15}N + ^{1}H \rightarrow ^{12}C + ^{4}He + 4.43\text{-MeV}$ gamma-ray reaction to probe for hydrogen. Because this is a resonance reaction, it occurs only when the ^{15}N is at a particular energy, the resonance energy. If a sample is bombarded with ^{15}N at the resonance energy, the yield of 4.43-MeV characteristic gamma rays is proportional to the hydrogen on the surface. If the ^{15}N energy is raised, there are no reactions with hydrogen on the surface because the ^{15}N is above the resonance energy. However, as the ^{15}N ions penetrate the sample, they lose energy and, as a result, reach the resonance energy at some depth. Now the yield of characteristic gamma rays is proportional to the hydrogen concentration at this depth. Thus, the method measures the hydrogen in a "window," about 3 nm wide, whose depth is determined by the beam energy. This is shown schematically in Fig. 3. In summary, by measuring the gamma-ray yield versus ^{15}N-beam energy, the hydrogen concentration versus depth is determined. Because this technique relies on a nuclear reaction which is independent of chemical bonding, it yields no information on the chemical form of the hydrogen but, as a result, the method is inherently quantitative.

The ^{15}N hydrogen profiling technique has been used to study hydrogen in a wide variety of high-technology thin-film materials such as: amorphous silicon (solar cells), where the presence of hydrogen is crucial to the usefulness of this material as a semiconductor; Nb_3Ge superconductor, which is the material with the highest known superconducting transition temperature (23 K) and where hydrogen as an impurity or as a deliberate dopant plays an important role in the transition from the normal to superconducting state; and insulators such as plasma-deposited silicon nitride or anodically grown Al_2O_3, where hydrogen as an impurity plays important roles in the chemical durability and electrical properties, respectively.

A different type of application, shown in Fig. 4, is an illustration of the use of hydrogen profiling in the study of a water-based chemical reaction. Because of the good sensitivity and depth resolution (approximately 3 nm), the kinetics of chemical reactions in which hydrogen is incorporated into a surface can be very efficiently studied by measuring hydrogen profiles of reacted surfaces. Figure 4a and b shows the time dependence of the penetration of hydrogen into a commercial soda-lime glass (SiO_2, Na_2O, and CaO) which has been exposed to water at 90°C for varying lengths of time. Figure 4c shows the superposition of hydrogen and sodium profiles, demonstrating that the Na is depleted wherever the hydrogen is incorporated. This sodium profile was also measured by using a resonant nuclear reaction.

Data such as those shown in Fig. 4 have been applied to settling long-standing arguments over the reaction mechanisms responsible for the weathering and corrosion of glass. In addition, this approach has been used to measure durability of both commercial glasses and new materials such as glasses designed to consolidate and isolate radioactive reactor wastes.

The example shown in Fig. 4 is just one illustration of the possible application of hydrogen profiling to the study of water-based reactions in a variety of materials.

Secondary ion mass spectrometry. Of all the ion-beam analysis techniques, perhaps the simplest conceptually is the direct observation of ions knocked (sputtered) off the sample by the bombarding beam. The difficulty with this technique is that the process by which ions are sputtered off surfaces is incompletely understood and, as a consequence, it is difficult to make secondary ion mass spectrometry (SIMS) quantitative.

Nevertheless, SIMS has become a widely used analytic tool because (1) it is useful for all elements, (2) it can have extreme sensitivity, and (3) since the sputtering process itself erodes away the sample, by measuring the yield of ions as a function of time, concentration depth profiles are deter-

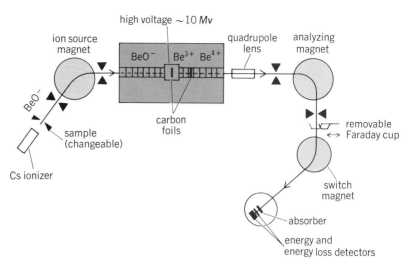

Fig. 5. Schematic illustration of the Yale ultrasensitive secondary ion mass spectrometry setup used to measure ^{10}Be/^{9}Be isotope ratios in geologic samples. (*From K. K. Turekian et al., Measurement of ^{10}Be in manganese nodules using a tandem Van de Graaff accelerator, Geophys. Res. Lett., 6:417–420, 1979*)

mined. Perhaps the most exciting recent development in this area has been the realization that by combining SIMS with a nuclear accelerator, the already good sensitivity can be improved by many orders of magnitude. In fact, the sensitivity of this method is now so great that for the first time ^{14}C dating can be accomplished by simply using SIMS to measure directly the ^{14}C/^{12}C ratio. This application requires measuring this ratio as small as 10^{-15}, which is impossible by all other known mass spectrometric methods. This method is revolutionizing radiocarbon dating because it gives rapid and accurate results for dates from only milligram samples of carbon.

The measurement of isotope ratios as small as 10^{-15} requires some special "tricks" to suppress backgrounds. The only way known to effectively suppress backgrounds to such a low level is to use a nuclear accelerator to accelerate the secondary ions to high energy. First of all, at high energy, background events caused by molecular ions of the same mass can be eliminated by passing the ion beam through a thin carbon foil, which causes the molecular ions to dissociate. Second, at high energy the rate of energy loss of an ion in a solid or gas is a function of its atomic number and mass, and these differences can be used further to filter out misidentified background events.

Figure 5 shows the experimental arrangement used at Yale to measure directly ^{10}Be in geologic samples by SIMS. ^{10}Be is a radioactive nucleus with a 1,500,000-year half-life which is continuously produced by cosmic rays interacting with the atmosphere. Because of its long half-life, it is useful in dating geological deposits and in tracing geophysical processes. While, in the past, assays for ^{10}Be were made by counting the decays of this nuclide, this method required large samples and very long counting times. The direct measurement of ^{10}Be using SIMS avoids these problems and greatly extends the range of geophysical questions which can be probed by using ^{10}Be.

In the apparatus in Fig. 5, the sample at left is bombarded with a Cs beam, and the BeO^- ions emitted are mass-analyzed and accelerated to the high-voltage terminal. There they pass through thin carbon foils, removing their electrons to form Be^{+4} ions which are further accelerated by the high voltage. After magnetic analysis, the ^{10}Be ions are detected in a solid-state detector.

Aside from the new uses of ultrasensitive SIMS for ^{14}C dating in archeology and for ^{10}Be dating in geophysics, it is clear that the remarkable sensitivity of this method is going to be important in answering more conventional materials science questions. For the first time, there now exists a general analytic method to assay for impurities at extremely low concentrations. It will be interesting to watch the application of this new method over the next few years.

For background information see CHANNELING IN SOLIDS; MICROPROBE, ION; NUCLEAR REACTION; TRACE ANALYSIS in the McGraw-Hill Encyclopedia of Science and Technology.

[WILLIAM A. LANFORD]

Bibliography:H. H. Anderson and J. Bottiger, (eds.), *Proceedings of the 4th International Conference on Ion Beam Analysis*, Aarhus, Denmark, June 25–29, 1979, publication pending; J. Duggan and I. L. Morgan (eds.), *Proceedings of the 5th Conference on Scientific and Industrial Applications of Small Accelerators*, Denton, TX, Nov. 6–8, 1978, *IEEE Trans. Nucl. Sci.*, MS–26, 1979.

Meteorology

Boundary-layer meteorology deals with the special phenomena and processes that are characteristic of the lowest kilometer of the atmosphere. The layer is distinguished by the fact that its properties are noticeably affected by the presence of the Earth's surface, and it is usually separated from the free atmosphere above it by a temperature inversion that inhibits upward transport of boundary-layer properties. The most important effects of the surface may be seen in the diurnal cycles of air temperature associated with heating and cooling of the ground, reduction of the wind speed from frictional losses of momentum at the surface, and varying degrees of prevalence of water vapor and pollutants that have their origin at the surface.

The overlying inversion is most pronounced in regions dominated by diverging anticyclonic flow at low levels. It is caused and maintained by the subsidence of potentially warm air at higher levels. Normally it is prevented from reaching all the way to the ground by convection, which is most pronounced during the daytime over land. Because of the convection, properties tend to become well mixed throughout the boundary layer. Occasionally, when anticyclonic flow stagnates for protracted periods, the mixed layer becomes extremely shallow, and dangerous concentrations of pollutants may result. In cyclonic regions and other disturbed weather areas, the inversion becomes higher and weaker, allowing the surface layer to penetrate upward into the main portion of the atmosphere. In this way the main body's heat and moisture are replenished, and suspended pollutants are washed out by precipitation.

Although the boundary later has been recognized as a special area of study for more than 50 years, the layer has recently come under more intense scrutiny for two reasons: transport and diffusion processes must be predicted quantitatively if ambient pollution standards are to be effectively controlled at the source; and limitations of longer-range prediction methods are believed to arise, in part, from deficiencies in the understanding and measurement of heat and water vapor sources at the surface.

Observational methods. Towers have been the most important platform for the study of the surface layer because they permit continuous recording of rapidly fluctuating data needed to monitor the turbulent transports and to better understand turbulence statistics related to diffusion. Within the last year, important tower facilities were established at Cabauw (200 m high) in the Netherlands and near Boulder, CO (300 m). At the Boulder site, sonic anemometers and thermocouples are mounted on booms at eight levels. The data that are continuously recorded there will permit the study of the vertical distribution of the spectra of vertical and horizontal velocity components, and their correlations with other meteorological quantities to be determined, through a substantial fraction of the total boundary layer under a large variety of meteorological conditions. The tower will also be used for comparison with aircraft measurements to extend the collection of turbulence data throughout the entire boundary layer for limited periods of time. The Boulder facility is operated jointly by the National Center for Atmospheric Research and the National Oceanic and Atmospheric Administration.

The gathering of useful information about the structure of the boundary layer usually involves large field observation programs with major investments in instrumentation and worker-power. The processes inherent in daytime convection, for example, involve significant variations of wind and temperature from time to time and point to point; for this reason, many observations scattered over space and time are needed to derive meaningful statistics. One of the most productive of such field programs was recently conducted in Minnesota. It was one of the first to make use of tethered barrage balloons as platforms for instrumental measurements.

The high cost of on-site field measurement programs has spurred the development of remote sensing methods. Radar is the best known of these methods and has been widely used for the mapping of precipitation. Multiple radar systems have been adapted to the determination of boundary-layer winds in certain meteorological situations. Less conventional methods based on light scattering (lidar) and sound scattering (sodar) have been receiving attention recently. Lidar has proved to be effective for continuous monitoring of the depth of the mixed layer, which is sensed by means of the light scattered from particles that have become dispersed throughout its depth. Sodar systems operate by emitting a short pulse of sound upward and receiving the scattered echo a short time later either at the point of emission (monostatic systems) or at other points deployed at selected surface locations (polystatic systems). Monostatic systems are able to monitor the vertical distribution of certain patterns of temperature inhomogeneities, and much effort is currently being expended on the interpretation of such structures. The method has been most useful in monitoring the growth of the nocturnal temperature inversion. Bistatic and tristatic systems have been useful in providing continuous measurements of the wind at levels up to several hundred meters.

Useful knowledge about boundary-layer processes has also been gained by the study of laboratory models. J. W. Deardorff and G. E. Willis have recently observed the structure of convection by experimenting with tanks of liquid heated from below. Other studies of convection and turbulent exchange have been conducted in low-speed wind tunnels that permit dynamic simulation of some atmospheric processes. Unfortunately, the extended atmospheric boundary layer is influenced by the Earth's rotation, and these effects have not been successfully simulated in wind tunnels.

Theoretical methods. Observational and theoretical studies of the boundary layer are greatly simplified by nondimensionalization. For example, similarity theory enables a single formula to describe all distributions of wind in the lowest 10 m by expressing wind speed and height in units of the friction velocity, related to the drag of the wind on the ground, and the Monin length derived from the heat transported vertically upward by the turbulence. For many years, it was thought that the length scale appropriate for the extended mixed layer should be one derived from the surface friction velocity and the angular velocity of the Earth's rotation. A significant step forward in the last 5 years has been the realization and proof that under convective conditions a new set of scales applies. The appropriate velocity scale is constructed from the vertical transport of buoyancy, and the length scale is the depth of the mixed layer. By using such scales, it is now becoming possible to combine data from many different sources into simple universal relationships.

Theoretical understanding and practical exploitation of boundary-layer processes have been advanced by the widespread application of computational models. More than 50 of these have been described in the past 3 years. The principal difficulty of all models has been the closure problem. Predictions, of necessity, deal with the average of an ensemble of fluctuating states, and since the equations of the mean state involve correlations between fluctuating quantities, new relationships must be formulated between the new variables and the mean variables originally contained in the equations. For example, the equation for the change of mean temperature contains contributions from the turbulent heat flux, and solution is possible only if the heat flux can be related in some way to the mean distributions of temperature and other mean properties. The traditional closure method known as K-theory relates the transport terms to the gradient of mean quantities. Its success has been limited. One new approach to this problem has been to simulate in detail nearly all of the fluctuations in a hypothetical convective boundary layer and to reconstruct the mean statistics from the model output. The simulation has been successful, but since it required 450 hr on one of the world's largest computers, the approach

is not likely to be widely adopted for practical use. Another avenue of model improvement currently receiving much attention is the so-called second-order closure approximation. In this approach, exact equations for the required statistics are derived, and approximations are then made on these equations rather than on the original set.

Air-pollution meteorology. The Clean Air Act and its amendments have required industrial emitters of pollution to use computational models in order to verify that their emissions will conform to acceptable ambient standards under the most severe meteorological conditions. Models based on theoretical and observational data are able to predict ground-level concentrations to about 10 km downwind of a stack with moderate success if the surface is flat and uniform. For greater distances and especially in heterogeneous terrain, three-dimensional mesoscale models are a necessity, and the range of uncertainty of predicted ground-level concentrations is still quite large. Much of this uncertainty arises from chemical reactions whose rates in the natural atmospheric environment are poorly known at best.

One of the most important of the poorly understood reactions is the transformation of sulfur dioxide (SO_2) to sulfate ions which, when assimilated in water droplets, result in acid precipitation. To obtain a better understanding of this problem, the Electric Power Research Institute recently conducted the Sulfate Regional Experiment (SURE). The region comprises roughly the northeastern quarter of the 48-state area. The study included as key elements the sampling of pollutants at more than 40 surface stations, extensive collection of samples by aircraft, chemical analysis of precipitation, and preparation of a detailed inventory of emissions in space and time over the area covered by the study. The combined cost of this study and a companion field program conducted by the Federal government exceeded $10,000,000. The ultimate goal of the study is to explain, by the use of computational models, the observed surface concentrations of various chemical species, given the previous source distribution history and knowledge of meteorological transports.

For background information *see* ATMOSPHERIC POLLUTION; METEOROLOGY; TEMPERATURE INVERSION in the McGraw-Hill Encyclopedia of Science and Technology.

[ALFRED K. BLACKADAR]

Bibliography: R. H. Clarke, *Quart. J. Roy. Meteorol. Soc.*, 96:91–114, 1970; D. A. Haugen (ed.), *Workshop on Micrometeorology*, American Meteorological Society, 1973; H. A. Panofsky, *Ann. Rev. Fluid Mech.*, 6:147–177, 1974; H. Tennekes, *J. Atmos. Sci.*, 27:1027–1034, 1970.

Microwave tube

Cyclotron resonance coupling between microwave fields and an electron beam in vacuum is the basis for a family of microwave generators called gyrotrons or cyclotron resonance masers. This type of coupling has the advantage that both the electron beam and the associated microwave structures can have dimensions which are large compared with a wavelength. Thus, cyclotron resonance masers should be greatly superior to conventional

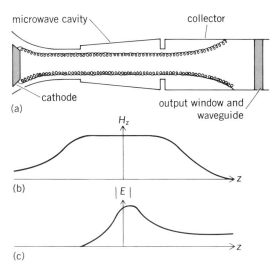

Fig. 1. Gyrotron. (a) Schematic diagram, showing elements. (b) Plot of typical dc magnetic field H_z as function of distance z along axis. (c) Plot of representative microwave electric field intensity $|E|$.

microwave tubes with respect to power capability at short wavelengths.

In 1973–1975, results fulfilling this promise were announced by Soviet publications involving about 12 authors. Continuous-wave outputs of 22 kW at 2.0-mm wavelength and 1.5 kW at 0.92 mm were announced. More recently, pulsed power output of 1250 kW at 6.7-mm wavelength and 1100 kW at 3.0 mm have been reported. The development of these power sources is particularly significant for magnetically confined plasma fusion experiments. Microwave heating has been considered an attractive method of supplying the energy needed to bring a reactor to ignition temperature. However, before the recent gyrotron achievements, it was not clear that sufficient microwave power could be produced at the very short wavelength required. Gyrotrons should also find appli-

cation in millimeter-wave radar and communications systems.

History. R. Q. Twiss recognized as early as 1958 that cyclotron resonance coupling could give rise to microwave amplification or generation. In 1959 J. Schneider and A. V. Gaponov published separate calculations of the effects, and R. H. Pantell reported experiments using this type of coupling. In the early 1960s there was considerable work by many groups on cyclotron resonance devices, but the results were not particularly impressive and most of this effort ended about 1965. One exception was the team of Soviet researchers, led by A. V. Gaponov, which produced the significant results quoted earlier. Another exception grew out of research on intense relativistic electron beams. V. L. Granatstein and others at the Naval Research Laboratory (NRL), using cyclotron resonance coupling to these beams, produced single-pulse, intense microwave output. Typical parameters were 350 MW output at 2-cm wavelength with 60-nanosecond pulse length, from an electron beam having energy of 2.6 MeV and current of 40 kA. Although efficiency in these experiments was low, the demonstration of very high peak power was significant.

Basic characteristics. The basic cyclotron resonance condition is given by Eq. (1), where ω is the operating frequency, n is an integer, and ω_c is the cyclotron frequency or angular velocity of the electron given by Eq. (2). Here, B is the dc magnetic

$$\omega = n\omega_c \qquad (1)$$

$$\omega_c = \frac{eB}{\gamma m_0} \qquad (2)$$

field, e is the electron charge, m_o is the rest mass, and γ is the relativistic mass factor. The fundamental cyclotron resonance occurs when $n = 1$. This is the strongest and most useful interaction. The resonance condition requires that very high magnetic fields be used for high-frequency devices. For example, a frequency of 120 GHz requires a magnetic field of about 45 teslas. Generally, the very-high-frequency gyrotrons have used superconducting magnets.

Larger values of n allow corresponding reductions in the required dc magnetic field. Practical devices have generally been limited to values of n no larger than 2 (second-harmonic operation).

The most important microwave field component in the gyrotron is the electric field tangential to the orbit of the electron. With the fundamental cyclotron resonance interaction, any spatial variation of the microwave fields is of little importance. It is this property that allows the gyrotron to use cross-section areas which are large compared with a wavelength.

Electron bunching in the gyrotron occurs by virtue of the relativistic mass effect included in Eq. (2). The transverse microwave electric field introduces a sinusoidal modulation of γ depending on the angular position of the electron in its orbit relative to the direction of the electric field. The modulation of γ results in a modulation of angular velocity as given by Eq. (2). As the beam drifts, this converts to angular bunching in the coordinate system centered on each electron orbit. By proper ad-

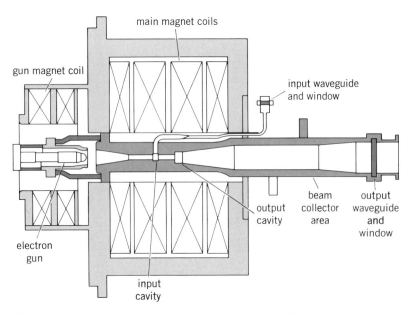

Fig. 2. Simplified cross section of pulsed gyro-klystron amplifier.

justment of phase conditions, the bunched beam can give up most of its energy to microwave energy.

A number of tube configurations are possible using the cyclotron resonance interaction. The simplest form, and that used for most practical gyrotrons to date, is an oscillator using a single resonant cavity. A gyroklystron amplifier employing two or more cavities is another alternative; and in a third variation a traveling-wave circuit, in analogy to a traveling-wave tube, is used.

A schematic representation of a gyrotron (single-cavity oscillator) is shown in Fig. 1, along with a typical dc magnetic field profile H_z and a representative microwave electric field distribution $|E|$. The electron beam is a hollow beam with all electrons having helical motion. For efficient operation, all electrons must have a large fraction of their total energy contained in motion transverse to the device axis.

Recent work in United States. In the United States, gyrotron experiments have recently been performed at Varian Associates, Inc., and at NRL. At Varian a pulsed gyroklystron amplifier has achieved 40 dB power gain at 28 GHz using an 80-kV, 8-A beam. A simplified cross section of this device is shown in Fig. 2. Peak power output was 65 kW for an efficiency of 9%. At Varian, a pulsed single-cavity oscillator, using the same beam, produced 250 kW peak output with an efficiency of 37%. A continuous-wave oscillator used the same beam parameters with a different microwave output coupling system. It has produced 105 kW continuous-wave output. At NRL a pulsed single-cavity oscillator has generated 100 kW peak at 35 GHz.

Gyro–traveling-wave tube experiments are in progress at both Varian and NRL. Results so far have been limited to a power gain of 18 dB and

efficiency of 6%. Calculations predict that gains of 30 dB or more with efficiency of 30–50% should be achievable.

A summary of gyrotron demonstrated power output versus frequency is shown in Fig. 3, along with power capabilities of conventional klystrons and traveling-wave tubes. In the case of the pulsed devices, it is the peak power output that is shown on the graph. The single-shot, short-pulse results refer to devices using intense relativistic beams which are not suitable for repetitive pulsing. The gyrotron results indicate a clear capability for producing higher power at high frequency. Additional practical engineering needs to be done to enable the full potential of the gyrotron to be realized, but its importance as a microwave generator is already obvious.

For background information *see* MICROWAVE TUBE; TRAVELING-WAVE TUBE in the McGraw-Hill Encyclopedia of Science and Technology.

[HOWARD R. JORY]

Bibliography: A. A. Andronov et al., *Infrared Phys.* 18:385–394, December 1978; V. A. Flyagin et al., *IEEE Trans.*, MTT-25:514–521, June 1977; J. L. Hirshfield and V. L. Granatstein, *IEEE Trans.*, MTT-25:522–527, June 1977; H. R. Jory, *Digest of the 1977 International Electron Devices Meeting*, IEEE, December 1977.

Miocene

This major epoch in earth history has been studied intensively for the past several years. The geological record suggests that major tectonic and biogeographic relationships were changing rapidly during the Miocene, accompanied by significant changes in the world oceans and their circulation. Intense volcanism on a global scale, worldwide fluctuations in sea level, and pulses of marked global cooling leading to the development of the permanent ice cap in Antarctica have been evidenced in terrestrial and marine sections. It is now known that these important happenings during the Miocene are interrelated. It will be the task of continuing research to disentangle the sequence of these events.

Deep Sea Drilling Program. The concentrated efforts of large groups of scientists of various disciplines working on the sample material recovered from the voyages of the research vessel *Glomar Challenger* have been largely responsible for the recent advances in knowledge of this epoch. In 1967 this ship embarked on the first leg in the first phase of a Deep Sea Drilling Program (DSDP) called the Joint Oceanographic Institutions for Deep Earth Sampling (JOIDES). Its well-defined objectives were to systematically core the world oceans. In 1975 the project entered its second phase with the cooperation of several foreign countries as the International Phase of Ocean Drilling (IPOD), and in the summer of 1979 embarked on the sixty-ninth and final leg of phase two in the Galapagos Trench. Fortunately for the scientists of the world, the project will be extended through 1980–1981, a third phase.

The enormous successes of this drilling project sponsored by the U.S. National Science Foundation and various nations have spurred many other oceanic projects. The wealth of information from

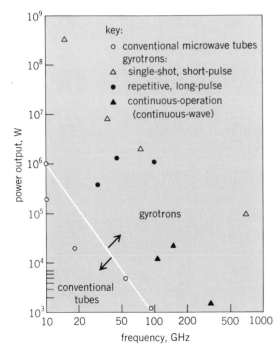

Fig. 3. Power output versus frequency of gyrotrons compared with that of conventional microwave tubes (data based on 1979 information).

these studies has revealed that dramatic events of global significance occurred during the Miocene. What is emerging from these studies is not only the history of past climates and ocean circulation patterns, but also new information on geochemistry, biological productivity, and other characteristics of the Earth's changing environment.

Miocene stages. The Miocene or lower Neogene (new Tertiary Period) spans the interval from the end of the Oligocene to the beginning of the Pliocene. The standard stages that are distinguished and the geochronologic ages are: early Miocene—Aquitanian, 24–22.5 Ma (mega-annum, or 10^6 years), Burdigalian, 22.5–16.5 Ma; middle Miocene—early-middle Langhian, 16.5–15.5 Ma, late-middle Serravallian, 15.5–11.5 Ma; late Miocene—early-late Tortonian, 11.5–6.6 Ma, late-late Messinian, 6.6–5.2 Ma. These are shown in Fig. 1 with some of the major geologic events which have occurred within the epoch.

The absolute age of 24± Ma for the lower boundary was conditionally accepted in 1975 at the 6th Congress on Mediterranean Neogene Stratigraphy, in Bratislava. The upper boundary of 5.2–5.4 Ma was established on direct paleontological evidence from land-based sections in Sicily which have been correlated with paleontological and pa-leomagnetic data from a deep-sea drilling site in the Tyrrhenian Sea.

Volcanism and sea-level changes. The Miocene was a time of intense volcanism on a global scale. Widespread volcanic ash deposits and lava flows on land and in the oceanic realm attest to the frequency and vigor of the igneous activity.

It has been suggested that volcanism is closely related to differential oceanic sea-floor spreading rates and subduction. Apparently there is also a correlation in orogenic belts between volcanism and global changes in sea level. Therefore, it has been suggested that changes in volcanism reflect fundamental tectonic processes on a worldwide scale. Consequently, the paleogeographic history of continents seems to be related to plate tectonics.

Eustatic sea-level changes may be controlled by plate movements that change the capacity of the ocean basins. Rapid spreading would cause the expansion of the mid-oceanic ridge and a concomitant reduction in the volumetric capacity of the ocean basins. This could result in a sea-level rise (transgression) of several hundred meters. Conversely, a contraction of the mid-oceanic ridge system would cause a significant lowering of sea level (regression). The continued continental emergence during the Cenozoic may be attributed to the reduction in sea-floor spreading rates.

Middle Miocene cooling. Worldwide sea-level lowering and colder climate seem to correlate well. Intervals of continental emergence correspond to times of low temperature. There is now strong geological and paleontological evidence which indicates that Quaternary glacial ages may have begun in the Miocene. It has also been proposed that this late Tertiary cooling might have been a result of extensive regional uplift which occurred during this epoch.

Within this framework, global transgressive and regressive events in the Miocene may be related to glacial buildup in Antarctica. Evidence of glaciation in the high latitudes of Alaska and Antarctica goes back approximately 14–16 Ma. It seems likely that by mid-Miocene time sufficient ice had accumulated in the area so that fluctuations in glacial mass could trigger major sea-level changes within the tectonic setting outlined above. P. Vail and coworkers recorded major sea-level falls occurring before Miocene and again before the Messinian (10.8–6.6 Ma). Minor dips in sea level are also recognized at 24, 22, 19, 15.5, and 13 Ma. These Miocene fluctuations are all charted in Fig. 2a.

Deep-sea drilling sites adjacent to Antarctica have recovered proof that there has been active glaciation on Antarctica since at least the beginning of the Miocene, 24± Ma. Isotope techniques have been used to determine when significant amounts of ice first formed at the poles.

Geological and paleontological evidence and $^{16}O/^{18}O$ measurements suggest major changes in the world oceans and their circulation patterns during the middle Miocene. Measurements of the $^{16}O/^{18}O$ ratio in fossil sediments disclose information about ice formation and water temperatures. ^{16}O, the lighter isotope, evaporates preferentially, and therefore the ice in glaciers is enriched in this isotope relative to sea water. The oxygen isotopic composition of calcareous benthic foraminifers

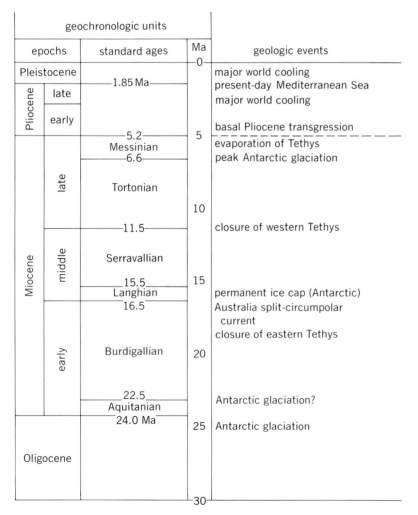

Fig. 1. Geological time scale and subdivisions of Miocene standard ages, with approximate times of important geological events.

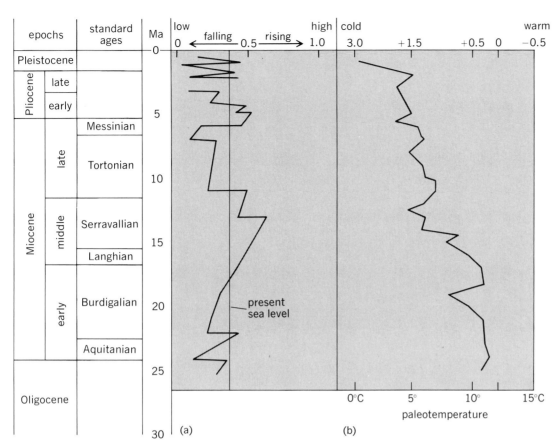

Fig. 2. Middle Miocene changes. (a) Global relative sea-level changes. (b) Oxygen isotopic composition of benthic foraminifers and paleotemperature trends from DSDP Site 289. (From data by P. Vail and co-workers, 1977, and T. Saito, 1977)

and the isotopic paleotemperatures from DSDP Site 289, western equatorial Pacific, are shown in Fig. 2b. These are charted next to the worldwide sea-level changes for comparison. During the middle Miocene, approximately 16 Ma, isotopic temperatures dropped sharply as a result of the major ice cap established in Antarctica and the subsequent development of deep-water circulation. At this time high-latitude temperatures dropped dramatically below the freezing point, but low-latitude temperatures remained the same or perhaps increased. It is postulated that the divergence between high- and low-latitude temperatures is related to the establishment of circum-Antarctic circulation due to the opening of the Australia-Antarctic seaway.

In the Northern Hemisphere the Atlantic-Arctic circulation was well established. This circulation pattern coupled with worldwide climatic deterioration and the location of the pole in the central Arctic resulted in the first freezing of the Arctic area. In southeastern Alaska major marine coolings dated about 13–14 Ma have also been recognized in association with nonmarine tillites.

In the Paratethys region evaporites were forming in the marginal basins. In the deep-ocean basins there was an increased formation of hiatuses and a marked rise in the calcium compensation depth (CCD). Maxima in hiatus abundance are found in the middle to late Miocene, minima in the early to middle Miocene.

T. C. Moore and coworkers in 1978 showed that peaks in hiatus abundance in the Miocene suggest the possibility of three episodes of erosion by bottom flow: at 18 Ma, associated with the formation of North Atlantic Deep Water; at 14 Ma, with the development of the east Antarctic ice cap; and at 12–10 Ma, when the Antarctic ice sheet reached the sea and Antarctic Bottom Water was formed. Permanent glaciation in Antarctica 16 Ma ago and the development of very cold corrosive bottom water increased calcium dissolution and elevated the CCD. Therefore, the dissolution pulse recognized between 15 and 10 Ma may be due to the onset of the flow of Antarctic Bottom Water.

Late Miocene–Messinian events. One of the most important paleoceanographic events in the Miocene is the marked increase in the volume of Antarctic ice during the late Miocene. This acme of Antarctic glaciation affected a broad region and may have produced the glacioeustatic lowering of sea level which led to the isolation of the Mediterranean Sea and the so-called crisis of salinity. The causal relationship of these events are the most important questions yet to be answered about the Miocene.

During the Messinian interval of the late Miocene (6.2 Ma) the connection between the western Tethys and the Atlantic closed completely, isolating the Mediterranean Sea from the world ocean. Deposits of evaporites 1.5–2 km thick from this period are found in the Mediterranean basins. The Messinian represents a very short interval, approximately 1 Ma, so that the extraction of such a great quantity of salt from the oceanic realm must have affected other parts of the world and a record of the effects should exist.

Several theories have been proposed to explain the desiccation which took place during the Messinian interval of the late Miocene. One group presents tectonic movements as the sole reason for the Messinian desiccation, whereas others believe a global sea-level change was the cause. More current theories, supported by an increasing amount of evidence, seem to suggest that the final separation of the Mediterranean from the Atlantic was due to a tectonic elevation of the area and a eustatic sea-level fall. The combination of a tectonic elevation of the land and a lowering of sea level had far-reaching effects in the world oceans. The Messinian crisis has a global significance.

A recent study by C. Adams and coworkers examined evidence for eustatic sea-level changes. Although they do not imply that this is the one and only cause for the Messinian crisis of salinity they present worldwide proof of both severe cooling and a major drop in sea level. Their evidence from widely separated areas is based on sediment facies, faunal and floral change, and isotopic compositions of microfossil assemblages.

As observed previously, cooling and accompanying regression may be attributed to the expansion of Antarctic glaciation. This could result in a sea-level drop of 50–70 m. Adams and coworkers propose that a fall in sea level of the magnitude of 50 m could convert shallow shelf areas into inland lakes. A drop of the magnitude of 70 m would isolate areas such as the Sea of Japan. It is of interest that subsequent studies of deep-sea cores from the Sea of Japan have resulted in discovery of a 3-m-thick sediment layer with abundant fresh-water diatoms with no admixture of marine forms.

This fresh-water sediment may be correlated with the latest Miocene Messinian Mediterranean desiccation. Therefore, it seems possible in the model of Adams and coworkers that a global sea-level drop might have been responsible for the final closing of the Iberian Portal and the isolation and desiccation of the Mediterranean. Perhaps, as they propose, the removal of the dissolved salts from the world oceans brought about by this desiccation might have caused the late Miocene glaciation by lowering salinity and raising the freezing point of sea water in high latitudes.

Or, on the other hand, was high-latitude glaciation completely independent of these Mediterranean events? The answers may well be found in the sediments and faunas from the continental shelves. Further investigation and continued research will add pieces to the puzzle, and eventually the sequence of events in this fascinating story may be reconstructed.

Thus the picture that emerges in the Miocene within the framework of gradual Cenozoic climatic deterioration is one of marked global cooling pulses. A strong equatorial polar thermal gradient, creation of a permanent Antarctic ice cap, and the establishment of deep-water circulation similar to that which exists today no doubt caused the sharp decline in isotopic temperatures near the beginning of the middle Miocene (Fig. 2).

Severe global cooling is also indicated about 6.2 Ma when the Antarctic ice sheet peaked in volume accompanied by rapid northward movement of the Antarctic Convergence, a worldwide sea-level fall, and desiccation of the Mediterranean—the Messinian crisis of salinity.

For background information *see* MIOCENE; PALEOCLIMATOLOGY in the McGraw-Hill Encyclopedia of Science and Technology.

[DOROTHY J. ECHOLS]

Bibliography: C. Adams et al., *Nature*, 269: 383–386, 1977; T. C. Moore, Jr., et al., *Micropaleontology*, 24(2):113–138, 1978; T. Saito, Late Cenozoic planktonic foraminiferal datum levels: The present state of knowledge toward accomplishing Pan-Pacific stratigraphic correlation, *Proceedings of the 1st International Congress on Pacific Neogene Stratigraphy, Tokyo, 1976*, 1977; P. R. Vail, R. M. Mitchum, Jr., and S. Thompson III, in C. Payton (ed.), *Seismic Stratigraphy and Global Changes of Sea Level*, Amer. Ass. Petrol. Geol. Memo. 26, 1977.

Molecular spectroscopy

High on the list of ways to determine the intimate details of molecular structure and behavior are the various interactions of molecules with light. From the wavelengths a molecule absorbs, one gains information about the energies involved in the rotational, vibrational, and electronic motions. From these, a quantum-mechanical picture of the molecule can be formed. Since the details of the electronic structure of molecules are often hidden behind congestion, diffuseness, and complexity in the absorption and emission spectra, spectroscopists are always looking for different ways of probing the molecule. One of the more useful new techniques is multiphoton ionization spectroscopy, whereby several photons are used to remove electrons from a molecule when the light is resonant with an excited electronic state.

Multiphoton spectroscopy. In a normal interaction of light with any system, a single quantum, or photon, is absorbed at a time, and the energy of this photon must be matched with the difference between two allowed energy states of the molecule. However, it turns out that the quantum-mechanical rule which restricts an absorption to a single photon is only approximate. It can be violated when enough photons are present simultaneously and the energies of two or more of the photons sum up to the energy between molecular states. It had been obvious that these multiphoton absorption events could occur since the early days of quantum mechanics, but only with the development of lasers have scientists had a light source powerful enough to use as a spectroscopic tool.

Advantages. The advantages of using multiple photons to cause a spectroscopic transition are:

(1) Selection rules are different so that the relative probabilities of transitions to the various excited electronic states are changed from those in one-photon spectra. This allows new electronic states to be seen and the spectral structure from overlapping states to be sorted out.

(2) Very energetic transitions can be studied by using the relatively low-energy photons of visible light. Tunable lasers emitting intense beams of ultraviolet light are not available, and conventional sources operating in the vacuum ultraviolet are weak and difficult to use.

(3) Lasers can be used to study almost any transition of the molecule because the photon energies are summed. A major advantage of a laser is that it can be easily made very monochromatic without losing much intensity. Thus a small laser can do the same job as a very large spectrograph.

(4) Electronic states can be identified by the relative absorption of linearly and circularly polarized light. In a single-photon transition there is no difference in the absorption of these two types of polarized light, but in a multiple quantum transition one type of light is caused to be absorbed more or less strongly than the other, depending upon the symmetries of the electronic states involved.

(5) The Doppler effect (spectral broadening caused by the random thermal motion of molecules in the gas phase) can be eliminated in a two-photon absorption by having one photon absorbed from each of two counterpropagating beams. The Doppler shift in the absorption of the first photon is opposite to that for the second, and the two exactly cancel.

(6) States with more than one excited electron can be accessed because each photon can interact with a different electron.

Detecting multiphoton transitions. Given these advantages to multiphoton spectroscopy, the main problem is to determine when a multiphoton transition has taken place, since it is such an improbable event. When using currently available dye lasers, it is typical that only 1 photon in 10^{10} will be absorbed even when the laser is tuned to one-half the energy of an electronic transition. Clearly, an experiment which tries to measure the small attenuation of the light will not succeed unless unreasonably intense lasers (an *n*-photon probability goes as the *n*th power of the light intensity) are used. However, the presence of an excited electronic state can be detected by virtue of one of the ways of getting rid of the absorbed energy: fluorescence, heating of the surroundings, photochemistry, or the loss of an electron. Alternatively, one can look for the enhancement of a two-photon scattering process that occurs when the sum of two photons equals a transition energy.

Multiphoton ionization technique. Perhaps the most versatile method of detecting when an excited molecular state has been formed is to look for electrons which are ejected from the excited molecule when it absorbs an additional one or two photons from the laser beam. If there is enough energy in the laser beam to cause a multiple-photon transition to an excited state, there is almost always enough light available to cause a much more probable single-photon transition to the continuum of energy levels associated with an electron leaving the molecule. Once an electron is free, it is very easy to detect. One can measure the number of photoelectrons as a current if there are enough of them, or increase their number by a gaseous cascade process, and then measure the current; or increase their number with an electron multiplier and then measure the current, but only when working in a good vacuum. To perform a spectroscopic experiment, then, one simply scans a tunable dye laser focused between charge-collecting plates and records the number of charges produced as a function of wavelength. The simplicity of the technique and the great amount of information obtained from the spectral details have contributed to the rapid spread of the use of multiphoton ionization spectroscopy to obtain quantum-mechanical descriptions, and should contribute to its development as an analytical technique.

Applications. Examples of the use of multiphoton ionization spectroscopy to gain information about molecular structure are the observation of previously undetected molecular states in benzene and butadiene by P. M. Johnson, in ammonia by G. Neiman and S. Colson, and in iodine by G. Petty, C. Tai, and F. W. Dalby and by K. Lehman, J. Somolarek, and L. Goodman. Many other molecules have been studied by various workers, adding to the knowledge of molecular excited states. Work by J. Berg, D. Parker, and M. El Sayed has demonstrated the use of polarization in multiphoton ionization spectroscopy for assigning states in the multiphoton ionization spectra. As demonstrated in the work of A. Williamson and R. Compton, the use of two independently tunable lasers is very helpful in unraveling complicated spectra where there are electronic states at two or more of the multiples of the photon energy. R. Turner and coworkers have shown that multiphoton ionization can also be used to study triplet states.

Although the probability of any one photon getting absorbed in a multiphoton experiment is vanishingly small, there are so many photons present in a typical laser pulse that the probability a molecule in the focus region of the laser beam is ionized is relatively high and can approach unity. This sensitivity leads to the possibility of using multiphoton ionization for spectroscopic experiments on very dilute gases. Any electrons that are formed can be easily collected and individually counted. Although obviously this presents possibilities as an analytical technique, specific molecular applications have not yet been developed.

Interpretation of larger molecules. The capabilities for the detection of gases at very low partial pressures has led to experiments conducted in supersonic free jets by Johnson and coworkers. In these experiments the gaseous molecules of interest are mixed in with a high pressure of helium and allowed to expand into a vacuum through a small orifice. The drastic expansion cools the molecules down to temperatures approaching absolute zero, simplifying the rotational and vibrational structure of the electronic transitions by orders of magnitude. In addition, the directional nature of the supersonic jet removes almost all of the Doppler component to the linewidths. With relief from spectral congestion and narrow inherent linewidths, rotational structure of larger molecules can be resolved and interpreted. Because of the extremely dilute nature of the gas in the supersonic expansion, only very sensitive techniques such as multiphoton ionization spectroscopy or fluorescence excitation can be used to observe spectroscopic transitions of molecules in a jet.

Selective ionization. One of the results of a multiphoton ionization is a molecular ion, and this has naturally led to the use of lasers as a source for mass spectrometry. E. Schlag and coworkers and R. Bernstein and coworkers have shown that the multiphoton ionization process can be a very effective way of cleanly producing ions. It is found that the degree of fragmentation depends very greatly upon the laser flux, with the predominant species being the parent ion at low powers, while fragmentation proceeds at high powers until the predominant species in the decomposition of a hydrocarbon is carbon atomic ion. Laser ionization may have some advantages in mass spectometry because one can selectively ionize various species by tuning the laser to hit or to avoid a given molecular transition. Some information may also be gained about the photodecomposition of molecular ions in intense light fields.

The kinetics of the multiphoton ionization process are greatly affected by molecular radiationless processes such as dissociation and relaxation to a lower electronic state. As lasers become more powerful and their characteristics become better defined, it will be possible to determine the rates of these radiationless processes as well as the cross sections of the radiational transitions, thus arriving at a clear overall understanding of the photophysics of molecules in intense light fields.

For background information *see* MOLECULAR STRUCTURE AND SPECTRA in the McGraw-Hill Encyclopedia of Science and Technology.

[PHILIP M. JOHNSON]

Bibliography: J. O. Berg, D. H. Parker, and M. A. El Sayed, *J. Chem. Phys.*, 68:5661–5662, 1978; P. M. Johnson et al., *J. Chem. Phys.*, 64:4143–4148, 1976; P. M. Johnson et al., *J. Chem. Phys.*, 68:3644–3653, 1978; A. D. Williamson and R. N. Compton, *Chem. Phys. Lett.*, 62:295–299, 1979; L. Zandee and R. B. Bernstein, *J. Chem. Phys.*, 70:2574–2575, 1979.

Neutrino

One of the most interesting developments in neutrino physics in recent years is the deep underwater muon and neutrino detector (DUMAND). For about 5 years, this project has been concerned with studying the problems encountered in designing, constructing, and operating an enormous neutrino detector at a depth of 2–3 mi (3–5 km) in the ocean. The primary purpose would be the detection and study of ultra-high-energy (UHE) neutrinos—those of 1–2 teraelectronvolts and above.

Such an undertaking is unique. The largest neutrino detector built to date is at the Super-Proton-Synchrotron (SPS) at the European Commission for Nuclear Research (CERN), Geneva, and is just over 1000 metric tons in mass. The DUMAND detector would employ a cubic kilometer of sea water—10^9 metric tons—as a detector for naturally occurring neutrinos. Its effective volume for extraterrestrial neutrinos would in fact be about 10 times greater, to 10^{10} metric tons. It would dwarf not only previous neutrino detectors but all previous long-term ocean installations, whether acoustic or oil-drilling equipment.

DUMAND installation. As presently envisaged, DUMAND will comprise an array of detector modules attached to cables or "strings" anchored to the bottom of the ocean, and stretched upward by flotation modules that provide buoyancy to hold the strings more or less vertical. The modules are distributed along the strings, 18 to a string; the strings are just under 800 m long. The array, hexagonal in shape, is shown in the illustration; it constitutes a hexagon 800 m on a side, covers an area

of 1.66 km², and occupies a volume just over 1 km³. It has three central "legs" in the shape of a Y, radiating from the center, to which are attached 60 rows or planes in three subarrays. To each of these rows is attached 21 sensor strings. With an additional central string, this makes a total of 1261 strings, each with 18 detector modules, for a total of 22,698 sensor modules. The modules are about 37 m apart along the strings, the strings are 40 m apart along the rows, and the rows 40 m apart along the legs.

The cabling terminates in a central junction box, which is connected by undersea cable to a shore station 40 to 55 km away, depending on where the site is chosen. The cable carries power from the shore to the array and collects data (preprocessed by undersea computers) from the array. Studies of possible sites in the Hawaiian Islands, financed by the State of Hawaii and the U.S. Office of Naval Research, are now being conducted by the Hawaiian Institute of Geophysics of the University of Hawaii.

The detector modules contain photomultiplier tubes that detect the weak Cerenkov light emitted by relativistic charged particles in a transparent medium like water. A major DUMAND site criterion is that the water be highly transparent, so that the detector spacing may be as large as possible, thus reducing the number of detectors required. It is expected that water with a 20-m attenuation length in the blue-green will be used, allowing spacings near 40 m for neutrinos in the teraelectronvolt range. For lower-energy neutrinos, such as those emitted in gravitational collapse, much closer detector spacings are needed.

Atmospheric neutrino detection. A detector for cosmic-ray and extraterrestrial neutrinos must be very massive because of the extremely small cross section of neutrinos, which are observable only through the weak interaction. Neutrinos are not uncommon in the universe; in fact, some cosmological theories postulate many more neutrinos in the universe than there are electrons and protons. They are simply inordinately difficult to detect. The Sun illuminates the Earth with about 13,000 MW of neutrinos, or 100 W/m²; they pass through the Earth almost unnoticed, depositing in it only a few watts. Solar neutrinos have been detected by R. Davis and collaborators, using several hundred tons of detector to produce a few radioactive atoms per month by neutrino collisions. *See* SUN.

Fortunately, the neutrino interaction cross section rises rapidly with energy—linearly above 10–20 MeV to energies at least in the teraelectronvolt region (no data are as yet available above a few hundred gigaelectronvolts). At 1 TeV the cross section is extrapolated to be 0.7×10^{-35} cm². Another increase by a factor of 30 would make the Earth opaque to neutrinos. It is generally believed that the cross section does not increase indefinitely with energy, but flattens off in the teraelectronvolt region because of the existence of the hypothetical intermediate vector boson $W^{\pm}$. According to current gauge theory, the mass of the W should be about 80 GeV/c^2—about 90 times that of the proton. If so, the neutrino-nucleon cross section and the distribution of energy among the products of the neutrino interaction will change markedly in the energy range of 3–30 TeV, a range beyond the

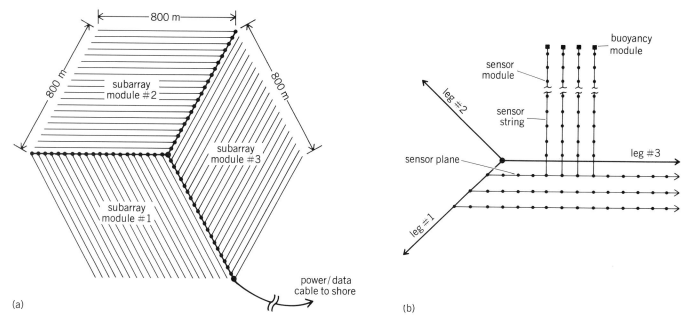

(a)

(b)

DUMAND 1978 Standard Array. (a) Plan view. (b) Detail, showing attachment of sensor strings and modules. (From A. Roberts and G. Wilkins, The DUMAND 1978 Stan- dard Array, in A. Roberts, ed., Proceedings of the 1978 DUMAND Summer Workshop, vol. 3, pp. 9–22, Scripps Institution of Oceanography, 1979)

capability of all existing or contemplated accelerators, but one which DUMAND is well suited to explore.

The only known source of neutrinos in this energy range is the Earth's atmosphere. Primary cosmic-ray protons (or other nuclei), striking atoms in the upper atmosphere, produce secondary pi mesons and K-mesons; these decay in the thin atmosphere, producing tertiary muons and neutrinos which can reach the surface of the Earth. In addition, a few muons and neutrinos are directly produced in the primary interaction.

The muons are highly penetrating particles, and energetic ones traverse the atmosphere; some will be energetic enough (a few teraelectronvolts) to reach the DUMAND detector through several miles of ocean. DUMAND will be able to cast light on some interesting cosmic-ray muon problems. One concerns the isotropy of high-energy cosmic-ray primaries, which depends on whether they are galactic or extragalactic in origin. Another concerns the changing primary composition as energy increases; the DUMAND experiment will help to determine whether the composition of cosmic-ray primaries shifts toward the heavy nuclei as energy increases, as some experiments indicate. *See* COSMIC RAYS.

Neutrino interactions. In the ocean a UHE neutrino produces a hadronic shower, and in most cases an energetic muon as well. By measuring the energy of both the shower and the muon, DUMAND will be able to detect the expected change in the energy distribution between muon and cascade predicted by theory if the W exists. It may also be able to observe the corresponding falloff from the linear cross-section rise. A search for the onset of opacity in neutrino transmission through the Earth would be equally revealing.

The angular distribution of atmospherically produced neutrinos will perhaps reveal the existence at high energies of the directly produced neutrinos; these would also be manifested by an increase to the ratio of electron neutrinos to muon neutrinos.

Detection of extraterrestrial neutrinos. As noted above, for this purpose the effective DUMAND volume is 10^{10} metric tons. Nothing is known from experiment concerning extraterrestrial neutrinos; they have never yet been observed, and the prospect of such observation is one of the more interesting aspects of DUMAND.

Two general types of extraterrestrial sources are to be expected. The UHE neutrinos which would be observed in the experiment must originate in processes similar to those in the Earth's atmosphere; the collision of high-energy charged particles with matter (or, at sufficiently high energies, with starlight or the 3° blackbody radiation). Any other source would be unexpected and highly exciting. A hypothetical source must first produce high-energy particles, which must then collide with matter. The Milky Way Galaxy is such a source, since most of the cosmic rays that reach the Earth are almost certainly of local origin (that is, within the Galaxy). Thus galaxies in general will provide a diffuse neutrino background, even if within individual galaxies the sources are concentrated. Some hypotheses estimate that in early stages of evolution, galaxies are much brighter neutrino sources than they are at present.

Since the DUMAND detector is directional, locating an incoming neutrino to 1° of arc or better, and since it can look simultaneously in all directions (remembering that the Earth is transparent), it is feasible to search for point neutrino sources. Experience with gamma-ray telescopes, which in the past have had resolution inferior to DUMAND's, indicates that point sources can be found and identified under such conditions. In fact, gamma-ray telescope results may indicate where to search for neutrinos; for wherever neutrinos are produced by meson decay, neutral pions that decay into gamma rays are being produced as well.

Among the point sources of neutrinos proposed to date, the following presently appear the most plausible:

1. Pulsars in binary stellar systems. Since pulsars are now believed to be the accelerating mechanism of cosmic rays, a pulsar with a stellar companion fulfills precisely the prescription of an accelerating source in conjunction with a target.

2. The same prescription is satisfied in the first few months after a supernova explosion, by the shell of expanding material around a new pulsar.

3. Quasar explosions. These events, of awesomely catastrophic intensity, may be interpreted as potential intense neutrino sources.

4. Steady-state emission from quasars and active galaxies. The cores of galaxies are regions of great activity that are still poorly understood; mechanisms can be suggested in which they would be copious neutrino sources (and gamma-ray sources as well, if not hidden from view by enough matter to absorb the gamma rays).

6. Known gamma-ray sources, in cases where the gamma rays are not due to synchrotron radiation; in most cases it is not known whether they are or not.

For background information *see* CERENKOV RADIATION; COSMIC RAYS; NEUTRINO; PULSAR; QUASARS in the McGraw-Hill Encyclopedia of Science and Technology.   [ARTHUR ROBERTS]

Bibliography: D. Eichler, Summary of Workshop on High Energy Neutrino Astronomy, in A. Roberts (ed.), *Proceedings of the 1978 DUMAND Summer Workshop*, vol. 2, pp. 353–359, 1978; A. Roberts and G. Wilkins, The 1978 DUMAND Standard Array, in A. Roberts (ed.), *Proceedings of the 1978 DUMAND Summer Workshop*, vol. 3, pp. 9–22, 1979.

Neutron

Ultracold neutrons (UCN), that is, neutrons with energies of order 10^{-7} eV or less, are totally reflected from most material surfaces, in contrast to the behavior of neutrons of higher energy (the average neutron energy in a thermal reactor is approxi-

mately 5×10^{-2} eV) which can penetrate significant distances through materials. The process of neutron reflection is similar to that of total reflection in ordinary optics and enables the storage of neutrons in material containers for significant periods of time. This in turn opens up the possibility of performing several fundamental experiments at increased levels of sensitivity.

Storage of UCN in material bottles. During the process of total reflection, UCN penetrate a small distance (approximately 10nm) into the reflecting surface. This means that there is a small probability that the UCN are absorbed or inelastically scattered by the nuclei in the material. Inelastic scattering results in the UCN gaining so much energy from the thermal vibrations of the nuclei that they can no longer be contained in the storage vessel. While the probability of a UCN being lost by these processes can be readily calculated by means of a straight forward theory, it soon became evident from experiments performed first by F. L. Shapiro and his group at Dubna and subsequently by A. Steyerl at Munich, as well as by groups at Leningrad and Risley (U.K.), that experimentally observed loss rates were much greater (sometimes by as much as $10 - 100$ times) than expected theoretically. Although this anomaly is still not completely understood, some recent experiments have resulted in significant progress toward achieving an understanding of the problem.

In an experiment performed at Yale University, W. A. Lanford and R. Golub measured the hydrogen content of some materials which had been used in UCN storage experiments, as a function of depth into the material. The measurements were carried out on samples which had been treated in the same way as the materials used in the UCN storage vessels, and were performed by bombarding the samples with ^{15}N ions. Because of the resonant reaction $^{15}N + H \rightarrow {}^{12}C + {}^{4}He + 4.43$ MeV γ-rays, the yield of 4.43-MeV γ-rays is a direct measure of the hydrogen content of the surface. The results showed that, in most of the samples studied, there was more than enough hydrogen present to account for the anomalous UCN loss rates. *See* MATERIALS ANALYSIS.

A. V. Strelkov and M. Hetzelt, working at Alma-Ata (Soviet Union), surrounded a thin-walled UCN storage chamber with a neutron detector (Fig. 1) in an attempt to discover if inelastic scattering was responsible for the large loss rates. If this were so, the neutrons, inelastically scattered from the UCN energy region to higher energies, would be expected to penetrate the walls of the storage chamber and so be registered in the surrounding detector. In fact, the external detector counted neutrons at a rate which fell off exponentially with time after the filling of the storage vessel, and the decay constant τ of the external detector counting rate was similar to that exhibited by the number of UCN in the vessel (Fig. 2), indicating that the external detector was, in fact, detecting neutrons which were escaping from the vessel. The experiment was successful; there were enough neutrons reaching the surrounding detector to account for all the anomalous losses. By varying the energy sensitivity of the detector, it was shown that the energy of the upscattered neutrons was close to the thermal energy range—consistent with the idea that the

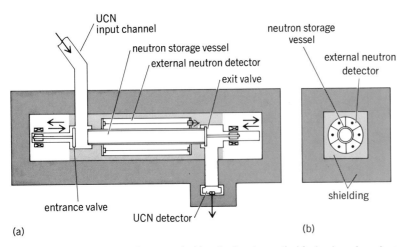

Fig. 1. UCN storage vessel surrounded by six-chamber cylindrical external neutron detector for escaping neutrons. (*a*) Section along length of vessel and detector. (*b*) Section across vessel and detector. (*From A. D. Stoika, A. V. Strelkov, and M. Hetzelt, Z. Phys., B29:349, 1977*).

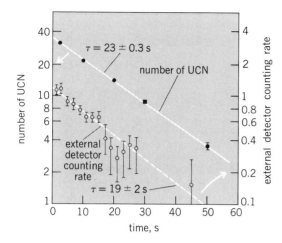

Fig. 2. Counting rate of external detector and number of UCN remaining in storage chamber (as measured by UCN detector) versus time after filling of the UCN storage vessel. *(From A. D. Stoika, A. V. Strelkov, and M. Hetzelt, Z. Phys., B29:349, 1977).*

anomalous losses are due to inelastic scattering on impurities.

An attempt to eliminate any possible impurities by measuring UCN loss rates on a surface which was continuously renewed by deposition from a metal vapor while the UCN loss rate measurements were in progress failed to show significant improvements in loss rate. This can be interpreted as evidence against the hypothesis that hydrogenous impurities are responsible for the anomalous loss rates. In these experiments, also performed by Strelkov and his collaborators at Alma-Ata, however, there was no attempt to monitor the hydrogen content of the freshly deposited surface. Preliminary measurements at Yale University indicate that the surface hydrogen content is not substantially reduced during evaporation — it seems that hydrogen continuously diffuses from the interior to the freshly formed surface as the deposition continues. A final resolution of the problem must await a new series of experiments, now in preparation, which will involve the measurement of loss rates on carefully prepared surfaces, monitored for their hydrogen content, as well as loss rate measurements at cryogenic temperatures.

Sources of UCN. Until now, UCN have been extracted from steady-state and pulsed reactors by means of either of two methods. One technique uses a "convertor" which restores the low-energy part of the neutron spectrum that cannot penetrate the walls of the UCN apparatus by slowing down, through inelastic scattering, faster neutrons that do penetrate the walls. The other method mechanically slows down faster neutrons after they leave the reactor. This latter objective has been accomplished by means of the Earth's gravitational field — neutrons traveling upward in a vertical direction lose energy to the Earth's gravitational field until they reach the UCN energy range — and also by a "neutron turbine" in which the neutrons are slowed down by reflection from a set of moving curved neutron reflectors. Both techniques have been pioneered by Steyerl at Munich.

Rotating crystal source. At the present time there are two new types of UCN sources under

development, both of which offer appreciably higher UCN densities than previously available. The first of these is similar to the neutron turbine in that it slows down faster neutrons by reflection from a moving surface, although in this case the moving surface will be that of a mica crystal which reflects those neutrons which satisfy the Bragg condition for diffraction from the crystal planes.

The use of Bragg reflection in place of the mirror reflection employed in the neutron turbine will allow the use of higher input velocities to the device. Production of UCN by this method has recently been demonstrated at the Argonne National Laboratory. It is planned to operate this "rotating crystal" UCN source in conjunction with a pulsed neutron source at Argonne. A pulsed neutron source is one which produces neutrons only during relatively short periods of time. This allows the neutron intensity at the peak of the pulse to be quite high while the average intensity and the dissipated power remain relatively low.

If a UCN storage chamber is equipped with a shutter which opens only when the neutrons arrive at the entrance to the chamber, it is possible for the UCN density to build up to a level corresponding to the peak of the pulse. Since pulsed sources have peak intensities greater than the intensities available from steady-state sources, this promises significantly larger UCN densities than are presently available.

Superthermal source. All the above methods of obtaining UCN are limited by the fact that one cannot increase the UCN density above that existing in the source, that is, in the final moderator or convertor in which the neutrons come to thermal equilibrium. However, this is not the case in the "superthermal" UCN source proposed by Golub and Pendlebury. In this source the neutrons are cooled outside the reactor in such a way that, in analogy to a gas-driven refrigerator, the final neutron "temperature" is lower than that of the surrounding medium — that is, one obtains a final UCN density higher than one would obtain if the neutrons were in thermal equilibrium with the surrounding material.

When neutrons scatter from superfluid helium, the helium can exchange energy and momentum

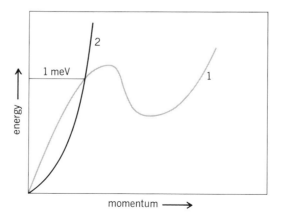

Fig. 3. Curve 1 shows the relation between the energy and momentum which liquid helium at low temperature can absorb from, or give up to, an external probe such as a neutron; and curve 2 the energy-momentum relation of a free neutron.

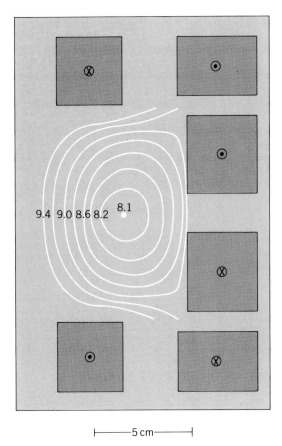

⊢————5 cm————⊣

Fig. 4. Cross section of a superconducting magnetic torus for storing neutrons. *(From K. J. Kugler, W. Paul, and U. Trinks, Phys. Lett., 72B:422, 1978)*

with the neutrons only if there is a strict relation (shown in Fig. 3) between the amount of energy and momentum transferred. Since there is a different relation between the energy and momentum that a neutron can transfer (governed by the energy-momentum relation of a free neutron, also shown in Fig. 3), it turns out that both relations can be satisfied only within certain limits of neutron energy. Thus only neutrons in a narrow energy range around 1 millielectronvolt (meV) can be transferred to the UCN energy range by collisions with the atoms in the liquid helium. (This energy range is near the intersection of the two curves in Fig. 3, since the energy and momentum of UCN are negligible on the scale of this figure.) Conversely, UCN can gain an energy of only 1 meV from the liquid helium. If the temperature of the helium is low enough (below 1 K), the latter process is extremely unlikely. Thus if a container of liquid helium whose walls are transparent to 1-meV neutrons is placed in a beam of such neutrons, there will be a constant production rate of UCN inside the liquid. If, in addition, the walls of the container are constructed of a good UCN reflector, the density of UCN in the liquid helium will build up to a value determined only by the loss rate from the UCN storage vessel. Such a source is now under construction at the Institut Laue-Langevin (ILL; Grenoble, France).

Magnetic storage of UCN. Because neutrons possess a magnetic dipole moment—and hence

behave like microscopic bar magnets—they can be contained in suitably constructed inhomogeneous magnetic fields. A group led by W. Paul at the University of Bonn has constructed two devices in which magnetic fields suitable for storing UCN are produced by superconducting coils. The first, in the form of a torus, has been tested at ILL and has demonstrated its ability to store neutrons for times comparable to the neutron beta-decay lifetime. A cross section of the device is shown in Fig. 4. The coils are bent in the form of a ring whose center is far off to the left of the figure. Centrifigual force prevents the neutrons, whose velocity normal to the plane of the page is about 15 m/s, from escaping to the left. The curves in the figure are the effective equipotential lines showing the combined effect of magnetic and centrifugal forces. The net force is everywhere perpendicular to the curves and directed toward their center. Since the device stores neutrons with energies above the UCN range, there is a possibility of the neutrons being lost due to resonances similar to those which occur in storage rings and accelerators for high-energy particles. Nevertheless, the experiment represents an exciting advance.

The group is attempting to apply the techniques developed with the torus to another device which will have a spherical storage region and which will store only true UCN so the problem with resonances should not arise. The magnetic bottle will be filled with liquid helium, and the UCN will be produced inside the storage region as in the superthermal source described above. After filling, the helium will be drained off, leaving the UCN trapped inside the magnetic field configuration and allowing the decay of the neutrons to be studied.

Neutron electric dipole moment. The first results on the use of UCN to search for the electric dipole moment (edm) of the neutron has been reported from Leningrad. The observation of an edm of the neutron or other elementary particle would be direct evidence for the violation of time reversal symmetry outside the system of K°-mesons where such symmetry violation has already been observed, and is of fundamental interest to physicists. A group under V. M. Lobashov obtained a preliminary value of $4.0 \pm 7.5 \times 10^{-25}$ e-cm (consistent with zero), where e is the electron charge, for the neutron edm. This is already an improvement by a factor of 2 over the best previous measurement and clearly demonstrates the potential of UCN for this fundamental experiment.

For background information *see* MESON; NEUTRON; NEUTRON DIFFRACTION; NEUTRON OPTICS; QUANTUM MECHANICS; REFRACTION OF WAVES; SYMMETRY LAWS (PHYSICS) in the McGraw-Hill Encyclopedia of Science and Technology.

[ROBERT GOLUB]

Bibliography: P. Ageron et al., *Phys. Lett.*, 66A: 469, 1978; I. S. Altarev et al., *Nuclear Physics*, 1979; T. W. Dombeck et al., *Nuclear Instruments and Methods*, 1980; R. Golub and J. M. Pendlebury, *Rep. Progr. Phys.*, 42:439, 1979.

Nobel prizes

The Swedish Royal Academy of Sciences announced 11 recipients of the Nobel prizes for 1979.

Medicine or physiology. Two men shared this prize for their independent work leading to the development of an x-ray technique known as computerized axial tomography (CAT): an American physicist, Allan McLeod Cormack, of Tufts University in Medford, MA, and a British electronics engineer, Godfrey Newbold Hounsfield, of EMI Ltd. The CAT scanner, in which x-rays are used to measure the density of body tissues and a computer helps display those measurements on a screen, greatly facilitates diagnosis of numerous ailments that formerly could be detected only by painful or dangerous tests.

Chemistry. This prize was awarded to Georg Wittig of Heidelberg University and Herbert C. Brown of Purdue University for their independent researches on organic compounds. Brown's development of organoboranes and Wittig's work, notably in the linking of carbon and phosphorus (the Wittig reaction), made possible the synthesis of numerous chemical products through the chemical joining of large molecules.

Physics. Sharing the award were two Americans, Steven Weinberg and Sheldon L. Glashow, of Harvard University, and a Pakistani, Abdus Salam, a professor at the University of London and director of the International Center for Theoretical Physics in Trieste. They were honored for their independent work leading to the unification hypothesis, or the Weinberg-Salam theory of weak interactions, whereby the relationship of electromagnetism to the nuclear force called the weak interaction could be explained. Glashow's research on the charmed quark contributed to the development of the theory.

Literature. The Greek lyric poet Odysseus Elytis was honored "for his poetry, which against the background of Greek tradition, depicts with sensuous strength and intellectual clearsightedness modern man's struggle for freedom and creativeness." The verses of Elytis, a resistance fighter during World War II, poignantly recall the period.

Economics. Sir Arthur Lewis, a British citizen and professor at Princeton University, and Theodore Schultz, an American and professor at the University of Chicago, were the recipients. Both were honored for their work on the economic problems of developing nations. Lewis is noted for his economic theory, the Lewis model, which describes how traditional societies evolve economically into modern ones. Schultz has been concerned with agricultural economic issues, particularly in low-income societies.

Peace. The 1979 Peace prize was awarded to Mother Teresa of Calcutta, founder of the Society of the Missionaries of Charity, for her lifetime of work aiding the destitute in Calcutta. Mother Teresa planned to use the stipend to build more houses, "especially for the lepers."

Nondestructive testing

In recent years it has been found that the radiation resulting from the annihilation of positrons with electrons in metal samples can be used to obtain information concerning atomic scale defects that are present in the specimens. Positron annihilation (PA) techniques have unusual sensitivity—in pure metals concentrations of vacancies as low as several atomic parts per million give rise to measurable effects. In some cases the techniques can also be utilized to determine what types of defects are dominant in the PA response, for example, simple vacancies, small clusters of vacancies, voids, and dislocations. Thus, PA has become a useful research tool in several areas of physical metallurgy.

Because of the paucity of adequate techniques for detecting mechanical damage in metals and alloys, especially that due to fatigue, increasing effort is being devoted to the development of PA as a nondestructive test technique. While improved

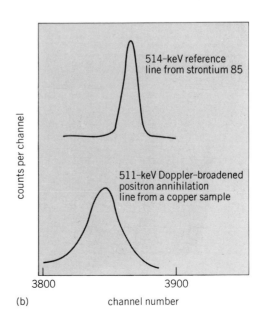

Fig. 1. Positron annihilation Doppler broadening. (a) Schematic representation of apparatus. (b) Plot of data. (From W. B. Gauster, Study of radiation damage in metals by positron annihilation, J. Vac. Sci. Technol., 15:688–696, March/April 1978)

understanding of the response of positrons to microstructural changes is still required, their sensitivity to the presence of defects on an atomic scale suggests that PA may have practical use where the production of such defects causes harmful changes in mechanical properties, such as in radiation damage in nuclear reactor materials.

Positrons in metals. Positrons have the same properties as electrons except for having the opposite electrical charge. They are obtained conveniently from certain radioactive isotopes, such as sodium-22 or germanium-68, that emit, as they decay, positrons with kinetic energies ranging up to 0.5 MeV (^{22}Na) or almost 2 MeV (^{68}Ge). After these particles are allowed to enter a solid target, they slow down (thermalize) very quickly, probably in a few times 10^{-12} s.

The positron and the electron are antiparticles of each other. When a positron encounters an electron, the two particles are annihilated and all their mass is converted into energy in the form of gamma rays according to the equation $E=mc^2$. Here E is the energy liberated in the interaction, m is the total mass of the particles, and c is the speed of light. About 1950 it was noticed that the characteristics of the gamma rays emitted in the annihilation of poritrons with electrons bound in a solid differed among various solids. More recently, sensitivity to structural imperfections in metals has been observed, and the systematics of these variations are a new tool for materials research.

Experimental techniques. Three widely used experimental PA techniques are determination of positron lifetime, angular correlation, and Doppler broadening. Each will be discussed in turn, together with the underlying phenomenology.

Lifetime. It is plausible that the rate at which positrons annihilate with electrons in a metal is proportional to the electron density at the site of decay. The reciprocal of that rate, the lifetime, is then inversely proportional to the electron density. For lifetime measurements, sodium-22 as a positron source has the advantage that together with each positron there is also emitted a 1.3-MeV gamma-ray photon. This gamma can be detected with a scintillation counter and can be discriminated from the annihilation photons. It is used to start an electronic "clock." One of the annihilation gammas is then detected by another scintillator and stops the clock. Many such readings are stored in a multichannel analyzer and processed by computer to determine a mean lifetime for the positrons. In a metal, the mean lifetime is typically in a few times 10^{-12} s.

Thus lifetime measurements yield information about electron densities at the sites at which positrons have come to rest. The two other experimental techniques both supply information about the distribution of momenta of the electrons with which positrons annihilate.

Angular correlation. In the most likely annihilation process, the total energy is carried off by two photons. If both the positron and the electron were at rest at the instant of annihilation, the two gamma quanta would leave in exactly opposite directions in order to conserve momentum. Also, each would have an energy of exactly 511 keV, corresponding to the rest mass of the electron or of the positron. In fact, however, the annihilating electron-positron pairs are not at rest at the time of the interactions, and the two annihilation gammas will deviate slightly from 180° by a few milliradians. The distribution of angles between the photons can be determined by registering the coincident counts in two detectors while one detector is held fixed and the other is moved slightly from the 180° direction. The shape of the resulting angular correlation curve reflects details of the electronic structure of the sample material.

Doppler broadening. The third technique involves measuring small energy shifts in the annihilation lines. Because the electron-positron pairs are not completely at rest, the energy of one annihilation photon may be larger than 511 keV by a small amount ΔE; and the energy of the other photon smaller by the same increment. Typically, ΔE is of the order of a few kiloelectronvolts out of the total 511 keV for positrons annihilating with electrons in a metal. The effect of these shifts is to broaden the spectral line. The broadening can be measured with an energy-resolving counter, such as a germanium detector, and yields the same in-

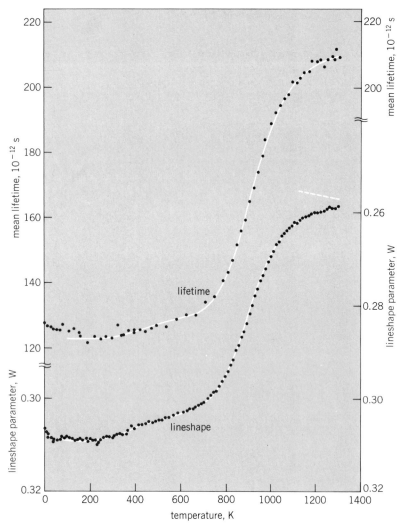

Fig. 2. The mean positron lifetime and a Doppler-broadening lineshape parameter *W* for gold as a function of sample temperature in kelvins. The lineshape parameter *W* is defined such that it decreases with increasing vacancy concentration. (*From D. Herlach et al., Positron lifetimes and annihilation lineshape in Cd and Au from 4.2 K to the melting points, Appl. Phys., 12:59–67, 1977*)

formation concerning electronic structure as the angular correlation curve.

Although angular correlation and Doppler broadening provide the same information, the two techniques have their respective advantages and disadvantages. Angular correlation provides high resolution, which is needed in some research problems. Doppler-broadening measurements, on the other hand, involve shorter counting times. In addition, they require only one detector, making the experimental geometry simpler. Of all three techniques, Doppler broadening is the most promising for nondestructive test purposes because of its ease of implementation and flexibility. A PA Doppler-broadening apparatus is represented schematically in Fig. 1a, and a plot of data in Fig. 1b. The positron source material is sandwiched between two samples. The germanium detector must be operated at low temperatures (cooled with the cryostat), and some stabilization technique is required in conjunction with an analog-to-digital converter (ADC) to achieve the desired resolution. The instrument resolution determines the broadening of the reference line in Fig. 1b, and the further broadening of the 511-keV PA line in the lower part of the figure is determined by the momentum distribution of the electrons in the sample. The data are usually represented by a lineshape parameter, such as the number of counts in the peak of the curve divided by the total number of counts.

Defects and positron annihilation. After a positron has slowed down in a metal sample, it moves about randomly for the remainder of its lifetime of 1 to 2×10^{-10} s. During that time it covers a distance of between 10^3 and 10^5 Å (10^2 and 10^4 nm). This, then, is a measure of the volume "sampled" before annihilation.

Open-volume defects in metals—vacancies, dislocation cores, voids, bubbles—are characterized by the absence of atoms from their regular lattice sites. At such a defect, there is a reduced electron density. Further, the electrons that remain are predominantly conduction electrons that are not bound closely to atomic cores.

If only a few atoms in a million are missing from the lattice, a measurable fraction of positrons can be trapped at the defects. Once trapped, the positrons see the reduced electron density at the defect, resulting in a longer positron lifetime. Further, the altered electron momentum distribution results in changes of the angular correlation curve and the Doppler-broadened lineshape. These changes can be measured accurately; thus positrons are an extremely sensitive probe for defects in metals.

Temperature dependence. An example of such changes, observed by the group of A. Seeger, is shown in Fig. 2. Both the mean lifetime of positrons and a Doppler-broadening lineshape parameter are seen to change with sample temperature in similar ways. It is known from other work that at any given temperature there is a certain concentration of vacancies in a metal sample, increasing with temperature up to a value of several hundred or a thousand parts per million at the melting point. The data of Fig. 2 are interpreted as follows: Below approximately 700 K the changes in lifetime and lineshape parameter are due to effects of

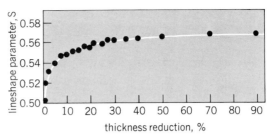

Fig. 3. Change of positron annihilation lineshape parameter S with thickness reduction produced by cold-rolling a nickel specimen. The lineshape parameter S is defined such that it increases with increasing defect concentration. *(From G. Dlubek, O. Brümmer, and E. Hensel, Positron annihilation investigation for an estimation of the dislocation density and vacancy concentration of plastically deformed polycrystalline Ni of different purity, Phys. Status Solidi, (A) 34:737–746, 1976)*

temperature on the behavior of positrons in a "perfect" metal lattice. Above 700 K there are sufficient vacancies to trap positrons before they annihilate with electrons, and a strong change due to temperature is observed. Finally, as the melting point is approached, the concentration of vacancies becomes large enough for every positron to be trapped at a defect during its lifetime, and a saturation of the effect sets in. From such careful measurements, important thermodynamic parameters relating to vacancy properties can be derived.

Metallurgical applications. Metallurgical applications began with the work of I. Ya. Dekhtyar in the Soviet Union in 1964. He observed that cold-working of iron-nickel alloys led to a narrowing of the angular correlation curve. Later S. Berko in the United States made similar measurements on aluminum and attributed the effect to trapping of positrons at dislocations introduced during the working of the metal. Since that time the range of applications has expanded steadily.

A specific example is given in Fig. 3. Here a lineshape parameter is plotted as a function of deformation at room temperature of a high-purity nickel sample. In this particular case, the lineshape parameter values were obtained from angular correlation measurements; but, as discussed above, the same information can be obtained from Doppler-broadening experiments. The PA effect sensitively monitors the changes in defect concentration caused by deformation up to 10% thickness reduction. The more gradual changes for greater deformations reflect saturation effects both in positron trapping and in the defect concentrations. From this and other work, it has been concluded that both the vacancies and the dislocations produced by plastic deformation of metals can act as positron traps.

Several further steps are required to assess the practicality of PA for nondestructive evaluation. From a fundamental viewpoint, additional work is required to correlate PA response with the microstructural changes that occur during deformation, as well as to link the latter with parameters of interest, such as remaining fatigue life of a component. Technological problems involve source-specimen-detector geometries, limitation of sample materials and environments, and so forth. With further progress in these areas, PA is likely to find application wherever rearrangements on an atomic scale lead to changes in the mechanical properties of metals.

For background information *see* CRYSTAL DEFECTS; NONDESTRUCTIVE TESTING; POSITRON; RADIATION DAMAGE TO MATERIALS in the McGraw-Hill Encyclopedia of Science and Technology. [W. B. GAUSTER]

Bibliography: W. Brandt, *Sci. Amer.*, 233(1): 34–42, July 1975; C. F. Coleman and A. E. Hughes, Positron Annihilation, in R. S. Sharpe (ed.), *Research Techniques in Nondestructive Testing*, vol. 3, pp. 355–394, 1977; W. B. Gauster, *J. Vac. Sci. Technol.*, 15:688–696, March/April 1978.

Nuclear magnetic resonance

Nuclear magnetic resonance (NMR) has, over the past 2 decades, become one of the sources of spectral data most commonly used for the structural and conformational analysis of organic molecules. Chemical shift, measured as a parts-per-million (ppm) alteration in an externally applied magnetic field, is widely accepted as an indication of electronic structure in the neighborhood of the nucleus observed, and spin-spin coupling constants are a widely recognized indicator of numbers of magnetic nuclei on directly bonded groups. Despite the analytic potential of these parameters, utilization has until recently been restricted largely to proton magnetic resonance of solution samples. The reasons have been the low sensitivity of most magnetic nuclei, and the high resolution required to extract chemical shift and coupling constant information. Recently the improvements in pulse Fourier transform instrumentation, the popularization of "other nuclei NMR," and the introduction of special techniques for the narrowing of lines in solids have extended applicability to other sample forms. These forms include crystalline and amorphous solids, cellular suspensions, perfused tissue samples, and whole organisms. Use of these alternate sample forms will be reviewed here.

Application to solids. NMR has never been strictly confined to the liquid phase. It is simply that factors other than chemical shift and through-bond spin-spin coupling dominate the spectrum unless they are explicitly removed. In most cases the dominant factor is through-space dipolar coupling of magnetic nuclei. For a pair of spin ½ nuclei, the interaction energy can be expressed in terms of their internuclear distance r, their gyromagnetic ratio γ, and the angle between the applied field and internuclear vector θ:

$$h\nu = \frac{3}{4}\hbar^2\gamma^2\frac{(1 - 3\cos^2\theta)}{r^3}$$

This energy adds or subtracts from the normal transition energy, depending on whether the adjacent nucleus is in a $+\frac{1}{2}$ or $-\frac{1}{2}$ spin state. A rigidly oriented molecule having two equivalent nuclei will therefore show two spectral lines whose separation is a function of molecular orientation in the externally applied magnetic field. In a single crystal having isolated pairs of nuclei with identical orientation, a pair of lines is seen for each distinct dipolar interaction. These discrete lines can be used to deduce structural information, as

was demonstrated by the very early work on gypsum, $CaSO_4 \cdot 2H_2O$. In solution, molecules rotate rapidly compared with the total splitting, and an average dipolar interaction would be observed. It is clear that the function in the equation, averaged over all angles, yields zero, and a single resonance with no through-space effects is observed. In powders or amorphous solids, there are often so many pairs of interactions and so many distinct values of θ that only a broad superposition of all possible splittings is observed. This broad pattern, which can be tens of kilohertz in width, has made extraction of normal high-resolution parameters from solid spectra difficult.

Radio-frequency (rf) decoupling. Several methods have been introduced which reduce dipolar broadening. The simplest involves the application of a strong (10 gauss or 1 millitesla) rf field at the frequency of an abundant nucleus causing dipolar broadening of a dilute spin. Precession of the abundant spins about the rf field effectively decouples them from the more dilute species. This is useful for observation of natural abundance ^{13}C or ^{31}P in samples where dipolar broadening due to protons is the major problem.

Magic angle sample spinning. A somewhat more complicated method is that of magic angle sample spinning (MASS). It is clear from an examination of the equation that if all nuclear pairs were at an angle of 54.7° (the magic angle) with respect to the magnetic field, dipolar interactions would disappear. Rapidly rotating the sample about an axis tilted to 54.7° has the effect of reducing all dipolar interactions to residuals along this axis. Rotation rates must be large compared with the total dipolar splitting. Since splittings are large, this is not practical as a sole means of resolution enhancement for all samples. However, MASS has the additional advantage that it removes a second source of broadening in solid spectra—that due to chemical shift anisotropy (CSA). Strong proton decoupling and MASS are therefore often used in combination. Rotors for MASS capable of holding powdered samples have been designed. Rotors can also be machined from samples of sufficient structural integrity.

Cross polarization. Observation of ^{13}C in solid samples offers an additional problem in that sensitivity is low. Superimposed on this is the fact that relaxation times are long in solids. This prevents efficient averaging because one must wait for recovery of spin magnetization after each pulse-observation cycle. A. Pines, M. Gibby, and J. Waugh have introduced a very effective procedure for improving sensitivity and efficiency of averaging in dilute spin samples. The procedure, now most frequently called cross polarization, relies on transfer of magnetization from an abundant, high-sensitivity nucleus such as the proton to the nucleus to be observed. For a brief period of time, between observations, rf fields are applied to both abundant (^1H) and dilute (^{13}C) spins at strengths such that both precess at equal frequencies in coordinate systems rotating with their respective rf fields. Exchange of magnetization is very effective under these circumstances, and magnetization of the less abundant nucleus (^{13}C) can be quickly restored to a value orders of magnitude larger than its normal equilibrium state. Cross polarization can also be used in combination with strong proton decoupling and MASS.

Polymers. An illustration of these techniques is given in Fig. 1. The spectra are on a sample of poly(phenylene oxide) machined in the form of a suitable rotor. Without MASS but in the presence of strong proton decoupling, spectra show resolution of aromatic and aliphatic resonances. Residual broadening is due to chemical shift anisotropy. With MASS, spectra approach the resolution of solution samples. Features are for the most part similar to solution spectra but do show some differ-

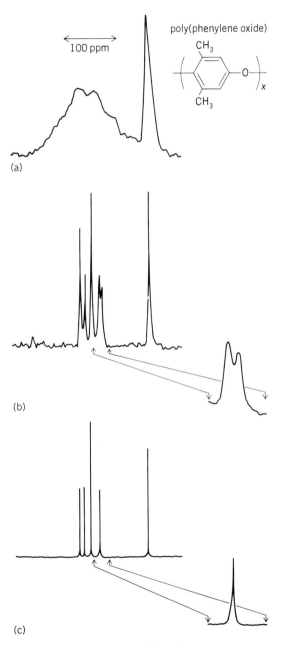

(a)

(b)

(c)

Fig. 1. Cross-polarization ^{13}C NMR spectra of poly-(phenylene oxide): (a) without spinning; (b) spinning at the magic angle; and (c) in solution. (*From J. Schaefer, E. O. Stejskal, and R. Buchdahl, Magic-angle ^{13}C NMR analysis of motion in solid glassy polymers, Macromolecules, 10:384–405, 1977*)

ences. The expanded peak due to the protonated aromatic carbon appears as a doublet in the solid but as a singlet in solution. This is attributed to loss of free rotation about the C-O bonds in the solid which can lead to nonequivalence of the two carbons. Applications to other solids, including fossil fuels and fibrous samples of biological origin, have also appeared in the literature.

Applications to tissue samples. Improved NMR sensitivity and the general impression that NMR need not be confined to solution samples have led to observation of high-resolution spectra of small molecules in heterogeneous semisolid phases. The quantitation of metabolites in cellular dispersions, tissue samples, and whole organs is a particularly interesting application in the biomedical field. Fibrous and membranous components of these samples yield spectra which are broadened by the same factors discussed for solid samples, but small metabolites dissolved in the cytosol or interstices of cells tumble freely and can give high-resolution spectra. The broadening of most resonances makes observation and identification of some metabolites possible by using a variety of nuclei, but ^{31}P has offered distinct advantages. First, ^{31}P is nearly 100% naturally abundant. Second, ^{31}P NMR is relatively sensitive, being about 7% as sensitive as proton NMR. And third, there are relatively few naturally occurring molecules that contain phosphorus, so that background and assignment problems are minimized.

The ^{31}P resonances observed in tissue samples belong to an important group of metabolites involved in energy storage and utilization. These include high-energy phosphates such as creatine phosphate (P-CR) and adenosinetriphosphate (ATP), as well as low-energy phosphates such as inorganic phosphate (P_i), sugar phosphates, and adenosinediphosphate (ADP). Relative levels of each of these substances can be used to assess the metabolic state of functioning tissue. In addition, chemical shifts of inorganic phosphate and ATP can be used as an indication of intracellular pH. These substances have pKa's near physiological pH, and their resonances undergo large chemical shift changes as they are titrated. ATP resonance shifts also respond to complexation with Mg^{++} and provide an indicator of Mg^{++} levels.

An example of an application to a perfused beating rabbit heart is presented in Fig. 2. A normal metabolic state is characterized by a high level of creatine phosphate (resonance at 3 ppm), a low level of inorganic phosphate (resonance at −1 ppm), an intermediate level of ATP (three resonances at 6, 12, and 19 ppm), and a relatively high pH (7.3). An ischemic condition, brought on by interruption of perfusate flow to the heart, is characterized by low creatine phosphate, high inorganic phosphate, and low pH. The ATP levels remain relatively constant at the expense of creatine phosphate, which acts as an energy reserve during brief periods of oxygen deficiency. It is clear from Fig. 2 that details of metabolic processes and the extent of recovery after ischemia can be followed by taking periodic spectra after restoration of perfusate flow. Applications in identification of damaged tissue areas or in evaluation of procedures for preserving organs for transplant are very promising.

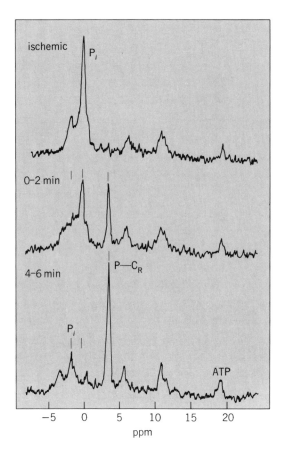

Fig. 2. ³¹P NMR spectra of a perfused rabbit heart. *(From W. E. Jacobus et al., Rapid ³¹P NMR of perfused hearts, in P. F. Agris, ed., Biomolecular Structure and Function, Academic Press, 1978)*

Applications to imaging. The analytical potential of NMR in medicine would be increased manyfold if signals arising from molecules within whole organs or organisms could be spatially resolved. Areas within the sample having abnormal metabolism or other abnormal physiological conditions could then be identified. Substantial progress toward this objective has been made. There are now several approaches to spatial resolution of NMR signals and display of data as an image of the sample. A method based on the application of linear static field gradients has been introduced by P. Lauterbur. Molecules in different parts of a sample have resonances at different frequencies as a result of the variation in magnetic field across the sample. Spectra collected in the pulse Fourier transform mode under the influence of a gradient represent a one-dimensional projection of the object. Several projections may be combined to form two- and three-dimensional images. R. Damadian has introduced a field-focusing technique which employs a magnetic field shaped such that only a small volume within the sample satisfies the resonance condition. The object of interest can be scanned by moving the object and accumulating a spectrum at each step. Images are produced by systematically scanning a two- or three-dimensional grid. W. Hinshaw has introduced a technique employing pulse Fourier transform averaging and time-dependent field gradients. Two orthogonal time-dependent gradients can be applied such that time-independent planes intersect to form a sensitive line. Signals arising along this line can be resolved in the third direction by application of a static gradient. Scanning of the line through the object produces an image.

P. Mansfield has employed selective rf pulses along with switched linear field gradients to achieve spatial resolution in a single pulse sequence and Fourier transformation. Time savings appear to be substantial, although the procedure may be technically more difficult. It appears that additional improvements in technology will be forthcoming.

In principle, many different nuclei and many different NMR parameters could be displayed as an image. The simplest applications, however, display amplitudes of a proton signal as a function of two spatial parameters by using variations in shading to indicate variations in signal amplitude. In biological samples, water dominates the signal amplitude, allowing distinction of fleshy and structural components. Figure 3 is an example taken from work by Hinshaw showing a cross section of a human wrist. Muscle and bone are clearly differentiated as dark and light areas. Although imaging is not unique to NMR, the promise of combining the analytical capabilities realized in solution with observation of structurally complex samples is an interesting prospect.

For background information *see* NUCLEAR MAGNETIC RESONANCE (NMR) in the McGraw-Hill Encyclopedia of Science and Technology.

[JAMES H. PRESTEGARD]

Bibliography: P. F. Agris (ed.), *Biomolecular Structure and Function*, 1978; W. S. Hinshaw, P. A. Bottomley, and G. N. Holland, *Nature*, 270:722–723, 1977; A. Pines, M. G. Gibby, and J. S. Waugh, *J. Chem. Phys.*, 59:569–590, 1973; J. Schaeffer, E. O. Stejskal, and R. Buchdahl, *Macromolecules*, 10:384–405, 1977.

Nuclear physics

In the study of nuclear reactions induced by heavy ions, a major area of research is concerned with the formation of nuclei in extreme conditions, far removed from the normal states of nuclear matter. Nuclei are produced in states of very high angular

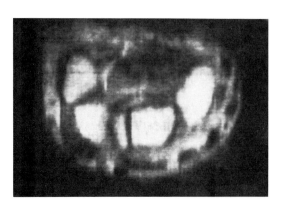

Fig. 3. NMR image of protons in a thin section through a human wrist. *(From W. S. Hinshaw, P. A. Bottomley, and G. N. Holland, Radiographic thin-section image of the human wrist by NMR, Nature, 270:722–723, 1977)*

momentum, up to the limit at which they fly apart in fission; the limiting states of nuclear temperature and density are also under extensive study. Determining another limit, the edge of nuclear stability, continues to pose a challenge to both experimentalists and theorists, who are studying exotic nuclei for this purpose. In recent years the main advances in creating these nuclei, far removed from the valley of stability, have exploited low-energy collisions of heavy ions or very-high-energy proton collisions. New experiments on reactions induced by relativistic heavy-ion beams now show great promise for reaching the limit of stability for light nuclei in the near future.

Exotic nuclei. Approximately 300 stable nuclei exist in nature. They constitute the stable elements and isotopes ranging from hydrogen to uranium (Fig. 1). Over the last half-century or so, another 1300 radioisotopes have been produced in a variety of nuclear reactions, and have been identified and studied. These nuclides are also stable against prompt decay by strong interactions into their constituent nucleons. Instead, they are forced to decay by emitting electrons or positrons (beta decay), which is such a weak process that the nuclei are long-lived on a nuclear time scale of the order of 10^{-22} s, and can be regarded as stable. From these 1600 nuclei, the present sophisticated knowledge of nuclear physics has been built; they are denoted in Fig. 1 by the irregular boundary on either side of the squares indicating the stable isotopes.

Theoretical considerations indicate the possible existence of another 6000 nuclei, yet to be discovered. These include completely new elements, in particular the superheavy elements which constitute a new island of stability beyond the present periodic table, but also new isotopes of the known elements, carrying very large neutron excess or neutron deficiency. The theoretical limits of stability of nuclei against strong interaction decay appear as the outer boundaries in Fig. 1, and indicate that a nucleus like $^{70}_{20}\text{Ca}^{50}$, containing 20 protons and 50 neutrons, might be stable. It is important to make experimental tests of the theoretical predictions, which are based on familiar nuclear models like the shell model and the liquid-drop model, developed for conventional nuclei close to the valley of stability. The experiments on exotic nuclei test scientists' ability to describe the general macroscopic properties of nuclear energy in terms of surface and volume effects, as well as the general symmetry features of the nuclear interaction.

Experimental methods. During 1979 developments in two experimental approaches using heavy-ion collisions showed great promise for future spectacular advances in the production of exotic nuclei.

Low-energy heavy-ion reactions. One method makes use of low-energy heavy-ion reactions in which two nuclei, for example, argon and thorium, are brought together to form a temporary association like a rotating dinuclear molecule. This reaction mechanism, known as deeply inelastic scattering, is illustrated schematically in the right-hand side of Fig. 2. While the two nuclei are in contact, various degrees of freedom are relaxed by the diffusion of nucleons through the area of contact.

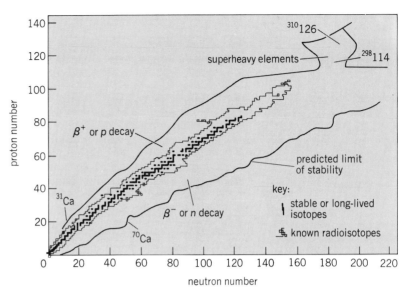

Fig. 1. Chart of the stable nuclei, displayed by atomic number defining the elements on the vertical axis, and by neutron number defining isotopes on the horizontal axis. (*Adapted from D. A. Bromley, Trends in nuclear physics: The challenge of precision, in J. de Boer and H. J. Mang, eds., Proceedings of the International Conference on Nuclear Physics, Munich 1973, North Holland/American Elsevier, 1973*)

Such a process occurs on a relatively slow nuclear time scale of 10^{-21} s, allowing sufficient time for the neutron-proton ratio (N/Z) to equilibrate to the value associated with the intermediate complex, rather than with the initial projectile and target values. As a result, when the molecule finally separates into fragments, for example, into the magnesium isotopes illustrated in Fig. 2, the most probable outcome (M_0) is close to the isotope ^{27}Mg (12 protons and 15 neutrons) reflecting the N/Z of

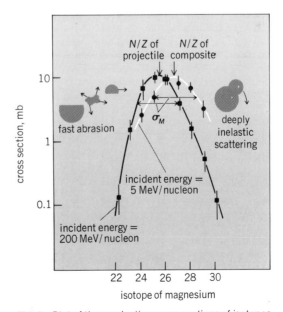

Fig. 2. Plot of the production cross sections of isotopes of magnesium in reactions of argon-40 on thorium-232, at incident energies of 5 MeV/nucleon and 200 MeV/nucleon. Deeply inelastic scattering (occurring at 5 MeV/nucleon) and fast abrasion (at 200 MeV/nucleon) are schematically illustrated. (1 mb = 10^{-31} m².)

the composite system, rather than the most stable isotope of magnesium, namely, ^{24}Mg (12 protons and 12 neutrons). The statistical fluctuations in the diffusion process ensure that there is a distribution of cross section for isotopes *(M)* about this mean value (M_0), according to the following expression,

$$\exp\left[-\frac{(M-M_0)^2}{2\sigma_M^2}\right]$$

where σ_M is the statistical dispersion. In the example of ^{40}Ar + ^{232}Th→Mg isotopes, at an incident energy of 5 MeV/nucleon, that is, a total energy of 200 MeV, a significant cross section for the highly neutron excess product ^{29}Mg is still observed. A recent application of this approach to the ^{40}Ar + ^{238}U reaction led to the production of four new nuclides, ^{37}Si, ^{40}P, ^{41}S, and ^{42}S, which are, in fact, the last known isotopes of silicon, phosphorus, and sulfur. During the past few years, experiments of this type have also established the present limits of stability for many of the light elements.

Spallation reaction. Another very successful technique is the spallation reaction, in which high-energy protons of typically 1000 MeV are used to shatter a heavy target like uranium. Occasionally, from the resultant nuclear debris, exotic nuclei emerge. In contrast to the deeply inelastic process described above, this type of reaction occurs on a very rapid time scale of 10^{-23} s. Both approaches, however, suffer from disadvantage that the exotic species are emitted with very low velocity in the laboratory, making them hard to identify in the face of competing reaction products, millions of times more intense. Nevertheless, prior to the use of the new approach discussed in the following section, deeply inelastic and spallation reactions have been the key to present knowledge of the stable and unstable nuclei.

Abrasion reaction. The abrasion reaction combines some aspects of the spallation and deeply inelastic reactions. As illustrated schematically in the left-hand side of Fig. 2, a heavy projectile is accelerated to very high energies and used to bombard the target. The case shown is ^{40}Ar on ^{232}Th at 200 MeV/nucleon, equivalent to a total incident energy of 40 times 200, that is, 8000 MeV. At this energy, there is insufficent time for the dinuclear molecule to be formed. Instead, the nuclear matter is rapidly abraded from the projectile and target, creating an intermediate zone of highly excited nuclear matter, or a fireball. The abraded projectile fragment continues with the full beam velocity in a direction close to that of the incident beam. Since the reaction takes place on a fast time scale of approximately 10^{-22} to 10^{-23} s, it gives an instantaneous snapshot of the ground-state motion of the projectile. One consequence is a shift of the most probable isotope of any element emitted to the N/Z values reflecting the original ratio in the incident productile, rather than in the intermediate dinuclear complex. The data in Fig. 2 illustrate this behavior, since the peak cross section for the magnesium isotopes appears at ^{25}Mg rather than at ^{27}Mg, corresponding to a shift downward of two neutrons. The distribution of isotopes about the mean value continues to have the same gaussian form of the deeply inelastic reaction, but the dispersion σ_M is determined by the fluctuations of neutrons and protons across the abraded interface, due to the quantal fluctuations of the zero point motion in the projectile ground state rather than to the statistical diffusion of excited nucleons in deeply inelastic scattering.

In spite of the shift of the isotope distribution to lower N/Z values, the abrasion reaction holds great promise for expanding knowledge of the exotic nuclei as they emerge from the peculiar shape of the sheared projectile. Whereas the reaction products in the previous techniques are distributed over all directions in space, the products from high-energy abrasion processes are concentrated into a narrow cone around the initial beam direction. Since practical limitations usually imply that the detection apparatus covers only a small solid angle, the resultant gain in efficiency is typically a factor of 10,000. This gain is of enormous advantage in detecting exotic events which are produced only rarely in any nuclear reaction. The other advantages stem from the high energy of the emerging and incident particles, both simplifying the problem of identifying exotic nuclei and permitting the use of targets a thousand times thicker than in deeply inelastic experiments at low energies. The overall boost in efficiency therefore approaches a factor of 10,000,000.

Experimental results. The first experiment to exploit the abrasion mechanism used an argon beam of 8000 MeV from the Berkeley Bevalac, in which the heavy-ion SuperHilac accelerator injected beam into the high-energy synchrotron. For the collision of an ^{40}Ar projectile containing 18 protons and 22 neutrons with a carbon target, protons and neutrons are sheared from the projectile. The removal of 8, 7, 6, 5, . . ., protons would result in nuclei with large neutron excess, namely, $^{32}_{10}$Ne. $^{33}_{11}$Na, $^{34}_{12}$Mg, and $^{35}_{13}$Al (the subscript is the number of protons, defining the element, and the superscript is the mass number, that is, the total number of neutrons and protons). The removal of protons alone is, however, relatively unlikely, because the motion of neutrons and protons is correlated in the nuclear ground state. This motion must be quantatively understood in order to predict the expected number of events leading to a particular nucleus in a given reaction, and to establish a criterion for the absence of an unstable nucleus.

In Fig. 3 the experimental results are shown as a chart of nuclides. The vertical axis denotes the number of protons (or, equivalently, the charge, or the atomic number) defining the different elements, whereas the horizontal axis represents the mass number (total number of protons and neutrons) of successive isotopes (more precisely, the nuclide mass in atomic mass units). Various consistency and redundancy requirements on the particle identification in an extensive off-line computer analysis of the data effectively eliminate all spurious background events, so that the individual nuclides appear as clearly defined groups at a given Z and N value. The boundary marks the limit of stability reached in previous experiments, and the three boxes denote the additional nuclei, ^{28}Ne, ^{33}Mg, and ^{35}Al, discovered by this first exploration of the abrasion mechanism.

Future prospects. The nuclide ^{29}Ne, one isotope beyond the presently determined limit, is predict-

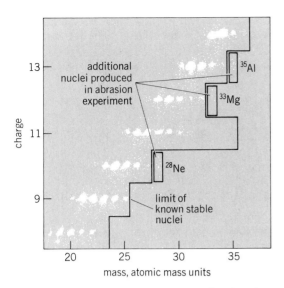

Fig. 3. Display of nuclei created by the abrasion of argon on a carbon target at an incident energy of 8000 MeV. (*From T. J. M. Symons et al., Observations of new neutron-rich isotopes by fragmentation of 205-MeV/nucleon ^{40}Ar ions, Phys. Rev. Lett., 42:40–43, 1979*)

ed to be unstable by many theoretical mass formulas. Likewise, the nuclide ^{34}Mg is predicted to be only barely bound. The presence or absence of these nuclei in an extension of the above experiment could therefore serve as a crucial test of present theories. Such extensions will become possible with the acceleration of relativistic beams of more neutron excess projectiles like ^{48}Ca (containing 20 protons and 28 neutrons) or even uranium. The expected production cross sections are more than an order of magnitude higher than in the experiment with argon, and future explorations with these beams may make it possible for the first time to map out the stable mass surface for a very wide range of elements.

The question of which nuclei are stable, and of how many neutrons a given number of protons can bind, is central to nuclear structure physics. It is possible, for example, that the nuclear density, which is surprisingly constant throughout the known periodic table, could be altered in very exotic nuclei, leading to a change in the binding energy. Present methods of estimating the stability of nuclei by means of theoretical mass formulas could then be seriously in error. The detailed understanding of an exotic nucleus like ^{70}Ca is also a challenge to existing models of nuclear structure, which have been evolved in the valley of stability.

Moving to new regions of the N, Z plane opens a rich field for speculation. It is conjectured that nuclei of very large neutron excess may exist in which the extra neutrons form an expanded cloud enveloping a conventional nucleus at the center. Such two-component nuclei have never been observed, nor is it clear what the optimum form of such differentiated nuclei would be. One possibility is that toroidal and bubble nuclei with a central depression of the density may become stable. These exciting speculations about nuclei at the limit of stability are likely to be pursued with in-

creasing vigor in the years ahead, as the improvement of relativistic heavy ion beams leads to more detailed studies of abrasion reactions.

For background information *see* ISOTOPE; NUCLEAR REACTION; NUCLEAR STRUCTURE in the McGraw-Hill Encyclopedia of Science and Technology.   [DAVID K. SCOTT]

Bibliography: P. Auger et al., *Z. Phys.*, A289: 255–259, 1979; G. W. Butler et al., *Phys., Rev. Lett.*, 38:1380–1382, 1977; J. Cerny and A. M. Poskanzer, *Sci. Amer.*, 238(6):60–72, 1978; T. J. M. Symons et al., *Phys. Rev. Lett.*, 42:40–43, 1979.

Nuclear power

Major events related to the development of nuclear power in 1979 included an accident at the Three Mile Island nuclear reactor and controversy over the health effects of low-level radiation.

Three Mile Island reactor accident. The accident at the Three Mile Island nuclear power plant near Harrisburg, PA, occurred on Mar. 28, 1979. Caused by a human error in leaving closed some valves it was compounded by a succession of mechanical failures intertwined with human errors which resulted in releases of radioactive gases, principally ^{133}Xe, into the atmosphere and serious overheating of the reactor fuel. The fuel overheating occurred when both the regular cooling water pumps and the emergency core-cooling systems were manually shut off by the operators, who failed to recognize the problem because they relied on a water level indicator which proved to be faulty rather than temperature indicators which showed serious temperature rises. As a result of the overheating, water reacted chemically with the zirconium fuel cladding to produce hydrogen gas. This formed a large bubble that posed a serious threat for several days. At first there were fears that the hydrogen might explode and break the reactor vessel, but it was soon determined that the oxygen was insufficient for such an explosion. It was then feared that the bubble might block the coolant flow and thereby cause a meltdown. Eventually, the hydrogen was bled off by locally heating the water in the "pressurizer" to drive hydrogen out of solution, and then releasing the gases into the containment. There was never a danger of a hydrogen explosion breaking the containment since there was not enough hydrogen generated for this under any circumstances. An indicated pressure spike in the containment was 14 psi (97 kPa), and this may be interpreted as a hydrogen explosion; the containment would rupture at about 100 psi (689 kPa).

Effect on population. Pregnant women and small children were advised to leave the area within a 5-mi (8-km) radius, and when the hydrogen bubble first appeared and seemed to be growing, a general evacuation of the area close to the plant was considered but was not implemented. A significant fraction of the local citizenry left of their own accord.

The total population dose within a 50-mi (80-km) radius was estimated as 3300 man-rem, an average of 1.7 millirem per person for the 2,000,000 people in the area. The highest hypothetical individual exposure was 83 mrem. These estimates are based on assuming that all people normally living there remained, with no credit for protection from being inside buildings or for absorption of gamma rays in passing through bodies; these factors represent a substantial conservatism. From 3300 man-rem one would expect 0.6 eventual excess fatal cancer, whereas the normal number of fatal cancers expected among the population of 2,000,000 is 325,000. An 83-mrem exposure to an individual gives a cancer risk of 15 chances in 1,000,000, and is about equal to the extra annual natural exposure from living in Colorado.

Response to the incident. A special Presidential commission was appointed to study the incident, and there were several government and industry investigations. It was widely noted that formation of a hydrogen bubble that might block coolant flow was not an accident sequence considered in the Reactor Safety Study directed by N. C. Rasmussen in 1975 (formation of hydrogen and hydrogen explosions were treated). The consensus was that there was a need for better operator training, for improved emeroency response, and for computer aid in interpreting varying and possibly disparate signals that form the basis for operator judgment in an emergency. Preliminary estimates were that the reactor could be brought back into operation in about 4 years at a cost close to $400,000,000. Perhaps the most important effect of the incident was a marked decrease in public confidence in the safety of nuclear power.

Health effects of low-level radiation. Considerable controversy arose over claims that low-level radiation is more dangerous than it was previously believed to be.

Industrial workers. The controversy was started by the study of T. Mancuso, A. Stewart, and G. Kneale on workers at the Hanford plant (a Federal facility near Richland, WA), and that group published two further "refinements" during the year (giving successively lower but still substantial effects). There were at least 20 critiques of their work, and their findings were explicitly rejected by the U.S. National Academy of Sciences Committee on Biological Effects of Ionizing Radiation (BEIR) and the U.K. National Radiological Protection Board. The only evidence from their study that has been generally accepted is that there were two extra cases each (3 observed versus 1 expected) of cancer of the pancreas and multiple myeloma among workers with about 30-rad exposure. Quite possibly these were caused by chemical exposures since those exposed to higher radiation levels were generally also exposed to more chemicals. No excess of these diseases at the same exposure level occurs among the Japanese atomic bomb survivors, and the number of people involved is an order of magnitude larger in the latter case.

More "new evidence" was an excess of leukemia reported among workers at the Portsmouth naval shipyard (although Mancuso and colleagues found no excess leukemia). According to the most recent report, there were 6 cases versus 1.1 expected among radiation workers, and their average dose was 1.3 rem. Since the natural plus medical x-ray exposure to the general population is about 10 rem (cumulative), it is difficult to understand how 10 rem plus all nonradiation factors can cause only 1.1 leukemias while an extra 1.3 rem causes 5

cases. One must suspect other causes—there are many carcinogens in a shipyard—or improper dosimetry; in any case, no information emerges on the dose-effect relationship.

Atomic bomb tests. Similar considerations apply to the 3000 soliders who in 1950 occupied an area in the Nevada test site shortly after the "smoky test" of a nuclear weapon. Among this group there have been 8 leukemias versus 4 expected, but the average dose was only 1 rem.

There was a report by J. L. Lyons and colleagues of 19 excess childhood leukemias (3 to 25 within two standard deviations) among the Utah population living downwind from the Nevada test site (although there was about an equal deficiency of other childhood cancers). Since there is no dosimetry data, no information emerges on the dose-effect relationship.

Other radiation sources. A reanalysis of a 1959–1962 study of effects of medical x-rays was reported by I. Bross and colleagues. It concludes that usual estimates of effects of low-level radiation should be multiplied by 10. The Bross paper was immediately followed by a critique, and it was rejected by the BEIR committee. It uses a complex theory with many more adjustable parameters than can be evaluated from the data with statistical significance.

Excess leukemia was reported in Grand Junction, CO, where uranium mill tailings were used in construction, but there was no excess of lung cancer, which would be the principal expected effect of the tailings (due to their radon emission). A subsequent study found that the leukemia excess was not correlated with living in a house built with tailings, or with length of residence in Grand Junction.

There were reports of excess cancer near the Rocky Flats, CO, plutonium plant, but there are actually less cases than expected there of lung cancer, which is the principal effect expected from plutonium.

Critiques and rejections. The BEIR report which became available in prelminary form at the end of 1978 gave essentially the same estimates of effects of low-level radiation as its 1972 report. The U.N. Scientific Committee on Effects of Atomic Radiation issued a report in early 1978 which gave the same estimate as its 1972 report. The International Commission on Radiological Protection meets periodically and has decided not to change its estimates and recommendations because of the "new evidence."

Other major events. In 1978, for the first time, nuclear power was legally stopped as a result of a referendum in a state (Montana) and in a foreign country (Austria). A referendum in Switzerland which would have made new plants impossible and required relicensing of existing plants was defeated by only a 51.2% vote.

The first two tests of the emergency core-cooling system were carried out on the LOFT test reactor in Idaho, with a simulated double-ended pipe break. In both cases, the system worked substantially better than predicted; in the full-power test the maximum cladding temperature reached 1100°F (593°C) versus 1365°F (741°C) predicted.

The U.S. Nuclear Regulatory Commission withdrew its support for the Rasmussen Reactor Safety Study on the grounds that the uncertainties there were underestimated. It was stated that this did not imply that the Rasmussen estimates of the dangers were too low.

Summary of 1978 operations. During 1978 approximately 12.5% of United States electricity and 10% of West European electricity were obtained from nuclear fission. Nuclear plants in the United States had a weighted average net capacity factor of 68.0% and an availability factor of 76.6% versus 55.1% and 77.8% for coal-fired plants. The cost of nuclear-generated power was 1.5 cents/kWh, about the same as in 1976 and 1977. A coal-generated kilowatt-hour cost 1.8 cents in 1976, 2.0 cents in 1977, and 2.3 cents in 1978, and for oil these costs were 3.5, 3.9, and 4.0 cents respectively.

For background information *see* NUCLEAR POWER; REACTOR, NUCLEAR in the McGraw-Hill Encyclopedia of Science and Technology.

[BERNARD L. COHEN]

Oceanography

The Deep Sea Drilling Project has generated a great deal of information about the Earth. Improved knowledge of the ocean floors is providing a broader understanding of the nature of the Earth's crust, as well as the potential resources beyond the continental shelf.

Deep-sea drilling. More than 100 years ago, the HMS *Challenger* set out from England on a scientific project to study the world's oceans. The *Challenger*'s findings symbolized the beginning of oceanography as a science. In the 1970s, the *Challenger*'s 20th-century namesake produced no less startling accomplishments. As of the sixty-sixth leg of the Deep Sea Drilling Project, the *Glomar Challenger* had drilled at 493 sites, taking more than 30 mi (48 km) of cores from the ocean's bottom (Fig. 1).

Outstanding among its many accomplishments are the following: (1) The extreme youth (150,000,000–200,000,000 years), in geologic terms, of the Earth's crust beneath the sea has been demonstrated. (2) It has been established that parts of the ocean crust have moved horizontally for great distances, away from zones of crust formation along mid-ocean ridges. (3) As it has moved away from the mid-ocean ridge, the sea floor has subsided relative to the ocean surface. (4) Salt domes, which may have acted as traps for hydrocarbons, exist in the bottom of the Gulf of Mexico. (5) The drying up and later refilling of the Mediterranean Sea has been ascertained. (6) The Atlantic Ocean has grown from a series of narrow stagnant basins into the broad ocean of today. (7) Past climatic and ecological changes in the ocean have been studied. (8) Resource-rich deposits have been discovered. (9) The age of the Antarctic ice cap has been established. (10) The fixed location of a "hot spot" in the mantle has been found.

History of deep-sea drilling. The Deep Sea Drilling Project was born after the cancellation of the Mohole project, to which it has superficial similarities. Mohole, so named because it planned to drill 6 km to the Mohorovičić discontinuity, was con-

Fig. 1. Locations of sites (white dots) drilled by the *Glomar Challenger*.

ceived in the late 1950s. This project was aborted by Congress in 1966. In 1964 the Lamont-Doherty Geological Observatory, Rosenstiel Institute of Marine Sciences, Woods Hole Oceanographic Institution, and Scripps Institution of Oceanography coalesced into the Joint Oceanographic Institution for Deep Earth Sampling (JOIDES). As an advisory group, JOIDES submitted a proposal for funding to the National Science Foundation, and in June 1966 the Scripps Institution of Oceanography received a $12,600,000 contract for an 18-month program in the Atlantic and Pacific oceans. In August 1968, built and operated under the subcontract to Global Marine Corporation of Los Angeles, the *Glomar Challenger* drilled its first hole in the Gulf of Mexico. Three extensions and nearly $150,000,000 later, JOIDES has grown to include nine United States and five foreign laboratories. Since late 1975, in the International Phase of Ocean Drilling, the *Glomar Challenger* has provided raw material for researchers in the United States, the Soviet Union, West Germany, Japan, France, and the United Kingdom, each of whom contribute about $1,000,000 annually to the Deep Sea Drilling Project.

Glomar Challenger. The *Glomar Challenger* is a scientific drilling vessel created especially to meet the goals of deep-sea drilling. Launched in Galveston, TX, in July 1968, the ship is 400 ft (122 m)

long, displaces 10,000 tons (10,160 metric tons), and bears a 142-ft-high (43-m) drilling tower amidships. Aside from an unusual appearance, the ship possesses remarkable operating capabilities. It can be positioned in water too deep for anchors, and thus remain at a drilling site as long as desired. A dynamic positioning system, which maintains the ship's position within a radius of a few hundred feet, makes this possible (Fig. 2).

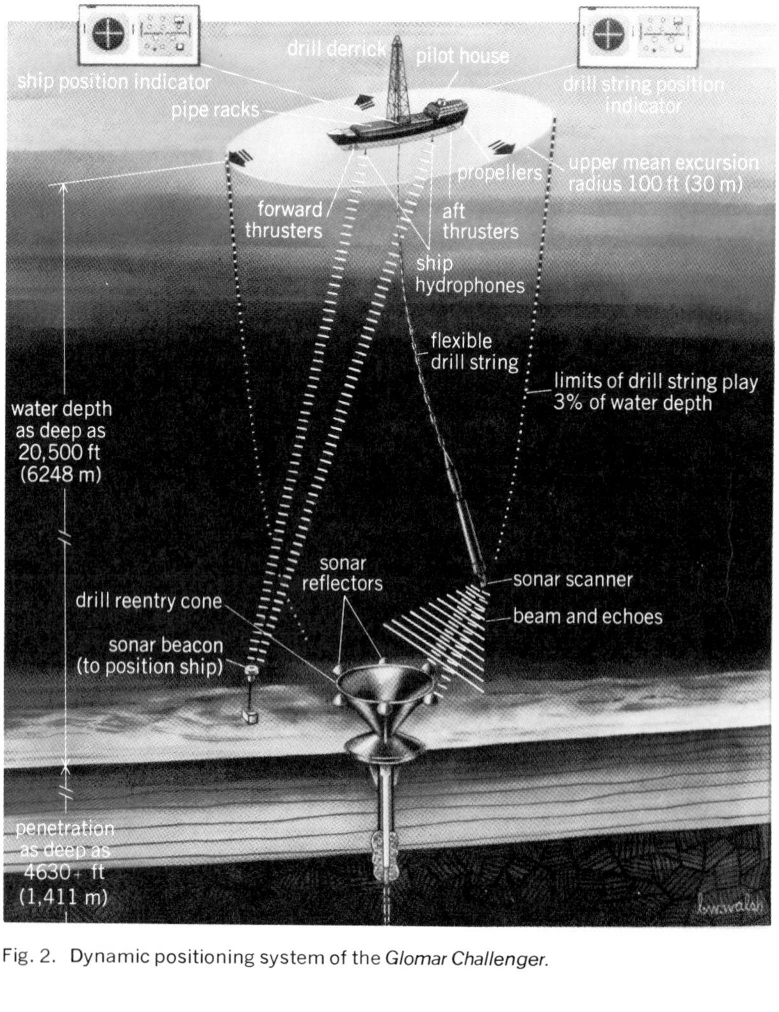

When the ship reaches a drilling site, a beacon that emits acoustic signals is dropped to the ocean floor. The pulses are received by hydrophones located in the hull. The ship's position is determined by triangulation from the difference in arrival times of the pulses at the individual hydrophones. A shipboard computer then calculates the amount of thrust required from each of the propulsion units to keep the ship precisely on station. These propulsion units consist of four thrusters (two forward and two aft) with propellers that provide lateral thrust to enable the ship to move sideways, and the more powerful main propulsion units.

So that drillers can effectively and rapidly control, on a rolling sea, the enormous lengths of heavy drill pipe needed to probe the ocean floor, the ship incorporates an automatic pipe-racking device that takes up most of the topside space forward of the derrick. More than 23,000 ft (7.0 km) of drill pipe are stacked in 90-ft (27-m) lengths in the pipe racker.

Beneath the derrick is an opening about 20×22 ft (6.1×6.7 m) extending through the bottom of the ship. The drill pipe is suspended from the derrick and passes through a stress-distributing structure shaped somewhat like the flared throat of a horn. This prevents the roll and pitch of the ship from bending the pipe sharply at any one point. The lengths of pipe in the drill string are raised and lowered by winches using huge pulleys with 1,000,000-lb (454-metric ton) capacity.

Glomar Challenger has living and storage facilities that permit the ship to remain at sea for 60 days at a time. There is berthing for 74 persons, including the ship's operating crew, the drill crew, and the scientists and technicians. Living spaces, the bridge, and laboratories dominate the upper decks aft of the drilling derrick. The ship is fitted out with one of the best laboratories ever designed for the study of geological materials at sea.

Satellite navigation permits positioning the ship accurately at any time and provides precise location of the drilling sites. The ship receives daily satellite weather photographs of cloud patterns, which are used in weather forecasts to increase the safety and effectiveness of the drilling operation and schedule.

Value of deep-sea drilling. Plate tectonics provides the now generally accepted conceptual model for the continuing evolution of the surface features of the Earth. The model was developed from geological and geophysical observations in the oceans and from earthquake seismology. Its validity has subsequently been strikingly demonstrated by evidence from samples recovered from deep beneath the ocean floor by the *Glomar Challenger*.

While providing a broad conceptual framework for understanding the evolution of the Earth, plate tectonics has also exposed many areas of igno-

Fig. 2. Dynamic positioning system of the *Glomar Challenger*.

rance and raised many questions. These questions are grouped under four broad headings: oceanic crustal studies, active margin studies, passive margin studies, and ocean paleoenvironment studies. Problems to be studied in the first two areas are oriented toward the study of processes going on within the solid Earth which are manifested at the active plate boundaries, such as mountain building.

Passive margin and ocean paleoenvironment studies relate to the processes associated with the birth of an ocean and the subsequent evolution of the ocean environment. The latter involves the interaction between climatology, geochemistry, oceanography, and the living earth.

Studies conducted in these four broad problem areas are of more than academic interest. Improved knowledge of the ocean floors will provide a sound scientific basis for the evaluation of potential hydrocarbon and other mineral resources beyond the continental shelf, both in passive margin and active margin areas, well in advance of any possible commercial activity.

Understanding of the nature and evolution of the Earth's crust and its state of stress in various tectonic settings will greatly clarify understanding of the mechanisms of plate tectonics and thus contribute to the development of a predictive understanding of major natural geologic hazards (earthquakes, volcanic eruptions).

The safe disposal of long-lived, human-generated poisonous waters (for example, nuclear waste) is a topic of growing concern. The deep oceans have been suggested as a possible repository, but until comprehensive information becomes available concerning sea-floor processes and rates of change, it is impossible to assess reliably the capability of the deep-sea floor to contain these substances for the necessary periods of time.

Finally, and also of global significance, is an assessment of the natural variability of long-term climatic changes. The most complete history of the Earth's changing climate, prior to the last few thousand years, is contained in the sediments and fossils of the ocean floor. Presently observed changes and the consequences forecast from these by mathematical models of the present climate regime can be tested against such observations.

Deep-sea drilling also contributes to a better understanding of the Earth processes in the following broad categories: (1) sediment diagenesis; (2) organic geochemistry; (3) paleoclimate and paleocirculation; (4) stratigraphic correlations; (5) ocean crust age distribution, chemical and petrologic nature, and structural evolution; (6) tectonics and sedimentation; and (7) resource evaluation of available mineral ores, reservoired hydrocarbons, source rock potential, reservoir rock potential, thermal history, geometry and chronological development.

Future of deep-sea drilling. In spite of its impressive accomplishments, the Deep Sea Drilling Project has scarcely gone farther than to scratch the surface of the ocean floor. Though much remains to be done within the limited capability of *Glomar Challenger*, drilling technology has now been developed to the point where a larger ship, capable of drilling in deeper water, penetrating

further beneath the sea bottom, and suspending much heavier loads from the drilling platform, could be effectively used. These greater drilling capabilities are needed to take advantage of the scientific opportunities revealed by the work already done.

By mid-1979, not much drilling had been accomplished in the continental margins, including the continental slope and the rise to seaward of the slope in the passive margins of the Atlantic, and the landward sides of trenches in the active margins that encircle the Pacific. Yet these are the regions that may contain many of the undiscovered keys to the dynamics of plate tectonics. The sediments of the continental margins are thick and complex, and deep penetration by drilling is necessary to understand their relationships.

Several practical reasons exist for drilling in the continental margins. The active margins are the loci of many destructive earthquakes and volcanic eruptions. Measurements of stress and strain at depth in boreholes may be important in the development of forecasting methods for these catastrophic phenomena. Enormous volumes of sediments have accumulated on the passive margins. Hydrocarbons may have been formed and trapped in these sediments on a large scale.

The project's contract is being amended to provide for two additional years of drilling during 1980 and 1981. The drilling plans for 1980 and 1981 have progressed through meetings and recommendations of working groups for the North and South Atlantic drilling. The basic program calls for five legs of ocean paleoenvironment drilling in the South Atlantic, six legs of passive margin drilling in the North Atlantic, and a commitment of one leg for Caribbean drilling.

A 1977 JOIDES report, *The Future of Scientific Ocean Drilling*, recommends an intense 10-year program for some continued shallow drilling as well as increased deep drilling, concentrating on the continental margins beyond 1981. The program (costed at about $500,000,000 – 600,000,000) calls for continued use of the *Glomar Challenger* until 1984, while a larger vessel, the *Glomar Explorer*, is being prepared to start deep drilling in 1981.

The *Challenger*'s drill string—a maximum of 25,000 ft (7.6 km)—is not adequate to reach the critical layer of volcanics or to drill in the deep trenches. Thus, many answers about heat flow, seismic activity, and the birth and destruction of the ocean crust elude the *Challenger*'s reach. The ship lacks other capabilities needed for such exploration. Its hull is not thick enough and its platform not stable enough to venture into the icy Antarctic seas in search of offshore resources. The worst problem of margin drilling is the risk of hitting oil and gas. To prevent disasters and control "gushers," should one occur, a ship must have a riser and blowout preventer, which the *Challenger* lacks. A riser, a casing like that used in land drilling, circulates drilling muds and equalizes the pressure if a gas chamber is struck. A blowout preventer seals a hole if a gusher occurs.

Though the *Challenger* could be upgraded, the ship lacks the necessary lifting capacity for a longer drill string and riser system. Changing priorities recommend the *Glomar Explorer*. Alternative

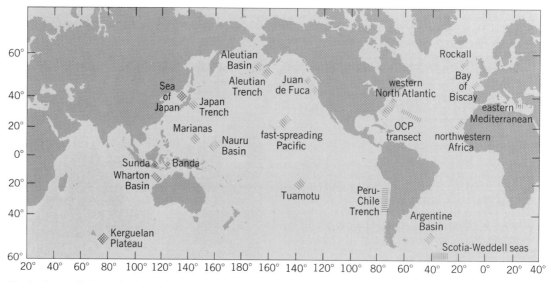

Fig. 3. Areas of interest for future *Glomar Explorer* drilling.

ships have been studied, but none has the capabilities of the *Explorer*—in particular, a lifting capacity sufficient for a 12,000-ft (3.7-km) riser system plus 30,000 ft (9.1 km) of drill string. The 618-ft (188-m) 25,000-hp (18.6-kW) *Explorer* has more than 10 times the lifting capability of the 400-ft (122-m) 10,000-hp (7.6-kW) *Challenger*. The *Explorer* has greater power, a more stable platform, and a thicker hull than the *Challenger*—everything needed to carry out JOIDE's program of increased margin and crustal drilling.

The entire future program of scientific ocean drilling is being carefully conceived, each portion to provide a conceptual and technical base for adjoining portions. During the remainder of the *Challenger* work, engineering studies, with special drilling, will provide important information for both technical development and environmental assessment and impact statements for future riser work. Similarly, a parallel geophysical program will provide for problem definition and for scientific site-specific drilling program development, including environmental safety (Fig. 3).

The possible advent of a scientific drilling vessel, having deep-water marine riser capability and a long drill string, and also the ability to operate in the adverse conditions of high latitudes, will open up profound opportunities in a field that is the very intersection of basic science and applied knowledge of the Earth.

This future program will provide information for resource assessment for fossil fuels within the largest accumulations of sediments on Earth; it will contribute to the technology necessary to recover them; it will provide information on the history, time scales, and processes of the oceans with which to more adequately judge environmental concerns. It will provide essential information for an energy and environment strategy for the future.

For background information *see* OCEANOGRAPHY; PLATE TECTONICS; SALT DOME in the McGraw-Hill Encyclopedia of Science and Technology.

[STAN M. WHITE]

Bibliography: *Initial Report Volumes of the Deep Sea Drilling Project*, vols. 1–49, U.S. Government Printing Office, 1968–1979; F. Press and R. Siever, *The Earth*, 2d ed., 1978; R. Revelle, *The Past and Future of Ocean Drilling*, pamphlet of JOI, Inc., Washington, DC, 1979; F. P. Shepard, *Submarine Geology*, 1979.

Oncology

Tumor promoters are a class of chemicals that complete the process of chemical carcinogenesis started by another group of chemicals known as initiators. This two-stage model of induction of cancer by chemicals was observed first in mouse skin, and now has been observed in other organs such as liver, prostate, lung, mammary gland, and urinary bladder. The initiation stage is accomplished by the exposure of the organ to a single small dose of a carcinogen. An initiating dose of a carcinogen per se will not lead to the development of visible tumors during the life-span of the animal. Visible tumors will result following prolonged exposure of the initiated organ to a second chemical, or tumor promoter, in the promotion stage. However, in most cases, initiators applied at higher doses are also able to cause cancer without the aid of a promoter. Tumor promoters are not carcinogenic by themselves; it is only their application following initiation that elicits tumors.

Initiation leads to a permanent change in the genetic machinery of the cell, perhaps by the interaction of the metabolite(s) of the initiator with deoxyribonucleic acid (DNA), and this process is thought to be irreversible. For instance, skin tumors will develop in mice even if 1 year (half the mouse's life-span) is allowed to elapse between initiation and application of a tumor promoter. In contrast, promotion is reversible; the promoter needs to be applied at frequent intervals for the development of tumors. Because initiation results from exposure to a single, noncarcinogenic dose of a carcinogen and because the process is irreversible, it is difficult to modify initiation to control human cancer. Because promotion is reversible and humans are constantly exposed to numerous promoters in the environment, the promotion aspect of chemical carcinogenesis is crucial in the etiology of human cancer. A list of some of the chemicals acting as initiators or promoters in various organs is given in Table 1.

The immediate aim of researchers studying promotion is to identify molecular event(s) that are unique and specific to tumor promotion. Such studies will facilitate understanding of the molecular mechanism of promotion and the development of drugs to prevent and treat cancer.

Among the various systems (both in living organisms and in laboratory cultures) developed to understand the biochemical mechanism of action of tumor promoters, mouse skin is the most extensively studied and is perhaps the most useful animal model (80% of human cancer is of epithelial origin). Various chemicals, including Tweens (nonionic surface active agents), α-limonene, iodoacetic acid, anthralin and other phenols, certain fatty acid esters, several surface active agents, and a series of phorbol esters (see illustration), have been shown to promote skin tumors. 12-O-tetradecanoyl-phorbol-13-acetate (TPA), a diterpene, is the most potent tumor promoter among a series of phorbol esters isolated from croton oil. Croton oil is obtained from the seeds of a leafy shrub, *Croton tiglium* (Euphorbiaceae), that is native to Southeast Asia.

Biochemical changes. TPA, either applied topically to mouse skin or added to cells in culture, elicits numerous biochemical and biological effects. Among the earliest effects of TPA that suggest it might interact with membranes are enhanced transport of small molecules such as 2-deoxyglucose and α-aminoisobutyric acid, stimulation of Na^+/K^+-dependent adenosinetriphosphatase and $5'$-nucleotidase activities, increased synthesis of phospholipids, altered membrane glycoproteins, and loss of the large-external-transformation-sensitive proteins. TPA treatment leads to sequential activation of ribonucleic acid (RNA), protein, and DNA synthesis. The other changes observed following TPA treatment include enhanced phosphorylation of nuclear histones, decreased histidase activity, and altered cyclic nucleotide metabolism. Many of these biochemical changes are observed in other cells and tissue systems stimulated to proliferate and, thus, cannot be causally related to skin tumor promotion. Among the well-characterized biochemical effects of TPA that are thought to be necessary for tumor promotion are the induction of epidermal ornithine decarboxylase activity and increased proteolytic activity.

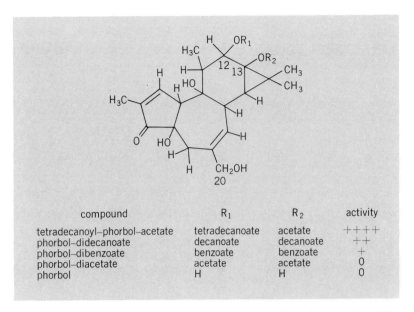

compound	R₁	R₂	activity
tetradecanoyl–phorbol–acetate	tetradecanoate	acetate	++++
phorbol–didecanoate	decanoate	decanoate	++
phorbol–dibenzoate	benzoate	benzoate	+
phorbol–diacetate	acetate	acetate	0
phorbol	H	H	0

Structure of phorbol and certain of its esters, together with their promoting ability. (*From R. K. Boutwell, The function and mechanism of promoters of carcinogenesis, CRC Critical Rev. Toxicol., 2:419–443, 1974*)

Ornithine decarboxylase activity. Ornithine decarboxylase, which forms putrescine by decarboxylation of ornithine, is the first and probably the rate-limiting enzyme in the biosynthesis of the polyamines, spermidine, and spermine, which are thought to be required for growth and malignant transformation. The role of TPA-induced mouse epidermal ornithine decarboxylase activity in skin tumor promotion was established by a number of observations. Application of TPA to mouse skin leads to a pronounced increase (about 200-fold) in ornithine decarboxylase activity at about 5 hr following treatment. The magnitude of the enzyme induction is dose-dependent and correlates with the ability of the dose to promote skin tumor formation. The degree of induction of ornithine decarboxylase activity by a number of structurally unrelated tumor promoters correlates with their tumor-promoting ability. Furthermore, the ability of a series of phorbol esters to induce ornithine decarboxylase activity correlates with their ability to promote skin tumor formation. Epidermal ornithine decarboxylase activity is induced following treatment of mouse skin with tumor promoters, and not after treatment with nonpromoting hyperplastic agents. The activity of ornithine decarboxylase is elevated in skin papillomas and carcinomas developed by the initiation-promotion procedure. Further support for the proposal that TPA-induced ornithine decarboxylase activity may be an essential component of the mechanism of skin tumor promotion was provided by the observations that vitamin A acid and its analogs and indomethacin (an inhibitor of prostaglandin synthesis) inhibit TPA-induced ornithine decarboxylase activity and formation of skin papillomas promoted by TPA (Table 2).

Role of prostaglandins. The suggestion that prostaglandins may be implicated in skin tumor promotion emerged from the finding that indomethacin inhibits both TPA-induced ornithine decarbox-

Table 1. Tumor initiators and promoters acting on various organs

Organ	Initiator	Promoter
Mouse skin	Dimethylbenz[a]anthracene	Phorbol diesters
Rat mammary gland	Dimethylbenz[a]anthracene	Prolactin
Rat liver	2-Acetylaminofluorene	Phenobarbital
Rat prostate	X-rays	Testosterone
Mouse lung	Urethan	Butylated hydroxytoluene
Rat urinary bladder	N-methyl-N-nitrosourea	Saccharin or cyclamate

Table 2. Inhibition of TPA-induced ornithine decarboxylase (ODC) activity and formation of skin papillomas by vitamin A acid and its analog and by indomethacin

Treatment	Dose, nanomoles	ODC activity, nanomoles CO_2/30 min/ mg protein	Papillomas per mouse
Acetone	—	3.5 ± 0.4	9.2
β-Retinoic acid	34	0.04	1.5
13-*cis*-Retinoic acid	140	0.5 ± 0.1	2.4
Trimethylmethoxyphenyl analog of ethyl retinoate	140	3.3 ± 0.6	9.3
Acetone	—	3.14 ± 0.4	12.2
Indomethacin	280	0.59 ± 0.05	3.6

ylase activity and the formation of skin papillomas. Inhibition of TPA-induced ornithine decarboxylase activity by indomethacin could be overcome by prostaglandin E_1 and E_2 applied concurrently with TPA. Furthermore, TPA treatment enhances the accumulation of prostaglandins in mouse epidermis; this accumulation can be blocked by treatment of mouse skin with indomethacin prior to TPA. Further support for the involvement of prostaglandins in the mechanism of action of TPA was provided by studies of cells in culture. Addition of phorbol diesters to dog kidney cells cultured in serum-depleted medium has been shown to stimulate prostaglandin production. In addition, release of prostaglandin E_2 in cultured peritoneal macrophages was observed as early as 1 hr following treatment of cells with a tumor promoter, and a correlation was found to exist between the ability of a compound to induce prostaglandin release in cell cultures and its ability to promote skin tumor formation in the living organism. The role of prostaglandins in promotion is still speculative; it remains to be seen whether prostaglandins can induce epidermal ornithine decarboxylase activity and can promote skin tumor formation.

Role of proteases. A possible role of proteases in skin tumor promotion is suggested by the data showing that protease inhibitors delayed the onset and reduced the number of tumors appearing on mouse skin. Recently, TPA has been found to increase a specific protease activity (plasminogen activator) in cells in culture. Increased plasminogen activator activity is dependent on the dose of TPA and correlates well with the dose that promotes skin tumor formation. Furthermore, nonpromoters failed to increase protease activity. The correlation between enhanced proteolytic activity and skin tumor promotion by TPA was further strengthened by the findings that the potent antipromoting steroids inhibit TPA-increased activity of plasminogen activator. The role of increased proteolytic activity in promotion is still unclear since the activity of plasminogen activator in mouse epidermis following TPA treatment has not been demonstrated.

Biological changes. Among the numerous biological effects of TPA, induction of sister chromatid exchange and inhibition of terminal differentiation have been shown to be specific for tumor promoters. TPA induces sister chromatid exchanges, whereas the nonpromoting derivative, 4-*O*-methyl

TPA, does not. Inhibitors of tumor promotion—antipain, leupeptin, and fluocinolone acetonide—inhibit formation of TPA-induced sister chromatid exchange in V79-4 Chinese hamster lung fibroblasts.

Under normal situations, a cell undergoes division after passing through a tightly coupled sequence of molecular changes; one of the resulting daughter cells is destined to perform a special function (differentiation), whereas the other daughter cell replaces the parent cell. Cancerous cells lose the ability to follow this highly organized pattern. Neither daughter cell performs specialized functions, but instead both continue to divide so that uncontrolled growth results. It is likely that tumor promoters may interfere with the normal differentiation of cells; this has been shown in a number of cell culture systems. Tumor promoters inhibit spontaneous and induced differentiation of Friend erythroleukemia cells in culture. Several morphological and biochemical changes suggestive of alteration of differentiation have been observed in adult mouse skin treated with TPA. For instance, TPA stimulates the synthesis of two proteins in epidermal cells in culture that are not otherwise found in adult skin.

Summary. The tumor-promoting phorbol diester TPA induces numerous biochemical and biological changes both in intact animals and in cultured cells, but the molecular mechanism of tumor promotion is still not completely understood. Various proposed mechanisms of skin tumor promotion by TPA include increased ornithine decarboxylase and proteolytic activities, induction of exchange of sister chromatids, and, more grossly, inhibition of differentiation. In order to fully understand the phenomenon of promotion, further study of the biochemical processes related to tumor promotion—in particular, the molecular events that lead to altered differentiation during promotion—is necessary. Furthermore, the identification of agents such as steroids, retinoids, and prostaglandin inhibitors that modify promotion will help to unravel the mystery of tumor promotion.

For background information *see* ONCOLOGY in the McGraw-Hill Encyclopedia of Science and Technology.

[AJIT K. VERMA; ROSWELL K. BOUTWELL]

Bibliography: R. K. Boutwell, *CRC Critical Rev. Toxicol.*, 2:419–443, 1974; A. K. Verma et al., *Cancer Res.*, 39:419–425, 1979; L. Diamond et al., *Life Sci.*, 23:1979–1988, 1978.

Optical phase conjugation

Optical phase conjugation involves the use of nonlinear optical effects to precisely reverse the direction of propagation of each plane wave in an arbitrary beam of light, thereby causing the return beam to exactly retrace the path of the incident beam. The process is also known as wavefront reversal or time-reversal reflection. The unique features of this recently discovered phenomenon suggest widespread application to the problems of optical beam transport through distorting or inhomogeneous media. Although closely related, the field of adaptive optics will not be discussed here.

Fundamental properties. Optical phase conjugation is a process by which a light beam interacting in a nonlinear material is reflected in such a manner as to retrace its optical path. As Fig. 1 shows, the image-transformation properties of this reflection are radically different from those of a conventional mirror. The incoming rays and those reflected by a conventional mirror (Fig. 1a) are related by inversion of the component of the k-vector or wave vector k normal to the mirror surface. Thus a light beam can be arbitrarily redirected by adjusting the orientation of a conventional mirror. In contrast, a phase-conjugate reflector (Fig. 1b) inverts the vector quantity $\vec{k}$ so that, regardless of the orientation of the device, the reflected conjugate light beam exactly retraces the path of the incident beam. This retracing occurs even though an aberrator (such as a piece of broken glass) may be in the path of the incident beam. Looking into a conventional mirror, one would see one's own face, whereas looking into a phase-conjugate mirror, one would see only the pupil of the eye. This is because any light emanating from, say, one's chin would be reversed by the phase conjugator and return to the chin, thereby missing the viewer's eye. A simple extension of the arrangement in Fig. 1b indicates that the phase conjugator will reflect a diverging beam as a converging one, and vice versa. These new and remarkable image-transformation properties (even in the presence of a distorting optical element) open the door to many potential applications in areas such as laser fusion, atmospheric propagation, fiber-optic propagation, image restoration, real-time holography, optical data processing, nonlinear microscopy, laser resonator design, and high-resolution nonlinear spectroscopy.

Optical conjugation techniques. Optical phase conjugation can be obtained in many materials whose properties are affected by strong applied optical fields. The response of the material may permit many beams to combine in such a way as to generate a new beam that is the phase-conjugate of one of the input beams. Processes associated with degenerate four-wave mixing, scattering from saturated resonances, stimulated Brillouin scattering, stimulated Raman scattering, photon echoes, and three-wave mixing have all been utilized to generate optical phase-conjugate reflections.

Degenerate four-wave mixing. As shown in Fig. 2, two strong counterpropagating (pump) beams with k-vectors $\vec{k_1}$ and $\vec{k_2}$ (at frequency ω) set up a standing wave in a clear material whose index of refraction varies linearly with intensity. This arrangement provides the conditions in which a third (probe) beam with k-vector $\vec{k_3}$, also at frequency ω, incident upon the material from any direction would result in a fourth beam with k-vector $\vec{k_4}$ being emitted in the sample precisely retracing the third one. (The term degenerate indicates that all beams have exactly the same frequency.) In this case, phase matching (even in birefringent materials) is obtained independent of the angle between $\vec{k_3}$ and $\vec{k_1}$. The electric field of the conjugate wave E_4 is given by the equation below, where

$$E_4 = E_3{}^* \tan\left(\frac{2\pi}{\lambda_0}\delta n\ell\right)$$

δn is the index change induced by one strong counterpropagating wave, λ_0 is the free-space optical wavelength, and ℓ is the length over which the probe beam overlaps the conjugation region. The

conjugate reflectivity is defined as the ratio of reflected and incident intensities, which is the square of the above tangent function. The essential feature of phase conjugation is that E_4 is proportional to the complex conjugate of E_3. Although degenerate four-wave mixing is a nonlinear optical effect, it is linear in the field one wishes to conjugate. This means that a superposition of E_3's will generate a corresponding superposition of E_4's; therefore faithful image reconstruction is possible.

To visualize the degenerate four-wave mixing effect, consider first the interaction of the weak probe wave with pump wave number two. The amount by which the index of refraction changes is proportional to the intensity $(E_3 + E_2)^2$, and the cross term corresponds to a phase grating (periodic phase disturbance) appropriately oriented to scatter the pump wave number one into the $\vec{k_4}$ direction. Similarly, the very same scattering process occurs with the roles of pump waves reversed. Creating the phase gratings can be thought of as "writing" of a hologram, and the subsequent scattering can be thought of as "reading" the hologram. Thus the four-wave mixing process is equivalent to volume holography in which the writing and reading are done simultaneously.

Conjugation using saturated resonances. Instead of using a clear material, as outlined above, the same beam geometry can be set up in an absorbing or amplifying medium partially (or totally) saturated by the pump waves. When the frequency of the light is equal to the resonance frequency of the transition, the induced disturbance corresponds to amplitude gratings which couple the four waves. Because of the complex nature of the resonant saturation process, one does not obtain the simple $\tan^2\left[(2\pi/\lambda_0)(\delta n\ell)\right]$ expression for the conjugate reflectivity. Instead, this effect is maximized when the intensities of the pump waves are about equal to the intensity which saturates the transition.

Stimulated Brillouin and Raman scattering. Earliest demonstrations of optical phase conjugation were performed by focusing an intense optical beam into a waveguide containing materials that exhibit backward stimulated Brillouin scattering. More recently, this technique has been extended to include the backward stimulated Raman effect. In both cases, the conjugate wave is downshifted by the frequencies characteristic of the effect.

Practical applications. Many practical applications of optical conjugators utilize their unusual image-transformation properties. Because the conjugation effect is not impaired by interposition of an aberrating material in the beam, the effect can be used to repair the damage done to the beam by otherwise unavoidable aberrations. This technique can be applied to improving the output beam quality of laser systems which contain optical phase inhomogeneities or imperfect optical components. In a laser, one of the two mirrors could be replaced by a phase-conjugating mirror, or in laser amplifier systems, a phase conjugate reflector could be used to reflect the beam back through the amplifier in a double-pass configuration. In both cases, the optical-beam quality would not be degraded by inhomogeneities in the amplifying medium, by deformations or imperfections in optical elements, windows, mirrors, and so forth, or by accidental misalignment of optical elements.

OPTICAL PHASE CONJUGATION

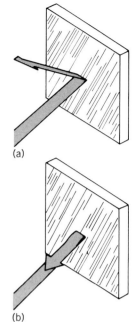

(a)

(b)

Fig. 1. Comparison of reflections (a) from a conventional mirror and (b) from an optical phase conjugator. (From I. J. Bigio et al., High efficiency phase-conjugate reflection in germanium and in inverted CO_2, in V. J. Corcoran, ed., Proceedings for the International Conference on Laser '78 for Optical and Quantum Electronics, STS Press, McLean, VA, 1979)

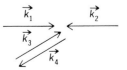
OPTICAL PHASE CONJUGATION

Fig. 2. Geometry of k-vectors for optical phase conjugation using degenerate four-wave mixing.

Aiming a laser beam through an imperfect medium to strike a distant target may be another application. The imperfect medium may be turbulent air, the air-water interface, or the focusing mirror in a laser-fusion experiment. Instead of conventional approaches, one could envision a phase-conjugation approach in which the target would be irradiated first with a weak diffuse probe beam. The glint returning from the target would pass through the imperfect medium, and through the laser system, and would then strike a conjugator. The conjugate beam would essentially contain all the information needed to strike the target after passing through both the amplifier and the imperfect medium a second time. Just as imperfections between the laser and the target would not impair the results, neither would problems associated with imperfections in the elements that constitute the laser amplifier.

Other applications are based upon the fact that four-wave mixing conjugation is a narrow-band mirror that is tunable by varying the frequency of the pump waves. There are also applications to fiber-optic communications. For example, a spatial image could be reconstructed by the conjugation process after having been "scrambled" during passage through a multimode fiber. Also, a fiber-optic communication network is limited in bandwidth by the available pulse rate; this rate is determined to a large extent by the dispersive temporal spreading of each pulse as it propagates down the fiber. The time-reversal aspect of phase conjugation could undo the spreading associated with linear dispersion and could therefore increase the possible data rate.

For background information *see* HOLOGRAPHY; LASER; OPTICAL COMMUNICATIONS; OPTICS, NONLINEAR; RAMAN EFFECT in the McGraw-Hill Encyclopedia of Science and Technology.

[ROBERT A. FISHER; BARRY J. FELDMAN]
Bibliography: A. Yariv, *IEEE J. Quantum Electron.*, QE(14):650–660, 1978; B. Ya. Zel'dovich et al., *Kvant. Elektron.*, 5:1800–1803, 1978 (transl., *Sov. J. Quantum Electron.*, 8:1021–1023, 1978).

Organic chemical synthesis

The future sources of raw materials for the manufacture of organic chemicals have been a problem of growing concern to the chemical industry. Before the petrochemical era began in the early 1930s, almost all organic chemicals were obtained as primary products or coproducts from the processing of coal, wood, and plant material. Today, most major organic chemicals (Table 1) are manufactured from petroleum and natural gas. Only a few building blocks are used to manufacture these chemicals. Interestingly, all but about 10% of the organic chemicals marketed today are made from synthesis gas, a mixture of carbon monoxide and hydrogen; the light olefins ethylene, propylene, and butadiene; and the aromatics benzene and *p*-xylene.

Despite the fact that the chemical industry's consumption of feedstocks amounts to only about 3% of the world's total demand for petroleum and natural gas, the uncertainties associated with petroleum and natural gas supplies and their increasing costs have led most manufacturers to begin searches for new raw materials. The interest in coal-derived chemicals has been renewed; and other fossil materials (such as oil shale, tar sands, and peat), nonfossil materials (such as organic wastes and land- and water-based vegetation), natural biochemical methods, and the indirect conversion of solar energy to organic chemicals via hydrogen and inorganic carbon resources are all under

Table 1. United States production of organic chemicals in 1978

Compound	10^9 kg	10^9 lb
Ethylene	12.76	28.133
Propylene[a]	6.52	14.375
Benzene[a, b]	5.05	11.13
Urea, primary solution	4.88	10.76
Ethylene dichloride	4.75	10.463
Toluene[a, b]	4.20	9.25
Ethylbenzene[c]	3.80	8.379
Vinyl chloride	3.15	6.955
Styrene	3.12	6.878
Formaldehyde, 37 wt %	2.92	6.433
Methanol, synthetic	2.88	6.36
Xylenes[a, b]	2.79	6.16
Terephthalic acid[d]	2.71	5.971
Ethylene oxide	2.19	4.838
Ethylene glycol	1.82	4.015
1,3-Butadiene, rubber grade	1.59	3.515
p-Xylene	1.58	3.493
Cumene	1.48	3.257
Acetic acid, synthetic	1.26	2.786
Phenol, synthetic[e]	1.24	2.726
Cyclohexane	1.06	2.335
Acetone	0.959	2.115
Propylene oxide	0.927	2.044
Acrylonitrile	0.795	1.752
Isopropyl alcohol	0.774	1.707
Adipic acid	0.767	1.69
Vinyl acetate	0.763	1.681
Acetic anhydride	0.680	1.50
Ethanol, synthetic	0.576	1.27
o-Xylene	0.460	1.014
Phthalic anhydride	0.423	0.933
Caprolactam	0.415	0.915
Carbon tetrachloride	0.334	0.737
Perchloroethylene	0.327	0.722
n-Butanol	0.293	0.646
Methyl chloroform	0.283	0.623
Aniline	0.275	0.606
Methyl ethyl ketone	0.271	0.598
Methylene chloride	0.257	0.567
Propylene glycol	0.250	0.551
Ethyl chloride	0.245	0.541
Dodecylbenzene[f]	0.238	0.525
Carbon disulfide	0.216	0.476
Bisphenol A	0.214	0.471
Methyl chloride	0.206	0.454
2-Ethylhexanol	0.188	0.415
Dioctyl phthalates	0.176	0.388
Ethanolamines[g]	0.165	0.364
Maleic anhydride	0.151	0.332
Ethyl acetate, 85%	0.103	0.227
Diisodecyl phthalate	0.075	0.165
Cresylic acid[c, h]	0.065	0.143
Pentaerythritol	0.055	0.121
Total	83.676	184.505

[a]All grades.

[b]Excludes tar distillers but includes material for blending in motor fuels.

[c]Excludes ethylbenzene produced and consumed in continuous-process styrene manufacture.

[d]Includes acid and dimethyl ester.

[e]Excludes coke and gas-retort ovens data.

[f]Includes straight-chain dodecylbenzene, tridecylbenzene, and other straight-chain alkylbenzenes.

[g]Includes mono-, di-, and triethanolamines.

[h]Includes mixed cresols.

SOURCE: Adapted from International Trade Commission, *Chem. Eng. News*, 57(19):24, May 7, 1979; 57(24):37, June 11, 1979.

study as potential replacements for oil and natural gas.

During the 1980s, petroleum and natural gas will continue to serve as the largest sources of feedstocks for organic chemicals manufacture. Only small contributions will be made by other materials to the feedstock pool. But beyond the late 1980s, increasing use of other materials is expected to occur. Coal is expected to make major contributions in the 1990s, and then renewable nonfossil materials and oil shale and possibly peat are projected to be used on a large scale.

Fossil raw materials. During the 1980s, increasing use of processes for the manufacture of the light olefins from crude oil and residuals is expected. Direct cracking of heavy stocks should help maintain olefin supplies, although processing costs will be higher. For example, ethylene and propylene, the top two building blocks of the organic chemicals industry, will continue to be manufactured by cracking of naphtha and of ethane and propane from natural gas and refinery gas. But the trend toward greater use of heavy stocks such as atmospheric and vacuum gas oils is expected to meet almost half the light olefin demand by 1990. In 1978 these stocks met less than 20% of demand. Also, direct crude oil cracking is being developed by several chemical companies. Large-scale commercialization of these processes should help free chemical companies from dependence on refiners and gas processors for supplies of ethane, propane, naphtha, and gas oil feedstocks.

Aromatics. Most of the primary aromatic chemicals, benzene, toluene, and the xylenes (BTX), will continue to be manufactured by petroleum refiners during the 1980s. Catalytic reformate will still be the main source of BTX, but the operation of cracking units for olefin production will supply larger amounts of benzene from the coproduct, pyrolysis gasoline. Catalytic reformate will decrease and pyrolysis gasoline will increase their respective percentage shares of the benzene market. A complicating factor that should begin to be felt in the 1980s is the use of ethanol, and perhaps methanol, as a motor fuel component in ethanol-gasoline blends (gasohol). Since these alcohols have relatively high octane blending values, the need for aromatics as octane improvers in unleaded gasolines may be lessened, and this may help relieve the aromatics shortages by increasing the BTX supplies available for chemical applications.

Synthesis gas. Although not listed in Table 1, synthesis gas is an important intermediate for the manufacture of methanol, ammonia, normal paraffins, olefins, oxo products, and other chemicals by established routes. In the United States, synthesis gas is used mainly for methanol and ammonia manufacture. It is currently made from natural gas, refinery gas, naphtha, and petroleum residuals by steam reforming and partial oxidation. It can also be made from other raw materials such as coal, shale oil, and peat by similar processes. In the 1980s natural gas and petroleum are expected to maintain their dominance in the United States as sources of synthesis gas, but the gradual introduction of large facilities for the conversion of coal to gaseous and liquid fuels, for which gasification

plays major roles, should have a large impact on the availability of synthesis gas for chemical applications. In addition, coal conversion plants can be operated to supply significant quantities of the light olefins, BTX, polynuclear aromatics, naphthenes, phenol, cresols, heterocyclic nitrogen bases, sulfur, ammonia, specialty solvents, and carbon black.

SNG and syncrude. Table 2 illustrates typical product distributions for typical plant sizes of some of the coal, oil shale, and peat conversion processes proposed for commercial use in the United States for the manufacture of substitute natural gas (SNG) and synthetic crude oil (syncrude). Methane, which is the dominant component in SNG, can be used in place of natural gas as a source of synthesis gas, and the liquid hydrocarbon products can serve as sources of organic chemicals in the same manner as petroleum liquids do. The light liquid hydrocarbon products usually contain high concentrations of BTX in the light oil fraction and provide a major source of aromatic chemicals. Light oils from coking operations have been a supplemental source of benzene, toluene, and the xylenes in recent years when conventional petroleum-derived BTX supplies were short. Relatively few SNG or syncrude plants of the capacities presently contemplated would be needed to supply enough BTX to meet all of the needs for these chemicals in the United States.

Table 2 also shows that salable by-products of ammonia, sulfur, and phenols are produced by several of the SNG and syncrude processes. Large quantities of ash also result from these processes. For some ash products, it may be feasible to recover salable quantities of inorganic chemicals. Other useful ash applications such as road aggregate may be feasible, too. Additionally, gasification and liquefaction processes yield low- or medium-Btu gases as intermediate process streams or as products. These gases are high in hydrogen and carbon monoxide and can serve as large sources of synthesis gas for organic chemical synthesis.

Acetylene. It should not be forgotten that acetylene derived from lime and coke via calcium carbide was the foundation of many synthetic organic chemicals until it was displaced by ethylene-based processes. Acetylene technology will probably not be used again on a large scale, but it could be employed if necessary.

Nonfossil raw materials. Simultaneously with the resurgence of a coal energy industry and the development of oil shale in the United States in the 1990s and beyond, nonfossil carbon sources such as trees, grasses, algae, and plants, and organic wastes such as municipal solid wastes, animal manures, agricultural and forestry residues, and industrial wastes—all of which can collectively be called biomass—are projected to serve as significant renewable energy supplies.

Biomass conversion. These materials will be converted to low-, intermediate-, and high-Btu gaseous fuels, liquid alcohol and hydrocarbon fuels, and densified solid fuels. The gasification and liquefaction processes used to manufacture these substitute fuels will in many cases yield organic chemical by-products or intermediate streams that

Table 2. Typical product distribution from modern coal, oil shale, and peat conversion processes

Process	Gasification				Liquefaction			Integrated process
	Lurgi[a]	HYGAS[b]	HYTORT[b]	PEATGAS[b]	H-COAL[c]	HYTORT[b]	SRC[d]	COG refinery[d]
Primary feed	Bituminous coal	Pittsburgh seam No. 8	Devonian shale	Minnesota peat	No. 6 Illinois coal	Devonian shale	Kentucky coal	Kentucky-Illinois coal
wt % S	1.07	4.5	5.73	0.27	5.0	5.57	3.4	–
Feed rate, metric tons/day	21,119	13,527	88,774	51,081[e]	25,200	86,740	19,061	52,345
short tons/day	23,280	14,911	97,857	56,307	27,778	92,614	21,011	57,700
SALABLE PRODUCTS:								
SNG, 10^6 m^3/day at STP	7.73	7.24	6.79	7.01	2.41	6.79	327[f]	8.92
10^6 SCF/day	288	270	253	261	89.9	253	–	332
LPG, metric tons/day	–	–	–	–	–	–	–	18,752
bbl/day	–	–	–	–	–	–	–	20,670
Liquid hydrocarbons, metric tons/day	218[g]	377[g]	110[g]	478[g]	9664[h]	7043[i]	2830[j]	13,299[g,k]
bbl/day	1430	2474	722	3239	67,466	46,214	18,607	100,000
Tars and tar oils, metric tons/day	1496	–	–	1018[l]	–	–	–	–
Solvent refined coal, metric tons/day	–	–	–	–	–	–	9026	2250
Phenolics, metric tons/day	123	–	–	–	–	–	25	142
Sulfur, metric tons/day	170	554	2073	53	992	2519	408	1637
Anhydrous ammonia, metric tons/day	233	84	465	519	186	413	34[m]	–

[a]Lurgi Mineralotechnik GmBH.
[b]Institute of Gas Technology.
[c]Hydrocarbon Research, Inc.
[d]Pittsburgh & Midway Coal Mining Co.
[e]50 wt % water.
[f]Medium-Btu gas in metric tons/day.
[g]Naphtha and BTX.
[h]About 84% has boiling range up to 343°C (650°F), and remainder has end point of 524°C (975°F).
[i]High-grade syncrude, contains BTX.
[j]About 25% light oil, 70% medium oil, 5% heavy oil.
[k]End point of 343°C (650°F).
[l]Mainly aromatic oil containing benzene, phenol, naphthalene.
[m]25% ammonia.

can be converted to organic chemicals. With the advent of modern renewable energy technology, biomass processes will also be developed specifically for the production of chemicals, several of which are still manufactured from biomass, especially wood (silvichemicals), by several different fermentation and thermochemical methods. Some of the biomass-derived products will be useful for both fuel and chemical applications.

Low-Btu fuel gases will be manufactured from biomass by thermochemical processes. Partial oxidation and steam-oxygen reforming processes, for example, yield gases that have heating values in the 3.9–17.7 megajoules/m^3 range at standard conditions (100–450 Btu/standard cubic foot), and that are high in hydrogen and carbon oxides. Subsequent processing of these gases in shift reactors to adjust the carbon monoxide–hydrogen ratio yields synthesis gases suitable for chemical applications. Short-residence-time–high-temperature pyrolysis and hydropyrolysis, in addition to forming synthesis gases, can also yield relatively high concentrations of ethylene. Processing of these gases to separate ethylene and other components will afford raw materials for organic chemical synthesis.

Intermediate-Btu gases that have heating values of 19.6–27.5 MJ/m^3 at standard conditions (500–700 Btu/SCF) are produced from aqueous slurries of biomass by anaerobic digestion. This gas is essentially a two-component gas containing methane and carbon dioxide. Removal of the carbon dioxide by the usual scrubbing or adsorption methods yields SNG, which has a heating value of about 39.3 MJ/m^3 at standard conditions (1000 Btu/SCF). Large-scale use of this technology would make the intermediate-Btu gas and methane available for chemical applications such as methanol and ammonia production via synthesis gas, although it appears more practical to manufacture synthesis gas directly from biomass by thermochemical conversion.

The momentum now building up for the large-scale production of fermentation alcohol from biomass for motor fuels in the United States may lead to substantial increases in the availability of ethanol, presumably at prices competitive with those of gasoline. Many researchers believe that this will occur in the 1980s, when low-cost methods, now in the laboratory, for the hydrolysis of cellulosics to high yields of glucose concentrates are commercialized, in which case an ethanol-based organic chemicals industry may become reality. Ethanol conversion to ethylene, acetaldehyde, and acetic acid by established methods would supply the building blocks needed to synthesize many organic chemicals. Some countries, notably Brazil, have already progressed a long way toward this goal. However, ethanol-based chemicals are possible in these countries at this time only because of ethanol price controls and government regulations which require ethanol as a motor fuel blending component in gasoline. The availability of low-cost glucose would also open the way to direct production of other chemicals by fermentation and chemical methods.

An important feature of essentially all biomass conversions, particularly the thermochemical processes, is that several products are formed. Most pyrolysis processes yield chars and liquids in addition to gaseous products. The proportion of each depends on the reaction time and temperature. Separation of the liquid products by extraction and distillation can often yield useful quantities of organic chemicals, especially the oxygenated aliphatics. Wood "distillation" was in fact a well-established commercial technology for the manufacture of chemicals until it was displaced by petrochemical processes in the 1930s. Table 3 illustrates the average product distribution and yields reported by the Ford Motor Company in 1929 for the continuous pyrolysis of hardwoods in its 363-ton/day commercial plant. The charcoal was converted to briquettes for use as fuel along with the wood gas, and the other products were separated by distillation.

The chemical composition of biomass lends itself to multiproduct chemical plants, and product selectivities can often be controlled by the process design. For example, mild acid hydrolysis of hardwood, which contains about 50% cellulose, 25% hemicelluloses, and 25% lignin, selectively converts the hemicelluloses and affords a predominantly xylose solution and a lignin-cellulose residue. Strong acid treatment of the residue yields a glucose solution and a lignin residue. The xylose and glucose solutions and the lignin residue can then be treated separately to produce other products with good selectivities. One design yields

furfural by strong acid treatment of the xylose solution, ethanol by alcoholic fermentation of the glucose solution, and phenols by hydrogenation of the lignin.

Nonenergy carbon forms. Another source of nonfossil raw materials that may be used for the manufacture of organic chemicals is the very large deposits and reservoirs of nonenergy carbon forms — carbon dioxide and the carbonates. Hydrogenation of inorganic carbon at elevated temperatures and pressures can yield methane and its derivatives for further synthesis. Hydrogen must of course be available in large quantities to facilitate the process. It can be produced by electrolysis and the thermochemical and photolytic splitting of water, and also by reforming biomass and fossil materials. The latter approach would tend to defeat the purpose of using renewable carbon sources for energy and chemical applications, but it is nevertheless feasible. Commercialization of this technology in future years will be highly dependent on the availability of low-cost hydrogen.

Natural biochemical methods. Specialty organic chemicals have long been obtained from biomass because they occur naturally in certain species and can be separated in pure form. Among the many examples are the alginic acids from the marine biomass *Macrocystis pyrifera* (giant brown kelp) and the saponin digitonin from the leaves and seeds of plants of the *Digitalis* species. Such complex compounds will continue to be derived from biomass as long as markets exist and the economics are favorable. These compounds, however, constitute only a small fraction of commercial organic chemicals, but research is in progress to develop new techniques that rely on natural biochemical production of heavy organic chemicals for large-scale systems.

The production of hydrocarbons in land-based biomass is a well-known phenomenon. Commercial production of natural rubber, *cis*-1,4-polyiso-

Table 3. Product yields from continuous pyrolysis of wood in commercial plant*

Product	Yield/metric ton dry wood	Composition	
		Component	mol %
Gas	156 m³ (5000 ft³)	Carbon dioxide	37.9
		Carbon monoxide	23.4
		Methane	16.8
		Nitrogen	16.0†
		Oxygen	2.4†
		Hydrogen	2.2
		Hydrocarbons	1.2
Charcoal	300 kg (600 lb)		
Ethyl acetate‡	61.12 liters (14.65 gal)		
Creosote oil	13.56 liters (3.25 gal)		
C.P. methanol	13.02 liters (3.12 gal)		
Ethyl formate	5.30 liters (1.27 gal)		
Methyl acetate	3.94 liters (0.945 gal)		
Methyl ethyl ketone	2.72 liters (0.653 gal)		
Other ketones	0.943 liter (0.226 gal)		
Allyl alcohol	0.20 liter (0.048 gal)		
Soluble tar	91.8 liters (22.0 gal)		
Pitch	33.0 kg (66.0 lb)		

*Feed: 70% maple, 25% birch, 5% ash, elm, and oak hogged to 20 × 5 × 2 cm. Average temperature: 515°C (959°F), reactor center.
†Wood gas probably contaminated with air.
‡From esterification of by-product acetic acid.

prene, from the heavea rubber tree *(Heavea brazil-iensis)* and terpene extraction from pine trees are established technologies. Recently, efforts have been directed to searches for plants native to North America that biochemically manufacture hydrocarbons within the plant. Many plants have been identified that produce isoprenoids of lower molecular weight than natural rubber, which suggests that standard refining procedures may be possible. Some plants in the Euphorbiaceae family, which includes the heavea rubber tree, and other families have been reported to be capable of producing up to 25 bbl/hectare/year (4.0 m³/ha/year) of hydrocarbons. Hydrocarbon plantations may thus be developed as future sources of fuels and organic chemicals.

The possibility of restoring the oleoresin industry as a large-scale source of hydrocarbons has received some attention. The pine oleoresin industry (gum naval stores) has decreased in importance as a source of organic chemicals since it was at its peak around 1900. Competition from petroleum substitutes and cheaper pine products, primarily from stumps and kraft pulping operations by extraction and steam distillation techniques, has resulted in current oleoresin production of about 3% of that in 1908. Research in progress on the stimulation of natural oleoresin formation by chemical injections into trees, and on the benefits of combined timber and oleoresin production in mixed stands, has provided a few leads to improved technology. Although the oleoresin industry will continue to supply modest amounts of specialty chemicals, its restoration in the 1980s and 1990s as a major supply of materials for the manufacture of other chemicals seems remote. Coal-derived substitutes are expected to be just as competitive as petroleum-derived substitutes.

The polysaccharide α-cellulose from trees and plants is one of the world's largest natural sources of organic polymers. It has been utilized as a raw material for the manufacture of many derivatives, including cellulosic ethers and esters as well as regenerated cellulose. The tonnage quantities of these products are small in comparison with those of most of the other organic chemicals listed in Table 1; 140,000 and 271,000 metric tons of cellulose acetates and rayon respectively were produced in 1978 in the United States, for example. Natural cellulose, however, is expected to continue as the only practical source of these and related polymers, unless new materials are developed as replacements for cellulose. This possibility is considered highly improbable in view of the renewable nature of cellulose and the well-established markets for cellulose derivatives.

For background information *see* NAVAL STORES; ORGANIC CHEMICAL SYNTHESIS in the McGraw-Hill Encyclopedia of Science and Technology.

[DONALD L. KLASS]

Bibliography: D. L. Klass, *Biosources Dig.*, 1: 47–78, 1979; D. L. Klass, *Chemtech*, 5:499–510, 1975; *Proceedings of Symposia on Clean Fuels from Biomass Sponsored by Institute of Gas Technology*, Jan. 27–30, 1976, Jan. 25–28, 1977, Aug. 14–18, 1978, Jan. 21–23, 1979; *Proceedings of Symposia on Clean Fuels from Coal Sponsored by Institute of Gas Technology*, Sept. 10–14, 1973, June 23–27, 1975, May 14–18, 1979.

Ozone, atmospheric

Ozone is a form of molecular oxygen that exists in minor quantities—a few parts per million—in the atmosphere. It is concentrated primarily in the stratosphere, the region between about 8 and 30 mi (12 and 50 km) above the Earth's surface. Ozone (O_3) is continuously being formed in the stratosphere by the action of sunlight on oxygen (O_2) and being destroyed by various chemical processes. The ozone layer in the stratosphere shields the surface of the Earth from high-energy ultraviolet (UV) radiation emitted by the Sun. The scientific community has warned in recent years that continued release of chlorofluorocarbon (CFC) gases could eventually cause a substantial depletion of stratospheric ozone with a consequent increase in biologically harmful UV radiation. The CFCs are used mainly as propellants in aerosol cans, as refrigerants, as solvents, and in the manufacture of plastic foam products. As a consequence of the threat to the ozone layer, the Food and Drug Administration and the Environmental Protection Agency in the United States in 1978 banned the manufacturing and processing of CFCs for use as aerosol propellants, and regulations on the other uses are now under consideration.

Ozone is also formed in photochemical smog in amounts which are small on a global scale, but rather significant on a local, urban scale due to the potentially harmful effects on human health which result from breathing air containing ozone concentrations above a few tenths of a part per million.

Atmospheric chemistry of CFCs. Nearly 10^6 metric tons of the CFCs trichlorofluoromethane ($CFCl_3$ or F-11) and dichlorodifluoromethane (CF_2Cl_2 or F-12) are being released every year to the atmosphere. These compounds are chemically very inert and are practically insoluble in water; thus, they are not destroyed in the lower atmosphere, but slowly diffuse upward into the stratosphere where they are decomposed by UV radiation from the Sun, releasing chlorine atoms (see illustration). The free chlorine atoms react with ozone through a catalytic chain mechanism, shown below. After these two reactions have occurred, the

$$Cl + O_3 \rightarrow ClO + O_2$$
$$ClO + O \rightarrow Cl + O_2$$

Cl atom is still present, but one molecule of O_3 and one oxygen atom O have been removed. (In the process of screening solar radiation, O_3 and O are rapidly interconverted to each other in the stratosphere.)

Chlorine-containing species are expected to participate in many other chemical and photochemical reactions in the stratosphere. Compounds such as methane (CH_4), nitrogen oxides (NO and NO_2), and the hydroperoxyl radical (HO_2) are natural constituents of the atmosphere, present in trace amounts. The Cl catalytic chain mechanism may be interrupted, for example, by reaction of the Cl atom with CH_4 to produce the relatively stable molecule HCl; or by reaction of ClO with NO_2 or HO_2 to produce $ClONO_2$ or HOCl, species which eventually break down through solar radiation, returning chlorine to its catalytically active form.

Mathematical models of the atmosphere. In order to estimate the magnitude of the ozone per-

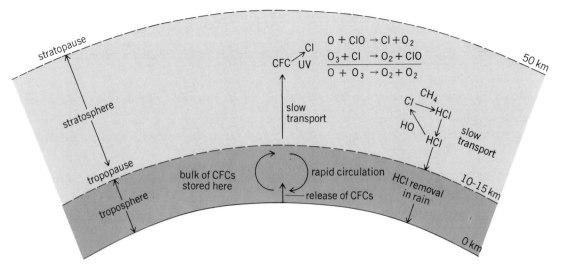

$$O + ClO \rightarrow Cl + O_2$$
$$O_3 + Cl \rightarrow O_2 + ClO$$
$$\overline{O + O_3 \rightarrow O_2 + O_2}$$

Simplified diagram of atmospheric behavior of chloro-fluorocarbons. (*From National Research Council, Panel on Atmospheric Chemistry, Halocarbons: Effects on Stratospheric Ozone, September 1976*)

turbations resulting from a given CFC release, atmospheric computer models are employed which incorporate information on atmospheric motions as well as information on the rates of over a hundred chemical and photochemical reactions; these chemical reaction rates are measured in the laboratory. Due to the complexity of atmospheric motions, the calculations have been made mainly with "one-dimensional" models, an approach which attempts to average these motions and the concentration of chemical species over latitude and longitude, leaving only their dependency on altitude and time. The results of actual measurements of these trace species in the atmosphere are then used to verify the models.

There is little question about the fundamental aspects of the problem: the measured CFC concentrations indicate that they do accumulate in the lower atmosphere and that they do reach the stratosphere. Chlorine atoms and ClO radicals are indeed found in the stratosphere together with other species, including O, OH, HO_2, NO, NO_2, and HCl. The observed concentrations are in reasonable agreement with theory if atmospheric variability, as well as the rather severe limitations of the models, is taken into account. However, there are discrepancies which are not understood at present, and additional experimental work in the laboratory as well as in the atmosphere is required to resolve them. For example, some measurements indicate that the ClO concentration above 35 km significantly exceeds the predicted values, while other ClO measurements yield concentrations in much closer agreement with theory: either some unknown chemistry is missing from the models or some of the measurements are in error.

Current results from one-dimensional models predict an eventual 15–18% depletion of stratospheric ozone if CFC release continues indefinitely at the present rate; it would take roughly 50 years to reach 8–10% depletion. If the CFC release were to be eliminated in the next decade, ozone would continue to decrease for at least another decade, and would then slowly increase, taking roughly a century to reach normal values again. The

models also indicate a present-day ozone depletion level of about 2% due to past CFC release. This value cannot be confirmed in the near future by direct observation of stratospheric ozone levels due to a poorly understood long-term variability of about ±5% in the natural ozone levels.

Other stratospheric pollutants. Besides F-11 and F-12, other CFCs are currently finding technological use, although not at the high tonnage level of F-11 or F-12. Among them, F-113 ($CFCl_2CF_2Cl$) and F-114 (CF_2ClCF_2Cl) are expected to affect stratospheric ozone in much the same manner as F-11 or F-12. F-22 (CHF_2Cl), used mainly as a refrigerant, is 10–15 times less effective than F-11 (on a molecule-per-molecule basis) in terms of stratospheric perturbations because some of the F-22 is destroyed by reaction with OH radicals in the lower atmosphere. This increased reactivity is due to the replacement in the CFC molecule of a chlorine or fluorine atom by a hydrogen atom. Similarly, methylchloroform (CH_3CCl_3)—used in rather large quantites as an industrial solvent—is partially destroyed in the lower atmosphere, but a significant amount does reach the stratosphere, releasing chlorine atoms. Methyl chloride (CH_3Cl) is also in technological use but in small quantities compared with the biological sources of CH_3Cl which provide the natural chlorine background to the stratosphere. (The CFCs now contribute a larger amount of chlorine to the stratosphere than all natural chlorine sources.)

Concern over pollution of the stratosphere originated in the early 1970s with the finding that nitrogen oxides emitted from high-flying aircraft—particularly the supersonic transport (SST)—could catalytically destroy ozone. The Climatic Impact Assessment Program (CIAP) of the Department of Transportation, formed in 1971, concluded in 1974 that fleets of several hundred SSTs would significantly reduce stratospheric ozone levels in the absence of adequate emission controls. At the end of CIAP, one major gap remained in terms of laboratory studies: the rates of chemical reactions involving HO_2 radicals were poorly known. In 1976 these rates were measured directly for the first time, and the $HO_2 + NO$ reaction turned out to be about 30 times faster than previously thought. This finding and other new results increased the predicted ozone depletion levels due to chlorine, and decreased the predicted effects due to nitrogen oxides to the extent that small SST fleets would now be expected to increase ozone levels slightly.

Other possible ways in which humans may contribute to pollution of the stratosphere include the use of nitrogen fertilizers, because of their contribution to the biological production of nitrous oxide, N_2O, which provides a stratospheric source of nitrogen oxides; and the space shuttle, because the combustion of the solid propellant will inject chlorine directly into the stratosphere (in amounts, however, that are very small compared with CFC emissions).

Environmental effects of CFC release. Recent experiments have shown that certain kinds of plankton, fish larvae, food crops, and other plant and animal species are extremely sensitive to UV radiation of the type that would penetrate the atmosphere and the surface layers of the oceans with increased intensity if the amount of ozone were to diminish: a 10% depletion of stratospheric ozone would lead roughly to a 20% increase in the intensity of biologically active ultraviolet radiation (UV-B). The overall impact on aquatic and terrestrial ecosystems of such a perturbation cannot be assessed at present — much research remains to be done. Nevertheless, it is a cause for great concern. The effects on humans are perhaps less serious, but can be much better quantified: there are extensive epidemiological studies which link the incidence of skin cancer to exposure to UV-B radiation. Damage to biological systems occurs because the genetic material deoxyribonucleic acid (DNA) and other nucleic acids break down upon absorption of UV-B light.

The large-scale release of CFCs also represents a threat to the world's climate. The CFCs absorb some of the infrared radiation emitted by the Earth in the wavelength region $8-15$ μm, thereby warming the atmosphere. This is the so-called greenhouse effect, which also occurs with carbon dioxide (at different infrared wavelengths). At present, the estimates of global temperature changes caused by the greenhouse effect are rather uncertain. Another mechanism for perturbing the climate involves the ozone layer: the temperature of the stratosphere would certainly change following a depletion of ozone, but it is very difficult to estimate with current models the effect that such a change would have on the lower atmosphere, close to the Earth's surface. The global weather is so complex that it is not yet possible to predict with any assurance the magnitude or even the type of climatic perturbations to be expected from the release of CFCs, but it is certainly another cause for serious concern.

In spite of the ban on CFCs for use as propellants in spray cans, the CFC emission levels in the United States are projected to increase at a rate of $5-8$% per year at least through the 1990s due to all other uses. The amount emitted by the rest of the world is expected to increase even faster — only Sweden has so far ordered a ban similar to the United States ban (and the Canadian government

has won agreement from manufacturers for a voluntary cutback in CFC use for aerosol sprays). Thus, unless further worldwide regulations are enforced, ozone depletion might reach the $20-30$% level toward the middle of next century, a level that is clearly unacceptable in view of its potential impact on biological systems and on the climate.

For background information *see* CLIMATE, MAN'S INFLUENCE ON; OZONE, ATMOSPHERIC in the McGraw-Hill Encyclopedia of Science and Technology.   [MARIO J. MOLINA]

Bibliography: R. D. Hudson (ed.), *Chlorofluoromethanes and the Stratosphere*, NASA Ref. Publ. no. 1010, August 1977; National Academy of Sciences, Committee on Impacts of Stratospheric Change, *Halocarbons: Environmental Effects of Chlorofluoromethane Release*, September 1976; National Research Council, Panel on Atmospheric Chemistry, *Halocarbons: Effects on Stratospheric Ozone*, September 1976.

Particle accelerator

The most important recent developments in particle accelerators center on colliding-beam machines. Greatly increased amounts of energy are available for studying the structure of elementary particles in such collisions. There are three kinds of colliding-beam machines in existence or being constructed, identified by the particles that participate in the collisions: electron-positron colliding beams, proton-antiproton colliding beams, and proton-proton colliding beams. Each type of machine poses different technological problems, and each has the potential of providing complementary insights into the fundamental structure of matter. The two most intriguing advances are the introduction of techniques to damp (or cool) the phase-space (temperature) population of particle beams, and the development of large superconducting dipole and quadrupole magnets. The future study of elementary particle physics is expected to be very closely tied to the success in developing and operating these types of machines.

Luminosity and center-of-mass energy. Before the different kinds of machines are discussed, it is valuable to describe the salient characteristics of these machines. The luminosity of a colliding-beam machine is defined by Eq. (1), where $\dot{N}$ is

$$\dot{N} = \mathscr{L}\sigma \tag{1}$$

the rate of collisions between the particles, σ is the cross section for collisions, and $\mathscr{L}$ is the luminosity expressed in units of cm^{-2} s^{-1}. The cross sections for observed elementary particle collisions range from 10^{-26} cm^2 to 10^{-39} cm^2, whereas attainable luminosities of 10^{33} cm^{-2} s^{-1} appear feasible. The luminosity of a given machine with bunched beams is given approximately by Eq. (2), where

$$\mathscr{L} = \frac{f}{2} \frac{N_1 N_2}{(r_1^2 + r_2^2)} \tag{2}$$

N_1, N_2 are the numbers of particles of type 1 and 2 in each bunch, f is the revolution frequency, and r_1, r_2 are the radii of the two beams assumed to be circular. There are fundamental limitations to the number of particles that can be stored in each bunch, and to the beam sizes, which in turn limit

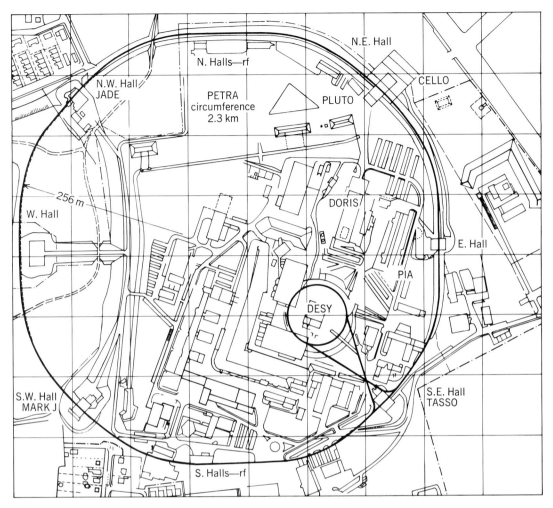

Fig. 1. Layout of PETRA, DESY and DORIS machines in Hamburg. The five experiments (PLUTO, CELLO, TASSO, MARK J, and JADE) under assembly in four halls are also indicated.

the maximum luminosity in any machine. However, for very-high-energy collisions of elementary particles, the most interesting processes are likely to have cross sections in excess of 10^{-33} cm^2, and are thus well matched to the limiting luminosities of these machines.

The center-of-mass energy E_{cm} for collisions of particles of equal energy $E_1 \cdot E_2 \cdot E$ is given by Eq. (3).

$$E_{cm} = (E_1 + E_2) = 2E \qquad (3)$$

The corresponding energy E_L of a projectile striking a resting proton mass (m_p) is given by Eq. 4.

$$E_L = \frac{E_{cm}^2}{2m_p} = \frac{2E^2}{m_p} \qquad (4)$$

For example, the proton-antiproton storage ring at Fermilab provides 2000 GeV in the center of mass. This requires a beam with about $2(1000)^2 = 2 \times 10^6$ GeV of energy in a conventional accelerator. It is clear that only colliding beams can provide such a large center-of-mass energy. However, cosmic-ray primaries of about 10^{11} GeV have occurred in extensive air shower observations. *See* COSMIC RAYS.

Electron-positron colliding beams. There are many electron-positron storage rings in Europe, the Soviet Union, and the United States. The fol-

lowing discussion is concerned with the new generation of storage rings at the Deutsches Elektronen Synchrotron (DESY) in Hamburg and the Stanford Linear Accelerator Center (SLAC) and the future Large European Project (LEP) proposed for the European Commission for Nuclear Research (CERN) in Geneva. Since the Positron Electron Tandem Ring Accelerator (PETRA) machine at DESY is operational, it will be described in some detail.

PETRA. When PETRA was designed, the aim was to build an electron-positron storage ring at DESY with the highest possible energy. The energy limit is essentially given by the energy loss due to synchrotron radiation, which is proportional to $E^4 R$, where E is the energy of the circulating electrons and R is the magnetic bending radius. To compensate for this loss, it is necessary to apply an accelerating field which is produced by an rf power P_{rf} proportional to E^8/R. For given R, the rf power increases with the eighth power of E; for this reason, technical and financial limits are quickly encountered.

The parameters for PETRA are given in Table 1, and its layout is shown in Fig. 1, including the machines DESY and DORIS (an earlier electron-positron storage ring) and the Positron Injection Area (PIA). Electrons and positrons are accelerat-

Table 1. Parameters of PETRA

Maximum energy of beam	19 GeV
Circumference	2.3 km
RF power	4.8 MW
RF cavities	64
Klystrons	8
Average magnetic radius	256 m
Magnetic field	≤ 0.4 T
One to four particle bunches feeding two to eight intersections	

ed in the DESY synchrotron to 5–7 GeV before being injected into PETRA. Positrons are preaccumulated in DORIS. PETRA has four long, straight sections, two of which house the rf accelerating structure. In the four short, straight sections the experiments are accommodated.

The construction of PETRA started in 1976 and was terminated by July 1978, a construction time of less than 2½ years. A first beam was stored on July 15, about 9 months earlier than foreseen in the original proposal. The capital investment was 100,000,000 German marks.

In the first few months of operation, several important steps were achieved. After the chromaticity had been corrected, beam lifetimes of several hours (limited by the vacuum) were obtained. The beams were accelereated to 14 GeV without any loss or other difficulties. However, the luminosity in mid-1979 was only 1% of the design value. Nevertheless, significant physics experiments had already started.

PEP. The Positron-Electron Project (PEP) machine at SLAC was expected to be completed in November 1979. The injection of electrons and positrons is accomplished by the 2-mi (3.2-km) linac (linear accelerator), which may give a more intense source. Otherwise, the PEP machine is very similar to the PETRA machine.

LEP. In 1977 the European Committee for Future Accelerators (ECFA) recommended that first priority be given to an electron-positron facility with beam energies of 100 GeV or more. As an intermediate step, energies of about 70 GeV per beam were thought to be useful.

On the basis of this recommendation, CERN carried out a conceptional study for a 100-GeV electron-positron ring. This study seemed to indicate that some serious problems might arise for such a machine. They would mainly stem from its large size. As explained above, in order to reduce synchrotron radiation losses, the radius of the ring must be increased. Since the necessary rf power goes as E^8, but only as $1/R$, a very large radius must be chosen. For 100 GeV per beam, a radius of

Table 2. Proposed parameters of the LEP70

Nominal energy per beam	70 GeV
Number of intersections	8
Ring circumference	22.2 km
Circulating current per beam	10.5 mA
Synchrotron radiation loss per turn	906 MeV
Synchrotron radiation loss (power two beams)	19.1 MW
Radio frequency	357 MHz
Length of active rf structure	1.34 km
Total rf generator power	74 MW
Number of klystrons	80

about 8 km seems necessary, resulting in a circumference close to 50 km. This leads to difficulties, for example, the need for precise positioning of magnets to avoid closed-orbit distortions, the occurrence of antidamped closed orbits, of low magnetic fields at injection, and problems in finding a site.

As a consequence, CERN decided to study in greater detail a ring optimized for an energy of 70 GeV per beam with an average radius of about 3.5 km. The main parameters of the proposed machine, the LEP70, are shown in Table 2.

At a later stage the energy of LEP70 could be increased if the rf cavities were replaced by superconducting ones. LEP could be constructed and operating by 1989. Operating results from PETRA and PEP will be crucial in the decision to build LEP, since unforeseen problems in these machines could affect LEP.

Proton-antiproton colliding beams. Since antiprotons do not occur naturally, it is necessary to provide an antiproton source that uses high-energy protons to produce the antiprotons. The protons (p) and antiprotons ($\bar{p}$) can circulate in the same storage rings just as electron-positron beams, but without the large synchrotron energy loss for such machines. The general scheme is outlined in Fig. 2.

The source intensity depends on two factors: the antiproton production characteristics, and the accumulation of antiprotons over many seconds. The realistic accumulation of large numbers of antiprotons requires phase-space compression. Two techniques have been devised to achieve this: electron cooling and stochastic cooling. These are discussed later. First, the plans for colliding beams at various laboratories will be outlined. These projects were initially discussed in 1976, and the machines are now under construction.

p̄p Collisons in the SPS. The general scheme for proton-antiproton collisions in the Super Proton Synchrotron (SPS) at CERN is as follows:

1. Antiprotons of momentum 3.5 GeV/c are produced by 27-GeV/c protons from the CPS (CERN Proton Synchrotron). (A momentum fraction $\delta\, p/p$ of $\pm\, 0.7 \times 10^{-2}$ is accepted.)

2. The antiprotons are transferred to a small storage ring (Antiproton Ring; APR) and cooled rapidly in momentum space (within a few seconds) and slowly in transverse phase space by stochastic cooling.

3. After approximately 10^{11} to 5×10^{11} antiprotons, are collected, they are transferred into the CPS and accelerated to 26 GeV/c, and bunched and injected into the SPS.

4. Protons are injected into the SPS at 26 GeV/c.

5. Protons and antiprotons are accelerated to 270 GeV/c and collide at two places where large experimental detectors will be located. The center-of-mass energy will be 540 GeV.

The proposed antiproton collector ring has an extremely large aperture-to-diameter ratio, which is needed in order to collect the large antiproton phase space.

p̄p Collisions at Fermilab. The basic scheme for proton-antiproton collisions at the Fermilab superconducting storage ring at Batavia, IL, is very different from that undertaken at CERN. It is shown schematically in Fig. 3. The Fermilab booster ac-

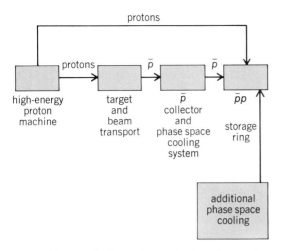

Fig. 2. Elements in the proton-antiproton storage ring scheme.

celerator provides protons that are accelerated to 80 GeV in the main ring and targeted to produce antiprotons. The antiprotons are in turn injected back into the booster or precooler, and decelerated and transferred to a cooling ring where electron cooling provides a rapid decrease in the beam phase space. The beam is then stacked in a small part of the orbit, and the electron cooling ring aperture is cleared for another injection of antiprotons. Ideally, the cooling time should match the repetition rate of the booster (13 Hz); this seems possible. The antiprotons are injected back into the booster or precooler after a sufficient intensity has been obtained, and subsequently both antiprotons and protons are injected into the superconducting energy doubler ring or teratron for 1000 × 1000 GeV collisions. A precooler or supercooler to increase the number of protons collected is to be added about 1982.

Phase-space compression. Techniques of phase-space compression or cooling of particle beams will now be discussed.

Electron cooling. Electron cooling at low energy will be considered since it is an example of phase-space cooling used in one of the actual schemes (Fermilab). Whenever protons stop in matter and give up their energy to electrons, the protons have been cooled by the cold electrons in matter. The central idea of beam cooling is to apply this cooling to finite-energy beams by accelerating the electrons to an energy such that the mean velocity of the electrons is the same as that of the protons or antiprotons. Then the cooling takes place in the co-moving system. For realistic electron currents, the cooling time can be as fast as $10-100$ ms for a proton beam of 200 MeV with an energy spread of $\pm$ 300 keV. For protons stopping in matter with 600 keV initial energy, the stopping time is about 10^{-14} s. The density of electrons in matter is about 10^{13} greater than that possible for an electron beam, and very crudely, the cooling time scales as the density of electrons. Thus it is not unreasonable that electron cooling times on the order of a fraction of a second are possible with energetic electron beams. The central problem of electron cooling is maintaining a cold electron beam for a finite distance. Space charge tends to cause the beam to diverge in the long straight section of the cooling ring, where cooling takes place. In order to overcome the beam blowup, a strong solenoid magnetic field was introduced in early experiments in Novosibirsk. A surprising result of this innovation was an enhanced cooling rate at low energy.

Stochastic cooling. Stochastic cooling of proton beams has been successfully proved at CERN by using both the Intersecting Storage Rings (ISR) and the small Initial Cooling Experiment (ICE) ring. When fluctuations in the beam are detected, correcting kicks are applied to the beam. It appears that momentum cooling can be made fast

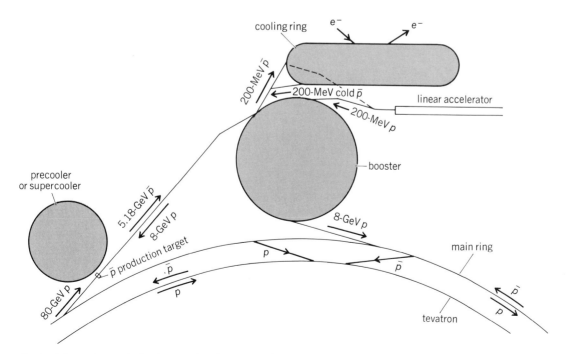

Fig. 3. Scheme for producing beams of antiprotons and protons at Fermilab.

(a few seconds), whereas the cooling of the transverse momentum is much slower. The cooling time is roughly independent of energy (if adequate power is available for the system), and thus this cooling is appropriate if the collector is to operate in the direct-current or steady-state mode, in which the protons are not accelerating.

In order to verify the behavior of stochastic cooling over a wide range of parameters, a special storage ring, ICE, was quickly brought into operation in 1977 at CERN. It has a diameter of 24 m and can store protons with momenta up to 2 GeV/c.

The pickup elements for cooling in vertical, horizontal, and longitudinal phase space are located in one straight section. The correction elements are essentially installed in another straight section. In the third straight section, equipment to study electron cooling is installed. This method will be useful for experiments using low-energy antiprotons.

ICE has been very successful in demonstrating that stochastic cooling is indeed a useful method for the main proton-antiproton projects. Cooling in each of the three phase planes operates independently. Initial cooling times as short as 15 s were achieved, and increases in momentum density up to a factor 22 were measured. With simultaneous cooling in the three phase planes, the beam lifetime increased by a factor of 40.

Cooling of high-energy beams. There is another interesting application of electron cooling. At high energy, proton and antiproton beams in a storage ring will naturally heat up (blow up) because of beam-beam interactions and multiple Coulomb scattering with the residual gas in the machine. In addition, the beam sizes may not be optimal to obtain the maximum luminosity for a given number of antiprotons or protons in the bunches. If the beams can be cooled by high-energy electrons, some of these problems may be overcome. The idea is to bring the electrons in the long straight section of an electron storage ring in conjunction with protons or antiprotons. The protons cool; the electrons heat up. The electrons are then cooled by synchrotron radiation.

Proton-proton colliding beams. A very large proton-proton storage ring called Isabelle is to be constructed at Brookhaven National Laboratory. This machine promises to provide high luminosity in contrast to the previously discussed proton-antiproton machines, which are really converted synchrotrons. (The energy doubler at Fermilab will be used as a high-performance proton-antiproton storage ring.) Also, there will be five interaction regions at Isabelle; only two or three at CERN and Fermilab.

Growth of available energy. The construction of these new generations of machines will occupy the 1980s. The available center-of-mass energy of a few hundred GeV to 2 TeV is likely to be adequate to observe the behavior of weak and electromagnetic interactions and to test for the unification of these forces. Even now there is discussion of very-high-energy storage rings of 5–20 TeV, very likely for proton-antiproton machines. These developments depend on the success of large superconducting rings of magnets. Such magnets have now been produced at both Fermilab and Brookhaven.

For background information *see* PARTICLE AC-CELERATOR in the McGraw-Hill Encyclopedia of Science and Technology.   [DAVID B. CLINE]

Bibliography: G. J. Budker, *At. Energy*, 22: 346–351, 1967; G. Carron et al., *Phys. Lett.*, 77B: 353–357, 1978; C. Rubbia, P. McIntyre, and D. Cline, in H. Faissner, H. Reithler, and P. Zerwas (eds.), *Proceedings of the International Neutrino Conference*, Aachen, pp. 683–687, 1976; H. Schopper, *Proceedings of the International Symposium on Perspectives and Future Projects in High Energy Physics*, Tokyo, pp. 279–295, 1978.

Petroleum

Current estimates of conventionally recoverable oil and gas indicate that an amount (Btu equivalent) approximately equal to present proved and produced reserves remain to be found. These potential resources probably will be found in many of the presently productive areas, and in unexplored or frontier basins which are located primarily in the arctic regions, offshore areas of the continental shelf, possibly the deep sea, and many areas which in the past have been relatively inaccessible. Present worldwide geologic distribution and per capita consumption of oil and gas reserves are not uniform geographically or politically. Current estimates indicate that future discoveries will include a greater percentage of relatively smaller accumulations and more gas than in the past, and will probably not be uniform in distribution. For the world in general, it appears that the addition of new reserves of oil is less than growth in current consumption, while new reserves of gas presently exceed consumption; that is, oil may be "topping out," whereas this point has not been reached by gas.

Resource estimates. By definition, most estimates of the world's ultimate conventionally recoverable petroleum resources involve: (1) primary recovery—on average, 25% of oil in place which may be recovered without an "external boost" (gas has a much higher conventional recovery, about 70%); and (2) conventional secondary recovery (that is, water flood or gas injection), which in many reservoirs may boost recovery to 30–40% of oil in place. Present world petroleum resource estimates do not include nonconventional petroleum resources which involve enhanced recovery (which may increase recovery of oil in place to 50% or more), tar sands, heavy oil belts, oil shales, gas in tight reservoirs, and geopressured gas. Generally included are those resources in polar regions (Antarctica) and deep-water zones. Nonconventional oil resources are estimated to represent over three times the ultimate conventionally recoverable resource base and, at present, only 10–15% of these resources can be considered as being potentially recoverable economically. Nonconventional gas resources may represent (dependent on the world's geopressured zones) 10–50 times the ultimate conventionally recoverable gas resource base, with less than 1% economically or physically recoverable currently.

Basins. Petroleum occurs in concentrated accumulations (fields) in depressed, sediment-filled areas (basins or provinces). Worldwide, over 600 basins and subbasins are known to occur (Fig. 1). Of these, about a quarter have production (two out of five producing basins have prolific giant fields), while 40% have had variable amounts of petroleum exploration. Over one-third of the world's basins may be considered as unexplored or as exploration frontier basins. Substantial reserves are still in the process of being discovered in many of the producing basins, and such basins are therefore considered development frontier basins.

Methods of estimation. Methods of estimating recoverable oil and gas range from rudimentary geologic appraisal in poorly known new exploration frontier basins to measurement of definable petroleum parameters in basins where fields have been established. Accuracy of estimates involves both high risk and a wide range of resource magnitude in exploration frontier basins, while both risk and resource estimates become more predictable in developing basins. Based upon past experience in new exploration frontier basins, one in two establishes significant commercial petroleum production, while one in four or five basins contains giant fields.

In estimating the potential resources of petroleum in new exploration frontier basins, the following methods are used: (1) volumetrics—some ratio between the volume of sediments in the basin being estimated, and the success ratios and recovery factors in generally similar established producing basins being considered; (2) analog, or "look-alike"—involving a comparison of geologic character of basin appraisal, a basin classification being used. Present classifications include six to eight basin types or classes based upon the basin's structural form or geologic architecture, relation to the underlying crust of the Earth, and genesis in relation to the phenomenon of sea-floor spreading and plate tectonics. Estimates are further modified or qualified by what is known of the basin's sedimentary fill. Although all producing basin types or classes have a range of examples from high recovery to low recovery, some basin types appear to represent a lower risk and a potential for higher reserves from giant fields, whereas other basin types appear to have lower reserves and a higher risk; these are known as petroleum characteristics. Considering the imperfect knowledge of new exploration frontier basins and the lack of uniform petroleum characteristics in basins of the same type, many geologists consider estimation of their potential so inaccurate as to be misleading; many other geologists believe that the above methods of appraisal, within broad limits, are generally feasible. Appraisal by these methods is often more acceptable if considered on a "no higher" basis (no higher reserves than the higher amounts in look-alike basins); also, the more basins being appraised, the more the risk can be averaged—that is, the estimated reserves in a single basin are compared to those of many basins.

In basins where considerable exploration and drilling has occurred, estimation of potential resources of petroleum often has sufficient information to apply a more sophisticated methodology. These include the petroleum zone concept, wherein a petroleum zone is a sedimentary volume which contains pools or petroleum accumulations showing common characteristics, and estimates are made by comparing the large set of established producing zones in the world through an extensive geologic parameter check list; and measurement of fundamental petroleum parameters, wherein the factors of reservoir rock, trap, source rock, and petroleum migration are measured or estimated for a more direct appraisal of the petroleum potential.

In development frontier basins where significant commercial discoveries have occurred, the "play concept" method of estimation is often applied, wherein look-alike plays with common characteristics are sought out for drilling. This essentially combines the petroleum zone concept and measurement of petroleum parameters.

In arriving at estimates of future conventionally recoverable petroleum resources, all of the above methods, along with risk or probability analysis, are used, in that future resources are anticipated to come from new frontier basins, partially explored basins, and both new and older producing basins.

Giant fields. Giant fields (over 5×10^8 bbl or 7.9×10^7 m^3, or Btu gas equivalent) have contributed substantially to world proved and produced reserves. Giant and supergiant (over 5×10^9 bbl or 7.9×10^8 m^3) fields represent slightly over 1% of the world's fields, yet they contribute to over 70% of all oil reserves and over 50% of all gas reserves found to date. The presence of giant fields in the future exploration frontier basins will have a strong influence on the magnitude, economics, and producibility of any estimates of future reserves. To date, giant fields (Figs. 2 and 3) continue to be discovered; however, their contribution to world reserve growth appears to be slackening. Fewer oil supergiants are being found, and in the past 15 years more gas has been found than oil, including the discovery of supergiant gas fields. To date, 30% of all giant field reserves are gas, while 37% of all reserves found are gas. From 1890 to 1950, gas averaged about 10% of all giant reserves. The drop in oil discovery rate of supergiants is related to the early phase of Mideast development, when the supergiants were found in the 1940s and 1950s, versus the supergiant gas discovered in the last 15 years in the Mideast and western Siberia. Supergiants in the Mideast and western Siberia have created *superbasins*, containing 57% of the world's reserves of oil and gas. At present, world consumption of oil exceeds yearly addition of oil reserves.

Basin location and characteristics. Classification and analysis of the world's 160 producing basins (Fig. 1) have suggested that certain basins may have a preference for giants and often have a variation in trap types, field size distribution (spread), age, depth, lithology of reservoirs, and maturation (geothermal) history of source beds. Most authors who have classified world basins generally agree on the classification of about three-quarters of the world's individual basins, and differ on about a quarter of those that are poorly known or are geologically controversial and therefore difficult to classify due to a geologic evolution which often includes a change in basin type through geologic time. The world's basins, as they are now known (Fig. 1), occupy about 25% of the Earth's surface, 15% on land and 10% offshore.

Offshore ocean basins with over 4500 ft (1372 m)

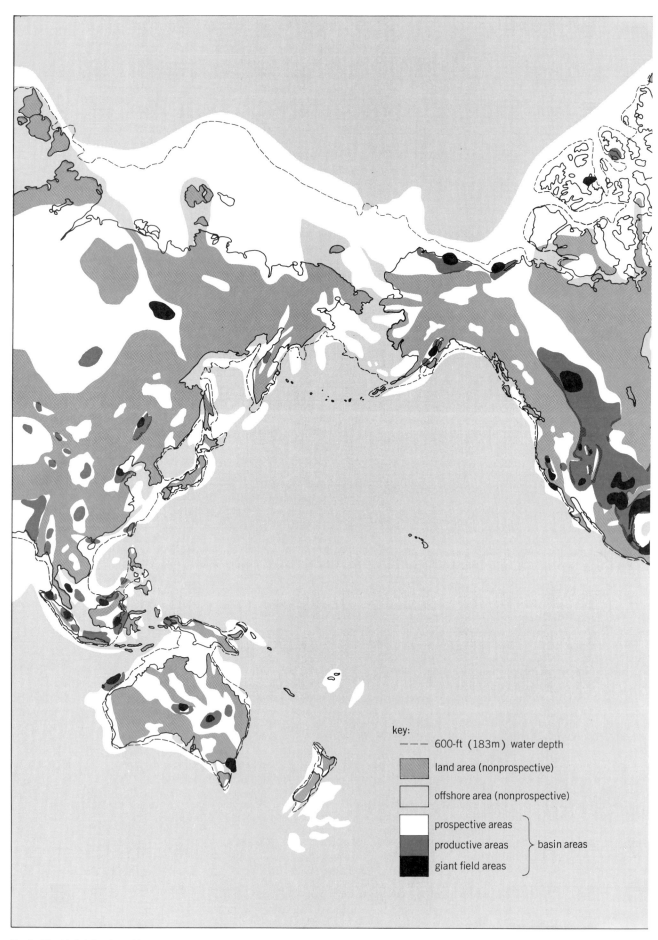

Fig. 1. World distribution of basin areas.

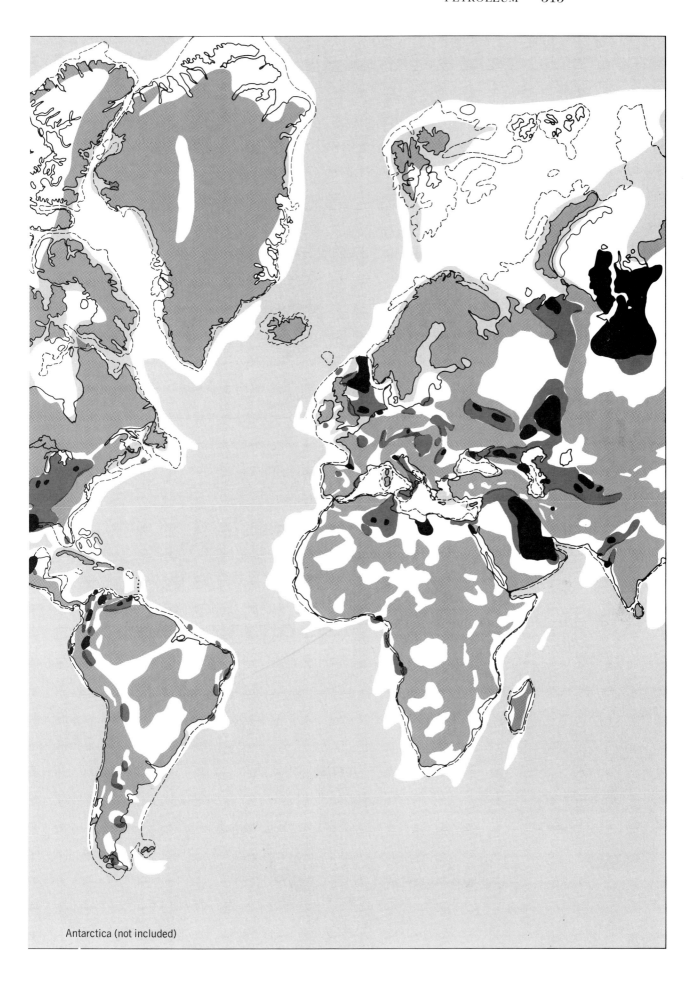

Antarctica (not included)

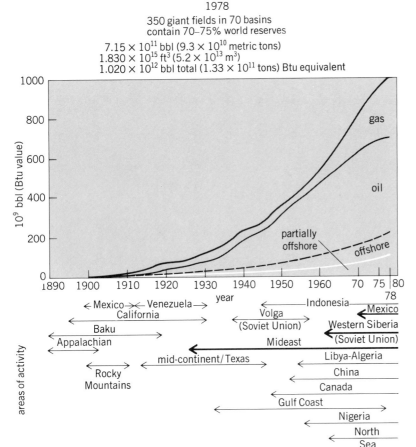

Fig. 2. Cumulative discovery of reserves in giant fields. Heavy arrows indicate super giant basins. (*From H. D. Klemme, Giant fields – update, Oil Gas J., May 7, 1977*)

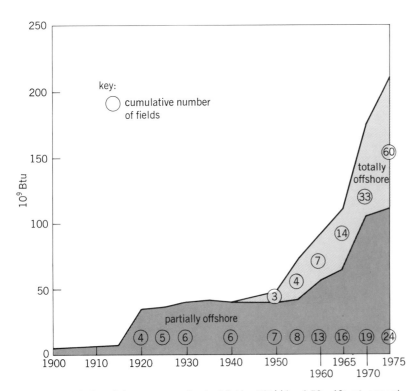

Fig. 3. Cumulative offshore reserves in giant fields—10^9 bbl or 1.59×10^8 m³, or equivalent (69% oil/31% gas). (*From H. D. Klemme, 200 billion bbl of offshore oil and gas, Oil Gas J., vol. 75, no. 35, 1977*)

of sediment represent 20% of the ocean area (with 7% of the area on the continental shelf, that is, at less than 600 ft (183 m) water depth; and 13% in deep-water areas, that is, at more than 600 ft water depth). Present offshore reserves are currently estimated at 21% of the world's proved and produced reserves, while estimates of ultimate resources of petroleum offshore range from 30 to 35%, indicating that 40–50% of all future resources will come from offshore. While the exploration potential of the continental shelf is fairly well established, the potential of deep-water areas is relatively unknown, has had little petroleum test drilling, and involves a wide range of estimates (3–25% of future resources). At present, deep-water areas have not been included in most estimates of future resources. In addition, petroleum data from offshore areas exhibit a change in the dominance of basin types from those onshore, changes in trap types and age and depth of reservoirs, and a reduction in supergiants and average size of giants—particularly from fields partially onshore to fields that are totally offshore (Figs. 2 and 3). Presently, many fields with low productivity over an extensive area which are commercial onshore might be uneconomic offshore, where a thick productive column and high productivity over a concentrated area are required.

A high percentage of future petroleum resources is expected to come from offshore, where fields smaller than those on land seem to be prevalent. In addition, productive basins on land seem to have smaller fields in the final stages of development. Therefore it is probable that future resources will involve a greater number of reserves from smaller fields than in the past.

Future resources. Future conventionally recoverable petroleum resources (Fig. 4) are expected to come from new exploration frontier basins (many offshore), new or developing frontier basins (offshore and difficult terrain on land), and older producing basins (on land extending offshore).

New exploration frontier basins. Frontier basins are present in onshore and offshore portions of the Arctic (IA and IB in Fig. 4), along the eastern Asia continental margin (II), along the margins of the Atlantic and Indian oceans (III), and in relatively inaccessible arctic and equatorial areas located in the interior of North and South America, Africa, and Asia (IVA, IVB, IVC, and IVD).

Three moderately large basins are located on the northeastern Siberia shelf (IA in Fig. 4). They are poorly known and have had little exploration appraisal. Their position, relative to the polar Arctic Ocean, is somewhat similar to the North American arctic area, where developing frontier basins of the North Slope, Beaufort Shelf, Mackenzie Delta, and Sverdrup basins are better known. These arctic basins appear to have a similar geologic setting and perhaps an initial genesis similar to many of those basins located between Eurasia and Africa, and the Gulf of Mexico–Caribbean zone between North and South America—basins which include the Mideast and Reforma-Campeche of Mexico. Basins of this type often have high reserves with giant and supergiant fields. In addition, several extensive drainage areas appear to have formed deltas along the northeastern margin of Siberia.

A large basin (IB in Fig. 4), poorly known and with the potential for several subbasins, lies north of the European Soviet Union and Norway. On-shore basins, extending offshore, suggest that this basin is a submerged complex cratonic basin (similar to most land basins which occupy central to marginal continental areas). Substantial gas production and some oil have been established on land in the Soviet Pechora basin, which may project offshore into the Barents Sea.

Over 30 basins and subbasins extend from western Alaska along the eastern Asia continental margin to the southeastern Asia archipelago of Indonesia (IIA in Fig. 4). Most of these basins appear to be in the same classification category as producing basins in Indonesia and California, where convergence of oceanic plates subducts continental plates, forming both island arcs and shear basins or rhombochasms. These basins have a higher risk than normal, often due to extreme deformation resulting in oil leakage and poor reservoirs due to an overabundance of volcanic sediments. However, those that are found productive have moderate to extremely high recovery per volume of sediment, with a potential for giant accumulations. The size of these frontier basins on average appears to be larger than the 12 presently producing basins which account for 5% of the world's established reserves. Minor production on land has been established along this trend (Kamchatka, Sakhalin, and Japan), and offshore discoveries have been reported.

Small basins, perhaps similar in origin to the eastern Asia basins, are located in northern and southern South America (IIB in Fig. 4). Basins in the northern coastal areas of Columbia and Venezuela, including the Gulf of Venezuela, have yet to be prospected. In the south the Malvinos basin may offer similar prospects. Both areas are currently involved in territorial disputes.

Over 50 basins and subbasins occur along the continental margins bordering the Atlantic and Indian oceans (III in Fig. 4). These basins are formed by the opening or "spreading" of their bounding ocean basins, literally pull-apart basins in zones of structural divergence. Because most of their sediments lie offshore on the continental shelf and bounding deep-water areas, they have only recently been prospected. More prospecting of these basins is required since many occupy positions near large centers of population. However, the results to date, with a few exceptions, indicate that recovery of petroleum from these basins is not providing the rich deposits associated with most other basin types (lack of supergiant accumulations or the concentrated accumulations required for offshore economics).

Relatively untested areas in the interior portions of the continents previously were often considered economically or operationally inaccessible (IVA, IVB, IVC, and IVD in Fig. 4). Most are simple cratonic basins. They occur in both arctic and equatorial regions. Limited exploratory drilling has occurred in most of these basins. Production from similar basins has resulted in relatively low recovery rates and does not represent a substantial contribution to world reserves. Recently, modest success in these basins has occurred in Australia and north-central Africa, where the simple nature of

these basins has been modified by both pre- and postbasin rift or graben development.

In addition, many delta areas around the world remain to be prospected. It has been estimated that one-third or more of new resources will come from new exploration frontier basins.

New or developing frontier basins. Basins which have recently established production are expected to continue for variable time periods to add substantial reserves. Some of those with significant reserves include the North Slope, Mackenzie-Beaufort, and Sverdrup basins of North America; the Andean trend of the Oriente, Maranon, Ucayali, Madre de Dios, Pirity, and Chaco basins of South America; the western Siberia–Kara Sea, Tarim, and Vilyuy–Lena basins of Asia; the Northwest Shelf subbasins of Australia; and the North Sea province in Europe.

Older producing basins or provinces. Basins which are expected to continue adding new reserves include the Mideast, southern Caspian Sea, north Black Sea, Pechora, Caucusus-Mangyshlak trend, and northern China basins of Eurasia; the Sirte basin of northern Africa; and the "thrust belt" play, Alberta basin, and Vera Cruz–Reforma-Campeche area of North America. In Mexico the Vera Cruz area production was established in 1908; however, it was not until the late 1960s and early 1970s that substantial reserves were found in the remote Reforma area and Campeche offshore, where more efficient and sophisticated exploration technology (similar to the discovery of new reserves in the United States thrust belt) contributed to success. Thus, although Vera Cruz–Reforma-Campeche is treated as a new frontier basin, it actually is an older producing basin.

Expectations. Present estimates of future oil and gas resources indicate a consensus in estimates which consider that an amount ranging from 60 to 140% of present proved and conventionally produced recoverable oil and gas (1600×10^9 bbl or 254×10^9 m^3, or Btu equivalent) remains to be found (see table). Estimates as low as 25% and as high as 600% present proved and produced reserves have been made. The general magnitude of the present consensus was expressed 20 years ago, when proved and produced reserves were about one-third of the present, although subsequent estimators have tended to increase their estimates to the magnitude of the present consensus or more. Continual update of new data and revision of forecasts are required.

With the world divided into three political, rather than geologic, sectors, namely OPEC areas, Communist areas, and the remaining areas, present proved reserves are distributed 57, 24, and 19% respectively (Figs. 2 and 3). A geologic appraisal of future petroleum resources indicates that each sector will share roughly an equal amount. The United States and Canada are expected to share less than one-third of what is available to the third group, or about 10% of the world's future oil and gas resources, yet they are presently consuming nearly 40% of the world's oil and gas production. Of the world's present oil reserves, 70% are in OPEC countries, and of this amount 80% are in the Arab countries and Iran—primarily in the Mideast basin. There are unequal ratios of

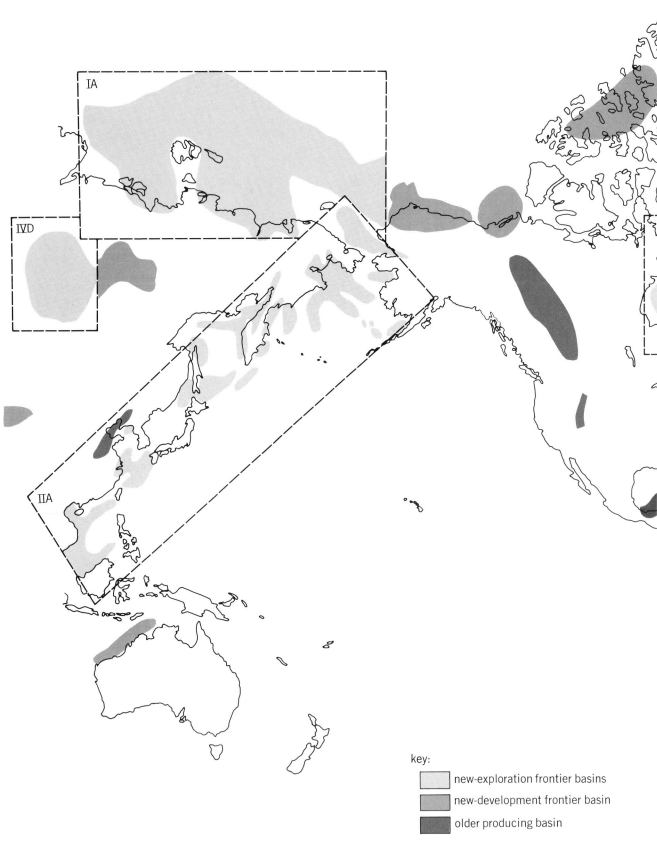

Fig. 4. World prospects for future oil and gas resources.

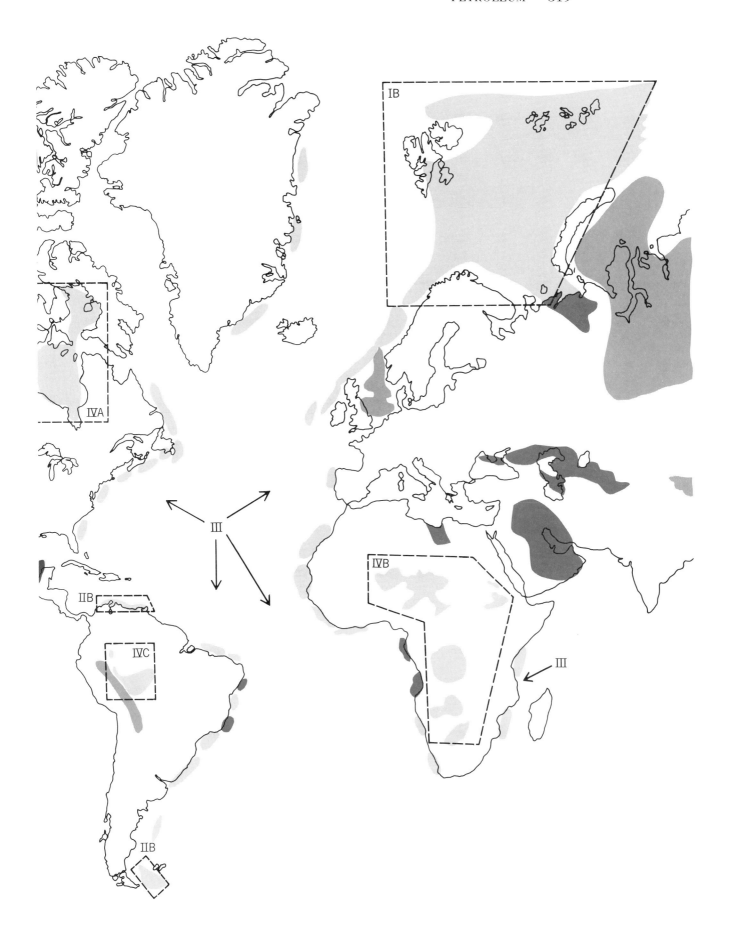

Conventionally recoverable oil and gas reserves (10⁹ bbl oil or gas Btu equivalent)*

Sector	Estimated proved reserves				Oil and gas consumption (1976)	Oil consumption, years to present reserves	Oil production, years to present reserves	Potential future reserves	
	Oil	Gas	Total	%				Low	High
OPEC	450	162	612	57%	1.1	766	40	333	666
Soviet Union, etc.	100	159	257	24%	5.1	25	20	333	666
Other areas	100	99	203	19%	22.6	6	18	333	666
			1072	100%					

*From H. D. Klemme, Worldwide petroleum exploration and prospects, Proceedings of the Southwestern Legal Foundation Exploration and Economics of Petroleum Industry, vol. 16, Matthew Bender, 1978.

oil to gas in each sector; that is, OPEC's reserves are 26% gas, Communist areas 61%, and the remaining sector 50%. With a continued ratio of 50:50 gas to oil in remaining countries and 40:60 gas to oil in the OPEC sector, the remaining countries are estimated to obtain from 33 to less than 75% of present proved oil reserves in the OPEC sector and 25 to 40% of ultimate OPEC oil resources. In addition, future resources in OPEC will involve more simple logistics since these resources will be in the general vicinity of established production facilities, whereas the greater portion of oil-producing areas in the remaining countries involves more difficult operating areas where production facilities will have to be established.

The increase in gas reserves (Fig. 2) noted in giant fields is expected to continue in the future discoveries associated with the addition of future oil and gas resources, approaching a ratio of 50:50 worldwide. This is expected as more discoveries in western Siberia continue to be added, as gas deposits bypassed in the past search for oil are exploited (as in the Mideast), and as deeper drilling in older basins encounters more thermally mature gas. Ultimate resources from giant fields are estimated to range from 75% to a much lower magnitude if smaller accumulations become economic or further development of smaller accumulations occurs in basins outside North America. Smaller field sizes, fewer supergiants from offshore areas, and the smaller accumulations associated with the late-phase development in producing basins suggest a future reduction in the reserve ratio of giant fields to nongiant fields. However, depending upon the economics of smaller accumulations, a figure of 50 to 60% for ultimate giant resources appears reasonable, particularly when some prolific basins like the Mideast, presently developing western Siberia, and Mexico's Reforma-Campeche basins have their largest fields as supergiants and their smaller fields as giants.

Constant updating and revision of petroleum resource information are necessary and are continually in progress. Substantial increases of reserves would require a discovery of another Mideast or western Siberia basin, the presence of large reserves from deep-water areas, or nonanalog accumulations of large reserves of petroleum (nonconventional accumulations).

While significant reserves have been projected for the Reforma-Campeche province of Mexico (up to 2 × 10¹¹ bbl or 3.18 × 10¹⁰ m³, or 6% of ultimate world resources), 57% of world reserves are located in the Mideast or western Siberia. It is believed that the world does not have other remaining basins of comparable size and geology. While deep-water areas are essentially untested and the extent of the reserves relatively unknown, available data suggest that the magnitude will not exceed the area of the presently known deep-water basins. Furthermore, a substantial drop in the success ratios in new exploration basins, new development frontier basins, and additional reserves in older producing basins would appreciably lower the forecasts of future petroleum resources. However, such a development would result in a reversal of historical data projections.

Most current estimates of ultimate conventionally recoverable oil and gas resources concede the possibility of substantial changes in resource magnitude. However, a consensus of current estimates using presently available data considers this unlikely. A 5–10-year lead time to bring on production and the possibility of slower exploration pace in difficult or remote new exploration frontiers are anticipated.

The time involved for depletion of present reserves and the addition of potential petroleum resources is dependent to a considerable degree upon political and economic factors.

For background information see MINERAL FUEL AREAS; OIL AND GAS, OFFSHORE; PETROLEUM GEOLOGY in the McGraw-Hill Encyclopedia of Science and Technology.   [H. DOUGLAS KLEMME]

Bibliography: T. D. Adams and M. A. Kirby, *Estimates of World Gas Reserves,* vol. 3 of the *Proceedings of the 9th World Petroleum Congress,* 1975; A. W. Bally, *A Geodynamic Scenario for Hydrocarbon Occurrences,* vol. 2 of the *Proceedings of the 9th World Petroleum Congress,* 1975; K. O. Emery, The potential and time frame for development of petroleum in the deep ocean, in R. F. Meyer (ed.), *The Future Supply of Nature Made Petroleum and Gas,* 1977; M. T. Halbouty et al., *Geology of Giant Petroleum Fields,* Amer. Ass. Petrol. Geol. Mem. no. 14, 1970; American Association of Petroleum Geologists, Reserves Symposium, in J. D. Haun (ed.), *Studies in Geology,* 1975; K. F. Huff, *Oil Gas J.,* 76(9):214, 1978; H. D. Klemme, *Bull. Can. Petrol. Geol.,* vol. 23, no. 1, 1975; H. D. Klemme, Giant fields—update, *Oil Gas J.,* May 7, 1977; H. D. Klemme, 200 billion bbl of offshore oil and gas, *Oil Gas J,* vol. 75, no. 35, 1977; H. D. Klemme, World's oil and gas reserves from an analysis of giant fields and basins, in R. F. Meyer (ed.), *The Future Supply of Nature Made*

Petroleum and Gas, 1977; H. D. Klemme, World-wide petroleum exploration and prospects, in *Proceedings of the Southwestern Legal Foundation Exploration and Economics of Petroleum Industry*, vol. 16, 1978; J. D. Moody and R. W. Esser, *An Estimate of World's Recoverable Crude Oil Resources*, vol. 3 of the *Proceedings of the 9th World Petroleum Congress*, 1975; R. Nerhing, *Giant Oil Fields and World Oil Resources*, Rand Corp. Rep. no. R-2284-CIA, 1978; H. R. Warman, *Geograph. J.*, 138(3): 287, 1972; L. G. Weeks, Where will energy come from in 2059, *Petrol. Eng.*, vol. 31, no. 9, 1959; P. W. J. Wood, *World Oil*, p. 141, June 1979; *World Energy Resources (1985–2020)*, World Energy Conference, 1978.

Petrology

One of the principal aims of petrologists is to determine the physical conditions (pressure, temperature, fluid-phase composition) under which rocks from the Earth's crust and upper mantle crystallized. Knowledge of the pressure-temperature conditions of crystallization of rocks of different geologic ages may then be used to deduce ancient geothermal gradients so as to lead to an understanding of the Earth's thermal history. In order to make accurate pressure-temperature (and fO_2, fH_2O, and so forth) estimates for a given phase assemblage, it is necessary to have precise data pertaining to the stabilities of the different phases under the conditions of interest. The problems involved in obtaining such stability data are exacerbated by the fact that all naturally occurring solid, liquid, and fluid phases are complex multicomponent solutions. Recent work has centered on an expansion of thermodynamic data on pure one-component minerals and simple solid solutions, and attempts to integrate calorimetric results with stability relations derived from phase equilibrium experiments. This has led to the derivation of a large body of precise internally consistent thermodynamic data for pure minerals and fluids and to an understanding of the mechanisms of solid solution in several complex mineral solutions.

It has been known for nearly 70 years that metamorphic rocks (recrystallized in the solid state) and slowly cooled igneous rocks (crystallized from melt) have mineral assemblages which formed under conditions approaching chemical equilibrium. This was originally deduced from the predictable relationships between chemical compositions of rocks and their mineralogic constitution and has been confirmed by more recent studies of element partitioning behavior between minerals. The observed approach to chemical equilibrium, although requiring confirmation for any particular suite of rocks, is the justification for applying equilibrium thermodynamic arguments or phase equilibrium experiments to geologic systems.

Thermodynamic data. Although the principles of chemical thermodynamics and their potential application to rocks have been long established, petrologists did not use thermodynamic data as a matter of routine until the early 1970s. This recent usage was stimulated by the publication of several extensive tables of thermodynamic data for minerals and related substances, the data set of Richard Robie and David Waldbaum being particularly

important. These workers tabulated 1-bar (10^5-Pa) enthalpies and free energies of formation for many pure minerals together with measured molar volumes and "third-law" entropies (S_i). As in Eq. (1),

$$S_i = \int_0^T \frac{C_{Pi}}{T}\, dT \qquad (1)$$

at temperature T the entropy is derived from the measured heat capacity of phase i (C_{Pi}).

Enthalpy (H_i), entropy, and volume (V_i) data for pure phases may be used to calculate the conditions of equilibrium for reaction (2), for example.

$$\mathrm{A} \rightleftharpoons \mathrm{B} \qquad (2)$$

If pure A coexists with pure B at temperature T and pressure P bars, then the free-energy change (ΔG) for the reaction is given by Eqs. (3) and (4).

$$\Delta G = (G_B - G_A) = 0 \qquad (3)$$

$$\Delta G = (H_B - H_A) - T(S_B - S_A)$$
$$+ \int_1^P (V_B - V_A)\, dP = 0 \qquad (4)$$

In Eq. (4) the enthalpy and entropy data refer to 1 bar and temperature T. Inspection of Eq. (4) demonstrates that, given H, S, and V [as $f(P)$], the equilibrium pressure may be calculated at any temperature. After publication of the (predominantly calorimetric) Robie-Waldbaum data set, it became apparent that there were substantial inconsistencies between the enthalpy of formation measurements from different laboratories, particularly for components containing aluminum. This was readily shown by calculating reactions in the manner just described and checking the results against direct phase equilibrium measurements of the reaction boundaries. Fortunately, these inconsistencies are now largely resolved, and a revised version of the calorimetric data set has recently been produced by Robie, B. Hemingway, and J. Fisher.

The process of checking calorimetric data against phase equilibrium experiments led a number of workers to believe that precise thermodynamic data were best derived indirectly from the latter rather than from calorimetric measurements. The procedure for determining the enthalpy of phase B at 1 bar and temperature T may be understood from consideration of Fig. 1. ΔG for the reaction is zero at P_1, T_1 and at P_2, T_2. If the volumes of both phases (generally assumed independent of pressure for solids) are known, average values of $(H_B - H_A)$ at 1 bar and $(S_B - S_A)$ in the temperature range T_2 to T_1 may be derived by applying Eq. (4) to each of the two points P_1, T_1 and P_2, T_2. Alternatively, if entropies and heat capacities are known or may be estimated, and the enthalpy of A is known, the two points give two values for H_B, one at T_1 (1 bar) and one at T_2 (1 bar). These are then generally referenced to a 298 K/1 bar value as in Eq. (5). Generally, the reference value at 298 K/1 bar is the enthalpy of formation from the elements, ΔH_f^0.

$$(H_B)_{298} = (H_B)_{T_1} - \int_{298}^{T_1} C_{pB}\, dT \qquad (5)$$

ally, the reference value at 298 K/1 bar is the enthalpy of formation from the elements, ΔH_f^0.

Having obtained enthalpy data for phase B, one can use this result to obtain information about other phases and hence build up an extensive data

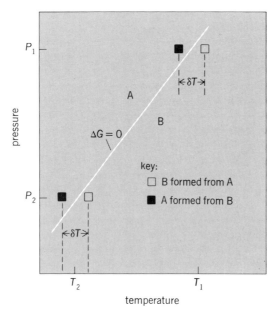

Fig. 1. Determination of free-energy change of reaction (ΔG) from phase equilibrium experiments.

set with a minimum of calorimetrically measured enthalpies. H. Helgeson, J. Delaney, H. Nesbitt, and D. Bird have used this approach to derive thermodynamic data for the pure end-members of most rock-forming minerals. They adopted calorimetric values of enthalpies of formation of 5 minerals and derived enthalpies of formation of 70 additional minerals in the system $Na_2O - K_2O - CaO - MgO - FeO - Fe_2O_3 - Al_2O_3 - SiO_2 - CO_2 - H_2O$. Wherever possible, they used measured heat capacities and third-law entropies of the minerals of interest. In many cases, however, it was necessary to estimate these parameters from measured values for minerals of a structural type similar to the mineral of interest. For example, "structural-analog" entropy for akermanite, $Ca_2MgSi_2O_7$, may be derived by considering reaction (6). With definitions (7), it is found that, to a very good approximation, Eq. (8) holds.

$$Ca_2MgSi_2O_7 + \alpha\text{-}Al_2O_3 \rightleftharpoons$$
Akermanite   Corundum   (6)
$$Ca_2Al_2SiO_7 + MgO + SiO_2$$
Gehlenite   Periclase   Quartz

$$S' = S_{gehlenite} + S_{periclase} + S_{quartz} - S_{corundum}$$
                                                            (7)
$$V' = V_{gehlenite} + V_{periclase} + V_{quartz} - V_{corundum}$$

$$S_{akermanite} = \frac{S'(V' + V_{akermanite})}{2V'}$$ (8)

This relationship enables calculation of the entropies of most silicates to within about 1% at 298 K and 1 bar, provided the required entropies are derived from a structurally analogous mineral (in this case akermanite from gehlenite). A similar approach enables accurate calculations of heat capacity.

Since considerably more than 70 reactions of the type shown in Fig. 1 were available for the estimation of thermodynamic data, it was possible for Helgeson and coworkers to make extensive checks on the consistency of values derived from different

sets of phase equilibrium experiments. Their final data set is internally consistent with relative uncertainties in enthalpies of formation on the order of 400 J mol⁻¹. This order of uncertainty is substantially less than can be obtained for standard calorimetric measurements of enthalpies of formation. The reason for the improved precision may be understood by considering Fig. 1. As a general rule, an equilibrium boundary may be "bracketed" in the manner shown to within ±10°C, that is, $\delta T = 20°C$. Uncertainties in calorimetric enthalpies of reaction ($H_B - H_A$) are of the order of 4000 J mol⁻¹ or more, and entropy changes ($S_B - S_A$) have uncertainties of the order of ±2 J mol⁻¹. At 1 bar the equilibrium temperature is given, from Eq. (4), by Eq. (9). Taking typical values of ΔH, ΔS, and T for

$$T = \frac{(H_B - H_A)}{(S_B - S_A)} = \frac{\Delta H}{\Delta S}$$ (9)

a reaction involving solids only as $\Delta H = 40,000$ J ± 4000, $\Delta S = 40$ J ±2, and $T = 1000$ K, it may readily be shown from standard error propagation that the uncertainty in temperature calculated from the calorimetric data is ±112°C. The phase equilibrium experiments are an order of magnitude more accurate. Thus, relative enthalpy data derived from experiments such as those shown in Fig. 1 are much more precise than the calorimetrically measured values.

It seems probable that this internally consistent set of tabulated thermodynamic data will form the basis of a considerable amount of petrological work in the next few years.

Solid solutions. The thermodynamic data discussed above relate to the stabilities of many of the pure minerals of geologic interest. In general, these data need to be modified to take account of the presence of other components in natural systems and activity-composition relationships obtained for complex phases.

There have been a number of recent phase equilibrium studies aimed at determining activity-composition relationships for binary and, in some cases, ternary solutions. The general method may be illustrated with respect to reaction (2). Suppose that B forms a stable solid solution with component C but that there is no solution between A and C. An experiment at some P and T fixes the activity of B in the B-C solid solution through the Van't Hoff equation (10). Taking the standard-state free-

$$\Delta G^0 = -RT \ln\left(\frac{a_B}{a_A}\right)$$ (10)

energy change of the reaction (ΔG^0) to be the value of ΔG for the pure phases at the pressure and temperature of the experiment, then activity data for the end members a_B/a_A may be calculated. Since, by definition with this standard state, activity is equal to 1 for pure phases, the ratio in Eq. (10) is simply equal to a_B (A is pure). The composition of the solution is measured after the experiment and the activity coefficient (γ_B) calculated from Eq. (11), where X_B is the measured mole frac-

$$\gamma_B = \frac{a_B}{X_B}$$ (11)

tion of B in the solid solution. The activity coefficient of C in the binary solution of composition X_C, X_B is then obtained from integration of the

Gibbs-Duhem equation at constant P and T and the excess free energy G^{XS} calculated according to Eq. (12). If G^{XS} is zero, the solution is ideal. The prob-

$$G^{XS} = RT \, X_B \ln \gamma_B + X_C \ln \gamma_C \qquad (12)$$

lem with the application of Eqs. (11) and (12) to minerals is that, in general, minerals have several sites on which mixing takes place and that the relationship between activity and composition depends sensitively on whether or not the site substitutions are coupled to one another. For example, in the binary pyroxene solution $CaMgSi_2O_6 -CaAl_2SiO_6$, the replacement of Mg^{2+} by Al^{3+} on octahedral sites is coupled to replacement of Si^{4+} on tetrahedral sites by Al^{3+}. If the replacing atoms are linked to one another in the structure (short-range-ordered), then the activity-composition relationships are of similar form to Eq. (11). However, if there is complete disorder on each of the different types of site regardless of the charge-balancing constraint, it may be shown that the relevant relationships are as in Eqs. (13).

$$a_{CaAl_2SiO_6} = X^2_{CaAl_2SiO_6}(2 - X_{CaAl_2SiO_6})\gamma_{CaAl_2SiO_6}$$

$$a_{CaMgSi_2O_6} = X_{CaMgSi_2O_6}\left(\frac{3X_{CaMgSi_2O_6} - 1}{2}\right)^2 \gamma_{CaMgSi_2O_6} \qquad (13)$$

Although either Eqs. (11) or (13) could satisfy the phase equilibrium data, vastly different values of G^{XS} are obtained if disorder is assumed. Fortunately, as in the case of pure minerals, calorimetric information is complementary to the phase equilibrium results. High-temperature-solution calorimetry by R. Newton, O. Kleppa, and coworkers has established the values of excess enthalpy for a number of binary pyroxene and garnet solutions. A general relationship of the same form as Eq. (4) (at 1 bar) between excess free energy, excess enthalpy (H^{XS}), and excess entropy S^{XS} may be written as Eq. (14). Since it is known that $CaMgSi_2O_6 - CaAl_2SiO_6$

$$(G^{XS})_{1bar} = (H^{XS})_{1bar} - TS^{XS} \qquad (14)$$

solid solutions have little or no excess entropy, possible order-disorder models for the coupled substitution can be tested against the constraint that G^{XS} and H^{XS} should be approximately equal. Figure 2 shows calorimetric excess enthalpies and phase equilibrium values of G^{XS} derived from the assumption of complete disorder on tetrahedral and on octahedral sites in the solution. Agreement is within experimental error and indicates that there is little or no short-range order. Thus the complementary G^{XS} and H^{XS} measurements enable the mechanism of solid solution to be determined.

Similar calorimetric and phase equilibrium measurements for Ca-Mg garnet solutions indicate that such garnets are markedly nonideal and that they have substantial excess entropies.

Applications. The thermodynamic data for pure phases and information on mixing properties which have been discussed above have many applications in the fields of igneous, metamorphic, and sedimentary petrology. One recent study of the stable assemblages in the Earth's upper mantle serves to illustrate their use.

The upper mantle is commonly believed to be of peridotitic composition, dominantly olivine with lesser amounts of pyroxenes, spinel, and garnet. One of the important phase transitions which is

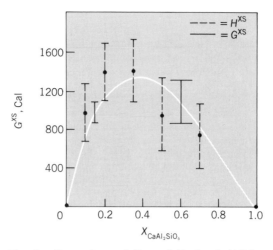

Fig. 2.   Comparison of G^{xs} and H^{xs} for $CaAl_2SiO_6-CaMgSi_2O_6$ pyroxene solid solutions. (*From B. J. Wood, Amer. J. Sci., 279:854-875, 1979*)

observed in such peridotitic material is the replacement of low-pressure spinel by high-pressure garnet through reaction (15). At temperatures of

Orthopyroxene + clinopyroxene + spinel $\rightleftharpoons$ garnet + olivine   (15)

$1300°$ C or so, several sets of phase equilibrium experiments indicate that the reaction has the relatively steep dP/dT slope of 25 bars/K (Fig. 3). The extrapolation of these data to lower pressures is, however, complicated by the varying compositions of orthopyroxene, clinopyroxene, and garnet solid solutions. Linear extrapolation of the high-temperature data suggests that garnet peridotite should be stable in the lowermost continental crust and that there is no field of spinel peridotite stability beneath the continents. D. Jenkins and R. Newton have shown, however, that the boundary slope flattens at low temperatures so that garnet peridotite would not be stable at the base of any but the very thickest continental crust. This was demonstrated from both phase equilibrium and calorimetric data

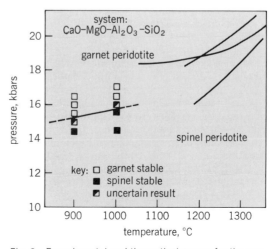

Fig. 3.   Experimental and theoretical curves for the garnet-spinel peridotite transition. Square symbols are experiments of D. Jenkins and R. Newton, and slope-fitted to them is the calculated calorimetric one.

with the result illustrated in Fig. 3. From calorimetric work on the end member and solid solution phases in reaction (15), they were able to calculate the slope of the reaction boundary at 16 kbar and 1000°C from the Clausius-Clapeyron equation (16).

$$\frac{dP}{dT} = \frac{\Delta H}{T \Delta V} = 4.5 \pm 2.7 \text{ bars/K} \qquad (16)$$

This calculated slope is in excellent agreement with their experimental "bracketing" of the equilibrium boundary.

For background information *see* ACTIVITY (THERMODYNAMICS); EQUILIBRIUM, PHASE; PETROLOGY: THERMODYNAMICS, CHEMICAL in the McGraw-Hill Encyclopedia of Science and Technology.

[BERNARD J. WOOD]

Bibliography: H. Helgeson et al., *Amer. J. Sci.*, 278(A):1–229, 1978; D. Jenkins and R. Newton, *Contrib. Mineral. Petrol.*, 68:407–419, 1979; R. Robie, B. Hemingway, and J. Fisher, *U.S. Geol. Surv. Bull.* no. 1452, 1978; B. Wood, *Amer. J. Sci.*, 279:854–875, 1979.

Photoacoustic spectroscopy

Recently several important advances were made in photoacoustic spectroscopy; these advances have further extended the capabilities of this new methodology.

Analytical applications. Photoacoustic spectroscopy is now routinely used to study all three phases of matter—gas, liquid, and solid. For gaseous samples and for most solid samples, a photoacoustic spectrometer with a gas-microphone mode of detection is usually used. Liquids can also be studied with such an apparatus, although the signal strength is considerably reduced for liquids with low optical absorptions. Recently, however, several researchers have performed photoacoustic studies on liquids by means of a piezoelectric mode of detection. In these experiments, the liquid sample completely fills a small chamber that is itself made of piezoelectric material, and the photoacoustically generated heat pulses within the liquid are detected as stress-strain signals by the piezoelectric chamber. With this method, all of the optical energy absorbed by the sample, which decays nonradiatively, can contribute to the photoacoustic signal, whereas in the gas-microphone method only that energy absorbed within a thermal diffusion length of the sample-gas boundary can contribute to the signal. The piezoelectric cell is thus better suited to the measurement of very small optical absorptions in liquid samples. Using such a method, researchers have recently been able to detect minute traces of metal ions in solution (0.02 ng/ml). This value is almost two orders of magnitude lower than that obtained from calorimeter analysis or from flame-absorption measurements.

Catalytic and surface studies. One of the major advantages of photoacoustic spectroscopy lies in its relative immunity to scattered light, and its consequent ability to provide absorption spectra of highly light-scattering materials such as powders. This capability is being put to good use in the study of catalytic compounds and catalytic reactions. For example, several recent studies of this nature involve the reactions of transition-metal complexes with polymeric ligands to form anchored catalysts. Photoacoustic spectroscopy has been used very effectively to investigate the electronic structures of these metal-polymer complexes in order to elucidate chemical processes and structure-reactivity relationships.

The ability of photoacoustics to detect absorption processes in the presence of strong light scattering and reflection has also made the technique useful in surface studies. Photoacoustic surface studies have been performed previously in the visible region, but more useful data can be obtained in the mid-infrared, since the molecular information there is more detailed and specific. Unfortunately, widely tunable infrared light sources are not available with sufficient intensity for a conventional photoacoustic spectrometer. Work is progressing at several laboratories to develop Fourier-transform photoacoustic spectrometers that will operate in the mid-infrared region. In the meantime, P.-E. Nordal and S. Kanstad have demonstrated the power of the photoacoustic method for infrared surface studies. Using a CO_2 laser, they demonstrated that it is possible to measure absorptions on surface layers that are only angstroms (10 A = 1 nm) thick. This is a sensitivity comparable to that attainable with electron-energy-loss spectroscopy, but with the advantage of optical spectral resolution.

Deexcitation. The photoacoustic effect measures the heat-producing deexcitation processes that occur in a system after it has been optically excited. This selective sensitivity of photoacoustics to the heat-producing deexcitation channel has been used to great advantage in the study of fluorescent materials, and in the study of photosensitive materials that exhibit photochemistry or photoconductivity.

These capabilities have been amply illustrated by recent experiments of D. Cahen and coworkers. In one set of experiments, they studied the mechanism of photosynthesis in both green plant matter and bacteria. They were able to obtain the activation spectrum for photosynthesis, and to study the role of inhibitors and the effects of intermediate storage states. In another experiment Cahen used photoacoustics to investigate the photovoltaic process in silicon, and to demonstrate how photoacoustics can be used to measure the photovoltaic efficiency of candidate materials for solar cells.

Phase transitions. The photoacoustic signal depends not only on the optical properties of the sample but also on its thermal properties. Since thermal properties generally undergo a change when the material undergoes a phase transition, monitoring the photoacoustic signal as a function of temperature can provide information about these phase transitions. This application of photoacoustics was recently demonstrated by R. Florian and coworkers. Using photoacoustics, they investigated the first-order liquid-solid transitions of gallium and of water, and the first-order structural phase transition of K_2SnCl_6. They found not only that the magnitude and phase of the photoacoustic signal change when the thermal parameters of the sample change, but that, in addition, a strong modification in both amplitude and phase signals occurs when a first-order transition is approached from lower temperatures. This is a result of the fact that, during the endothermic cycle of the phase transition, the periodic heat generated by the photoacoustic process is itself absorbed by the endothermic event, and is therefore lost, resulting in large changes in both the amplitude and the phase of the photoacoustic signal. The application of photoacoustics to phase-transition studies should constitute a useful complementary technique to the conventional calorimeter methodology.

For background information *see* SPECTROPHOTOMETRIC ANALYSIS in the McGraw-Hill Encyclopedia of Science and Technology.

[ALLAN ROSENCWAIG]

Bibliography: D. Cahen, S. Malkin, and E. I. Lerner, *FEBS Lett.*, 91:339, 1978; R. Florian et al., *Phys. Status Solidi*, 48:K35, 1978; P.-E. Nordal and S. O. Kanstad, *Opt. Commun.*, 24:95, 1978; A. Rosencwaig, *Adv. Electron. Electron Phys.*, 46:208, 1978.

Picosecond spectroscopy

The past few years have seen exciting developments in ultrashort laser pulses and their applications to picosecond and, now, subpicosecond laser spectroscopy. Meanwhile, molecular descriptions of the forces and torques which govern the dynamical structure of fluids have begun to dominate contemporary discussions of the liquid state, which was traditionally viewed in the formalism of continuum hydrodynamics. Electrons in liquids have proved to be sensitive microscopic probes of short-range liquid structure. Thus, it is not entirely a coincidence that the recent advances in understanding the dynamics of electron localization and solvation in liquids have arisen through picosecond spectroscopy, where the probe light pulses and the molecular dynamics of liquids share the same temporal domain.

Electrons trapped in polar liquids (e_s^-) have a fleeting chemical existence, seldom more than microseconds. The electron resides in discrete clusters of molecules whose rotational, vibrational, and configurational relaxations prove to be intimately involved in the electron localization mechanism and subsequent laser-induced electron-transfer processes. Picosecond absorption studies of electron solvation in fluids of different density and composition also reveal detailed aspects of the local solute-solvent interactions, which are unanticipated by the macroscopic liquid properties, and provide an excellent prototype for solvation phenomena in general. These results are of significance to the theory of excess electron states and electron transfer in disordered systems, and also to the more practical areas of dielectric breakdown and electric conduction in liquids, formation of atmospheric ion clusters, laser-induced nucleation phenomena, and fast optical switching.

Electrons in fluids. When a low-energy electron is injected adiabatically into a liquid, the initial electronic state is an extended or conduction state in which the quasi-free electron moves with high mobility $(\mu > 10^2 \text{ cm}^2 \text{ V}^{-1} \text{ s}^{-1})$, without modifying the liquid structure, until a strong electron-medium interaction blocks the quantum transport and the electron becomes localized in a shallow potential at times 10^{-12} s. Thermal fluctuations can pro-

mote the electron back into the continuum state. Alternatively, if the residence time in this quasi-localized state is comparable to the time scale of rotational diffusion in the liquid $(>10^{-12}$ s), then the adjacent molecules will be polarized and will orientationally relax in the field to form a cluster of molecules about the electron. (In helium, repulsive interactions dominate and the electron is trapped in a bubble.) In its configurationally relaxed ground electronic state, of typically 1–2-eV binding energy, e_s^- exhibits strong optical absorptions which are a signature of the trapping potential experienced by the electron. As Fig. 1 shows, the absorption maxima for e_s^- appear in spectral groupings according to the short-range electron-medium interactions and the molecular structure of the liquid host. Through picosecond spectroscopy it is possible to observe the transition between the extended and localized electronic states by directly monitoring the temporal evolution of the optical absorption of e_s^- following electron injection. The evidence shows that the dynamics of electron solvation can be correlated to the statistical dynamical structure of the fluid.

Techniques. The orientational and vibrational (population and phase) relaxations of molecules in liquids can be interrogated through many different picosecond techniques, including absorption and emission spectroscopy, fluorescence depolarization, transient-induced dichroism, and the nonlinear optical techniques of conjugate wave formation and optical Kerr spectroscopy. The time correlation functions derived from the results yield complementary information on the short-range dynamical correlations vis-à-vis the hydrodynamic contributions to the liquid state.

The electron localization in liquids at 298 K has been studied by means of two different pump-probe techniques. In one technique an intense

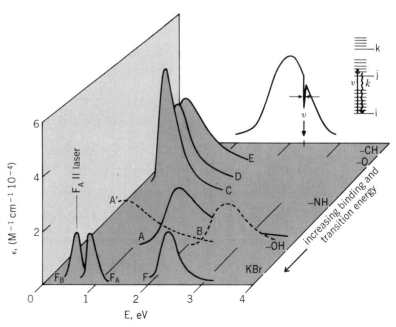

Fig. 1. Ensemble of electrons in fluids: optical absorptions in *n*-alcohols (A), diols (B), amines (C), ethers (D), and alkanes (E); for comparison, the color centers (F, F_A) and KBr color center laser emission (F_B) are shown. (*From G. A. Kenney-Wallace, Electrons in molecular clusters: Microscopic probes of fluids, Acc. Chem. Res., 11:433–439, 1978*)

picosecond pulse triggers electron ejection and a delayed, attenuated picosecond continuum probe pulse monitors the transient absorptions in the liquid. Single pulses of about 6 ps duration can be extracted from the mode-locked pulse train of a neodymium glass laser, amplified, and used to generate harmonics of the 1.06-μm fundamental pulse through successive frequency mixing in nonlinear crystals. The ultraviolet pulse is directed to the sample, where n-photon photoionization of an impurity molecule can occur in the host liquid under study, and the probe pulses are directed along an optical delay line and interrogate the developing e_s^- absorption at a known time interval after the photoionization. The repetition rate of many gigawatt mode-locked Nd glass or ruby solid-state lasers is typically 10^{-2} Hz, and thus this technique is most suited to experiments in which strong e_s^- absorption signals are observed and all the data can be reliably recorded in a single shot. For very weak signals in low-density fluids, the second technique must be used. That technique is based on direct ionization of the liquid with a picosecond pump pulse of high-energy electrons, and the probe is obtained from a synchronously pulsed Cerenkov continuum or as a delayed laser pulse. The unique, single-pulse stroboscopic electron beam facility at the Argonne National Laboratories can be operated at $\leq$60 Hz, thus permitting the use of signal averaging and sampling techniques to extract the weak transient absorption from the signals. However, the latest developments in subpicosecond laser technology promise to combine the gigawatt powers required for higher-order nonlinear processes and temporal resolution below 10^{-12} s with the wavelength tunability and repetition rates necessary to obtain the optimum signal-to-noise over a wide dynamic range.

Pure liquids. The picosecond absorption signal observed immediately after electrons are ejected (via photoionization, photodetachment, or direct ionization) into pure alcohols $CH_3(CH_2)_nOH$ corresponds to a low-intensity, structureless infrared spectrum, depicted in Fig. 1 as A′. This absorption gradually shifts to the characteristic nanosecond spectrum (A), with a time constant that depends on n, the length of the alkyl chains attached to the $-OH$ dipoles, which eventually compose the inner

structure of the cluster of molecules about the electron. The uppermost trace in Fig. 2 illustrates the temporal evolution of the signal in pure n-butanol and shows the fast and slow, exponential component from which the time constant is derived. In a series of alcohols, methanol to decanol, these times increase systematically from 11 to 50 ps, and can be correlated with the rotational diffusion times of the monomer alcohol molecules (from 12 to 48 ps) and liquid viscosity (from 0.5 to 14 cp) in a Stokes-Einstein-Debye related formalism, which includes the orientational polarizability of the molecule. Unexpectedly, electron solvation in glycol (B in Fig. 1) is too fast to resolve, despite the high viscosity of 20 cp, and the visible spectrum attributed to e_s^- is already present within 6 ps. Although the apparent hydrodynamic response of the liquid to the sudden perturbation presented by the excess electron masks the details at the molecular level, these responses are dramatically revealed in studies of alcohols diluted to very low number densities in alkane matrices of similar molecular structure. Upper limits of 2–4 ps and 5 ps have also been placed on electron solvation in water and methylamine, respectively, by P. M. Rentzepis and coworkers.

Dilute fluids. When the picosecond dynamics of electron localization and solvation are studied in dilute alcohol–alkane systems, three patterns of different kinetic and spectral behavior can be observed. Each pattern is a function of the average number density (ρ molecules cm^{-3}) of alcohol in the alkane. The patterns are seen in Fig. 2 for n-butanol in n-hexane, where the time-dependent absorptions at 514 nm during and after the ionizing pulse (represented by a horizontal bar) are given for a range of alcohol mole fractions, $6 \times 10^{-3} < \chi_4 < 1$.

In pure n-hexane there is again an instantaneous but weak infrared absorption, but the signal rapidly decays after the pulse. Addition of a small quantity of alcohol to the system does not affect this signal until a critical number density of alcohol is reached: $\rho_c = 2.7 \times 10^{19}$ cm^{-3}, or $\chi_4 = 0.006$. Above ρ_c, increasing ρ slows down the decay until, at 2×10^{20} cm^{-3}, a plateau is seen. No spectral shifts or significant changes in amplitude can be detected until a higher density is attained, whereupon the absorption signal continues to grow after the pulse. By $\chi_4 = 0.16$ the kinetics of the transient absorption and spectral shifts are identical to those observed at $\chi_4 = 0.58$, although the solvation time is about 51 ps in comparison to the 30 ps in pure n-butanol.

Alcohols are associated fluids, although the extent of hydrogen bonding and the distribution of the cluster sizes and molecular geometry vary from liquid to liquid. The clusters coexist in a dynamic equilibrium, and dielectric dispersion studies place relaxation times of 10^{-10} to 10^{-9} s on the aggregation cycles. Upon twofold dilution of a long-chain alcohol in an alkane, there is a significant drop in viscosity. This does not herald the breaking up of the clusters but reflects instead the loss of long-range order through the disentanglement of the alkyl chains. Recent ^{13}C nuclear magnetic resonance studies of the internal mobility of the carbon atom attached to the OH group, which

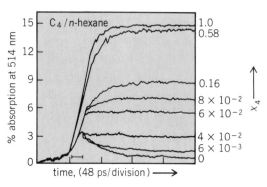

Fig. 2. Profile of electron solvation as a function of alcohol density in the n-butanol–n-hexane system. (*From G. A. Kenney-Wallace and C. D. Jonah, Picosecond spectroscopy of the dynamics of electron solvation in polar fluids, submitted for publication in J. Chem. Phys., 1979*)

is anchored in a hydrogen bond, clearly indicate that small clusters persist until mole fractions $\chi_{ROH} < 0.05$, whereas at $\chi_{ROH} = 10^{-3}$, the systems appear to comprise essentially monomeric alcohol.

Interpretation. The picosecond spectroscopy of electron solvation in these dilute fluids can thus be interpreted in the following way: In pure alkane the quasi-free electron scatters through the liquid, encountering density and configurational fluctuations that are "frozen-in" on this subpicosecond time scale. In sampling the statistical structure of the liquid, the electron can become localized, and this weakly bound state is the origin of the infrared absorption. The subsequent rapid decay is largely the result of ion recombination or electron attachment to impurity molecules. The presence of a very low number density of alcohol molecules does not affect this neutralization channel until the critical density is reached. This threshold ρ_c also marks the onset of clustering in the alcohols to form dimers and trimers, whose capture cross section for low-energy electrons is such as to introduce an effective competition for ion recombination. As the alcohol concentration increases further, the size and local structure of the alcohol clusters change, and since translational diffusion of alcohol molecules between isolated clusters is too slow to play a role at times much less than 10^{-9} s, the cluster must eventually grow to the state at which (at $\chi_{ROH} = 0.16$ in n-butanol) the electrons become localized and solvated in a microscopic structure that simulates the dynamics of a pure alcohol liquid. By estimating the magnitude of the solute-solvent interactions in these systems prior to electron injection, it is possible to predict qualitatively both the critical ρ value and the number density at which the cluster of alcohol molecules appears to behave like a liquid, at least from the microscopic perspective of an electron. Once trapped, the electron induces dipolar polarization and configurational relaxation of the unaligned molecules in the cluster, and this gives rise to the spectral shifts that correlate with rotational diffusion in the liquid. In ethylene glycol, twice the number density of —OH dipoles are available to form structures in the fluid, and the ultrafast solvation time reflects the availability of a chelating type of local structure requiring minimum reorganization upon electron solvation.

Picosecond spectroscopy thus reveals microscopic details of molecular structure and dynamics in liquids which are of fundamental significance to a quantitative understanding of fluids, to the dynamics of chemical reactions in liquids, and to the many processes of scientific and technical importance that occur in the liquid state.

For background information see SPECTROSCOPY in the McGraw-Hill Encyclopedia of Science and Technology.

[G. A. KENNEY-WALLACE]

Bibliography: G. A. Kenney-Wallace, *Acc. Chem. Res.*, 11:433–439, 1978; G. A. Kenney-Wallace and C. D. Jonah, *Chem. Phys. Lett.*, 47: 362–366, 1977; *Newer Aspects of Molecular Relaxation Processes*, Faraday Symposium of the Chemical Society (London), 1977; C. V. Shank, E. P. Ippen, and S. L. Shapiro (eds.), *Picosecond Phenomena*, 1978.

Plant cell

Higher-plant cells are made up of a nonliving portion called the cell wall and a living portion called the protoplast (from the Greek, "first form"). In 1892 J. von Klercker observed that it was possible to separate these two portions. Since that time, botanists have been exploring the possibilities of using isolated protoplasts (Fig. 1) to increase understanding of cellular processes. Isolated protoplasts are now being used in the manipulation of the genetic information of higher plants. This manipulation can be brought about by the fusion of dissimilar isolated protoplasts, by transplantation into protoplasts of cell organelles, and by the incorporation of different kinds of exogenous deoxyribonucleic acid (DNA) into the cytoplasm (the cell body excluding the nucleus).

Mechanical isolation of protoplasts. The isolation of protoplasts requires that the wall be removed from the cell and that the protoplast be stabilized against osmotic rupture. In normal plant cells, the wall not only serves as a protective barrier, but also pushes against the protoplast with a pressure to counteract the outward turgor pressure of the protoplast. If the protoplast is to be isolated from the wall, this turgor pressure must be reduced by placing the cell in an environment with an osmotic pressure approximately equal to that of the cell cytoplasm and sap. Sugars, salts, and sugar alcohols such as mannitol and sorbitol have been used to provide this pressure.

Until 1960 the techniques used to isolate living plant protoplasts were purely mechanical. The tissue was first incubated in a solution of high osmotic concentration. When the protoplasts had plasmolyzed away from the cell walls, the tissue was sliced or chopped with sharp blades to open the cells and release the protoplasts (Fig. 2). Unfortunately, this cutting also produced many small

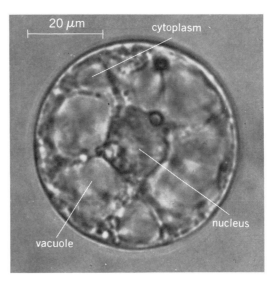

Fig. 1. A single enzymically isolated protoplast from the cortex of a pea root. The spherical structure is characteristic of healthy protoplasts. The nucleus is connected to the bulk of the cytoplasm by strands which traverse the vacuole.

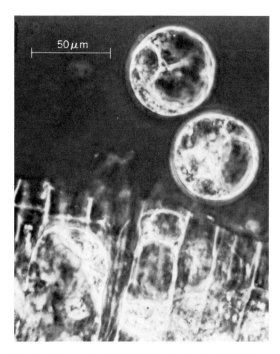

Fig. 2. Two mechanically isolated protoplasts float beside the piece of onion bulb epidermis from which they were isolated. Protoplasts can still be seen in the epidermis, while the empty walls of two cells are visible at the top of the tissue fragment.

fragments of cell wall and large numbers of damaged cells which contaminated the isolated protoplasts. The number of healthy protoplasts which could be isolated by these techniques was very small.

Enzymic isolation of protoplasts. In 1960 E. C. Cocking developed techniques which depended on the action of enzymes for the isolation of the protoplasts of cells from tomato roots. Some mechanical disruption was still necessary to release the protoplasts, but these techniques allowed the isolation of large populations of healthy protoplasts with little debris from the isolation treatment. The enzymes used to isolate the protoplasts are derived from snail intestine or fungal mycelium. The enzymes which are most effective are very impure, containing protein- and nucleic acid-degrading components which are detrimental to the long-term survival of the protoplasts. Most procedures which decrease the concentrations of these detrimental enzymes also decrease the effectiveness of the preparation in cell wall degradation.

Protoplast studies. Early work with protoplasts focused on the synthesis of cell wall material. It is believed that the wall which is synthesized initially in a regenerating protoplast is structurally or chemically different from that of a normal cell. Nonetheless, the large amount of cellulose synthesis triggered by the removal of the original cell wall makes these isolated protoplasts a unique tool in the study of an important biochemical synthesis. The isolated protoplasts have also been used to study the plant cell membranes. It is also possible to rupture the protoplasts by osmotic stress or gentle grinding to release the organelles of the plant cells. This technique is thought by some to be

less damaging to the organelles than other, more mechanical isolation procedures. The isolated organelles can be used to study the physiology or chemistry of these vital cell parts. Many researchers have studied the course of viral infection in isolated plant protoplasts; the technique has been used extensively in the study of tobacco mosaic virus, cowpea chlorotic mottle virus, potato virus X, and cucumber mosaic virus.

Source of isolated protoplasts. Much of the research work with protoplasts has focused on their use as a source of tissue for culture and for genetic manipulation. Since 1960, protoplasts from 21 species of flowering plants have been isolated and successfully cultured. Protoplasts have been isolated from many more species where culture has not yet been successful. Protoplasts have been isolated from every plant organ, from plant tissue cultures, and even from such specialized cell types as microspores. In addition, protoplasts have been isolated from cells of algae, fungi, mosses, and ferns.

Hybrids and cybrids. Although most of the genetic information in a plant cell is found in the nucleus, some is also found in mitochondria and chloroplasts. It is also possible that viral DNA or the isolated strands of genetic information called plasmids can be translated into protein by the synthetic machinery of the cell. Until it was possible to isolate the protoplast from the cell wall, the nonsexual manipulation of this genetic information was only rarely reported. A few experiments have indicated that the wall is not always a barrier to the manipulation of the genetic information of the plant. On the other hand, with the cell wall removed from the protoplast, it has been possible to transfer chloroplasts, mitochondria, viral DNA, isolated higher-plant DNA, and even whole nuclei of higher-plant cells into the cytoplasm of the protoplasts. Such transfers have been used to produce hybrids (called parasexual hybrids because of the mechanisms used in the mixing of the nuclear genetic information) in a number of plant species and cybrids (plants in which the cytoplasmic genetic information is derived from two parents or in which cytoplasmic genetic information is associated with a nucleus with which it is not found in nature).

Selection techniques. The greatest barrier to the wide-scale study of cybrids and parasexual hybrids is the selection of these cells from large populations of cells growing in cultures. Biochemical selection techniques have been used to increase the survival of the hybrids and cybrids in these cultures. Although these techniques have been used to isolate hybrid plants in tobacco, *Petunia*, and *Daucus* spp., as well as cybrid plants of tobacco, they have limited applications because the biochemical traits used in these selection techniques are carried in only a few cell lines. Other selection techniques for these genetically engineered plants are under investigation; it has recently been reported that tobacco-tomato and tobacco-potato hybrid plants have been created with these techniques.

Transfer of genetic information. The mechanisms used to transfer genetic information include the fusion of protoplasts and the stimulation of organelle uptake. Fusions are triggered by agents which act on the membrane of the cell. The most commonly used agents are polyethylene glycol (PEG) and sodium nitrate. PEG has also been used to induce the transplantation of chloroplasts. Liposomes (lipid-bound vesicles) and the plasmolysis of protoplasts themselves have been used to induce organelle transplantation. It has been shown that plasmolyzed cells take up small particles during the shrinking process. Latex spheres, virus particles, bacteria, and organelles have all been incorporated into protoplasts by this technique.

Limitations. Although it is clear that protoplasts can be isolated from any living plant cells, it must be borne in mind that the protoplasts from each tissue will carry with them the genetic information characteristic of their source. Diploid cells will yield protoplasts with the diploid amount of genetic information, while polyploid cells will yield protoplasts with multiple copies of each genetic message. Polyploid protoplasts are usually unsuitable for experiments where growth responses are to be studied or where manipulation of the genetic information is involved.

The engineering of genetic information in the protoplast often results in abnormal developmental responses. These include the rejection of transplanted organelles, chromosomes, or even specific genetic messages. The further development of the protoplasts in culture will be influenced by these responses.

To date, the most successful genetic engineering using protoplasts has been in experiments which attempt to mimic sexual hybridization. This may imply a limit to the results which can be expected from efforts to manipulate the genetic information in a higher plant. The isolation process does not stabilize the genetic composition of the cell. In fact, there is evidence that the protoplast isolation procedures result in gross chromosomal damage in some of the cells. Aneuploid metaphase figures are frequently observed in cultures of cells derived from protoplasts. Nonetheless, the isolation of higher-plant protoplasts provides a powerful tool in the study of many areas of plant genetics and physiology.

For background information *see* CULTURE, TISSUE; CTYOPLASM; CYTOPLASMIC INHERITANCE; PLANT CELL in the McGraw-Hill Encyclopedia of Science and Technology. [C. RANDALL LANDGREN]

Bibliography: E. C. Cocking and J. F. Peberdy (eds.), *The Use of Protoplasts from Fungi and Higher Plants as Genetic Systems*, 1974; J. Reinert and Y. P. S. Bajaj (eds.), *Plant Cell, Tissue, and Organ Culture*, 1977; H. E. Street (ed.), *Plant Tissue and Cell Culture*, 1977; T. A. Thorpe (ed.), *Frontiers of Plant Tissue Culture 1978*, 1978.

Plant disease control

Fungi may become resistant to chemicals used in agriculture to control fungal plant diseases, just as bacteria may become resistant to antibiotics used for control of bacterial diseases in humans. A decade ago, after the introduction of a new type of fungicide, the occurrence of fungicide resistance became a problem in practice. Recent work has contributed to an understanding of the nature of this phenomenon and to the development of methods to cope with it.

Fungicides. Fungicides are chemicals that kill fungi. They have been used on a large scale for about a century for control of fungal plant pathogens that threaten crops. Until recently, only conventional fungicides were used, which provide only superficial protection against attack by pathogenic organisms. These fungicides have caused hardly any problems with respect to development of fungicide resistance. About 1969, however, a new type of fungicide became available. Such a chemical is taken up by the plant and transported in the plant system, and is therefore called a systemic fungicide. Since introduction of the type in agriculture, failure to control disease with compounds that were originally very effective has been reported frequently. It appeared that in these instances the fungi had become less sensitive or resistant to the fungicide.

Origin and nature of resistance. A reduction in sensitivity of a fungus to a fungicide may be caused by genetic or nongenetic changes in the fungal cell. Since a decrease in sensitivity due to nongenetic changes (adaptation) is not stable and is lost again in the absence of the fungicide, it is of little practical importance. Therefore only the reduction in sensitivity due to genetic changes, for which the term resistance is reserved, will be discussed here.

Development of resistance may be the result of: a decreased permeability of the fungal membrane to the fungicide, so that entrance into the cell becomes more difficult; an increased detoxification of the fungicide in the fungal cell before the sites of action are reached; or a change at the sites of action themselves, so that affinity to the fungicide is lost or reduced.

Mutants. In almost all instances where resistance to agricultural fungicides has been analyzed, the changes responsible for resistance appeared to be due to gene mutation. These mutations may occur spontaneously, thus being independent of the presence of the fungicide and not deliberately induced by treatment with a mutagenic agent. The frequency with which resistant mutants appear usually lies between 10^{-4} and 10^{-9}. In some instances, however, resistant cells may be present continuously in the fungal population. They occur at low frequencies, but may be detected even before the fungicide has been introduced. In all of these instances, the effect of the fungicide is mainly selective. Elimination of the sensitive individuals favors development and multiplication of resistant strains. The possibility that the fungicide itself may exert a mutagenic effect, however, cannot be excluded a priori, although it is certainly not the rule.

Type of fungicide. The chance that fungicide-resistant mutants may emerge depends primarily on the type of fungicide. It is very small for the so-called multisite inhibitors, that is, those fungicides which interfere with fungal metabolism at many sites in the cell. Most conventional fungicides, such as the dithiocarbamates and metal toxicants, belong to this type of fungicide. With these, mutation leading to changes at all involved sites is unlikely or impossible, and a significant increase in tolerance due to a change in permeability of the cell membrane or to increased detoxification does not seem to occur often. The situation is, however, different for fungicides that act primarily at a specific site in fungal metabolism. In this instance, mutation in a single gene may already result in resistance. As systemic fungicides probably all are specific-site inhibitors, it is not surprising that it was possible to obtain resistant mutants to most systemic fungicides when tested in laboratory culture experiments.

The development of resistance to a specific-site inhibitor might be compared with the emergence of a new physiological race of a pathogen, which may break the so-called vertical resistance in a crop variety. In both instances, the fungus expresses its ability to adapt to adverse conditions by mutation in a single gene. In the same way, development of resistance to a multisite inhibitor seems just as difficult as breakdown of horizontal resistance of a plant by a pathogen. The latter, considered to be polygenic, is efficient against all physiological races of the pathogen.

The possibility of obtaining fungicide-resistant mutants in the laboratory does not necessarily imply that the use of such a fungicide will lead also to problems in practice. This will depend on the chance that a resistant fungal population can be built up in the field.

Buildup of a resistant population. Disease control with an initially effective fungicide will fail only when the rare fungicide-resistant cells can multiply to such an extent that a large part of the fungal population becomes resistant. Whether this will happen, and how soon, depends on various factors, the most important of which are the fitness of the resistant mutants, the degree of pressure exerted by the fungicide on the fungal population, and the type of disease.

Fitness of resistant mutants. When resistant cells are just as fit and virulent as the original sensitive pathogen, the shift toward a resistant fungal population may occur rapidly because of selection pressure by the fungicide. However, if increased resistance is genetically linked to a decreased fitness and virulence so that resistant strains are always weaker than sensitive ones, the buildup of a resistant population will be seriously hampered or even impossible.

Selection pressure exerted by fungicide. A high and continuous selection pressure upon a fungal population that contains resistant individuals will enhance the shift toward a resistant population. Important in this respect are the frequency of application, the doses used, the persistence of the chemical, and the completeness with which the leaves are covered. Also, the method of application may influence the result; for example, application of a chemical to the soil may result in a continuous uptake and selection pressure.

Type of disease. The type of disease and the nature of the pathogen may influence the speed with which a resistant population may build up. Resistant mutants of a pathogen that sporulates abundantly on aerial parts of a crop may spread more rapidly than those of a slowly growing pathogen with few or no aerial spores. Moreover, if a type of disease is concerned where the pathogen is difficult to eliminate completely by application of the fungicide, competition by the sensitive fungus may counteract the spread of a resistant mutant.

These principles may be illustrated with a few examples. Before the introduction of systemic fungicides, sugarbeet leaf spot, caused by *Cercospora beticola*, was controlled in Greece by conventional fungicides. After the introduction of the systemic fungicide benomyl, almost the complete beet crop there was sprayed frequently with this new fungicide, which caused a high selection pressure. The resistant mutants that appeared possessed a fitness equal or almost equal to that of the sensitive pathogen, and a rapid shift to a completely resistant pathogen population occurred. For that reason, benomyl was replaced by a conventional fungicide. On the other hand, no problems have been encountered with application of benomyl for the control of a foot disease in wheat, caused by *Cerosporella herpotrichoides*. This slow-spreading disease could be controlled with only one or two sprays per season, which sprays exerted a relatively low selection pressure. Another systemic fungicide which caused problems in practice was dimethirimol, which initially provided excellent control of cucumber powdery mildew in commercial greenhouses in the Netherlands. Application to the soil around the plant allowed a continuous uptake by the roots and resulted in a lasting, high selection pressure. Within a year, disease control failed because the fungus developed resistance to the fungicide.

Several other systemic fungicides did not create resistance problems in the field, although resistant mutants were easily obtained in the laboratory. This was the case with some fungicides that interfere with sterol biosynthesis in fungi. It seems that increased resistance to these chemicals is linked to a decreased fitness and virulence.

Detecting and coping with fungicide resistance. When disease control fails with a fungicide that was originally very effective, the possibility should be considered that the pathogen has developed a resistance to the fungicide. In order to prove that this is the case, and that the failure is not due to an inadequate application technique or to other factors, samples of diseased plants should be collected and investigated in laboratory or greenhouse experiments.

Detection methods. Resistance can be detected by growing the fungus on nutrient media containing normally lethal doses of the fungicide, or by inoculation of fungicide-treated plants under controlled conditions. It may then be too late, however, to avoid crop losses in that particular season. It is therefore important to detect an increase in the percentage of resistant fungal cells in a much earlier phase, so that control measures can be changed in time. A regular control (monitoring) for the presence of resistant cells is advisable.

Changing fungicides. When failure of disease control by a fungicide due to development of resistance becomes evident, the user may consider abandoning this fungicide and replacing it with another one. This strategy, which is used by entomologists when insects become resistant to a particular insecticide, can be followed only when other effective chemicals are available. This is mostly not yet the case with respect to systemic fungicides. Moreover, the lifespan of a newly developed fungicide may then become too short in relation to the costs of its development. Therefore, it is important to consider continuation of the use of a fungicide, to which resistance may develop, in such a way that buildup of a resistant population is avoided or drastically reduced, such as by application of two different chemicals, alternately or as a mixture.

Management measures. On the basis of the principles mentioned above, the following management measures may be considered:

1. The use of a chemical to which resistance may develop should be restricted to carefully selected diseases, for example, where buildup of a resistant pathogen population may be expected to be slow or absent in view of the type of disease. The chemical should be used very selectively or not at all against diseases that can be controlled by other chemicals or methods.

2. Use of the fungicide in such a manner that the pathogen is under a continuous high selection pressure should be avoided. Its use should be restricted to critical periods; the doses applied and the frequency of application should be limited to what is strictly necessary; and use of such a fungicide should be avoided if it is the sole compound against a particular disease over a vast area. Further, the method of application should be carefully considered, as some methods may result in a continuous high selection pressure, for example, application of a systemic fungicide to seed, roots, or the soil.

3. Application of two fungicides with a different mechanism of action, used alternatively or in combination, may delay or even prevent the buildup of a resistant population.

These recommendations indicate that it is better to reduce the disease to an acceptable level by restricted use of the chemical than to try to eliminate the pathogen completely. The latter procedure requires a high selection pressure and will, therefore, favor an increase in the number of resistant individuals.

For background information *see* FUNGISTAT AND FUNGICIDE; PLANT DISEASE CONTROL in the McGraw-Hill Encyclopedia of Science and Technology.

[JOHAN DEKKER]

Bibliography: J. Dekker, in R. W. Marsh (ed.), *Systemic Fungicides*, pp. 176–197, 1977.

Plant growth

Trees differ from other woody perennial plants in their larger size at maturity. Research on tree growth is predominantly centered on increasing the yield of stem wood from forests to supply lumber, paper, fuel, and other wood products. Recent research has investigated development of tissues from apical and lateral meristems of trees.

Growth in plants generally means the development of organized structures. Longitudinal growth depends upon cell division in apical meristems and upon subsequent elongation of those cells. Cell enlongation occurs by the incorporation of new matter into the primary cell wall and by resulting adjustments in the water status of the cell. If external water potential is not equal to the water potential of the cell, then water flow occurs until equilibrium is reached. The water potential of a cell is equal to the turgor pressure, that is, the elastic resistance of the wall to expansion, minus the osmotic potential of the cytoplasm. Turgor pressure itself cannot by definition expand the cell. If bondings in the cell wall are loosened, such as occurs in the incorporation of new cell wall material, then the resistance of the cell wall to expansion is decreased, and water is taken up. Cell wall expansion (growth) will occur until new bonds reduce the elasticity of the wall, turgor pressure rises, and water uptake ceases.

Predetermined and free growth. Spring shoot extension in mature conifers is the result of elongation of needles and their associated stem units which were formed the previous year and overwintered as the bud (Fig. 1); this is called predetermined growth. In trees less than 10 years old, free growth, the initiation of needles and stem units and their elongation in the same growing season also occurs. Increased height growth can occur due to free growth (Fig. 1), and can result in differences that determine the relative heights of individual trees for several decades. Free-growth capacity, although often different from the inherent potential for predetermined growth, could determine which individuals attain dominant height as tree crowns grow and begin to compete for light, water, and nutrients. Work on Douglas-fir and western hemlock indicates that controlled-release fertilization at time of planting increases the amount of free growth as well as the number and length of stem units overwintered in the bud.

Annual growth and dormancy. Research has described the annual growth and dormancy cycle in north temperate conifers (Fig. 2, inner ring). Dormancy has often been described in terms of the growth response of the terminal bud to favorable environments (Fig. 2, middle ring), and may be more precisely defined as the period when no cell division occurs in the stem apex. Development of Douglas-fir buds using this criterion is described by the middle ring in Fig. 2. In older trees shoot extension occurs from mid-May until July. In young trees this extends into late August or early September, when environmental conditions allow free growth. Planting of tree seedlings during the late, slow leaf-initiation phase has usually resulted in low survival and poor height growth the following year. Cold storage of seedlings in late October, which reduces their rate of cell division to dormancy conditions a few weeks early, can reduce height growth by 23% the following year. It is apparent that many physiological events occur during this period that affect growth. Soluble sugar levels and the ratio of the hormones auxin and abscisic acid drop sharply during the late leaf-initiation stage— this hormone ratio reaching a low in early December at the onset of dormancy.

Once buds are dormant, they must be chilled (either naturally over the winter or in cold storage) before they will break bud and exhibit normal growth. The chilling effect is not well understood, but some evidence indicates that chilling causes changes in abscisic acid and auxin levels which affect dormancy. Douglas-fir requires an average of 2000 hr and western hemlock 660 hr between 0 and 5°C to "condition" the buds so that they will break normally. Recent research indicates that warming of the soil in the spring could cause increased transport of gibberellic acid from the root to the

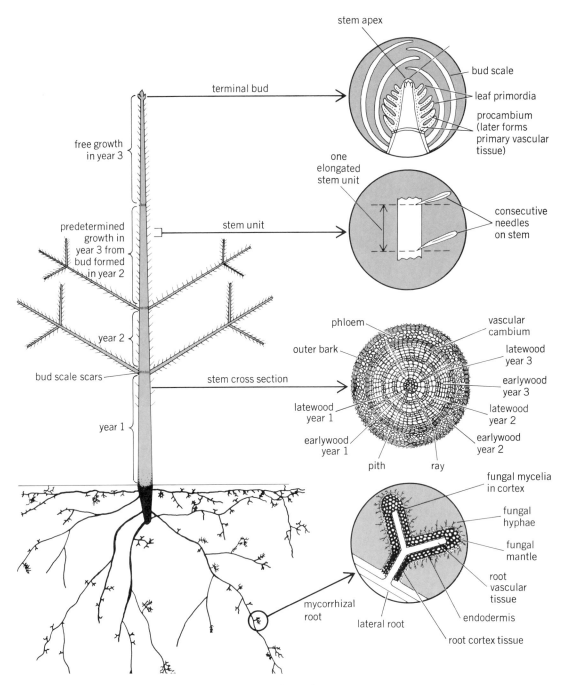

Fig. 1. A 3-year-old Douglas-fir 2 years after outplanting in the forest.

bud, initiating bud break in Douglas-fir and also in eastern cottonwood.

Cambium and vascular tissue. Transport of materials between the root and the shoot occurs upward in the xylem, the water-conducting vascular tissue, and downward in the phloem, the sugar-conducting vascular tissue. Radial transport of materials occurs mostly in ray tissue (Fig. 1). If the bark is removed from an actively growing tree, it usually separates from the wood (secondary xylem) at the vascular cambium. This cambium produces secondary xylem to the inside and secondary phloem to the outside. The cambium encircles the tree just under the bark on the bole, branches, and

main root and lateral roots except just behind the growing tips.

Primary xylem and phloem. New leaves are connected to the secondary xylem and phloem by primary xylem and phloem. Research indicates that in eastern cottonwood certain substances transported from leaves through the vascular system affect the development of new leaves and the primary vascular tissues serving them.

Secondary xylem and phloem. New concepts of growth and development of secondary xylem tissue from the vascular cambium have recently developed. The rate of cell division in the vascular cambium largely determines the width of the annu-

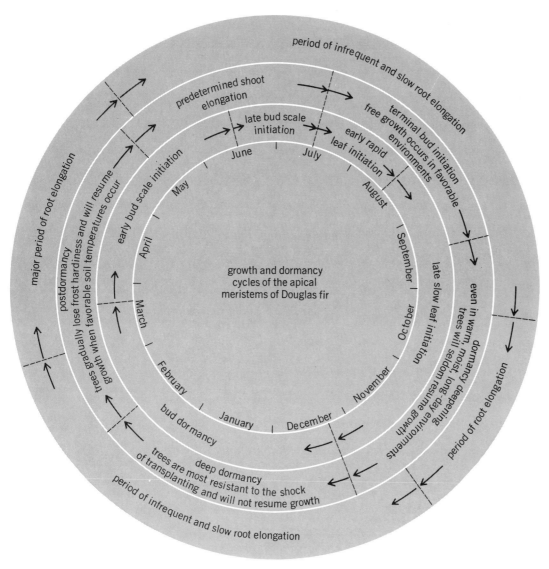

Fig. 2. Phases of tree growth. The inner ring describes vegetative bud development; the middle ring describes growth and dormancy cycles in terms of shoot growth response in a favorable environment; and the outer ring describes periods of root elongation.

al growth ring. Fast-growing trees have higher rates of cell division, and the rate increases considerably more than in less vigorous trees over the growing season. The rates of cell elongation, deposition of the secondary cell wall, and lysis of the cytoplasm to form the mature, dead, water-conducting tracheid or vessel do not vary with time of season or tree vigor. Late in the growing season, increased levels of auxin in the area where cell maturation is occurring may cause a delay in the onset of lysis of the cytoplasm, resulting in a longer period of secondary cell wall formation. The cells which mature during this period have comparatively thick secondary walls and are termed latewood (Fig. 1).

Auxin. In 1936 British researchers proposed that gradients in cambial cell division rate from the top to the base of the tree are caused by parallel gradients of auxin. It was recently demonstrated that while auxin must be present, its concentration does not determine the rate of cambial cell divi-

sion. A possible explanation of earlier observations is that the cambium at different locations responds differently to auxin, depending upon cell division inhibitor levels which vary with the season and distance from the stem apex.

Roots. Radial growth of the vascular cambium of tree roots has been another subject of interest. Reduced xylem development observed above root branching points in red pine could, according to recent Canadian work, be due to an interaction between hormones and assimilates moving in the phloem, and to unequal distribution of nutrients, hormones, or water itself moving upward toward the shoot in the main and lateral root xylem. Work on roots of Sitka spruce in Scotland indicates that nutrients moving in the xylem could cause these effects through control of cambial growth rate. Canadian research on red pine roots indicates that the cambium is complete in the stem but in the root shows an increasing occurrence of discontinuities with age. Distance of a root from the stem has a greater effect on occurrence of discontinuities than depth in the soil.

In Sitka spruce seedlings, root growth is stimulated only in roots to which nutrients are applied. Internal nutrient concentrations increase in unfertilized roots until they are only slightly reduced relative to those of fertilized roots. This indicates that nutrients which may be internally translocated to unfertilized roots have little effect on growth.

Root extension growth of Douglas-fir is usually most active in the weeks just prior to bud burst, then sharply curtailed while shoot elongation occurs (Fig. 2, outer ring). In north temperate conifers, root growth is inhibited if the chilling requirement of the buds is not fulfilled.

Excavation of the roots of trees in forests has shown frequent grafting between roots of neighboring trees. Callusing over of the cambial area on freshly cut stumps in a forest is attributed to these root connections to intact trees. Several substances have been shown to be transported through such grafts. The effect of grafting on competition for water and nutrients within a stand of trees is not well understood.

The roots of many plants, including trees, form mycorrhizae, symbiotic relationships with specific fungi. Mycorrhizal fungi increase the ability of the tree to take up phosphorus. Some mycorrhizal fungi allow tree seedlings to withstand highly acid or alkaline soils, drought, and high surface temperatures.

There is a strong interaction between root and shoot growth of trees. This could be due to the root surface area affecting water and nutrient uptake and the shoot and leaf surface area affecting water loss, nutrient usage, and photosythate production. Recent research indicates that hormones produced in the root could control the development of the shoot, and as discussed above, the converse could also be true.

Conclusions. There have been substantial advances in knowledge of the nature of tree growth. Such work will continue to be the basis for the development of cultural practices necessary to increase the rate of utilizable wood production from intensively managed forests.

For background information *see* PLANT GROWTH in the McGraw-Hill Encyclopedia of Science and Technology.   [WILLIAM C. CARLSON]

Bibliography: H. G. Burstrom, *Amer. J. Bot.*, 66(1):98–104, 1979; M. G. R. Cannell and F. T. Last (eds.), *Tree Physiology and Yield Improvement*, 1976; P. R. Larson, *Amer. J. Bot.*, 66(1): 1332–1348, 1979; J. N. Owens, M. Molder, and H. Langer, *Can. J. Bot.*, 55(21):2728–2745, 1977.

Pluto

During the morning of June 22, 1978, a satellite of Pluto was discovered by James W. Christy on photographs of the planet at the U.S. Naval Observatory in Washington, DC. The discovery was unexpected because previous observations had led to the conclusion that a satellite of detectable size around Pluto was extremely unlikely. On June 23 the orbital period of the satellite was deduced after examination of older photographs in the Naval Observatory's collection. Knowledge of the satellite's orbital period allowed the first accurate determination of Pluto's mass. The result obtained that same day by Robert S. Harrington of the Naval Observatory was that the total mass of the Pluto/satellite system is 0.002 times the mass of the Earth.

The discovery was announced by Naval Observatory Superintendent, Captain Joseph C. Smith, on July 6 after two postdiscovery observations of the satellite verified the orbital period and after intense analysis of all previous Pluto observations had rendered alternative explanations highly improbable. The name selected for the new satellite by its discoverer was Charon.

The consequences of the satellite's discovery for scientific research can be placed in three categories. First, the knowledge of the mass and orbit of Pluto/Charon allows a more specific description of the properties of both objects to be deduced from existing observations. Second, intensive new observations will be made during the next few years. Third, because Pluto/Charon's extremely low mass cannot explain suspected deviations in the orbits of Uranus and Neptune, speculation is growing about the possible existence of a tenth planet.

Prediscovery observations. Pluto itself was discovered in 1930 by Clyde Tombaugh at the Lowell Observatory in Flagstaff, AZ. Observations since then have established that its orbital period around the Sun is 248 years and that its distance from the Sun varies from 30 to 49 astronomical units. (The Earth's orbit has a radius of 1 AU = 1.496×10^{11} m.) It is also known that Pluto's brightness varies by 10–20% in a regular way with a period of 6.387 days, which is due to rotation of the planet. The spectrum of the light from Pluto contains an absorption feature at a wavelength of 1.7 μm which, along with other information about the outer solar system, suggests a surface of methane ice.

Beyond these observations, data have been difficult to obtain. Because of its great distance from Earth and Sun, never less than 29 AU, Pluto is more than a million times fainter than the planet Mars. The disk of Pluto has never been seen, being about 10 times smaller than the blur image that always occurs in the light which passes through the atmosphere.

Difficulty of discovery. Since 1965, approximately 50 photographs of Pluto have been taken with the 155-cm Astrometric Reflector at the Naval Observatory's Flagstaff Station (only a few miles from the discovery site of Pluto at Lowell Observatory). These photographs are obtained to measure the position of Pluto in its orbit with improved accuracy. (Such photographs, always on glass plates instead of film, are usually referred to simply as plates.) The presence of Charon is obvious in a postdiscovery photograph of July 2, 1978, and shows faintly on the discovery plates of April 13 and May 12, 1978. Of the remaining 50 plates, 40 reveal nothing, because the blur image caused by the light's passage through the atmosphere is larger than the Pluto-to-Charon separation. Five plates taken in better observing conditions in 1965 and 5 more in 1970 reveal only indicative elongations of the blur image which can be used to estimate the orientation of the underlying Pluto-to-Charon separation. Thus, less than 1 plate in 10 was taken when the atmosphere was still enough to reveal Pluto's close companion.

Interpretation of orbit. Although the image of Charon is always blended with that of Pluto, approximate measurements of the angular separation are possible. Harold D. Ables of the Flagstaff Station measured the separation to be 0.84 arc-second; this observation was made at a time of maximum separation in the Pluto/Charon orbit. At Pluto's distance of 30 AU, this angle corresponds to a separation of 18,000 km. On the day following his discovery, Christy measured the position angle of the elongation of Pluto/Charon on plates exposed in 1965 and 1970. Because the angle of elongation progressed through a full rotation in about 6 days in 1970, he hypothesized that the orbital period is identical to the known brightness variation period of Pluto's light, which is 6.387 days.

Harrington then derived orbital elements and computer-processed the position angle measures, verifying the 6.387-day period to be exactly that which was required to explain the measures. Harrington went one step further. The 10–20% light variation could not be explained by the presence of Charon. In fact, the original interpretation had to be retained: Pluto's light variation is due primarily to the rotation of Pluto itself, that is, is due simply to brightness variations on Pluto's surface. Why is the rotation period of Pluto identical to the orbital period of Charon? Harrington knew that the closeness of the two bodies, only six Pluto diameters, was more than sufficient to produce strong tidal interaction. He thus hypothesized two probable results of a billion years of tidal friction: the orbit of Charon is circular, and the orbit and the rotation of both bodies are synchronous. Thus, Pluto and Charon are thought always to keep their same faces toward each other in their mutual revolution.

Deduced characteristics of Pluto/Charon. Although the disks of Pluto/Charon can not be resolved through the Earth's atmosphere, analysis of the light variations yields further information. Astronomers D. P. Cruikshank, C. B. Pilcher, and D. Morrison have determined that the spectrum of the light from Pluto (to which Charon contributes 20%) indicates a surface of methane ice which is expected to reflect roughly 50% of the Sun's light. This albedo of 0.5, and Pluto/Charon's apparent brightness, leads to a diameter of 3000 km (similar to that of the Earth's Moon) for Pluto, and a diameter of about 1500 km for Charon. The orbital period of 6.387 days and separation of 18,000 km require a total of 0.002 Earth mass. These sizes and mass imply an average density for Pluto/Charon of close to 1.0 g/cm³ (like that of water). This density is similar to that of the other satellites and planets in the outer solar system.

The uniqueness of Pluto/Charon lies in the large relative size of the satellite (one-half the diameter of the primary body), in the small relative size of the separation (6 Pluto diameters and 12 Charon diameters), in the exact synchronicity of revolution and rotation, and in the fact that Pluto/Charon is, in both size and mass, the smallest planet in the solar system.

Future observations. Although resolution of the disks of Pluto and Charon may be accomplished by the Large Space Telescope in 1984, much is to be gained by intense observation of Pluto/Charon prior to that date. In particular, because the orbit lies almost in the line of sight to Earth, eclipses are expected to occur within the next few years. Such eclipses would take place every 3 days for about 5 years. Duration of eclipse in the center of that period would be about 5 hr, depending on the exact size of both bodies. Studies of the light variations during eclipse and of the length of each eclipse will allow detailed mapping of both surfaces.

Pluto was discovered in 1930 near a position predicted by Percival Lowell from an analysis of alleged residuals between the predicted and observed positions of Uranus, after allowing for perturbations of other planets, including Neptune. Pluto's mass, however, is now known to be more than a hundred times too small to have produced those residuals. Consequently astronomers now regard the discovery of Pluto to be an accidental result of the prediction. Some astronomers now suspect such residuals in the orbits of both Uranus and Neptune. This raises the possibility that the real cause – "Planet X" – of the deviation remains to be discovered.

For background information see PLUTO; SATELLITE (ASTRONOMY) in the McGraw-Hill Encyclopedia of Science and Technology.

[JAMES W. CHRISTY]

Bibliography: L. E. Andersson and J. D. Fix, *Icarus*, 20:279–283, 1973; J. W. Christy and R. S. Harrington, *Astron. J.*, 83:1005–1008, 1978; D. P. Cruikshank, C. B. Pilcher, and D. Morrison, *Science*, 194:835–837, 1976.

Polar-coordinate navigation systems

A new system called wide-aperture digital VOR has been developed which provides bearing indications 10 times more accurate than present VOR (very-high-frequency omnidirectional range) facilities. This accuracy would enhance the practicality of using the VOR for area navigation techniques in high-density terminal areas.

Objectives. Aircraft navigation throughout the United States and in many parts of the world is based on guidance information supplied by VOR facilities, usually in conjunction with co-located distance-measuring equipment (DME). The primary use of VOR/DME has been in providing guidance along designated routes for aircraft flying along radials from one VOR to the next. This practice is not desirable because the resulting flight path is seldom a direct path from takeoff point to destination. Furthermore, aircraft tend to bunch together over the VOR sites.

Area navigation (RNAV) techniques can eliminate these disadvantages by using the area coverage of the VOR/DME station. The standard VOR/DME is adequate for RNAV operation for enroute and low-density terminal areas. VOR system accuracy is considered to be ±4.5° on a 95% probability (2-sigma deviation) basis, but the accuracy requirements for high-density terminal areas are more severe and a reduction in the total error of two or three times is considered necessary to achieve the needed capability.

The objective of the wide-aperture digital VOR is to provide a VOR, compatible with present VOR facilities, that would reduce siting errors to one-tenth of those commonly encountered at present, provide 10 times more accuracy, and utilize digital techniques to simplify signal processing. Development system tests, indicating that the system ac-

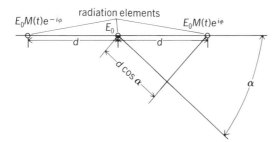

Fig. 1. Basic three-element interferometer. *(From A. S. Palatnick, Wide-aperture digital VOR, IEEE Trans. Aerosp. Electron. Syst., AES-14:853–865, November 1978)*

curacy is 0.22° (2-sigma deviation), show that these goals were achieved.

Principles of operation. Consider a transmitting station consisting of three radiation elements equally spaced on a straight line as shown in Fig. 1. The center element radiates a constant-frequency carrier wave, while the outer two elements radiate a series of pulses at the same frequency. The radio-frequency phase of the signal applied to the right element is advanced by an amount ϕ, and the phase of the signal applied to the left element is retarded by the same amount. Thus, if E_0 is the carrier wave radiated by the center element, the signals radiated by the outer elements are $E_0 M(t) \exp(i\phi)$ and $E_0 M(t) \exp(-i\phi)$, where $M(t)$, the modulating function, is a string of pulses. Also, assume that the receiver is located at a distance that is large compared to the baseline dimensions. The angle between the line of sight and the baseline is α, the distance between the center element and each of the outer elements is d, and the wavelength of the carrier frequency is λ. Then, at the receiver, the signal from the right element will be advanced by an amount Θ, given by Eq. (1), due to

$$\Theta = (2\pi d/\lambda) \cos \alpha \qquad (1)$$

geometry, and further advanced ϕ, due to the electrical phasing. The phase of the signal from the left element will be equal and opposite. The total signal received is given by Eq. (2).

$$\begin{aligned} E_R(t) &= E_0[1 + M(t)(\exp i(\Theta + \phi) \\ &\qquad + \exp - i(\Theta + \phi))] \\ &= E_0[1 + 2M(t)\cos(\Theta + \phi)] \end{aligned} \qquad (2)$$

If the sum $\Theta + \phi$ is between -90 and $+90°$, the cosine is positive and the envelope will be modulated by a series of positive pulses. On the other hand, if the sum $\Theta + \phi$ is between 90 and 270°, the envelope will be modulated by a series of negative pulses.

If ϕ is varied from zero to π radians, depending on the value of Θ, a series of positive pulses followed by a series of negative pulses, or the reverse sequence, may be received. When positive pulses are received first, at the instant the positive pulses vanish, the sum $\Theta + \phi$ is 90° plus or minus some multiple of 360°. Similarly, if negative pulses are received first, the sum $\Theta + \phi$ is 270° plus or minus some multiple of 360° at the instant the modulation disappears. If ϕ is known as a function of time, then Θ can be determined with a possible ambiguity of a multiple of 360°. By varying both ϕ and the

pulses at a uniform rate, the measurement of Θ, and thus $\cos \alpha$, can be made by merely counting the number of positive and negative pulses and noting which came first.

System of interferometers. The wide-aperture digital VOR is based on a system of crossed wide-baseline interferometers that measure the angles between the line of sight and each of the two baselines. Interferometer accuracy is directly related to the baseline spacing. If the baseline exceeds one wavelength, ambiguities exist which must be resolved through the use of additional interferometers, including one with a baseline of less than one wavelength. To eliminate the possibility of ambiguity resolution errors, due to multipath or instrumentation errors, the system uses four interferometer pairs, with a spacing ratio of 4 to 1 between adjacent pairs, on each baseline. The inner pair spacing is 0.5λ and, therefore, the outermost pair spacing is 32λ.

Bearing computation. Because the geometry of a pair of crossed interferometers does not have radial symmetry, the basic system measurements of the direction cosines must be further processed to obtain the required azimuth angle.

The system antennas are placed along two orthogonal baselines, with one in a north-south direction as shown in Fig. 2. The coordinates are R, A, and E, respectively slant range, the azimuth measured clockwise from north, and the elevation above the horizon. Thus, the following relations hold: $X = R \sin A \cos E$, $Y = R \cos A \cos E$, and $Z = R \sin E$.

By definition, the direction cosines are: $X/R = \cos \alpha$, $Y/R = \cos \beta$, and $Z/R = \cos \gamma$, where α and β are the angles between the line of sight and each of the two baselines, and γ is the angle between the line of sight and the vertical.

The azimuth angle A is obtained as follows: $\tan A = X/Y = \cos \alpha/\cos \beta$ and the azimuth angle $A = \tan^{-1}(\cos \alpha/\cos \beta)$. The final calculation for azimuth angle is independent of station constants, since these terms appear in both the numerator and denominator and cancel.

Ground station equipment. The ground station (Fig. 3a) contains a transmitting antenna array, a radio-frequency generator unit, and a control unit. The radio-frequency transmitter provides a continuous-wave signal that is split into two lines. One of these represents the carrier and the other the

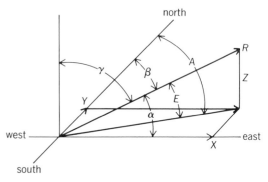

Fig. 2. Geometry of wide-aperture digital VOR system. *(From A. S. Palatnick, Wide-aperture digital VOR, IEEE Trans. Aerosp. Electron. Syst., AES-14:853–865, November 1978)*

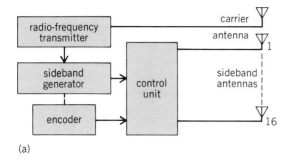

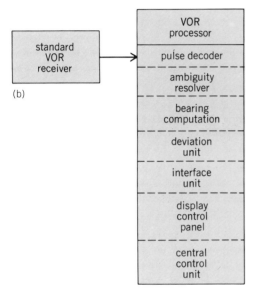

Fig. 3. Overall block diagram of wide-aperture digital VOR system. *(a)* Ground station. *(b)* Airborne equipment.

sideband. The carrier feeds the center antenna, while the sideband signal feeds a goniometer in the sideband generator. The goniometer is driven at a constant speed and provides the sideband signals that rotate in phase opposition. The sideband signals are routed to the control unit, which pulse-modulates the radio-frequency signal and sends it to the appropriate interferometer antenna pair. A digital shaft encoder, driven in synchronism with the goniometer, generates the related pulse-modulation control signals.

The transmitting antenna array contains 17 Alford loop antennas, the central carrier antenna, and the 16 sideband antennas. Each of the eight interferometers is energized successively by a set of pulse and phase shifts. The four transmissions on each axis permit complete resolution of ambiguities and the precise calculation of the azimuth angle with respect to the station.

Aircraft equipment. The airborne units (Fig. 3*b*) are the standard VOR receiver and the VOR processor. The detected audio output of the conventional VOR receiver feeds the pulse decoder, which determines the presence of pulses and their sign, positive or negative. The ambiguity resolver board counts the number of positive and negative pulses and temporarily stores the counts. Upon the completion of the transmissions from all interferometer pairs, the ambiguity resolver extracts the stored counts and processes them to obtain the

direction cosines for each baseline. The direction cosines are fed to the bearing computation control units, which compute the tangent of the azimuth angle, and finally the azimuth angle. This angle is presented as a direct numeric readout.

The deviation module obtains the difference between the omnibearing selector (OBS) command and the actual bearing. The interface provides the necessary operations to tie in the display control panel and convert digital data to analog format for all normal VOR display requirements. The central-control timing module generates the control signals for the transfer and computation cycles in the processor.

For background information *see* POLAR-COORDINATE NAVIGATION SYSTEMS; RADIO RANGE in the McGraw-Hill Encyclopedia of Science and Technology.   [ALBERT S. PALATNICK]

Bibliography: AIL, Cutler-Hammer, *Wide Aperture (Digital) VOR*, Fed. Aviat. Admin. Rep. FAA-RD-76-224, August 1976; A. S. Palatnick, *IEEE Trans. Aerosp. Electron Syst.*, AES-14:853–865, November 1978.

Polarization of nuclei

The properties of nuclei are influenced by the fact that nuclear interactions depend strongly on angular momentum and spin. It is therefore experimentally useful to control the spins of nuclei and to study this dependence. Basic research involving polarized nuclei advanced considerably in the 1970s.

If the quantized magnetic substates (spin orientations) of a collection of nuclei are predominantly in a given direction, the nuclei are said to be polarized. Polarization observables of a beam or target can be characterized by several parameters: vector polarization, tensor polarization, and analyzing powers. Vector polarization has three components; for example, along a quantization axis z it has a component P_z given by Eq. (1), where F_+ and F_-

$$P_z = F_+ - F_- \qquad (1)$$

are the fraction of the particles in magnetic substates parallel and antiparallel to that axis. For spin $\frac{1}{2}$ particles, such as protons (p), neutrons (n), tritons (^{3}H or t), and helions (^{3}He), P_z completely describes the polarization with respect to the z axis. Tensor polarization is required for particles with spin $s > \frac{1}{2}$. For $s = 1$, the simplest tensor component P_{zz} is given by Eq. (2), with F_0 the frac-

$$P_{zz} = 1 - 3F_0 \qquad (2)$$

tion of particles in magnetic substate zero (spin orientation transverse to the z axis). Analyzing powers indicate the effect on scattering cross sections of changes in the polarization of the beam or target.

Polarized beams. Charged-particle polarized beams are produced in ion sources and then accelerated to the required energy. Atomic-beam sources use Stern-Gerlach separation in an inhomogeneous magnetic field of magnetic substates in a neutral atomic beam. The atoms emerge polarized in electron spin, then undergo hyperfine transitions in a radio-frequency field to produce the desired nuclear polarization. Finally, the atoms are ionized and injected into an accelerator. Beams of protons, deuterons (d), ^{6}Li, ^{7}Li, and ^{23}Na have been

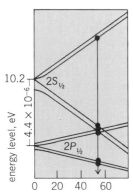

Fig. 1. Curves of energy levels for the eight nuclear magnetic substates belonging to the $2S_{1/2}$ and $2P_{1/2}$ atomic states of hydrogen. The arrow joins those magnetic substates which can be depopulated to produce a polarized beam of atoms in the uppermost state.

polarized in atomic-beam sources. Lamb shift sources produce polarized beams from one-electron systems in their $2S_{1/2}$ excited state. In zero magnetic field the $2S_{1/2}$ and $2P_{1/2}$ states are separated by the Lamb shift. In an external magnetic field which alters the energy levels of these states through the Zeeman effect (Fig. 1), the states can be mixed by magnetic and electric fields, and selected nuclear magnetic substates can be depopulated, leaving the remaining atoms in a state with large nuclear polarization. The Lamb shift scheme is used for beams of p, d, t, and ^{3}He nuclei.

Polarized neutron beams can be produced from nuclei by using spin-dependent nuclear reactions. Those reactions commonly used for production of energetic neutrons, $d(d,n)^3$He and $t(d,n)^4$He, are very strongly spin-dependent, so that emerging neutron beams are often significantly polarized, and enhancement of neutron polarization is achieved by initiating these reactions with polarized deuteron beams.

Polarized targets. For nuclear interactions that depend on the spin of the target nucleus, a polarized target greatly enhances the effects of spin dependence. Adaptations of room-temperature polarized-beam techniques are being used at the University of Wisconsin for gas-jet targets of polarized hydrogen, while at the University of Hamburg very thin targets of polarized ^{6}Li are produced similarly. Other researchers are using optical pumping to polarize ^{3}He nuclei.

Cryogenic target-polarization schemes are based on the fact that in thermal equilibrium at absolute temperature T, the populations $N(m)$ of allowed magnetic substates m are related by Eq. (3), where

$$N(m+1)/N(m) = \exp{(\mu B/skT)} \qquad (3)$$

μ is the nuclear magnetic moment, s the nuclear spin, B the magnetic field at the nucleus, and k Boltzmann's constant. Typically $B > 30$ kilogauss (3 tesla), $T < 50$ millikelvins is required to produce significantly different populations, and therefore polarization. This is achieved by using superconducting magnets and dilution refrigerators. In high-energy physics, polarized proton and deuteron targets are prepared by exposing the target material in an organic host to microwave radiation at an electron-spin resonance frequency, thus producing the desired polarization. The polarization is frozen in by using temperatures of about 50 mK. Polarization can also be transferred to nuclei in a spin refrigerator which uses rotation of magnetically highly anisotropic materials at about 1 K.

Symmetry principles. Polarized nuclei are being used in elegant experiments testing the nuclear interaction for invariance under the basic symmetry operations of parity (reversal of space coordinates) and time reversal, and for charge symmetry. Parity conservation has been tested at the Los Alamos and Argonne national laboratories and at the Swiss Institute for Nuclear Research by placing limits of order parts per million or less on the difference between total cross sections in proton-proton elastic scattering for beams polarized parallel and antiparallel to the direction of motion. The small differences are consistent with theories in which parity nonconservation arises from the presence of weak-interaction currents in nuclei.

Time-reversal symmetry in the weak interaction has been studied at Princeton for the β-decay of polarized ^{19}Ne. The very few events which appear to violate time-reversal symmetry can be attributed to additional electromagnetic effects which, however, are invariant under time reversal.

Charge symmetry in the nuclear interaction can be investigated only in the presence of the symmetry-breaking Coulomb interaction, except for neutron-proton scattering. If charge symmetry holds, the analyzing powers for protons and neutrons in the reaction $p(n,n)p$ with polarized beam and target should be equal. Experiments to test this are under development at Indiana University and at Tri-Universities Meson Facility in Vancouver.

Spin dependence in nuclear interactions. Polarization data greatly aid in interpreting the angular momentum dependence of the nucleon-nucleon interaction. Very accurate data from Triangle Universities Nuclear Laboratory for neutron-proton scattering with polarized neutrons, using the apparatus shown in Fig. 2, show that the nucleon-nucleon interaction is still not known definitively.

Inside a nucleus, the potential in which a nucleon moves contains a spin-orbit part, written as $V_{so}(r)$l·s, where r is the nucleon-nucleus separation, l is the relative orbital angular momentum, and s is the nucleon spin. Spin-orbit coupling has been elucidated by use of polarized beams of neutrons and protons in elastic-scattering experiments. Spin-spin terms in the nucleon-nucleus interaction are studied by scattering polarized nucleon beams from polarized targets and measuring the change in scattering cross section with rever-

Fig. 2. Apparatus for measuring the analyzing power for neutron-proton scattering. Analyzing power is determined from the left-right asymmetry in polarized neutron scattering from the cylindrical target (center) into the scintillation detectors at left and right. (*W. Tornow et al., Phys. Rev. Lett., 39:915–918, 1977; photograph by E. N. Mitchell, University of North Carolina at Chapel Hill*)

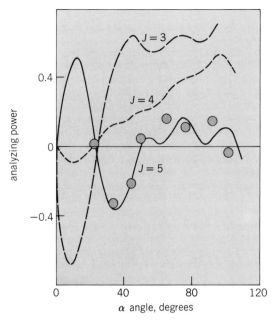

Fig. 3. Vector analyzing power as a function of α angle relative to the incident deuteron beam direction for a ^{28}Si $(d,\alpha)^{26}$Al reaction with $L = 4$. Curves give theoretical predictions for the three possible values of J. Circles give experimentally observed values, indicating that $J = 5$. (From E. J. Ludwig et al., Phys. Rev. Lett., 40:441–444, 1978)

sal of spin orientation. Nuclear research groups at the universities of Stanford, Tokyo, Groningen, and Hamburg have made such studies for targets of ^{59}Co and ^{165}Ho. Since the spin-spin interaction is apparently about 100 times weaker than the average potential in the nucleus, the experiments are technically very challenging.

Nuclear shapes. A nucleus with spin $s > \frac{1}{2}$ usually has a nonspherical shape. For the deuteron, one can investigate those parts of the neutron-proton relative motion that are nonspherical, the D state, by studying tensor analyzing powers for (d,p) reactions at bombarding energies well below the Coulomb barrier. If the deuteron spin is perpendicular to the direction of motion, the neutron in the deuteron approaches the nucleus more closely and is more likely to be captured by it than when the spin is along the direction of motion. Groups at the universities of Wisconsin and Birmingham have used this to give new information on the deuteron. Similar studies of (d,t) and $(d,^3\text{He})$ reactions with polarized deuteron beams have revealed nonspherical components of t and ^{3}He.

Tensor-polarized ^{6}Li and ^{7}Li have been shown at the Max Planck Institute for Nuclear Physics in Heidelberg to have much different analyzing powers at bombarding energies near the Coulomb barrier, since ^{7}Li is very nonspherical and its scattering is therefore sensitive to spin alignment, but ^{6}Li is nearly spherical and is therefore insensitive to alignment. Fusion and fission of heavy nuclei are expected to be sensitive to their orientation, so that use of polarized heavy nuclei will produce better understanding of the fission process.

Nuclear structure. An important quantum number of a nuclear energy level is the total angu-

lar momentum J. For example, the (d,α) reaction removes a neutron-proton pair from the target nucleus. For a given orbital angular momentum L of this pair about the rest of the nucleus, J usually has three possible values: $J = L - 1, L$, or $L + 1$. The analyzing power of a vector-polarized deuteron beam shows a clear distinction between the three possibilities (Fig. 3). Such studies are important in understanding correlations between nucleon motions in nuclei.

For background information see ANGULAR MOMENTUM (QUANTUM THEORY); ION SOURCES; NUCLEAR REACTION; NUCLEAR STRUCTURE; NUCLEAR TARGETS, POLARIZED; QUANTUM THEORY, NONRELATIVISTIC; SCATTERING EXPERIMENTS, NUCLEAR; SPIN (QUANTUM MECHANICS); SYMMETRY LAWS (PHYSICS) in the McGraw-Hill Encyclopedia of Science and Technology.

[W. J. THOMPSON]

Bibliography: A. D. Krisch, Sci. Amer., 240: 68–80, 1979; W. J. Thompson and T. B. Clegg, Phys. Today, 32:32–39, 1979; W. Haeberli (pp. 151–191) and by P. Catillon (pp. 193–212) in Nuclear Spectroscopy and Reactions, Part A, 1974.

Printing

Although many new products and developments were introduced during 1979, the areas of most active interest were: (1) small, lower-cost electronic scanners for color separation and correction, with much of the sophistication of a large, higher-cost scanners; (2) page makeup with advanced photocomposition systems and electronic scanners; (3) dry-output phototypesetters; (4) emulsion-type inks for simplified conversion of letterpress newspaper presses to lithographic operation; and (5) systems for ink setting and drying. These developments are described in this article.

Electronic scanners for color reproduction. Making color separations using electronic scanning techniques has been commercially acceptable since 1950, when Printing Developments Inc. introduced the first commercial electronic color-separating scanner. Many improvements in electronic scanning have been successfully introduced to the printing industry since that time.

The more sophisticated, contemporary electronic scanners are capable of producing, in one operation, completely color-corrected, enlarged or reduced, cropped, and properly screened sets of separations from original copy in a matter of minutes. In many cases, the quality of these separations far exceeds that achieved by conventional photomechanical techniques. This capability, coupled with speed, makes electronic scanners highly cost-effective pieces of production equipment, even though their current price can reach as high as $375,000.

Because of this high initial cost, two areas of research have received much attention from scanner manufacturers in recent years. It has become obvious that a less expensive but still productive electronic scanner would find excellent market acceptance. Second, it has long been known that one of the largest expenses in the preparation of complete pages, including color copy, is incurred in assembling the page so that it is ready to go to the plate or cylinder maker.

In 1979 several low-cost ($59,000 to $150,000) scanning systems were introduced, as were some electronic composition systems which allow the electronic makeup of full-color pages. These systems employ electronic scanners, color video terminals, and contemporary interactive computer technology. In addition, Printing Developments Inc. introduced a unique modular scanning system.

Four scanning systems which fall into the low-cost category are discussed here: Dainippon Screen of Kyoto, Japan, introduced two scanners—the SG-100 and the SG-601; K. S. Paul of London, the Lino Scan 3040; and R. Hell of Kiel, Germany, the model DC-299. The specifications of the DC-299 (Hell) are shown in Table 1. Each of these machines has interesting and distinct features.

Dainippon Screen scanners have been produced and sold in Japan for some years and were only recently introduced into the United States. The specifications for both the SG-100 and the SG-601 are given in Table 1. The SG-100 is a desktop design, whereas the SG-601 is a conventional design. Like the Hell DC-299, the SG-601 can produce either continuous-tone or contact-screened separations.

The SG-601 uses a HeNe laser as an exposing light source and is the only low-cost scanner in which a laser is used for writing. The laser light is red, and therefore requires panchromatic film. This film requirement means loading the scanner in total darkness—with no safe light. However, there are two possible solutions to this problem: Film makers may develop film which could be used with something besides a conventional red safe light. A more logical solution would be to replace the HeNe laser with an argon ion laser (blue light).

K. S. Paul has been associated with scanning technology almost since the beginning of electronic scanning. Its Lino Scan 3040 is described in Table 1. This scanner has a special color-analyzing read head which uses only one photoreceptor, whereas other scanners use one for each of the additive primary colors of red, green, and blue. The most innovative aspect of this scanner is its color computer, which has five preset color computations. If it is successful, this will be a very useful method of increasing scanner productivity by eliminating some of the scanner set-up time.

Electronic page composition. With the rapid advance of digital computer technology, and in particular, the rapidly decreasing cost of electronic data storage, it has become economically and technologically feasible for graphic arts scientists to attempt high-quality color page composition.

Several systems for newspaper-quality graphics and text have been available for some time. Similar machines are also in use for real estate listing systems. These should not be confused with the new high-quality color composition systems. Another very important aspect of these composition systems is their potential for greatly reducing, and in some cases completely eliminating, the need for intermediate photographic films. With the increasing cost of silver halide film, this aspect of these systems will be of growing importance in the future.

By 1979 at least four color composition systems had been demonstrated: one system from the Hell organization, one from Sci-Tex, an Israeli company, one from Crosfield in Great Britain, and one from the Comtal Corporation of Pasadena, CA. All have significant differences. Three of the systems are totally interactive; that is, the operator can manipulate images on a computer-controlled color video terminal, in real time, changing position, cropping, and modifying colors. The next decade will reveal the full extent to which this technology will be accepted by the printing world.

It should be kept in mind that because of the ongoing rapid development of these page composition systems, it is possible and even probable that some of the characteristics of the systems described herein will be changed by 1980. This discussion is, nevertheless, representative of the state of affairs in this rapidly advancing technology during 1979.

Crosfield Magnascan 570 System. This is a very interesting system, although it is not truly interactive. It consists of a modified Crosfield Magnascan 550 color scanner, a digital computer, magnetic disk storage units, an electronic layout table, and a monochromatic video terminal.

Its operation is simple. Color work, as well as text, is scanned into the system via the scanner. This inputs the material into the system and produces a film to final size. This film, along with any other film that may have been produced, is placed on the electronic layout table in its proper position on a page layout, which is also on the table. The

Table 1. Specifications for four low-cost color scanning systems

Function	DC−299	SG−100	SG−601	Lino Scan 3040
Scanner type	Rotary drum−helical scan, input and output on a common axis	Rotary drum−helical scan, input and output on a common axis	Read and write on common-axis rotary drum−helical scan	Rotary drum−helical scan, read and write on common axis
Input format	20 × 24 in. (508 × 610 mm)	10 × 14 in. (254 × 356 mm)	11 × 14 in. (279 × 356 mm)	8 × 10 in. (203 × 254 mm)
Output format	20 × 24 in. (508 × 610 mm)	14 × 18 in. (356 × 457 mm)	14 × 20 in. (356 × 508 mm)	13 × 15³⁄₄ in. (330 × 400 mm)
Enlargement	Electronic, 10−2500%	1:1 or 2:1 only	50−1200%−1% increments	10−1000% in 1% increments
Read light source	Quartz halogen	Quartz halogen	Quartz halogen	Xenon arc
Write light source	Glow lamp	Glow lamp	HeNe laser	Xenon arc
Type of output	Continuous-tone or contact screened separations produced serially	Continuous-tone separations, serially produced	Continuous-tone or contact screened separations serially produced	Continuous-tone or contact screened separations serially produced
Scanning speed	2 in./min at 500 lines/in. (50.8 mm/min at 20 lines/mm)	1 in./min at 500 lines/in. scan rate (25.4 mm/min at 20 lines/mm)	2 in./min at 500 lines/in. scan rate (50.8 mm/min at 20 lines/mm	1½ in./min at 500 lines/in. scanning rate (38.1 mm/min at 20 lines/mm)

operator is able to call electronically for special borders, croppings, tints, and cuts until the page layout is completed. The operator's feedback is a monochromatic line representation of the layout on the video terminal.

Once the operator is satisfied with the layout, the separations to be made are called for. The computer, using the information from the electronic layout, now assembles the pictures stored on the magnetic disks exactly as the operator inidcated on the layout table. The final, completely assembled pages are produced by the system's scanner in the form of completely finished separation films. The assembled pages can be viewed for the first time in color after the output separations are proofed.

Sci-Tex. Sci-Tex developed an interactive composition system called Response 300. The system consists of a conventional color scanner, a series of magnetic disk memories, a computer system, a color video terminal, and finally, the firm's own output scanner, which exposes the final separation films in a combination of electronically generated halftones and line exposures.

The process begins when the operator scans into the memory all graphics via the imput scanner. Once this has been accomplished, the operator calls up the various images on the color video terminal and can then position the various elements via the control panel to create the desired page layout. The operator may add various tints, vignettes, outlines, and other common layout effects. Furthermore, local as well as global color corrections can be made. When the operator is satisfied that the layout is correct, the system is directed to produce the final, fully composed, plate-ready films.

Hell Chromacom. The Hell composition system, scheduled to be ready for commercial introduction before the end of 1980, has operating features similar to those of the Sci-Tex systems with some notable exceptions. This system is supposed to have a greater interactive capability than any of the systems discussed so far. It will be the first to allow the electronically composed pages to be used to drive an electronic gravure cylinder engraver, thereby completely eliminating the use of intermediate films of any kind.

Comtal. The Comtal Corporation has been developing an electronic page composition system for several years. Original efforts were in the designing, manufacturing, and marketing of interactive computer-controlled color video display systems for the aerospace industry. Then the firm developed its composition system known as the Vision One/20 system. When this system is coupled to a scanner, electronic composition and production of plate-ready color pages can be carried out. The system features a twin video display. One display is an extremely high-resolution black-and-white terminal, and the other is a high-resolution (800-line) color video display. Work such as critical montaging and cuts can be done in extremely high resolution on the monochrome display. Color correction and retouching, both local and global, can be carried out on the color display. The system can perform other composition functions similar to those of the Hell and Sci-Tex systems.

PDI Compuscale/Compudot Scanning System. This system is the only one besides the Hell DC-300 which is capable of purely electronic halftone screening, sometimes referred to as dot generation. The PDI scanner is unique in that it is a purely modular system. The read or input scanner is a stand-alone unit. Its only connection with the write or output scanner is purely electronic; there is no mechanical connection. All the main modules are also stand-alone units—the color computer, the electronic screener, and the central control computer system, as well as the previously mentioned input and output units. Because of its modular design, it is particularly suited as an input or output scanner, or both, for electronic page-composition systems. The main specifications for this system are given in Table 2. The system is available with a four-up 12 × 15-in. (305 × 381-mm) format output scanner or an interchangeable four-up 20 × 24-in. (508 × 610-mm) format output scanner.

[BRIAN A. CHAPMAN]

Typesetting. Typesetting is the art, practice, and business of setting type for the purpose of printing, making it possible by means of the printing press or other methods of replication to provide identical copies of books, magazines, brochures, and other printed pieces. Typesetting requires the sequential assembly of characters from a repertoire (consisting of extended alphabets, figures, and other signs) into an attractive layout. Hence, typographer is a term describing one who is skilled in composition, whereas a type designer is concerned with the shapes and features of characters.

Since 1960 typesetting has undergone a revolution that is comparable in its impact to that which occurred when movable, interchangeable types were introduced by Johann Gutenberg, (about 1397-1468) or when mechanical methods of typesetting were developed just before the turn of the century. Today, typesetting occurs at very high speeds, images are created in a variety of ways, and the process is usually output from some sort of electronic text-editing system.

Mechanical methods. During the first half of the 20th century, the principal methods of setting type made use of the Linotype, or slug-casting machine, and the Monotype. Out of many hundreds of inven-

Table 2. Specifications for the PDI scanning system

Function	Description
Scanner type	Rotary drum—helical scan, modular design with separate input and output scanners
Input format	12 × 18 in. (305 × 457 mm)
Output format	12 × 15-in. (305 × 381-mm) films or 20 × 24-in. (508 × 610-mm) films, all four separations produced simultaneously
Enlargement	Electronic, 20–2000% in 0.1% increments
Read light source	Quartz halogen
Write light source	Quartz halogen
Type of output	Continuous-tone or halftone separations at conventional screen angles (operator-selectable)
Scanning speed	Four 20 × 24-in. (508 × 610-mm) halftone separations in 25 min or four 12 × 15-in. (305 × 381-mm) halftone separations in 10 min

tions for the composing of type, these two, which gradually found their way into common use around the turn of the century, soon prevailed throughout the Western world, representing the first significant technological innovations since Gutenberg's movable types (cast in a foundry and set by hand).

These mechanical methods of setting type involved the use of hot metal, since the individual lines (slugs)—or characters (in the case of Monotype)—were cast from molten metal after the molds or matrices were assembled into justified lines, that is, lines with even left- and right-hand margins.

Cold type. After about 1950 cold type was introduced, appearing in two forms. One form is called strike-on, since the image is created by the use of typewriterlike devices, offering a repertoire of proportionally spaced rather than monospaced characters, much as hot-metal types vary in width from character to character. The other method of setting cold type is by photocomposition. Both processes were made feasible partly because of a gradual shift in the printing process from letterpress (which required raised characters to carry ink) to offset lithography (which involved the use of printing plates created by photographic or chemical processes).

Photocomposition. Several generations of photocomposition technology have so far been developed. First generation devices (the Monophoto and the Intertype Fotosetter) used the same principles employed in hot-metal setting, merely adapting them to the photographic process. Second-generation machines were based on different principles of character selection and laydown, involving electronic circuitry and xenon flash elements, and significantly increased setting speeds (Fig. 1). Louis M. Moyroud and René A. Higonnet are generally identified with the first successful second-generation devices, introduced by a company called Photon.

Third-generation typesetters came into being in the late 1960s with the advent of such products as the Hell (German) Digiset, the RCA VideoComp, and the Harris (Intertype) Fototronic CRT (cathode-ray tube). These devices created characters on the face of a CRT, exposing very fine strokes or slices of stored character masters (usually retained in memory in digital format) which can be sized and even slanted (obliqued) by electronic manipulation (Fig. 2). Since the late 1960s the price of third-generation devices has been falling; speed, quality, and reliability have increased; and these machines, capable of composing type at the rate of from 600 to several thousand newspaper lines per minute, are widely used.

Computer technology. Of equal importance is the software which facilitates the assembly of the desired characters, and formats them into lines and even pages. Computers take the original text stream, add up character width values, subtract this total from an overall line length allowance, and decide how to break lines (hyphenation and justification). They then tell the typesetting machine how to find and set these characters in the desired type faces and sizes.

Coincident with this development has been the application of computer technology (especially for minicomputers and microcomputers) for text input and editing, and the introduction (in newspapers and some magazines) of electronic newsrooms, where writers enter their text directly into the system and forward it to editors, who review it on their own video terminals and make desired changes, and in turn command the computer to hyphenate and justify and to typeset the textual material, even in formatted pages (Fig. 3).

Office equipment. Successive models of second- and third-generation typesetting machines experienced such drastic price reductions, as new technology (the "computer in a chip") led the way, that it became possible for many firms that formerly purchased typesetting from trade shops to install their own equipment and to use it in an office, much as a typewriter might be used. In 1979 this trend was accelerated by the introduction of new photographic processes which provided dry chemistry, so that photographic image-carrying material no longer required the application of fluids, often messy in an office setting.

And simultaneously, office automation continued, with the introduction of microcomputer-driven word processors. Gradually the input, storage, and editing of text became dissociated from its output, so that the user could determine whether to output by means of a strike-on or typewriterlike process or by phototypesetting. *See* WORD PROCESSING.

Improvements in the quality of output of office printers are also occurring, so that a computer line printer can form high-quality character images of varying widths by the use of dot matrix patterns. Even office copying machines can be used as relatively high-quality output devices. Laser technology, xerography, and ink-jet printing have come into play, so that the revolution in typesetting has now merged with the revolution in word processing and office automation.

Bypassing the typesetter. Not only can type now be composed by means of high-speed third-gener-

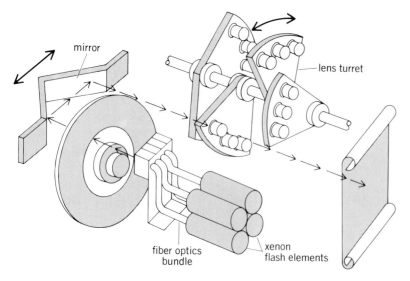

Fig. 1. Photon (Dymo) Pacesetter. Once exposed, the character is bounced against twin mirrors, which minimize the escapement distance. The escapement takes place before the images are sized. The lens turret could offer a total of 16 sizes, but focusing and sizing are accomplished in part by locating the lenses at different distances from the mirror assembly. (*From J. W. Seybold, Fundamentals of Modern Composition, Seybold Publications, 1979*)

Fig. 2. Mergenthaler Linotron 202, a third-generation typesetter. Floppy diskettes store fonts and font width information, as well as text input, if the machine is not driven on-line from a text-editing system. The cathode-ray tube, which is at the right in the cabinet and points toward the ceiling, has a fiber optic faceplate which comes into direct contact with the photographic paper or film contained in the casette directly above. (*Mergenthaler*)

ation devices, and with newer (fourth-generation) laser technology, but typesetting machines as such can be bypassed altogether. Type images, including such graphics as line drawings and halftones, can be written directly onto the printing plate or even onto the printing (paper) surface itself.

[JOHN W. SEYBOLD]

Delphi system. Stereotype plates are commonly used on newspaper rotary letterpress machines, but recently photopolymer relief plates have been adopted (to take advantage of the new phototypesetting) which have faster information processing, less environmental pollution, and improved work efficiency. New systems use a combination of photosensitive resin and phototypesetting. To print newspapers with these, the web offset printing process is most effective, but very expensive. Therefore, various types of direct lithographic

printing systems (DLPS) have been developed. DLPS can use the lithographic printing system by modifying rotary letterpress machines. They can shorten the plate-making time by using lithographic plates and can also avoid the large investment required to buy a web offset press. However, since DLPS need the same type of dampening unit used in conventional lithographic presses, many problems (such as stripping, scumming, or tinting) can be caused by troublesome water control, and they affect the printed result (Fig. 4).

In order to overcome these difficulties, the direct emulsion lithography (Delphi) system was developed. It requires a specially prepared emulsion ink which consists of oil-based ink and dampening water. The emulsion ink is separated into the oil ink and the dampening water on the press when it is subjected to roller pressure at a cool

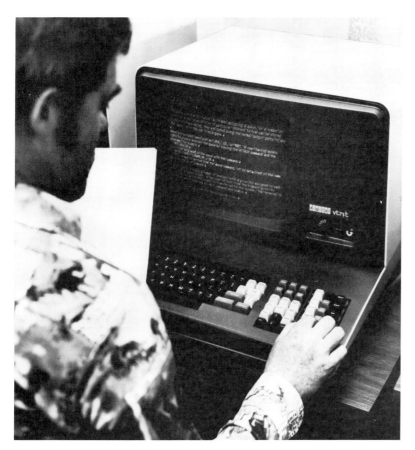

Fig. 3. A reporter writes a story directly on a video terminal. After completion, it is routed electronically to a similar terminal where an editor reviews the article, makes changes, and sends it on for composition.

temperature just before the ink reaches the printing plates. Thus, the dampening roller that is required for normal lithographic printing processes can be eliminated.

Printing mechanisms. To print with this system, the press must be equipped with a roller cooling device (which requires a simple and not-too-costly modification of the press). The emulsion ink is supplied to the press in the usual manner, and is broken into its oil component of the absorption-drying type and its aqueous component (dampening solution) just before reaching the plate. When the ink reaches the cooled roller, its viscosity is increased and the emulsion is broken into the two components by a shearing force between the rollers. Thus lithographic printing is achieved with a good printing result very close to that of web offset printing.

The emulsion does not freeze or decompose during storage or transportation at temperatures as low as −5°C (24°F), and it is of the water-in-oil type so that there is no danger of corrosion of ink tanks, printing press, or ink supply pumps during storage. There is also no possibility of the ink catching fire.

Other features. Any conventional lithographic plate can be used with the Delphi system. The emulsion has little misting tendency, and although the paper dust is more than with the web offset press, it is less than with the rotary letterpress.

[SABURO HASHIMOTO]

Ink drying. During the 1970s, research and development effort in the printing ink industry was concentrated on novel methods for drying inks at high rates of speed. There are several reasons for this activity: energy shortages, particularly fossil fuels; environmental, health, and safety regulations; and increasing worldwide inflation.

Liquid inks. Effective in 1980, new United States government regulations regarding the use of volatile organic compounds will put tremendous pressure on printers using both flexographic and rotogravure printing methods. Traditionally, in both these methods highly volatile solvents such as alcohols, esters, and low-boiling hydrocarbon solvents are used. The regulations will severely limit the use of these solvents unless some form of control technology, such as incineration or solvent recovery, is utilized.

There is thus a strong trend toward the use of water-based inks, particularly in flexography and packaging gravure on paper and board. One of the traditionally high-volume markets for flexographic inks is in the printing of polyethylene film. Even in this difficult area, in which alcohol-based polyamide inks are now used, ink companies are actively researching the use of water-based systems containing less than 20% of alcohol in the solvent.

In publication gravure, in which the normal formulations use mainly simple hydrocarbon solvent systems, solvent recovery is the method of choice to control emissions. Water-based inks, while available, are not a cost-effective solution in this marketplace. In addition, there has been an increasing demand for high-gloss, excellent scuff-resistant publication gravure inks in the American market. These have been developed, but generally use an all-toluene-based ink. Again this means that solvent recovery is a preferred method for controlling emissions from these inks, since the solvent can be recycled.

In order to successfully use water-based inks at the high speeds of publication printing, improvements must be made in the ink properties in press drying systems, and in the surfaces of coated papers which have a tendency to pucker when printed with water-based inks.

Paste inks. These inks are utilized for the high-volume, high-quality magazine, book, catalog, and packaging printing done in the United States. The majority of the inks are printed by offset lithography, although letterpress is still being used for some magazine and packaging printing.

The traditional method of drying such web-printed heat-set inks has been with gas-fired thermal ovens, and their use continues to dominate in this marketplace. These inks dry by evaporation of the hydrocarbon solvent used in their formulation.

The problems that have arisen because of air pollution regulations, particularly those for visible smoke and odor, are being solved in most installations not by the use of control equipment but by reformulation of the heat-set inks.

One trend is to use low-energy inks, which require 20–30% less gas because of the reduced web temperatures required for drying them. This, in turn, tends to reduce both smoke and odor, which are generally accentuated by high temperatures. In another effort to reduce these emissions,

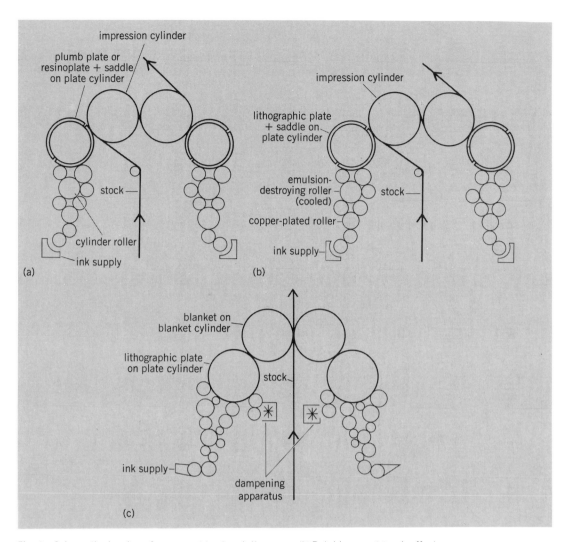

Fig. 4.   Schematic drawing of presses: (a) rotary letterpress; (b) Delphi press; (c) web offset press.

a high content of solids has been used in the ink. Conventional heat-set inks contain as much as 45% of high-boiling kerosine fractions as solvents. The goal is to reduce these fractions to as little as 15%, while maintaining the proper rheology and printing properties. Reducing the quantity of evaporated solvent in high-solids inks should help to reduce smoke and odor to very low levels.

1. Radiation-curing inks. The use of ultraviolet (uv) light for drying printing inks is now nearly a decade old, and it has grown in a number of markets where the economics are favorable. Currently, these are in metal decorating, sheet-fed lithography (particularly in carton and package printing), and the printing of plastic containers.

Ultraviolet dryers use less than half the energy used by conventional gas-fired thermal dryers. The ink formulation technology is now well established, with the inks fully equivalent to conventional heat-set inks in performance.

Ultraviolet inks have not grown as strongly in the web offset field, mainly because of the economics of the inks themselves, which still are several times the price of conventional web heat-set inks.

There is a strong trend toward the use of uv-cured clear varnishes and overcoatings, both press-applied and roller-coated, on sheet-fed

equipment. These are used to impart product resistance and high gloss to packaging materials, record albums, and the covers of paperback books. Currently they are applied over dry conventional inks, but active research is concerned with finding ways to apply these coatings over wet conventional inks, so that there will be further economic incentive for their use.

The use of electron-beam-cured coatings and inks is presently at a low level, mainly because the high capital cost of curing equipment has limited sales. The availability of the linear cathode electron-beam generator is likely to accelerate interest in this method of curing, and four units were to be in operation by the end of 1979. Inks have been made for letterpress, lithography, and gravure which cure well with low dose rates of $1-2$ megarads. They should be less expensive than their uv counterparts. In addition, the electron beam equipment is highly energy-efficient, using much less than half the energy of an equivalent uv dryer.

2. Infrared setting inks. There is much current interest in the use of infrared radiation (peak wavelengths from 1.2 to 3.5 μm) on sheet-fed lithographic equipment to accelerate the setting of specially made inks. The tackfree state achieved by the nearly instantaneous setting of the ink permits

sharp reductions in the amount of anti-setoff spray needed to prevent smearing and sticking in the pile. The infrared inks which have been developed set within a matter of seconds when the ink film is brought to a temperature of about 110–120°F (43–49°C) by an infrared radiator. Some of these inks also contain active catalysts to promote the rapid reaction, and others use special polymers and solvents to achieve the rapid setting.

In web printing, the use of infrared dryers and inks has been confined mainly to light-coverage forms printed with black ink only. Infrared setting is not expected to be a viable drying method for multicolor web printing on coated papers.

3. Chemically reactive inks. Among other methods of drying ink which are being researched are some which do not use external energy input but, rather, depend on a chemical reaction to achieve the solidification of the ink. In one such process, which originated in Australia, a chemical vapor bath is used; an ink containing a coreactant chemical is passed through the bath. The action of the vapor causes an instantaneous solidification of the ink film. The process is being used for several sheet-fed operations on nonporous substrates. It seems unlikely that it can be applied in its present form to high-speed web printing because of the difficulty of containing chemical leaks from the vapor bath under such conditions.

There are, however, many other reactive chemical materials which may possibly be used in a two-part ink system so that chemical energy, rather than externally applied energy, can be used to dry the ink. The feasibility of this approach is being actively investigated. Sprays, fountain solution additives, and paper-coating additives are possible ways whereby catalysts can be applied.

The current level of research in the printing ink industry is certain to lead to higher-technology inks utilizing a variety of unique drying methods, so that cost-effective options are available to all segments of the printing marketplace.

For background information *see* INK; PRINTING in the McGraw-Hill Encyclopedia of Science and Technology.

[ROBERT W. BASSEMIR]

Bibliography: M. H. Bruno, *Status of Printing in the U.S.A.—1979*, 15th International Conference of the International Association of Research Institutes for the Graphic Arts Industry, Randalls Road, Leatherbend, England, 1979; A. H. Phillips, *Computer Peripherals and Typesetting*, 1968; J. W. Seybold, *Fundamentals of Modern Photocomposition*, 1979; V. Strauss, *The Printing Industry*, 1967; U.S. Department of Commerce, *U.S. Industrial Outlook*, 1979.

Quantum mechanics

Whether there are "hidden variables" not accounted for in quantum-mechanical descriptions was long regarded to be empirically undecidable. New light was thrown upon this question by J. S. Bell's theorem in 1964 that no local hidden-variables theory can agree with all statistical predictions of quantum mechanics. Subsequent experiments concerning polarization correlations of appropriate pairs of particles mainly favored quantum mechanics and disconfirmed the entire family of local hidden-variables theories. The

philosophical implication is that either relativistic space-time structure or traditional physical realism is incorrect.

Einstein-Podolsky-Rosen argument. A modern version of the argument of A. Einstein, B. Podolsky, and N. Rosen (EPR) that quantum-mechanical descriptions of physical systems are incomplete considers photons γ_1 and γ_2 propagating respectively along the $\hat{z}$ and $-\hat{z}$ axes in the polarization state Ψ of total spin 0 and parity 1. This state can be expressed by Eq. (1)

$$\Psi = (1/\sqrt{2})\left[|a>_1|a>_2 + |a+\tfrac{\pi}{2}>_1|a+\tfrac{\pi}{2}>_2\right] \quad (1)$$

for any angle a, where $|a>_i$ ($i = 1, 2$) says that γ_i is linearly polarized at an angle a with the $\hat{x}$ axis. The argument proceeds from three premises:

(i) Predictions with probability 1 on the basis of Ψ are correct.

(ii) A sufficient condition for the existence of an "element of physical reality" corresponding to a physical quantity of a system is predictability with certainty (probability 1) of the value of the quantity without disturbing the system.

(iii) There is no action at a distance.

If γ_1 impinges upon an ideal analyzer oriented at angle a, it will exhibit linear polarization either along a or along $a + (\pi/2)$. In either case, the linear polarization of γ_2 with respect to these two directions is predictable with certainty from Eq. (1). Because of premise iii, the measurement performed upon γ_1 in no way disturbs γ_2, and hence by premise ii an "element of physical reality" corresponds to linear polarization of γ_2 with respect to directions a and $a + (\pi/2)$. This conclusion is true for every angle a, because a is arbitrary in Eq. (1). But no quantum-mechanical description specifies linear polarization along all directions, and therefore quantum mechanics is incomplete. The part of the putative total state not contained in the quantum-mechanical description is commonly called the hidden variables.

Bell's theorem. Bell's theorem concerns only those hidden-variables theories which he called local, one reasonable definition of which is given in Eq. (2).

$$p_{12}(\lambda,a,b) = p_1(\lambda,a) \cdot p_2(\lambda,b) \quad (2)$$

Among the many variants of his theorem (by E. P. Wigner, F. J. Belinfante, H. P. Stapp, B. d'Espagnat, F. Selleri, and others), the following by J. F. Clauser and M. A. Horne, which concerns both deterministic and stochastic local theories, is especially clear.

Let systems 1 and 2 impinge upon two spatially well-separated assemblies of apparatus, each consisting of an analyzer, with an adjustable parameter, and a detector (Fig. 1). Let λ, a point in some space Λ on which a normalized probability distribution μ is defined, be the total state of $1 + 2$ at some initial time. Suppose that there is a well-defined probability $p_1(\lambda,a)$ of detecting 1 when λ and parameter a are fixed, an analogously well-defined probability $p_2(\lambda,b)$ of detecting 2, and a well-defined probability $p_{12}(\lambda,a,b)$ of detecting both when the parameters are respectively a and b. (The character of the parameters is arbitrary.) Finally, assume the locality condition given by Eq. (2). If

$p_1(a)$, $p_2(b)$, and $p_{12}(a,b)$ are defined as the integrals over Λ with measure μ of $p_1(\lambda,a)$, $p_2(\lambda,b)$, and $p_{12}(\lambda,a,b)$ respectively, then inequality (3)

$$p_{12}(a,b) + p_{12}(a,b') + p_{12}(a',b) - p_{12}(a',b')$$
$$- p_1(a) - p_2(b) \leq 0 \quad (3)$$

is valid. This is an immediate consequence of Eq. (2) and the straightforward mathematical lemma, inequality (4),

$$xy + xy' + x'y - x'y' - x - y \leq 0 \quad (4)$$

where x, x', y, y' are real numbers between 0 and 1. Some quantum-mechanical predictions based upon Ψ of Eq. (1) violate inequality (3). Let a and a' be orientations of the analyzer of γ_1, and b and b' be orientations of the analyzer of γ_2. The detection probabilities implied by Ψ if analyzers and detectors are ideal are given by Eqs. (5).

$$p_1^{qm}(a) = p_2^{qm}(b) = \tfrac{1}{2}$$
$$p_{12}^{qm}(a,b) = \tfrac{1}{2}\cos^2(a-b) \quad (5)$$

Therefore if a, a', b, b' are chosen respectively to be $\pi/4, 0, \pi/8, 3\pi/8$, then Eq. (6) holds,

$$p_{12}^{qm}(a,b) + p_{12}^{qm}(a,b') + p_{12}^{qm}(a',b)$$
$$- p_{12}^{qm}(a',b') - p_1^{qm}(a) - p_2^{qm}(b) = 0.207 \quad (6)$$

contrary to inequality (3) and hence to any local hidden-variables theory.

Experimental tests. In principle, Bell's theorem permits a decisive test between quantum mechanics and all local hidden-variables theories. In prac-

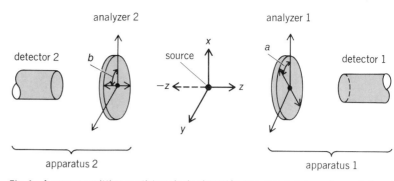

Fig. 1. A source emitting particle pairs is viewed by two apparatus, each consisting of an analyzer and an associated detector. The analyzers have parameters a and b respectively, here shown as angles, which are externally adjustable. (*From J. F. Clauser and A. Shimony, Bell's theorem: Experimental tests and implications, Rep. Prog. Phys., 41: 1881–1927, 1978*)

tice, it is difficult to prepare an appropriate quantum state and to find sufficiently good analyzers and detectors. An experiment proposed by Clauser, Horne, A. Shimony, and R. A. Holt in 1969, using photon pairs emitted in an atomic cascade, overcame two of these difficulties. The quantum polarization state of the pair is close enough to Ψ to imply sharp discrepancy with inequality (3) were ideal apparatus available; and low-frequency photons permit better than 90% efficient polarization analysis by calcite prisms or piles of glass plates. Unfortunately, the best detectors available for these photons are only about 20% efficient, so that

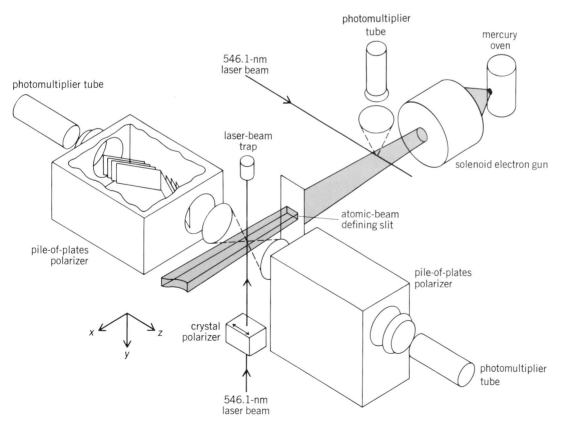

Fig. 2. Schematic diagram of an experiment to test between quantum mechanics and local hidden-variables theories, using a cascade in mercury-200 and employing a dye laser. The polarizer plate arrangement

is indicated, but actual analyzers have 14 plates each. (*From E. S. Fry and R. C. Thompson, An experimental test of local hidden variable theories using an atomic cascade in Hg²⁰⁰, Phys. Rev. Lett., 37:465–468, 1976*) quantum predictions with actual apparatus do not violate inequality (3). However, supplementing a local hidden-variables theory by either of two reasonable assumptions about detection probability yields testable inequality (7),

$$K \equiv |R(\pi/8) - R(3\pi/8)|/R_0 \le \tfrac{1}{4} \qquad (7)$$

where $R(s)$ is the coincidence counting rate when the relative orientation of the analyzers is s, and R_0 is the coincidence counting rate with both analyzers removed.

Results of atomic cascade experiment. This experiment has been performed four times.

(1) S. Freedman and Clauser (1972) used the $4p^2\ {}^1S_0 \to 4p4s\ {}^1P_1 \to 4s^2\ {}^1S_0$ cascade in calcium and found $K_{exp} = 0.300 \pm 0.008$, in comparison with $K_{qm} = 0.301 \pm 0.007$ predicted by quantum mechanics.

(2) Holt and F. M. Pipkin (1973) used the $9\ {}^1P_1 \to 7\ {}^3S_1 \to 6\ {}^3P_0$ cascade in mercury-198 and found $K_{exp} = 0.216 \pm 0.013$ in comparison with $K_{qm} = 0.266$.

(3) Clauser (1976) repeated the experiment of Holt and Pipkin with different analyzers and found $K_{exp} = 0.2885 \pm 0.0093$ in comparison with $K_{qm} = 0.2841$.

(4) E. S. Fry and R. C. Thompson (1976) used the $7\ {}^3S_1 \to 6\ {}^3P_1 \to 6\ {}^1S_0$ cascade in mercury-200 and found $K_{exp} = 0.296 \pm 0.014$ in comparison with $K_{qm} = 0.294 \pm 0.007$.

All results except that of Holt and Pipkin agreed with the quantum-mechanical predictions and disagreed with inequality (7). The anomalous result may have resulted from incomplete correction for optical activity in tube windows and lenses. The result of Fry and Thompson is considered particularly reliable, because high pumping rate from a dye laser (Fig. 2) permitted data collection in 80 min as compared with 154.5 hr of Holt and Pipkin.

Positronium annihilation and proton-proton scattering experiments. Quantum-mechanical results were also obtained by L. R. Kasday, J. D. Ullman, and C. S. Wu (1970), who observed the polarization correlation of photon pairs produced by positronium annihilation. Three groups repeated this experiment, twice with the same result and once not. Finally, M. Lamehi-Rachti and W. Mittig (1976) found quantum-mechanical results in measurements of spin correlations in proton pairs produced by low-energy S-wave scattering.

Future experiments. Two improvements over extant experiments are either in process or envisaged. A. Aspect and C. Imbert are using an acoustooptical device effectively to switch the orientation of polarization analyzers while the photons are in flight. This procedure makes it impossible for Bell's locality condition to be violated without also violating relativistic causality. Several groups are studying the feasibility of Stern-Gerlach measurements of the spins of two atoms produced in a singlet state by molecular dissociation. Highly efficient spin analysis and detection may both be possible, thus avoiding the necessity of resorting to auxiliary assumptions as in the photon experiments.

Implications. The preponderant disconfirmation of local hidden-variables theories (expected also of the new experiments) has momentous philosophi-cal implications. The abandonment of either premise ii or premise iii of EPR seems unavoidable. The latter alternative implies a radical revision of current theories of space-time structure. The former implies the renunciation of traditional physical realism, as N. Bohr proposed in his answer to EPR.

For background information *see* QUANTUM MECHANICS; QUANTUM THEORY, NONRELATIVIS-TIC in the McGraw-Hill Encyclopedia of Science and Technology.

[ABNER SHIMONY]

Bibliography: J. S. Bell, *Physics*, 1:195–200, 1964; J. S. Bell et al., Experimental quantum mechanics, *Progress in Scientific Culture*, Ettore Majorana Centre, pp. 439–460, winter 1976; J. F. Clauser and A. Shimony, *Rep. Prog. Phys.*, 41:1881–1927, 1978; B. d'Espagnat (ed.), *Foundations of Quantum Mechanics*, 1971.

Rhynchocoela

The Rhynchocoela are mainly free-living predators, and the pattern of their nutritional physiology has become apparent in recent years from studies by J. B. Jennings, Ray Gibson, Pamela Roe, and others. This pattern is of twofold interest, first as an extension of knowledge of the biology of these widely distributed, largely marine littoral, acoelomate invertebrates, and second because of its significance in the context of the evolution of animal alimentary systems.

Anatomy and physiology. The rhynchocoelans are the most primitive animals possessing an alimentary tract with both mouth and anus; they also show the beginnings, within the tract, of those cellular and regional specializations which dominate the organization of the gut in all higher animals. The cellular specializations consist of the separation of the digestive epithelium, or gastrodermis, into secretory and phagocytic components; at the regional level the tract is divided into a foregut (sometimes further subdivided into esophagus and stomach) used for reception and killing of the prey, and an intestine in which digestion and assimilation occur. Physiologically, too, the rhynchocoelan gut shows (as an emerging feature) well-marked temporal, and less distinct spatial, separations of the acidic and alkaline phases of digestion characteristic of higher forms. Within these phases, endopeptidases, exopeptidases, carbohydrases, lipases, and acid and alkaline phosphatases act in concert to achieve complete digestion of the food; digestion is largely intracellular, but there is extensive extracellular hydrolysis in the intestinal lumen before uptake of partly digested food by the gastrodermal phagocytes.

Diet and food detection. The diet of carnivorous rhynchocoelans includes protozoans, turbellarians, nematodes, annelids, crustaceans, mollusks, insect larvae, and, often, any material of animal origin that is not too badly decomposed. Rhynchocoelans are predominantly active carnivores but can be of some importance in the ecosystem as scavengers. Food may be encountered directly during searching movements, or detected from distances of up to 20 cm; many rhynchocoelans respond to mechanical disturbances of the water in their vicinity and to chemical stimuli emanating from damaged or even intact prey organisms. The chemosensory organs and other sensory structures used in food detection are situated in a pair of anterior lateral cephalic grooves and possibly also in the two cerebral organs associated with the cerebral ganglia.

Modes of nutrition. Feeding is effected by the proboscis, which is everted extremely rapidly from the dorsal rhynchocoel and used to seize motile living prey and draw it back to the anteriorly (or subterminally) situated mouth for ingestion. The proboscis may be unarmed, as in the orders Palaeonemertini, Heteronemertini and Bdellonemertini, or armed with one or more stylets (Hoplonemertini) which pierce the prey's body and often allow injection of toxic or paralyzing secretions produced within the proboscis. Ingestion occurs by the anterior end of the rhynchocoelan arching dorsally to expose the mouth; the lips and buccal cavity are distended as they are applied to the prey; and vermiform organisms are drawn intact into the foregut, usually bent into a U or J shape depending upon where they have been grasped by the proboscis. In the foregut, acidic secretions from epithelial glands rapidly kill the prey unless death has already occurred during capture, and mucus is secreted from other glands to facilitate onward passage into the intestine. In the palaeonemertine *Cephalothrix* and the heteronemertine *Lineus*, and probably in most other members of the Palaeonemertini and Heteronemertini, production of acid is mediated by the enzyme carbonic anhydrase in a process paralleling that used for hydrochloric acid secretion in the oxyntic (parietal) cells of the stomach in vertebrates.

Soft parts of larger, nonvermiform prey, and portions of decaying carrion, are sucked up in a similar fashion, but generally pass directly to the intestine without pausing in the foregut. Ingestion normally occupies only a few minutes, but a large organism such as a polychaete may take up to 2 hr. If the prey is significantly larger than its captor and the ingested portion fills the foregut and intestine, the uningested remainder is nipped off by oral contraction and enzymic digestion, and abandoned.

The Bdellonemertini are unusual among the Rhynchocoela in being entosymbiotic in habit; three species of the single genus of the order, *Malacobdella*, live entocommensally in the mantle cavity of marine bivalve mollusks, and one other occurs in the lung of a fresh-water gastropod. *Malacobdella grossa*, in the bivalve *Zirfaea crispata*, uses its proboscis in the normal manner to capture larger dietary constituents (crustacean larvae and the like), but it feeds principally as an unselective microphagous omnivore. Bacteria, diatoms and other unicellular algae, protozoans, and microcrustaceans form the bulk of the diet; these enter the host's mantle cavity with the feeding current and are captured by the rhynchocoelan by a simple filtration mechanism based on the modified foregut, which in the genus *Malacobdella* is termed a pharynx. The pharyngeal wall is produced into numerous ciliated filiform papillae; when the mouth and pharynx are dilated, the papillae are separated and water fills the pharyngeal chamber. Constriction of the esophagus prevents the water from entering the intestine; contraction of the pharynx then expels the water back through the mouth, but

during this process the papillae intermesh and suspended organisms are retained. Trapped particles are moved posteriorly by ciliary action, aided by bendings and contractions of the papillae, and pass through the now open esophagus into the intestine.

Absorption of nutrients. These heterotrophic modes of nutrition in carnivorous and microphagous rhynchocoelans clearly supply the bulk of the organism's dietary requirements. There is, however, some evidence that dissolved organic nutrients can be taken directly into the body, across the epidermis, from the surrounding aqueous medium. In *Lineus ruber* F. M. Fisher and J. A. Oaks have shown that radiolabeled glucose and amino acids are absorbed along this route and quickly incorporated into more complex substances within the body. Glucose, for example, showed within minutes of exposure a 17-fold concentration within the tissues as compared with the external concentration. Microvilli on the epidermal cells are probably involved in this absorption; in other species epidermal enzymes are believed either to prepare external dissolved nutrients for absorption or to facilitate their passage across the epidermis.

Digestion. Digestion in the Rhynchocoela is rapid and generally completed within 24 hr of ingestion. In carnivorous species, food entering the intestine is of a low pH (4.0–5.0) due to the acidic foregut or proboscis secretions; in the intestinal lumen it is attacked by endopeptidases working optimally at this pH value, and rapidly reduced to a semifluid heterogeneous mass. In Palaeonemertini and Heteronemertini the endopeptidases come from the gastrodermal gland cells, but in the Hoplonemertini they are released from the columnar phagocytes by distal abstriction of enzyme-loaded spheres. The hoplonemertine gland cells also discharge at this time, but their secretions have not been identified.

The extracellular acidic proteolysis is followed by phagocytosis of partly digested food by the gastrodermal columnar cells. These are ciliated, presumably to facilitate mixing and movement of the gut contents since the alimentary system (as in all acoelomates) lacks the muscular layers present in other animals. This phenomenon of phagocytosis by ciliated gut cells appears to be unique to the rhynchocoelans, as it has not been described in any other group. In *L. ruber* J. B. Jennings showed that pseudopodialike extensions develop on the distal surface of the phagocytes between the cilia and engulf food from the intestinal lumen (see illustration). Engulfed material passes into the cell as a classic food vacuole or phagosome (as in protozoans and cnidarians), in which digestion is completed and from which soluble products pass into the cytoplasm of the phagocyte and thence to the rest of the body. Enzymes responsible for intracellular digestion have been identified in place histochemically; they are produced by the normal intracellular mechanisms involving the rough endoplasmic reticulum, and are packaged by the Golgi apparatus into membrane-bound vesicles or lysosomes. The lysosomes fuse with newly formed phagosomes to form heterolysosomes within which digestion proceeds.

The acidic endopeptic digestion initiated in the intestinal lumen continues for a short time in the

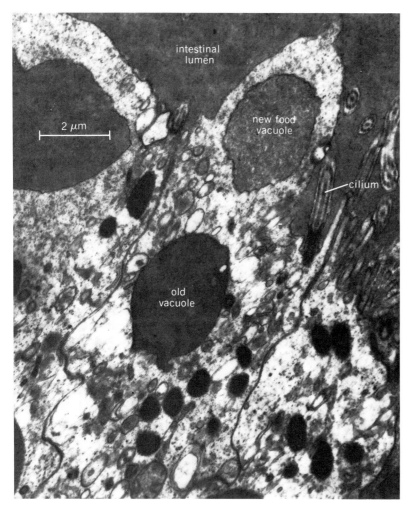

Electron micrograph showing the distal border of three intestinal cells of *Lineus ruber*. The central cell shows a food vacuole newly formed by fusion of outpushings from the cell which have engulfed material lying in the intestinal lumen. An older vacuole lies within the cell, and a cilium can be seen to the right of the newly formed vacuole. (*Courtesy of J. B. Jennings*)

heterolysosomes, the enzymes responsible being those engulfed with the food plus others of lysosomal origin. At this time there is much acid phosphatase activity around and within the heterolysosomes. These acid hydrolases eventually disappear and are replaced by alkaline phosphatase, arylamidases (indicative of exopeptic activity), lipases, and carbohydrases which complete digestion. There is thus only a partial spatial separation of initial acidic proteolysis (gut lumen + intracellular) from the terminal alkaline proteolysis, lipolysis, and carbohydrate hydrolysis (intracellular), but a complete temporal separation.

In the microphagous bdellonemertine *M. grossa*, digestion follows the same sequence as in carnivorous species, but with a marked decrease in emphasis on proteolysis. The gastrodermal gland cells secrete an α-amylase, not an endopeptidase, and the phagocytes produce only weak esterases (not characterized as endopeptidases) and no arylamidases. These variations presumably reflect the difference in diet between *Malacobdella* and carnivorous species, the former ingesting a greater proportion of carbohydrate-rich foods such as algae and bacteria.

Vascular system. The Rhynchocoela are the simplest animals possessing a vascular system. This is of simple construction but, curiously, shows a physiological feature not reported from any other group. The vessel walls and plasma contain at all times arylamidases broadly similar to those of the gut; their function is unknown, but possibly related to peptide circulation in relation to nutrition, reproduction, or osmoregulation.

For background information *see* DIGESTIVE SYSTEM (INVERTEBRATE); ENZYME; FEEDING MECHANISMS (INVERTEBRATE); LYSOSOME; PHAGOCYTOSIS; RHYNCHOCOELA; VACUOLE in the McGraw-Hill Encyclopedia of Science and Technology.

[J. B. JENNINGS]

Bibliography: F. M. Fisher and J. A. Oaks, *Biol. Bull.*, 154:213–225, 1978; R. Gibson, *Nemerteans*, 1972; R. Gibson and J. B. Jennings, *J. Mar. Biol. Ass. U.K.*, 49:17–32, 1969; J. B. Jennings, *Biol. Bull.*, 137:476–485, 1969; J. B. Jennings and R. Gibson, *Biol. Bull.*, 136:405–433, 1969; P. Roe, *Biol. Bull.*, 139:80–91, 1970.

Rock

Recent work in the areas of structural geology, metamorphism, sedimentary diagenesis, and lithification indicates that pressure solution and related mechanisms are of major importance in all these fields. As a result, research in pressure solution has been renewed after a long period of inactivity. The principal areas of investigation are: transport mechanisms and the role of water; thermodynamics of bodies under nonhydrostatic stress; flow laws for pressure solution; type and magnitude of strain; and the role of pressure solution in dolomitization and cementation. This article outlines current knowledge on the nature of pressure solution in terms of these problems, highlighting areas of particular interest.

Pressure solution. Pressure solution is a process of textural change in rocks resulting from the transfer of material from surfaces with high normal pressures to surfaces with low normal pressures where the transferred phases are deposited. Pressure solution is the dominant mode of deformation at low strain rates ($\approx 10^{-14}$/s) where temperatures and pressures also are low ($\approx 50 - 200°C$ and $\approx 50 - 100$ bars, or $5 - 10$ megapascals).

The concept of pressure solution was first developed by H. C. Sorbey following observation of the effects of pressure on the solubility of various salts. Sorbey found that the solubility of salts under pressure was directly related to the condition that the crystalline volume of a given salt be greater than its partial volume in solution. The solubility of salts meeting this condition was increased, whereas those with a greater partial molar volume in solution showed a decrease in solubility under increased pressure. Since that time, the process of pressure solution has been suggested for a variety of phenomena in a wide range of rock types. These phenomena include diagenesis and dolomitization in sedimentary rocks, formation of cleavage and folds, and metamorphic differentiation.

Pressure solution mechanism. Although petrographic and experimental evidence for pressure solution has been generally accepted, the theoretical basis for the process is uncertain. Outstanding problems involve two aspects of the pressure solution process: the role of pressure in controlling solubility; and the transport mechanism.

Questions regarding the role of pressure arise since the partial molar volume of the dissolved phase $(\bar{V}_f)$ is generally expected to be larger than in the crystalline state. This difficulty has been overcome by a series of theoretical treatments which show that the migration of the dissolved phases occurs down chemical potential gradients which are primarily a function of stress differences along the contact boundary, and not a function of $\bar{V}_f$. At present the mechanism for pressure solution proposed by P. F. Weyl has received greatest support from experimental and theoretical work. According to this model, dissolution occurs at the grain-to-grain contacts, with material diffusing away along an adsorbed water layer.

The necessity for a water layer has been questioned by D. Elliott, who proposed that the fluid phase along the grain boundary could result from lattice misfit between adjacent grains, with diffusion occurring by vacancy migration along the grain contacts. Elliott pointed out that this type of deformation was analogous to Coble creep or grain boundary sliding. Materials which deform by this mechanism follow a linear flow law.

Although Coble creep is indicated for rocks deforming in the range 200–350°C, this may not be true for rocks deforming at lower temperatures. W. Alvarez and coworkers and T. Engelder and R. Engelder have described deformation in sedimentary terranes at temperatures of less than 100°C. The deformation occurs in such a manner that almost all of the strain (70–100%) is accommodated along discrete cleavage surfaces without any evidence of grain boundary sliding. Moreover, there is evidence that in these terranes large amounts of solute (continuing up to 50% of the initial solid volume) have moved over distances on the order of tens and hundreds of meters. Since in many localities no sinks for the material have been found, the system is apparently open rather than closed as postulated in diffusional models. P. Robin has suggested that the deformation under these very-low-temperature conditions still follows a Coble-creep mechanism — a suggestion based on a Weyl-type model where a network of clays and micas act to enhance diffusion transfer. To date, little evidence of such a network has been found. An alternative model has been suggested by Y. Mimram in which the transfer mechanism is by bulk flow of meteoric waters circulating along systems of joints and microfractures. Recent studies in such very-low-grade terranes show that cleavage nucleates on preexisting joint surfaces, strongly supporting the mechanism proposed by Mimram.

Deformation features. Pressure solution results in two types of secondary fabrics: shape changes of individual grains associated with grain boundary sliding; and the development of discrete cleavage surfaces whose spacing may range from that of the individual grains to a meter or more. In general, both types of fabric develop during deformation. The shape change induced by pressure solution can simply be the result of the loss of material from surfaces normal to the greatest compressive stress operating on the grain. This material can be either removed entirely or deposited on the grain. The deposits may take the form of growths which parallel the local direction of least stress, or may become pore space cement. This latter response is of major importance in diagenesis.

Development of cleavage by pressure solution has been extensively documented in both sedimentary and metamorphic terranes. Grain boundary sliding apparently dominates at the higher temperature and pressures of metamorphic terranes, whereas the formation of discrete spaced cleavage surfaces with little or no grain boundary sliding characterizes sedimentary terranes. These discrete surfaces are readily recognized by the insoluble residues present on them. In the past, spaced cleavages have been mistaken as "fracture cleavage" or, in some rocks, joints.

Pressure solution is now recognized as the dominant mechanism in both folding and the layer parallel shortening which precedes folding. R. Groshong has demonstrated that, during pure bending of a limestone, removal of material from the region of compression beneath the neutral surfaces is effected by pressure solution along cleavage surfaces. S. Mitra has shown that grain boundary sliding controlled by pressure solution operated to produce the large-scale (wavelength about 10 km) folding of quartzites in the South Mountain anticline of Maryland.

Other geologic processes. Pressure solution has long been recognized as an important factor in the formation of cements and the reduction of porosity during the lithification of sedimentary rocks. Recently B. Logan and V. Seminiuk have documented a spectrum of diagenetic textures developed in carbonate rocks in western Australia during low-grade metamorphism. They also present evidence that pressure solution plays a major role in dolomitization by a process which is as yet poorly understood but may involve the interchange of Ca^{++}-with Mg^{++}-rich fluids across the pressure solution interface, with the subsequent deposition of the more stable dolomite. H. Wanless has suggested that pressure solution may induce dolomitization by homogeneous dissolution (that is, a process which occurs uniformly throughout the rock and not at a discrete interface) of Mg-rich calcite, such that the more mobile Ca^{++} leaves behind an Mg-enriched carbonate which becomes dolomite.

Pressure solution has also been suggested by D. Gray and D. Durney as a mechanism for metamorphic differentiation. Gray has shown that the limbs of microscopic kink folds are regions of extensive loss of silica which results in bands with alumino-silicate enrichment. This process is envisioned as continuing until widespread metamorphic segregations are produced.

For background information *see* Cleavage, rock; High-pressure phenomena: Metamorphic rocks; Petrology; Rock in the McGraw-Hill Encyclopedia of Science and Technology.

[PETER A. GEISER]

Bibliography: D. W. Durney, *Geology*, 6: 369–372, 1978; M. S. Paterson, *Rev. Geophys. Space Phys.* 11:355–389, 1973; H. C. Sorbey, *Roy. Soc. London Proc.*, 12:538–550, 1963.

Roof construction

During the last 10 years tents and long-span air-supported structures have found applications in permanent construction. This has been made possible by the development of a fiber-glass fabric coated with Teflon. This material is inert, incombustible, and resistant to moisture vapor and ultraviolet light. Consequently, the long life required for normal building financing and the fire safety requirements of local building codes can be readily met.

Benefits of fiber-glass fabric. The benefits resulting from this type of construction material are many:

Roof weights are between 1 and 2 lb/ft² (48–96 newtons/m²) in comparison to weights of 15 to 30 lb/ft² (0.7–1.4 kilonewtons/m²) for conventional construction. This leads to significant savings in the cost of supporting columns and foundation.

Large prefabricated components ready for erection are shipped to the site. This results in significant savings in time and cost of construction.

Fig. 1. Bullock's of Northern California Department Store, San Jose. (*Bill Apton, Photographer*)

Assembly of other components can occur simultaneously; and actual roof erection times are of the order of one-tenth of the time required for conventional construction.

The roof is translucent, allowing natural light to flood the interior space. Translucency can be varied from 4 to 18%, yielding light levels from 300 to 1500 foot-candles (3200–16,000 lux). Since the Teflon coating makes the fabric self-cleaning, translucency levels can be maintained.

The fabric bleaches to a pure white, and with self-cleaning, a highly reflective surface results with significant reduction in air-conditioning loads.

The high performance of these new materials is the consequence of a unit price about six times more expensive then conventional roofing materials. The total roof structure cost, though, is less expensive because the roofing material is more than a climatic barrier. It is a primary structural element spanning as much as 60 ft (18 m) between cables, arches, or masts that serve as the secondary structural elements.

Prestressing. Fabric can perform this load-carrying function only by being prestressed, either by being engineered and fabricated into a tension structure similar to a tent or Conestoga wagon or by being prestressed by air pressure as in a balloon.

Fabricated tension structure. Since the fiber-glass fabric is largely inextensible (that is, it deforms very little under load), it will not find its proper shape by being stretched into position. Rather, the proper shape must be established mathematically by using modern high-speed computers. Even methods of model analysis which were in vogue for polyester fabrics as little as 10 years ago are no longer viable for these new permanent fabrics. Mathematically, one must first determine the membrane shape for a given set of prestress conditions, namely fabric stress and loading. This load case requires no information as to the load deformation characteristics of the membrane. Only conditions of static equilibrium need to be satisfied.

Once the membrane shape is established, it must be checked for conditions of superimposed loads, which occur for example under wind or snow. These new loads will be carried by internal forces changing in magnitude and the membrane configuration changing in shpae.

In addition to satisfying conditions of static equilibrium, it is necessary to consider how the membrane deforms under load (material stress-strain properties) and how these changes occur so that the surface remains continuous (that is, conditions of compatibility must be met). A nonlinear and generally iterative mathematical procedure is required in the analysis. Once the behavior under load has been established, it is necessary to check that the maximum stresses do not cause overstress in the fabric membrane and that the minimum stresses do not result in the fabric membrane losing stress and going slack. If this latter condition exists, higher levels of prestress are required, and a new initial configuration needs to be established.

Once the initial configuration is established and checked, it is necessary to find the cutting shapes for the fabric by considering the deformation of the

Fig. 2.   The Silverdome, Pontiac, MI.

Fig. 3.   Haj Terminal roof structure, Jeddah, Saudi Arabia.

material in going from the cut shape to the pre-stress case. This compensation for stretch is rather complex as the material is nonhomogeneous and nonelastic, and is variable depending on the variables of material production. Compensation will even vary from roll to roll of the same type of fabric.

The secondary elements of the structure—the cables, arches, or masts to which the fabric spans—are also included in the structural analysis. For the tensile structures, it is always necessary to consider that in the eventuality the fabric rips, the collapse of these elements will be prevented.

Air-supported structure. Air-supported roofs, on the other hand, are prestressed by air pressure that is mechanically induced. Air pressure can be lost due to a number of reasons, such as loss of power, failure of emergency generators, openings resulting from structural failure of non-roof-related components, and failure of fabric. These new air structures are designed so that loss of prestress would cause the cables and fabric to hang freely, or to be supported by a frame or other structures. For large-span air structures, the possibility of this deflation is remote since the requirements for mechanical ventilation results in such large quantities of air being moved that openings on the order of 2000 ft² (186 m²) can be maintained. Since these roofs cannot collapse, repair is quickly implemented and the roof reinflated much as a car tire once the source of the leak is found.

Applications. This new form of construction has found applications from small-span canopies for department stores (Fig. 1) to stadiums spanning more than 700 ft (213 m) and covering 10 acres (4 square hectometers; Fig. 2). A modular roof consisting of 150-ft-square (46-m-square) modules repeating 210 times and covering more than 100 acres (40 hm²) is presently under construction as an airlines terminal (Fig. 3).

As new materials are developed, tensile strengths will increase. Fabric roofs weighing less than 1 lb/ft² (48 N/m²) that span stadiums today will weigh less than ¼ lb/ft² (12 N/m²) and may well span cities tomorrow.

For background information *see* ROOF CONSTRUCTION in the McGraw-Hill Encyclopedia of Science and Technology.   [DAVID H. GEIGER]

Bibliography: H. Berger, *Bull. Int. Ass. Shell Spatial Structures,* XVII-3(62):23–30, December 1976; D. Geiger, *Bull. Int. Ass. Shell Spatial Structures,* XVIII-2(64):15–24, August 1977; F. Otto, *Tensile Structures,* 1969.

Satellite communications

A new generation of spacecraft for communications is being developed to expand traffic capacity and to lower costs for the near future, while even more advanced technology is emerging from initial design and testing of spacecraft designed to provide new and improved communications services in the late 1980s. The principal advances are in spacecraft communications antennas. The new space systems will be capable of aiming small, shaped radio beams to only the desired areas where the system's ground stations or other satellites are located, and will be able to actively track spacecraft that carry instruments to observe and sense atmospheric, water, and land-resource conditions from space, reporting the data in real time. Other new satellite systems using today's techniques will employ technology reflecting the higher reliability and ready fabrication of complex devices brought about by recent advances in solid-state electronics. Intelsat V, the Tracking and Data Relay Satellite System (TDRSS), and examples of new technology under development are described below.

Geosynchronous satellites. The geosynchronous orbit—a path 36,000 km high, where an object over the Equator orbiting eastward at 10,860 km/hr is synchronized with the Earth's rotation, appearing stationary in the sky to a surface observer—is the prime orbital location for communications satellites. Although the path for radio signals transmitted by satellites in geosynchronous orbit is longer than that for low-orbiting satellites, advantages of the simpler systems used with geosynchronous satellites and the reliability of transmissions relayed by them have proved these spacecraft to be feasible and economical. Relatively small-diameter (4.5-m) ground antennas without tracking capability are in use today for commercial reception of a television signal relayed by a communications satellite for rebroadcast, or for over a thousand telephone circuits. The largest antennas, over 33 m in diameter, provide gateway service for many thousands of two-way voice circuits between continents. The geosynchronous round-trip time for the signal—about a quarter of a second—causes delays in the inherent echo of the speaker's voice and listener's response which have brought some complaints from users. However, as advanced echo cancelers emerge from current laboratory testing into full use, they will quiet the speaker's earpiece during speech, and this short delay will be even less noticeable. More evident to the user of communication satellites is the reduction in cost by a factor of 4 for a transatlantic phone call since satellite service and advanced cables were introduced. The satellite techniques are now well proved, as a result of the efforts by the National Aeronautics and Space Administration (NASA) 15 years ago. Research in private industry since then has rapidly brought basic communications technology and ground systems into very advanced stages, and United States industry is supplying most of the communications satellites for the world today.

Intelsat V. The first organization for commercial relaying of color television and telephone traffic, Intelsat was formed in 1964. Using an updated fourth-generation system, Intelsat IVA, this 104-country consortium is now providing better and more wide-reaching communications than were imagined just 20 years ago. Five of these satellites are now in use in the system. Three span the Atlantic, one the Indian Ocean, and one the Pacific, with several on-orbit spares available. Each generation has increased the traffic-handling capacity of the system, which has risen from the original 240 one-way calls for each satellite to 6000 today. Expected design lifetimes of 7 years necessitate the preparation of replacements with expanded capability to meet needs of the next period, and the fifth generation is now under development.

The capacity of the Intelsat V spacecraft (Fig. 1), designed and built by Ford Aerospace, will be

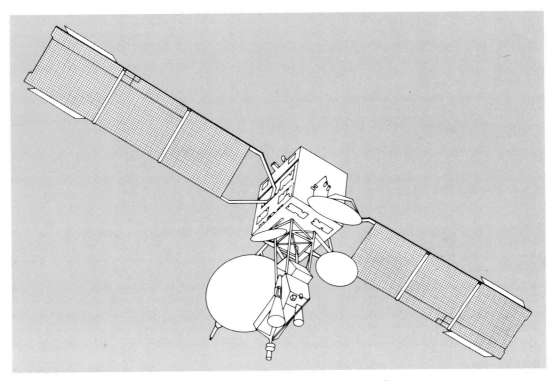

Fig. 1. Intelsat V communications satellite. (*Ford Aerospace and Communications Corp.*)

12,000 simultaneous phone circuits. As in all commercial communication satellites, high-quality color television can replace part or all of the phone channels.

Stabilization of spacecraft. Intelsat V is the first satellite in the series to use body stabilization instead of the spin-despin system pioneered by Hughes Aircraft. Although a spinning drum covered with solar cells and a motor-driven, Earth-pointed, "despun" antenna platform can be operated reliably and many are currently in use, the solar-generated electric power for such a spacecraft is limited. In contrast, body-stabilized spacecraft are almost unlimited in generation of power, because long booms with solar panels can be extended to generate the kilowatts needed for transmitters of the largest spacecraft.

Gyroscopic sensors and cold-gas-thruster attitude-control systems provide orientation for the basic body within one-tenth of a degree, so that the communications antennas can be stably pointed toward the ground station areas for the life of the spacecraft. Because the spacecraft body and the antennas must rotate daily with the Earth, the angle to the Sun is constantly changing. To provide maximum power, the solar panels are motor-driven to maintain a position normal to the Sun line. Overall, this system is more complicated than the spin-despin system, although many spacecraft have operated in orbit for 7 years using the spin-despin system; but the power advantage provides for more channel capacity per launch, which reduces per-circuit costs. As more body-stabilized systems are proved to have long lifetimes, this design is likely to pervade the geosynchronous orbit in the future.

Frequency allocation. For the first time in the system, Intelsat V will carry communications equipment to use the Ku-band part of the radio spectrum, where 14.0–14.5 GHz is allocated for uplinks and 11.7–12.2 GHz is allocated for downlinks. Each 500-MHz segment provides for twelve 36-MHz transponders, or receiver-transmitter combinations. These segments are separated to avoid jamming by the downlink transmitter of the sensitive uplink receivers. This standard technique for communications spacecraft is repeated in the satellite's additional use of the C-band at 5.9–6.4 GHz up and 3.9–4.4 GHz down.

Frequency reuse. To increase even further the capacity of the system, the concept of frequency reuse is employed to increase the number of channels without requiring added allocations of frequency spectrum. Two methods are used. In spatial reuse, the antenna beams are shaped so that the ground coverage area is contoured to permit communications within desired areas, while areas where other system users or other channels may use the same frequencies are sharply cut off. This technique has been used for years in radial patterns on terrestrial systems, but its implementation on spacecraft has awaited greater capacity in weight, size, power, and stabilization, and only recently has the technique come into popular use.

The other method of reuse, pioneered by Intelsat IVA and RCA's Satcom in 1975, discriminates between same-frequency signals by transmitting (and receiving) each signal polarized orthogonally to the other, generally using linear polarization. In this method, adequate separation of signals depends also on the receiver systems' tolerance for coherent interference, but a difference in power level of signals at the receiver of 33 dB (1/2000) is adequate and has been proved effective and practical to achieve. Exceptional design and care in execution are required to achieve signal separation over the whole frequency band and the entire antenna pattern. Another system using this concept effectively is the domestic satellite system, Comstar, operated for American Telephone and Telegraph and General Telephone and Electronics by the COMSAT Corporation. On all these satellites, the 500-MHz spectrum allocation can be used for 1000 MHz of traffic, reducing the potential for congestion, which could raise costs and cause interference.

TDRSS—Advanced Westar. Another body-stabilized satellite system used for communications is in the final integration stage preparing for launch in 1980. This satellite combines replacement of the current Western Union domestic satellites (Westars) with a new and unique capability for replacing NASA tracking stations around the world by relaying signals to and from other spacecraft to a single ground station at White Sands, NM. This Tracking and Data Relay Satellite System will be in use with commercial Western Union ground systems for telephone, television, and data relay under the service name Advanced Westar. Initiated by a need to reduce the costs of operating the many tracking stations that receive data from the low-orbiting NASA and National Oceanic and Atmospheric Administration (NOAA) spacecraft on weather, atmospheric, and oceanic conditions, as well as from the space shuttle orbiter, the TDRSS/Advanced Westar spacecraft (Fig. 2) may be the most sophisticated satellite ever constructed for commercial service.

Previous relay experiments. The concept of relaying data between spacecraft is not new. It was used during the Apollo-Soyuz rendezvous in space over Europe, when these satellites relayed live television signals to the NASA experimental satellite *ATS 6*, which in turn sent the signals back to a ground station in Madrid, where they were again turned around and sent via Intelsats for broadcast use around the world. Another successful satellite relay system is in use between the Lincoln Experimental Satellites *LES 8* and *9*, operating a millimeter-wave cross-link at 36 and 38 GHz.

Implementation. These experiments brought confidence to the relay system designers, but the commercial implementation of the TDRSS is on a grander scale. NASA has contracted with Western Union for 10 years of continuous, wideband, multiuse service. The spacecraft designer-builder, TRW Systems, is taking advantage of the similar needs of NASA and of Western Union's commercial service. Both need spot-beam coverage of heavy traffic areas—large cities for Western Union, and wideband data satellites for NASA. The frequencies for both systems will be in the same general spectrum region in the Ku-band—11.2–15.225 GHz. However, each user also has a special frequency requirement: S-band (2.2 GHz) for links to and from NASA's spacecraft and C-band for standard Westar service replacement. By innovative design, the six satellites can all be the same, and will share the command and telemetry facility at White Sands. The crucial need for relay of data and voice to and from the crewed space shuttle orbiter places a high reliability requirement on the TDRSS for continuous worldwide coverage, which in turn means that an on-orbit spare is required. This spare, plus the service needs of

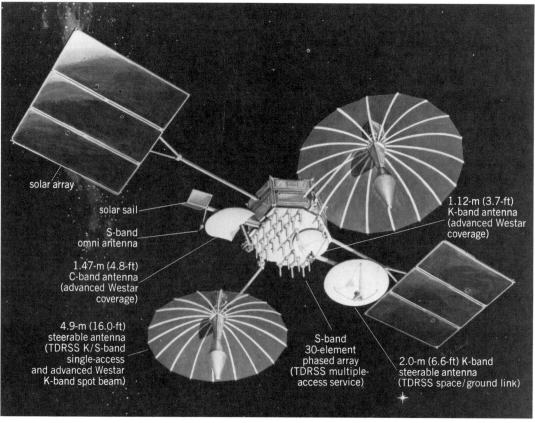

Fig. 2. TDRSS/Advanced Westar shared satellite. (*TRW Systems*)

Western Union, leads to four of the satellites being in orbit at one time.

Spacecraft positioning. Located at 42° and 172° west longitude, the TDRSS craft will be able to communicate with any spacecraft in Earth orbit except for a limited area—below 1000 km altitude between 45° and 102° east longitude (the Indian Ocean area). The satellite used for Advanced Westar service will be located at 103°W, and the shared spare will be placed initially at 99°W, midway between the TDRSS spacecraft to minimize orbital movement time to replace either one. Thruster-jet firing to remove orbital energy will lower the spacecraft, moving it eastward, and another firing to restore that energy will make it again geostationary in the new position. Similar actions to add energy first will relocate it westward. Perturbations in gravitational forces caused by the Sun and Moon, unevenness in the Earth's own gravity, and solar pressure on the spacecraft require that station-keeping fuel be on board at launch for the entire 10-year life, assuring users that the spacecraft are truly geostationary within 0.1° of the correct latitude and longitude. Although these techniques are not new, the TDRSS dependence of expensive, wideband, complex satellite systems—some carrying astronauts—is stressing the need for high reliability of the entire complex, including the ground hardware and software. NASA is depending on this system for expediting the flow of data from such systems as Landsat.

Satellite control and data management. These user satellites will be controlled from Goddard Space Flight Center by yet another satellite-relayed link through commercial "domsats" to White Sands. Then, after the collected scientific data about the Earth are selected, processed, and structured for analysts' use, the data can be sent again by satellite, presenting to the ultimate users the sensed information about their selected area of interest. For the space shuttle, data will be relayed to TDRSS to and from Houston's Johnson Space Center.

NASA research. While this seemingly easy access between spacecraft and user is a tribute to technology of the past few years, even further gains are required to assure that this access continues so that the pressures of the information generation are met. NASA has reestablished its program to provide for the needed advances in space communications by planning for the development of spacecraft antennas that can produce multiple simultaneous directional beams, and the natural companion on-board switching systems that will allow signal path routing between desired users at high rates. In its optimum form, an advanced frequency-reuse system would have perhaps hundreds of small, shaped beams that cover areas only where there are potential communicators. The system would achieve the highest isolation between beams, which is the most difficult design parameter for multibeam antennas. Then it would be possible to use the entire allocated spectrum in each beam. Recently completed tests on a 15-beam antenna designed by TRW Systems under NASA contract for coverage of the United States show that at least 7.5 times frequency reuse can be cost-effectively built in a contiguous-beam system.

In another technique, which is called Satellite-Switched Time-Division Multiple Access (SSTDMA), short bursts of very wideband data are sent from the spacecraft by a highly agile, beam-forming phased array antenna. Each burst is sent to a prescribed location, and uplink bursts are received, processed, sorted, and returned on the downlink to the intended addressee. During the next 10 years, various experimental systems will test these concepts, probably adding a giant switchboard and computer control to make effective use of satellites for high-speed traffic. Developments by NASA and efforts by Bell Labs should make substantial contributions in these techniques. The expansion of satellite traffic capacity must continue in attempts to keep ahead of ever-increasing needs for information transfer. *See* COMMUNICATIONS SATELLITE.

For background information *see* COMMUNICATIONS SATELLITE; SPACECRAFT GROUND INSTRUMENTATION in the McGraw-Hill Encyclopedia of Science and Technology. [DONALD K. DEMENT]

Bibliography: P. Barghellini, Evolution of U.S. domestic satellite communications, *3d Jerusalem Conference on Information Technology*, 1978; D. K. Dement, An overview of the NASA Satellite Communications Program, *ICC '79 Conference Record*, vol. 1, IEEE, pp. 15.1.1–15.1.4, June 1979; P. Kaul et al., Advanced WESTAR SS/TDMA System, *4th International Conference on Digital Satellite Communication Systems*, Montreal, pp. 36–43, October 1978; E. A. Ohm, System aspects of a multibeam antenna for full U.S. coverage, *ICC '79 Conference Record*, vol. 3, pp. 49.2.1–49.2.5, 1979; J. Ramasastry et al., Western Union's satellite-switched TDMA Advanced WESTAR System, *AIAA 7th Communications Satellite Conference*, San Diego, pp. 497–506, April 1978; D. O. Reudink and Y. S. Leh, A scanning spot beam satellite system, *Bell Sys. Tech. J.*, 56(8):1549–1560, October 1977; R. J. Rusch et al., INTELSAT V spacecraft design summary, *AIAA 7th Communications Satellite Conference*, San Diego, pp. 8–20, April 1978.

Satellites, navigation by

The Global Positioning System (GPS) when fully implemented will be a universal positioning or navigation system that will provide for the first time three-dimensional position accuracies to 10 m, velocity to an accuracy of 0.03 m/s, and time to atomic clock accuracy. This information will be available anywhere on Earth. The system is composed of satellites, control stations, and user equipment. A total of 24 satellites will be placed in orbit (Fig. 1) in the fully implemented system. Four are already in orbit, and the initial tests indicate that all the goal specifications have been achieved. Master stations and satellite monitor stations control the system and transmit data to the satellites. User equipment consists of a variety of receivers which are under development to cover both military and civilian requirements. Very sophisticated receivers have been designed and built for use in checking of the GPS in the early deployment stages. Some of the receivers will be lightweight, small, and inexpensive. It is anticipated that user equipment will be installed on airplanes, ships, and ground vehicles and will also be available as a portable navigation set.

Fig. 1. Orbits of satellites in the Global Positioning System.

$$(X_1 - \boxed{U_x})^2 + (Y_1 - \boxed{U_y})^2 + (Z_1 - \boxed{U_z})^2 = (R_1 - \boxed{C_B})^2 \qquad R_1 = c(\Delta T_1) \quad (1)$$

$$(X_2 - \boxed{U_x})^2 + (Y_2 - \boxed{U_y})^2 + (Z_2 - \boxed{U_z})^2 = (R_2 - \boxed{C_B})^2 \qquad R_2 = c(\Delta T_2) \quad (2)$$

$$(X_3 - \boxed{U_x})^2 + (Y_3 - \boxed{U_y})^2 + (Z_3 - \boxed{U_z})^2 = (R_3 - \boxed{C_B})^2 \qquad R_3 = c(\Delta T_3) \quad (3)$$

$$(X_4 - \boxed{U_x})^2 + (Y_4 - \boxed{U_y})^2 + (Z_4 - \boxed{U_z})^2 = (R_4 - \boxed{C_B})^2 \qquad R_4 = c(\Delta T_4) \quad (4)$$

Basis of operation. The GPS satellite is called NAVSTAR and has eight major subsystems and more than 33,000 parts, each of which is designed to last 5 years. The long service life is a result of a very conservative design coupled with a total redundancy in all assemblies. The heart of the GPS system is the accurate timekeeping subassembly carried in each satellite.

The basis for the GPS accurate position (and time) determination is the precise measurement of the transit time of the radiated signals from 4 satellites from the constellation of 24. Since accurate time is essential, the satellites carry atomic clocks. To prevent a catastrophic failure during the satellite service life, three clocks are carried in each satellite. These clocks will lose or gain an average of only 1 s in 30,000 years. The signals are transmitted at two frequencies, 1227 MHz (designated L_2) and 1575 MHz (designated L_1) to permit modeling and corrections to be made for variations of time delay of the signal passing through the ionosphere.

To use the GPS and obtain the highest accuracy, great care must be taken to model and correct any errors in the received time delay such as clock drift and propagation delay. The satellites are continuously monitored, and error corrections are transmitted to the individual NAVSTAR satellites for inclusion in the satellite-generated signal so that the receiver can automatically correct for position and time.

Calculation of position and time. The receiver must detect the satellite signals and then perform a set of mathematical calculations that are based upon the model shown in Fig. 2. The satellite signals give the ephemeres positions (X_1,Y_1,Z_1), (X_2,Y_2,Z_2), . . . , of the satellites, and the time cor-

rection factor for an individual satellite. If four satellite signals are received, Eqs. (1)–(4) must be solved simultaneously for four unknowns: the user's position coordinates (U_x,U_y,U_z), and the user's clock bias C_B. In these equations, R_1, R_2, . . . , are the apparent ranges from the user to the satellites, given by the auxiliary equations where c is the speed of light, and ΔT_1, ΔT_2, . . . , are the differences between the broadcast times and the user's time.

If the user receiver had a very accurate clock, only three satellites would be required. If only two-dimensional position (such as latitude and longitude) and time are required, three satellites would give real-time measurements.

Satellite signals. The signal that the satellite transmits must be easy to receive, contain the navigation message in a form that is readily usable, and not be subject to errors due to anomalies such as multipath. It must contain time information, data for clock correction, space vehicle status data, synchronization information, ephemeris of the NAVSTAR satellite, corrections for delays in signal propagation, and the means to convey special messages.

All of this information is formatted and transmitted as pseudo-random-noise signals. Two signal codes are used: one called the Precision (P) code has a reset every 7 days and a frequency of 10.23 MHz; the other is called the Clear-Access (C/A) and has a 1-ms epoch and a frequency of 1.023 MHz. Both the P and C/A signals carry the same information with a data bit frequency of 50 bits/s. The navigation message consists of five subframes of 6-s length making a data frame of 30 s and 1500 bits. The data are NRZ (nonreturn-to-zero).

Velocity determination. The GPS is designed to provide accurate position and time, but it also provides precision velocity measurements to the user by measuring the Doppler shift in the satellite-transmitted carrier frequency. The accuracy of this measure is primarily influenced by the dynamics of the host vehicle of the receiver.

Satellite characteristics. Twenty-four satellites will be used, and these will be in 12 hr orbits as illustrated in Fig. 1. The orbital inclination is 63°, and the satellites are divided into three groups of eight, each group in a separate orbital plane, staggered by 120°. This arrangement permits continuous four-satellite coverage of the entire Earth. New satellites will be added as older satellites wear out after 5–7 years of service life.

The power of the transmitted navigation signal is given in Table 1. The initial satellites carry rubidium frequency standards, but future satellites will carry even more accurate cesium standards. The satellites are three-axis-stabilized and carry skewed reaction wheels. The antennas always point to the Earth, and the solar panels rotate so the solar cell surface is always aimed at the Sun.

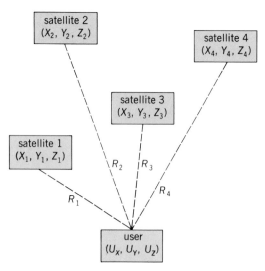

Fig. 2. Model for mathematical calculations in the Global Positioning System.

Table 1. Power of transmitted navigation signal

Signal	Power
L_1 C/A signal	+26.8 dBW
L_1 P signal	+23.8 dBW
L_2 P or C/A signal	+19.1 dBW

Table 2. Minimum signal at receiver

Frequency	P signal	C/A signal
L_1 (1575 MHz)	−163 dBW	−160 dBW
L_2 (1227 MHz)	−166 dBW	−166 dBW

Control station characteristics. Monitor stations containing GPS user receivers monitor the position and status of the satellite on a continuous basis. These data are sent to a master station. Here the ephemeris coefficients and time and satellite data are calculated and processed. At regular intervals, the master station updates the satellite memory to permit an accurate navigation message to be transmitted. It essentially keeps track of and fine-tunes the satellites as they pass each day or as required.

Signal format. Each of the 6-s subframes contains ten 30-bit words. All subframes are preceded by a telemetry (TLM) message and a HOW (Hand Over Word). The TLM word contains an 8-bit Barker synchronization word. The HOW permits the user receivers to rapidly synchronize to the satellite. The remainder of the subframe contains the satellite ephemeris, the satellite clock correction, the ionospheric model, and the almanac data.

User equipment characteristics. If the satellite is at an elevation angle of 5°, a 0-dB right-hand (RH) circular polarized antenna on Earth will have a minimum signal as given by Table 2. All receivers must use the message format described above. However, many different receivers can be configured for different user applications. The receivers may receive the satellites simultaneously or sequentially, may use all of the received information or selected parts of the transmitted information,

and may use L_1 and L_2 or a single frequency. The choice will depend on the vehicle dynamics, desires for rapid readout, required accuracy, cost, and so on. In any case, all receivers must be capable of satellite selection, signal acquisition, tracking and measurement and data recovery, some correction for propagation effects, and calculation of the navigation parameters from the time-of-arrival measurements.

As a minimum, an antenna, receiver, computer, and output or display device is required in each assembly of user equipment. The antenna must have a clear look at the sky. The basic receiver must rapidly lock only five or less signals (five if four satellites are used, since one satellite may sink below the horizon) and must use tracking loops with narrow bandwidth to keep the noise low.

The output of the receiver must be fed to a computer. The computer may be more or less elaborate depending on the navigation requirements. The output devices will vary depending upon the applications. In some, the output will be a direct digital readout, and in other applications the output will go directly to control computers, either as the sole input or with other navigational data to obtain a more significant output. The navigational output may be in the form of cartesian earth coordinates, latitude, longitude, and altitude, or corrections to other navigational signals; and in the form of a wired signal in some form of direct readout depending on the user needs.

For background information *see* SATELLITES, NAVIGATION BY in the McGraw-Hill Encyclopedia of Science and Technology.   [LOUIS FEIT]

Bibliography: B. G. Glazer, *J. Inst. Navig.*, 25(2): 173–178, 1978; J. J. Spilker, *J. Inst. Navig.*, 25(2): 121–146, 1979.

Scanning proton microprobe

The scanning proton microprobe (SPM) is a recent addition to the arsenal of analytic tools for determination of the spatial distribution of trace elements in samples of interest to many sciences. In its

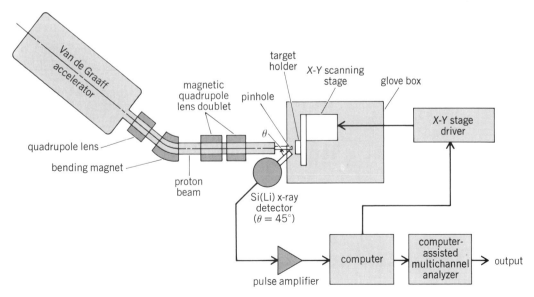

Fig. 1. Schematic of scanning proton microprobe facility.

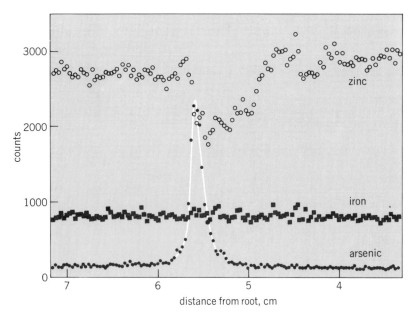

Fig. 2. Portion of SPM scan of hair pulled from worker 6 months after poisoning by arsene.

earliest and still commonest mode, the SPM marries the techniques of beam-scanning microscopy with those of proton-induced x-ray emission (PIXE); in this mode it is the proton analog of the scanning electron microprobe (SEMX). The first SPM was developed in the early 1970s by J. Cook-son and colleagues, who focused a beam of protons from a 4-MeV Van de Graaff accelerator, at the Harwell Laboratories in England, to spots as small as 4 μm in diameter and measured the characteristic x-rays induced in corresponding areas of the target by these energetic protons. They demonstrated that the well-documented advantage of PIXE over electron techniques in terms of measuring trace concentrations of many elements at levels at and below 1 μg/g of specimen was retained when focused beams were used. L. Grodzins and P. Horowitz demonstrated the added utility and uniqueness of SPM for studies in controlled environments when they showed that the proton beam could be brought out of the normal acclerator vacuum system, with little loss of its microbeam quality.

Advantages and limitations. The advantages of protons (or heavier-ion beams) over electrons in microbeam work derives from the proton's heavier mass, which is reflected in much reduced scattering as the protons (or heavier ions) traverse matter. The negligible scattering allows the proton beam to be brought into the atmosphere, where it diverges little over the millimeter distance to the target (the advantages of keeping the target at normal room conditions are hard to overemphasize); it results in excellent spatial resolution even for thick targets since the integrity of the microbeam is retained over the effective path length in the sample; and, most importantly, it means that the background of continuous radiation of x-rays, emitted when the charged particles are deflected by the nuclei in the sample, is orders of magnitude smaller than that produced by the easily deflected electrons used in the SEMX, and hence the ratio of signal to background is one to several orders greater.

These advantages have proved useful for many applications in which the disadvantages of radiation damage and beam heating characteristic of the protons are unimportant, and spatial resolution below a micrometer is unnecessary. Each proton knocks several atoms from their lattice position on traversing the effective path of the target, and the radiation damage can be severe even though there is no visible destruction. Beam heating is generally an order of magnitude greater than encountered with SEMX and becomes a serious problem at high resolution. At this time the spatial resolution of energetic proton beams has not been less than about 1 μm, but advances in this new field are rapid and smaller-diameter beams will be available in the future.

Apparatus. A schematic drawing of the facility developed at the Lincoln Laboratories of the Massachusetts Institute of Technology by a team of scientists from there and Harvard is shown in Fig. 1. The proton beam from the 4-MV Van de Graaff accelerator is focused by appropriate lenses to a windowless exit hole which is differentially pumped to preserve a good vacuum on the accelerator side and atmospheric pressure on the sample side. The beam, whose diameter is controllable, is fixed in position; the target is moved across the beam in a raster pattern by stepping motors which are under computer control; at each resolved area of the target the spectrum of x-rays is sorted and stored in appropriate memory bins of the computer

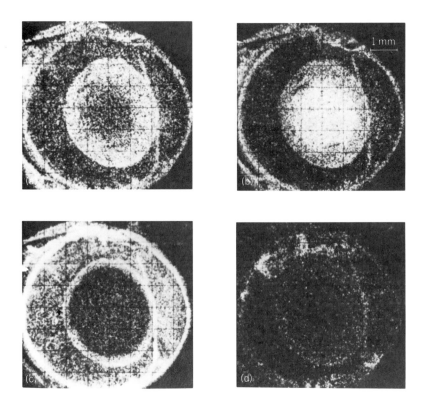

Fig. 3. SPM scans showing distribution of minor elements in lens from eye of a rat. (a) Potassium. (b) Sulfur. (c) Chlorine. (d) Calcium. (From P. Horowitz et al., Elemental analysis of biological specimens in air with a proton microprobe, Science, 194:1162–1165, 1976)

to provide a two-dimensional map of the distribution of elements in the specimen.

Applications. One of the simplest but important uses of the SPM has been in the study of the elemental distribution along the length of a single hair. Since a hair, typically 100 μm in diameter, grows about 300 μm per day, a 12-cm scan gives a 1-year history of the elements in the hair with a time resolution which can be less than a day. There is considerable questioning as to whether the elemental distribution of hair of a healthy subject reflects the nutritional uptake of elements such as zinc and iron, which are required by the body, but there is no doubt that hair is one of the purge systems the body uses for toxic elements. Figure 2 shows a portion of the scan of a hair pulled 6 months after a worker was poisoned by arsene. The sharp spike of 200 μg/g of arsenic in the hair was not accompanied by a change in any other element but zinc, which had been depleted, possibly by the body's mechanism fighting the toxic invasion.

The ability of the SPM to study tissues in controlled environments is illustrated by the photographs in Fig. 3, where the distributions of minor elements (potassium, sulfur, chlorine, and calcium) are shown for a hemisected frozen hydrated rat lens which was examined at low resolution in a dry nitrogen environment at about 100 K. The mobile electrolytes are caught without distortion; the high sulfur content in the center of the lens shows the high concentration of peptides and proteins there, while the potassium ions concentrate in the aqueous phase. (Certain cataract lenses, a principal focus of this study, show quite different distributions.) It is difficult to obtain the quantitative results on such aqueous tissues, easily obtained from the computer output of the SPM, when the lens is studied in vacuum since the dehydration techniques can alter both the morphological integrity and the density distribution of the elements in the lens.

For background information *see* MICROPROBE, ION; MICROPROBE, X-RAY; TRACE ANALYSIS in the McGraw-Hill Encyclopedia of Science and Technology.   [LEE GRODZINS]

Bibliography: J. A. Cookson, A. T. G. Ferguson, and F. D. Pilling, *J. Radioanal. Chem.*, 12:39, 1972; P. Horowitz et al., *Science*, 194:1162–1165, 1976; P. Horowitz and L. Grodzins, *Science*, 189:795–797, 1975.

Seismology

In recent years there have been a number of advances in the use of seismology for studying the Earth's structure. Among these are the use of deep crustal seismic reflection profiling and the study of seismic-wave velocity.

Seismic sounding of the crust. New and exciting information on the detailed structure of the Earth's continental crust is being obtained in the United States through the technique of deep crustal seismic reflection profiling. The information includes the behavior at depth of major faults; the variable character of the crust-mantle transition; the response of the crust to compression, extension, and strike-slip motions; and the variation of crustal structure between different geological provinces.

Refraction versus reflection profiling. Gross characteristics of the continental crust, such as its thickness and generalized seismic velocity structures, have been determined in many parts of the world by means of the seismic refraction technique, in which seismic waves are generated and recorded in such a way that they propagate laterally through the Earth's crust. This method therefore results in models of the crust described in terms of layers of different seismic velocities, with lateral small-scale variations averaged out.

Outcrops of rock representative of the deep crust (such as in Precambrian shield areas and deeply eroded mountain belts) suggest that the detailed structure of the crust at depth is much more complex. A horizontally layered crust is also difficult to reconcile with ideas of crustal genesis in the framework of plate tectonics. These expected complexities have been observed at depth by using the technique of seismic reflection profiling.

With this method, seismic-wave generating and receiving stations are arranged so that waves propagate in a predominantly vertical direction and lateral differences in crustal structure are better re-

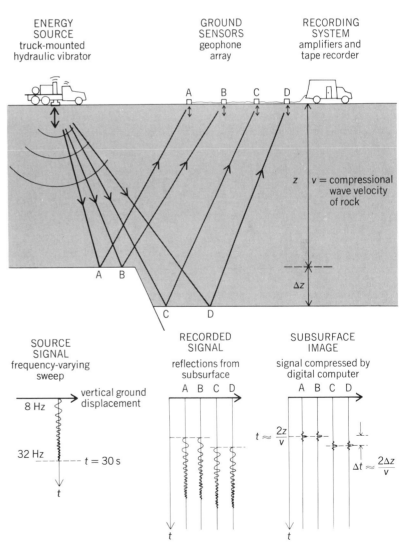

Fig. 1.   The Vibroseis (registered trademark of Continental Oil Company) method of crustal reflection profiling. The object is to map the structure of the Earth's crust with high resolution.

solved. An advantage of this method is that data are processed into a form closely imaging geological structures; also, data acquisition and processing techniques have been developed to a high level of sophistication, largely because of exploration for economical hydrocarbons in the shallowest parts of the crust.

Deep crustal reflection profiling has been carried out over the last 15 years in Germany, Canada, Australia, and the Soviet Union. As of 1979, the most intensive effort was being made in the United States, under the auspices of the Consortium for Continental Reflection Profiling (COCORP).

COCORP. The Consortium was formed in 1975 to carry out reflection profiling across areas of geological interest in the United States where the acquisition of high-resolution subsurface data might help solve a given problem or increase understanding of a particular geological province. COCORP uses sophisticated oil industry equipment and techniques, and maintains a contract seismic crew continually in the field. The Vibroseis technique (Fig. 1; Vibroseis is a registered trademark of Continental Oil Company) is used because it is most economical in terms of flexibility, rate of data collection, and quality of data. The final seismic sections represent cross sections of

the crust with the depth scale in two-way travel time (in seconds); to convert to approximate depth in kilometers, multiply by 3. A typical seismic section is shown in Fig. 2.

COCORP study areas. The Rio Grande Rift in New Mexico, interpreted as being the locus of present crustal extension, is flanked by normal faults. COCORP seismic profiles show fault offsets of up to 4 km, and the eastern boundary fault may be traced as a zone lacking coherent reflected energy that dips steeply into the rift to depths of at least 11 seconds. The western rift boundary near the surface appears to be a series of smaller step faults, and becomes seismically indistinguishable below about 3 s. The geometry of these bounding faults closely constrains theories of formation of this rift. The most spectacular feature of the COCORP profiles is a distinct complex group of seismic events at about 7 s in the middle of the rift (Fig. 3). This event corresponds in dip and depth to a midcrustal magma body, the presence of which had previously been suggested from the study of microearthquake data. These COCORP profiles are the first detailed studies to be made of a crustal magma body.

During the Laramide Orogeny, large uplifts of basement rock occurred along reverse faults in Wyoming. Whether these uplifts represent compressional or vertical tectonics has been much debated. COCORP profiles across the largest of these, the Wind River Mountains, brought to the surface about 6×10^7 years ago, revealed that they are underlain by a profound thrust fault with an average dip of about 35° (Fig. 2). This thrust can be traced on the seismic sections to at least 24 km depth (about 8 s), and at least 21 km of crustal shortening has occurred along it. The thrust zone generates a strong acoustic contrast, enabling thrusting processes to be studied in detail. The geometry of the fault shows that compressional forces were responsible for the formation of the Wind River Mountains, and were probably the dominant tectonic element of the Laramide Orogeny.

Profiling in Georgia and Tennessee has revealed that the major tectonic structure in the Southern Appalachians is the Blue Ridge Thrust Sheet, which may be traced from its toe at the Cartersville–Great Smoky Fault in a southeasterly direction underlying the Piedmont and rooting east of the Carolina Slate Belt. The thrust sheet is 6–15 km thick and has moved at least 220–260 km to the northwest. Valley and Ridge sediments outcropping in Tennessee appear to underlie the thrust sheet and are relatively undeformed. These sediments represent an enormous, unexplored possible source region for hydrocarbons. Other faults in this area, such as the Brevard and the Hayesville, appear to be minor splays from the main Blue Ridge thrust. This COCORP profile has been extremely useful in unraveling the complexities of Southern Appalachians geology.

Other COCORP studies have revealed: (1) The structure of the San Andreas Fault; brittle deformation near the surface apparently passes below a depth of 12–14 km into a zone about 4 km wide of possibly ductile flow. (2) The detailed structure of a Precambrian rift below the Michigan Basin. (3) An extensive, faulted layer of basalt underlying the

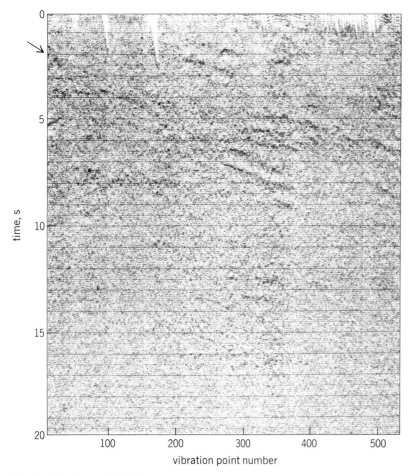

Fig. 2. Unmigrated COCORP seismic section (Wyoming Line 1A) recorded in the Wind River Mountains of Wyoming. The horizontal length of the section is about 50 km. The arrow marks the trace of the Wind River thrust which reaches the surface to the left of the section.

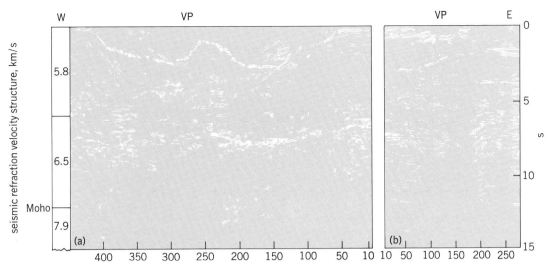

seismic refraction velocity structure, km/s

W   VP

5.8

6.5

Moho

7.9

(a)

400 350 300 250 200 150 100 50 10   10 50 100 150 200 250

VP   E
                          0

                       — 5

                       s

                       —10

(b)   —15

Fig. 3. Drawings of unmigrated COCORP seismic sections across the Rio Grande Rift. (a) Socorro Line 1A. (b) Abo Pass Line 1. The western rift boundary is at about vibration point (VP) 400 on Socorro Line 1A; the eastern boundary is at about VP 200 on Abo Pass Line 1. The inferred magma body can be seen at about 7 s between VP 250, Line 1A, and VP 50, Line 1. Horizontal length is about 80 km. (*From L. D. Brown et al., Deep structure of the Rio Grande Rift from seismic reflection profiling, in R. E. Riecker, ed., Rio Grande Rift: Tectonism and Magmatism, Amer. Geophys. Union Spec. Publ., pp. 169–184, 1979*)

Atlantic coastal plain near Charleston SC; movements on these faults may have been responsible for the 1886 Charleston earthquake and the continuing minor seismic activity in this area. (4) The crust-mantle transition (typically at a depth of 40–50 km) has distinct seismic signatures in different areas; preliminary interpretations of these signatures suggest three types of lower boundary to the crust: a complexly layered transition, a simple discontinuity, and a boundary that gives rise to a strong velocity gradient.

New ideas about continental crust. High-resolution seismic profiling studies of the deep crust in the United States and elsewhere are thus providing images of structures which approach the complexities seen at the surface of the Earth in Precambrian shields and eroded mountain belts. Major faults and folded structures can be identified at depth. Tentative interpretation of the varying seismic character of different geological provinces suggest that it may be possible to map deep into the crust such features as homogeneous plutons or metamorphic terrains characterized by differing degrees and attitudes of layering. These data on the fine structure may be combined with chemical and mineralogical data from kimberlite pipes and with laboratory data on rock mechanical properties to increase understanding of the effects of plate tectonic processes on the continental crust.

[JONATHAN BREWER]

Seismic waves. Seismology is the most powerful method for determining the structure and elastic properties of the interior of the Earth. Seismic waves attenuate as they propagate; the study of this attenuation provides information about the nonelastic properties of the Earth's crust, mantle, and core. The physics of the attenuation, however, has been obscure. Recent research has contributed to an understanding of both the phenomenology and physics and the effect which absorption has on seismic wave velocities.

The elastic moduli of an absorbing medium increase with frequency. This fact has been used by R. Hart, D. Anderson, and H. Kanamori to reconcile Earth models obtained by body-wave and free-oscillation techniques. Previously, shear-wave travel times predicted by models obtained from free-oscillation inversion were as much as 5 s slower than the observations.

Seismic quality factor (Q). Recent work on the attenuation of seismic waves by Anderson, Hart, R. Sailor, and A. Dziewonski indicates that there is a low-Q layer in the upper mantle in the vicinity of the asthenosphere. The lithosphere and lower mantle are relatively high-Q. The inner core of the Earth is also highly attenuating. The damping of the radial modes can be interpreted in terms of bulk attenuation in the core. This is the first evidence of a bulk viscosity mechanism in the Earth and makes it possible to place bounds on core viscosity.

Attention is now being focused on the frequency dependence of seismic-wave attenuation. Theoretically, the seismic quality factor should increase linearly with frequency at high frequency. Such an effect has been found from a study of the core reflected shear phase ScS by S. Sipkin and T. Jordan. The theory of the spectral behavior of Q for relaxation mechanisms, such as dislocation relaxation, has been worked out by H. Liu, Anderson, Kanamori, and J. Minster. In general, at geophysically important frequencies, Q decreases with period, and it is to be expected that tides and Chandler wobble have a much smaller Q than shorter-period seismic waves. At very long periods, Q should increase linearly with period. The location of the low-frequency and high-frequency cutoffs depends on temperature and pressure and therefore depth.

Dislocation relaxation. The upper-mantle low-velocity zone has often been equated with the asthenosphere. It has not been obvious, however, that the same mechanism which leads to low viscosity should also be responsible for the attenuation of seismic waves. The low velocity and high attenuation are now believed to result from dislocation relaxation, a mechanism most effective at temperatures near the melting point. The same dislocations, relaxing further by self-diffusion and then multiplication, are responsible for steady-state creep. This mechanism is also most effective near the melting point. Thus the two different types of nonelastic behavior are linked through the relative diffusivities of the controlling point defects and the temperature. The effects of temperature and pressure are less on attenuation than on creep rate, since the controlling ion is smaller.

Seismic mechanisms. Attenuation in the outer core is very low, as expected for fluids in general and molten metals in particular. The mechanism at seismic frequencies is shear and volume viscosity rather than thermal conductivity. The high attenuation in the inner core is possibly a result of pressure increasing the characteristic relaxation time so that it enters the seismic band. If so, the inner core can be treated as a viscous fluid rather than a crystalline solid.

It now appears that dislocations are responsible for both the long-term rheology of the Earth's mantle and the removal of energy from a seismic wave. Diffusion of either heat or point defects across grains is much too slow a process. Relaxation of grain boundaries by partial melts is too fast. Stress-induced migration of dislocations has the appropriate characteristics to explain both the anelasticity associated with the small-amplitude, short-duration stresses of seismic waves and the nonreversible creep associated with large-stress, long-duration phenomena such as postglacial rebound and plate tectonics.

Seismic-wave interpretation. In the small-stress situation, dislocations bow in their glide planes. The anelastic strain is proportional to the area swept out. For a solid containing a Frank network of completely relaxed dislocations, the shear modulus is decreased by about 16% compared with the same solid with no dislocations or unrelaxed dislocations. Measurements made at high temperature, or low frequency, therefore yield smaller velocities than those made, for example, in the laboratory at the other extreme. In the general case, the elastic properties of a solid are frequency-dependent. Seismic measurements are made over the frequency range of several hertz to 54 min. Laboratory measurements, with which seismic data are compared to infer composition, crystal structure, and temperature, are commonly made at frequencies greater than a kilohertz. Attenuation, therefore, must be understood not only for its own sake but in order to interpret seismic velocities and to compare data taken by different techniques at different frequencies.

The actual reduction of shear velocity in the upper-mantle low-velocity zone is almost exactly that predicted by dislocation relaxation. The relaxation time is comparable to seismic frequencies for diffusion of interstitial point defects at upper-mantle temperatures. Therefore partial melting is not required in order to explain the low velocities and high attenuation found in the upper mantle.

There are two time constants associated with dislocation motions. Stress-induced glide is controlled by the diffusion of interstitials and is relatively rapid. The associated time constant at high temperature falls in the seismic band. Climb of dislocations requires self-diffusion and is much slower. It is responsible for transient and steady-state creep. In principle, seismic-wave attenuation can be used to estimate viscosities appropriate for long-term deformation. In addition, it is sensitive to such geophysically important parameters as dislocation density and temperature.

Minster and Anderson showed that the "high-temperature background," an exponential increase of attenuation with temperature that occurs above one-half of the melting point, is due to dislocation bowing. The frequency dependence can be explained by a hyperbolic distribution of active dislocation lengths. This gives the Jeffreys-Lomnitz transient creep relation and a mildly frequency-dependent Q in the absorption band. The absorption band is bounded by the longest and shortest relaxation times, which in turn are controlled by the longest and shortest active dislocations. These times depend exponentially on temperature and pressure. In the mantle the absorption band shifts with respect to the seismic band. This gives a variation of Q with depth for a given frequency.

For background information *see* EARTH, INTERIOR OF; SEISMOLOGY in the McGraw-Hill Encyclopedia of Science and Technology.

[DON L. ANDERSON]

Bibliography: D. L. Anderson and R. S. Hart, *J. Geophys. Res.*, 83:5869–5882, 1978; D. L. Anderson and J. B. Minster, *EOS, Trans. Amer. Geophys. Union*, 59:1182, 1978; J. A. Brewer et al., *Tectonophysics*, in press; L. D. Brown et al., in R. E. Riecker (ed.), *Rio Grande Rift: Tectonism and Magmatism*, Amer. Geophys. Union Spec. Publ., pp. 169–184, 1979; R. S. Hart, D. L. Anderson, and H. Kanamori, *J. Geophys. Res.*, 82:1647–1654, 1977; H. Kanamori and D. L. Anderson, *Rev. Geophys. Space Phys.*, 15:105–112, 1977; H. P. Liu, D. L. Anderson, and H. Kanamori, *R. Astron. Soc., Geophys. J.*, 47:41, 1976; J. B. Minster and D. L. Anderson, *EOS, Trans. Amer. Geophys. Union*, 59:1183, 1978; J. Oliver et al., *Geol. Soc. Amer. Bull.*, 87:1537–1546, 1976; S. Mueller, *Amer. Geophys. Union Monogr.*, 20:289–318, 1977; R. V. Sailor and A. M. Dziewonski, *R. Astron. Soc., Geophys. J.*, 53:559–581, 1978; S. A. Sipkin and T. H. Jordan, *EOS, Trans. Amer. Geophys. Union*, 59:1142, 1978.

Shale

Recent studies of shales have focused on sedimentology and depositional environments rather than mineralogy. This shift in emphasis parallels an earlier shift in the study of sandstones and suggests by analogy that a major new field of inquiry is being opened.

Depositional environments. Shale deposition requires relatively quiet water. Quiet conditions may be found in water below the zone of wave mixing (this depth is variable, dependent on the overall energy of the coastline, but normally is at least

20 m) or in shallower water protected from wave and current action. Most shales in the stratigraphic record are marine or marginal-marine, but a few deposits formed in lakes are known. One of these, the Green River Shale of the Rocky Mountain region, is important as an oil-bearing shale and a probable future supplier of petroleum.

Modern areas of mud deposition include open shelves below the wave zone, coastal mudbanks, delta foreslopes, lagoons, estuaries, marshes, and tidal flats. Shales deposited in similar environments are common in the stratigraphic record on the continents.

Much thicker masses of muddy sediment accumulate on the continental rise, in water several thousand meters deep. Sediments derived from the adjacent continent are probably delivered to the rise by episodic downslope turbidity-current flows. Muds can then be suspended and redeposited by the permanent geostrophic bottom currents. In addition, there is a constant rain of fine particles from the overlying water mass (pelagic sediment), including clays carried by surface currents, wind-blown particles, and biogenic silica and calcium carbonate fragments. On the deep-ocean floor, far from continental influence, the pelagic component dominates; these muds are finer-grained and accumulate more slowly than those on the continental rise. Both rise and pelagic sediments are preservable and recognizable in the stratigraphic record of mountain belts, where sediments deposited in deep water have been uplifted.

Depositional processes. The actual mechanics of deposition of fine-grained particles is not well understood. Calculations using Stokes' law show that clay particles have extremely long settling times, even in absolutely still water. Given the widespread presence of turbulence in the seas and oceans, it is somewhat mysterious that clays accumulate at all. Flocculation of clays into aggregates with much shorter settling times could account for rapid deposition in estuarine and deltaic zones. W. A. Pryor has shown that biogenic pelletization has a similar effect. Clay particles are molded into fecal pellets in the guts of invertebrate detritus feeders, and these pellets then function as sedimentary grains. Upon deposition and burial the pellets may disintegrate, producing a structureless clay.

Such cryptic depositional processes may be further masked by the extensive postdepositional bioturbation which muddy sediments undergo. Under normal marine conditions the annual sedimentation layer is probably ingested and redeposited by burrowers at least once and perhaps several times per year. By the time an individual clay particle has been buried below the zone of burrowing (less than 10 cm below the sediment-water interface), it may have passed through the guts of a multitude of deposit feeders. Often, the resulting mudstone displays little trace of depositional structure.

Black shales. Recent interest in black shales is primarily due to their potential for supplying hydrocarbons, either indirectly, as source beds for petroleum reservoirs, or directly, as gas shales. Black shales contain large amounts of organic carbon (3–15%) and are commonly pyritic and phosphatic, all indicative of an environment lacking oxygen. Dissolved oxygen content depends on the

rate of supply by addition from the atmosphere or from the oxygenated open ocean versus the rate of uptake by respiration, bacterial decomposition, and chemical oxidation. Even in very shallow water (1–10 m) oxygenation may be low due to stagnant conditions (lack of mixing) or high organic productivity. Quiet-water swamps, marshes, and lagoons can thus be environments of black mud deposition.

Deep-water deposition of black muds requires low-oxygen conditions at depth, in which planktonic carbonaceous material can accumulate. The surface waters of a large basin will be oxygenated through direct contact (wave-mixing and thermal overturn) with the atmosphere. Below the surface zone, oxygenation diminishes because of respiration, and oxygen can be replenished only by lateral influx of normally oxygenated open ocean water. If influx is blocked by a sill, oxygenation in the basin will decrease to zero at some depth, below which truly anaerobic conditions will prevail.

Silled basin model. D. C. Rhoads and C. W. Byers have elaborated a model for the facies sequence to be expected in a transect down the slope of a silled basin (see illustration). The oxygenated (aerobic) surface zone will support abundant benthic life, so muddy sediments of this zone should contain shelly fossils and thoroughly bioturbated sediments. In the marginally oxygenated (dysaerobic) zone below, calcified epifaunal animals will be eliminated, whereas infauna tolerant of low-oxygen conditions will persist. Hence the sediments will be bioturbated, but lacking in fossils. In fully anaerobic conditions at depth, all metazoan animals will be absent, and the sediments will lack bioturbate texture as well as benthic fossils.

Application of this sedimentary model to shales can provide the direction of slope of the ancient basin, through recognition of a progressively deeper-water facies sequence. By analogy with the depth zonation in modern basins, actual paleowater depths can be approximated. Also, by comparing the depth zonation with thickness trends, the

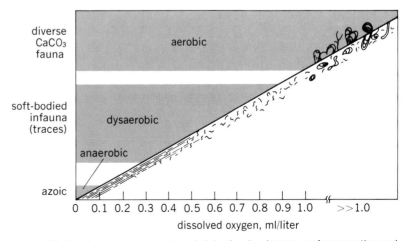

Model of facies change in a stagnant euxinic basin, showing zones of oxygenation and distribution of fossils and bioturbated sediment. (*After D. C. Rhoads and J. W. Morse, Evolutionary and ecologic significance of oxygen-deficient marine basins, Lethaia, 4:413–428,1971*)

relationship between sedimentation and subsidence can be clarified. This anaerobic basin model has been advanced to explain several shales of Devonian and Cretaceous age.

For example, in the Early Cretaceous Mowry Shale of Wyoming, a west-to-east transect reveals changes in the sedimentary fabric. Shale in the west contains abundant evidence of burrowing animals. Eastward, bioturbate texture gives way to laminated silt and clay. This facies sequence implies an eastward-dipping submarine slope, and the point of transition records the dysaerobic-anaerobic boundary's impingement on that slope. In a modern anaerobic basin, the Black Sea, the top of the anaerobic water mass lies at about 150 m depth. By analogy, depths in the Mowry Sea must have reached at least 150 m. Because the Mowry Shale also thins eastward across Wyoming, it is clear that the deepest water accumulated the least sediment. Sedimentation proceeded most rapidly in the western shallow zone, implying contemporaneous subsidence of the sea floor there.

Oxygen-minimum zone model. Another model for deep-water deposition, for widespread black shales of Pennsylvanian age, has been advanced by P. H. Heckel. As observed previously, oxygenation at depth depends on lateral influx of normally oxygenated ocean water. However, in parts of the world ocean, oxygen is strongly depleted in a zone a few hundred meters thick which lies below the surface mixed layer. This oxygen-minimum zone is produced by oxidation of settling organic carbon generated in the surface waters above. An epicontinental sea which received influx at depth only from the oxygen-minimum zone would remain dysaerobic even in the absence of a sill. Thus if the Pennsylvanian seas which flooded the midcontinent were deep enough to keep surface oxygen from reaching the bottom and had quasi-estuarine bottom currents which tapped the oxygen-minimum zone, then black muds would be deposited over broad areas.

Records of ocean basin sediments. Evidence for the importance of the oxygen-minimum layer also appears in the sedimentary record of the ocean basins. Deep-ocean drilling has revealed the presence of black mud zones in the normally well-oxidized sediment column of the ocean floor. A. G. Fischer and M. A. Arthur have detected a cyclicity of 32,000,000 years between periods of black mud deposition, with the most pronounced episode coming at the end of the Early Cretaceous (about 100,000,000 years before present). During that time the oxygen-minimum layer apparently expanded vertically due to sluggish global circulation, leading to dysaerobic-anaerobic conditions of deposition on the ocean floor. Epicontinental seas of Early Cretaceous age also deposited black shales, possibly because of the upward expansion of the oxygen-minimum zone to a point where dysaerobic water lapped onto the continental surface. Because there is no preserved oceanic sedimentary record of Paleozoic time, it is not possible to correlate the Pennsylvanian and Devonian shales discussed previously with events in the world ocean. However, W. B. N. Berry and P. Wilde have noted the progressive decrease in abundance of black shales through the Paleozoic Era and have proposed a gradual increase in ventilation of the ocean to account for it. Early in the Precambrian the oceans and atmosphere were anaerobic. With the development of oxygen in the atmosphere through photosynthesis, the surface ocean waters became aerobic, but the deep water may have remained anaerobic until global deep circulation was initiated by glaciation in the late Precambrian. Berry and Wilde's model is similar to that of Fischer and Arthur: both postulate an expanded oxygen-minimum layer to produce widespread black shales. However, this expanded layer may have been a permanent feature in early Paleozoic time, gradually diminishing by the late Paleozoic, and only episodic in the Mesozoic and Cenozoic.

For background information *see* MARINE GEOLOGY; SHALE in the McGraw-Hill Encyclopedia of Science and Technology.   [CHARLES W. BYERS]

Bibliography: C. W. Byers, in H. E. Cook and P. Enos (eds.), *Deep-Water Carbonate Environments*, SEPM Spec. Pub. 25, pp. 5–17, 1977; A. G. Fischer and M. A. Arthur, in H. E. Cook and P. Enos (eds.), *Deep-Water Carbonate Environments*, SEPM Spec. Pub. 25, pp. 19–50, 1977; P. H. Heckel, *Amer. Ass. Petrol. Geol. Bull.*, 61:1045–1068, 1977; W. A. Pryor, *Geol. Soc. Amer. Bull.*, 86:1244–1254, 1975.

Ship, merchant

In 1978 an international conference adopted important new regulations governing the design and operation of tankers in order to prevent oil pollution at sea. Another major conference adopted regulations on the training and certification of seafarers. Both sets of regulations were in the process of being ratified as of mid-1979, but the tanker regulations were already being brought into force in the United States.

IMCO. Almost all significant worldwide regulation changes in merchant ship design are created at the United Nations special agency for marine safety, the International Maritime Consultative Organization (IMCO). IMCO has a secretary general, a council, and an assembly just like the parent organization, and is located in London. It was quite natural for the UN marine safety organization to be thus located since under the old system of international law England was the official host country of both of the original major maritime safety conventions: the Safety of Life at Sea Convention of 1930 and the International Convention on Load Lines of 1930.

Since IMCO was created in 1960, it has been host in all international maritime safety matters. Moreover, marine law has become an ongoing process. Rather than waiting for the host country to call a new conference, delegations from all member nations meet at least once a year to discuss improvements in each safety area for existing conventions, proposed recommendations, voluntary codes of safe design practice for new ship types, and so on. The technical groups meet separately on each functional topic such as fire protection, lifesaving apparatus, subdivision and stability, bulk chemicals, and containers and bulk cargoes.

In addition to the functional subcommittee meetings, there has been at least one full-scale conference each year for the last 10 years in maritime safety.

Tanker safety and pollution prevention. Of particular interest was the Marine Pollution Protocol of 1978 on Tanker Safety and Pollution Prevention, which is commonly called TSPP-1978. This was a special convention called on short notice and completed within 1 year to consider safety measures for tankers in addition to those already in the Marine Pollution Convention of 1973. The reason a new agreement was necessary became evident during the winter of 1975/76 when over a dozen tanker accidents occurred within about 2 months, including the *Argo Merchant* off Nantucket Island.

These accidents caused President Jimmy Carter to send initiatives to Congress on Mar. 17, 1977, to help prevent pollution at sea by a series of immediate actions. He asked that: Congress ratify the 1973 International Convention for the Prevention of Pollution from Ships (MARPOL 73); international ship construction standards of tankers over 20,000 deadweight tons (DWT; 20,321 metric tons deadweight) be reformed by requiring double bottoms, segregated ballast, inert-gas systems, extra radar equipment, collision avoidance equipment, and improved steering standards; international inspection and certification of tankers be improved worldwide, including a new safety information system; both American and international crew training and license standards be improved; oil pollution liability and compensation legislation be approved in the United States; and Federal ability to respond to oil pollution emergencies be improved.

During the summer and fall of 1977, Americans in government and in industry joined in implementing the Presidential initiatives in many ways. The United States requested a special conference from IMCO as soon as possible, and one was set up for February 1978 in London (normally, several years of preparatory work are needed for a full diplomatic conference). Visits to many nations were undertaken to explain the United States' ideas for change and to gain support. Finally, the conference met for 2 weeks in February 1978 and adopted greatly expanded inspection and certification requirements.

In particular, the TSPP protocol produced agreements in six areas of design and construction, and strengthened inspection and survey requirements and recommendations for improved crew standards. The design, construction, and equipment areas are: (1) segregated ballast on new tankers; (2) segregated ballast or equivalent in existing tankers; (3) drainage and discharge arrangements; (4) inert-gas systems for protection of cargo tanks; (5) steering gear improvements; and (6) radar and collision avoidance aids. Each of these design improvements will be described separately.

Segregated ballast on new tankers. There are two ways that tankers may contribute to the pollution of the oceans: by a catastrophic accident such as a stranding, structural failure, or collision; and by putting sea water in cargo tanks for ballast on return trips and for ordinary cleaning of cargo tanks between trips. This second type of pollution has been accused of causing more pollution than catastrophic accidents. Since pollution from this source is more controllable than that from a navigation accident, a strong effort was made in the 1970s to reduce or eliminate pumping oily water overboard. One of the best ways to accomplish this is by placing as much ballast as possible in separate tanks. In MARPOL 73, agreement to require segregated ballast was limited to very large tankers—those of 70,000 DWT (71,123 metric tons deadweight) and above. TSPP-78 moved this lower limit down to 20,000 DWT (20,321 metric tons deadweight) for all new oil tankers, and 30,000 DWT (30,481 metric tons deadweight) for all new "product" tankers. A product tanker carries refined oil products such as gasoline, kerosine, or aviation fuel.

The conference also agreed that the segregated ballast tankage in each ship should be located along the sides and bottom of the ship so as to protect the oil cargo from loss in a minor accident. Formulas for the best location of this protection are currently being considered by the technical groups which meet continuously at IMCO.

Segregated ballast on existing tankers. The conference decided that the tankers already sailing the oceans should also be required to reduce their operational pollution by one or more equivalent methods. If they could change cargo tanks to segregated ballast tanks, it was considered the best approach. The second-best solution was a rededication of a few cargo tanks to clean ballast tanks. Both systems reduce the amount of oil that a tanker can carry by varying amounts according to the design of the ship.

Drainage and discharge arrangements. These are requirements to minimize cross-connections in cargo pumping systems which could inadvertently cause oil to contaminate ballast water.

Since MARPOL 73, the tanker industry has been experimenting with a new way to clean tanks between loaded trips. This method uses the oil itself in a high-pressure washdown arrangement to scour cargo tank bulkheads and wash the heavier sediments down to the tank bottoms so that the residues can be pumped to the cargo terminal along with the oil. The system is used only while discharging the cargo.

Inert-gas system. A development of the tanker industry to reduce the possibility of explosions in tankers has been the use of the exhaust gas from a ship's boiler or main engine which is low in oxygen content for venting main cargo tanks, instead of allowing air to enter the cargo tanks while offloading oil at the terminal. The crude-oil washing system combined with the inert-gas system was recognized as having a certain equivalency by the TSPP convention. Further, the timetable for both crude-oil and product tankers requires that most will be equipped with an inert-gas system by 1983.

Steering gear systems. All tankers above 100,000 gross tons (entire capacity of 283,168 m³) are to have two remote steering gear control systems operating separately from the navigating bridge. Additionally, the main steering gear of such tankers must comprise at least two identical power units, each of which is capable of operating the rudder by itself.

Radar. The conference agreed that all ships (not just tankers) of 1600 gross tons (4531 m³) or more should be equipped with radar, and ships of 10,000 gross tons (28,317 m³) or more should have two radars, each capable of independent operation.

Training and certification of seafarers. The second major conference in 1978, the International Conference on Training and Certification of Seafarers (Standards of Training and Watchkeeping— 1978), took place in London in June. Unlike the TSPP conference, this one had been planned since 1972 and a draft convention had been produced. As a result of the Presidential initiatives of March 1977, this conference was convened 5 months before it was originally scheduled, and it contained requirements for masters, mates, and engineers to handle tankers in order to reduce the human failure aspect of tanker accidents.

The Standards of Training and Watchkeeping conference adopted minimum requirements for both deck and engineer officers and ratings (seamen deck and engine departments). It agreed upon minimum knowledge and sea time requirements for qualifying for each level and for renewing the certificate every 5 years. It adopted a special chapter of extra requirements for masters, officers, and ratings of oil tankers, chemical tankers, and liquefied-gas tankers. Finally, it adopted rules for knowledge of survival craft (which includes lifeboats).

The basic requirement for experience at sea before getting a master's or other officer's license is now 3 years of sea time. For deck officers, 2 years of this may be obtained at special merchant marine school. For engineering officers, all of it may be obtained at a nautical school. A qualified deck watch officer may obtain a chief mate's license after 18 months' sea experience and a master's license after 3 years. Engineering watch officers may obtain a second engineer's license after 12 months and a chief engineer's license after 3 years. Finally, the conference spelled out a list of technical subjects in which each watch officer is expected to be knowledgeable no matter what country the license is issued in.

Although both of these conferences were attended by many nations, the regulations must now be ratified by enough countries to bring them into force in international law. Meanwhile Congress has required the U.S. Coast Guard, which is responsible for marine regulations in the United States, to place the new TSPP regulations into effect in several steps beginning in 1979.

For background information *see* SHIP, MERCHANT in the McGraw-Hill Encyclopedia of Science and Technology.

[WILLIAM A. CLEARY, JR.]

Bibliography: Code of Federal Regulations, Department of Transportation, *Tank Vessels of 10,000 Gross Tons or More and Tank Vessels of 20,000 DWT or More Carrying Oil in Bulk*, Feb. 12, 1979; Code of Federal Regulations, *Tanker Safety and Pollution Prevention, Information and Regulatory Implementation Plan*, Apr. 20, 1978: Exxon Corporation, *Exxon Marine*, vol. 23, no. 2, summer 1978; International Maritime Consultative Organization, *Final Act of the International Conference on Training and Certification of Seafarers*, STW/CONF/12, 1978; *IMCO News*, no. 1, 1979; W. D. Snider, USCG Cmdr., in *IMCO Conference on Tanker Safety and Pollution Prevention*, Society of Naval Architects and Marine Engineers, San Diego Section, 1978.

Ship design

The past few decades have witnessed many significant developments in the maritime industry. Moreover, advancements in hull and machinery design, such as container ships, VLCCs (very large crude carriers), controllable-pitch propellers, offshore drilling structures, transverse thrusters, roll-on/roll-off ships, and stabilizers, to name but a few, have contributed significantly to international commerce by increasing the efficiency of seaborne trade.

That such innovations have proved so successful is due in part to the work of classification societies, whose major function is to assure the fitness of merchant ships and other marine structures for their intended service. Classification societies review and analyze design plans and survey construction to assure that vessels which have been presented for classification are structurally and mechanically safe.

Similarly, classification societies are currently hard at work to assure that designs now under development will result in safe vessels to serve society tomorrow. This is accomplished through a systematic design review procedure by which the societies verify that a vessel's plans adhere to recognized design standards. If a design is so novel that it is not covered by a codified set of standards, the society performs a rigorous analysis using proved engineering principles and practices to ascertain if the design is acceptable and capable of producing a safe ship. The effect, then, is that classification societies act as a safeguard in assuring that designs submitted to them are acceptable.

Rules. The standards used for review, known as Rules, which are integral to the classification societies in their duties of design review and other functions, are promulgated and annually updated through committees composed of owners, builders, naval architects, underwriters, marine engineers, and individuals from other allied fields who are eminent in the marine industry. The committee arrangement allows the societies to maintain close contact with developments and interests in various geographical regions as well as various technological and scientific disciplines. In addition, the committee arrangement has the advantage of allowing all segments of the industry to participate in producing the societies' Rules. As a result, the Rules are recognized as authoritative and impartial, and serve as a self-regulatory mechanism for the marine industry.

Review of design. The actual classification procedure begins with the genesis of the vessel—its design. Vessel design plans are formally reviewed by technical surveyors of the various societies to verify that the plans adhere to accepted standards of good design practice as embodied in the Rules. As explained, the Rules are predicated upon principles of naval architecture, marine engineering, and other engineering and scientific disciplines that have proved satisfactory through service experience and practice.

So in reviewing a given set of design plans, a society compares them with a compendium of experience factors and scientific principles. In this way a society is able to determine if the design is adequate in its structural and mechanical concept and, therefore, acceptable to be translated into an actual marine vessel. However, it is important to understand that the Rules do permit a certain latitude of ideas without which vessels could only follos the designs of predecessors.

Specifically in regard to the Rules of the American Bureau of Shipping (ABS), variations in the stated design requirements are sanctioned under a general proviso that, in essence, permits alternative arrangements to those specified if they can be demonstrated as sound and acceptable. In so demonstrating, the owner and the designer must apply a recognized scientific basis such as valid theoretical approaches, authoritative test data, or proved engineering principles to support the design.

If the design is found acceptable, it is approved and certified. On the other hand, if a society finds a design unacceptable, the underlying reasons are discussed with the owner, thereby affording an opportunity to resolve the deficiency and institute corrective measures.

Construction. Once a plan has been formally approved by a classification society, construction starts at the shipyard. Thus begins the second phase of classification, during which time society field surveyors "live" with the vessel; that is, they survey its construction to certify that the approved plans are followed and the Rules are adhered to in all respects. Surveyors also witness the testing of material, machinery, and components at the manufacturers' plants to determine that they also comply with the Rules.

Sea trials. When completed, the vessel is given sea trials, marking the third phase of the classification procedure. Society field surveyors attend the trials to verify that the vessel performs according to specifications. The vessel is then officially classed, that is, approved, by the society.

However, in order to retain its classification status the vessel must undergo periodic surveys, as detailed in the Rules. In this way it can be determined that the vessel continues to be maintained in a structurally and mechanically fit condition for service.

Function of classification. The decision as to classification rests with the owner. The owner selects the society to class the vessel, and, indeed, decides whether the vessel is to be classed. However, it strongly behooves an owner to have the vessel classed due to the magnitude of the investment it represents, the lives and cargo it will carry, and the environment in which it will operate. Classification provides the owner with evidence that due diligence has been exercised in making the vessel sound. This is important for the owner's peace of mind, and also because there are many interested parties such as underwriters, financial institutions, government bodies, and charters that may wish this assurance.

The influence of the classification society on the development of ship design, then, is to provide a mechanism by which a given design, be it standard or innovative, can be demonstrated as structurally and mechanically safe. In this way the revolutionary design concepts of today can be developed into realities tomorrow.

For background information *see* SHIP DESIGN; SHIPBUILDING in the McGraw-Hill Encyclopedia of Science and Technology. [ROBERT F. VOLLACK]

Bibliography: American Bureau of Shipping: *Rules for Building and Classing Offshore Mobile Drilling Units*, 1973; *Rules for Building and Classing Steel Vessels*, published annually; *Rules for Building and Classing Steel Vessels Under 61 Meters (200 Feet) in Length*, 1973; *Rules for Building and Classing Underwater Systems and Vehicles*, 1979.

Skeletal system disorders

Such diverse diseases as psoriatic arthritis, Reiter's syndrome, and the arthritis associated with granulomatous or ulcerative colitis have in common with ankylosing spondylitis a similar pattern of inflammation of various tissues. While the spine and related structures are the major targets, the anterior chamber of the eye and the media of the aorta are also involved. Thus, on clinicopathologic grounds a kinship among these diseases was recognized even before it was learned that a unique genetic trait, the cellular imprint of the HLA-B27 (formerly called HLA-W27) antigen, was found to underlie this association.

Inheritance of HLA-B27. In the human, the chromosome pair C6 bears a segment called the HLA region. Within this region, a series of loci, in close proximity to each other, carry genes which code for the production of proteins involved in the host's histocompatability and immune responses. Genes of the four loci of the HLA region, namely A, B, C, and D, code for distinct cell membrane products. Each of the four loci consists of a large series of allelic genes. An individual would inherit one allelic gene for each of the A, B, C, and D loci from the maternal parent and the other HLA allelic gene for each locus from the paternal parent. Thus, in the absence of genetic crossing-over, the alleles of the HLA region are inherited as a common unit on one autosomal chromosome 6, from each parent. This unit is called a haplotype (from haploid gene type).

Thus, every individual has as many as eight gene products coded for by eight alleles, four derived from each patient. Each of these allelic genes governs the production of a separate membrane protein, called the HLA antigen, that appears on most nucleated cells of the individual.

Should an individual have only one parent with the B27 allele at the B locus, he or she would have a 50% chance of inheriting this allele, and would then be a heterozygote for B27. In certain cases, a person might inherit the B27 allele from each of the two parents, thus making him or her a homozygote for B27. Current evidence suggests that homozygosity confers a somewhat higher degree of disease susceptibility on an individual who bears a double dose of these alleles compared with the individual who is a heterozygote. Since these HLA proteins in an incompatible tissue allograft can induce a strong immune response in the host against the transplant, such as a kidney, they are known as histocompatibility or tissue antigens.

Testing for HLA-B27. Lymphocytes and other nucleated cells, but not red cells, carry the HLA proteins on their membranes. Because of easy accessibility, lymphocytes are used as the test cell and, after separation from other cellular elements in the blood specimen, are tested against a large number of human antisera representing about 45 or more different HLA antigens corresponding to the various cell surface antigens. By using a large battery of antisera, each of the HLA antigens in an individual can be determined. Occasionally, only one antigen of a locus is recognized, implying either that one of the antigens may be present in a double dose, that is, the person is a homozygote for the HLA allele, or that the individual possesses an unknown antigen against which a typing antiserum has not yet been found. Family studies would help resolve whether the person is a homozygote, since each parent would have to carry the same allele.

Ankylosing spondylitis. Several large series of patients with ankylosing spondylitis have been examined and compared with control individuals. The frequency of the HLA-B27 antigen in the normal Caucasian population in the United States is approximately 8%, while among patients with ankylosing spondylitis the frequency runs well over 90%. In black Africans the frequency of HLA-B27 is almost 0%, but in black Americans the frequency is approximately 4%, a finding that corresponds to a lesser prevalence but not total absence of this disease in American blacks. More recent studies indicate, however, that only about half of those blacks who do have spondylitis carry the B27 antigen. Such B27-negative persons frequently possess a B-locus gene coding for an HLA antigen that is immunologically cross-reactive with B27 as B7, BW22, or BW42.

When one compares the incidence of various HLA antigens in a group of Caucasian patients who have ankylosing spondylitis with the incidence in a control group consisting of Caucasian individuals who may have gout or rheumatoid arthritis or who may be healthy individuals, one finds that the distribution of all antigens other than B27 is more or less the same in all groups. However, B27 is uniquely associated with spondylitis.

Psoriatic arthritis. In the psoriatic patient, the presence of B27 antigen corresponds more closely with the arthritic component of the disease than with skin changes alone. By inference, patients with skin psoriasis who lack B27 may be at less risk of arthritis. When the type of arthritis that exists in such patients is examined more closely, one finds that the association of B27 corresponds more closely with those psoriatic arthritic patients who have spinal involvement than with those patients who have only peripheral arthritis without spondylitis (20% or more may be HLA-B27-positive).

Reiter's disease. In several series, an even stronger association was noted between the B27 antigen and Reiter's disease than reported above for patients with psoriasis. Carriage of the B27 antigen confers upon the patient with Reiter's disease a propensity toward a more severe form of ill-ness. Uveitis, a relapsing course rather than an isolated event, and signs of toxicity, as weight loss and fever, were more common in those individuals with the B27 marker. Furthermore, sacroiliac disease is almost exclusively reserved for this group with B27, while peripheral arthritis occurs both with and without B27.

Colitic arthritis. Patients with ulcerative colitis and a peripheral arthritis that usually presents as an asymmetrical oligoarthritis of the lower extremities do not share in the tendency to carry the B27 antigen. On the other hand, patients who have an associated spondylitis are usually B27-positive. A similar pattern marks the association of spondylitis and Crohn's disease.

Yersinia arthritis. Species of *Yersinia* (formerly identified as *Pasteurella*) cause an acute febrile and diarrheal illness, occasionally severe enough to present as an abdominal crisis. Some patients also develop erythema nodosum and arthritis. Most instances of this type of infection have been reported from Scandinavian countries. Among this group of patients, those who have arthritis are B27-positive, whereas those who are ill but do not have arthritis lack B27. The genetic carriage of B27 appears to condition the person for the development of arthritis after exposure to an exogenous agent such as *Yersinia*. As noted previously in patients with Reiter's disease, the B27 state is related to the development of arthritis which is more likely to involve a greater number of joints for a longer period of time and, in particular, to have a close association with sacroiliitis and eye findings.

Risk of spondylitic disease in B27 population. Although a very close association with B27 has been found for patients who already have ankylosing spondylitis or one of the variants, the risk of developing future disease if one inherits the B27 antigen has not previously been known. Certainly, the prevalence of spondylitis or its variants is far less than the known number of B27 males or females in the population. For example, of the total population of approximately 200,000,000 in the United States, 8,000,000 males and an equal number of females are B27-positive. Even for the males with a much higher prevalence of spondylitis than females, overt disease is found in only 5% of the B27-positive group. Recent studies indicate that the remaining B27 individuals, that is, the other 95% of the population, share an approximate 20% risk of spondylitic disease. It should be emphasized, however, that the degree of actual illness in these latter persons was mild, so that the concern about the magnitude of the health problem that might be involved in this sizable fraction of the population should not be overemphasized.

Significance of association. All these studies document a clear relationship between spondylitis and the B27 antigen regardless of whether the disease is primarily ankylosing spondylitis or the spondylitic element in the variant diseases. Earlier clinical studies that demonstrated a hereditary tendency are now supported by the inheritance of the allelic gene coding for the B27 antigen. Moreover, the marked vulnerability of about one-quarter of the carriers of B27 for symptoms of disease further strengthens the suggestion that the B27 state is a necessary prerequisite for its development. Still unexplained is the overwhelming preponderance of definite disease in males, although females might share equally in the risk for subclinical disease.

In addition to a genetic component, the necessity for some environmental factor seems apparent. The demonstration that B27-positive patients with *Yersinia* infection are the patients who also develop arthritis links an external factor such as a bacterium with a genetic vulnerability. In the example of infection by *Yersinia* a bacterium may be considered as the triggering agent. In Reiter's syndrome one might postulate that some type of infectious process occurs, although at present no infectious agent has been definitely incriminated. Nevertheless, epidemiological data would lend support to such a view since the disease often appears after venereal contact or diarrheal epidemics. In other conditions, such as psoriasis or perhaps some forms of colitis, inflammation of the skin or the bowel may cause the B27-positive individual to develop spinal disease. In ankylosing spondylitis the external stimulus remains to be determined. For all of these conditions, a viable hypothesis can be constructed that some environmental factor may initiate the development of an inflammatory process in a B27 individual. When this happens, selected tissues come under attack, as the spine, the anterior chamber of the eye, or the elastic tissue of the aorta.

The exact manner by which the B27-positive individual is made susceptible to disease still remains obscure. According to one hypothesis, the offending microorganism mimics the antigenic characteristics of the B27 antigen so that the host does not recognize the foreign invader. Thus disguised, the agent gets a foothold to cause disease. Another hypothesis suggests that a close linkage exists between the gene that codes for B27 antigen and another more important gene, such as the immune response gene, at a nearby site that governs the host's ability to respond to a given antigen. If a particularly "poor" immune response gene(s) is more often linked to the B27 gene, the B27-positive individual would not be able to form a strong immune response to an offending agent; whereas other individuals without B27, who more frequently may have "strong" immune response gene(s), would have an effective immune response. This phenomenon of high association of certain types of linked genes is known as gametic association.

Whatever the ultimate explanation, the syndrome of the spondylitic variants is closely associated with both the genetic factor, B27, and an environmental factor, presumably a microbial agent, a circumstance that creates a unique opportunity in medicine to analyze and study the interplay between these forces.

For background information *see* HUMAN GENETICS; IMMUNOLOGY; SKELETAL SYSTEM DISORDERS in the McGraw-Hill Encyclopedia of Science and Technology.

[FRANK R. SCHMID]

Bibliography: D. A. Brewerton et al., *Lancet*, 1: 904, 1973; R. H. Goldin and R. Bluestone, *Clin. Rheum. Dis.*, 2:231, 1976; L. Schlosstein et al., *N. Engl. J. Med.*, 288:704, 1973.

Soil

Nitrification inhibitors are used in agriculture to delay transformation of ammonia to nitrate by soil microorganisms. Recent work has focused on understanding the mode of action of nitrification inhibitors, reducing the potential for nitrogen loss when inhibitors are used, and investigating factors that affect the persistence and bioactivity of these inhibitors.

Mode of action. The biological oxidation of ammonia-N (ammonium-N) to nitrate-N is carried out in soil by chemoautotrophic bacteria which derive from this process (nitrification) the energy needed for their metabolic activities. Many chemicals inhibit the growth of soil microorganisms, including the nitrifying bacteria. Chemicals which are nonspecific in their action affect nitrifier growth and proliferation by creating unfavorable microenvironments, by disrupting cell membranes or otherwise changing cell ultrastructure, or by interfering with cell respiratory or intermediary metabolism. Many chemicals which specifically inhibit nitrification interfere with energy-transfer systems in metabolic pathways. They include compounds which bind enzymes or heme proteins, chelate metals, affect the formation and release of high-energy phosphate bonds, or act as trappers of free radicals involved in the nitrification process.

Biochemistry. Ammonia is oxidized to nitrite by *Nitrosomonas* and *Nitrosocystis* by the probable reaction sequence: ammonia → hydroxylamine → (nitroxyl)? → (nitrohydroxylamine)? → nitrite. The overall reaction involves a net transfer of seven electrons and produces about 65 kilocalories (270 kilojoules) of energy per mole of ammonia-N oxidized. The final oxidation to nitrate is carried out mainly by *Nitrobacter*, involving transfer of one electron for each nitrite ion oxidized and producing 15 to 20 kcal/mole (63 to 84 kJ/mole). The energy balance in the nitro and nitroso groups of microorganisms is critical and, when upset, severely inhibits growth and activity.

Ammonia oxidation to hydroxylamine probably involves a metal such as copper as a component of a cytochrome or other oxidase enzyme system. There is evidence also for the involvement of molecular oxygen, an enzyme which binds carbon monoxide, one or more free radicals, a catalaselike enzyme, and some factor which first activates ammonia before its oxidation. Among the chemicals which chelate metals and inhibit ammonia oxidation to hydroxylamine are allylthiourea, diethyldithiocarbamate, ethyl xanthate, salicylaldoxime, cyanate, 8-quinolinol, and *o*-phenanthroline. Thiourea apparently impedes ammonia transfer into the *Nitrosomonas* cell, thereby inhibiting ammonia oxidation. Methylamine may compete with ammonia for permease, an enzyme which facilitates the penetration of cell membranes. The short-chain primary alcohols (methanol, ethanol, *n*-propanol, and *n*-butanol) inhibit ammonia oxidation, perhaps by trapping free radicals.

Allylthiourea has no effect, and the other chemicals mentioned have only slight effect on the oxidation of hydroxylamine. This process involves a dehydrogenation, with electron transfer through a flavoprotein-cytochrome electron transfer chain to form a product believed to be nitroxyl, NOH. Cytochrome inhibitors such as atabrine, dicumarol, cyanide, and hydrazine block the oxidation of hydroxylamine.

Little is known about the postulated intermediates, nitroxyl and nitrohydroxylamine. There is evidence that nitroxyl oxidation also involves a flavoprotein-cytochrome electron transfer system.

The "nitrite oxidase" of *Nitrobacter* is characterized by a unique reverse flow of electrons through a cytochrome system, yielding one high-energy phosphate bond per nitrite. It is also involved in respiration. Cyanate, certain isothiocyanates, and mercuric chloride are highly toxic to nitrite oxidase. Chlorate suppresses *Nitrobacter* growth, but does not directly inhibit nitrite oxidation. Although flavoproteins are not directly involved in nitrite oxidation, quinacrine, an inhibitor of flavoprotein activity, also inhibits nitrite oxidation. Chemicals such as citrate and amines which chelate iron and molybdenum also interfere with the flavoprotein-cytochrome and respiratory systems, thereby indirectly inhibiting nitrite oxidation. Several nitrophenols inhibit nitrite oxidation by interfering with the production of energy-rich phosphate bonds.

Specific inhibitors. The ideal nitrification inhibitor for use in agriculture specifically blocks ammonia but not nitrite oxidation, does not adversely affect other beneficial soil microorganisms and higher plants, and is not toxic to humans and animals in amounts needed to inhibit nitrification effectively. Of the chemicals available for use as nitrification inhibitors in agriculture, nitrapyrin, or 2-chloro-6-(trichloromethyl)pyridine, has received the most study. It acts by chelating copper in the cytochrome component of the system involved in ammonia oxidation to hydroxylamine. In solution cultures, addition of cupric ion will reverse the inhibition. As already noted, allylthiourea also blocks ammonia oxidation, but this blockage can only partly be reversed by addition of copper. Thiourea, used commercially in Japan as a nitrification inhibitor, acts by retarding ammonia transfer into the bacterial cell. Dicyandiamide, currently under market development, apparently inhibits the cytochrome oxidase involved in ammonia oxidation. Little is known about the mode of action of other chemicals developed as nitrification inhibitors. The substituted triazines perhaps retard nitrifier growth by interfering with carbon dioxide assimilation. The thiazoles have chemical structures that may substitute in part for essential growth factors, such as *p*-aminobenzoic acid. Triazoles may interfere with cataselike enzymes, believed to be involved in ammonia oxidation.

Pesticides. Many pesticides affect the activity of the nitrifiers to some degree. For example, eptam (a herbicide), maneb (a fungicide), and vapam (a fumigant) are thio- or dithiocarbamates which inhibit nitrite, ammonia, and ammonia plus nitrite oxidation, respectively. Other pesticides which inhibit ammonia oxidation are dyrene, terrazole, and chloropicrin. Monuron (CMU), PCP, telone, BHC, and vorlex affect ammonia and nitrite oxidation. Lindane, chlordane, DDD, CIPC, and heptachlor retard nitrite oxidation. The limited information available suggests that one mode of action, for example, that of CIPC and eptam, is inhibition of energy transfer from energy-rich phosphate bonds. The development of chemicals for use in agriculture proceeded independently of biochemical studies of their effect on nitrification, leaving a large gap in knowledge of their mode of action as nitrification inhibitors.   [ROLAND D. HAUCK]

Potential for nitrogen loss. The potential for loss of nitrogen from cultivated and natural soil-plant systems is the subject of broad concern. Efficient utilization of fertilizer nitrogen by crops is desirable from the standpoint of water purity since unused nitrates (NO_3^-) from fertilizer may leach from the soil and appear in groundwaters and streams. At high levels, nitrates in drinking water are considered a health hazard. Furthermore, atmospheric scientists suspect that nitrous oxide, N_2O, from the decomposition of nitrate in the soil may react to destroy the ozone layer in the stratosphere. A final and relatively unimportant concern is the cost of the lost nitrogen to the farm operator.

Fate of fertilizer nitrogen. There are several things that can happen to fertilizer nitrogen after it is added to the soil. The most desirable is absorption and removal in the crop. In general, about half of the nitrogen fertilizer is removed in the crop; some of it remains in the roots and the unharvested portions of the crop, and some is assimilated by soil microorganisms. The latter portion becomes part of the soil organic matter which decays through a period of years and returns the nitrogen to the soil solution. Soil erosion brings about a loss of nitrogen by physical removal of the soil. Since the smaller soil particles are higher in soil organic matter, the loss of nitrogen during erosion is disproportionately greater than the loss of whole soil. Uptake of nitrogen by plants is not entirely a one-way process. Small amounts of nitrogen are returned to the plant environment in the water of guttation excreted from leaves and also in root exudates. Fragrances and aromas from plants contain traces of nitrogen. Gases involved in plant biochemical reactions, such as N_2, N_2O, and NH_3, are often emitted in small amounts from plant leaves.

Some fertilizers, such as urea, ammonia, or ammonium salts, when added to the soil can lose nitrogen directly to the atmosphere through volatilization of ammonia. This type of loss is more likely in calcareous and alkaline soils and can be largely prevented by careful incorporation of the fertilizer into the soil.

Leaching and denitrification are usually the most serious loss processes. Although these two processes are entirely different, both bring about the loss of nitrogen in the nitrate form and both occur more extensively in excessively wet soils. Leaching is a physical process whereby nitrates dissolved in the drainage water are moved down, through, and out of the soil. Denitrification is a biochemical, microbial process which converts nitrates to nitrogen gases that may then escape from the surface of the soil.

Sources of nitrates in soils. Fertilizers are the main source of nitrogen in most highly productive agricultural soils. Nitrates are also released by decomposition of organic nitrogen compounds in the soil organic matter. Small amounts of nitrate come from atmospheric pollution and also from nitrogen fixation by lightning.

Fertilizer nitrogen is applied to the soil in several different forms. Anhydrous ammonia and aqueous solutions of ammonia often containing urea and ammonium nitrate are common. Also, ammonium salts, such as ammonium sulfate, are often used. Regardless of the form applied, virtually all nonnitrate fertilizer nitrogen not immediately taken up by the plant is transformed by soil bacteria to nitrates. Since the nitrate form of nitrogen is subject to both denitrification and leaching, this transformation means that the fertilizer nitrogen has been put into an unstable condition.

Ammonium nitrogen, on the other hand, is not readily leached from soils, and it is not converted directly to nitrogen gases. Until it undergoes the transformation to nitrate, it is a plant-usable, stable form of nitrogen. However, since the transformation of ammonium to nitrate usually takes less than a few weeks in warm moist soils, the advantage of using ammonium nitrogen over nitrate nitrogen for the prevention of leaching and denitrification is real but usually of relatively short duration.

Since the ammonium is neither leached nor subject to denitrification, nitrification inhibitors are sometimes added with the fertilizer. Presumably, if the nitrification (transformation of ammonium to nitrate) could be inhibited or delayed, the potential for nitrogen loss by denitrification and leaching could be reduced.

At present, the only nitrification inhibitor approved for sale in the United States is nitrapyrin, or 2-chloro-6(trichloromethyl)pyridine, a product of the Dow Company marketed under the trademark N-Serve. This chemical acts on the soil bacteria that transform ammonium to nitrite and thus prolongs the time the fertilizer nitrogen will remain in the soil. Yield responses from nitrification inhibitors have been variable, but their use is increasing because farm operators want to distribute their peak labor load over a longer period and use N-Serve to permit earlier application of fertilizer.

Amounts of nitrogen lost. Amounts of nitrogen lost by leaching and denitrification may be a very few pounds per acre under dry-land agriculture to hundreds of pounds per acre under irrigated agriculture in warm, humid regions. Since denitrification is a microbial process, it takes place slowly in cold weather. Similarly, leaching does not occur during the winter in cold regions when the soil is frozen. Losses from agricultural soils have been estimated at around 75 lb/acre (84 kg/ha) in the eastern United States, 20 lb/acre (22 kg/ha) in the central cornbelt, and as much as 900 lb/acre (1000 kg/ha) in some irrigated areas of California that are cropped, fertilized, and irrigated continuously throughout the year.

Estimates of losses of fertilizer nitrogen by denitrification range from 15 to 25% but may be greater in warm, wet soils.

Several practices may be utilized to reduce the potential for nitrogen loss. Most of these are already followed by farm operators because they are generally beneficial in addition to their effect in conserving nitrogen. These practices include erosion control, good drainage, use of cover crops,

delayed and proper fertilizer applications, fertilizer rates adjusted to expected yields, and perhaps the use of a nitrification inhibitor.

[L. T. KURTZ]

Persistence and bioactivity. There are a number of soil properties and environmental variables which are known to influence the persistence and bioactivity of nitrification inhibitors. Virtually all of the studies have been with nitrapyrin, and the discussion will be largely oriented to this compound.

Factors of potential importance. The table lists the factors potentially important in the persistence or bioactivity of nitrification inhibitors. Bioactivity and persistence are closely related. Bioactivity is considered here to be the relative inhibition of nitrification. A compound may persist for a long time but not be bioactive due, for example, to adsorption on soil organic matter; yet in another situation the compound may degrade rapidly but have a high bioactivity while it is present.

A number of factors, including soil pH, temperature, and organic matter content, are linked with many of the processes that affect bioactivity, indicating that complex interactions are involved. Soil pH is particularly hard to evaluate since it fluctuates widely, especially in zones where high pH−forming fertilizers are being nitrified. The intrinsic nitrifying ability of a soil is another dynamic factor that is difficult to quantify.

Persistence. Cleve Goring conducted most of the initial research on nitrapyrin. He demonstrated that it was tightly adsorbed on soil organic matter, and that as the organic matter of a soil increased, more nitrapyrin was required to give the same degree of inhibition of nitrification. This has been confirmed in several other studies, including that by Geoffry Briggs and coworkers, who found that nitrapyrin degradation was slowed by sorption on organic matter. Since nitrapyrin is so tightly adsorbed on organic matter, it does not leach through the soil. It does have a fairly high vapor pressure, and can readily volatilize from the soil surface. However, as Briggs demonstrated, it does not move far from the zone of placement when incorporated in the soil. This can limit its effectiveness

Potentially important factors in persistence and bioactivity of nitrification inhibitors

	Persistence
Sorption	Surface area (clay, organic matter), temperature, pH
Degradation	Temperature, sorption, pH
Volatilization	Sorption, degradation, placement, pH
Movement	Volatilization, sorption, leaching, pH
	Bioactivity
Persistence	All of above
Sorption	Activity of sorbed phase
Nitrifier activity	Species population, genera, pH, organic matter, aeration, crop, climate, temperature, water, salinity, sample pretreatment, degree of inhibition
Nitrifier recovery	Degree of inhibition, pH increase or decrease due to N fertilizer, salts, Cu availability
Form of fertilizer	Rate of hydrolysis, pH effects
Mode of fertilizer application	Band and broadcast, rate of diffusion and leaching of ammonium versus that of inhibitor

since the fertilizer nitrogen may move away from the nitrapyrin-affected zone and then be nitrified.

Nitrapyrin degrades rapidly by a chemical hydrolysis reaction to a much less active compound, 6-chloropicolinic acid. The rate of this reaction is primarily a function of temperature, and it occurs much faster in warm than in cold soils.

Bioactivity. The bioactivity of the inhibitor is related to the activity of the nitrifier population. Nitrification proceeds most rapidly in soils with near-neutral pH and increases at about a twofold rate for each 10°C rise from about 5 to 30°C. Thus more of the inhibitor would be required when nitrification is rapid than when it is slow. Also, when conditions for nitrification are ideal, the nitrifier population can reinfest the treated zone following degradation of the compound. The form of nitrogen fertilizer also can affect bioactivity. For example, urea, when it hydrolyzes to ammonium, increases the soil pH markedly, oftentimes giving conditions more favorable for nitrification.

Disulfides. Another group of compounds that are attracting interest as nitrification inhibitors are the disulfides. These compounds degrade to carbon disulfide, CS_2, which is quite volatile, and not nearly as tightly adsorbed on soil components as nitrapyrin and is shorter-lived. Thus it controls nitrification in a large volume of soil, but for a shorter time.

For background information *see* FERTILIZER; NITROGEN CYCLE in the McGraw-Hill Encyclopedia of Science and Technology.

[DENNIS R. KEENEY]

Bibliography: M. I. H. Aleem, *Annu. Rev. Plant Physiol.*, 21:67–90, 1970; J. Ashworth et al., *J. Sci. Food Agric.*, 28:673–683, 1977; G. G. Briggs, *J. Sci. Food Agric.*, 26:1083–1092, 1975; J. K. R. Gaser, *Soils Fertil.*, 33:542–554, 1970; C. A. I. Goring, *Soil Sci.*, 93:211–218, 1962; R. D. Hauck, in *Nitrification Inhibitors: Potential and Limitations*, Amer. Soc. Agron. Spec. Publ., 1979; D. R. Nielsen and J. G. MacDonald (eds.), *Nitrogen in the Environment*, vols. 1 and 2, 1978; L. F. Welch, *Nitrogen Use and Behavior in Crop Production*, Univ. Ill. Agric. Exper. Sta. Bull. no. 761, 1979.

Soil chemistry

One of the most important properties of soils is the ability to retain metal cations in forms available to plants. This property is almost wholly attributable to clay minerals and organics, both of which are known to adsorb metals in forms that may or may not be exchangeable by other metal ions. The recent use of electron spin resonance (ESR) spectroscopy as a means of investigating these metal-surface interactions has improved understanding of the mechanisms of soil reactions with metals.

Advantages of ESR. Although ESR has not often been applied to natural clay or organic systems, it has a number of advantages over more conventional research techniques. The method is nondestructive, allowing the investigation of systems under near-ambient conditions with little disruption of surface structures or chemical equilibria. Also, ESR is inherently sensitive, with a detection limit of observable species in solution on the order of 10^{-9} molar. This feature permits studies of trace metals in natural biological and mineral systems. In addition, samples can be analyzed in the gaseous, liquid, or solid state, and the chemical complexity of most natural sample matrices does not generally interfere with the technique since the spectrometer can detect only unpaired electrons in the sample.

Principle of ESR. The ESR experiment is based upon the principle that the spin of any unpaired electron in an atom or molecule, when placed in a magnetic field, becomes quantized and must accept one of two possible values, $+1/2$ or $-1/2$. These quantum states correspond to two allowed orientations of the electron spin, parallel or antiparallel to the applied magnetic field direction. The two orientations have slightly different energies, and an electromagnetic field of the proper frequency can be applied perpendicular to the magnetic field, providing the energy for an electron to "flip" from the lower- to the higher-energy spin state. The frequency of radiation required when the applied magnetic field is about 3400 gauss (as in most ESR instruments) is about 10^{10} Hz, corresponding to microwave radiation with a wavelength of about 3 cm. Measurement of the exact magnetic field required for the unpaired electrons in a sample to absorb radiation of a given frequency provides information about the type of atomic or molecular orbitals containing these electrons. In addition, the amount of radiation absorbed is a measure of the number of unpaired electrons in the sample. An ESR spectrum generally represents the first derivative of microwave absorption as the magnetic field is scanned.

Spectra of soil colloids. Most atoms and molecules have electrons which are paired in orbitals, so that the total electron spin is zero. As a result, these species are termed diamagnetic and do not possess ESR spectra. However, many transition metals in their common oxidation states have unpaired electrons in d-orbitals. Divalent manganese (Mn^{2+}) has five unpaired electrons, a total spin of 5/2, and is therefore paramagnetic. Soil colloids commonly demonstrate a very broad resonance attributable to Fe^{3+} oxide particles or coatings, a sharper low-field signal due to Fe^{3+} in structural sites of clay minerals, and a six-line spectrum arising from exchangeable Mn^{2+} (Fig. 1). The addition of Cu^{2+} to a soil produces a signal that can be assigned to adsorbed but hydrated and freely rotating $Cu(H_2O)_6^{2+}$. However, the added Cu^{2+} also enhances the structural Fe^{3+} and exchangeable Mn^{2+} signals as the time of reaction is increased (Fig. 1). It is likely that structural Fe^{2+}, which cannot be observed by ESR at room temperature, is oxidized to Fe^{3+} by Cu^{2+}. The appearance of the Mn^{2+} signal on addition of Cu^{2+} to the soil colloids probably results from displacement of exchangeable Mn^{2+} from sites on clays or organics. The signal is composed of six resonance lines as a result of the magnetic interaction between the electron spin and the nuclear spin of 5/2. Since the Mn^{2+} ESR spectrum is highly sensitive to the symmetry of the immediate chemical environment, signal broadening probably explains the inability to detect exchangeable Mn^{2+} until it is displaced from sites by another metal ion. These example experiments on a very complex and heterogeneous soil material demonstrate the value of ESR as an investigative tool.

Metal-clay systems. Experiments with much simpler metal-colloid systems have elucidated the surface bonding mechanisms that are likely to control metal availability in soils. A large fraction of many soil clays is composed of layer silicates, whose platelike layers are made up of one or two silica sheets bonded to an octahedral sheet of alumina or magnesia. Ionic substitutions within these layers result in a net negative charge that must be balanced by exchangeable cations positioned between the layers. Many of the metal ions present in soils are electrostatically attracted to clays by this type of permanent charge. The ESR spectra of paramagnetic ions adsorbed on clays which are capable of expanding in water indicate that the degree of exchange ion mobility is highly dependent on the degree of expansion of the clay. Thus, ESR investigations of the vanadyl (VO^{2+}) and copper (Cu^{2+}) ions have indicated a nearly rigid alignment of the square pyramidal $VO(H_2O)_5^{2+}$ and square planar $Cu(H_2O)_4^{2+}$ species in the interlayer regions of air-dry montmorillonite (an expanding clay mineral common in soils). A narrow spacing allows only a single monolayer of water between the clay platelets, preventing rotational motion of these ions and forcing a high degree of orientation (Fig. 2). However, most clays are much more highly hydrated in natural soil environments. Upon addition of water to montmorillonite, the spacing between clay layers increases to about 10 angstrom units (1 nm) when divalent ions occupy the interlayers, and ESR spectra of the ions show a very great increase in rotational mobility (Fig. 3). In fact, an estimate of the rotational correlation time of VO^{2+}, which is a measurement of the time required for the ion to reorient by random thermal tumbling, is only about 50% longer in the adsorbed

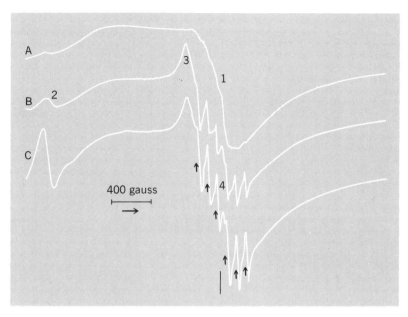

Fig. 1. ESR spectra of the colloidal fraction ($<2\mu$m) of an acid mineral soil: A, wetted with water; B, wetted with $CuCl_2$ solution for several minutes; and C, the same as B after several hours of reaction time. Signals 1, 2, 3, and 4 are assigned to free iron oxides, structural Fe^{3+}, adsorbed Cu^{2+} and exchangeable Mn^{2+}, respectively. The last-mentioned signal is composed of six resonance lines, denoted by arrows. The center field position (3300 gauss) is marked by a vertical line.

state than in bulk solution, where it has been estimated to be about 5×10^{-11} s. A similar ESR experiment with Mn^{2+} suggests a 30% longer correlation time for the adsorbed ion. Evidently, the inter-

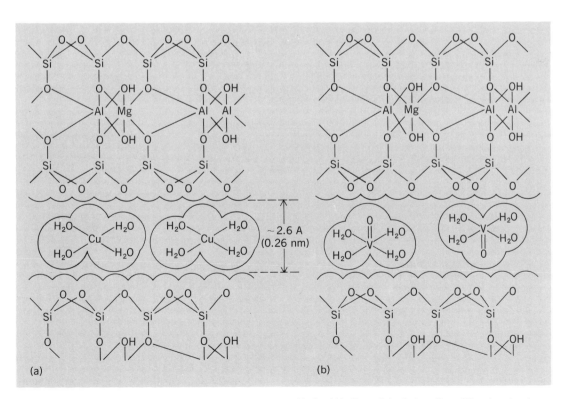

(a)

(b)

Fig. 2. Cross-sectional diagram of montmorillonite clay platelets when air-dried, with (a) Cu^{2+} and (b) VO^{2+} as interlayer cations. Not the isomorphous substitution of Mg for Al in the octahedral portion of the clay structure, producing a net negative charge in the layers.

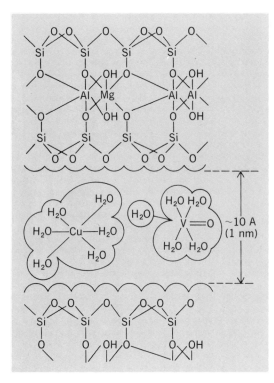

Fig. 3. Cross-sectional diagram of montmorillonite clay platelets when fully wetted, with VO^{2+} and Cu^{2+} as fully hydrated and mobile interlayer cations.

layer "solution" of fully hydrated expanding clays is not greatly different in terms of viscosity than bulk solution, and the divalent exchange ions are rather loosely held in a hydrated form in the vicinity of the clay exchange sites. With this picture of ions on clays, one should not be surprised that ion exchange in expanding clays is a very rapid reaction, with little preference of one divalent metal ion over another. For example, Cu^{2+} and Ca^{2+} have very similar affinities for the surfaces of clays bearing permanent charge.

Other types of layer silicate do not fully expand in water. For example, vermiculite tends to maintain two molecular layers of water between the platelets when divalent ions largely occupy the exchange sites. ESR experiments show that an octahedral arrangement of water molecules surrounds each exchangeable cation in such a way that half of the molecules can hydrogen-bond with one layer, while the other half bonds with the adjacent layer. The resulting arrangement produces a rigid orientation of exchange ions even when excess water is present. Rates of metal ion displacement from vermiculite clays are therefore much slower than from montmorillonite. In contrast, kaolinite is a type of clay with layers that do not separate at all in water, so that the only available sites for metal adsorption are on the outer planar surfaces and edges. Studies using ESR suggest that metal adsorption on the planar surfaces predominates, and that metals on these types of sites are as loosely held by electrostatic forces as they are on expanded montmorillonite.

Specific adsorption. This is a term often used in soil chemistry when a metal ion adsorbs on soil materials in a nonexchangeable form, or shows a strong preference for the adsorbed state when compared with other metals. There is little evidence that the electrostatic adsorption of metals by layer silicate clays can produce specific adsorption. However, amorphous oxides of aluminum and iron strongly chemisorb certain metals, notably Cu^{2+}, Zn^{2+}, and Pb^{2+}. The ESR spectrum of Cu^{2+} chemisorbed on aluminum oxide clearly indicates that it is immobilized at well-defined octahedral sites, forming Cu-O-Al bonds. The specific affinity of Cu^{2+} for oxide surfaces can be explained on the basis of ionic size and bonding nature. It is highly likely that chemisorption of certain trace metals occurs on soil oxide surfaces. Unlike the layer silicates, oxides adsorb metals such as Cu^{2+} and Zn^{2+} in a nonexchangeable form.

Organometal complexes. Another process of metal ion adsorption in soils, the formation of complexes with organic matter, has been known to be important for a long time. However, only recently the use of ESR has provided a more detailed picture of the types of organometal complexes formed. B. Lakatos and coworkers studied Mn^{2+}, VO^{2+}, and Cu^{2+} complexation with organic matter extracted from soils, concluding from the ESR spectra that these ions form inner-sphere complexes with organic functional groups; that is, there is a direct bond formed between carboxylic or phenolic oxygens of the soil organics and the metals. However, other ESR studies suggest that, at least at low pH, Mn^{2+} maintains a sphere of hydration water around it when adsorbed at the negatively charged carboxylate sites in soil organics and is therefore relatively mobile in the adsorbed state. This result is consistent with the observation that Cu^{2+} and VO^{2+} are much more strongly adsorbed on soil organics than Mn^{2+}. In fact, the similarity of the affinities of Mn^{2+}, Ca^{2+}, and Mg^{2+} for soil organics suggests that all of these ions form outer-sphere complexes, and should be readily exchangeable. Some metals such as Ni^{2+} and Zn^{2+} may represent a borderline state between inner- and outer-sphere complexation, since they are more strongly adsorbed than Mn^{2+} but more weakly adsorbed than Cu^{2+}. In general, the metal-surface association at organic exchange sites is stronger than that of layer silicate clays, but not all metals are adsorbed by the same mechanism. ESR has clearly established certain facts about natural metal-organic complexes, including the dominance of oxygen ligands (probably carboxylate), the fact that Mn^{2+} and numerous other metals are unlikely to form chelates with soil organics at common soil pH values, and the lack of strongly covalent bonds between the metal and organic oxygen atoms.

Organic radicals. Structural and exchangeable transition metals are not the only paramagnetic species to be found in soils. Soil organics generally produce a narrow signal that is due to organic radicals, species containing unpaired electrons in molecular orbitals. S. A. Wilson and J. H. Weber found that this signal increased in intensity as the pH was raised, and was greater for the acid-insoluble (humic acid) than the acid-soluble (fulvic acid) fraction. The results are consistent with the existence of quinones, especially in the less soluble humic acids, which can be converted to semiquinone radicals at high pH. The presence of organic radicals may significantly influence oxidation-reduction processes in soils.

The use of stable organic radicals as ESR "spin probes" of colloid surfaces may provide another area of useful research, since these organic probes can be synthesized with molecular properties that imitate compounds of interest. In this way, probes may report the fate of organics added to soil colloids, supplying details of surface reactions at the molecular scale.

For background information *see* CLAY MINERALS; SOIL CHEMISTRY in the McGraw-Hill Encyclopedia of Science and Technology.

[MURRAY B. MC BRIDE]

Bibliography: B. Lakatos, T. Tibai, and J. Meisel, *Geoderma*, 19:319–338, 1977; M. B. McBride, *Soil Sci.*, 126:200–209, 1978; M. B. McBride, *Soil Sci. Soc. Amer. J.*, 42:27–31, 1978; S. A. Wilson and J. H. Weber, *Analyt. Lett.*, 10:75–84, 1977.

Solar energy

Because solar energy is both ubiquitous and diffuse, its direct conversion into optical or mechanical energy differs profoundly from conventional processes. With the exception of nuclear, geothermal, and tidal energy, conventional sources are also derived from the Sun through biochemistry and the circulation of the atmosphere and hydrosphere. Humans can mine the accumulated deposits of primeval biochemical reactions—coal and oil—and can harvest the products of ongoing photosynthesis—wood and crops. Mining and harvesting are concentration processes in that fuels are originally dispersed over spaces much larger than the engine or animal consumer. When humans extract power from moving air and water, they do so at points where it is naturally concentrated. Examples include coastlines for wind and wave power, warm ocean currents over colder water for ocean thermal energy, and waterfalls for hydroelectric power. The direct conversion of solar energy, to be economical, must take place within spaces and times much smaller than those available to conventional, indirect solar conversion. Hence, the concentration and conversion of solar photons into useful energy must be much more efficient.

Advances in solar energy conversion must be measured relative to both conversion efficiency and fabrication costs. Essentially, the diffuse, limited nature of the solar resource demands high efficiency in the conversion system, and the increasing scarcity of conventional energy resources requires the economic production of these systems. Both constraints place a great emphasis on materials research. To illustrate, consider the problem of concentrating the solar flux, common to most solar technologies. Lenses and mirrors with high transmittance and reflectance, respectively, must be produced at low cost from inexpensive materials. These concentrators should maintain their qualities for years despite exposure to the elements. In many projected systems, concentrator production, maintenance, and replacement costs dominate the economics.

Three basic technologies can be defined with reference to the energy end product: photothermal, in which heat is produced; photovoltaic or photoelectrochemical, in which electricity is made; and photochemical, in which a fuel is generated. Hybrid systems are also possible: photovoltaic systems which require cooling can exploit the waste heat.

Photothermal conversion. Photothermal conversion is the most familiar and tangible of the solar technologies; it has influenced the design of housing and clothing for millennia. While many of the advances in photothermal conversion involve active, concentrating systems with specific optical properties, passive, nonconcentrating systems, such as buildings designed to behave as collectors and provide their own heating, are also being improved.

Passive systems. Modern passive systems use, among other techniques: (1) direct gain from south-facing windows; (2) thermal storage, which involves massive masonry (Trombe) walls or water-filled drum structures that absorb solar energy during the day and release it at night; (3) roof ponds, which operate like storage walls; (4) thermosyphon devices, also used for water heating, in which a fluid heated in a flat plate collector moves by convection through a structure or to a tank; and (5) solar greenhouses, which are built next to a house and act like large collectors to provide warm air.

Active systems. Basically, devices are defined as active if they have moving parts, like mirrors that track the Sun or pumps that circulate the heat transfer fluid. Since the modern era of cheap energy ended in the early 1970s innovations have engendered devices that range from low-temperature pool heaters to high-temperature receivers for electrical power generation.

Two collector geometries with little or no concentration are the flat plate collector, in which an absorber plate, in contact with a heat transfer fluid, is covered by a light-transmitting plate and insulated from the ambient air, and the evacuated tube collector, in which a transparent, evacuated outer tube insulates an absorbing inner tube containing a heat transfer fluid. To achieve temperatures much above the boiling point of water, concentration of the photon flux on the absorber is required. Concentrator geometries can be classified by their effect on sunlight; the radiation is directed to either a plane, a line, or a point. Effective concentration for a flat plate collector or a solar oven can be achieved with flat reflectors adjacent to the absorber surface. Higher temperatures are possible in line-focus concentrators, which generally use troughlike geometries to focus sunlight to a line at which a long tube containing a fluid is heated. Still greater heat demands point-focus concentrators. A conceptually simple approach employs a single large paraboloidal dish which tracks the Sun; a more complex design amasses large arrays of nearly flat tracking mirrors to direct sunlight onto a single central boiler. Demonstration projects using line-focus concentrators already exist; notable successes include an irrigation pump system near Phoenix, AZ, that produces 38 kW of electrical power, and a 20,000-gal (75,600-liter) system for heating water to 88°C at a soup factory near Sacramento, CA. A prototype for a point-focus facility is located at Sandia Laboratories in Albuquerque, NM; 5550 16-ft² (1.5-m²) mirrors produce 5 MW of thermal power atop a 200-ft (61-m) tower. Proposals for high-temperature central electrical power stations applying both point- and line-focus geometries have been advanced. A point-focus pilot plant that produces 10 MW of electrical power was begun in California's Mojave Desert in 1978.

Absorber surfaces. The demand for better absorber coatings has motivated considerable research into their optical properties. Lying between the more massive concentrators and heat transfer arrangements in a collector system, a small amount of material in an absorber film has a major impact on overall efficiency. As seen in the illustration, the solar spectrum and the thermal spectrum of a heated collector (whose general form is indicated by the blackbody spectra) do not overlap at typical collector temperatures. This separation can be exploited. Clearly, maximum absorption of sunlight is desirable. To attain higher collector temperatures, however, the surface must be made reflective in the thermal infrared. To illustrate, ignore other losses and assume that the power loss P_l from the surface is equal to thermal reradiation, which can be expressed as $\epsilon\sigma T^4$, where ϵ is the emittance, σ is the Stefan-Boltzmann constant, and T is the absorber temperature. If in equilibrium with P_g, the power gain from sunlight, P_l is constant, and a decrease in ϵ will force T to increase. Under normally encountered conditions, one can set the emittance equal to the absorptance (Kirchhoff's law), and for opaque materials, transmittance can be ignored and the reflectance will equal the absorptance subtracted from unity. Hence, a low emittance implies a high reflectance. As the wavelengths increase past a point around a few micrometers, then, a good selective absorber changes abruptly from absorber to reflector, as

shown in the illustration by the step function for the optimum spectral reflectance. (The exact wavelength at which the abrupt change in optimum spectral reflectance occurs depends on the temperature of the radiating surface.)

Typically, a number of physical effects are used, often in concert, to achieve the desired spectral selectivity: (1) Though no single, homogeneous material exhibits the ideal spectral profile, some are close enough to be improved by other effects. (2) Interference effects in a surface layer can trap solar radiation. Similarly, interference effects underlie antireflection coatings, which permit the initial penetration of solar photons. (3) Internal scattering phenomena from small particles embedded in the coating material can absorb sunlight while ignoring longer-wavelength infrared radiation. (4) Tandem stacks employ two films: the upper absorbs well across the solar spectrum and transmits in the thermal infrared, while the lower emits poorly in the infrared. (5) Surface textures, such as dendrites, can scatter and eventually absorb sunlight.

Absorber surfaces are usually produced by applying the selective material as a thin film to a heat-conducting substrate, such as copper. A number of techniques are available, including electroplating, physical or chemical vapor deposition, painting, bonding of foils, and physical or chemical treatment of the surface of the substrate material. Low-temperature commercial surfaces include black chrome, a mixture of chrome particles in a matrix of chrome oxides, and Tabor black, a mixture of zinc and nickel sulfides. For higher-temperature selectivity, few coatings have reached the commercial stage, but among the promising candidates are tandem stacks based on silicon absorbers and refractory metal reflectors, and cermets containing metals in matrices of alumina.

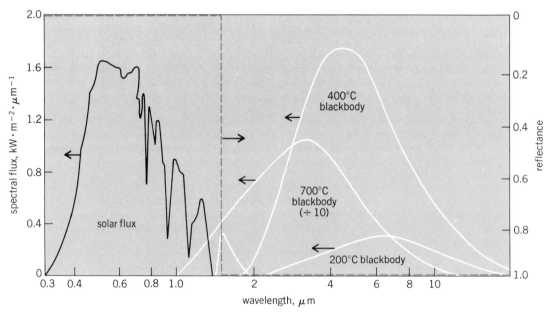

Spectral flux of solar radiation and of blackbodies at various temperatures as a function of wavelength, and optimum spectral reflectance of a selective absorber surface (broken line), also as a function of wavelength.

(From P. J. Call, National Absorber Program Plan for Absorber Surfaces R & D, Solar Energy Research Institute, 1979)

Photovoltaic conversion. Because of the widespread dependence on electricity, devices that convert the energy of solar photons directly into electrical current could fill a critical role in the energy economy. Strategies that improve efficiency include: (1) concentration of solar photons on solar cells; (2) increasing the absorption of photons at the cell junction; (3) matching the photon energy to the band gap of the solar cell; and (4) reducing losses of mobile electrons between the absorption locus and the surface contacts.

High-efficiency solar cells can handle concentrated sunlight, and since inexpensive concentrators, such as Fresnel lenses or parabolic mirrors made of molded plastic, can be cheaper than solar cells of equal area, the first approach is attractive economically. Prototype gallium arsenide cells, which perform better than silicon cells at high temperatures, have shown efficiencies around 20% at high concentration ratios. Secondly, increased photon absorption at the junction can be arranged by moving both contacts to one side by interdigitation, by stacking or superimposing two junctions, and by arranging the junctions vertically, parallel to the incoming light. Since energy beyond that necessary to bridge the band gap is normally wasted in solar cells, the third approach employs different materials to capture photons of various energies. Either silicon, germanium, or other cell materials could be stacked vertically, with the upper layers absorbing high-energy photons, allowing the other photons to cascade, or selective mirrors could direct the photons to the proper cells. Finally, losses of electron hole pairs can be minimized by making the cell material as orderly as possible. Originally, this problem was solved by using very pure, large silicon crystals, but cost considerations have motivated a search for other solutions, such as suitable polycrystalline and amorphous silicon structures. Since the photon penetration depth is a basic material property, and since longer diffusion lengths, equivalent to lower losses, are a result of expensive crystalline order and purity, thin films of highly absorptive materials are a promising option. In this instance, the trade-offs between crystal and thin-film techniques for making cells must include economics, as very cheap thin-film cells may win out, despite lower efficiencies.

Photochemical conversion. The most attractive goal for this conversion process has been the production of hydrogen and oxygen from water through the absorption of solar photons. However, despite demonstrations in a number of laboratories of the feasibility of the process, efficiencies remain low, long-term stabilities of the systems are unknown, and cost studies are very few. Of course, photosynthesis is a photochemical process, and its better understanding may lead to the conception of more efficient reactions. At present, photochemical conversion is much less developed than photothermal or photovoltaic technologies.

For background information *see* SOLAR BATTERY; SOLAR ENERGY; SOLAR HEATING in the McGraw-Hill Encyclopedia of Science and Technology   [MICHAEL R. JACOBSON]

Bibliography: P. J. Call, *National Absorber Program Plan for Absorber Surfaces R & D*, 1979; A. D. Meinel and M. P. Meinel, *Applied Solar Energy*, 1976; W. D. Metz and A. L. Hammond, *Solar Energy in America*, 1978; Office of Technology Assessment, *Application of Solar Technology to Today's Energy Needs*, June 1978; B. O. Seraphin, *Solar Energy Conversion*, vol. 31 of *Topics in Applied Physics*, 1979.

Solvent extraction

Recently developed separation processes are based on the application of supercritical gases. In simplified terms, this new method, often described as destraction, is characterized by the following features: (1) Destraction is a high-pressure technique, necessarily requiring special equipment. (2) High-boiling or even nonvolatile material can be dissolved by supercritical gases, whereby a loaded supercritical phase is formed. (3) The ability of a supercritical gas to take up other substances generally increases with increasing density or, at constant density, with increasing temperature. (4) Substances belonging to the same chemical class are taken up into the supercritical phase in the order of their increasing boiling points (for example, olefins in ethylene). (5) The take-up according to boiling points can be overlapped by the selective affinity of the individual substance for the supercritical gas (for example, caffeine in CO_2). (6) The material taken up can be recovered by decreasing the density of the supercritical phase, either by reducing the pressure at constant temperature or by raising the temperature at constant pressure. (7) The phenomena of distillation and extraction are utilized simultaneously; that is, enhancement of vapor pressure and phase separation both play a role. (8) The method can be used in fractionation and is particularly suitable for the isolation of thermally labile substances.

Phenomenological considerations. Distillation and extraction are among the most important separation procedures. Whereas separation by distillation is based upon the different vapor pressures of the components, separation by extraction is based on the properties of the material which determine the intermolecular interaction with the molecules of the extraction agent. Both of these effects are, to a certain extent, united in the general process of separation with supercritical gases. A supercritical gas is one which is above its critical pressure P_c and temperature T_c. Under these conditions the gas cannot be liquefied.

Effective destraction can be obtained in a temperature range of $10-100°C$ above the critical temperature of the gas being used and in a pressure range of $50-300$ atm ($5-30$ MPa). Gases which are particularly effective are those whose critical temperatures are neither extremely high nor extremely low. Satisfactory results are obtained with the following gases: ethane (T_c 32°C, P_c 48 atm = 4.9 MPa), ethylene (T_c 9°C, P_c 50 atm = 5.1 MPa), propane (T_c 97°C, P_c 42 atm = 4.3 MPa), propene (T_c 92°C, P_c 46 atm = 4.7 MPa), CO_2 (T_c 31°C, P_c 73 atm = 7.4 MPa), and NH_3 (T_c 132°C, P_c 111 atm = 11.2 MPa). NO and N_2O can also be used, but care must be taken since explosions can occur, particularly in the case of N_2O.

The following example illustrates the phe-

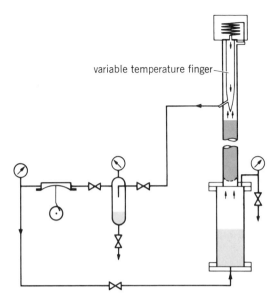

Fig. 1. Apparatus for fractional destraction. *(From Ange-wandt Chemie, Internat. Ed., 17(10):701–784, Verlag Chemie, Weinheim)*

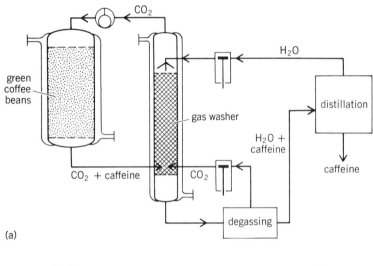

(a)

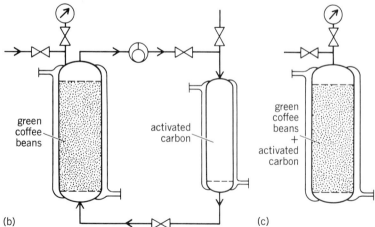

(b)   (c)

Fig. 2. Decaffeination of green coffee beans with supercritical CO_2. Removal of caffeine from the CO_2 is accomplished by: (a) washing with water; (b) adsorption on activated charcoal; or (c) without recycling, by adsorption on activated charcoal admixed with the coffee beans. *(From Angewandt Chemie, Internat. Ed., 17(10):701–784, Verlag Chemie, Weinheim)*

nomena: If ethylene is bubbled through paraffin oil at room temperature (23°C, that is, 14°C above T_c) and atmospheric pressure ($\rho \sim 0.0013$ g cm^{-3}), 100 g of ethylene transports about 0.1 g of oil. When the pressure is raised, the amount of oil transported per 100 g of ethylene hardly changes until the critical pressure ($P_c = 50$ atm = 5.1 MPa, $\rho \sim 0.22$ g cm^{-3}) is reached; at this point a dramatic increase in the amount of oil transported is observed, and at 200 atm or 20 MPa ($\rho \sim 0.4$ g cm^{-3}) 100 g of ethylene transports about 25 g of paraffin oil. At 200 atm and 70°C ($\rho \sim 0.32$ g cm^{-3}) the value falls to about 7 g. These effects can be interpreted to a first approximation as follows: The amount transported at atmospheric pressure corresponds roughly to that expected from the partial pressure of the paraffin oil. As the pressure rises, the solubility of ethylene in the oil increases, and is accompanied, at most, by an insignificant increase in the partial pressure as a result of changes in the intermolecular forces. At and just above the critical pressure, the density of the gas increases dramatically as a result of the increased compressibility (the pressure in this case says little about the degree of packing, that is, the density). The high-density gas is now able to take up large quantities of oil, and a loaded supercritical gas phase is formed. Raising the temperature from 23 to 70°C at constant pressure (200 atm) leads to a decrease both in density and in the ability of the gas to transport oil. If, however, the temperature is increased while the density is kept constant by increasing the pressure, then the ability of the gas to take up oil also increases. The large increase in the density above the critical pressure is also accompanied by a dramatic increase in the solubility of the gas in the liquid phase. This facilitates transfer of the oil into the supercritical gas phase.

Fractionation. The observation that an increase in temperature at constant pressure is accompanied by a decrease in density, and hence in the ability to transport material, forms the basis of an important application: mixtures of high-boiling material can be separated. This possibility is illustrated in Fig. 1. The apparatus, which consists of a still (25 liters), a column, and a heated finger, is charged with a mixture of α-olefins (traces C_{14}, 2 liters C_{16}, 2 liters C_{18}, and 7 liters C_{20}), and fractionated with ethane ($T_c = 32$°C, $P_c = 48$ atm = 4.9 MPa) at 45°C and an initial pressure of 60 atm (6.1 MPa) ($\rho \sim 0.2$ g cm^{-3}) with a finger temperature of 85°C ($\rho \sim 0.08$ g cm^{-3}). The supercritical ethane-olefin phase passes through the column, which is filled with copper rings, and reaches the hot finger, where refluxing occurs. The supercritical gas is bled off and partially depressurized to 30 atm or 3.0 MPa ($\rho \sim 0.03$–0.04 g cm^{-3}). The product separates out and is collected (60 ml/hr). The gas is then repressurized. The pressure is slowly increased during the destraction from 60 to 110 atm or 6.1 to 11.1 MPa ($\rho \sim 0.35$ g cm^{-3}). (Increasing the pressure in a destraction is analogous to raising the temperature during a distillation.) A gas chromatographic analysis of the various fractions shows that a successful separation has been accomplished; for example, a fraction containing about 25% of the C_{16} olefin has a purity of 95.5%.

Since the components and the gas are aliphatic hydrocarbons, this separation is based primarily on the differences in vapor pressure of the components and thus resembles a distillation; that is, the components are taken up into the gas phase in the order of their increasing boiling points. This effect can be overlapped by a separation according to the class of compound in those cases in which certain components in the mixture have, for example, little affinity for the supercritical gas and others are taken up preferentially. This separation process can also be applied to materials which are not amenable to fractional distillation.

Practical applications. The fact that destraction can be carried out at relatively low temperatures makes the method particulary appropriate for the separation of thermally labile substances. This is why destraction has received considerable attention in the last few years from the food industries, for it is possible, for example, to remove fats and oils from vegetable and animal matter under mild conditions without the necessity of a final step in which solvent is removed. A variation of the procedure is to extract the desired product under subcritical conditions, that is, using a liquid gas, and then to separate the extract into its components under supercritical conditions. It has proved possible to remove cocoa butter from cocoa beans, soybean oil from soybeans, the essential oils from spices, as well as the valuable constituents from the hop resins.

The decaffeination of coffee is a process that is due to go into commercial operation shortly. In this process green coffee beans having a certain water content are treated with supercritical carbon dioxide. It seems obvious that CO_2 does not pose a physiological problem. Of particular interest is the fact that the caffeine is selectively removed by the supercritical CO_2; that is, no substances are lost that contribute to the aroma formed on roasting. Decaffeinated coffee and untreated coffee should then differ only in their pharmacologic effects.

The decaffeination can be carried out in three ways (Fig. 2). In the first variation, presoaked green coffee beans are treated in a pressure vessel with CO_2 at $160-220$ atm or $16.2-22.3$ MPa ($\rho \sim 0.4-0.65$ g cm^{-3}), the CO_2 being continuously recycled (Fig. 2a). The caffeine diffuses out of the beans into the supercritical CO_2 and is then carried out of the pressure vessel into a washing tower where it is washed out with water at $70-90°C$. Three to five liters of wash water is needed to treat 1 kg of raw coffee. After 10 hr (at $90°C$) all of the caffeine is in the wash water, which is then degassed; the caffeine is recovered by distillation. The caffeine content in the bean is decreased from an initial value of between 3 and 0.7% to one as low as 0.02%, that is, lower than that stipulated (0.08%).

The separation can be carried out with the help of active charcoal instead of water (Fig. 2b). However, in this case the caffeine must then be extracted from the charcoal.

The third variation (Fig. 2c) has certain advantages. The pressure vessel is charged with a mixture of coffee beans and active charcoal pellets, the diameter of which is such that they fill the space between the beans; that is, the packing situation for the beans is practically identical to that in the absence of active charcoal. One kilogram of active charcoal is sufficient for 3 kg of coffee. A CO_2 pressure of 220 atm (22.3 MPa) at $90°C$ is used. The caffeine in the supercritical CO_2 diffuses directly into the active charcoal; there is no need to recycle the gas. The required degree of decaffeination is reached after 5 hr. The mixture is then separated into its components by being passed over a vibrating sieve.

The decaffeination of green coffee illustrates well the advantages associated with destraction, and it can be anticipated that many other separation problems can be solved using this technique.

For background information *see* SEPARATION, CHEMICAL AND PHYSICAL; SOLVENT EXTRACTION in the McGraw-Hill Encyclopedia of Science and Technology.   [GÜNTHER WILKE]

Bibliography: *Angew. Chem. Int. Ed. Engl.*, 17: 701–754, 1978.

Space flight

The year ending Sept. 1, 1979, was notable for Soviet accomplishments in crewed space flight and for American achievements in planetary exploration and astrophysics. Other nations increased the rate and extent of their involvement in space flight activities, and so did commercial interests. International cooperation in space-based weather and environment observations continued to develop. The Soviet Union's *Salyut 6* space station was used extensively for materials-processing experiments. Major space missions are listed in the table.

Public concern rose after the nuclear debris–scattering impact of *Cosmos 954* in Canada upon reentry on Jan. 24, 1979. Concern grew to anxiety awaiting the reentry of the massive 70-metric-ton Skylab station, which disintegrated over the Indian Ocean on July 11, scattering debris harmlessly in a remote area of Australia.

Space shuttle. Development difficulties compelled postponements of the anticipated launch date of the space shuttle orbiter *Columbia*. Problems included hydrogen piping failure, high-pressure-fuel turbine malfunction, and the application of thermal protection system tiles. President Jimmy Carter in August 1979 requested a personal briefing because of the importance of the shuttle to budget and defense space requirements, and a special panel was created to examine the shuttle's status and problems. The orbiter *Enterprise* was delivered atop NASA's 747 carrier aircraft to Kennedy Space Center on Apr. 10, 1979. It was mated on May 1 with the external tank and two inert solid rocket boosters and was moved to the Complex 39 launch pad in the first checkout of the full space shuttle configuration. The *Columbia* had arrived on March 24 for preparations prior to launch. Industry observers estimated a launch date for midsummer 1980.

Soviet launch activity. In calendar year 1978, there were 88 Soviet launches to orbit or beyond (in 1977, 98; in 1976, 99, the highest year). In the first 6 months of 1979, there were 49 launches, and in the year ending September 1979, there were 99 launches.

The 88 launches of calendar year 1978 produced

Major space missions from September 1978 to September 1979

Payload name	Launch date	Payload country or organization	Purpose and comments
Tiros-N	10/12/78	U.S	Polar orbiting meteorological research satellite for NOAA (National Oceanic and Atmospheric Administration)
Intercosmos 18	10/24/78	Soviet Union	Magnetospheric-ionospheric spacecraft; joint effort with East Germany, Hungary, Poland, Romania, and Czechoslovakia
Nimbus 7	10/24/78	U.S.	Testing advanced sensors and technology for atmospheric, meteorological, and pollution studies
Cosmos 1045	10/26/78	Soviet Union	Carried Radio 1 and 2 piggyback; the first amateur radio relay spacecraft orbited by the Soviet Union
Prognoz 7	10/31/78	Soviet Union	Scientific mission with experiments from France, Hungary, Sweden, and Czechoslovakia
HEAO-B	11/13/78	U.S.	Second astronomical observatory to study highly energetic radiation from space
Telesat-D	12/14/78	Canada	U.S. launch of Canadian domestic communications satellite
SCATHA	1/30/79	U.S.	USAF research on preventing harmful electronic discharges caused by Spacecraft Charging At High Altitudes (SCATHA); geosynchronous orbit
Cosmos 1076	2/12/79	Soviet Union	Mission with two "ocean viewing" telescopes and other oceanographic instruments
Sage-A	2/18/79	U.S.	Stratospheric aerosol and gas experiment
Soyuz 33	4/10/79	Soviet Union	Carried a Bulgarian and a Soviet cosmonaut; failed to dock with *Salyut 6*; problems forced 8-*g* ballistic reentry profile
UK 6	6/2/79	U.K.	U.S. launch of U.K. scientific research payload
Soyuz 34	6/6/79	Soviet Union	Uncrewed supply-ferry vessel; used for reentry by *Salyut 6* crew, which went to orbit in *Soyuz 32*
Bhaskara	6/7/79	India	Soviet launch of second satellite built in India; carried two television cameras for Earth resources mission plus microwave radiometers for moisture data
NOAA-A	6/27/79	U.S.	First operational spacecraft of NOAA's *Tiros-N* series
Gorizont 2	7/6/79	Soviet Union	Positioned over Atlantic Ocean for Olympic Games television relay

127 payloads, grouped generally as follows: science, 9; communications satellites — regular, 10; radio store and transmit, tactical military, 35; electronic ferret, 6; weather, 1; navigation, 8; geodesy, 1; Earth resources, 2; military photographic – recoverable, 33; early warning, 2; ocean surveillance, 2; antisatellite interceptor, 1 (prior to American-Soviet negotiations on this weapon); Earth-oriented crewed, 5; Earth-oriented, crew-related, but uncrewed, 7; Venus, 4; miscellaneous, 1.

Soviet crewed flight. The Soviet Union moved briskly toward its goal of large crewed space stations by a sequence of successes with its *Salyut 6* spacecraft, launched Sept. 29, 1977. The three-man American flight of 84 days aboard Skylab in 1974 has been exceeded by three successive occupations of *Salyut 6*: a two-man 96-day stay, ending March 1978; a 140-day period by Alexander Ivanchenkov and Vladimir Kovalenok, ending Nov. 2, 1978; and a 175-day flight by Valeriy Ryumin and Vladimir Lyakhov, ending Aug. 19, 1979. The Soviets have stated that the long-term goal of this work is a flight to the planets. Closed environmental cycles will be needed for such long-distance flights, and the Soviets have reported a ground-laboratory successful run of 1 year of a closed cycle capable of supporting a human crew.

Salyut's extended missions are made possible by visits of automated supply vessels of the Progress class, and by visits of crewed or automated Soyuz class spacecraft. Station crews consume between 20 and 66 lb (9 and 30 kg) of supplies daily. Visiting cosmonauts bring supplies, deposit and retrieve equipment and experiments, and use the previously docked Soyuz ferry for return to Earth.

Left behind is a more reliable Soyuz reentry vehicle, less degraded by the space environment. Ivanchenkov and Kovalenok, who were launched in *Soyuz 29*, returned in *Soyuz 31*, after transferring propellants from *Progress 4* to *Salyut 6*, leaving the station usable for further occupancy, which began 115 days later. Salyut's propellants are said to be unsymmetrical dimethylhydrazine and nitrogen tetroxide oxidizer. The Progress tankers are commanded into a destructive reentry after unloading and separation. Air Force Lt. Col. Lyakhov and civilian Ryumin, a civil engineer, entered orbit Feb. 25, 1979, aboard *Soyuz 32*, and returned aboard *Soyuz 34*.

Key research activities in *Salyut 6* include astronomical observations; Earth photography by multispectral and topographical cameras; scores of biological experiments, with the most important being human reactions to space flight; plant growth and sensory system experiments; and the processing of materials under variable gravity conditions.

Progress 5 docked at Salyut's aft port March 14 with modularized components for replacement of major system elements in a preventive maintenance program for the aging station. Thirty-eight repair operations were completed. *Progress 5* delivered also a Yelena telescope for detecting gamma rays and electrons in low Earth orbit, a receiver for the first transmission of television pictures to the spacecraft, and French-supplied samples for materials-processing experiments.

Progress 5 maneuvered the station into an improved rendezvous orbit for the Apr. 11, 1979, arrival of *Soyuz 33* bearing Bulgarian Georgi Ivanov and Russian Nikolay Rukavishnikov. The approach correction motors of their ship malfunctioned during final rendezvous maneuvers. They were forced to use the backup main propulsion system engines for an emergency return to Earth, on April 12, in a ballistic reentry that placed an acceleration of 8g on the crew, twice the Soyuz norm.

This failure threatened the endurance mission by inadequate supplies and by the need to return the station crew before their *Soyuz 32* ferry exceeded its time limit in orbit. The Soviets coped by orbiting *Progress 6* on May 13. It docked 2 days later at the aft port for unloading, and was undocked on June 8 to allow that day's arrival of *Soyuz 34*, launched uncrewed on June 6. Operating uncrewed, the now surplus *Soyuz 32* was undocked from the forward Salyut port, and on June 13 was commanded to a recoverable reentry, returning film, biological samples, and other data. In a continuing display of capability the Soviets then undocked *Soyuz 34* from the space station's aft port on June 14, maneuvered 180°, and Salyut redocked it on the forward port. This allowed *Progress 7* to dock on June 30, leaving the crew well supplied.

In the first deployment of a large structure from a crewed orbiting vehicle, the cosmonauts, using components delivered by *Progress 7*, erected on *Salyut 6* the first large radio telescope in space. Experiments with the 33-ft-diameter (10-m) KRT-10 radio telescope were conducted daily, observing stars and the Sun; microwave observations of Earth provided data on humidity and surface water content.

Soviet spokesmen regard these operations as a rehearsal for continuous space research in Earth orbit, and feel that the Salyut stations are now clearly established as orbital laboratories, the development of which is the main trend in Soviet cosmonautics.

Materials processing in space. Several nations conducted materials-processing research in space. Benefits anticipated from experiments under variable gravity include: ultrapure materials, avoidance of wall contamination through "containerless" processing, composite materials unobtainable on Earth, bioseparation through electrophoresis, compositional uniformity in semiconductor crystals, and the opportunity to study fluid dynamics in a regime where surface tension and viscosity forces dominate behavior.

The United States' *SPAR-V* (space processing applications rocket) carried four successful experiments in solidification mechanics and immiscible alloys processing. West German activity continued in the Texus series of rocket flights, with *Texus 2* launched from the Kiruna Range in northern Sweden on Nov. 16, 1978. The experiments focused on melting, solidification, and fluid effects of materials. American and European experiments continue to be prepared for flight on Spacelab, aboard the shuttle.

The *Salyut 6* station's two furnaces supported dozens of experiments. The Splav (alloy) furnace heats entire samples and cools them at controlled rates. The Kristall (crystal) furnace permits directional solidification when the sample is frozen by progressively extracting heat from one end. Other *Salyut 6* apparatus can place metallic coatings on surfaces by vaporization and condensation pro-

cesses. Soviet processed materials included cadmium sulfide, germanium crystals, aluminum-germanium compounds, and indium antimonide and gallium arsenide alloyed with zinc and tellurium. French-supplied samples were processed in the redesigned Kristall furnace brought up by the *Progress 5* supply ship. Alloys of tin, lead, aluminum, and copper were prepared. *Progress 5* delivered also a centrifuge which provided artificial gravity for plant growth experiments. Metallurgical experiments necessary for space welding and construction were conducted, showing the Soviet appreciation of the industrial potential of space and of large space structures. For some experiments, *Salyut 6* was placed into a drift mode with thrusters shut down so that small acceleration forces would not affect the materials being processed.

Earth resources observations. Satellite observation of renewable resources (crops, forests, and so forth) gained further credibility with the successful completion of the 3-year Large Area Crop Inventory Experiment (LACIE). Using *Landsat 3* data, an Agriculture Department and NASA team estimated 1977 Soviet wheat production at 91,400,000 metric tons. Later, the Soviet Union announced it had been 92,000,000 tons, compared to a predicted 102,000,000 tons. For spring wheat the inventory was successful only in areas where fields are very large, and crop discrimination techniques are not expected to be adequate without the improved resolution of the *Landsat-D* to be launched in late 1981.

Remote-sensing initiatives by the Agriculture Department have set priority on early warning of changes which will affect productivity, and the ability to make good commodity production forecasts. The stakes are high. The department estimates, for example, that earlier information on the 1977 Soviet shortfall would have made farmers' selling prices higher, saving American taxpayers about $500,000,000 in wheat deficiency payments. The Landsat system demonstrated capabilities also in providing meaningful data on such diverse problems as land use change, monitoring forest clear-cutting practices, the status of irrigated lands, Corps of Engineers watershed runoff management, gypsy moth defoliation extent, wildland capacities for supporting animal life, population growth on the urban fringe, and snow-cover area measurements.

Nonrenewable resources observations focused on improving exploration strategies for energy and mineral deposits. The Heat Capacity Mapping Mission satellite continued experiments for identifying types of rocks through their thermal inertia signatures by measuring the rate of cooling from day to night temperatures.

Unexpected benefit came from Seasat and *GEOS 3* by their mapping of Earth terrain contours by microwave to an accuracy comparable with ground and aircraft surveys. *GEOS 3*–derived contours agreed within 1 m with Geological Survey small-scale topographic maps. The spacecraft was able to measure the distance to Earth features within 50 cm (20 in.).

Salyut 6 cosmonauts conducted Earth resources–oriented observations of Soviet rangelands for comparison with complementary ground measurements taken simultaneously in the areas observed.

Environmental observations. The Stratospheric Aerosol and Gas Experiment satellite (SAGE) was launched Feb. 18, 1979, to globally map vertical profiles of ozone, aerosols, nitrogen dioxide, and molecular extinction in the stratosphere. Aerosols (particles or droplets under a millionth of a meter in diameter) filter and reduce the amount of sunlight reaching and reflecting from the Earth's surface, thus affecting temperature and perhaps climate. On April 23, SAGE detected a veil of stratospheric debris from the eruption of a Caribbean volcano, providing the first opportunity for satellite measurement and tracking of aerosols distributing globally from a point source. *Nimbus 7* was orbited October 24 to monitor pollution in the atmosphere and oceans, and to help determine if Earth's ozone layer is being depleted by aircraft exhausts or fluorocarbons from aerosol sprays. Its LIMS (Limb Infrared Monitor of the Stratosphere) experiment, monitoring seasonal and daily changes in concentrations of nitrogen oxides, is leading toward altitude profiles of nitrogen compounds. *See* OZONE, ATMOSPHERIC.

In January began the Global Weather Experiment, the largest international scientific experiment ever conducted. The existing World Weather Watch was supplemented by 10 spacecraft in this effort by the 147 members of the World Meteorological Organization. The satellites included five in geosynchronous orbit: three American (GOES), one Japanese (GMS), and one European (ESA's Meteosat). Also participating were polar orbiting weather satellites such as *Tiros-N* and *NOAA-A* of the United States, and the Meteor class of the Soviet Union. Satellites, balloons, aircraft, buoys, and ships generated data for the 85% of the Earth's area not previously covered by the World Weather Watch.

In October, Bangladesh installed a ground station for data from weather satellites of the United States, Soviet Union, and Japan. One day after starting operations, it tracked a storm in the Bay of Bengal toward a predictable landfall. The populace was evacuated and not one life was lost, although tens of thousands have died in the past when such storms struck without warning.

The *GEOS 3* satellite altimeter mapped the southern portion of the Greenland ice sheet to an accuracy of 2 m in ice sheet mass balance monitoring studies. It also defined ocean geoids that helped refine gravity field models and calibrate the Seasat altimeter. Seasat demonstrated the ability to measure ocean surface winds, sea state, major ocean currents, ocean circulation patterns, Gulf Stream shear, sea ice conditions, and orographic steering of on-shore winds by terrain features. Covering 95% of the oceans every 36 hr, Seasat carried five sensors—three types of radar and two different radiometers—for its oceanographic measurements. The most important instrument, the synthetic aperture radar, with a resolution of 25 m, obtained all-weather pictures of sea and ice conditions. The mission achieved 80% of its technical objectives.

The depth of clear, shallow ocean waters proved measureable from Landsat images, prompting the U.S. Defense Mapping Agency to request 12,000 images on a global basis in the next 5 years for updating charts, especially in areas of high tanker traffic.

The Soviet Union reported repeated success in visual and photographic environmental observations from *Salyut 6*, detecting forest fires, tracking sandstorm migrations across oceans, accurately locating large schools of fish, documenting coverage for a snow and ice atlas of the world, and utilizing the cryogenically cooled BST-1M submillimeter telescope to obtain data on cyclonic activity.

Astrophysics. The second High Energy Astronomical Observatory, *HEAO 2*, was orbited Nov. 13, 1978, to do intensive studies of selected x-ray sources identified by *HEAO 1*, used for x-ray sky survey. Discovering sometimes three new x-ray sources a day, *HEAO 1* had brought the known number from about 300 to over 1000. Gas depletion ended the observatory's 17-month life on Jan. 9, 1979. *HEAO 1* permitted a definitive study of the "hard x-ray" space background glow from a very thin, ultrahot electrified gas generating copius x-rays, but no visible light. If this plasma were distributed evenly in the universe, its inferred density would provide half the mass required to slow and reverse celestial expansion (causing a closed universe). The *HEAO 2*, however, found that much of the diffuse x-ray background radiation originates from quasars. Coupled with consideration of the mass in quasars, the *HEAO 2* findings favor an open universe. The satellite discovered also that x-ray emissions from some globular star clusters switch on and off. Prior to *HEAO 2*, x-ray spectra of supernovas had shown only silicon and iron. Its data, however, showed highly ionized magnesium, sulfur, argon, and calcium as well. The findings support a concept that the Sun and planets formed from a supernova's explosive debris.

The first image of an x-ray "burster" came from *HEAO 2*. These bizarre celestial events are rare, apparently originating from compact objects of less than 50 km diameter, and they release more x-ray energy in 10 s than the Sun does in a week. Previous spacecraft x-ray instruments returned only numerical data. The *HEAO 2* also recorded x-ray sources 100 times weaker than seen before, including some in close proximity to a strong source, and some with no counterparts in visible light. *HEAO 2* images disclosed the most powerful x-ray producers known, quasars estimated to be over 10×10^9 light-years (10^{26} m) away. *Pioneer 10*, observing cosmic radiation, determined its behavior to be highly erratic at distances from the Sun beyond 16 astronomical units (2.4×10^{12} m), suggesting proximity of the spacecraft to the heliospheric boundary.

Satellite observatories have located several candidates for black holes, collapsed stars whose great mass prevents light escaping. Two candidates have very good supporting evidence provided by ultraviolet and x-ray instruments. One is Cygnus X-1, and the latest is Scorpius V-861. The x-ray instrument on NASA's 6-year-old astronomy satellite, Copernicus, located the latter. Supporting evidence came from observations in ultraviolet by the American-British-European satellite, International Ultraviolet Explorer (IUE). Both candidates are orbiting large, visible stars and gravitationally siphoning their matter which emits detectable radiation as it streams into the dense companion. The IUE has allowed direct measurement of the flow of matter (as a carbon plasma) out of the giant blue star of Cygnus X-1. Other IUE data suggest the possibility of a massive black hole at the core of some globular star clusters. The IUE's shortwave ultraviolet detection apparatus is capable of observing the core of a million-star cluster for the first time. What shows up is radiation from a group of 10 to 20 bright blue stars that orbit the core of the cluster. They may be circling a black hole with a mass on the order of 10^3 solar masses.

The 4-year-old Small Astronomy Satellite (SAS 3) reentered in April. It discovered half the known (35) x-ray bursters, all of which are at least a billion times farther away than the Sun. SAS 3 data support the idea that the bursts are due to thermonuclear reactions on neutron stars. SAS 3 discovered "rapid bursters" which may occur several times a minute.

Jupiter flybys. *Voyager 1* began regular, systematic color imaging of Jupiter in early January 1979, when 60,000,000 km from the planet. Its closest approach (174,000 mi or 280,000 km) occurred on March 5, and that of *Voyager 2* (404,000 mi or 650,000 km) on July 9. *Voyager 1* flew by just south of the equator, and *Voyager 2* passed across the southern hemisphere, each surviving the punishing Jovian radiation. They will arrive at Saturn in November 1980 and August 1981, respectively. An option exists to target *Voyager 2* past Saturn for a January 1986 encounter with Uranus.

A rocky debris ring was discovered, with bright and dark concentric structure, 5000 mi (8000 km) wide and probably less than 18 mi (29 km) thick (inside the orbit of the nearest satellite, Amalthea). Images of the dark side of Jupiter indicated lightning bolts uniformly distributed globally and an aurora perhaps 18,000 mi (29,000 km) long occurring in at least three layers up to 1400 mi (2250 km) above the cloud tops. The Great Red Spot was found to be rotating counterclockwise in the southern hemisphere, and is therefore an anticyclonic "high-pressure" center. Jupiter is seen to be essentially a large ball of gas, composed primarily of hydrogen, with a ratio like the Sun's of approximately 1 atom of helium for 10 molecules of hydrogen. Also confirmed was the presence of ammonia, methane, acetylene, ethane, and water vapor. Photographs disclosed almost a mirror image in the multicolored parallel atmospheric stripes flowing in the north and south hemispheres, with a regular pattern of alternating eastward and westward velocities.

Photographs of the satellite Io from 8000 mi (13,000 km) showed the first volcanic activity on a body other than Earth. Enormous, simultaneous eruptions indicate that Io has the most active surface in the solar system, probably due to tidal heating. Clouds of ejecta, primarily sulfur dioxide, moving perhaps 2000 mph (900 m/s), rose as high as 150 mi (240 km). Some particulates apparently escape Io's gravity to form a doughnut-shaped cloud (torus) of heavy ions surrounding Jupiter in Io's plane. This may be the source of low-frequency radio emissions discovered by the Voyagers. Jupiter's wobbling rotation apparently causes oscillations of the dense torus plasma, generating the radio waves.

Moving within Jupiter's magnetosphere, Io generates a current flow of about 5,000,000 A, which moves from Io in a "flux tube" along a Jovian magnetic field line to the planet's ionosphere, over to another field line, and back to the other side of Io, closing a direct-current circuit.

The Voyagers disclosed a stunning variety in Jupiter's satellites. Only Io appears to have an "atmosphere," a haze of gas perhaps 100 mi (160 km) high with a density millionths that of Earth's atmosphere. Fresh cinder cones dot its surface. The satellites' densities decrease outward from Jupiter while the spacing of their orbits decreases. *Voyager 2*, passing within 78,000 mi (125,000 km) of Ganymede and Callisto, revealed their surfaces riddled with craters. Neither showed evidence of major surface relief, possibly due to creep or flow in their icy crusts. Europa looks flat and without large craters, but it is marked with complex, intersecting arcuate features that may be cracks or large fractures, possibly occurring in a thin ice crust. The density of Ganymede, the largest satellite, suggests a bulk composition of about 50% water by weight.

Previous spacecraft to fly by Jupiter are *Pioneer 10*, which crossed the orbit of Uranus on July 11, and *Pioneer 11*, which encountered Saturn in September 1979. *Pioneer 10*, launched in March 1972, will cross Pluto's orbit in 1987, and will be the first spacecraft to leave the solar system. Its nuclear power source will thereafter be insufficient to send radio signals receivable on Earth, as it heads for the constellation Taurus. After Saturn encounter, *Pioneer 11* will also head out of the solar system, but in a direction opposite to *Pioneer 10*. Each spacecraft carries a message plaque to explain its origin for any intelligent species it may encounter in its endless journey.

Venus explorations. A December 1978 scientific assault on Venus involved 10 spacecraft, 6 from the United States and 4 from the Soviet Union. The American Pioneer *Venus 1* began on December 5 an 8-month, highly elliptical, atmosphere-skimming orbit of the cloud-covered planet. Four days later arrived Pioneer *Venus 2*, composed of five probes for globally dispersed parachute descent. The Soviet *Venera 12* arrived on December 21, and *Venera 11* on December 25, each dropping a lander which sent data to Earth (for 110 and 95 min, respectively) via its flyby bus until occultation by the planet. American and Soviet scientists coordinated experiments for correlation and later exchanged data. Both nations confirmed an unexpectedly high concentration of the primordial isotope argon-36.

The Pioneer probes passed through three distinct layers of clouds above 40 km altitude and found the atmosphere clear below. The lead-melting surface temperature (455°C) has been considered excessive if due to entrapment of solar radiation (the greenhouse effect) by only the dense (96 ±1%) carbon dioxide atmosphere. The 1979 findings showed two other contributors: water vapor occurring below the clouds (0.1–0.5%), plus an 85% sulfuric acid solution in the cloud layers.

Below the clouds, Pioneer found 60 times the level of molecular oxygen (60 parts per million) expected from measurements from Earth. The surface pressure was reported at 91.5 bars (9.15 MPa) from the day probe. Venous apparently has no ozone.

Almost three-fourths of the incoming solar radiation is reflected away by the clouds and atmosphere. About 60% of the remainder is absorbed in the clouds, and somewhat over 2% of the original incident radiation is finally absorbed at the surface. The topmost cloud layer is believed to consist chiefly of sulfuric acid particles of the $1-3$-μm size. There is some evidence for a fourth cloud layer, thin and stratified and probably not global, just below the main cloud deck. Below 33 km the atmosphere did not have particulate matter, but is attenuating because of the high concentration of CO_2.

Strong, impulsive signals received by the orbiter's electric field detector suggest lightning activity in the lower atmosphere. Lightning apparently occurs as often as 25 times per second in relatively small areas, in contrast with Earth lightning occurrences of 100 times a second globally. The orbiter's cloud photopolarimeter acquired ultraviolet maps of Venus for the first time. Its radar mapper discovered a mountain range higher than Mount Everest, plus a vast chasm extending hundreds of miles across the equatorial region, suggesting that crustal movements like those on Earth may be at work. The orbiter instruments also determined that Venus's ionosphere is significantly weaker on the night side from the absence of solar ultraviolet radiation in a darkness period equal to 58 Earth nights. Both ion density and the extent of the ionosphere vary widely with solar wind pressure, largely due to the absence of a deflecting planetary magnetic field. Atmospheric circulation appears to be the cause of day and night side temperatures being very close. Radiometer data suggest that temperatures are 10°C hotter above the polar cloud tops than at the equator and that polar temperatures are affected by a downward-moving polar vortex. *See* VENUS.

Sun-Earth relations. *OSO 8*, designed for measurements in the Sun's quiet period (solar minimum), was turned off 2 years after an expected 1-year lifetime. Its pointed experiments obtained the most accurate observations to date of the solar chromosphere and transition regions. *OSO 8* results influenced the planning for observations to be made by the Solar Maximum Mission satellite (SMM), due for launch early in 1980. The SMM will emphasize solar flare mechanisms studies. *OSO 8* did not detect the mechanical waves theorized as the cause of coronal heating, and this spurred the current belief that magnetic field dissipation may supply the needed energy.

Because the International Sun Earth Explorers 1 and 2 (*ISEE 1* and 2) fly in tandem, they have recorded significant differences in the Earth's magnetic field in the same regions only minutes apart. With *ISEE 3*, these spacecraft have elevated the investigation of space plasma from a level of exploration to a level of highly coordinated, fine-resolution science. With *ISEE 3* in its heliocentric

orbit at the libration point since Nov. 28, 1978, early continuous measurements of the fluctuating solar wind are being obtained about an hour before the particles reach the magnetosphere. The trio of spacecraft have revealed the complex and dynamic nature of the sunward boundary between the magnetosphere and the solar wind.

GEOS 1 discovered in the outer magnetosphere large quantities of thermal O^{++} and He^{++} ions from the ionosphere. It gathered evidence that perhaps half the hot 0.1–16-keV plasma in the outer magnetosphere comes from the ionosphere rather than from the Sun, as previously thought.

The Small Scientific Satellite *(S3-3)* showed that the polar ionospheric acceleration regions lie generally in the auroral zones, occur at all local times, contain high electric fields, extend at least from 1500 to 8000 km altitude, often extend hundreds of kilometers in latitude, and inject ionospheric oxygen, helium, and hydrogen ions upward into the outer magnetosphere with energies up to 16 keV.

Continuing analysis of Skylab solar data has produced a nearly "standard" model of a flare (its temperature, density, and magnetic field). Of particular importance are the facts that the hottest flare kernel was always found to be confined to a small magnetic, plasma-filled flux tube which connects regions of strong magnetic field on the Sun's surface; and that a continuous energy release must be occurring in the secondary thermal phase of the flare. It is still not clear, however, just what mechanism originally triggers the flare.

The Atmosphere Explorers (*AE-C* and -*E*), investigated the detailed interactions between solar ultraviolet radiation, the neutral atmosphere, and the resulting ionized products. Reaction rates are being systematically worked out between atmospheric constituents, many of which are also important in the production and loss of ozone. The global distribution of atomic nitrogen has been measured down to 140 km.

Another important result is that the increase in solar ultraviolet activity has been found to be quite different from the last solar cycle. In particular, it has increased more rapidly than would be deduced from the ground-based measurements at the 10.7-cm radio frequency, a standard index of solar activity. The higher ultraviolet solar activity observed from *AE-C* and -*E* explains the rapid decay of the Skylab orbit, owing to increased heating of the Earth's upper atmosphere, whereas the other solar indices had predicted a slower decay.

It appears that the so-called solar constant is not constant at all, but may vary up to 0.3%, with important implications for the Earth's climate. Measurements of solar irradiance continued, using rocket calibration flights and instruments on *Nimbus 6* and *7*, and will later utilize instruments on SMM and *Spacelab 1*.

Mars observations. Functioning since August 1976, *Viking Orbiter 2*. designed for 5 months of operation, was shut down due to a leak of attitude control jet gas. *Orbiter 1* continued to function, as did the two Viking Landers. Ultraviolet spectra data on the atmosphere confirmed the virtual lack of ozone in some seasons.

A February 1979 colloquium on Mars generally agreed that the Landers failed to find signs of life, but debate continues on whether it ever existed or now exists within rocks in a Martian analog of the endolithic organisms found in porous rocks of Antarctic dry valleys. Only the Landers' labeled release experiment showed results suggesting a metabolizing system, but these are now believed due to a strong oxidant, possibly hydrogen peroxide in the Martian soil.

Four groups of material were found by the Landers: (1) easily penetrated drift material (in a 23-cm penetration by the backhoe) at Chryse Planitia; (2) blocky, strong, soillike material at Chryse, requiring great force to break through; (3) crusty, cloddy material in the Utopian Planitia with a thin veneer of fine-grained soil, variously lumpy and smooth; (4) rocks in abundance at both sites. The magnets on the backhoes attracted strongly magnetic material from the soil, and those on the Lander tops accumulated magnetic material from the atmosphere. *Lander 2* photographed what appeared to be surface frost that lasted 100 days in the 1977 northern winter. A confirming high-resolution picture taken in May 1979 showed what scientists believe is a new layer of water frost about one-thousandth of an inch (25 μm) thick. Apparently, frozen CO_2 and water in the atmosphere settle on dust particles and fall. When the Sun evaporates the CO_2, the water frost is left behind.

For background information *see* JUPITER; MANNED SPACE FLIGHT; MARS; SATELLITES, APPLICATIONS; SATELLITES, SCIENTIFIC; SPACE FLIGHT; SPACE PROBE; VENUS in the McGraw-Hill Encyclopedia of Science and Technology.

[CHARLES BOYLE]

Bibliography: *Astronaut. Aeronaut.*, 1978 Highlights, December 1978; *Aviat. Week Space Technol.*, issues from Sept. 4, 1978, through Sept. 3, 1979; *Defense/Space Business Daily* issues from Sept. 1, 1978, through Sept. 4, 1979; *Goddard Space Flight Center News*, vol. 25, no. 16, December 1978; N. Hinners (9/26/78) and A. Calio (9/27/78 and 2/28/79), Statements before the Subcommittee on Space Science and Applications, Committee on Science and Technology, U.S. House of Representatives; N. Hinners (2/28/79), Statement before the Subcommittee on Science, Technology and Space, Committee on Commerce, Science and Transportation, U.S. Senate; *NASA Activities*, monthly issues from September 1978 through August 1979; NASA News Releases 79–13, 79–84, 1979; *Science*, 203:743–745, Feb. 23, 1979.

Space telescope

A long-sought aim of astronomers should be realized in late 1983 with the scheduled launching of the space telescope (ST). Actually a multi-purpose observatory, the ST (Fig. 1) is furnished with an array of instruments to investigate light gathered by its 2.4-m (94.5-in.) Cassegrain reflector. The mirrors are of unprecedented precision and provide an order of magnitude increase in observing capabilities. Thus, the ST is confidently expected to revolutionize optical and ultraviolet astronomy in the 1980s. This project is a joint venture of the National Aeronautics and Space Administration and the European Space Agency, and also involves many universities, scientific laboratories, and aerospace corporations.

Rationale. The advantages of observation with a large telescope above the Earth's atmosphere are many.

1. Resolution. The Earth's atmosphere smears images of astronomical objects, and as a result fine detail of 1 arc-second is all that usually can be recorded with the best Earth-based telescopes (1 arc-second corresponds to approximately 2 km at the distance of the Moon). The instruments on the ST routinely will receive an image good to 0.1 arc-second—fully a factor-of-10 improvement. Concentration of the light into a much smaller area greatly enhances the ability to detect faint objects.

2. Continuous operation. At the 500-km altitude of the ST, observations can be carried out night and day.

3. Atmospheric penetration. Some kinds of electromagnetic radiation are absorbed by atmospheric gases, and thus the radiation is not observable even from the highest mountaintops. This includes portions of the ultraviolet and infrared spectra.

4. Sky background. The Earth's atmosphere is a source of weak light that contaminates photographs of very faint objects. This interference is largely avoided by the ST.

Major systems. The ST observatory (Fig. 2) is composed of four major systems. (1) The telescope itself is called the optical telescope assembly. It gathers light from astronomical objects and focuses it on the scientific instruments. (2) The scientific instruments, normally five in number, analyze the light from the optical telescope assembly. The first complement of instruments and their scientific programs are discussed below. (3) The solar array generates electrical power for the observatory. (4) The support systems module supplies the other necessary observatory functions, such as electrical and thermal control, and receives commands from the control center and returns data to the ground. It has a computer that controls telescope pointing with high accuracy, using reference signals from the fine-guidance system, itself part of the optical telescope assembly.

The ST, which will be placed into Earth orbit by the space shuttle, has an estimated operating life-

Fig. 1. Space telescope in orbit. *(NASA)*

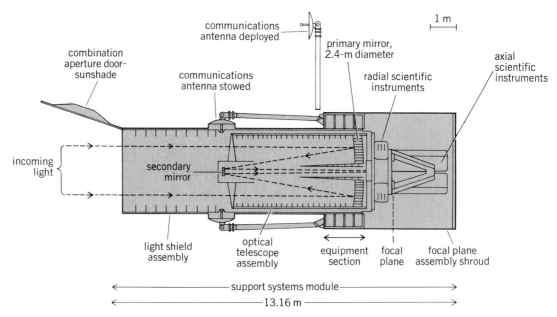

Fig. 2. Cross-sectional view of the space telescope showing some of its components.

time of 15 years. It is designed to be visited by astronauts at intervals as short as 2½ years, for maintenance, repairs, and perhaps replacement of a scientific instrument. The ST may even be returned to Earth at longer intervals if more extensive refurbishment is required.

Scientific instruments. The initial complement of five instruments is listed in the table.

Wide-field/planetary camera. This is an imaging device sensitive to radiation in the broad wavelength range from 115 to 1100 nm. In the wide-field mode, it has a field of view measuring 2.7 by 2.7 arc-minutes; sharper images are recorded in the planetary mode with a field of 1.2 by 1.2 arc-minutes. Different parts of the spectrum are isolated

by choosing from a large number of filters and transmission gratings. This instrument is unique because it alone is mounted in the radial instrument bay shown in Fig. 2; light reaches the camera by means of a pick-off mirror located on the optical axis of the optical telescope assembly. The other scientific instruments (described below) are mounted in the axial bays.

This instrument will provide photographs needed to determine the cosmic distance scale up to 10 times farther than possible from the ground. Thus the volume of space surveyed by astronomers will be dramatically increased. Planetary photography with this camera will produce detailed photographs of Jupiter with resolution comparable to those obtained by *Voyagers 1* and *2*, except at their closest approach to the planet.

Faint-object camera. This instrument is designed to fully exploit the resolution of the ST on the faintest detectable astronomical objects. Its resolution will be at least 0.1 arc-second. Filters will isolate chosen wavelengths of light in the range from 120 to 700 nm. High-quality observations of stars as faint as twenty-eight visual magnitude (invisible with the most powerful Earth-based telescopes) can be obtained in exposures of 10 hr. The field of view is quite small, only 11 by 11 arcseconds.

The faint-object camera will permit the optical identification of very faint sources known previously only through their emission of x-rays or radio waves. This instrument will also excel in observing individual stars (useful as distance indicators) in other galaxies out to the Virgo cluster.

Faint-object spectrograph. This is a versatile instrument operating in the wavelength range of 115 to 700 nm. Two spectral resolution modes are provided. These have chromatic resolving powers $\lambda/\delta\lambda$ of approximately 100 and 1000. In the former mode the spectrograph will obtain spectra of stars as faint as twenty-sixth visual magnitude.

This instrument will probe physical conditions

Science instruments for the space telescope

Lead institution	Instrument	Sample major science programs
California Institute of Technology	Wide-field/ planetary camera	Determination of the cosmic distance scale. synoptic studies of planets
European Space Agency	Faint-object camera	Optical identification of faint x-ray and radio sources, observation of stellar distance indicators
University of California, San Diego	Faint-object spectrograph	Physical conditions in active galactic nuclei, chemical evolution of galaxies, determination of material motions in comets
Goddard Space Flight Center	High-resolution spectrograph	Composition and physical conditions in the interstellar gas, studies of stellar mass loss and mass transfer
University of Wisconsin	High-speed photometer	Studies of variability of compact sources, brightness measurements of the zodiacal light and diffuse galactic light

in quasars and in the nuclei of Seyfert galaxies. It will be used to record spectra of stars and nebulae in other galaxies in order to study the chemical evolution of galaxies. Also, this spectrograph would permit spectra of comets to be obtained much farther from the Sun than before, and would probe motions in the cometary plasma tail.

High-resolution spectrograph. This instrument provides the capability for observations in two high-spectral-resolution modes. These are λ/δλ of 20,000 and 100,000, the latter being equal to the resolution of the largest ground-based Coudé spectrographs. The wavelength range is 110 to 320 nm.

The high spectral resolution and relatively high sensitivity permit the accurate measurement of absorption lines produced by the interstellar gas in the spectra of stars. These lines can be analyzed to determine chemical abundances and physical conditions. Analysis of stellar line profiles allows studies of mass loss from red giant stars, which may be the main source of the interstellar gas from which stars are formed. Spectra of close binary systems having a very hot star as one component allow studies of mass transfer.

High-speed photometer. This instrument is designed to produce very accurate, time-resolved observations over the wavelength range of 120 to 800 nm. Filters isolate different wavelength bands. This instrument will explore routinely the important domain of rapid cosmic phenomena in the range from 1 s to 1/1000 of a second; this range of time variations is particularly hard to observe from the ground because of atmospheric noise. Even faster variations can be observed when necessary.

The high temporal resolution of this instrument makes it ideal for studies of the very compact objects such as pulsars and black holes, currently thought to be terminal stages of stellar evolution. The time scale on which the energy output of these objects is observed to undergo significant variations can reveal the object's size and physical conditions. The high accuracy makes possible precise measurements of the zodiacal light and of the diffuse galactic light.

Astrometry. The ST will make major contributions to astrometry, the precise measurement of star positions, through use of the fine-guidance system. Repeated measurements can determine stellar positions to 0.002 arc-second. The useful faint limit could be as low as the twentieth magnitude. The astrometry effort is led by astronomers from the University of Texas.

Measurements can be applied to the standard problems of astrometry but with the advantage of a factor-of-approximately-10 greater accuracy. Several instruments on the ST have distance determinations as a major program. These include distances to the farthest galaxies and to bright stars in relatively nearby galaxies. It is well to remember that indirect distance measurements rest on the direct determinations from trigonometric parallaxes and moving clusters. The ST should greatly expand the base of fundamental parallax determinations. Other applications for astrometry include the search for extrasolar planets.

Observing with the ST. The operation of the ST will be the joint responsibility of a Space Telescope Science Institute (at a site to be determined)

and the Space Telescope Operations Control Center, with its Science Support Center, at the Goddard Space Flight Center in Greenbelt, MD. Communication with the observatory will be through the Tracking and Data Relay Satellite System. Real-time interaction with the ST is necessary for some observations. These could include modification of an observing program on a target of unknown properties on the basis of quick-look data or acquisition of a target under difficult circumstances, for example, an object in a dense star cluster or a rapidly moving object such as a comet. However, the usual mode of observation will involve preplanned sequences stored electronically on board the spacecraft. Two computers on board the ST will supervise its performance and regulate the receipt of commands and the transmission of data. *See* SATELLITE COMMUNICATIONS.

Prospects. The ST is clearly a powerful assembly of instruments that holds the promise of huge advances in many branches of astronomy. Some of these advances are straightforward and have been described. However, there is a tradition, going back to Galileo, of spectacular, unexpected discoveries whenever an observing facility with a major increase in capabilities is placed into operation.

For background information *see* ASTRONOMICAL INSTRUMENTS; TELESCOPE in the McGraw-Hill Encyclopedia of Science and Technology.

[JOHN C. BRANDT]

Bibliography: D. S. Leckrone, *Proc. Soc. Photo-Opt. Instrum. Eng.*, 183:56–73, 1979; M. S. Longair, *Quart. J. Roy. Astron. Soc.*, 20:5–28, 1979.

Sputtering

Whenever an energetic ion or atom impinges upon a solid surface, the phenomenon of sputtering may occur. Energy deposited in the target by collisions between the ion and a target atom is shared by nearby atoms or molecules, with the subsequent development of a recoil cascade, as shown schematically in the illustration. As a consequence, a recoiling target atom may be able to overcome chemical binding forces and escape from the surface. The sputtering yield S is the average number of atoms which are ejected from the target per incident ion. Typical values of S range from around 0.01 for proton impact to 50, or larger, for the impact of a heavy ion such as Au. Sputtering may thus be characterized as the erosion of a surface on the atomic scale.

Since fluxes of energetic particles are common in the solar system, and elsewhere in space, it is to be expected that the erosive effects of these parti-

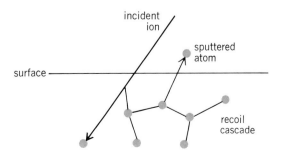

Schematic picture of recoil cascade and sputtered atom.

cles are also common. Sites where sputtering is thought to occur as a natural phenomenon include the surface of the Moon, interplanetary and circumstellar dust grains, the atmosphere of Mars, the surfaces of some of the satellites of Jupiter, and the rings of Saturn and Uranus. The mass loss associated with sputter erosion provides one limit to the lifetime of small grains in space, may cause erasure or modification of geological features on planets and satellites subjected to intense charged-particle bombardment, and may contribute to the loss of planetary atmospheres. Moreover, laboratory experiments have shown conclusively that sputtering of a chemically complex surface often leads to preferential emission of some elements (or isotopes). In the solar system the conclusion is that bombarded surfaces may be both chemically and isotopically differentiated.

One of the most important charged-particle fluxes in the solar system is the solar wind (SW), consisting of 1-keV/amu (atomic mass unit) protons and alpha particles (helium nuclei) streaming out from the Sun. At the Earth's orbit, the proton flux is 2×10^8 cm^{-2} s^{-1}, and the alpha flux about a factor of 20 smaller. The sputtering yield of 1-keV protons on mineral targets composed of light elements is on the order of 0.01, and for 4-keV alpha particles, S is about an order of magnitude larger. Thus the proton and alpha components are nearly equivalent in their erosive capability. Other, heavier constituents of the SW, although having much higher sputtering yields, are not sufficiently abundant to be important in this respect.

Sputter-induced erosion. Interplanetary dust grains are subject to SW bombardment, and as a consequence are eroded away at a rate of about 5×10^{-6} μm y^{-1}. Thus a grain of 1-μm diameter orbiting the Sun at a radius of 1 astronomical unit (au) has a lifetime against sputtering of about 2×10^5 y. Although such small objects are expected to be swept up by radiation forces before their complete sputter destruction can occur, nevertheless, these grains, as well as larger debris fragments, will experience extensive radiation damage of their surfaces owing to SW bombardment.

Outside the solar system, grain bombardment is also important. Thus the survival of grains which have condensed around a supernova is determined in part by the sputtering they suffer from the stellar atmosphere as radiation pressure hurls them outward at high velocity. Some laboratory experiments indicate that sputter-processed surfaces are highly activated, in a chemical sense, so that random grain collisions may result in spontaneous welding to form a larger aggregate. It has been suggested that this process is important in the amalgamation of grains into more massive planetary objects.

The lunar surface has also been extensively sputtered by SW bombardment. An inferred sputtering rate of 5×10^{-6} μm y^{-1} corresponds to the removal of a few centimeters of material over the course of lunar history. This is a tiny mass loss in terms of modification of geological features, but more significant erosion effects may occur on the interior satellites of Jupiter. Here the charged-particle radiation is associated with the Jovian magnetosphere, rather than the SW. The net flux of protons more energetic than 100 keV might be as large as 10^{11} cm^{-2} s^{-1}, several orders of magnitude greater than that of the SW. Furthermore, there may be a heavy-ion component to the radiation field surrounding the satellite, and any such component would be a very effective sputtering agent. The recently discovered sodium cloud associated with the Jovian satellite Io is attributed to the magnetospheric sputtering of Na-containing compounds on the satellite's surface. The typical energy of a sputtered particle is several electronvolts, which is of the same order of magnitude required for an atom to escape from the gravitational field of Io (or any similar-size object, such as the Moon). If all sputtered atoms were to escape Io, a 1% surface Na concentration and maintenance of the cloud in a steady-state condition would imply the loss of about 100 m of surface material over the age of the satellite.

Erosion of icy surfaces. Because of the very high erosion rate accompanying the ion bombardment of water ice (S is approximately 0.4 for 500-keV proton bombardment), sputter-induced erosion effects on the icy Jovian satellites Europa, Ganymede, and Callisto may be especially important. Voyager probe experiments in 1979 gave the first close look at these new worlds, where erasure of major geological features by ion erosion has probably proceeded considerably further than on the Moon. Up to 500 m of relief may have been removed over the lifetime of the solar system. On Io, volcanism is probably the major agent of geological change, but ion smoothing of surface features might play a more significant role on the icy satellites, or on the tiny interior satellite Amalthea.

It is suggested that, farther out in the solar system, sputtering by ions in the Saturnian radiation belts causes substantial erosion of ice particles in the outer rings, with a net erosion rate as high at 10^{-6} cm y^{-1} (2000 times faster than the corresponding erosion of the Moon's surface). The lifetime of these particles may be set by sputtering to be less than the age of the solar system.

The fact that the sputtering yield of ice can be as much as a thousand times greater than that of silicate and metallic mineral fragments, such as are thought to be embedded in the icy crust of the three largest Galilean satellites of Jupiter, can have an important effect on the evolution of the surface. After sufficient sputtering, a thin mineral armor may form which will be of lower albedo than pure ice or frost, and which would serve to protect the planet from subsequent rapid erosion. Effects of this nature may also be responsible for the very low albedo of the recently discovered rings of Uranus. It has been suggested that differences in brightness between the leading and trailing hemispheres of some Jovian and Saturnian satellites have their origins in asymmetries in the bombarding ion flux.

Generation and erosion of atmospheres. The high sputtering yields associated with the icy surface of Ganymede may lead to the existence of a tenuous O_2 atmosphere on that satellite. Sputtered water molecules are dissociated by sunlight to form H and OH, and subsequent reactions form molecular oxygen and hydrogen gas. Because of its low mass, H_2 undergoes thermal escape. This avenue is unopen to the heavier O_2 component, which must escape via such mechanisms as photochemical reactions or sputtering. The O_2 equilibrium pressure, which may be as high as 10^{-3} millibar or 0.1 pascal (10^{-6} Earth's atmospheric pressure), is determined by the competition between these loss processes and the rate of production of water vapor via sputtering and sublimation.

Magnetospheric particles around Jupiter may also cause atmospheric loss from its larger satellites, and the SW can have a similar effect on Mars. If the present SW flux were directly incident on the Martian atmosphere for 4×10^9 y, it would lead through sputtering to a loss of mass comparable to the mass of the present atmosphere. The sputtering yield of a specified gravitationally bound gas under charged-particle impact is not much different from the yield of a solid substance composed of similar elements. The chemical binding energies which must be supplied to remove an atom or molecule from a solid surface are of the same order of magnitude as the gravitational potential energy of such a particle in the atmosphere of a planet. Also, although the matter at the upper surface of the atmosphere, or exobase, where sputtering occurs, is some 10^{14} times more rarefied than a typical solid, there is no large density effect on the yield itself. The reason is that, although an incident ion must travel considerably farther before it suffers a collision in a gas, the subsequent recoils find it correspondingly easier to make their way to the exobase and escape. Although the value of S does not reflect variations in target density, the size of the collision cascade is strongly affected; a cascade diameter of 1 nm in a solid becomes 10 km at the exobase.

Elemental and isotopic fractionation. Evidence for elemental and isotopic fractionation of solid surfaces comes from the analysis of returned lunar samples. The outer 10 nm of 1-μm grains are found to be isotopically enriched at the level of a few percent in the heavier isotopes of the pairs $^{30}Si/^{28}Si$, $^{18}O/^{16}O$, and $^{34}S/^{32}S$. In this same layer, silicon (with atomic mass number A = 28) is also enriched with respect to oxygen (A = 16), compared with the bulk value, by about 40%. The thickness of this layer corresponds roughly to the range of the incident SW particles, and evidently sputter fractionation plays a role. Here the Moon's gravitational field functions as a giant isotope separator, heavier atoms falling back preferentially to the surface to build up the enriched layer. Detailed calculations of this process are complicated by the very high roughness of the lunar surface; a certain fraction of the sputtered atoms will strike an immediately adjoining surface. Mass or chemical dependence of the sticking probability can also lead to fractionation.

One of the surprising results arising from the Martian Viking Lander experiments was that the atmospheric $^{15}N/^{14}N$ isotope ratio has a value nearly double the terrestrial value. If the terrestrial abundance of these isotopes is assumed to be representative of the solar system abundance, then a specific mechanism is necessary to explain the anomalous Martian result. Sputtering is one mechanism which can contribute to atmospheric enrichments of heavy isotopes. Sputtering occurs at the very top of the atmosphere and, therefore, will preferentially remove the lighter elements and lighter isotopes, which are relatively more abundant in this region than in the bulk atmosphere, by virtue of diffusive separation. Further, S increases as the molecular mass decreases because of an inverse dependence of the sputtering yield on the gravitational potential energy. Such sputter mass loss and fractionation, of course, are effective only to the extent that the SW actually impinges upon the atmosphere. Although, in contrast to the Earth, Mars has no significant magnetic field to deflect the SW, the Martian ionosphere is expected to divert the SW around the planet. If this flow is completely collisionless, with respect to exospheric molecules, no direct sputtering will occur. At the present time, the exact nature of the SW flow is not known, so it is difficult to assess the overall effect of sputtering on the evolution of the Martian atmosphere. At maximum effectiveness, SW sputtering could easily account for the $^{15}N/^{14}N$ enrichment, as well as for the preferential removal of N_2 with respect to CO_2, which has presumably led to the present CO_2- rich, N_2-poor atmosphere. Photochemical reactions are also significant sources of atmospheric fractionation, but the relative importance of these two mechanisms remains to be assessed.

There are other sputtering mechanisms which may operate in space. Fast (megaelectronvolt) electrons are thought to sputter circumstellar dust grains. The small electron mass leads to a small value for the sputtering yield S of approximately 3×10^{-7} for 1-MeV electrons, but these particles can pass successively through many 1-μm grains, so that their sputter efficiency is much higher than the yield value alone indicates. Photons, neutrons, and cosmic rays may all also cause sputtering, but the yields are very small and would be important only in special circumstances.

For background information *see* JUPITER; MARS; MOON; SATURN; SOLAR WIND in the McGraw-Hill Encyclopedia of Science and Technology.

[PETER K. HAFF]

Bibliography: W. L. Brown et al., *Phys. Rev. Lett.*, 40:1027–1030, 1978; A. F. Cheng and L. J. Lanzerotti, *J. Geophys. Res.*, 83:2597–2602, 1978; D. L. Matson, T. V. Johnson, and F. P. Fanale, *Astrophys. J. Lett.*, 192:L43–L46, 1974; Z. E. Switkowski et al., *J. Geophys. Res.*, 82:3797–3804, 1977.

Stem

Many plants are constructurally highly organized. This is particularly true of rhizomatous plants — their regular branching patterns allow computer-aided predictions to be made of their growth potential.

It is conventional to consider the structure or morphology of a flowering plant in terms of stem, leaf, and root, although these categories are not mutually exclusive. The stem bears leaves which may be green photosynthetic organs or may be reduced to scales or otherwise modified. The stem also bears buds and sometimes roots. The combined structure of stem, leaves, and buds can be referred to as the shoot. The stems of a great many plants are not self-supporting; for example, climbing plants have narrow stems, and their weight is borne by other vegetation. Rhizomatous plants are

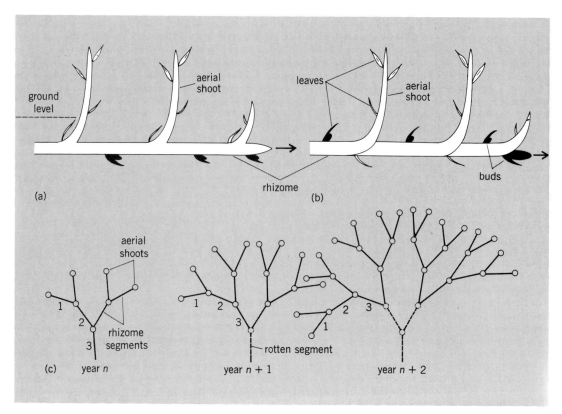

Fig. 1. Growth features of a rhizomatous plant. Side views of rhizome showing (a) monopodial growth and (b) sympodial growth. (c) Plan view of Y-shaped sympodial branching in three successive seasons: a 3-year-old plant with seven aerial shoots in year n has developed into four 3-year-old plants still having seven aerial shoots by year n + 2.

supported by the ground, the bulk of the stem lying horizontally just below the soil level.

Features of rhizomatous plants. The prostrate habit of rhizomatous plants results in a number of distinctive ecological and morphological features. The rhizome may be monopodial or, more frequently, sympodial in its development. A monopodial stem is derived from a single apical meristem; a sympodial stem is derived from a successive series of apical meristems (Fig. 1). Growth

Fig. 2. An excavated plant of *Alpinia speciosa* demonstrating a hexagonal branching pattern. (*From A. D. Bell, Computerized Vegetative Mobility in Rhizomatous Plants, in A. Lindenmayer and G. Rozenberg, eds., Automata, Languages, Development, North Holland Publishing Co., 1976*)

is commonly confined to the youngest, distal end. The proximal, and therefore oldest, end eventually dies and usually rots away. In many rhizomatous plants, side branches are located in a regular manner on the parent stem and grow out at fixed angles, thus developing an organized pattern of two-dimensional branching.

There are two major consequences of this mode of growth. The first is that the process of extension at one end, coupled with rotting at the other, causes the rhizome to "move" steadily through the soil; new ground is always being explored and exploited. The second is that, each time the rotting reaches a branch junction, the plant is diverted in two directions, the two daughter plants moving away from one another.

Growth pattern. The individual members (ramets) of such a plant population (a genet) developing in this manner are potentially immortal, although paradoxically, all may always be the same fixed age and even the same size (Fig. 1). Rhizomatous plants thus form clones in which all individuals are genetically the same and are constantly moving about while multiplying in number. The movement is not necessarily random but, rather, may be highly organized because of the conservative details of the branching patterns. This regularity of pattern may be considered in terms of a selected response resulting in economical spread in two dimensions: the sections of stem between aerial leaves or shoots function in the manner of spacers, locating the aerial shoots and associated roots at regular and systematic positions in the substrate. The patterns found in rhizomatous plants are often the patterns recognized by geographers and mathematicians as the most economical means of "tessellating the plane." In other words, the plant manufactures a minimum amount of stem during the location of the aerial shoot-root complexes, and lays down the shortest possible transport route, linking up these sites of food synthesis and water-nutrient uptake.

A particularly striking example of this system is shown by *Alpinia speciosa* (Fig. 2). Excavations and detailed measurements reveal that the sympodial rhizome system develops basically as a hexagonal grid, that is, much like the most perfect form of two-dimensional spread. It is tempting to suppose that this type of rhizome stem configuration is the consequence of competitive selection. However, the situation is undoubtedly complex, with factors in addition to spread influencing the pattern. Nevertheless, different patterns and different scales of patterns must affect the performance of various plants, and it is reasonable to

suppose that specific organizations are in some way related to specific habits. In a tentative manner, A. D. Bell and P. B. Tomlinson recognize three basic rhizome branching types: a linear system with infrequent branching, a system based approximately on 45° angles, and a Y-shaped branching giving rise to hexagonal patterns, as exemplified by *Alpinia*.

The hexagonal pattern, at least potentially, has a built-in problem. If such a pattern of growth were complete, it would result in many aerial shoot-root complexes being located directly on top of one another. This eventuality does not occur in *Alpinia*, for two reasons. A proportion of side buds fails to develop fully so that the grid is incomplete but uncongested, and the angles of the grid are consistently distorted from 120°, conferring a random twist to the pattern.

Seasonal growth. The sympodial units of the *Alpinia* rhizome are produced in seasonal flushes, this observation having been confirmed by tagging rhizome buds and shoots and monitoring their development over a 12-month period. Once the

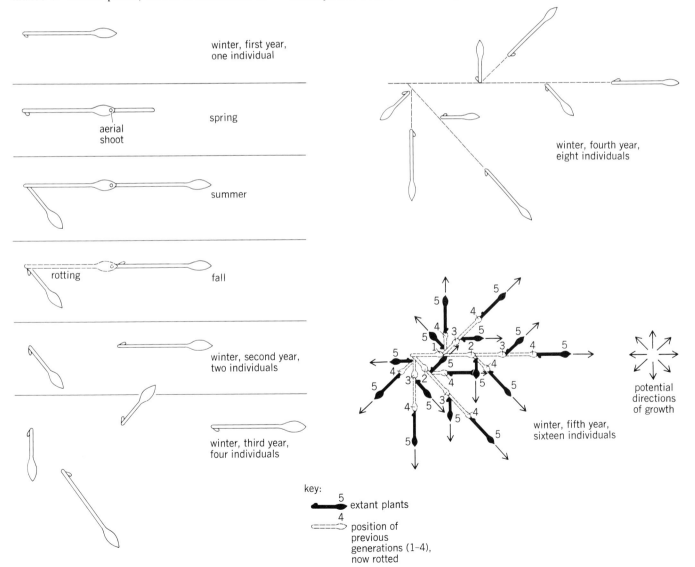

Fig. 3. Plan view of the development of a clone of *Medeola virginiana* rhizomes. The repetitive pattern allows extrapolation into the future. (*From A. D. Bell, J. Arnold Arbor., 55:458–468, 1974*)

rate of development is known and can be identified morphologically, it is a simple matter to age excavated plants. It has been found, for example, that a segment of the *Alpinia* rhizome persists in the soil for 10 years on average before it dies and rots (by which time it is the oldest segment on the plant). Each ramet of *Alpinia* is thus approximately 10 years old. The aerial shoots (the vertical ends of the horizontal rhizome segment) survive for only 3 years, after which they are shed by the formation of a well-defined abscission zone. It is usually possible to age rhizomatous plants having seasonal growth by paying careful attention to morphological details, confirmed by the observation of tagged material during a year's growth. Such information can be provided by the annual extension of units as in *Alpinia*, the presence of annual growth scars due to a periodic congestion of nodes (such growth pauses may be discernible in continuously growing plants only by charting the length of each successive internode), by recognizing that a consistent number of leaves (possibly scale leaves or scars) are produced per year, or by the consistent production of lateral shoots each year in specific numbers or at specific locations.

Predictive studies. The conservation of pattern and regularity of growth of many rhizomatous plants allow predictive studies to be made. A simple example is shown in Fig. 3. The Indian cucumber, *Medeola virginiana*, grows in woods in New England and develops a sympodially branching rhizome system. Excavation and tagging reveal that the orientation and position of the two active buds borne on any one sympodial unit are extremely precise and must follow a set pattern from one year to the next. Also, sympodial units persist only for one season in this species, and then rot away. Since the stem lengths and branch angles are relatively constant, it is possible to extrapolate the lines of future growth of the expanding population and to gain information on the future area, density, and size of the clone. More elaborate rhizome patterns, particularly when considered over a longer time period, require the use of a computer. A visual display unit allows the year-by-year development of a clone to be analyzed and recorded, thus giving valuable information in temporal, spatial, and numerical form on the population dynamics of the individual. Such information can be retained in the form of printed maps, photographs, or movie film of the developmental display itself, and tables of birth and death rates of constituents of the rhizome pattern.

Stoloniferous plants. The facets of rhizome growth discussed above apply equally well to the growth of stoloniferous plants, which may be loosely described as plants that spread horizontally over the surface of the ground and also ultimately form clones. A substantial part of the world's flora develops in this manner. Of immediate consequence to humans are many of the herbage grasses, bare land stabilizers, crop plants, and weeds. A detailed study of the form of growth and organization in rhizomatous plants is thus of considerable current interest. The consideration of plant growth architecture in terms of apical meristem (or bud) numbers, and with the use of modern methods, is a valuable step toward understanding plant productivity.

For background information *see* STEM (BOTANY) in the McGraw-Hill Encyclopedia of Science and Technology.   [ADRIAN D. BELL]

Bibliography: A. D. Bell, *Ann. Bot.*, 43:209–223, 1979; A. D. Bell, *J. Arnold Arbor.*, 55:458–468, 1974; A. D. Bell, D. Roberts, and A. Smith, *J. Theor. Biol.*, vol. 81, 1979; A. D. Bell and P. B. Tomlinson, *Bot. J. Linn. Soc.*, vol. 80, 1980.

Sun

For many years the Sun has been considered as a main-sequence star, 4.6×10^9 years old, that changes its luminosity with time consistent with its slow evolutionary development. This attitude has prevailed because broadly based theoretical understanding of stellar evolution forecasts that the Sun should be dynamically stable and constant in its luminosity for very long periods of time. It is known from the biological record that the Earth's climate has remained hospitable for most species for over 2×10^9 years, indicating little or no change in the Sun. However, in recent years a number of observations have shown that the Sun is a dynamic star producing intense solar flares and exhibiting surface oscillations, with dramatic long-term changes in its activity cycle. Some of these recent findings will be reviewed that relate changes in the solar activity in the past to terrestrial climate. The Sun produces energy by the nuclear burning of hydrogen to form helium. These nuclear processes can be observed directly by measuring the neutrino radiation from the Sun. The present status of these experiments and ideas for new solar neutrino detectors will be discussed.

Solar luminosity. In recent years there has been a resurgence of interest in monitoring the Sun's behavior to search for variations in its luminosity, rotation, radius, and activity. The solar luminosity has an important bearing on the Earth's climate. Changes in the solar luminosity of a few tenths of a percent will have important effects on the climate. Some climatologists have claimed that a 3% change in the solar luninosity would cause the oceans to freeze in a period of 10^7 years. It is indeed fortunate that the Earth's distance from the Sun and the Earth's atmospheric pressure and composition are nearly perfectly adjusted to allow the development of life.

Study of the present and past history of the solar luminosity has been a topic of great interest for hundreds of years. The longest series of measurements monitoring changes in the solar constant are those performed under the direction of C. G. Abbot of the Smithsonian Institution. He measured the solar radiation reaching ground levels for the long period 1908 to 1952. These measurements were limited in accuracy by atmospheric absorption corrections, and subsequent interpretations of these data concluded that the solar constant has changed less than 1% over this 44-year period. To minimize atmospheric effects, more recent measurements were made at higher altitude with aircraft, balloons, and rocket-borne instruments. These measurements resulted in a value of the solar constant of 1373 ± 20 W/m² at the Earth's mean distance from the Sun.

In the summer of 1975 the *Nimbus 6* spacecraft was launched to begin monitoring the solar luminosity. Its instrument is capable of measuring the solar radiation at several wavelengths. The first 6 months of operation showed that the solar constant varied less than 0.2%. These direct measurements have so far not found a clearly observable change in the solar luminosity. Measurements capable of observing long-term changes in the solar constant of 0.1% from a space platform are needed to search for secular changes in the solar luminosity associated with the sunspot cycle and for long-period climatic changes. Surprisingly, the best evidence for change in the solar activity and the related solar luminosity in the recent past has come from the study of sunspots.

Sunspots and solar activity. Sunspots occur on the solar surface, and the number of spots varies with time in a nearly regular manner. The periodicity in the spot numbers follows an 11-year cycle. Sunspots contain an intense and nearly vertical magnetic field. When a pair of spots occurs, it is found that in a given cycle the magnetic field orientation in the leading spot is opposite to that in the following spot. This polarity difference persists in a given 11-year cycle. However, in the next 11-year cycle the relative magnetic field orientation in the leading spot compared with the following spot reverses. Because of this alternation of magnetic fields, the complete sunspot cycle is regarded as a 22-year cycle.

Figure 1 shows a plot of the recorded sunspot numbers versus time from the earliest observations in the early 17th century to the present time. The 11-year cycle can clearly be seen, and it is readily noticed, that there is a variation in the height of the cycles. It was pointed out by F. G. W. Spörer and E. W. Maunder in 1887–1890 that there was a period from 1645 to 1710 that the Sun was essentially devoid of sunspots. Their papers were ignored by modern astronomers, and their conclusions were regarded as being based upon faulty and incomplete data. However, the subject was recently carefully reexamined by John Eddy of the High Altitude Observatory and Smithsonian Astrophysical Observatory. His modern studies showed this important feature in the history of solar activity was indeed correct. Figure 1 shows his reconstruction of the sunspot data. The period from 1645 to 1710 was after the development of the telescope, and many excellent astronomers clearly capable of observing even small spots in the Sun were making systematic observations. It is now abundantly clear that this was a period in which the solar activity as measured by the sunspots was extremely low, even lower than during the present-day solar minimum. This period, from 1645 to 1710, is now called the Maunder minimum.

Eddy's historical studies revealed that observers reporting on numerous solar eclipses during the period of the minimum did not observe the beautiful corona that one sees today. Even at solar minimum today, one observes coronal streamers at the solar equator. The lack of visible corona was of course directly related to the greatly diminished solar activity during the Maunder minimum. Another piece of evidence consistent with a greatly diminished solar activity during the Maunder minimum was that auroral displays were not described or reported during this period, though they were reported before and afterward. Auroral displays are an easily visible feature associated with an active Sun and attributed to solar flare particles penetrating the Earth's magnetic field at the poles.

Solar rotation. The Sun's rotation can be measured by observing sunspots on the surface or by measuring the small shift in the wavelength of characteristic lines in the solar spectrum (Doppler shift). Measurements show that the Sun does not rotate uniformly as a rigid body. The surface of the Sun rotates faster at the equator (25-day period) than it does at the poles (31-day period). Eddy and his associates Dorothy Trotter and Peter Gilman have examined the early observations to look for past changes in the solar rotation, as observed by sunspot movements. Fortunately, two 17th-century solar astronomers made careful drawings of directly projected images of the Sun and recorded the data and time. Christoph Scheiner made drawings in the 2-year period 1625–1626 before the Maunder minimum, and Johannes Hevelius made similar drawings in 1642–1644 at the start of the period of minimum sunspot activity. The analysis of these early observations revealed that in 1625–1626 the Sun rotated with the same equatorial and polar velocity distribution as it does today, but in 1642–1644 the Sun rotated more rapidly (24-day period) in its equatorial regions than it does today. There are, in addition, sunspot observations by Thomas Harriot from 1611 to 1613. These observations were recorded in his notebooks without a clear explanation of his technique. Richard Herr of the University of Delaware has reanalyzed these records and found that the solar rotation was even slower according to Harriot's

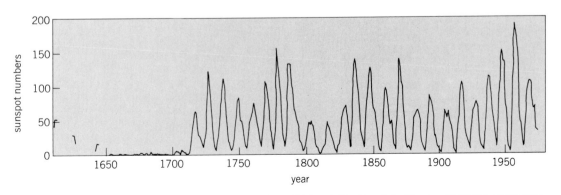

Fig. 1. Sunspot number observations from the 17th century to the present time. The Maunder minimum in the solar activity from 1645 to 1710 is evident. (*From J. A. Eddy, ed., The New Solar Physics, AAAS Selected Symposium 17, Westview Press, 1978*)

data; the period at the equator was 26 days. These changes in the rotational rate of the Sun as observed by sunspot motion are not understood, but do appear to be associated with the Maunder minimum in the solar activity. It must be understood that the total angular momentum of the Sun has not changed in this relatively brief period; it is only the surface regions and associated magnetic fields that apparently changed.

Solar activity and climate changes. The connection between solar activity changes and variations in the Earth's climate was deduced rather indirectly by relating these phenomena to the ^{14}C radioactivity in plants. Solar activity affects directly the intensity of cosmic radiation reaching the Earth, because the electric and magnetic fields associated with the plasma leaving the Sun partially shield the inner regions of the solar system from the incoming cosmic-ray particles. The solar plasma affects mainly the low-energy cosmic rays, those having an energy less than 1 GeV. When the Sun is active, the cosmic-ray intensity is reduced. On the other hand, a quiet inactive Sun allows the cosmic-ray intensity to increase. The variation in the cosmic-ray intensity in turn affects the production of ^{14}C in the Earth's atmosphere. This radioactive isotope of carbon with a half-life of 5600 years is incorporated in all living plants and animals.

The ^{14}C activity in growth rings in very old trees, the bristlecone pines of California, has been measured, giving a record of the amount of ^{14}C radioactivity present in the Earth's atmosphere over the last 7000 years. Paul Damon and his associates at the University of Arizona have made measurements accurate to 0.1% that enable one to see small variations in ^{14}C levels. Figure 2 shows the ^{14}C record measured by this group. Because ^{14}C has a very long half-life, only a long period of diminished or increased cosmic-ray intensity would be noticed in the record. There are several effects

that must be taken into account to interpret the changes observed. There is a smooth variation that arises from the slow change in the magnetic field of the Earth. It can be noticed that there is a small increase in ^{14}C activity during the Maunder minimum, and another dip earlier that would correspond to another minimum in the solar activity called the Spörer minimum. This period covered by the Spörer and Maunder minima corresponds to the Little Ice Age, a period when Europe suffered a period of intense cold and glaciers advanced in Europe, Alaska, and Greenland. If one accepts this connection between solar activity and the Earth's climate as Eddy suggests, then the other increases and decreases in ^{14}C activity marked with arrows in Fig. 2 should correspond to climate changes. The record of the climate in the past 7000 years does fit the pattern of ^{14}C variations reasonably well, giving a connection between solar activity and climate.

The major glaciations of the Earth could also be explained by small changes in the solar luminosity. For many years there has been an active discussion among scientists about the cause of glaciation. In recent years the debate has centered on two main causes: solar variations, and changes in the amount of radiation reaching the Earth as a result of changes in orbital motion and precession of the polar axis—the so-called Milankovitch theory. Recently a strong correlation has been obtained between the ocean water temperature derived from oxygen isotope measurements and orbital variations favoring this explanation of periodic glaciation. However, theories of the structure and evolution of the Sun predict that the Sun should gradually increase its luminosity as it ages. The increase expected from these theories is 5% per 10^9 years. If these calculations are correct, the solar luminosity 2×10^9 years ago would be 10% lower than its present value and life in its present form could not have existed on the Earth. Climatologists who have considered this problem believe that there were corresponding changes in the composition of the Earth's atmosphere to counteract the greatly diminished solar luminosity.

Energy production and neutrinos. The Sun is generally believed to be producing energy by nuclear processes that occur in the central regions where the temperature is nearly 15,000,000°C. Nuclear processes are the only source of energy that are capable of maintaining the solar luminosity at essentially its present value for 4.6×10^9 years. Knowledge of the detailed processes and the energy production rates is derived from theoretical models of the internal structure of the Sun and from understanding of nuclear processes. The internal structure is deduced from a set of equations expressing: (1) the rate of production of nuclear energy; (2) the rate of transfer of energy throughout the mass of the Sun by various photon interaction processes; and (3) the internal mass distribution determined by the hydrostatic balance between the inward-directed gravitational forces and the outward-directed kinetic and radiation pressures.

The nuclear processes have been selected on the basis of laboratory measurements of nuclear reaction rates. There are two sets of possible nuclear reactions; one is the proton-proton chain and

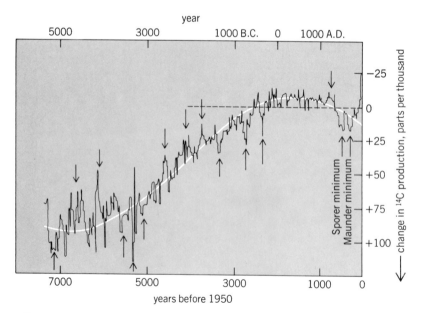

Fig. 2. Carbon-14 production derived from tree ring data. Fluctuations attributed to solar activity changes are marked with an arrow. (*From J. A. Eddy, ed., The New Solar Physics, AAAS Selected Symposium 17, Westview Press, 1978*)

the other is the carbon-nitrogen cycle. In the Sun most of the energy is generated by the proton-proton chain since this chain will operate at lower temperatures. Less than 2% of the Sun's energy is generated by the carbon-nitrogen cycle. The detailed processes are shown in Fig. 3. Both of these sets of reactions correspond to the overall burning of four atoms of hydrogen to produce helium, two positrons (e^+) and two neutrinos (ν). The positrons and gamma rays are absorbed in the Sun and their energy converted to thermal energy. The neutrinos are weakly interacting neutral, massless particles that escape from the Sun.

Solar neutrino flux experiment. An experiment to measure the solar neutrino flux has been carried out by Brookhaven National Laboratory. The goal of this experiment is to test the theory quantitatively. The experiment depends upon the neutrino capture reaction below, which is the inverse of the

$$\nu + {}^{37}Cl \underset{\text{Decay}}{\overset{\text{Capture}}{\rightleftharpoons}} {}^{37}Ar + e^-$$

normal radioactive decay of ${}^{37}Ar$ (half-life 35 days). Since the probability of capturing a neutrino is exceedingly small, a very large mass of ${}^{37}Cl$ is needed. The experiment uses a tank containing 380,000 liters (380 m³ or 615 metric tons) of a chlorine-containing compound, perchloroethylene (C_2Cl_4). The tank is shielded from cosmic rays by placing it nearly a mile underground in the Homestake Gold Mine at Lead, SD (Fig. 4). The operation of the Brookhaven solar neutrino detector is very simple. Helium is passed through the liquid perchloroethylene, removing the few atoms of ${}^{37}Ar$ formed. The ${}^{37}Ar$ is recovered from the helium gas stream by a charcoal filter, purified, and placed in a very small proportional counter to measure the ${}^{37}Ar$ radioactivity. Experiments are performed four or five times per year.

Observations over a period of 7 years from 1972 to 1979 have revealed a very low signal above background effects. The low rate, if attributed to solar neutrinos, would correspond to a neutrino capture rate in the tank of 0.41 ± 0.07 per day. To compare this rate to the theory, one expresses the neutrino capture rate in solar neutrino units (SNU, defined as 10^{-36} capture per second per ${}^{37}Cl$ atom). The rate corresponds to 2.2 ± 0.4 SNU.

Comparison of results with standard theory. The solar neutrino capture rate expected from the most recent (1979) solar model calculations of John Bahcall and Roger Ulrich is $5.5 \pm 30\%$ SNU, a factor of 2.5 above the experimental value. There has been a long-standing discrepancy between the observations of this single experiment and the standard theory of the Sun. The ${}^{37}Cl$ experiment is especially sensitive to the energy spectrum and fluxes of the solar neutrinos. The table lists the neutrino processes in the Sun, the individual neutrino energies, and the theoretical fluxes. One can consider the energy-flux spectrum to be in three broad ranges: a high flux of very low-energy neutrinos from the chain-initiating proton-proton reaction, a modest flux of medium-energy neutrinos from a number of processes, and a very low flux of high-energy neutrinos from ${}^8B$ decay. The ${}^{37}Cl$ experiment has an energy threshold of 0.814 MeV and therefore would not observe the low-energy

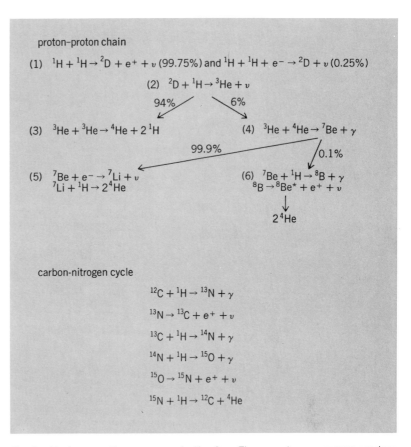

Fig. 3. Nuclear reaction processes in the Sun. These nuclear processes produce gamma radiation (γ), positrons (e^+), and neutrinos (ν).

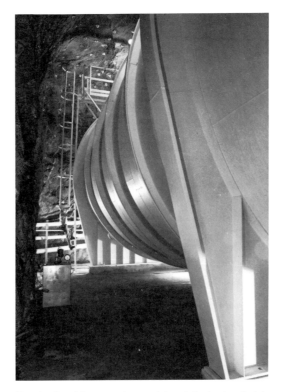

Fig. 4. Brookhaven solar neutrino detector located in the Homestake Gold Mine at Lead, SD.

Solar neutrino energies and fluxes at the Earth

Energy range	Solar process	Neutrino energies, MeV	Flux, neutrinos per cm²
Low	$^1H + {}^1H \rightarrow {}^1D + e^+ + \nu$	Spectrum, 0–0.42	6.1×10^{10}
Medium	^7Be decay	Line, 0.861	3.4×10^9
	$^1H + {}^1H + e^- \rightarrow {}^2D + \nu$	Line, 1.44	1.5×10^8
	^{13}N decay	Spectrum, 0–1.7	2.6×10^8
	^{15}O decay	Spectrum, 0–1.2	1.8×10^8
High	^8B decay	Spectrum, 0–14	3.2×10^6

neutrinos. The neutrino capture cross sections increase with the square of the energy, and furthermore depend upon the particular nuclear characteristics of the capturing nucleus. For these rather technical reasons the ^{37}Cl experiment is most sensitive to the high-energy ^8B neutrinos; approximately 75% of the expected solar neutrino capture rate should arise from this single source. It will be noticed from Fig. 3 that ^8B is produced by a small branch in the energy chain. This branch is very temperature-sensitive.

Explanations for discrepancy. This feature of the ^{37}Cl experiment has stimulated astrophysicists to propose various changes in the standard solar model that would lead to lower central temperatures. For example, it has been suggested that the internal regions of the Sun are devoid of elements heavier than hydrogen and helium. In the standard model the Sun is presumed to have 1.5% of elements heavier than helium throughout, exactly the same concentration as observed in its photosphere. The presence of heavy elements in the interior decreases the rate of energy transport and thereby increases the interior temperatures. This solar model with a reduced heavy-element concentration does yield a much lower ^8B neutrino flux in agreement with the ^{37}Cl experiment. If this model is valid, however, one must invoke a mechanism for adding heavy elements to the Sun's photosphere after the Sun has settled down and become a stable main-sequence star. Possibly material has been added by the infall of material in the solar system or by the collection of material by the Sun as it travels around the Galaxy.

Another explanation that has been considered is that there is a periodic variation in the energy generation processes, and at the present time the energy production is low. The standard theory would say that if the Sun's luminosity did vary, the period would be expected to be longer than 10,000 years. If the nuclear processes are indeed varying with time, there is no theoretical understanding of the process. However, the ^{37}Cl experiment with its high sensitivity to the temperature-sensitive ^8B production is an ideal monitor for variations in the solar energy production mechanisms. For this reason the operation of the ^{37}Cl experiment is continuing to monitor the solar neutrino flux.

Proposed experiments. In recent years a number of new methods for observing the solar neutrino flux have been proposed. Many astrophysicists feel that an experiment capable of observing the abundant low-energy neutrinos from the proton-proton reaction should be built. The flux of these low-energy neutrinos can be calculated accurately, and the flux is essentially independent of various assumptions and input data used in the solar model calculations. Observing these low-energy neutrinos would be a direct test of neutrino theory. Solar neutrino experiments depend heavily on understanding of the properties of neutrinos. If, contrary to present understanding of neutrino physics, the neutrino decayed, oscillated into a neutrino of a different character, or suffered energy loss by some unknown process, the solar neutrino flux could be greatly altered. The effects could be so small that they would not be observed in terrestrially based laboratory neutrino experiments, but could still greatly affect the solar neutrino flux.

Two experiments have been proposed to observe the low-energy proton-proton neutrinos. One is a radiochemical experiment based upon the neutrino capture reaction $\nu + {}^{71}Ga \rightarrow {}^{71}Ge + e^-$, to form radioactive ^{71}Ge with a half-life of 11.4 days. Using this reaction for observing solar neutrinos was originally suggested by V. Kuzmin of the Soviet Union. The detector would be based upon a radiochemical process in which ^{71}Ge is extracted from 50 tons of gallium and placed in a small proportional counter for observing the ^{71}Ge radioactivity. Experimental techniques are being developed, and a 1.5-ton pilot experiment is being built as a collaborative effort by scientists from Brookhaven National Laboratory, Max Planck Institut-Heidelberg, Weizmann Institute, Institute for Advanced Study, and the University of Pennsylvania.

Another low-energy solar neutrino detector, suggested by R. S. Raghaven of the Bell Laboratories, is based upon the neutrino capture reaction, $\nu + {}^{115}In \rightarrow {}^{115}Sn^* + e^-$. The product $^{115}Sn^*$ is formed in an excited state which decays rapidly with the emission of gamma rays. Techniques of directly observing the neutrino capture event are being developed that would incorporate indium in a scintillation counter or some other particle detector. To observe one neutrino capture a day, more than 3 tons of indium would be required.

Another radiochemical solar neutrino detector system is being developed at Brookhaven that depends upon the neutrino capture reaction, $\nu + {}^7Li \rightarrow {}^7Be + e^-$. This reaction has a higher threshold energy than the ^{37}Cl reaction, and therefore it would not be capable of observing the low-energy proton-proton reaction. However, a detector based upon this reaction would have a higher sensitivity to the medium-energy neutrinos than the ^{37}Cl experiment. For this detector only 5 tons of lithium would be needed to observe the theoretically calculated solar neutrino flux. A second solar neutrino detector is urgently needed to verify the results of the ^{37}Cl experiment and give further information on the spectrum of neutrinos from the Sun.

For background information *see* NEUTRINO; RADIOCARBON DATING: SUN in the McGraw-Hill Encyclopedia of Science and Technology.

[RAYMOND DAVIS, JR.]

Bibliography: J. N. Bahcall, *Rev. Mod. Phys.*, 50:881–903, 1978: J. A. Eddy (ed.), *The New Solar Physics*, AAAS Selected Symposium 17, 1978; J. A. Eddy, *Sci. Amer.*, 236(5):80–92, May 1977; O. R. White (ed.), *The Solar Output and Its Variation*, 1977.

Television

In recent years, techniques have been developed for reducing the high transmission rate required for digital transmission of television signals. These techniques, known as bandwidth reduction or bit-rate compression, involve elimination of redundant information from the signals.

Digital transmission. Digital transmission of signals is accomplished by sending on-off pulses, in contrast to analog transmission, where a continuous waveform of the signal, such as the picture or speech intensity, is transmitted. Digital transmission has a number of advantages: it offers flexibility; digital signals of different types can be easily multiplexed or encrypted; and transmission over long distances can be achieved with easy regeneration of the signal, without adding additional noise or distortion to the signal. Digitization of the analog signal is done by first sampling the signal and then representing each sample by a string of binary digits (bits), 1 or 0, specifying the on or off nature of the pulse. The transmission cost would be proportional to the product of the number of samples that are generated per second and the number of bits required to specify a sample.

The television signal contains a sequence of "snapshots" (called television frames) taken from a scene at a rate of 30 times a second. Within each television frame, sampling is done horizontally to generate 525 scan lines, each of which contains approximately 500 samples (monochrome television); for color television a larger number of samples (between 700 and 900) is required. Each sample is represented by 8 bits, giving a total of about 64,000,000 bits per second for monochrome television and 80,000,000 to 100,000,000 bits per second for color television. This transmission rate is rather high, but it can be reduced by techniques of bandwidth compression.

Bandwidth compression. Bandwidth compression, or more correctly bit-rate reduction, is accomplished by eliminating from transmission redundant information normally contained in the signals. Picture signals contain a significant amount of redundancy. Although there are approximately 8,000,000 samples generated per second, their intensities are not independent of one another. For example, the intensity of one picture sample contains considerable information about the intensity of a sample that is next to it, either horizontally or vertically in the same television frame or at the same spatial location in the previous television frame. Thus, in the illustration, intensities of picture elements A, B, D, E, F, G are close to each other most of the time. Bandwidth reduction techniques have been devised to take advantage of the similarity of the intensities of spatially as well as temporally adjacent picture samples. These techniques can be generally classified into two categories: differential pulse-code modulation (DPCM), in which intensity differences are sent; and transform coding, in which linear combinations of intensities in a block of samples are taken and only some of the combinations are selected for transmission. Practical systems based on transform coding have not yet proliferated due to their complexity.

Conditional replenishment. One popular application of DPCM is conditional replenishment, in which each television frame is divided into two parts: one part which is practically the same as the previous frame, and the other part (called the moving area) which has changed since the previous frame. Two types of information are transmitted about the moving area: addresses specifying the location of the picture samples in the moving area, and information by which the intensities of the moving area picture samples can be reconstructed at the receiver. Comparison with the previous frame intensities requires storage of an entire television frame (about 2,000,000 bits for black and white television and 4,000,000 bits for color television), both at the transmitter and at the receiver.

Since the motion in a real television scene occurs randomly and in bursts, the amount of information about the moving area will change as a function of time. To transmit it over a channel, which works at a constant bit rate, the output of the encoder has to be smoothed out by storing it in a storage buffer prior to transmission. The encoded data enter the buffer at an irregular rate, but they exit the buffer and enter the channel at the constant bit rate of the channel. If the buffer gets nearly full, then certain samples are deleted from transmission, thereby reducing the resolution. Some of the strategies for deleting samples from transmission are: (1) transmitting 1 in every n ($n = 2, 4, \ldots$) samples along a scan line — this reduces horizontal resolution; (2) transmitting 1 in every n ($n = 2, 4, \ldots$) scan lines — this reduces vertical resolution; (3) transmitting 1 in every n frames — this reduces temporal resolution. The blur introduced by these strategies is less visible since it is generally introduced when there is large motion in the scene, in which case the human visual acuity is low.

In the last 5 – 10 years several improvements of the conditional replenishment technique have been made, resulting in commercially available encoders. For television signals generated from video conferences, where the camera motion is limited and the scenes do not contain a large amount of movement, reasonable-quality pictures can be obtained at a transmission rate of 3 – 6 megabits per second, which is a reduction over the uncompressed bit rate by a factor of about 10 – 20. For broadcast television, on the other hand, where there can be large changes from one frame to the next and where there is a stricter picture quality requirement, reductions in the bit rate can be made only by a factor of 3 – 4.

Motion-compensated coding schemes. In television scenes which contain moving objects, a more efficient encoding can be performed by estimating the motion of objects and then using the motion to compare intensities in successive frames which are spatially displaced by an amount equal to the motion of the object. Such schemes are called motion-compensated coding schemes. The operation of such a scheme is shown in the illustration. If a point on an object moves from location C to location B in a frame time, then, instead of comparing intensities A and B having the same spatial location, as in conditional replenishment, intensities A and C are compared. If the estimate of dis-

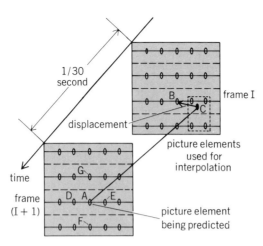

key: --- scan line from alternate field

Diagram of the sampled nature of the television signal, showing the operation of a motion-compensated coding scheme.

placement is accurate, the intensity difference between A and C is much lower than the intensity difference between A and B, resulting in a lower amount of information to be transmitted. The intensity at point C is computed by interpolating from the intensities of the picture elements in the small box in frame I in the illustration. Procedures for estimating translations of moving objects from frame to frame by recursive adjustments have been devised recently. These are simple to implement in hardware. Computer simulations show that, for many video conferencing and broadcast types of scenes, the bit rate can be reduced by a factor of 1.5–3 over the conditional replenishment. The resulting bit rate for monochrome television signals from video conferencing scenes would then be 1–2 megabits per second and for broadcast television 10–20 megabits per second. Hardware implementation of these schemes is expected in the new few years with many modifications to improve the efficiency and decrease the complexity.

For background information *see* COMMUNICATIONS, ELECTRICAL; PULSE MODULATION; TELEVISION; TELEVISION SCANNING in the McGraw-Hill Encyclopedia of Science and Technology.

ARUN N. NETRAVALI

Bibliography: B. G. Haskell, F. W. Mounts, and J. C. Candy, *Proc. IEEE*, 60(7):792–800, July 1972; T. Ishiguro et al., *National Telecommunications Conference Record*, Dallas, pp. 6.4-1 to 6.4-5, November 1976; J. O. Limb, R. F. W. Pease, and K. A. Walsh, *Bell Sys. Tech. J.*, 53(6):1137–1173, August 1974; F. W. Mounts, *Bell Sys. Tech. J.*, 48(7):2545–2554, September 1969; A. N. Netravali and J. D. Robbins, *Bell Sys. Tech. J.*, 58(3):631–670, March 1979.

Terrain sensing, remote

Remote sensing is the process of obtaining information about features on the surface of the Earth through the use of cameras and other sensing devices that are operated from aircraft and space-craft. In recent years this process has become increasingly useful to agriculturalists, foresters, and range managers. Special emphasis has been placed on detecting insect and disease infestations as early as possible so that control measures can be implemented most effectively.

Basic considerations. The rapidly increasing demand for food and fiber, whether it be at the regional, national, or global level, recently has created a greater need to manage the vegetation resources from which these products are derived. Hence, it has become increasingly important to obtain accurate, timely inventories of these vegetation resources so that resource managers will know how much of each kind of vegetation is present within their area of responsibility. In many instances the manager also must have timely, reliable information about the condition of the vegetation resources, so that prompt action can be taken to remedy developing problems from insects, diseases, or other plant-damaging agents. Prompt detection of the nature and location of each problem allows timely remedial action to be taken, thereby minimizing losses.

Foliar analysis. Ordinarily, differences in the species composition or vigor of vegetation can be detected by remote sensing only by differentiation of distinct photographic tones or colors. As seen from overhead, leaves usually are far more conspicuous than stems, flowers, or fruit. Consequently, on aerial or space photographs, the tones or colors of vegetated areas are governed primarily by foliar reflectances, wavelength by wavelength, within that part of the spectrum being examined. A spectrophotometer is used to select the parts of the spectrum that contain the largest spectral differences among the vegetation types or conditions to be distinguished. Then a photographic film can be chosen which is highly sensitive to these parts of the spectrum. If the film responds to unwanted parts of the spectrum, it is usually possible to employ a filter.

Interaction with sunlight. Most of the recent progress relative to remote sensing of agriculture, forestry, and range resources has stemmed from a clear understanding of the following phenomena.

Some wavelengths of a beam of sunlight falling on a healthy green leaf will be reflected to a far greater extent than others (Figs. 1 and 2). With respect to the visible part of the spectrum (0.4 to 0.7 μm), the solar reflectance curve of a healthy broad-leafed plant results primarily from the interaction of light with chlorophyll. Chlorophyll preferentially absorbs energy in the blue and red parts of the visible spectrum and preferentially reflects green wavelengths. These properties account for the green coloration of chlorophyll itself and of the plants which are dependent upon it for food manufacture. Recent research has shown, however, that to understand the spectral reflectance of foliage in the near-infrared part of the spectrum, a more overall look must be taken at the cellular structure of a leaf.

In the middle and lower portions of a typical leaf (Fig. 1), known as spongy mesophyll, individual cells are widely spaced and interspersed with air spaces. "Spongy" structures, such as this tissue, are excellent reflectors in the 0.4–0.9-μm range,

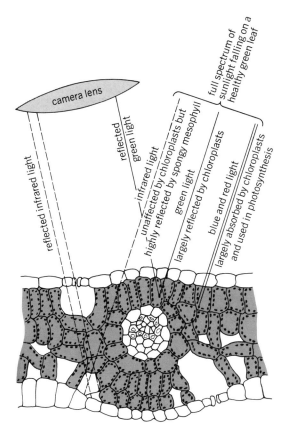

Fig. 1. Cross section of a normal, healthy oats leaf with annotations designed to explain the shape of the spectral reflectance curve for such a leaf.

within which virtually all aerial and space photography is done. Herein lies the explanation for the uniformly high infrared reflectance of healthy, green plants. Specifically, when a beam of sunlight falls on a leaf, the spongy mesophyll is "illuminated" primarily with infrared radiant energy because the visible wavelengths have been largely absorbed or reflected by chlorophyll; hence, they are not available to illuminate, and be reflected from, the mesophyll tissue.

Reflectance curves. The three curves in Fig. 2 provide a comparison of the typical foliar reflectances of a plant at three stages of vigor. It will be noted that, typically, loss of infrared reflectance is the first indication of loss of vigor. Since this early evidence of loss of vigor is in wavelengths beyond the visible red, the change cannot be seen directly by the human eye or on a conventional photograph, both of which are insensitive to infrared radiation. It can be seen, however, through the use of either a black-and-white infrared-sensitive film (which records, in shades of gray, the differences in infrared reflectance) or on an infrared-sensitive color film. Infrared photography is therefore said to provide previsual symptoms of plant stress.

Curve C of Fig. 2 indicates that there is little or no recovery from the earlier loss in infrared reflectance; and because of a gradual decay of chlorophyll (and a consequent unmasking of the leaf's yellow pigments) there is an increased reflectance of yellow light and a consequent visible change in the leaf's appearance from green to yellow.

Although by this time the somewhat higher reflectance of yellow light causes the foliage to photograph in a lighter tone than previously on panchromatic film, this is of little practical value because by then it usually is too late to employ measures that would reduce or eliminate the damage.

Further study of Fig. 1 facilitates the understanding of why a typical leaf can exhibit, in sequence, the three spectral reflectances that have just been described. Key to this understanding are the following facts: The spongy mesophyll can remain spongy (and thus highly infrared-reflective) only when there is an adequate supply of water reaching the leaf. Under this normal, healthy circumstance the mesophyll cells are full of water and are in a distended state, thus forming a spongy structure composed of rounded cells and intervening air spaces. In virtually every instance in which a plant undergoes stress (whether due to some pathogen, insect, or other damaging agent), the supply of water to the mesophyll tissue of the leaf soon becomes restricted. This restricted water supply results in a collapse of the spongy mesophyll (and consequent loss in the leaf's infrared reflectance) quite some time before there is any noticeable decay of the chlorophyll. Eventually, the inadequate nutrition of the leaf causes a general loss of vigor, including an accelerated decay of the chlorophyll and finally chlorosis. The chlorotic foliage then appears in light tones in panchromatic photography.

Color infrared photography. In recent years an improved type of color film known as color infrared or infrared Ektachrome has been developed. Using this film, agriculturists, foresters, and range managers are better able to determine the vigor, yield, production, and distribution of most of the crop types, timber types, and browse species.

Color infrared film is a subtractive reversal film, in which the responses of the three dye layers during processing are inversely proportional to the

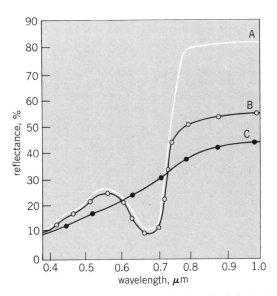

Fig. 2. Foliar reflectance curves of a broad-leafed plant at three states of vigor: curve A, in the healthy green state; B, shortly after the plant has started to lose vigor; and C, after the plant has been more severely damaged.

exposures received by the layers (see table). For example, if this film is used to photograph some highly infrared-reflective feature, such as healthy vegetation, the cyan-coupled dye, which is linked to the infrared-sensitive layer, will be largely eliminated at the time of film processing. As shown by curve A of Fig. 1, however, the reflectance of healthy vegetation in the green and red portions of the spectrum is sufficiently low to permit the correspondingly linked yellow and magenta dyes to be retained at the time of film processing. Normally, in order to extract information from this processed color infrared transparency, the photo interpreter places it over a light table which emits white light. In areas of the transparency where healthy vegetation is imaged, the high concentration of yellow dye almost completely absorbs blue light, and the high concentration of magenta dye almost completely absorbs green light. Because of the low concentration of cyan dye in such areas, however, there is little absorption of red light, that is, it is fully transmitted from the light source, through the transparency, and to the viewer. Consequently the images of healthy vegetation appear red.

Color infrared simulations from space. Space photographs have been used increasingly since the launching (1972) of the world's first Earth resources technology satellite *(ERTS 1)*. To date, there have been three spacecraft of this type (now designated as *Landsat 1, 2* and *3*) launched in a near-polar, Sun-synchronous circular Earth orbit at an altitude of 570 mi (917 km). Each Landsat color infrared simulation is made from three black-and-white space photographs. These in turn are made from records which the Landsat multispectral scanner system has telemetered to the ground of the Earth's "scene brightness" in the same green, red, and near-infrared bands as are employed when color infrared film is used. Each picture element (pixel) for which such a Landsat color infrared simulation is obtained corresoponds to an area on the ground of about 1.1 acres (0.45 ha), with a spatial resolution of approximately 250 ft (76 m). Similar color infrared simulations of agricultural, forest, and rangeland areas, as well as space photographs taken directly with color infrared film, have been acquired by the Skylab astronauts from an altitude of 270 mi (435 km) and often with resolutions of bettter than 100 ft (30 m). Because of the demonstrated high value of that photography to farmers, foresters, and range managers, space photographs of even higher resolution will soon be acquired on many of NASA's space shuttle flights.

Color infrared applications. Aerial or space photography has recently been used with great success in each of the following areas: (1) mapping citrus crop diseases in California, Arizona, and Florida; (2) mapping forest habitat types in the Rocky Mountains; (3) assessing timber losses due to bark beetle attacks in the southern pine region; (4) monitoring air pollution damage to vegetation in Switzerland; (5) monitoring the vigor of agricultural crops in Nebraska; (6) mapping wetlands vegetation along the Atlantic seaboard from New Jersey to Florida and also in Texas and California; (7) mapping desert vegetation in Arizona, central Australia, and Africa's Sahel; (8) determining irrigated land acreage in Nevada; (9) mapping frost damage to eucalyptus forests in California; (10) monitoring wildlife habitats in Colorado; (11) mapping redwood stands in the Coast Range of California; (12) mapping lightning-induced fire damage to sugarcane in Florida; (13) monitoring rotten neck disease in Louisiana's rice fields; (14) determining the prevalence of cereal crop diseases in California and Texas; and (15) monitoring algal blooms in the fresh-water lakes of Nevada and California.

Of particular importance has been the success achieved in a multiagency research program known as the Large Area Crop Inventory Experiment (LACIE). This program seeks first to monitor the global production of wheat, and eventually of other major food and fiber crops as well, primarily from the color infrared coverage provided by Landsat vehicles, at 9-day intervals (weather permitting), throughout each growing season. Successful use of such coverage to help provide increasingly accurate forecasts of crop production already has been made.

For background information *see* AERIAL PHOTOGRAPH; TERRAIN SENSING, REMOTE in the McGraw-Hill Encyclopedia of Science and Technology.   [ROBERT N. COLWELL]

Bibliography: W. Ciesla et al., *Proceedings of the 7th Biennial Workshop on Color Aerial Photography in the Plant Sciences, Davis, CA.*, American Society of Photogrammetry, May 1979; R. N. Colwell et al., *Monitoring Earth Resources from Aircraft and Spacecraft*, NASA SP–275, 1971; *Skylab EREP Investigations Summary*, NASA SP–399, 1978.

Transistor

The static induction transistor (SIT) is a new device under development for use at high current and voltage. The name was introduced in 1972 by J. Nishizawa, T. Terasaki, and J. Shibata. The SIT is similar in structure to a short-channel vertical-junction field-effect transistor (JFET), which has received intermittent attention since it was introduced in 1950 by Nishizawa, and in a somewhat different form by W. Shockley in 1952. Unlike the conventional JFET, the current-voltage characteristics do not saturate. They are very similar in form to those of a vacuum triode and, for this reason, SITs are sometimes referred to as "triodelike" FETs. Present devices can handle 20 A and 500 V at 50 MHz, or 1000 W at 1 GHz. Developemets are under way on devices to deliver 100 W at 3–4 GHz, and 10 W at 9–10 GHz. The SIT is currently

Spectral characteristics of color infrared film

Characteristic	Blue	Green	Red	Infrared
			Spectral region	
Sensitivity with a yellow filter	*	Green	Red	Infrared
Color of dye layers	*	Yellow	Magenta	Cyan
Resulting color in photography	*	Blue	Green	Red

*All three layers of color infrared film are sensitive to blue light, but the yellow ("minus blue") filter prohibits blue light from sensitizing any of the three layers.

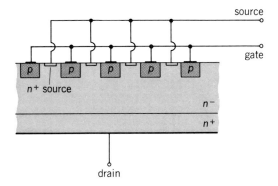

Fig. 1. Cross section of SIT showing physical structure.

used as a low-distortion audio power amplifier and as a fast high-current switch in switched dc power supplies. Developments will provide devices for power amplifiers and drivers in microwave applications. The devices may eventually replace magnetrons in microwave ovens.

Device physics. Figure 1 shows a cross section of the SIT. The applied voltages are such that the space charge region around the p-type gate regions extends throughout the n^- region. Under these conditions, the electrostatic potential along a line extending from source to drain midway between the p-type gates has the shape shown schematically in Fig. 2. The SIT is a majority carrier device. The drain current is carried by electrons injected from the n^+ source into the space charge region, where they move principally by drift to the n^+ drain region. As shown in Fig. 2, these electrons must surmount a potential barrier ϕ_B in order to be injected into the space charge region. The number of source electrons that have sufficient energy to cross the barrier is an exponential function of the barrier height, which in turn is a function of the drain voltage, gate voltage, and device geometry. As a result, the drain current varies exponentially with changes in either gate or drain voltage. A detailed analysis of the operation must include other effects such as the saturation velocity of the electrons and space charge limits on the current, but the exponential nature of the cur-

rent-voltage characteristics remains the same. A typical set of characteristics is shown in Fig. 3.

The temperature dependence of the drain current is found to be positive at low currents but negative at high currents, a fact that eliminates the problem of thermal runaway in high-power applications.

A related device is the field terminated diode (FTD), sometimes referred to as the field controlled thyristor (FCT). It differs from the SIT in that the n^+ substrate is replaced by a heavily doped p^+ substrate. The general features of the rest of the structure remain the same. Inclusion of the n-p^+ junction in the structure gives it a reverse-voltage-blocking capability that the SIT does not have. In the forward direction, FTD devices have the capability of blocking more than 1000 V with an applied gate bias of about 30 V, and simultaneously exhibiting a low forward voltage drop in the "on" state. The gate structure allows the forward current to be turned off in less than 1 μs.

Applications. SITs find application where high voltage and high current are simultaneously required. They are now commercially used as audio amplifiers with minimum breakdown voltages of 40 V between the source and gate, and 200 V between the gate and drain. In the low-to-medium frequency range, they are finding application as the switch in switched dc power supplies. For this application, devices that can handle 500 V and 20 A and be switched on or off in less than 1 μs are used. Use of the SIT should allow the switching frequency of the supply to be raised, thereby reducing the size of the inductance and capacitance required, and hence the cost of the supply. At microwave frequencies, the combination of high voltage at high current is the SIT's most attractive feature. Bipolar transistors can operate at high current but low voltages, while other devices, such as traveling-wave tubes, operate at high voltage but low current. It is the possibility of an inexpensive high-voltage high-current device that will operate with 60 – 100 V at up to 6 GHz and above that motivates

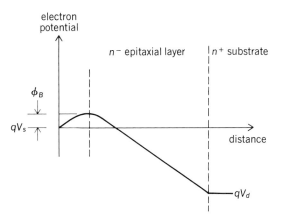

Fig. 2. One-dimensional potential distribution in SIT. V_s = source voltage; V_d = drain voltage; q = electron charge.

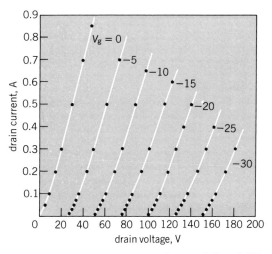

Fig. 3. Typical current voltage characteristics of SIT. Numbers next to curves give gate voltage V_g in volts. (From J. I. Nishizawa and K. Yamamoto, High-frequency high-power static induction transistor, IEEE Trans. Electron Devices, ED-25(3):314–323, 1978)

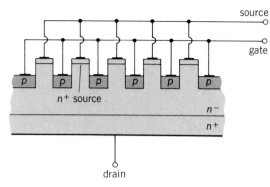

Fig. 4. Cross section of high-frequency SIT showing physical separation of source and gate.

the present developments. For applications at frequencies above 6 GHz, the GaAs field-effect transistor is a more likely candidate at this time. One very large commercial use of the SIT may be as a power source for microwave ovens, where it would represent a significant cost advantage over the magnetrons presently used. The requirement of 700 W at 2.45 GHz appears to be within reach of the SIT.

Fabrication. The fabrication of the SIT begins with a heavily doped n^+ silicon substrate. The lightly doped high-resistivity n^- layer is epitaxially grown on the n^+ substrate. The thickness of this layer determines the maximum gate-to-drain operating voltage and is about 40 μm for a 500-V device. The p gate regions are formed by diffusion, as are the n^+ source regions. The fabrication is completed by depositing and etching a layer of metal to provide electrical contact to the source and gate regions. Contact to the drain is made via the back side of the n^+ substrate when the chip is bonded into the final package. The p-type gate regions must be close enough together that the space between them can be totally depleted of movable carriers when the normal gate and drain voltages are applied. Center-to-center gate spacing in some devices is as low as 5 μm, which means that high-resolution fine-line lithography and other recently developed processing methods must be used in fabrication. These methods include shallow ion implants, self-aligned diffusion processes, and x-ray lithography. In fact, it is the development of these methods that led to the renewed interest in the device.

The high capacitance between the gate and source is one problem of the structure which must be held to a minimum for high-frequency devices. Much of the present development effort is devoted to this problem. The input capacitance is most directly reduced by increasing the physical separation between the p-type gates and the n^+ source regions. This is usually done by placing the gate in the bottom of a narrow recessed area as shown in Fig. 4. The fabrication of such a structure with center-to-center gate spacing of a few micrometers requires very sophisticated processing methods.

For background information *see* TRANSISTOR in the McGraw-Hill Encyclopedia of Science and Technology.

[ROBERT J. HUBER]

Bibliography: J. I. Morenza and D. Esteve, *Solid State Electron.*, 21:739–746, 1978; J. I. Nishizawa, T. Terasaki, and J. Shibata, *IEEE Trans. Electron Devices*, ED-22(4):185–197, April 1975; J. I. Nishizawa and K. Yamamoto, *IEEE Trans. Electron Devices*, ED-25(3):314–322, March 1978.

Tropical meteorology

The most significant new developments in tropical meteorology have concerned analysis of data from the recent GATE experiment and the discovery of the importance of direct momentum field perturbations. It has been shown that large-scale waves in the easterly trade winds modulate precipitation in the tropics. These waves derive much of their energy from condensation heating, but this mode of heating is quite inefficient at producing long-term changes in the pressure field. Direct perturbations of the rotation field are energetically weaker than condensation heating events, but they are much more efficient producers of large-scale circulation changes.

GATE. For the last 5 years, tropical meteorologists have concentrated their efforts on the processing and analysis of data from the GARP Atlantic Tropical Experiment (GATE) of the Global Atmospheric Research Program (GARP). GATE was a major international field experiment held in the late summer of 1974 over the tropical Atlantic west of Africa. Hundreds of scientists from throughout the world's meteorological community gathered for 3 months to accumulate data from rawinsonde balloons, satellites, weather radars, tethered balloons, shipboard instrument systems, and aircraft sensors.

GATE grew out of an increasing awareness that the fundamental behavior of tropical weather systems is quite different from that of the familiar and more analyzed middle-latitude systems. Poleward of approximately 30° latitude, the weather is governed by traveling-wave disturbances on the scale of thousands of kilometers. These waves convert the enormous potential energy of the Earth's Equator-to-pole temperature gradient into kinetic or motion energy on the scale of the waves. Horizontal and vertical motion fields associated with the waves can be adequately described in terms of smoothly varying averages over large areas. For many purposes, such as short-term weather prediction, the dynamics and energetics of middle-latitude weather systems can be analyzed and modeled adequately without regard to physical processes occurring on smaller scales. In the tropics, however, horizontal temperature gradients are much weaker. Large-scale disturbances such as traveling waves in the easterly trade winds and the rarer but much more destructive hurricanes derive most of their energy from the release of latent heat of condensation in convective clouds. The clouds are phenomena operating on spatial scales of no more than a few kilometers—orders of magnitude smaller than the weather systems themselves. The crucial interactions which occur between these phenomena of vastly different time and space scales are complex and poorly understood.

It will not be possible in the foreseeable future to describe or simulate the characteristics of every cloud in a tropical weather system. Hope for improved operational analysis and modeling of tropical weather depends upon the ability to find methods of representing the effects of clouds upon larger scales, and vice versa, by studying variations in large-scale, and hence observable, parameters. This was the problem addressed by GATE. The tropical atmosphere was observed simultaneously on scales ranging from hemispherical satellite observations to cloud microphysical measurements of particles just fractions of a millimeter in diameter. Enormous amounts of data were collected, processed, validated and cross-checked, and research results are now beginning to emerge.

Easterly waves. It has been known since the 1930s that there are large-scale waves in the easterly trade winds of the West Indies and central Atlantic. Analysis of GATE data has greatly improved the understanding of these easterly waves. The waves tend to stimulate the clouds and rainfall to the west of the low-pressure line (trough) and suppress convection near the high-pressure axis (ridge). This results from large-scale vertical motion patterns associated with the wind field. The waves form due to instabilities in the flow field of the easterly jet stream at 10,000 ft (3000 m) over western Africa. Once formed, the wave energy comes from a combination of latent heat release in clouds and kinetic (motion) energy drawn from the jet. The waves tend to weaken as they cross the central Atlantic, but they can be traced eastward as far as the eastern Pacific. A few of the wave troughs move into favorable regions and intensify into hurricanes, although half of the Atlantic hurricanes form from other types of disturbances.

Cloud clusters. Individual convective systems, or cloud clusters, typically extend over areas of 200–400 km and exist for an average of about 24 hr. Twelve of these clusters were analyzed during GATE with radar and balloon measurements. Data from all 12 systems were averaged together. It was shown that the buildup of rainfall measured by the radar lagged observed large-scale convergence of mass in the lower levels of the atmosphere by 4–6 hr. Low-level mass convergence (net inward flow) implies upward vertical motion, and it was originally assumed that the ascending air cooled and moistened the middle atmospheric levels, enhancing cloud development. However, recent analysis suggests that clouds form very rapidly in the presence of large-scale upward motion. Most of the lag between mass convergence and radar rainfall estimates reflects the storage of liquid water in clouds as the convective system intensifies. Similar results were obtained from several detailed studies of individual cloud clusters.

Since synoptic weather systems in the tropics draw much of their energy from latent heat release, it is of interest to examine the time and space scales on which the air is warmed. One might expect to find anomalously warm air in a region containing above-normal rainfall. This proves to be incorrect. Examination of the rawinsonde data showed that mean warming of the GATE ship array (see illustration) was not correlated with latent heat release on the observable 3-hr time scale despite rates of latent heat release as high as 15° C per hour (equivalent warming averaged horizontally over the ship array and vertically from the sur-

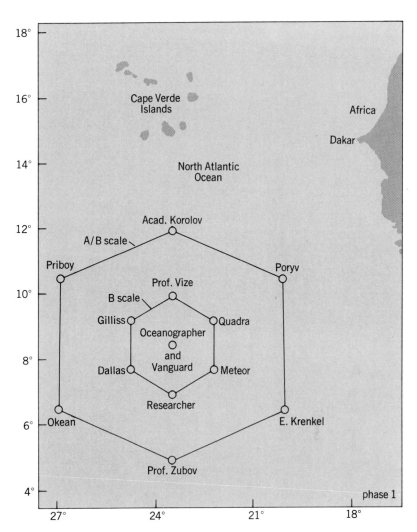

Ship array during the 1974 GATE experiment.

face to 16 km). Correlation coefficients between the time rate of change of temperature of this volume and the rate of condensation in the volume were actually slightly negative. There was some tendency, however, for temperatures to increase slightly at upper levels and decrease below during strong convective periods. It is clear that the tropical atmosphere rapidly disperses released latent heat laterally to large scales.

Diurnal variations. Several diurnal variations were documented during GATE. Radar-estimated rainfall and satellite cloud cover were at a maximum in the late afternoon. This differed from rainfall measurements in most other tropical oceanic regions. Temperatures underwent a rather regular diurnal cycle, showing an average maximum near 1:30 P.M. and a minimum at about 4:30 A.M. The temperature cycle resulted from a combination of direct solar heating of the atmosphere and very-large-scale circulation features—probably related to the atmospheric tide. Vertical motion also varied diurnally with a late-afternoon upward maximum corresponding to the rainfall maximum as expected.

Momentum perturbations. Since the early discovery that latent heat was fundamental in tropical energetics, studies of the role of clouds have concentrated on their warming properties. The more complex direct effects of clouds on the momentum field have generally been neglected. However, the finding that condensation heating is rapidly dispersed away from the convection implies that this energy source is inefficient with regard to causing lasting perturbations of the large-scale flow and pressure fields. Special circulation configurations are necessary to accumulate significant warming due to latent heat release. (This accounts for the relative rarity of tropical cyclones.) Although cloud momentum fluxes involve less energy than the condensation processes, the former effect may be important if it constitutes a more efficient mode of altering large-scale circulations. Recent analysis using numerical computer models suggests that small perturbations of the rotation field on the scale of a cloud cluster (200 to 400 km) in the tropics result in long-lasting changes in the pressure and temperature fields. Conversely, changes in the direct warming or the closely related vertical motion fields are rapidly dissipated with very little long-term effect. Studies of the direct effects of clouds on the momentum field have acquired great importance.

Cloud momentum fluxes are not the only process which can perturb the rotation patterns of the tropical atmosphere. When strong horizontal wind gradients exist, advection can cause changes in the effective rotation rates of air about a point. Local pressure patterns induced by lines of deep convection, such as squall lines, can also alter larger-scale rotation. All of these mechanisms are fundamentally similar. The resulting momentum perturbations are quite efficient modulators of tropical pressure patterns. Although latent heat release supplies much larger amounts of energy to the tropical atmosphere, the relative inefficiency of this process in altering large-scale pressure fields raises important questions for the future as to the relative importance of momentum and latent heat processes.

For background information see METEOROLOGY; TROPICAL METEOROLOGY in the McGraw-Hill Encyclopedia of Science and Technology.

[WILLIAM M. FRANK]
Bibliography: W. Frank, J. Atmos. Sci., 35: 1256–1264, 1978; W. Gray and R. Jacobson, Mon. Wea. Rev., 105:1171–1188, 1977; R. Reed, D. C. Norquist, and E. E. Recker, Mon. Wea. Rev., 105: 317–333, 1977.

Vegetation management

To early humans vegetation provided a source of food, shelter, and fuel for heating and cooking. Modern humans have the same basic uses for vegetation. Yet the addition of advanced technology has extended vegetation management far beyond the use of fire by primitive peoples. Vegetation managers seek to create a particular mixture of plants that meets human needs. The science of vegetation management recognizes and works within the continual processes of change in plant species and life-form. The major challenge is proper application of management techniques, including fire, to a variable resource.

Vegetation change. Vegetation is constantly changing. From an available species pool, plants interact with their environment and each other.

Those plants best suited to that environment grow and flourish. In relatively constant climates, orderly and predictable patterns of change or succession from early colonizing plants to steady-state species can be discerned. The understanding of successional patterns associated with environmental changes underlies modern vegetation management. The ecological and economic costs of maintaining or creating a particular vegetation stage correlate with how closely human goals align with natural change. In general, the longer-lasting a desired stage of succession, the lower the maintenance cost.

West Coast brush fields. To control large wildfires in extensive flammable brush fields along the West Coast, government land management agencies have embarked on a vegetation management program that seeks to reduce wildfire damage through the reduction of vegetative fuels. The major objective is to alter the vegetation to maintain a fuel distribution that would retard the spread and intensity of wildfire. Currently the most important aspect of such management is the construction of fuel-breaks, which are wide strips or blocks of land on which natural vegetation has been partially or totally removed and replaced by vegetation of lower volume and hazard.

A management objective including fire is important in preserving and maintaining natural ecosystems. In the western United States many types of vegetation have evolved, with the occurrence of fire as natural as rain, wind, and sunlight. Prior to humans, the high frequency of lightning fire prevented the formation of large accumulations of fuel. Some forests types were burned as frequently as every 4 to 10 years by low-intensity fire.

In recent years strong fire suppression programs have created fuel buildup conditions that virtually ensure the total destruction of the very resources they were designed to protect. A policy of fire exclusion is a strong vegetation management policy, although it is seldom considered to be such. Plants adapted to periodic disturbance by fire are rapidly replaced by more shade-tolerant ones. Grasslands are invaded by woody plants. Forest shrublands are shaded out by trees. Open parklike forests are choked with suppressed understory trees. The National Park Service of the U.S. Department of the Interior is charged with the objective of preserving and maintaining natural ecosystems, and is currently reintroducing fire as a natural process to the large national parks in the western United States. In high-elevation areas, where fuel concentrations have not been significantly increased by years of fire suppression, naturally ignited fires are allowed to burn unchecked. In this way fire-adapted plants and animals are favored, fuels are reduced, and the landscape is maintained for future generations.

Management techniques. The available vegetation management techniques depend on funding, equipment, climate, topography, soil characteristics, and vegetation structure and density. Mechanical, chemical, and prescribed burning techniques are frequently used in combination to maintain a particular vegetation state. Often plant seeding is done in conjunction with manipulation to control initial vegetation stages on a site. Grazing animals represent an important management tool.

Grazing is selective for certain species, and the season, intensity, and distribution of grazing can be controlled by humans. Social acceptability of a management technique increasingly dominates technical effectiveness in the selection of management methods.

Mechanical manipulation is a common technique used to alter large woody vegetation forms. Large crawler-type tractors are employed to clear the vegetation using a variety of implements. On projects involving small acreages, a bulldozer with its blade set approximately 30 cm above the ground is used to push over woody plants. Special toothed bulldozer blades called brush rakes or rock rakes are also used. These blades allow the surface soil to pass through them. With such implements, plants can be uprooted and piled without large-scale soil redistribution. Various brush cutters and large agricultural disk implements are employed where terrain and fuel type permit. On larger projects, naval surplus anchor chains are pulled between two large crawler-type tractors to break the vegetation off at ground level. An individual link from such a chain may weigh as much as 45 kg. Such chains have also been modified by adding steel bars running across the width of each link. These bars allow the chain to dig into the soil and uproot many plants.

Chemical herbicides, such as the phenoxy compounds 2,4-dichlorophenoxyacetic acid (2,4–D) and 2,4,5-trichlorophenoxyacetic acid (2,4,5–T), have been used extensively in the past 3 decades to control vegetation sprouts and seedlings on areas originally cleared by fire or mechanical techniques. These and other chemicals have also been used for direct control of certain highly susceptible species such as hardwoods in logged areas and sagebrush on rangelands. Problems with social acceptability and possible environmental costs of using chemicals in vegetation management may limit their use in future years.

Prescribed burning. The use of fire in wildlands to maintain a vegetation type, manage fuel loading, or manipulate wildlife habitat involves many considerations. The "ideal" fire is one that achieves the stated objective with the minimum of fire intensity. Such a fire will be easy to manage and will produce a minimum of undesirable biological, physical, and social effects.

The most recent technological advances concern aerial incendiary devices and dispensing techniques for carrying out large-scale prescribed burning. There are presently three basic aerial incendiary ignition systems used in North America. The first consists of the dropping of burning hydrocarbon mixtures from helicopters. Gasoline-diesel mixtures or jelled fuels are poured or pumped out of a large drum through a long spout. At the end of the spout is a large industrial sparkplug for fuel ignition. This arrangement is suspended below a helicopter, and the pilot can regulate the flow rate of fuel and time of ignition by way of electrical controls. This system, developed in the Pacific Northwest for burning large areas of forest harvest residues, is currently being used in wildlife habitat manipulation and fuel management work in California.

A second system employs the dropping or launching of pyrotechnical devices. A common delayed-action incendiary device (DAID) is simply a large form of a safety match. Like a safety match, a special striker pad must be used to obtain ignition. This pad and a metal box containing the DAIDs are mounted on a helicopter in such a position that the devices can be ignited and dropped with one sweeping arm motion.

The third system employs a DAID that utilizes an exothermic chemical reaction to create enough heat for ignition. For eucalyptus forest fuel management in Australia, incendiaries have been developed that use the heat from the reaction of potassium permanganate and ethylene glycol to bring about ignition. A small plastic pill bottle containing powdered potassium permanganate is injected with a measured amount of ethylene glycol just as the bottle is being dropped from an airplane. The quantity of ethylene glycol injected controls the length of time before ignition. Specially constructed dispensing equipment gives control over the rate at which the bottles are dropped and the spacing between ignition centers. More recently, the Canadian Forestry Service has developed a similar ignition system using the same chemicals dropped in a high-impact polystyrene ball similar to a ping-pong ball.

Summary. Advancement in the technology of vegetation management depends on scientific understanding of rates of change in many vegetation types. As the energy used for manipulation becomes more costly, vegetation management must become more attuned to natural processes.

For background information *see* VEGETATION MANAGEMENT in the McGraw-Hill Encyclopedia of Science and Technology. [RONALD H. WAKIMOTO]

Bibliography: J. R. Baxter, D. R. Packham, and G. B. Peet, *Control Burning from Aircraft*, CSIRO, Melbourne, 1966; J. M. Dodge, *Science*, 177(4044): 139–142, 1972; G. A. Roby and L. R. Green, *Mechanical Methods of Chaparral Modification*, USDA Agric. Handb. no. 487, 1976.

Venus

In December 1978 the U.S. National Aeronautics and Space Administration (NASA) carried out a comprehensive exploration of the atmosphere of Venus, using simultaneous observations from five entry vehicles and an orbiter. Although the Soviet Union has previously conducted a long series of successful missions to the Earth's sister planet, including the landing of spacecraft on its surface, this was the first effort by the United States to obtain data from either an orbiter or an atmospheric entry probe. The spacecraft that carried out these measurements were part of the Pioneer Venus Project, operated for NASA at the Ames Research Center in California.

Orbiter observations. The Pioneer Venus orbiter, launched on May 20, 1978, arrived at Venus on December 4 and was inserted into a highly eccentric, nearly polar orbit with an initial period of 24 hr. During its planned lifetime of 243 days, the orbiter maintained an apoapsis altitude of 67,000 km and a periapsis altitude of 150–260 km—low enough to permit direct sampling of the upper atmosphere and ionosphere of the planet. The orbiter is a spin-stabilized spacecraft with a 5 revolutions-per-minute spin rate. Its scientific payload includes neutral and ion mass spectrometers, a photopolarimeter, an infrared radiometer, an ultraviolet spectrometer, a surface radar mapper, and several instruments designed to measure magnetic and electric fields and properties of the plasma through which the spacecraft moves.

In its initial orbits, the Pioneer orbiter revealed a number of dramatic results. The absence of an intrinsic magnetic field was demonstrated; the magnetic fields present are due to the interaction of the solar wind plasma with the ionosphere. The radar mapper showed large-scale differences in elevation on the surface, but cast some doubt on previous claims from ground-based radar that the surface of Venus is very cratered. Thermal measurements indicated an enhanced emission near the poles, suggesting a lower cloud opacity at high latitudes, and ultraviolet spin-scan images displayed a rich cloud structure in the upper atmosphere. These investigations were continued throughout a full Venus day (= 243 Earth days).

Multiprobe observations. The Pioneer multiprobe spacecraft, a companion to the orbiter, was launched on August 8, 1978, and followed a faster trajectory, arriving at the planet on December 9, four days after insertion of the orbiter. Within an 11-minute period, five independent vehicles deployed by the multiprobe entered the atmosphere of Venus: a large probe, three small probes (called the North, Day, and Night probes), and the spacecraft "bus" that had supported the probes until their separation 3 weeks before entry. All four probes began to gather data at an altitude of 70 km. The large probe, which carried seven instruments, deployed a parachute for a 17-min descent through the middle atmosphere; it then jettisoned the chute at an altitude of 47 km (pressure about 2 bars or 200 kPa) and continued in free fall for an additional 36 min. The smaller probes descended without parachutes, requiring about 55 min each to fall 200 km from entry altitude to the surface. Although none of the probes was designed to survive impact on the oven-hot surface of the planet, the Day Probe continued to transmit for 67 min after landing. Some of the instruments on each of the three small probes, however, ceased functioning at 13 km altitude, for reasons that remain mysterious but may be related to electrical discharges induced in the probes.

Atmospheric structure. The structure of the atmosphere of Venus from the surface up to 110 km was derived from the probes' measurements of pressure and temperature, together with their aerodynamic behavior as deduced from on-board accelerometers. At the surface, all four probes measured temperatures between 721 and 732 K and pressures between 86 and 95 bars (8.6 and 9.5 MPa), as expected from previous calculations and the Soviet Venera measurements. From the surface to 40 km height, the temperature gradient does not favor vertical circulation and the atmosphere is highly stable. Only near the dense clouds, at about 45-50 km altitude, does significant convection occur.

Cloud structure. Direct measurements of clouds were provided by two instruments. Each of the probes carried a nephelometer (cloud meter), which emitted a light beam and measured the light backscattered by cloud particles adjacent to the

probe. A more elaborate instrument designed to measure the numbers of particles in size ranges from a micrometer to a millimeter—fine dust to large raindrops—was carried in the large probe.

Data from these instruments showed that the main cloud deck on Venus is 3–4 km thick and is centered at an altitude of 49 km. Here, the aerosol sizes are greatest; most of the particles are on the order of 10 μm across. Although no direct compositional measurements were made, the Pioneer scientists believe that these clouds are composed primarily of solid sulfur. About half the total mass of the clouds is in this layer, which cannot be observed from the Earth.

Above 51 km, there are fewer cloud particles per unit volume, and the average size appears to decrease. The probe data extend up to about 62 km altitude, a region where the particle size is typically about 2 μm. This altitude probably is close to the deepest that can be seen from above, and it is generally agreed that the clouds are composed of sulfuric acid, identified several years ago from Earth-based polarimetric observations. Within these acid clouds, the typical visibility is about 1 km, characteristic of a light haze or smog rather than a conventional terrestrial cloud. Below the main cloud deck, haze layers extend down to about 32 km, but the atmosphere is completely clear from this altitude to the surface.

Absorption and emission of radiation. An important question for the dynamics of a planetary atmosphere is which regions are heated strongly by the absorption of sunlight. In the case of Venus, the amount of light absorbed by the surface and reemitted as infrared radiation is also a critical quantity needed to evaluate the efficiency of the greenhouse effect that maintains the high surface temperature. All four probes carried radiometers to measure the deposition of sunlight and the emission of thermal energy by the atmosphere and clouds. Of the sunlight striking Venus, about 70% is reflected back to space from the upper clouds. About half of the remaining energy is absorbed in haze above 60 km. Additional absorption takes place in the main cloud deck, but about 3% of the original incident sunlight emerges into the clear atmosphere below 32 km, and more than 2% is finally absorbed at the surface, which is relatively dark (about 15% reflectivity). This deposition of energy at the surface is greater than has sometimes been assumed, and it appears ample to maintain the greenhouse effect.

Atmospheric stability. Overall, the atmosphere of Venus is extremely stable, and few differences appeared from probe to probe, even in the fine details of cloud structure. There appears to be no "weather" at the surface and probably little geographical climatic variation.

Chemical composition of atmosphere. One of the most important objectives of the Pioneer Venus mission was the accurate determination of the chemical composition of the atmosphere. The large probe carried two instruments to measure composition: a mass spectrometer and a gas chromatograph. Preliminary Pioneer data showed that the predominant gas in the atmosphere of Venus is carbon dioxide. Previous indications were that the concentration is 96–98%; the Pioneer gas chromatograph gave $96 \pm 1\%$. The second-ranking gas is nitrogen, at 3.4%; it had not been reliably measured before. Third is water vapor, at 0.1–0.5% (below the clouds). This concentration is slightly larger than had been expected, and it further supports the greenhouse explanation for the high surface temperature of Venus.

Other gases detected in Venus's atmosphere, all at trace levels, are oxygen, argon, neon, and sulfur dioxide. The sulfur dioxide is further evidence that sulfur compounds are the main constituents of the clouds. The most controversial results, however, are those for argon. The gas chromatograph found a concentration of 20–30 parts per million, while the mass spectrometer found about twice as much, based on a preliminary calibration. The mass spectrometer, which could distinguish among the isotopes of argon, found that the total amount of argon-36, which is thought to be primordial, is more than a hundred times greater for Venus than for Earth, and its abundance relative to argon-40 is also much greater. Since argon is too heavy to escape from the atmosphere of either planet, it appears that Venus was endowed with more argon-36 from the beginning, or else that additional argon is locked up in the interior of the Earth. Either interpretation would be unexpected, requiring important revisions in the use of the argon abundance as an indicator of planetary histories.

Conclusions. The Pioneer Venus results do not seem likely to generate any major upheavals in scientists' views of Venus. The massive, stagnant atmosphere of carbon dioxide, the planetwide surface temperature of 740 K, and the deep cloudy hazes of sulfur and sulfuric acid appear to be confirmed. The new information on cloud opacities and deposition of sunlight in the atmosphere and at the surface supports the greenhouse explanation for the high surface temperature. The biggest surprises arise from the high level of argon-36, and of noble gases generally, in the atmosphere. In addition, the orbital radar data are pointing out the extremely primitive state of knowledge of the geology of Venus and of the processes that shape its surface.

For background information *see* VENUS in the McGraw-Hill Encyclopedia of Science and Technology.   [DAVID MORRISON]

Bibliography: Special issue, *Science*, 203:743–808, Feb. 23, 1979; Special issue, *Space Sci. Rev.*, vol. 20, no. 4, June 1977.

Water treatment

Improved analytical technology has allowed identification of organic chemicals in drinking water that have not been previously reported. Many of these organic substances are suspected of having adverse public health effects when consumed, even at low concentrations. Of the many substances identified, chlorinated organic substances have generally been found in the highest concentrations. Studies are being conducted to provide water treatment technology capable of producing a water quality that minimizes the potential for organic contaminants to reach the public through their drinking water.

No single technique will provide removal of the myriad of organic contaminants that might be present in any given water supply. Therefore, many techniques are under investigation to provide the technology for removing a broad spectrum of organic substances.

The most promising treatment approaches have received fairly intensive investigations and can be discussed in terms of reducing relatively low-molecular-weight chlorinated organics such as chloroform, as one category, and reducing other chlorinated and nonchlorinated organics, as a second category.

Reduction of trihalomethanes. Chlorinated organics, such as chloroform and other trihalomethanes, are formed during the chlorination process as practiced at many locations across the United States. Studies by the Environmental Protection Agency (EPA) have shown that changes in disinfection procedures can result in reductions in the levels of chloroform produced.

Use of chlorine. Chloroform concentrations can be lowered if chlorine is applied to the water with the lowest possible organic content. Therefore, in locations where the procedure is feasible, a utility should consider moving the point of application of chlorine to the stage in the treatment process where the water should have the lowest organic content: after filtration, or after coagulation and settling, if these unit processes are employed. These practices have reduced, but not eliminated, trihalomethane concentration of the finished water at many locations. Utilities making such a change in disinfection practice should carefully monitor the microbiological quality of their drinking water to make sure it has not deteriorated.

Other disinfectants. Further reduction in chloroform concentration can be obtained if a disinfectant, such as ozone, chlorine dioxide, or chloramine, is used instead of chlorine. These three disinfectants do not produce trihalomethanes, although they may produce other organic or inorganic byproducts that have yet to be identified or evaluated for toxicity. Furthermore, chloramine is a weak disinfectant, and microbiological quality control therefore must be more carefully monitored. Finally, ozone does not produce a disinfectant residual, and thus the addition of chlorine or some other residual-producing disinfectant may also be necessary. If chlorine is used, some chloroform may be formed during passage through the distribution system. Other oxidants (disinfectants), such as hydrogen peroxide and potassium permanganate, will also be evaluated as alternatives to chlorine.

Use of activated carbon. Water containing very little organic matter can be produced when fresh granular activated carbon is used as a medium for the adsorption of organic compounds. This water can then be disinfected with chlorine, ozone, or chlorine dioxide, and little chloroform or other organic by-products will be produced because of the small quantity of organic matter available for reaction with the disinfectant. This treatment technique has the additional benefit of removing many organic raw-water contaminants, other than precursors, thereby providing consumers with an additional margin of safety. Other adsorbents may also produce the same effect.

The chief disadvantage of adsorption on granular activated carbon as a treatment technique is that the adsorption capacity of the material is limited. In general, the use of granular activated carbon for the control of chloroform precursors means that the frequency of reactivation will have to be increased over that commonly used when taste and odor control is the only objective. Other adsorbents are being evaluated for their ability to control disinfection by-product precursors, as well as removing raw-water contaminants.

The techniques described above are all preferable to attempting to remove chloroform once it has been formed, as no unit process has yet been demonstrated to be very effective for chloroform removal. Although all the information concerning these processes is not known and an extensive research program is refining the information, results of field investigations support the recommended use of any of these treatment processes in certain circumstances at this time (spring 1979).

Reduction of other organics. Although many techniques are being investigated for removing organic contaminants, adsorption provides a potential for removing a broad spectrum of substances with a single unit process. A number of major investigations using activated carbon and resins as adsorbents are being conducted. The table is a list of some of the major studies being conducted by the EPA. The projects at all of the locations cited have different specific objectives, but all have the common goal of removing organic contaminants.

Adsorbents. Projects completed at Miami, FL, and Kansas City, MO, have shown that activated carbon can operate as an effective adsorbent for a broad spectrum of organic contaminants under the conditions at those locations. These projects have also shown that various resins can be used to remove certain specific types of organics even more effectively than activated carbon, but the resins do not appear to possess the same broad-spectrum adsorptive properties as activated carbon. Both studies have included investigations for removing chlorinated organics, including trihalomethanes, and other organic substances. The water used in the studies was the water processed in the standard treatment system at each location. Thus, the results for removal of organic substances are considered similar to what can be achieved at those locations under the conditions evaluated.

A recently initiated study of a groundwater at Glen Cove, NY, will assess the ability of synthetic adsorbents and aeration to remove chlorinated organics in the groundwater at that location. In-place reactivation of the synthetic adsorbent will also be investigated.

Other projects now operational, such as those at Cincinnati, OH, Manchester, NH, and Little Falls, NJ, are providing additional data relative to on-site granular carbon reactivation and the subsequent adsorptive capacity of the reactivated material in full-scale treatment units.

Studies conducted at Jefferson Parish, LA, have provided data on full-scale units that show the ability of activated carbon for removing both relatively low- and high-molecular-weight organic substances with and without prior sand filtration. A series of 6-month investigations using different plant operating conditions has been conducted to assess the effects of changing ambient conditions on activated-carbon-bed operation.

A 100-gallons-per-minute (6.3 dm^3/sec) pilot plant project at Evansville, IN, is using chlorine dioxide disinfection to minimize the formation of trihalomethanes. Granular activated carbon is used to remove organic substances that are normally found in the source water and any organics that might result by disinfection with chlorine dioxide. The combination of these techniques could thus provide a prevention and removal process resulting in a relatively organic-free water.

Investigations at Beaver Falls, PA, and Huntington, WV, have demonstrated the performance of activated carbon as a replacement for sand in existing filter beds. The ability of activated carbon to adsorb a broad spectrum of organics at various efficiencies has been demonstrated during long-term studies of full-scale units at these locations.

Ozonation. All of the previously discussed projects dealing with adsorption have attempted to use the adsorbent primarily without any enhancement of biological activity. Recent European developments have cited the potential for increased removal for some classes of substances by incorporating an ozonation or oxygenation process prior to granular activated carbon adsorption. Projects are being started at Philadelphia, PA, Shreveport, LA, and Miami, FL, to assess the advantages and disadvantages of the system using ozone plus granular activated carbon. The concept involves the use

Studies of organic removal involving adsorption

Location	Primary area of interest	Source water
Miami, FL	Synthetic resins and granular activated carbon adsorption	Groundwater
Kansas City, MO	Synthetic resins, granular activated carbon adsorption; in-place resin reactivation	River
Glen Cove, NY	Synthetic resins adsorption and aeration; in-place reactivation	Groundwater
Manchester, NH	Granular activated carbon adsorption with fluidized-bed carbon reactivation	Lake
Little Falls, NJ	Granular activated carbon adsorption with infrared carbon reactivation	River
Cincinnati, OH	Granular activated carbon adsorption with fluidized-bed carbon reactivation	River
Jefferson Parish, LA	Granular activated carbon adsorption with and without prior sand filtration	River
Evansville, IN	Granular activated carbon adsorption and chlorine dioxide disinfection	River
Beaver Falls, PA	Granular activated carbon adsorption	River
Huntington, WV	Granular activated carbon adsorption	River
Philadelphia, PA	Granular activated carbon adsorption preceded by ozonation	River
Shreveport, LA	Granular activated carbon adsorption preceded by ozonation	Surface impoundment
Miami, FL	Granular activated carbon adsorption preceded by ozonation	Groundwater

of ozone as an oxidant to cause some substances to be more biodegradable, and also to provide a source of oxygen to maintain a biologically active system in the activated carbon bed. The primary objectives of these studies are to assess the potential for significant increases in the carbon reactivation cycle as compared with carbon systems without enhanced biological activity.

Although the ozonation procedure may also result in creation of some substances that are less adsorbable by carbon, the combined result of bacterial degradation of organics on the carbon surface and adsorption of the resulting organic mixture will be evaluated.

Acceptance of new technology. Cost effectiveness is also being studied as an integral part of assessing the ultimate acceptance of improved technology. Essentially, water treatment technology is available to minimize the threat of adverse public health effects posed by organic chemicals in drinking water. The eventual acceptance of improved technology in the United States will likely depend on current activities related to the types of studies briefly described.

For background information *see* WATER TREATMENT in the McGraw-Hill Encyclopedia of Science and Technology.   [JACK DE MARCO]

Bibliography: J. M. Symons, *Interim Guide for the Control of Chloroform and Other Trihalomethanes*, Water Supply Research Division, EPA, Cincinnati, June 1976.

Wave motion in fluids

The careful observer of the Earth's oceans and atmosphere is struck by their rich display of wavelike features—periodic, undulatory, propagating disturbances of airflow and ocean current. For most of history people have taken notice only of those waves discernible to the naked eye, producing obvious structure in the ocean surface and well-defined patterns in cloud. In the last several decades, however, waves and their effects have become increasingly manifest through sensitive instruments that detect their unseen influence on fluid pressure, motion, temperature, and turbulence fields. Ground-based remote sensors such as lidars, radars, and acoustic echo sounders now routinely monitor invisible wave motions in atmospheric and oceanic flows, while the commanding view from satellite platforms has revealed wave organization of the atmosphere and oceans on the larger scales unperceived by the earthbound observer. At the same time, it has become increasingly apparent that waves are not mere curiosities of atmospheric and oceanic structure. In many respects, they constitute the very warp and woof of geophysical fluid dynamics—transporting momentum and energy, organizing the weather into its familiar patterns, and triggering convection and mixing. For these reasons, geophysical wave studies have enjoyed steadily increasing impetus in recent years.

Wave types. An orderly discussion of the many wave types in the ocean and atmosphere first requires some sort of classification. Only recently has wave taxonomy seemed in any sense complete. The variety of wave types now distinguished stems from a corresponding diversity in the nature of the forces—inertial, Coriolis, pressure-gradient, surface-tension, buoyancy, electromagnetic, and viscous—governing the motions. The relative importance of these forces depends mainly on the spatial and temporal scales, that is, the wavelengths and periods, of the motions in question. It is therefore convenient to catalog the major wave types according to the characteristic scale of their motions.

On the very smallest spatial scales, surface-tension effects, which depend upon the radius of curvature of the air-water interface, produce surface waves on bodies of water known as capillary waves or "cat's paws." On similarly small scales, near fluid boundaries, and in the rarefied upper atmosphere, viscosity and thermal conductivity dominate, introducing viscous and thermal-conduction wave modes.

Because air and water are compressible, they sustain acoustic waves at frequencies greater than about 10 mHz whose wavelengths range from millimeters (ultrasound) to tens of kilometers (infrasound). Somewhat overlapping this frequency range, but extending to longer periods (as great as several hours) and larger wavelengths (as great as several thousand kilometers), are the atmospheric and oceanic gravity waves, which depend upon buoyancy for their existence. Unlike the acoustic modes, which propagate at the speed of sound, the gravity-wave modes typically propagate at speeds comparable to atmospheric wind speeds, some tens of meters per second or less. At the higher frequencies, these waves are evanescent; that is, their energy is trapped in the vertical and propagates only horizontally. Wind waves on the ocean and shear waves in the atmosphere known as Kelvin-Helmholtz waves are examples (Figs. 1 and 2). At the lower frequencies, the waves are able to propagate their energy vertically; in atmo-

Fig. 1. Cloud photograph revealing Kelvin-Helmholtz shear waves of the type responsible for much "turbulence" encountered by aircraft. (*Courtesy of W. Carroll Campbell, Boulder, CO*)

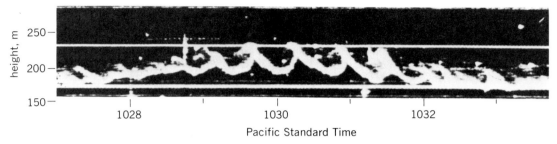

height, m
250 —
200 —
150 —

1028   1030   1032

Pacific Standard Time

Fig. 2. Kelvin-Helmholtz wave motions revealed by frequency-modulation–continuous-wave radar data. (*From E. E. Gossard, D. R. Jensen, and J. H. Richter, An* *analytical study of tropospheric structure as seen by high-resolution radar, J. Atmos. Sci., 28:794–807, 1971*)

spheric parlance, these are the internal gravity waves. In equatorial latitudes, there exist gravity waves trapped by latitudinal variation of the Coriolis parameter.

At temperate latitudes, for still larger wavelengths (greater than several thousand kilometers) and periods (the order of a day or more), the effects of the Earth's sphericity and rotation dominate. Atmospheric and oceanic tides forced by solar heating and solar and lunar gravitational attraction provide one example. Far more important are the so-called Rossby waves and planetary waves of several days' period, which are responsible for the familiar pressure highs and lows constituting temperate-latitude weather, and for major gyres observed in oceanic circulation patterns. These arise as a result of latitudinal variation of the Coriolis force on the rotating Earth. They propagate at speeds that depend upon wavelength and background flow velocity, but are typically the order of the background flow speed itself. Thus the typical Rossby-wave weather pattern (Fig. 3) may take several days to cross the continental United States. At the very longest periods, the Earth's atmosphere supports a semiannual and even a quasi-biennial oscillation, both readily identifiable components of climatological data.

Wave sources. In the Earth's atmosphere and oceans, waves constitute free modes of oscillation of the system; hence they require only slight excitation to reach noticeable amplitudes. For the most part, they are generated by instability in the background flow and by simple forcing. Barotropic (shear) and baroclinic (gravitational) instabilities in the global zonal flow field at temperate latitudes are major sources of Rossby waves. The instability of the gravity-stratified shear flows of the lower troposphere gives rise to much of the observed gravity-wave energy; depending upon the application, these mechanisms are referred to variously as Ekman-layer instability (governing boundary-layer flows), shear-flow instability, or Kelvin-Helmholtz instability. Forcing by bottom topography (continental land masses, mountain ranges, and even individual mountains and hills, as well as underwater features such as the continental shelves) excites planetary-, Rossby-, and gravity-wave modes. Other forcing includes surface-heating effects associated with continental and smaller-scale topography, the generation of acoustic- and gravity-wave modes by the development of thermals and convective clouds, and the wind generation of waves on the surface of the ocean. Atmospheric infrasound is generated by thunderstorms, volcanic eruptions, the airflow over mountains, nonlinear ocean-wave interactions near storms at sea, and even auroras. Just a very few years ago, many of these excitation processes were only qualitatively understood; now they are understood even in quantitative respects.

Wave propagation and dissipation. Once generated, all these wave types propagate in a manner determined by the dispersive properties of the background medium. Because this background varies spatially and temporally, waves experience refraction, scattering, partial and total reflection, and energy trapping. One of the more dramatic interactions occurs at wave "critical levels"—regions in the atmosphere and ocean where the wave-phase speed happens to match the fluid

Fig. 3. Breaking Rossby wave producing a storm in the Gulf of Alaska, shown in *NOAA* 5 satellite photograph, Sept. 28, 1978. (*NOAA*)

speed component in the same direction. Here the wave energy approaches the level asymptotically, so that little is allowed to penetrate, while at the same time wave amplitudes tend to grow without limit. As a result, nonlinear effects come strongly into play, producing significant spectral energy transfer and wave–mean flow interaction. Thus such regions act as sinks for wave energy and sources for clear-air turbulence (CAT). In the lower troposphere, CAT poses a significant hazard to aircraft.

Ultimately, all waves dissipate, either gradually in the course of horizontal propagation over many wave cycles or dramatically as in wave-critical-level encounters. The important dissipative processes include molecular viscosity and thermal conductivity (which erode the wave-associated fluctuations in velocity and temperature through diffusion), radiative damping (which smooths out temperature fluctuations through radiative transfer), molecular relaxation effects, and, in the ionosphere, ion drag (resulting from constraints placed on ionospheric motions by the Earth's magnetic field). Most of these become increasingly severe with altitude, and have their greatest influence in the upper atmosphere. In consequence, wave dissipation is a significant energy input to the upper atmosphere.

Importance of waves. In recent years scientists have discovered that waves exert a profound influence on weather and sea state on all scales from the minute to the large. On the local scale, waves determine surfing conditions along a beach, generate patches of CAT disturbing otherwise smooth aircraft flights, or change sea-state conditions affecting both commercial shipping and pleasure boating. Atmospheric gravity waves of somewhat larger dimension trigger convective clouds and severe storms, and subsequently organize these, producing cloud clusters, rain bands, and squall lines (Figs. 4 and 5). On still larger scales, Rossby waves produce the familiar pressure patterns of alternating highs and lows so prominent on conventional temperate-latitude weather maps. And at the global extreme, Rossby waves and planetary waves can induce short-term climatological variability through processes such as wave blocking, which may determine jet-stream and cold-front location and movement for months at a time. Such blocking has played a role in the severe winters experienced by the northeastern United States in 1976–1977 and 1977–1978.

Waves play a significant part in atmospheric and oceanic dynamics, through their major contributions to global energy and momentum budgets. For example, in this manner, Rossby waves produce the bulk of the meridional energy and momentum transports associated with the global atmospheric and oceanic circulations of temperate latitudes. Similarly, gravity waves produce significant vertical transports, particularly in the tropospheric jet stream. Without such transports, jet-stream winds might well be two times faster than observed values.

Even more dramatic are the effects of upward momentum and energy transports from the dense, energy-rich lower atmosphere to the highly rarefied upper atmosphere. There, where densities are

Fig. 4. Apollo spacecraft photograph of cloud streets off the Georgia coast, Apr. 8, 1968. *(NASA)*

some 10^{-6} or less of the surface values, a little energy input goes a long way. Scientists now know that gravity waves, planetary waves, and tides contribute some 20% of the total energy supply re-

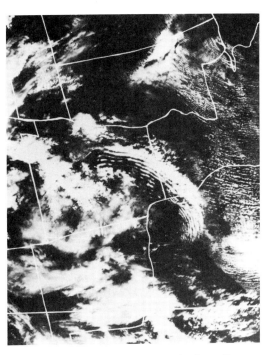

Fig. 5. *NOAA 2* satellite photograph of gravity-wave squall-line bands launched and propagating radially from a severe storm in Oklahoma, May 22, 1973. *(NOAA)*

ceived by the thermosphere (the upper atmosphere above about 140 km), accounting in part for its high temperature (some 1000–2000 K). In fact, if winds and radiative damping did not confine Rossby-wave energies to the middle and lower atmosphere, the Earth's thermospheric temperature might be as much as 100 times greater than the observed value. The result would be a steady boil-off of the planet's atmosphere, rendering it too rarefied to sustain life as now known. In the course of transport, waves often execute important spectral energy transfers, exchanging energy and momentum from large-scale modes to small and back again. Such momentum and energy exchanges are a vital component of atmospheric and oceanic dynamics, controlling the evolution of the circulation over time. The poor understanding of these processes currently poses a significant obstacle to the successful parameterization of these effects in global circulation models, and hence limits ability to forecast the weather.

By contrast with turbulent motions, waves in the atmosphere and ocean are relatively forecastable. Once generated, they tend to propagate with phase speeds, wavelengths, and periods that are well defined in terms of the properties of the background medium. Although they evolve, through instability, nonlinear interactions, and dissipative processes, this evolution is often relatively slow. As a result, the discovery by C. G. Rossby in 1939 that temperate-latitude weather patterns could be described in terms of waves produced significant improvements in synoptic-scale forecasting. In a similar way, gravity waves are currently believed to hold the key to many forecasting problems in mesometeorology. While the wave approach to forecasting has been supplanted operationally by numerical solution to the equations governing atmospheric motion, wave physics is required if there is to be some insight and understanding of the atmospheric processes behind the forecast.

For these reasons, wave studies have attracted the interest of theorist and experimenter alike during the past several decades. Kinematic wave studies have reached maturity, while dynamic studies remain in their infancy. Ahead lies the challenging tasks of putting wave processes into the proper perspective relative to atmospheric physics as a whole.

For background information *see* WAVE MOTION IN FLUIDS in the McGraw-Hill Encyclopedia of Science and Technology.   [WILLIAM H. HOOKE]

Bibliography: E. E. Gossard and W. H. Hooke, *Waves in the Atmosphere*, 1975; B. Kinsman, *Wind Waves*, 1965; J. Lighthill, *Waves in Fluids*, 1978; J. S. Turner, *Buoyancy Effects in Fluids*, 1973.

Wheat

Wheat *(Triticum)* can be hybridized with rye *(Secale)* to form an intergeneric hybrid called triticale. Triticale is a new crop species, known technically as × *Triticosecale*. It gets its name from the parent species, with the X denoting its hybrid origin. Although triticale plants were first described in 1876, only in the mid-1960s did they begin to receive great interest from scientists and farmers. Recent improvements in the hybrid have resulted in favorable grain yields and a good potential for

use as a food and feed grain. The amount of triticale grown throughout the world is not well documented, but estimates indicate that about 2,000,000 acres (810,000 ha) were grown in 1979. Some estimates indicate that about 200,000 acres (81,000 ha) were in the United States, while the largest area devoted to triticale is in the Soviet Union.

Origin. Triticale can be produced by hybridizing bread wheat *(T. aestivum)* or durum wheat *(T. turgidum)* with rye *(S. cereale)*. Rye has 14 chromosomes, with seven pairs designated as R chromosomes. Bread wheat has 42 chromosomes, which occur as seven pairs in each of three groups (genomes), designated A, B, and D. This species can be hybridized with rye quite easily, without any special treatment. On the other hand, durum wheat, with 28 chromosomes (seven pairs in each of the A and B genomes), is more difficult to cross with rye. The hybrid is made by transferring rye pollen to the stigmas of wheat, and after a few days it is necessary to excise the young embryo and grow it on nutrient medium.

The hybrid plant is treated with colchicine in the seedling stage, causing the chromosome number to double. The resulting plant, called an amphiploid, is octoploid and has 56 chromosomes (A, B, D, and R genomes) if the cross was with bread wheat, and is hexaploid and has 42 chromosomes (A, B, and R genomes) if the cross was with durum wheat. Figure 1 shows the scheme for producing triticale.

Types. The triticales mentioned above are called primary types because they are derived from the direct cross of wheat and rye. Such triticales can be hybridized among themselves or hybridized to wheat and other related species. Plants

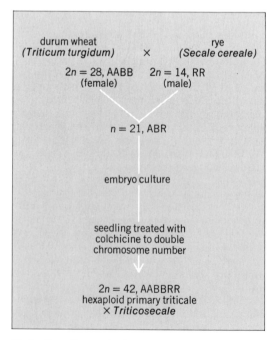

Fig. 1. Flow diagram for hexaploid triticale development showing chromosome numbers and genome identifications. The octoploid form $(2n = 56)$ is produced in a similar way, but starting with bread wheat *(Triticum aestivum, $2n = 42$, AABBDD)* as the female parent.

obtained from such crosses are called secondary triticales. Some of the plants from these crosses are not pure triticales, because they may have fewer than all seven pairs of the rye chromosomes and more than seven pairs of the chromosomes from a wheat genome; for example, a plant may have 7A, 7B, 3D, and 4R chromosome pairs. These are called secondary substitional triticales.

Triticale, like wheat, is mainly self-pollinating, in contrast to rye, which is predominantly wind-pollinated. The amount of natural outcrossing is slightly higher in triticale than in wheat, and so fields for producing seed should be isolated from other triticales and from wheat fields to retain cultivar purity.

Adaptation. Triticale can be grown under production-management practices commonly used for other cereal grains, including barley, wheat, rye, and oats. As with other cereal grains, the triticale cultivars must have growth and agronomic characteristics compatible with the intended production area. For example, triticales with spring growth habit are used for spring planting in the northern United States or for fall planting in areas with mild winters, whereas winter growth habit is required for fall planting in areas where winter killing by cold temperatures is common. Triticale performance can be improved by modifying management techniques appropriately. For example, earlier planting in the fall or spring in California was found to improve triticale performance relative to wheat. The most recently developed cultivars can equal or surpass wheat in grain yields in some locations but are substantially lower in other areas. Triticale varieties resemble other grains in adaptation responses, and cultivars have yet to be developed for many regions where triticale could be grown.

Usage. The first triticales had two major defects—poor spike fertility and wrinkled grains. Poor fertility contributed to low grain yields and increased the chance for infection by ergot (*Claviceps purpurea*). The most recently developed triticales have very good fertility. Despite major improvement in kernel characteristics, further improvement is still needed. Early triticales had a volume weight of less than 50 lb per bushel (64 kg/hl), but this has been increased to about 57 lb per bushel (74 kg/hl), compared to wheat at 60 lb per bushel (77 kg/hl). Figure 2 compares wheat and triticale kernels. These defects of triticale grain have an important bearing on its use, because ergot-infested grain is not suitable for human food or livestock feed, and the wrinkled grain affects the way triticale can be processed for food or feed uses. Triticale grain has been generally superior in protein and lysine content. Lysine, an amino acid essential for human and nonruminant animal nutrition, is deficient in most cereal grains.

Triticale grain has been evaluated for most applications already used with wheat and other grains. It can be used in pastry, bread, and extruded breakfast food or snack products. Such adaptations involve some modifications of the processing methods developed for wheat. A very acceptable bread made with 30% whole-grain triticale meal and 70% wheat flour is marketed in the

Fig. 2. Comparison of grains of (a) wheat with (b) triticale.

United States. Experimental studies have shown that triticale grain can be used as a starch source in beer making, and there has been interest in Canada in using triticale in the distilling industry.

Feeding trials with poultry, swine, and cattle have had variable results, but when there is no ergot infestation, triticale can be used by animals at about the same efficiency as with other grains. It is also used successfully as a fall grazing crop for cattle in some areas.

Future. Genetic improvement in triticale performance and quality has been rapid, and many scientists are optimistic that triticale will make a major contribution to increased world food production. Its use in specialty foods is very possible, although the impetus for increased production will likely depend on its acceptance as livestock feed.

For background information *see* BREEDING (PLANT); WHEAT in the McGraw-Hill Encyclopedia of Science and Technology. [CALVIN O. QUALSET]

Bibliography: J. H. Hulse and E. M. Laing, *Nutritive Value of Triticale Protein (and the Proteins of Wheat and Rye)*, International Development Research Centre, Ottawa, 1974; K. Lorenz, *CRC Crit. Rev. Food Technol.*, pp. 175–280, 1974; A. Müntzing, *Advances in Plant Breeding No. 10*, supplement to *J. Plant Breeding*, Verlag Paul Parey, Berlin, 1979; Y. V. Wu et al., *J. Agricul. Food Chem.*, 26:1039–1048, 1978; F. J. Zillinsky (ed.), *Triticale Breeding and Research at CIMMYT*, Research Bull. no. 24, International Maize and Wheat Improvement Center, Mexico City, 1973.

Winged bean

The winged bean (*Psophocarpus tetragonolobus*) is cultivated in the high-rainfall tropics and subtropics of Southeast Asia, Oceania, and Africa. Its green pods, leaves, seeds, tubers, flowers, and young shoots (see illustration) are all edible and form a minor part of the human diet, although all parts are not utilized in every region. The plant is a climbing perennial, usually grown as an annual on a trellis. It is easy to grow and yields abundantly.

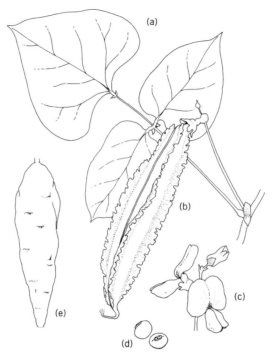

The winged bean: (a) leaf; (b) pod (about one-half life size); (c) flower; (d) seed; and (e) tuber (about one-half life size).

Other members of the genus *Psophocarpus* are indigenous to Africa, but the winged bean itself has not been found in the wild state. The very early history of the winged bean's dispersal and cultivation is not known. When the first Europeans reached Papua, New Guinea, they found the crop already growing in many parts of the island.

Although the most common name for the plant in English is winged bean, it is also known as asparagus bean, four-angled bean, Goa bean, and Manila bean. Native names include tua-pu (Thailand), pai-myeet (Burma), dara-dhambala (Sri Lanka), calamismis, seguidilla (Philippines), katjang botor (Indonesia), and kadjang-outan (Papua).

Morphology. Recent collections made in Papua, the Philippines, Indonesia, and elsewhere have revealed that a great deal of morphological variation exists in the crop. Mature pods range in length from 12 to 70 cm and contain from 3 to 20 seeds each. The majority of pods show a rectangular shape in cross section with four wings extending from the angles. The wings themselves may be undulate, dentate, serrate, or lobed. At maturity, the wings collapse and the pods become flattened. Other pod shapes include a semi-flattened and flat type. The pod surface may be smooth or roughened. Pods exhibit color variation, with a background of green, pink, or pale yellow and various intensities of purple coloration in the wings, resulting in wholly dark purple pods in some lines. The globular shiny seeds may be creamy white, yellow, brown, black, or mottled, and vary in weight from 0.06 to 0.42 g. Flower colors are basically blue or purple, although some lines have flowers that are almost white or deep reddish purple. Stems range in color from green to deep purple. The trifoliolate leaves are typically broadly rhomboid to ovate and vary in size.

Nutritional value. Chemical analysis of winged bean seed, leaves, immature pods, flowers, and tubers reveals that the crop is an excellent source of protein, oil, vitamins, and minerals. Protein and oil contents in mature raw seed range from 29 to 42 and 13 to 19 g per 100 g of seed (fresh-weight basis), respectively. On the average, winged bean seeds have only slightly less protein and oil than soybeans. Protein content of the tubers ranges from 5 to 20 g per 100 g (fresh-weight basis). These protein values are from 5 to 15 times higher than those of the staple root crops of the humid tropics. The amino acid profile of the winged bean also compares favorably with that of the soybean. However, as in the soybean, the sulfur-containing amino acids, methionine and cystine, are the main limiting amino acids in winged bean seeds. The fatty acid profile of the winged bean is quite similar in quality to that of the soybean.

Use as a food. The winged bean is cultivated usually as an annual backyard or horticultural crop on a small scale throughout Southeast Asia and to a lesser extent in parts of Oceania and Africa. At present, the plant is grown mainly for the tender young pods, although all parts of the plant are edible and highly nutritious. The use of different plant organs and methods of their consumption vary from location to location.

In Papua the green pods are the most popular edible part, with tuber consumption confined to relatively few tribes. Young pods and tubers are traditionally cooked in "mumu" fashion, a method of steam cooking, either in a pit dug in the ground or on the surface of the ground. Leaves and flowers may be boiled or fried, and eaten alone or as a supplement to sweet potatoes or cooked bananas. Ripe seeds may also be steamed, boiled, fried, or roasted within the pod.

In Indonesia, young leaves, shoots, and immature pods are eaten raw as leafy vegetables, or steamed or cooked with other vegetables to make side dishes and stews. Ripe seeds may be fried, roasted, or boiled and eaten as a snack, or fermented to make "tempeh."

Burma appears to be the only country where winged bean tubers are used on a large scale. The tubers are slightly sweet with the firmness of an apple, and may be boiled or eaten raw. Peeled small tubers dipped in oil and salted are popular as snacks.

Other uses. The winged bean has potential as a cover crop or fallow restorative crop because of its highly effective nodulation system and tolerance of poor soils. The feasibility of using it as a multirole second crop in rubber plantations and small holdings is being assessed in Malaysia and Sri Lanka.

Cultivation. The time taken for the winged bean to flower after sowing depends upon the season and location, as the plant is day length–sensitive. Time from planting to flowering may vary from 4 to 36 weeks. Young green pods are ready for harvesting about 2 to 3 weeks after fertilization. Mature fruits and ripe seeds can be picked 3 to 7 months after planting, while tubers are harvested 5 to 12 months after planting. In the 1978–1979 crop year, international trials were conducted to evaluate yield potential of the various parts of the winged bean that are utilized.

For background information *see* LEGUME in the McGraw-Hill Encyclopedia of Science and Technology.

[THEODORE HYMOWITZ; CHRISTINE A. NEWELL]

Bibliography: A. Claydon, A review of the nutritional value of the winged bean *Psophocarpus tetragonolobus* (L.) DC. with special reference to Papua New Guinea, *Sci. New Guinea*, 3(2): 103–114, 1975; T. N. Khan, Papua New Guinea: A centre of genetic diversity in winged bean (*Psophocarpus tetragonolobus* (L.) DC.), *Euphytica*, 25:693–706, 1976; C. A. Newell and T. Hymowitz, The potential of the winged bean—*Psophocarpus tetragonolobus* (L.) DC.—as an agricultural crop, in G. A. Ritchie (ed.), *New Agricultural Crops*, 1979; S. Sastrapradja et al., A survey of variation in *Psophocarpus tetragonolobus* (L.) DC. with reference to the Javanese samples, *Ann. Bogorienses*, 6(4):221–230, 1978.

Word processing

Word processing (WP) is a term commonly used to describe a system for accomplishing office work through people, procedures, and technologically advanced equipment. Originally limited to improved production of written communications generated from dictation in an office, WP has expanded to encompass all business communications generated by an organization, including the entire spectrum of office work. Although still commonly used, the term, which in the United States dates back to the early 1960s, fails to reflect this comprehensive expansion into the storage, retrieval, manipulation, and distribution of information and, most recently, a growing interdependence with electronic data processing. As a result, new descriptive terms such as information processing, administrative systems, and office systems are being used.

WP represents a further stage in modern society's application of automation, reaching beyond manufacturing and production lines into the office. "Automatic typewriters"—the generic predecessors of multifunction information processors—were the starting point. Now, a wide range of advanced equipment is available to electronically record keystrokes in electronic memories and on a variety of magnetic storage media—tape, card, cassette, or diskette—and electronically assist in correcting, editing, revising, and manipulating text material (Fig. 1). Retyping is limited to changes and corrections; perfect copy is automatically produced by pressing a button. Once stored, words, sentences, and paragraphs can be automatically rearranged, reassembled, and retyped repeatedly, and customized versions produced. Distribution of the finished work has also been affected. The effect in the office has become analogous to automation in the factory: higher production, lower costs, higher quality, and more efficient use of people.

More than equipment helps achieve these goals. A division of labor and specialization, particularly among secretaries, is generally evident in the use of support personnel. In addition, revised and updated procedures are required to take full advantage of the capabilities of increasingly sophisticated equipment, from initial dictation and keyboarding through the final stages of distribution

and transmission. New management structures are developing to integrate this advanced equipment with both procedures and personnel.

WP's basic goal is to provide an economic and efficient solution to the paperwork explosion characterizing the 1960s and 1970s and to streamline the flow of information and knowledge in an organization. WP modernizes the office, which is becoming the dominant work environment for the American labor force. (The U.S. Department of Labor estimates that by 1985 white-collar workers will outnumber blue-collar workers by a ratio of 3 to 2.) Rising costs, increasing amounts of information, and growing numbers of office workers are major factors in the spread of WP, especially in business organizations and government units heavily involved in paperwork.

HISTORY AND BACKGROUND

In the late 1950s, Ulrich Steinhilper, of the International Business Machines Corporation, originated the term *textverarbeitung*, in Germany, to convey the notion of processing words systematically in a manner analogous to the systematic way electronic data processing handles numbers. In the United States, the term (translated as "word processing" rather than "text processing") was introduced as a marketing technique, first to sell dictation equipment, then in 1967 as an element of a sales campaign for the IBM Magnetic Tape/Selectric Typewriter.

The new automatic typewriters (introduced in 1964) increased secretarial production by recording typed material on a magnetic tape cartridge and limiting retyping to changes and corrections. In addition, form letters were customized by using stock codes which made it possible to insert personalized information. Not only was the secretary freed from the necessity of completely retyping material, but the final copy was automatically turned out in perfect form at four to five times the speed of an average typist.

Initially, organizations either bought or rented

Fig. 1. The IBM 6/450 Information Processor includes diskette storage, mag card reader/recorder, multilingual keyboard, and functional display. The print station at right contains a high-speed ink jet printer, automatic paper and envelope feeder, and stackers. (*IBM*)

the new typewriters, but did not utilize them fully since only one-third of the traditional secretary's time is devoted to typing. In the late 1960s, as a response to emerging user requirements, a systems approach was advanced, concentrating typing and correspondence in centers where the machines were operated all day. The system was called word processing.

In the early 1970s, the concept was further refined. Adjustments were made for the kind of typing work (production or custom) and for the various ways of arranging work groups. As electronically stored memory capability increased through technological advances, the information processor emerged, comprising keyboard, video display, high-speed printer, and extensive storage capacity for keyboarded material.

The next evolutionary step during the mid-1970s involved the two-thirds of secretarial time spent on nontyping work. WP reached into daily record processing—file updating, list preparation, selection, scheduling, formatting, and printout of stored material—as information recorded on diskettes was called upon as needed. Word processors extracted selectively and rearranged stored information on demand. In 1976 communication features were added: word processors were directly linked so that copy could be printed out at the receiving end, thereby bypassing mail routing. In addition, the laser was used to speed printing.

CURRENT APPLICATION

WP's objectives in any situation are specifically: increased productivity of managers and professionals; more responsive secretarial support for principals (managers and other administrators who originate written material); more efficient output of paper work; reduced costs; improved communications; and a better end product.

The application of WP varies in its specific characteristics as the office work in an organization varies. An integrated office system is designed in terms of how text matter is generated (longhand, shorthand, or machine dictation), length of documents, volume, and variety. A crucial variable is the type of work, whether it is oriented to production or custom environments. Production work is routine, predictable, and unvaried, whereas custom work is complex, unpredictable, constantly changing, and flexible. Typically, a tendency toward either production or custom work will represent most of an organization's information handling.

WP's application can be viewed in terms of the three basic elements of equipment, people, and procedures. Cumulative experience has demonstrated that neglect of any one of these elements defeats the purposes of WP.

Equipment. Generally speaking, the office is the last place where organizations have applied new technology to improve production and lower costs. This is widely cited as a major factor in low office productivity. While industrial productivity jumped by almost 90% in the past decade, productivity increased less than 5% in the office, the most labor-intensive segment of the American economy. Capital investment reflects this gap: the capital investment supporting the typical manufacturing

worker is reported to be eight to ten times the investment supporting the typical office worker. In recent years, growth in sales of WP equipment reflects the increased interest in improving office productivity as labor costs have mounted steadily. Studies by the Dartnell Institute of Business Research, for example, report that the average cost of a business letter increased from $2.54 in 1968 to $4.77 in 1978, with more than half of the latter amount attributed to the costs of originator's and secretary's time.

WP equipment aims at improving productivity in these two fundamental phases of office work: input and output. Input is provided by the principal who originates the ideas, words, and information for transmittal by word in written or spoken form. Output involves the production of written material and its distribution.

Input. Input originates by hand-held microphone, by built-in microphone, and by telephone activation from an outside location. Dictation equipment ranges from battery-powered portable units to centralized systems serving a large number of originators. Cartridges or tape cassettes have a capacity ranging from 6 min to more than 2 hr to record material for playback and transcription. Department of Labor statistics show that dictating to a machine is two to three time faster than dictating to a secretary and six times faster than writing longhand. Yet, many offices still commonly use face-to-face dictation and longhand to originate material. The reluctance of originators to use dictation equipment is a significant barrier to full achievement of WP's productivity potential. Overall, dictation equipment can cut by more than 25% the time needed to get material out compared with face-to-face dictation or writing in longhand.

Output. WP equipment, which is manufactured by many companies, has had its greatest impact on output, the preparation and distribution of the entire range of typed and printed communication in an organization. Electronic storage capability makes possible the storage of keyboarded material for correction and change, without time-consuming retyping. Word spacing, margins, and formatting are done automatically. Using the capability to compare, information processors select, qualify, rearrange, or place stored material in sequence, then automatically turn out printed material. The increasingly popular feature of keyboard displays enables the user to make corrections on a video screen without the use of any paper until a final draft emerges.

Equipment varies according to the amount of material which is stored magnetically and the speed with which it is automatically played out. With a printer (either as a component of an information processor or operating separately), the speed of printing, for example, can reach 1800 characters per second compared with 15.5 characters per second provided by a magnetic card typewriter. With optical character recognition (OCR) equipment, higher-function devices can store and edit typewritten material for high-speed output by specialized printers.

Whether producing individual copies, producing multiple print sets, or distributing material electronically, WP is characterized by increasing speed of production. Versatility, as well, is exemplified by an office information distributor, sometimes referred to as an "intelligent" copier, which prints with a laser and receives and transmits documents electronically over ordinary telephone lines. It also links WP with data processing, using customized formats to print typewriterlike originals of computer-based information.

Copiers routinely handle a variety of documents, drawings, photographs, and texts of thick books, and can reproduce color. Copier output has reached a speed of a page per half second, with costs reduced to pennies per page. Facsimile machines can operate as remote output printers of information transmitted electronically. In addition, WP equipment anywhere can be linked directly for electronic communication and can be linked to central computers for access to their data bases. Meanwhile, WP has begun to go beyond paper into micrographics, by which information is stored on microfilm and microfiche for automatic retrieval.

People. Secretaries are directly affected by WP as their work is divided, specialization introduced, and their job status raised. A noteworthy result is the opening up of managerial opportunities in supervising WP. For the organization, these changes in secretarial assignments result in time, labor, and money savings, as well as increased opportunities for the individual.

A common way of reorganizing an office to achieve full WP potential involves the division of secretaries into correspondence or administrative specialists. This constitutes a move away from the traditional system in which the secretary reports directly to one or more principals for whom the secretary performs a full range of duties from typing to making travel plans. In the shift from such an arrangement, an adjustment period has been necessary for both principals and secretaries. While the particular forms of specialization vary considerably in organizations, the development of WP management is universally required. Not only does equipment change; so do work relationships,

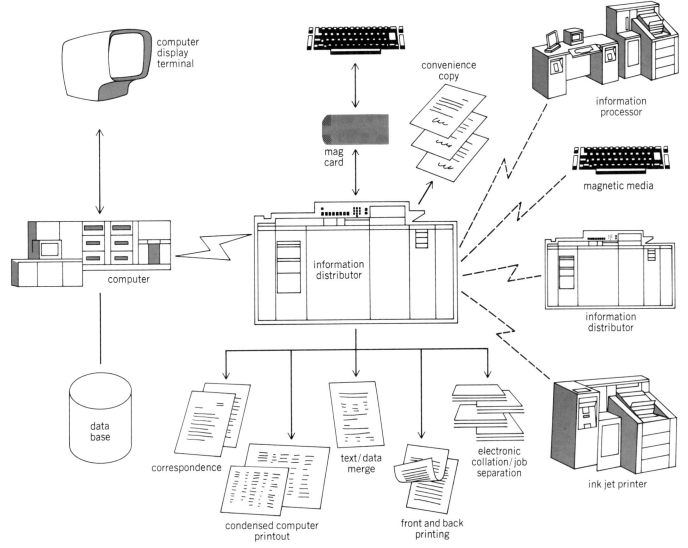

Fig. 2.   The multipurpose IBM 6670 Information Distributor represents a new step toward the office of the future through its ability to access and process a wide variety of information. The unit prints with a laser, receives and transmits documents electronically, processes text and data, and can also make convenience copies. (*IBM*)

duties, secretarial responsibilites, and chains of command.

Secretarial specialization in correspondence and administration is only one of many possibilities. In a consolidated system, administrative secretaries can specialize even further by concentrating on specific office duties, such as answering the phone, scheduling, or making travel plans. In such an arrangement, all secretaries report to secretarial managers rather than to the principals they support. Typically, correspondence secretaries work in WP centers concentrated in one or more office locations. In a further variation, a widely dispersed secretarial force with their WP equipment distributed near end-user locations is coordinated by a computer to manage, control, and handle work flow. This arrangement eliminates the need for a WP center. As new equipment and more integrated applications become available, even more variations will develop.

Procedures. WP required procedures to determine the way office work is most efficiently produced by people interfacing with equipment. The step-by-step movement of information from input through output must be handled in uniform ways. This encompasses formatting, labeling, storing, and retrieving, as well as manipulation of material that is generated. Since procedures must be tailor-made for specific situations, no particular procedure is standard, but the need for these procedures is universal.

TRENDS

"Office of the future," a term being used with increasing frequency, reflects the broad implications and potentials of WP. It takes into account the full range of functions in the office and of technology currently available, as well as that anticipated in the near future. It points up the disappearance of the boundary between WP and data processing. The office of the future will link computers, information processors, work groups in correspondence and administration, varieties of printers, copiers, and telephone equipment (Fig. 2).

The immediate outlook is for greater use of WP equipment and its systems approach. Independent market research companies uniformly predict substantial increases in equipment use and sales during the 1980s. A key to further expansion will be the opportunities for low-cost electronic document distribution via broadband satellite communication, which will provide inexpensive transmission of the electrically coded information generated by WP.

Meanwhile, rapidly improving technology continues to narrow the gap between originators and equipment, and increase the flexibility and capacity of WP equipment. Managers already can use a WP video display to review calendars and schedules, to file references and correspondence, and to issue instructions by using keyboard on their own desk. Tomorrow, letters may be dictated directly to a voice-activated typing device, while intelligent copiers produce hard copy output directly from information processors and computers. In the projected mixture of technology and systems in the office, both WP and data processing may well blend into information processing.

[EDWARD W. GORE, JR.]

Xenon compounds

Many types of xenon compounds have been prepared in recent years: halides, oxides, oxyfluorides, xenates, perxenates, perchlorates, fluorosulfates, molecular adducts, and complex fluorometallate salts. The pattern of bonding in these compounds is fairly uniform; the xenon is bonded only to electronegative elements, such as fluorine, chlorine, and oxygen. Very recently, however, new methods of synthesis have yielded compounds that contain xenon-xenon and xenon-nitrogen bonds.

Dixenon cation. Xenon reacts at room temperature with the dioxygenyl salt $O_2^+SbF_6^-$ to form $XeF^+Sb_2F_{11}^-$, a yellow crystalline product. Each time that a portion of gas is added, a bright green color can be seen very briefly in the solid. The source of this color is dixenon cation, Xe_2^+, an intermediate, paramagnetic species, which is formed by displacement of oxygen in the first stage of reaction, as shown in Eq. (1).

$$2Xe + O_2^+SbF_6^- \rightarrow Xe_2^+SbF_6^- + O_2 \qquad (1)$$

Spectra. Stable solutions of the cation can be prepared in antimony pentafluoride solvent by a reverse technique—reduction of $XeF^+Sb_2F_{11}^-$. Metals (Hg, Pb, Sn, Bi, Zn), metal oxides (PbO, As_2O_3), water, and gases (SO_2, PF_3) are suitable reducing agents. When a small amount of reducing agent is used, part of the divalent xenon is reduced to Xe_2^+ and part is reduced to elemental xenon, as shown in Eqs. (2) and (3); when an excess is used, all of the divalent xenon is reduced to elemental xenon. The ultraviolet-visible spectrum of the original yellow solution contains only one band at 285

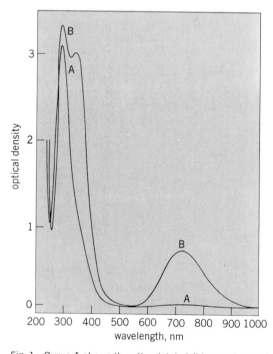

Fig. 1. Curve A shows the ultraviolet-visible spectrum of a solution of $XeF^+Sb_2F_{11}^-$ in antimony pentafluoride (yellow); curve B, of the same solution after partial reduction of the xenon (green). (*From L. Stein et al., Formation of the dixenon cation, Xe_2^+, in fluorantimonate (V) media by oxidation-reduction methods: Spectroscopic properties of the ion, J. Chem. Soc. Chem. Comm., pp. 502–504, 1978*)

$$2XeF^+Sb_2F_{11}^- + {}^3\!/_2 SO_2 \rightarrow$$
$$Xe_2^+Sb_2F_{11}^- + {}^3\!/_2 SO_2F_2 + 2SbF_5 \quad (2)$$

$$XeF^+Sb_2F_{11}^- + SO_2 \rightarrow Xe + SO_2F_2 + 2SbF_5 \quad (3)$$

nm, corresponding to an electronic transition of XeF^+ cation; that of the partly reduced green solution contains two additional bands at 335 and 710 nm, corresponding to electronic transitions of Xe_2^+ cation (Fig. 1). In Raman spectra, molecular vibrations of these ions can be discerned. A band at 619 cm^{-1}, generated by the xenon-fluorine stretching vibration of XeF^+ cation, grows weaker as this ion is reduced, and a band at 123 cm^{-1}, generated by the xenon-xenon stretching vibration of Xe_2^+ cation, grows stronger. Other bands originating in $Sb_2F_{11}^-$ anion and antimony pentafluoride solvent remain unchanged.

When $Xe_2^+Sb_2F_{11}^-$ is prepared from the natural mixture of xenon isotopes and frozen in a matrix of antimony pentafluoride at 4.5 K, it yields the complex electron paramagnetic resonance (EPR) spectrum shown in Fig. 2a; however, when it is prepared from a single isotope, ^{136}Xe, which has zero nuclear spin, it yields the simple spectrum shown in Fig. 2b. The latter spectrum has only two lines and can be readily interpreted. The unpaired electron that gives rise to the EPR signal occurs in a xenon species that has axial symmetry; when the axis is aligned parallel to the imposed magnetic field, one line is produced, and when it is aligned perpendicular to the field, the second line is produced. Hence this spectrum conforms to the pattern which is expected for the diatomic ion Xe_2^+. The first spectrum contains many additional lines (hyperfine lines) because of the presence of xenon isotopes that have nonzero nuclear spins. It can be reproduced almost perfectly with a computer by superimposing lines that arise from interaction of the electron and nuclear spins, and by adjusting the intensities of the lines to conform to the abundances of the different types of isotopes.

Bonding. An unusual type of chemical bond, called a three-electron bond, accounts for the stability of the ion. The unpaired electron occupies one of the overlapping orbitals that form the bond, as shown in Fig. 3, and two electrons with opposed spins occupy the second orbital. By interchanging the positions of the paired and unpaired electrons, a second structure similar to the first one is obtained. Resonance occurs between these two structures and provides the binding energy of the ion. Three-electron bonds also occur in molecular oxygen, nitric oxide, nitrogen dioxide, the superoxide ion O_2^-, and the ozonide ion O_3^-.

Iodine ions. Dixenon cation bears a striking resemblance to I_2^- anion, which has been characterized by spectroscopic methods in hydroxylic solvents and matrix-isolated salts like $K^+I_2^-$, $Rb^+I_2^-$, and $Cs^+I_2^-$. This similarity is not surprising, inasmuch as iodine and xenon are adjoining elements in the periodic table and the two ions are isoelectronic. The iodine anion is dark green and paramagnetic and has absorption bands at 370–400 and 737–800 nm, close to those of Xe_2^+. It also has a nearly identical vibration frequency, 115 cm^{-1}. In addition to I_2^- anion, iodine forms the polyatomic ions I_3^-, I_4^-, I_2^+, I_3^+, and I_5^+. Since the chemistry of xenon parallels that of iodine in many ways, it seems probable that xenon will also form

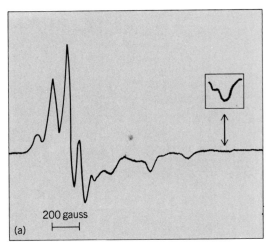

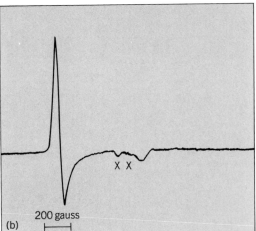

Fig. 2. EPR spectra of $Xe_2^+Sb_2F_{11}^-$: *(a)* derived from the natural mixture of xenon isotopes, with the insert showing the weakest line at increased gain; *(b)* derived from ^{136}Xe, X denoting an impurity band. Both samples were frozen in antimony pentafluoride at 4.5 K. (*From L. Stein et al., Formation of the dixenon cation, Xe₂⁺, in fluorantimonate (V) media by oxidation-reduction methods: Spectroscopic properties of the ion, J. Chem. Soc. Chem. Comm., pp. 502–504, 1978*)

more than one polyatomic ion. Efforts are currently being made to prepare other polyatomic xenon species by combining xenon with Xe_2^+ at high pressures. [LAWRENCE STEIN]

Xenon-nitrogen bonds. Since the preparation of the first xenon compound by Neil Bartlett in 1962, a considerable number of compounds containing xenon bonded to oxygen and fluorine have been prepared. These include the binary fluorides and oxides, the oxyfluorides, xenon esters of several

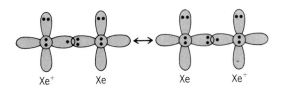

Fig. 3. Resonance structures which lead to the formation of the three-electron bond in Xe_2^+ cation. Only *p* orbitals are shown.

strong acids, and a variety of anions and cations of these compounds in complex salts. In spite of the impressive progress made in the chemistry of xenon, the number of different xenon-element bonds is very small. Nearly all compounds that can be isolated under normal laboratory conditions contain only xenon-fluorine or xenon-oxygen bonds.

Xenon-fluorine and xenon-oxygen bonds. These bond types fall into four classes: Xe-F, as in XeF_2; M-F-Xe, as in $Xe_2F_3^+$; Xe-O, as in $OXeF_4$; and Xe-O, as in $Xe(OSeF_5)_2$. Only two other bond types have been found in isolable compounds. These are two examples of Xe-Cl bonds in $M_9(XeO_3Cl_2)_4Cl$, where M represents Cs or Rb, and Xe-N bonds in $FXeN(SO_2F)_2$ and $[XeN(SO_2F)_2]_2F^+AsF_6^-$.

The existence of Xe-N bonds was first demonstrated in 1974 by R. LeBlond and D. DesMarteau. They prepared $FXeN(SO_2F)_2$ by reaction of XeF_2 with the nitrogen acid, $HN(SO_2F)_2$, as shown in reaction (4).

$$XeF_2 + HN(SO_2F)_2 \xrightarrow[0°C]{CF_2Cl_2} FXeN(SO_2F)_2 + HF \quad (4)$$

The ^{19}F nuclear magnetic resonance (NMR) and Raman spectra in Figs. 4a and 5a show the covalent nature of the compound and eliminate alternative bonding schemes such as $FXe^+N(SO_2F)_2$ or formula (5). They also suggest that the Xe-F

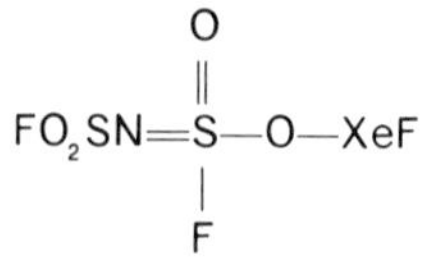

$$\begin{array}{c} O \\ \| \\ FO_2SN{=}S{-}O{-}XeF \\ | \\ F \end{array} \quad (5)$$

bond is remarkably similar to that in XeF_2. Xenon-fluorine bonds typically exhibit a three-line multiplet in the ^{19}F NMR spectrum, with the central line corresponding to the fluorine bound to all xenon isotopes with nuclear spins other than $I = \frac{1}{2}$. The two outer lines correspond to fluorine bound to ^{129}Xe ($I = \frac{1}{2}$), and their separation is due to the spin-spin coupling between the two different nuclei of spin $I = \frac{1}{2}$. The ^{129}Xe-F coupling constant is typically very large, in the range of 5000 to 6000 Hz for most xenon(II) compounds. The Raman spectra of compounds containing terminal Xe-F bonds show very intense Raman emission in the 500 to 650 cm^{-1} region. In XeF_2 this is found at 496 cm^{-1} and in $XeF^+Sb_2F_{11}^-$ at 621 cm^{-1}. The position of this $\nu(XeF)$ stretching frequency can be taken as a measure of importance of FXe^+ in the bonding. The value of 504 cm^{-1} for $\nu(XeF)$ in $FXeN(SO_2F)_2$ supports the similarity of the Xe-F bond to that in XeF_2. The compound is moderately stable at 22°C and decomposes slowly to XeF_2, Xe, and $[N(SO_2F)_2]_2$ as shown in Eq. (6).

$$2FXeN(SO_2F)_2 \longrightarrow XeF_2 + Xe + [N(SO_2F)_2]_2 \quad (6)$$

This decomposition is quite analogous to that of xenon esters, such as $FXeOSO_2F$, as is shown in Eq. (7).

$$2FXeOSO_2F \longrightarrow XeF_2 + Xe + (OSO_2F)_2 \quad (7)$$

Criteria for bonding. The discovery of $FXeN(SO_2F)_2$ clearly showed that the possibilities for making additional xenon-element bond types are good. It is, however, necessary that the ligand possess some special characteristics, which at the moment are not precisely defined. The ligand must be very electronegative, must be capable of

existing as a stable anion, and should form a chlorine(I) derivative. If all these criteria are met, the chances are good that a Xe(II) compound can be obtained. In essentially all Xe(II) esters of the type of FXeOR and Xe(OR)$_2$, RO satisfies these criteria. There have been many attempts to form compounds that do not meet these criteria, and essentially all have failed. The high electronegativity is necessary, because all descriptions of bonding in xenon compounds require considerable transfer of charge from xenon to the ligand. The chlorine(I) requirement serves to define qualitatively the acceptable electronegativity of the ligand, that is, it must be greater than chlorine. The stable anion requirement is necessary because the compounds have considerable ionic character, as shown in the canonical forms for $FXeOSO_2F$, $FXe^+OSO_2F^- \leftrightarrow FO_2SOXe^+F^-$.

Unusual compounds. With the synthesis of $FXeN(SO_2F)_2$, it might have been thought that a variety of other compounds would have soon followed. This has not been the case, and only very recently has the second xenon-nitrogen compound been obtained. It is a rather unusual compound and is more stable than the first example. As mentioned earlier, the Xe-F bond in $FXeN(SO_2F)_2$ is quite similar to that in XeF_2. Xenon difluoride undergoes reaction with AsF_5 to form $FXe^+AsF_6^-$ and $Xe_2F_3^+AsF_6^-$, as shown in reactions (8) and (9).

$$XeF_2 + AsF_5 \longrightarrow FXe^+AsF_6^- \quad (8)$$

$$2XeF_2 + AsF_5 \longrightarrow Xe_2F_3^+AsF_6^- \quad (9)$$

The 1:1 compound with AsF_5 is unstable with regard to the 2:1 compound, as shown in reaction (10). The reaction of $FXeN(SO_2F)_2$ with AsF_5 is

$$2FXe^+AsF_6^- \xrightarrow{22°C} Xe_2F_3^+AsF_6^- + AsF_5 \quad (10)$$

indeed similar to XeF_2, as shown in reactions (11) and (12). The 1:1 compound has never been adequately characterized, and its structure is unknown.

The Raman spectrum of the salt is shown in Fig. 5b. It is clear by comparison with Fig. 5a that there are no terminal Xe-F bonds in the compound.

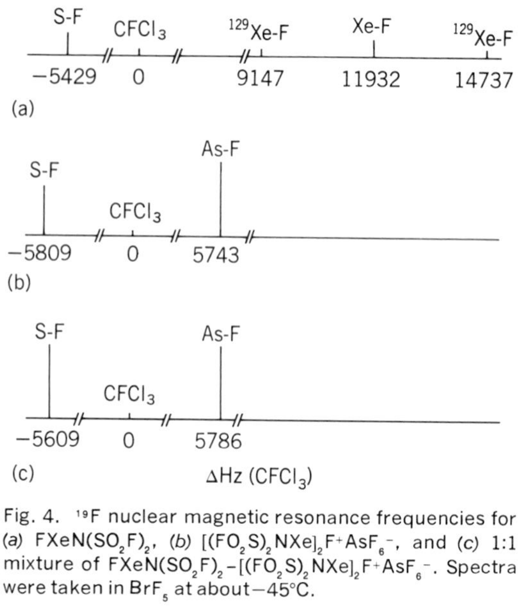

Fig. 4. ^{19}F nuclear magnetic resonance frequencies for (a) $FXeN(SO_2F)_2$, (b) $[(FO_2S)_2NXe]_2F^+AsF_6^-$, and (c) 1:1 mixture of $FXeN(SO_2F)_2$-$[(FO_2S)_2NXe]_2F^+AsF_6^-$. Spectra were taken in BrF_5 at about $-45°C$.

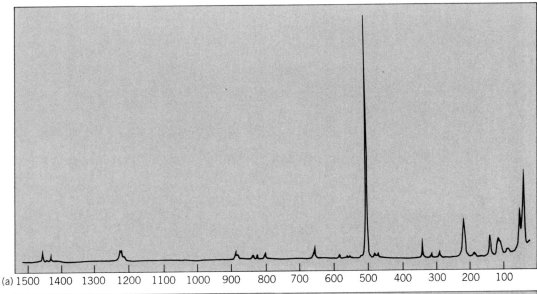

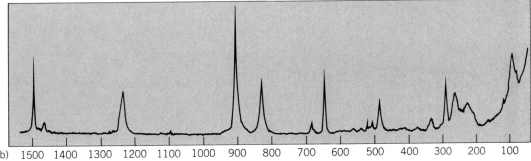

Fig. 5. Raman spectra of (a) solid $FXeN(SO_2F)_2$ and (b) $[(FO_2S)_2NXe]_2F^+AsF_6^-$, recorded at about $-100°C$.

$$FXeN(SO_2F)_2 + AsF_5 \xrightarrow{-78 \text{ to } -15°C}$$
$$[(FO_2S)_2 NXe^+AsF_6^-] \quad (11)$$

$$[(FO_2S)_2 NXe^+AsF_6^-] \xrightarrow[\text{Vacuum}]{22°C}$$
$$[(FOO_2S)_2 NXe]_2F^+AsF_6^- + AsF_5 \quad (12)$$

Furthermore, the simplicity of the spectrum reflects the greater covalency in the cation compared to $FXeN(SO_2F)_2$. This is exactly as expected, since the vibrational spectra of $N(SO_2F)_2^-$ are considerably more complex than in a compound such as $ClN(SO_2F)_2$. In FXe^+ and $Xe_2F_3^+$, the XeF bonds are considerably strengthened. In $[(FO_2S)_2NXe]F^+$, the Xe-N bonds should be similarly strengthened and the $N(SO_2F)_2$ groups less ionic.

The ^{19}F NMR spectrum of the compound in BrF_5 is shown in Fig. 4b. No resonance due to the bridging fluorine was observed. This was thought to be due to some exchange process involving $FXeN(SO_2F)_2$ and $XeN(SO_2F)_2^+$, as in Eq. (13).

$$[(FO_2S)_2NXe]_2F^+$$
$$= (FO_2S)_2 NXe^+ + FXeN(SO_2F)_2 \quad (13)$$

Good evidence for this was obtained by adding $FXeN(SO_2F)_2$ to the solution. The spectrum of $FXeN(SO_2F)_2$ shown in Fig. 4a clearly shows the F-Xe resonance. In Fig. 4c the spectrum of the mixture is shown. The Xe-F resonance is no longer observable and the F-S resonances merge to a single peak. The 1:1 mixture requires the total area of the resonances of the fluorine atoms bound to sul-

fur to be equal to that of the AsF_6^- as observed. This evidence, while indirect, strongly supports the postulated exchange or some related exchange phenomenon. The NMR and Raman evidence, along with data from synthesis, offers convincing proof for the existence of the novel cation $[(FO_2S)_2NXe]_2F^+$. This compound is the only example where the terminal Xe-F bonds in $Xe_2F_3^+$ have been replaced by another group. Other related cations may be discovered in the future, and continued synthetic work will undoubtedly result in additional xenon-nitrogen compounds and possibly other new xenon-element bonds. As the variety of xenon compounds increases, a better understanding of the nature of the bonding in these compounds will be possible.

For background information *see* XENON COMPOUNDS in the McGraw-Hill Encyclopedia of Science and Technology. [DARRYL D. DES MARTEAU]

Bibliography: N. Bartlett and F. O. Sladky, in A. F. Trotman-Dickenson (ed.), *Comprehensive Inorganic Chemistry*, vol. 1, pp. 213–330, 1973; D. D. DesMarteau, *J. Amer. Chem. Soc.*, 100:6270, 1978; D. T. Hawkins, W. E. Falconer, and N. Bartlett, *Noble Gas Compounds: A Bibliography, 1962–1976*, 1978; W. F. Howard, Jr., and L. Andrews, *J. Amer. Chem. Soc.*, 97:2956–2959, 1975; R. D. LeBlond and D. D. DesMarteau, *J. Chem. Soc. Chem. Commun.*, pp. 555–556, 1974; L. Stein et al., *J. Chem. Soc. Chem. Commun.*, pp. 502–504, 1978; R. D. Willet, S. W. Petersen, and B. A. Coyle, *J. Amer. Chem. Soc.*, 99:8202–8207, 1977.

McGRAW-HILL YEARBOOK OF SCIENCE AND TECHNOLOGY

List of Contributors

List of Contributors

A

Ackerman, Dr. Ralph A. *Scripps Institution of Oceanography, La Jolla, CA.* EGG (FOWL).

Anderson, Dr. Don L. *Seismology Laboratory, California Institute of Technology, Pasadena.* SEISMOLOGY (in part).

B

Baker, Dr. Victor R. *Department of Geological Sciences, University of Texas.* FLUVIAL EROSION LANDFORMS.

Bassemir, Robert W. *Graphic Arts Laboratories, Sun Chemical Corporation, Carlstadt, NJ.* PRINTING (in part).

Beaty, Dr. Chester B. *Department of Geography, University of Lethbridge, Alberta, Canada.* GLACIATION.

Bell, Dr. Adrian. *School of Plant Biology, University College of North Wales, Bangor, Gwynedd.* STEM.

Birdsall, Dr. Blair. *Steinman, Boynton, Gronquist & Birdsall, New York.* BRIDGE.

Blackadar, Prof. Alfred K. *Department of Meteorology, Pennsylvania State University.* METEOROLOGY.

Blanden, Dr. R. V. *Department of Microbiology, John Curtin School of Medical Research, Canberra City, Australia.* IMMUNOLOGY, CELLULAR.

Blumenfeld, Prof. L. A. *Institute of Chemical Physics, Academy of Sciences of the U.S.S.R., Moscow.* BIOENERGETICS.

Boutwell, Dr. Roswell K. *Professor of Oncology, McArdle Laboratory for Cancer Research, University of Wisconsin Medical Center, Madison.* ONCOLOGY (coauthored).

Boxer, Emanuel. *NASA Langley Research Center, Hampton, VA.* JET PROPULSION.

Boyer, Dr. Keith. *Los Alamos Scientific Laboratory.* FUSION, NUCLEAR.

Boyle, Charles. *NASA Goddard Space Flight Center, Greenbelt, MD.* SPACE FLIGHT.

Brandt, Dr. John C. *Chief, Laboratory for Astronomy and Solar Physics, NASA Goddard Space Flight Center, Greenbelt, MD.* SPACE TELESCOPE.

Bresler, Dr. Eshel. *Department of Soil Physics, Agricultural Research Organization, Bet Dagan, Israel.* IRRIGATION OF CROPS (in part).

Brewer, Dr. Jon. *Department of Geological Sciences, Cornell University.* SEISMOLOGY (in part).

Brown, Dr. Stephen C. *Department of Biological Sciences, State University of New York, Albany.* INVERTEBRATE ARCHITECTURE.

Bruck, Dr. David K. *Department of Botany, University of California, Berkeley.* LEAF (BOTANY) (in part).

Byers, Dr. C. W. *Department of Geology and Geophysics, University of Wisconsin.* SHALE.

C

Carlson, Dr. William C. *Department of Forestry, University of Kentucky.* PLANT GROWTH.

Casten, Thomas R. *General Manager, Cummins Cogeneration, Cummins Corporation, Columbus, IN.* ELECTRIC POWER GENERATION.

Changnon, Dr. Stanley A., Jr. *Illinois State Water Survey, Urbana.* HAIL.

Chapman, Brian. *Printing Developments, Inc., East Norwalk, CT.* PRINTING (in part).

Chapman, Prof. David S. *Department of Geology and Geophysics, College of Mines and Mineral Industries, University of Utah.* LITHOSPHERE.

Christy, Dr. James W. *U.S. Naval Observatory, Washington, DC.* PLUTO.

Clark, Dr. Nolan. *Agricultural Engineer, USDA, Southwestern Great Plains Research Center, Bushland, TX.* IRRIGATION OF CROPS (in part).

Cleary, William A., Jr. *Chief, Ship Characteristics Branch, Office of Merchant Marine Safety, U.S. Coast Guard Headquarters, Washington, DC.* SHIP, MERCHANT.

Cline, Prof. David. *Department of Physics, University of Wisconsin.* PARTICLE ACCELERATOR.

Cohan, Christopher S. *Graduate Student, Case Western Reserve University.* GASTROPODA (coauthored).

Cohen, Prof. Bernard. *Department of Physics, University of Pittsburgh.* NUCLEAR POWER.

Colwell, Prof. Robert N. *Department of Forestry, University of California, Berkeley.* TERRAIN SENSING, REMOTE.

D

Davis, Dr. Raymond, Jr. *Brookhaven National Laboratory, Upton, NY.* SUN.

Dekker, Dr. Johan. *Department of Phytopathology, Agricultural University, Wageningen, The Netherlands.* PLANT DISEASE CONTROL.

DeMarco, Dr. Jack. *Municipal Environmental Research Laboratory, U.S. Environmental Protection Agency, Cincinnati.* WATER TREATMENT.

Dement, Dr. Donald. *NASA Headquarters, Washington, DC.* SATELLITE COMMUNICATIONS.

Des Marteau, Prof. Darryl D. *Department of Chemistry, Kansas State University.* XENON COMPOUNDS (in part).

Dickinson, Dr. Dale F. *Jet Propulsion Laboratory, Pasadena, CA.* MASER.

Duguay, Dr. Linda E. *School of Marine and Atmospheric Science, University of Miami.* FORAMINIFERA (coauthored).

E

Echols, Dr. Dorothy. *Department of Earth and Planetary Sciences, Washington University, St Louis, MO.* MIOCENE.

Edelson, Dr. Edward. *Department of Chemistry, University of Southern California.* LIFE, ORIGIN OF.

Edelstein, Dr. Norman M. *Materials and Molecular Research Division, Lawrence Berkeley Laboratory, Berkeley, CA.* ACTINIDE ELEMENTS.

Edmonds, Dr. Robert L. *College of Forest Resources, University of Washington, Seattle.* AEROBIOLOGY (feature).

Epel, Dr. David. *Professor of Biological Sciences, Hopkins Marine Station, Stanford University, Pacific Grove, CA.* FERTILIZATION.

Evenson, Dr. Kenneth M. *Time and Frequency Division, National Bureau of Standards, Boulder, CO.* LIGHT.

F

Feit, Louis. *ITT Defense Communication Division, Nutley, NJ.* SATELLITES, NAVIGATION BY.

Feldman, Dr. Barry J. *Los Alamos Scientific Laboratory.* OPTICAL PHASE CONJUGATION (coathored).

428   LIST OF CONTRIBUTORS

Fernsler, Dr. R. F. *Naval Research Laboratory, Washington, DC.* ELECTRON BEAM CHANNELING (coauthored).

Fisher, Dr. Robert, *Los Alamos Scientific Laboratory.* OPTICAL PHASE CONJUGATION (coauthored).

Flavell, Dr. R. A. *Division of Gene Structure and Expression, National Institute for Medical Research, London.* EUKARYOTIC GENES (feature).

Franc, Dr. Jean-Marie. *Laboratoire d'Histologie et Biologie Tissulaire, Université Claude Bernard, Villeurbanne, France.* CTENOPHORA.

Frank, Dr. William. *Department of Environmental Science, University of Virginia.* TROPICAL METEOROLOGY.

G

Gamborg, Dr. Oluf L. *International Plant Research Institute, San Carlos, CA.* BREEDING, PLANT.

Gammon, Dr. Richard H. *Department of Atmospheric Sciences, University of Washington, Seattle.* INTERSTELLAR MATTER.

Garrett, Dr. C. J. R. *Department of Oceanography, Dalhousie University, Halifax, Nova Scotia.* DEEP-OCEAN MIXING PROCESSES.

Gartner, Dr. Stefan. *Department of Oceanography, Texas A & M University.* EXTINCTION (BIOLOGY).

Gauster, Dr. W. B. *Sandia Laboratories, Livermore, CA.* NONDESTRUCTIVE TESTING.

Geiger, Dr. David H. *Geiger-Berger Associates, New York.* ROOF CONSTRUCTION.

Geiser, Dr. Peter A. *Department of Geology, University of Connecticut.* ROCK.

Gerola, Dr. Humberto. *T. J. Watson Research Center, IBM Corporation, Yorktown Heights, NY.* GALAXY, EXTERNAL.

Giannasi, Dr. David E. *Department of Botany, University of Georgia.* CHEMOTAXONOMY.

Goldfine, Dr. Ira D. *Director, Cell Biology Research Laboratory, Mount Zion Hospital and Medical Center; Assistant Professor of Medicine, University of California, San Francisco.* HORMONE.

Golub, Prof. Robert. *School of Mathematical and Physical Sciences, University of Sussex, Brighton, England.* NEUTRON.

Gore, Edward W., Jr. *Vice President, International Market Requirements, Office Products Division, IBM Corporation, Franklin Lakes, NJ.* WORD PROCESSING.

Gossard, Dr. Arthur. *Bell Laboratories, Murray Hill, NJ.* CRYSTAL GROWTH.

Gott, Prof. J. Richard, III. *Department of Astrophysical Sciences, Princeton University.* COSMOLOGY (in part).

Greig, Dr. J. R. *Supervisory Research Physicist, Physics Division, Naval Research Laboratory, Washington, DC.* ELECTRON BEAM CHANNELING (coauthored).

Grodzins, Dr. Lee. *Department of Physics, Massachusetts Institute of Technology.* SCANNING PROTON MICROPROBE.

H

Haff, Prof. Peter. *Kellogg Radiation Laboratory, California Institute of Technology, Pasadena.* SPUTTERING.

Hashimoto, Saburo. *Tokyo Ink Manufacturing, Ltd.* PRINTING (in part).

Hauck, Dr. R. D. *Soil and Fertilizer Research Branch, Tennessee Valley Authority, Muscle Shoals, AL.* SOIL (in part).

Haugh, Dr. Bruce. *American Museum of Natural History, New York.* ECHINODERMATA.

Hayes, William C. *Editor in Chief, "Electrical World," McGraw-Hill Publications Company, New York.* ELECTRICAL UTILITY INDUSTRY.

Hébant, Dr. Charles. *Laboratoire de Paleobotanique et Evolution des Vegetaux, Academie de Montpellier, France.* BRYOPHYTA.

Heilmann, Dr. I. U. *Brookhaven National Laboratory, Upton, NY.* MAGNETISM (coauthored).

Hensel, Dr. John. *Bell Laboratories, Murray Hill, NJ.* EXCITON.

Hills, Dr. Graham W. *Department of Chemistry, University of North Carolina, Chapel Hill.* INFRARED SPECTROSCOPY.

Hooke, Dr. William H. *Environmental Research Laboratories, National Oceanic and Atmospheric Administration, Boulder, CO.* WAVE MOTION IN FLUIDS.

Hopwood, Prof. David A. *Department of Genetics, John Innes Institute, Norwich, England.* CROWN GALL.

Huber, Dr. Robert J. *Electrical Engineering Department, University of Utah.* TRANSISTOR.

Hymowitz, Prof. Theodore. *Department of Agronomy, College of Agriculture, University of Illinois.* WINGED BEAN (coauthored).

J

Jacobson, Dr. Michael. *Optical Sciences Center, University of Arizona.* SOLAR ENERGY.

Jenike, Dr. Andrew W. *Jenike and Johanson, Inc., North Billerica, MA.* FLOW OF SOLIDS.

Jenkins, Dr. C. B. *Geothermal Resources Division, Aminoil, USA, Inc., Santa Rose, CA.* GEOTHERMAL ENERGY.

Jennings, Dr. J. B. *Department of Zoology, University of Leeds, England.* RHYNCHOCOELA.

Jensen, Dr. Clayton E. *Cape Coral, FL.* ENVIRONMENT.

Johnson, Dr. Philip. *Department of Chemistry, State University of New York, Stony Brook.* MOLECULAR SPECTROSCOPY.

Jonscher, Prof. A. K. *Chelsea Dielectrics Group, Chelsea College, University of London.* DIELECTRICS.

Jory, Howard R. *Varian Associates, Palo Alto, CA.* MICROWAVE TUBE.

K

Kaplan, Dr. Donald R. *Department of Botany, University of California, Berkeley.* LEAF (BOTANY) (in part).

Keeney, Dr. D. R. *Department of Soil Science, College of Agriculture and Life Sciences, University of Wisconsin.* SOIL (in part).

Kenney-Wallace, Dr. Geraldine A. *Associate Professor of Chemistry, Lash Miller Chemical Laboratories, University of Toronto.* PICOSECOND SPECTROSCOPY.

Klass, Dr. Donald. *Institute of Gas Technology, Chicago.* ORGANIC CHEMICAL SYNTHESIS.

Klemme, Dr. H. D. *Weeks Petroleum Corporation, Westport, CT.* PETROLEUM.

Kurtz, Dr. L. T. *Professor of Soil Fertility, Department of Agronomy, University of Illinois.* SOIL (in part).

L

Landgren, Dr. C. R. *Biology Department, Middlebury College, Middlebury, VT.* PLANT CELL.

Lanford, Prof. William A. *Department of Physics, State University of New York at Albany.* MATERIALS ANALYSIS (in part).

Laurence, Prof. K. M. *Child Health Laboratories, Welsh National School of Medicine, Cardiff.* CLINICAL PATHOLOGY.

Lechene, Dr. Claude. *Director, National Biotechnology Resource in Electron Probe Microanalysis; Visiting Professor of Physiology, Harvard Medical School, Boston.* ELECTRON PROBE MICROANALYSIS IN HEALTH RESEARCH (feature—coauthored).

Lee, Dr. John. *Department of Biology, City College of New York.* FORAMINIFERA (in part).

Linsley, Prof. John. *Physics Department, University of New Mexico.* COSMIC RAYS.

Lucas, Dr. J. G. *Air Navigation Research Group, University of Sydney, Australia.* INSTRUMENT LANDING SYSTEM.

Luyten, Dr. J. *Department of Physical Oceanography, Woods Hole Oceanographic Institution, Woods Hole, MA.* INDIAN OCEAN (in part).

M

McBride, Dr. Murray B. *Agronomy Department, Cornell University.* SOIL CHEMISTRY.

Mackie, Dr. George O. *Department of Biology, University of Victoria, British Columbia, Canada.* EPITHELIUM.

Matyas, Dr. Steven M. *IBM Systems Communications, Kingston, NY.* CRYPTOGRAPHY (coauthored).

Meek, Prof. Devon W. *Department of Chemistry, Ohio State University.* COORDINATION CHEMISTRY.

Mellichamp, Prof. Duncan A. *Department of Chemical and Nuclear Engineering, University of California, Santa Barbara.* CONTROL SYSTEMS.

Meyer, Dr. Carl H. *Advisory Engineer, IBM Systems Communications, Kingston, NY.* CRYPTOGRAPHY (coauthored).

Molina, Dr. Mario J. *Department of Chemistry, University of California, Irvine.* OZONE, ATMOSPHERIC.

Morley, Dr. Joseph J. *Lamont-Doherty Geological Observatory, Palisades, NY.* PALEOCLIMATOLOGY (feature).

Morrison, Dr. David D. *Institute for Astronomy, University of Hawaii.* VENUS.

Mpitsos, Dr. George J. *Department of Anatomy, Case Western Reserve University School of Medicine, Cleveland.* GASTROPODA (coauthored).

Muller, Dr. Richard A. *Lawrence Berkeley Laboratory, University of California, Berkeley.* COSMOLOGY (in part).

N

Natesan, Dr. K. *Materials Science Division, Argonne National Laboratories, Argonne, IL.* CORROSION.

Netravali, A. N. *Bell Laboratories, Holmdel, NJ.* TELEVISION.

Newell, Dr. Christine. *Department of Agronomy, College of Agriculture, University of Illinois.* WINGED BEAN (coauthored).

O

Obrist, Dr. Paul A. *Department of Psychiatry, University of North Carolina Medical School, Chapel Hill.* BIOFEEDBACK.

Olson, Dr. Roger. *Research Associate, Aspen Institute for Humanistic Studies, Boulder, CO.* CLIMATIC CHANGE.

Olton, Dr. D. S. *Department of Psychology, Johns Hopkins University.* BEHAVIORAL NEUROSCIENCE.

P

Palatnick, A. S. *Group Leader, A.I.L. Division, Cutler-Hammer, Deer Park, NY.* POLAR-COORDINATE NAVIGATION SYSTEMS.

Parkinson, Dr. Dennis. *Department of Microbiology, University of Calgary, Alberta, Canada.* FOREST SOIL.

Paulus, Dr. Jean-Michel. *Département de Clinique et de Pathologie Médicales, Institut de Médecine, Hôpital Universitaire de Bavière, Liège, Belgium; Institut de Pathologie Cellulaire, Hôpital de Bicêtre, Le Kremlin-Bicêtre, France.* BLOOD.

Peck, Dr. Eugene L. *Hydrological Research Laboratory, National Weather Service, National Oceanic and Atmospheric Administration, Silver Spring, MD.* FLASH FLOODS.

Prestegard, Dr. James. *Department of Chemistry, Yale University.* NUCLEAR MAGNETIC RESONANCE.

Q

Qualset, Prof. Calvin O. *Chairman, Department of Agronomy and Range Science, College of Agricultural and Environmental Sciences, University of California, Davis.* WHEAT.

R

Repetski, Dr. John E. *U.S. National Museum, U.S. Department of the Interior, Washington, DC.* FOSSIL.

Reudink, Dr. D. O. *Bell Laboratories, Holmdel, NJ.* COMMUNICATIONS SATELLITE.

Roberts, Dr. Arthur. *Fermi National Accelerator Laboratory, Batavia, IL.* NEUTRINO.

Robertson, Dr. Gordon. *Head, Advanced Automation Department, Engineering Research Center, Western Electric Corporation, Princeton, NJ.* INDUSTRIAL ROBOTS (feature).

Roman, Dr. Richard. *Research Fellow, National Biotechnology Resource in Electron Probe Microanalysis, Department of Physiology, Harvard Medical School, Boston.* ELECTRON PROBE MICROANALYSIS IN HEALTH RESEARCH (feature — coauthored).

Rosencwaig, Dr. Allen. *Lawrence Livermore Laboratory, University of California, Livermore.* PHOTOACOUSTIC SPECTROSCOPY.

S

Sassaman, Dr. Clay. *Department of Biology, University of California, Riverside.* ENZYME.

Schaffner, Charles E. *Syska & Hennessy, Inc., Engineers, New York.* BUILDINGS (coauthored).

Scheer, Dr. Bradley T. *Professor Emeritus of Biology, University of Oregon.* CELL MEMBRANES; ENDOCRINE MECHANISMS.

Schippers, Dr. P. Anne. *Long Island Horticultural Research Laboratory, Riverhead, NY.* HYDROPONICS.

Schmid, Dr. Frank R. *Northwestern University Medical School, Chicago.* SKELETAL SYSTEM DISORDERS.

Schneider, Dr. Edward L. *National Institute of Aging Center, Bethesda, MD.* CELL (BIOLOGY).

Schoenman, Richard L. *Flight Controls Technology, Boeing Commercial Airplane Company, Seattle.* FLIGHT CONTROLS (coauthored).

Scott, Dr. David. *Physics Division, Lawrence Berkeley Laboratory, University of California, Berkeley.* NUCLEAR PHYSICS.

Sepkoski, Dr. John. *Department of the Geophysical Sciences, University of Chicago.* ANIMAL EVOLUTION.

Seybold, John. *Media, PA.* PRINTING (in part).

Shimony, Prof. Abner. *Physics Department, Boston University.* QUANTUM MECHANICS.

Shirane, Dr. Gen. *Brookhaven National Laboratory, Upton, NY.* MAGNETISM (coauthored).

Shomber, Henry A. *Flight Controls Technology, Boeing Commercial Airplane Company, Seattle.* FLIGHT CONTROLS (coauthored).

Shugart, Dr. Herman H. *Senior Ecologist, Environmental Sciences Division, Oak Ridge National Laboratory.* THE ECOLOGICAL NICHE (feature).

Silverman, Irving. *Syska & Hennessy, Inc., Engineers, New York.* BUILDINGS (coauthored).

Stankey, Dr. George H. *Forestry Sciences Laboratory, Missoula, MT.* FOREST AND FORESTRY.

Starke, Prof. Edgar A., Jr. *Metallurgy Department, Georgia Institute of Technology.* HEAT TREATMENT (METALLURGY).

Stein, Dr. Lawrence. *Chemistry Division, Argonne National Laboratory.* XENON COMPOUNDS (in part).

Stringfellow, Dr. Dale A. *Department of Experimental Biology, Upjohn Company, Kalamazoo, MI.* INTERFERON.

Sumner, Dr. Malcolm E. *Department of Agronomy, University of Georgia.* AGRICULTURE, SOIL AND CROP PRACTICES IN.

T

Taylor, Dr. Dennis. *School of Marine and Atmospheric Science, University of Miami.* FORAMINIFERA (coauthored).

Taylor, Dr. Joseph H., Jr. *Department of Physics and Astronomy, University of Massachusetts.* GRAVITATION.

Teng, Dr. P. S. *Department of Plant Pathology, University of Minnesota.* COMPUTER SIMULATION OF PLANT DISEASE EPIDEMICS (feature—coauthored).

Thompson, Richard. *Vice President of Commodities, Hills Brothers Coffee Company, San Francisco.* COFFEE.

Thompson, Prof. W. J. *Physics Department, University of North Carolina, Chapel Hill.* POLARIZATION OF NUCLEI.

Tilton, Dr. Richard. *Department of Laboratory Medicine, University of Connecticut Health Center, Farmington.* CLINICAL MICROBIOLOGY.

V

Verma, Dr. A. K. *Research Associate, McArdle Laboratory for Cancer Research, University of Wisconsin Medical Center, Madison.* ONCOLOGY (coauthored).

Vollack, Dr. Robert F. *Vice President, American Bureau of Shipping, New York.* SHIP DESIGN.

W

Wakimoto, Prof. Ronald H. *Department of Forestry and Conservation, University of California, Berkeley.* VEGETATION MANAGEMENT.

Wallen, Dr. Lowell L. *Northern Regional Research Center, U.S. Department of Agriculture, Peoria, IL.* AFLATOXIN.

Warren, Dr. Bruce A. *Department of Physical Oceanography, Woods Hole Oceanographic Institution, Woods Hole, MA.* INDIAN OCEAN (in part).

Wells, Dr. Carol L. *Department of Bacteriology, Virginia Polytechnic Institute.* BACTERIOLOGY, MEDICAL (coauthored).

White, Dr. Stan M. *Robertson Research, Houston.* OCEANOGRAPHY.

Wilke, Dr. G. *Director, Max-Planck Instituts für Kohlenforschung, Mülheim, Federal Republic of Germany.* SOLVENT EXTRACTION.

Wilkins, Dr. Tracy. *Department of Bacteriology, Virginia Polytechnic Institute.* BACTERIOLOGY, MEDICAL (coauthored).

Wood, Dr. Bernard J. *Department of the Geophysical Sciences, University of Chicago.* PETROLOGY.

Y

Yankee, Prof. Herbert W. *Mechanical Engineering Department, Worcester Polytechnic Institute.* INDUSTRIAL ENGINEERING.

Z

Zadoks, Dr. J. C. *Laboratory for Phytopathology, Agricultural University, Wageningen, The Netherlands.* COMPUTER SIMULATION OF PLANT DISEASE EPIDEMICS (feature—coauthored).

Zeiger, Dr. Eduardo. *Department of Biological Sciences, Stanford University.* EPIDERMIS (PLANT).

McGRAW-HILL YEARBOOK OF SCIENCE AND TECHNOLOGY

Index

Index